HILBERT TRANSFORMS

The Hilbert transform arises widely in a variety of applications, including problems in aerodynamics, condensed matter physics, optics, fluids, and engineering. This work, written in an easy-to-use style, is destined to become the definitive reference on the subject. It contains a thorough discussion of all the common Hilbert transforms, mathematical techniques for evaluating them, and a detailed discussion of their application. Especially valuable features are the tabulation of analytically evaluated Hilbert transforms, and an atlas that immediately illustrates how the Hilbert transform alters a function. These will provide useful and convenient resources for researchers.

A collection of exercises is provided for the reader to test comprehension of the material in each chapter. The bibliography is an extensive collection of references to both the classical mathematical papers, and to a diverse array of applications.

FREDERICK W. KING is a Professor in the Department of Chemistry at the University of Wisconsin-Eau Claire.

All the titles listed below can be obtained from good booksellers or from Cambridge University Press. For a complete series listing visit

`http://www.cambridge.org/uk/series/sSeries.asp?code=EOM`

67 F. Bergeron, G. Labelle and P. Leroux *Combinatorial Species and Tree-Like Structures*
68 R. Goodman and N. Wallach *Representations and Invariants of the Classical Groups*
69 T. Beth, D. Jungnickel, and H. Lenz *Design Theory 1, 2nd edn*
70 A. Pietsch and J. Wenzel *Orthonormal Systems for Banach Space Geometry*
71 G. E. Andrews, R. Askey and R. Roy *Special Functions*
72 R. Ticciati *Quantum Field Theory for Mathematicians*
73 M. Stern *Semimodular Lattices*
74 I. Lasiecka and R. Triggiani *Control Theory for Partial Differential Equations I*
75 I. Lasiecka and R. Triggiani *Control Theory for Partial Differential Equations II*
76 A. A. Ivanov *Geometry of Sporadic Groups I*
77 A. Schinzel *Polynomials with Special Regard to Reducibility*
78 H. Lenz, T. Beth, and D. Jungnickel *Design Theory II, 2nd edn*
79 T. Palmer *Banach Algebras and the General Theory of *-Albegras II*
80 O. Stormark *Lie's Structural Approach to PDE Systems*
81 C. F. Dunkl and Y. Xu *Orthogonal Polynomials of Several Variables*
82 J. P. Mayberry *The Foundations of Mathematics in the Theory of Sets*
83 C. Foias, O. Manley, R. Rosa and R. Temam *Navier–Stokes Equations and Turbulence*
84 B. Polster and G. Steinke *Geometries on Surfaces*
85 R. B. Paris and D. Kaminski *Asymptotics and Mellin–Barnes Integrals*
86 R. McEliece *The Theory of Information and Coding, 2nd edn*
87 B. Magurn *Algebraic Introduction to K-Theory*
88 T. Mora *Solving Polynomial Equation Systems I*
89 K. Bichteler *Stochastic Integration with Jumps*
90 M. Lothaire *Algebraic Combinatorics on Words*
91 A. A. Ivanov and S. V. Shpectorov *Geometry of Sporadic Groups II*
92 P. McMullen and E. Schulte *Abstract Regular Polytopes*
93 G. Gierz et al. *Continuous Lattices and Domains*
94 S. Finch *Mathematical Constants*
95 Y. Jabri *The Mountain Pass Theorem*
96 G. Gasper and M. Rahman *Basic Hypergeometric Series, 2nd edn*
97 M. C. Pedicchio and W. Tholen (eds.) *Categorical Foundations*
98 M. E. H. Ismail *Classical and Quantum Orthogonal Polynomials in One Variable*
99 T. Mora *Solving Polynomial Equation Systems II*
100 E. Olivieri and M. Eulália Vares *Large Deviations and Metastability*
101 A. Kushner, V. Lychagin and V. Rubtsov *Contact Geometry and Nonlinear Differential Equations*
102 L. W. Beineke, R. J. Wilson, P. J. Cameron. (eds.) *Topics in Algebraic Graph Theory*
103 O. Staffans *Well-Posed Linear Systems*
104 J. M. Lewis, S. Lakshmivarahan and S. Dhall *Dynamic Data Assimilation*
105 M. Lothaire *Applied Combinatorics on Words*
106 A. Markoe *Analytic Tomography*
107 P. A. Martin *Multiple Scattering*
108 R. A. Brualdi *Combinatorial Matrix Classes*
110 M.-J. Lai and L. L. Schumaker *Spline Functions on Triangulations*
111 R. T. Curtis *Symmetric Generation of Groups*
112 H. Salzmann, T. Grundhöfer, H. Hähl and R. Löwen *The Classical Fields*
113 S. Peszat and J. Zabczyk *Stochastic Partial Differential Equations with Lévy Noise*
114 J. Beck *Combinatorial Games*
116 D. Z. Arov and H. Dym *J-Contractive Matrix Valued Functions and Related Topics*
117 R. Glowinski, J.-L. Lions and J. He *Exact and Approximate Controllability for Distributed Parameter Systems*
118 A. A. Borovkov and K. A. Borovkov *Asymptotic Analysis of Random Walks*
119 M. Deza and M. Dutour Sikirić *Geometry of Chemical Graphs*
120 T. Nishiura *Absolute Measurable Spaces*
121 M. Prest *Purity, Spectra and Localisation*
122 S. Khrushchev *Orthogonal Polynomials and Continued Fractions: From Euler's Point of View*
123 H. Nagamochi and T. Ibaraki *Algorithmic Aspects of Graph Connectivity*
124 F. W. King *Hilbert Transforms 1*
125 F. W. King *Hilbert Transforms 2*

ENCYCLOPEDIA OF MATHEMATICS AND ITS APPLICATIONS

Hilbert transforms

Volume 2

FREDERICK W. KING

University of Wisconsin-Eau Claire

CAMBRIDGE
UNIVERSITY PRESS

CAMBRIDGE UNIVERSITY PRESS
Cambridge, New York, Melbourne, Madrid, Cape Town, Singapore, São Paulo, Delhi

Cambridge University Press
The Edinburgh Building, Cambridge CB2 8RU, UK

Published in the United States of America by Cambridge University Press, New York

www.cambridge.org
Information on this title: www.cambridge.org/9780521517201

First published 2009

Printed in the United Kingdom at the University Press, Cambridge

A catalogue record for this publication is available from the British Library

Library of Congress Cataloguing in Publication data
King, Frederick W., 1947–
Hilbert transforms / Frederick W. King.
p. cm.
Includes bibliographical references and index.
ISBN 978-0-521-51720-1 (hardback)
1. Hilbert transform. I. Title.
QA432.K56 2008
515′.723–dc22 2008013534

ISBN 978-0-521-51720-1 hardback

To the memory of my mother

Contents

Volume I

Preface

My objective in this book is to present an elementary introduction to the theory of the Hilbert transform and a selection of applications where this transform is applied. The treatment is directed primarily at mathematically well prepared upper division undergraduates in physics and related sciences, as well as engineering, and first-year graduate students in these areas. Undergraduate students with a major in applied mathematics will find material of interest in this work.

I have attempted to make the treatment self-contained. To that end, I have collected a number of topics for review in Chapter 2. A reader with a good undergraduate mathematics background could possibly skip over much of this chapter. For others, it might serve as a highly condensed review of material used later in the text. The principal background mathematics assumed of the reader is a solid foundation in basic calculus, including introductory differential equations, a course in linear and abstract algebra, some exposure to operator theory basics, and an introductory knowledge of complex variables. Readers with a few deficiencies in these areas will find a number of recommendations for further reading at the end of Chapter 2. Some of the applications discussed require the reader to be familiar with basic electrodynamics.

A focus of the book is on problem solving rather than on proving theorems. Theorems are, for the most part, not stated or proved in the most general form possible. The end-notes will typically provide additional reference sources of more detailed discussions about the various theorems presented. I have not attempted to sketch the proof of every theorem stated, but for the key results connected to the Hilbert transform, at minimum an outline of the essential elements is usually presented. Consistent with the problem-solving emphasis is that all the different techniques that I know for evaluating Hilbert transforms are displayed in the book.

I take the opportunity to introduce special functions in a number of settings. I do this for two reasons. Special functions occur widely in problems of great importance in many areas of physics and engineering, and, accordingly, it is essential that students gain exposure to this important area of mathematics. Since many Hilbert transforms evaluate to special functions, it is imperative that the reader know when to stop doing algebraic manipulations. I have incorporated several mathematical topics for which few or no applications are known to the writer. The selection process was governed in part by the potential that I thought a particular area might have in problem solving, and I have done this with the full knowledge that crystal-ball reading is an art rather than a science!

The exercises are intended as a means for the reader to test his/her comprehension of the material in each chapter. The vast majority of the problems are by design routine applications of ideas discussed in the text. A small percentage of the problems are likely to be fairly challenging for an undergraduate reader, and a few problems could be labeled rather difficult. Most readers will have no trouble deciding when they have encountered an example of this latter group.

I have compiled an extensive table of Hilbert transforms of common mathematical functions. I hope this table will be useful in three ways. First, it serves as the answer key for a number of exercises that are placed throughout the text. Since many additional Hilbert transform pairs can be established by differentiation, or by appropriate multiplicative operations, etc., this table can be used to generate a great number of exercises, much to the delight of the reader. Second, I hope it will provide a useful reference source for those looking for the Hilbert transform of a particular function. Finally, for those searching for a particular Hilbert transform not present in the table, finding related transforms may give an idea on how to approach the evaluation, and give some clues as to whether a closed form expression in terms of standard functions is likely to be possible. In several sections the table includes a few specific cases followed by the general formula. This has been done to allow the reader to access the Hilbert transform of some of the simpler special cases as quickly as possible, rather than reducing a more complicated general formula.

The mini atlas of functions and the associated Hilbert transforms given in Appendix 2 is intended to provide a visual representation for a selection of Hilbert transform pairs. I hope this will be valuable for students in the applied sciences and engineering.

The reference list is rather extensive, but is not intended to be exhaustive. There are far too many published articles on Hilbert transforms to provide a complete set of references. I have attempted to give a generous number of references to applications. Many citations are given to the classical mathematical papers on the topics of the book, and for the serious student these works can be read with great profit. The Notes section at the end of each chapter gives a guide as to where to start reading for further information on topics discussed in the chapter. Elaborations and further details on the proofs of different theorems will often be located in the references cited in the end-notes.

My final task is to thank those who have helped. Logan Ausman, Dr. Matt Feldmann, Geir Helleloid, Dr. Kai-Erik Peiponen, Dr. Ignacio Porras, Dr. Jarkko Saarinen, and Corey Schuster read various chapters and made a number of useful suggestions to improve the presentation. Dr. Walter Reid and Dr. Jim Walker gave me some helpful comments on a preliminary draft of the first three chapters. Julia Boryskina and Hristina Ninova assisted with the translation of a number of technical papers. Several other students did translations and I offer a collective thanks to them.

Julia also provided assistance in the construction of the atlas of Hilbert transforms and with a number of the figures. Ali Elgindi did some numerical checking on the table of Hilbert transforms and Julia also did a few preliminary tests. Thanks are extended to Irene Pizzie for her efforts to improve the presentation.

The author would greatly appreciate if readers would bring to his attention any errors that escaped detection. The URL http://www.chem.uwec.edu/king/forward is the web address where corrections will be posted. It is the author's intention to maintain this site actively.

Symbols

The first occurrence or a definition is indicated by a section reference or an equation number. HT is an abbreviation for the tables of Hilbert transforms given in the Appendixes (Table A.1).

$\|a\|$	sum of the components of the multi-index a; §15.5
$\arg z$	argument of a complex number; Eq. (2.69)
A_p	the A_p condition for $1 \leq p \leq \infty$; Eq. (7.377), §7.12
$b\Omega$	boundary of a bounded domain Ω; §3.1
B	Boas transform operator; §16.4
B_n	generalization of the Boas transform operator; Eq. (16.84)
$\mathbf{B}(t)$	magnetic induction; §17.9
$\mathcal{B}$	generalization of the Boas transform operator; Eq. (16.80)
$B(a,b)$	beta function (Euler's integral of the first kind); Eq. (5.112), HT-01
$BV([a,b])$	class of functions that have bounded variation on the interval $[a,b]$; §4.25
C	designation for a contour (usually closed); §2.8.1.
C	a positive (often unspecified) constant; in derivations such a constant need not be the same at each occurrence, even though the same symbol is employed.
C	SI unit for charge, the coulomb, §19.1
$\mathbb{C}$	the set of complex numbers; §2.10
C_n	symmetry operation such that rotation by $2\pi/n$ leaves the system invariant; §21.3
C^∞	infinitely differentiable function for all points of $\mathbb{R}$; §2.15.2
C_0^∞	class of functions that are infinitely differentiable with compact support; §2.15.2
C^k	class of functions that are continuously differentiable up to order k; §2.15.2

C_0^k	class of functions that are continuously differentiable up to order k and have compact support; §2.15.2.
C_p	positive constant depending on the parameter p; often not the same at each occurrence in the sequence of steps of a proof
cas	Hartley cas function; Eq. (5.59)
$\mathscr{C}f$	Cauchy transform of the function f; Eq. (3.19)
$C(z)$	Fresnel cosine integral; Eq. (14.171), HT-01
$\mathrm{chirp}(x)$	chirp function, Exercise 18.13
$\mathrm{Ci}(x)$	cosine integral; Eq. (8.78), HT-01
$\mathrm{ci}(x)$	cosine integral; HT-01
$\mathrm{cie}(\alpha, \beta)$	cosine-exponential integral; HT-01
$\mathrm{Cie}(\alpha, \beta)$	cosine-exponential integral; HT-01
$\mathrm{Cl}_2(x)$	Clausen function; HT-01
$C_n^\lambda(x)$	Gegenbauer polynomials (ultraspherical polynomials); §9.1, Eq. (11.298), HT-01

$\mathbf{D}$	electric displacement; §19.1
$\mathcal{D}$	space of all C^∞ functions with compact support; §2.15.2, §10.2
$\mathcal{D}'$	space of all distributions on $\mathcal{D}$; §10.2
$\mathcal{D}'_+$	space of distributions with support on the right of some point; §10.2
$\mathcal{D}_{L^p}$	space of test functions; §10.2
$\mathcal{D}'_{L^q}$	space of distributions; §10.2
$\mathcal{D}'_{\mathscr{L}}$	space of distributions; Eq. (17.240)
$D(x)$	Dawson's integral; Eq. (5.32)
$D_n(\theta)$	Dirichlet kernel; Eq. (6.56)
$D_n^\lambda(x)$	ultraspherical function of the second kind; Eq. (11.299)

$-e$	electronic charge
E	identity element; §2.10, Eq. (2.150)
E	energy of a signal; Eq. (18.1)
E	one-dimensional Euclidean space
E^1	one-dimensional Euclidean space; §2.11.1
E^n	n-dimensional Euclidean space; §2.11.1
E^σ	class of entire functions of exponential type; §2.8.7, §7.4
$\mathcal{E}$	space of all C^∞ functions with arbitrary support on $\mathbb{R}$; §10.2
$\mathcal{E}(t)$	envelope function; Eq. (18.76)
$\mathcal{E}'$	space of distributions having compact support; §10.2
$E_1(x)$	exponential integral; Eq. (5.98)
E_n	eigenvalues of the unperturbed Hamiltonian; §22.4, Eq. (22.57)
$E_n(z)$	exponential integral; Eq. (14.200), HT-01
$\mathbf{E}(t)$	electric field; §17.9
$\mathbf{E_i}(\omega)$	incident electric field; Eq. (20.3)
$\mathbf{E_r}(\omega)$	reflected electric field; Eq. (20.3)

$\mathbf{E_L}$	left circularly polarized electric wave; Eq. (21.16)
$\mathbf{E_R}$	right circularly polarized electric wave; Eq. (21.17)
$\mathrm{Ei}(x)$	exponential integral function; Eq. (5.101), HT-01
$\mathrm{erf}(z)$	error function; Eq. (5.27), HT-01
$\mathrm{erfc}(z)$	complementary error function; Eq. (5.141)
$\mathbf{E}_v(z)$	Weber's function; HT-01
$\mathbf{F}$	Lorentz force; Eq. (21.50).
$f()$ or f	function (at no particular specified point); §1.2
$f(x)$	function f evaluated at the point x; §1.2
f_j	oscillator strength; §19.2
$f[n]$	element of a discrete sequence; §13.2, §13.6
$\{f[n]\}$	discrete sequence; §13.6
$f_e(x)$	even function; Eq. (4.8)
$f_o(x)$	odd function; Eq. (4.9)
$f_\downarrow(c)$	limit approaching c from $c+0$; Eq. (2.22)
$f_\uparrow(c)$	limit approaching c from $c-0$; Eq. (2.23)
$\mathcal{F}f$	fourier transform of the function f; §2.6, Eq. (2.46)
$\mathcal{F}_n f$	n-dimensional Fourier transform of the function f; §15.6
$\mathcal{F}^{-1}f$	inverse Fourier transform of the function f; §2.6, Eq. (2.47)
$\mathcal{F}_c f$	Fourier cosine transform of the function f; Eq. (5.41)
$\mathcal{F}_s f$	Fourier sine transform of the function f; Eq. (5.40)
$\mathcal{F}_N$	N-point DFT operator; §13.4
$\mathcal{F}_Q$	fractional Fourier transform; §18.10, Eq. (18.147)
$\mathcal{F}_\alpha$	discrete fractional Fourier transform; §18.13, Eq. (18.240)
$\hat{f}$	Fourier transform of the function f; §2.6
$\tilde{f}$	conjugate series of f; Eq. (6.118); alternative notation for $\mathcal{H}f$; §6.1
f'	derivative of the function f
$f^+(z)$	function f evaluated at an interior point to a contour; Eq. (3.152)
$f^-(z)$	function f evaluated at an exterior point to a contour; Eq. (3.153)
$F_n(\theta)$	Fejér kernel; Eq. (6.63)
${}_1F_1(\alpha;\beta;x)$	Kummer's confluent hypergeometric function; Eq. (5.30), HT-01
${}_2F_1(a,b;c;z)$	hypergeometric function (or Gauss' hypergeometric function); HT-01
$\boldsymbol{F}_{\mathrm{ext}}$	external force; §17.9
$\boldsymbol{F}_{\mathrm{rad}}$	radiative reaction force; §17.9
$f(\omega, 0)$	scattering amplitude at $\theta = 0$; Eq. (19.309)
$F_{\mathbf{h}}$	scattering factor; §23.4
$\mathrm{floor}[x]$	the greatest integer $\leq x$
$G(a,x)$	Hilbert transform of the Gaussian function; §4.7. The abbreviation $G(1,x) \equiv G(x)$ is employed; §9.3
$G^{(n)}_{kl_1\cdots l_2}(t_1,t_2,\ldots,t_n)$	tensor components of the nth-order response function; §22.1

$\hbar$	Planck's constant divided by 2π
H	Hilbert transform operator on $\mathbb{R}$; Eqs. (1.2) and (1.1.4)
$\mathcal{H}$	Hilbert transform operator for the disc; Eq. (3.202)
$\mathbf{H}$	magnetic field; Eq. (19.7)
$\mathcal{H}_\tau$	Hilbert transform operator for period 2τ; Eq. (3.286)
Hf	Hilbert transform of the function f; §1.2
$(Hf)(x)$	Hilbert transform of the function f on the real line evaluated at the point x; Eq. (1.2)
$H_e f$	Hilbert transform of the even function f on $\mathbb{R}^+$; Eq. (4.11)
$H_o f$	Hilbert transform of the odd function f on $\mathbb{R}^+$; Eq. (4.12)
$H_1 f$	one-sided Hilbert transform of the function f; Eq. (8.18)
$H_I f$	Hilbert's integral of the function f; Eq. (7.33)
$H_n f$	n-dimensional Hilbert transform of the function f; Eq. (15.26)
$\mathscr{H}_n f$	general n-dimensional Hilbert transform of the function f in E^n; Eq. (15.2)
$\mathscr{H}_{n,\varepsilon} f$	general n-dimensional truncated Hilbert transform of the function f in E^n; Eq. (15.7)
$H_{(k)} f$	Hilbert transform of the function $f(x_1, x_2, \ldots, x_k, \ldots, x_n)$ in the variable x_k; Eq. (15.36)
H^{-1}	inverse Hilbert transform operator; Eq. (4.26)
H^+	adjoint of the Hilbert transform operator; Eq. (4.194)
H_α	fractional Hilbert transform operator; Eqs. (18.209) and (18.216)
$\mathscr{H}$	Hamiltonian for an electronic system; §22.4, Eq. (22.54)
$\mathscr{H}_0$	unperturbed Hamiltonian for an electronic system; §22.4, Eq. (22.57)
$\mathscr{H}$	space of test functions; §10.14
$\mathfrak{H}$	inner product space; §2.10.1
$\mathfrak{H}$	Hilbert space; §2.10
$H_n f$	n-dimensional Hilbert transform of the function f, for $n \geq 2$; Eq. (15.26)
$H_1(f,g)(x)$	bilinear Hilbert transform; §16.5
$H_a(f,g)(x)$	bilinear singular integral operator; Eq. (16.85)
$H_n(x)$	Hermite polynomials; §9.3, Eq. (9.39), HT-01
$\mathrm{H}(x)$	Heaviside step function; Eqs. (10.54) and (18.116)
$\mathsf{H}(\omega)$	response function for a linear system; Eq. (13.1), §18.2
$\mathsf{H}_p(\omega)$	fractional Hilbert filter; §18.9, Eq. (18.142)
$H^p(D)$	Hardy space for the unit disc; §2.10.2
H^p	Hardy space for the upper half complex plane; §2.10.2
H_ν	response function at the frequency ν; Eq. (13.3)
$H_\varepsilon f$	truncated Hilbert transform; Eq. (3.3)
$H_E f$	truncated Hilbert transform; Eq. (4.507)

$H_{\mathrm{M}}f$	maximal Hilbert transform function; Eq. (7.280)
$\mathcal{H}_{\mathrm{M}}f$	maximal Hilbert transform function; Eq. (7.282)
$H_{\mathrm{S}}F$	Hilbert–Stieltjes transform of the function F; Eq. (4.551)
H_{K}	Kober's extension of the Hilbert transform operator; §16.3
H_{R_m}	Redheffer's extension of the Hilbert transform operator; §16.2
$\mathbf{H}_j$	vectorial Hilbert transform operator; Eq. (16.100)
$H_{\theta,\varepsilon}$	truncated directional Hilbert transform operator; Eq. (16.103)
H_θ	directional Hilbert transform operator; Eq. (16.104)
H_θ	helical Hilbert transform operator; Eq. (16.131)
H_{M_θ}	directional maximal Hilbert transform operator; Eq. (16.105)
H_{M_θ}	maximal helical Hilbert transform operator; Eq. (16.132)
$H_{\mathrm{M}_{n\theta}}$	double maximal helical Hilbert transform operator; Eq. (16.136)
$H_\Gamma f$	Hilbert transform of f along the curve Γ; Eq. (16.109)
$\bar{H}_\Gamma f$	modified Hilbert transform of f along the curve Γ; Eq. (16.113)
$\mathrm{H}_{\mathrm{A}}f$	Hartley transform of a function f; Eq. (5.58)
$\mathrm{H}_{\mathrm{A}}^{-1}$	inverse Hartley transform operator; Eq. (5.60)
$H^{(1)}_{\pm v}(z)$, $H^{(2)}_{\pm v}(z)$	Bessel functions of the third kind (Hankel functions of the first kind and second kind, respectively); §9.9
$h^{(1)}_n(z)$	spherical Bessel functions of the third kind; Eq. (9.131) (spherical Hankel functions of the first kind)
$h^{(2)}_n(z)$	spherical Bessel functions of the third kind; Eq. (9.132) (spherical Hankel functions of the second kind)
$h_n(x)$	Hermite–Gaussian functions; Eq. (18.179)
$\mathbf{h}_k$	discrete Hermite–Gaussian vector functions; Eqs. (18.254) and (18.255)
$\mathbf{H}_v(z)$	Struve's function; Eq. (9.77), HT-01
$H_{\mathrm{D}}\{f[n]\}$	discrete Hilbert transform of the sequence $\{f[n]\}$; Eq. (13.127)
$\{H_{\mathrm{sD}}f\}(x)$	semi-discrete Hilbert transform of the sequence $\{f[]\}$; Eq. (13.133)
$\mathcal{H}_{\mathrm{D}}\{f[n]\}$	alternative definition of the discrete Hilbert transform; Eq. (13.158)
$(\mathcal{H}_{\mathrm{D}_{pq}}x)[n]$	discrete fractional Hilbert transform; Eq. (18.269)

i	imaginary unit (engineers typically use j); §2.8
I	identity operator; §4.4
I	interval; §7.9
$\lvert I\rvert$	length of an interval; §7.9
$I_n(x)$	modified Bessel function of the first kind; HT-01
$i(t)$	input (time-dependent in general) to a system; §17.1–17.2
iff	if and only if
Im	imaginary part of a complex function
inf	infimum, the greatest lower bound of a set; §2.8

Ind f	index of a function; Eq. (11.179)
J	SI unit of energy, the joule; §19.1
$J_{\pm v}(z)$	Bessel function of the first kind; §9.6, HT-01
$\mathbf{J}_v(z)$	Anger's function; $9.12, HT-01
$j_n(z)$	spherical Bessel function of the first kind; Eq. (9.115)
k	wave number; Eq. (19.87)
$\mathbf{k}$	wave vector; §20.7
$k(x,y)$	Kernel function; §1.2, Eq. (1.3)
$K(x)$	Calderón–Zygmund kernel function; §15.1
$K_n(x)$	modified Bessel function of the third kind; HT-01
$l(I)$	length of an interval I; §2.11.1
$\mathscr{L}f$	Laplace transform of the function f; Eq. (5.91)
$\mathscr{L}_2 f$	bilateral (or two-sided) Laplace transform of the function f; Eq. (5.92)
L	class of functions that are Lebesgue integrable on a given interval; §2.11.1
$L(a,b)$	class of functions that are Lebesgue integrable on the interval (a,b); 2.11.1
L^1_{loc}	class of functions that are Lebesgue integrable on every subinterval of a given interval; Eq. (4.121)
L^2	class of functions that are Lebesgue square integrable on a given interval; §2.11.1
L^p	class of functions f such that $\|f\|^p$ is Lebesgue integrable on a given interval; §2.11.1
$L^p(\mathbb{R})$	class of functions f such that $\|f\|^p$ is Lebesgue integrable on the real line; §2.11.1
l^p	§13.11
$l^p(\mathbb{Z})$	§13.11
L^∞	class of essentially bounded functions; §2.11.1
$L^p_{2\pi}$	class of periodic functions f such that $\|f\|^p$is Lebesgue integrable on the interval (0, 2π). $L^p_{2\tau}$ has a similar meaning for periodic functions with period 2τ.
$L^{\alpha,p}$	class of functions f such that $\|x\|^\alpha \|f(x)\|^p$ is Lebesgue integrable on a particular interval; Eq. (7.186)
$L^p(\mu)$	class of μ-measurable functions; §7.12
$L_n(x)$	Laguerre polynomials; §9.4, Eq. (9.60)
$\mathbf{L}_v(z)$	modified Struve function; HT-01
$\text{Li}_n(z)$	polylogarithm function; HT-01
$\text{Li}_2(z)$	dilogarithm function; HT-01
Lip m	Lipschitz condition of order m; §2.3
log	logarithm to the base e; the alternative notation ln is also common usage
$\log^+ f$	maximum of $\{\log\|f\|, 0\}$; Eq. (7.74)

$\mathbf{M}$	magnetization; Eq. (20.111)
$\operatorname{mod} z$	modulus of a complex number; Eq. (2.68)
$m(E)$	measure of the set E; §2.11.1
Mf	Hardy–Littlewood maximal function; §7.9
$\mathcal{M}f$	Mellin transform of f; Eq. (5.102)
$\mathcal{M}^{-1}$	inverse Mellin transform operator; Eq. (5.107)
m	SI unit for length, the meter; §19.1
$m\{g(\lambda)\}$	distribution function of g; §4.25, §7.2, Eqs. (4.556), (7.55)
$m_{X,Y}(\omega)$	relative multiplier connecting $X(\omega)$ and $Y(\omega)$; Eq. (18.60)
$\mathbb{N}$	set of positive integers; $1, 2, 3, \ldots$
N	complex refractive index; Eq. (19.90)
N^{NL}	nonlinear complex refractive index; §22.13
$\mathcal{N}$	number of molecules per unit volume; §19.2
$n(\omega)$	angular frequency-dependent refractive index; Eq. (19.91)
$n^{\mathrm{NL}}(\omega, E)$	nonlinear refractive index; §22.13, Eq. (22.238)
$N_{\pm}(\omega)$	complex refractive indices for circularly polarized modes; §21.3, Eq. (21.47)
$n_{\pm}(\omega)$	real parts of $N_{\pm}(\omega)$; Eqs. (21.79) and (21.80)
$\mathcal{O}$	linear operator on a vector space; §2.10
$\mathcal{O}^{+}$	*adjoint* operator to $\mathcal{O}$; §2.10
$\mathcal{O}^{-1}$	inverse of an operator $\mathcal{O}$; §2.10
$O(\,)$	Bachmann order notation, of the order of; Eq. (2.1)
$o(\,)$	Landau order notation, of the order of; Eq. (2.6)
$\mathcal{O}'_C$	space of distributions that decrease rapidly at infinity; §10.2
$P\int$	Cauchy principal value; §2.4, Eq. (2.18)
$P(r, \theta)$	Poisson kernel for the disc; Eq. (3.49)
$P(x, y)$	Poisson kernel for the half plane; Eq. (3.31)
P_{ε}	Poisson operator; §7.10, Eq. (7.290)
P_{+}	projection operator; Eq. (4.352)
P_{-}	projection operator; Eq. (4.353)
$Pf(x^{-1})$	pseudofunction; §10.1
$\mathbf{P}(\mathbf{x})$	electric polarization of a medium; Eq. (19.1)
$\mathrm{P}_n(x)$	one of the orthogonal polynomials; §9.1
$P_n(x)$	Legendre polynomials; §9.2, Eqs. (9.10) and (9.27)
$P_{\nu}^{m}(x)$	associated Legendre function of the first kind; HT-01
$P_n^{(\alpha,\beta)}(x)$	Jacobi polynomials; §9.1
$\mathscr{S}_T$	space of periodic testing functions of period T; §10.2
$\mathscr{S}'_T$	space of periodic distributions of period T; §10.2
$p.v.\frac{1}{x}$	distribution; §10.1

$\mathcal{P}\frac{1}{x}$	distribution; §10.1
$Q(r,\theta)$	conjugate Poisson kernel for the disc; Eq. (3.50)
$Q(x,y)$	conjugate Poisson kernel for the half plane; Eq. (3.32)
Q_ε	conjugate Poisson operator; §7.10, Eq. (7.291)
$Q_n(x)$	Legendre function of the second kind; Eq. (11.263)
$Q_\nu^m(x)$	associated Legendre function of the second kind; HT-01
$Q_n^{(\alpha,\beta)}(x)$	Jacobi function of the second kind; HT-01
R	reflection operator; Eq. (4.73)
R	radius for a semicircular contour
A	Radon transform; §5.10, Eqs. (5.152) and (5.155)
$R_i(z_i)$	residue corresponding to the pole at $z = z_i$; §2.8.5
$R_j f$	Riesz transform of the function f; §15.12
$\mathbb{R}$	real line; the set of real numbers
$\mathbb{R}^+$	positive real axis interval; §3.4
$\mathbb{R} \times \mathbb{R}$	Euclidean plane
$\mathbb{R}^n$	n-dimensional Euclidean space; §2.15.2
$\mathcal{R}$	simply connected region; §2.8.1
$\mathcal{R}$	radius for a semicircular contour
$\mathfrak{R}_p$	Riesz constant; §4.20, Eqs. (4.382) and (4.384)
$\tilde{r}(\omega)$	generalized or complex reflectivity; Eq. (20.1)
$\tilde{r}_\pm(\omega)$	generalized reflectivity for circularly polarized modes; Eq. (21.132)
$r(\omega)$	reflectivity amplitude; Eq. (20.1)
$r(t)$	response (time-dependent) from a system; §17.1, §17.2
R_{n0}	rotational strength; Eq. (21.233)
$R(\omega)$	reflectivity; Eq. (20.2)
$re^{i\theta}$	polar form of the complex number z
Re	real part of a complex number
rect(x)	reactangular pulse function; §18.7.3
$\text{Res}\{g(z)\}_{z=z_0}$	residue at the pole $z = z_0$ of the function g; §2.8.5, Eq. (2.93)
$\mathcal{S}f$	Stieltjes transform of the function f; Eqs. (5.77), and (8.6)
S_a	dilation operator (homothetic operator); Eqs. (4.70) and (15.68)
sgn	signum function (sign function); Eqs. (1.14) and (18.120)
S^{n-1}	locus of points $x \in \mathbb{R}^n$ for which $\|x\| = 1$; §16.6
$S(z)$	Fresnel sine integral; Eq. (14.172), HT-01
$S(E)$	S-function (S-matrix); §17.12
$S(\omega)$	Fourier transform of a signal $s(t)$ in the frequency domain; §18.1, Eq. (18.2)
$s(t)$	signal in the time domain; §18.1
Shi(z)	hyperbolic sine integral function; Eq. (14.201), HT-01
Si(x)	sine integral; Eq. (8.79), HT-01

$\mathrm{si}(x)$	integral; Eq. (9.170), HT-01
$\mathrm{sie}\,(\alpha,\ \beta)$	sine-exponential integral; HT-01
$\mathrm{sinc}\,x$	sinc function; Eq. (4.260), HT-01
sup	supremum, the least upper bound
supp	support of the function; §2.15.2
T	finite Hilbert transform operator; chap. 11, Eq. (11.2)
T	used to denote a distribution; §2.15.2
T_{ab}	finite Hilbert transform operator on the interval (a, b); Eq. (12.98)
$T_n(x)$	Chebyshev polynomials of the first kind; §9.1, HT-01
$\mathbb{T}$	circle group; §3.10
Tr	trace; §22.4, Eq. (22.63)
$U_n(x)$	Chebyshev polynomials of the second kind; §9.1, HT-01
$u[n]$	unit step sequence; Eq. (13.91)
V	total variation of a function; Eq. (4.554)
V	SI unit for potential, the volt; §19.1
$w(x)$	weight function; §9.1
$W(x)$	weight function; §14.4
w_i	weight points in a quadrature scheme; Eq. (14.15)
$W^{p,m}$	Sobolev space; §10.2
$\bar{x}_j$	any value in the interval $[x_{j-1}, x_j]$; §2.11
x_i	sampling points in a quadrature scheme; Eq. (14.15)
$\lvert x\rvert$	norm of x in E^n; §15.1
x	vector cross product
$\times$	direct product; §10.6. Also used for Cartesian product of Euclidean spaces; §15.13
$\boldsymbol{x}(t)$	time-dependent particle displacement; §17.2, §17.9
$X(z)$	Z transform (one-sided or two-sided); Eqs. (13.38) and (13.39)
$Y_\upsilon(z)$	Bessel function of the second kind (Weber's function, Neumann's function); §9.6, 9.8, HT-01
$y_n(z)$	spherical Bessel function of the second kind; Eq. (9.116)
z	complex variable, $z = x + \mathrm{i}y$; Eq. (2.67)
$\bar{z}$	complex conjugate of z
z^*	complex conjugate of z
z_I	inverse point (or image point) of z; Eq. (3.35)
$\mathbb{Z}$	set of integers $0, \pm 1, \pm 2, \ldots$
$\mathbb{Z}^+$	set of non-negative integers $0, 1, 2, \ldots$
$Z\{x_n\}$	Z transform of the sequence $\{x_n\}$; §13.6, Eq. (13.38)

$\mathcal{Z}$	space of test functions whose Fourier transforms belong to $\mathcal{D}$; §10.2
$\mathcal{Z}'$	space of ultradistributions; §10.2
$\mathcal{Z}_1$	space of Fourier transforms of test functions belonging to $\mathscr{S}_1$; §10.14
$Z_s(\omega)$	surface impedance function; Eq. (20.129)

Greek Letters

α	polarizability § 19.1, Eq. (19.6)
$\alpha(\omega)$	absorption coefficient of a medium; Eq. (19. 92)
$\beta(2)$	Catalan's constant (0.915 965 594 177 219 015 1 ...); HT-01
γ	Euler's constant (0.577 215 664 9 ...)
Γ	contour in the complex plane (frequently used to signify a non-closed contour)
$\Gamma(z)$	gamma function; Eq. (4.118), HT-01
$\Gamma(a,z)$	incomplete gamma function; Eq. (8.38), HT-01
Γ_{mn}	damping constant for a transition between the levels *m* and *n*; §22.4, Eq. (22.71)
$\delta(x)$	Dirac delta distribution; §2.15, §10.3, Eq. (10.1)
$\delta[n]$	unit sample sequence; Eq. (13.92)
$\delta^+(x)$	Heisenberg delta function; Eq. (10.9)
$\delta^-(x)$	Heisenberg delta function; Eq. (10.10)
δ_{nm}	Kronecker delta; Eq. (2.38). When one of the subscript indices appears with a negative sign, the notation $\delta_{n,-m}$ is employed
Δ	difference operator; Eq. (2.305)
Δ	Laplacian operator; Eq. (15.161)
Δ_τ	Dirac comb distribution; Eq. (10.212)
$\Delta R(\omega)$	magnetoreflection; Eq. (21.125)
Δx	length of an interval; §2.11, Eq. (2.166)
ε	permittivity of the medium; Eq. (19.5)
$\varepsilon^{\mathrm{NL}}$	nonlinear dielectric permittivity; §22.13
ε_0	vacuum permittivity; §19.1
ϵ_{ijk}	Levi-Civita pseudotensor; Eq. (21.212)
$\boldsymbol{\varepsilon}(\mathbf{k}, \omega)$	spatial-dependent electric permittivity; Eq. (20.177)
$\varsigma(n)$	Riemann zeta function; Eq. (2.288)
$\theta(\omega)$	phase; Eq. (20.1)
$\theta(\omega)$	ellipticity per unit length; Eq. (21.195)
$\theta_{\mathrm{F}}(\omega)$	ellipticity function; Eq. (21.95)
$\kappa(\omega)$	measure of the absorption of a propagating wave in a medium; Eq. (19.92)
$\kappa_\pm(\omega)$	imaginary parts of $N_\pm(\omega)$; Eqs. (21.79) and (21.80)
$\kappa^{\mathrm{NL}}(\omega, E)$	imaginary part of the nonlinear complex refractive index; §22.13, Eq. (22.238)
$\Lambda(x)$	unit triangular function; Eq. (4.265)

Λ_α	space of Lipschitz continuous functions; §6.16, §15.1
μ	continuous Borel measure; §7.12
μ	magnetic permeability; §19.1
$\mu_o(A)$	Lebesgue outer measure of a set A; Eq. (2.183)
μ_0	permeability of the vacuum; §19.1
$\boldsymbol{\mu}$	electric dipole operator; Eq. (22.56)
$\Pi_{2a}(x)$	unit rectangular step function; Eqs. (9.19) and (18.122). For $a = 1/2$, the abbreviation $\Pi(x) \equiv \Pi_1(x)$ is employed
ρ	density operator; §22.4, Eq. (22.51)
$\rho(\mathbf{r})$	electronic density; §23.4, Eq. (23.75)
ρ_{mn}	matrix element of the density operator; §22.4
$\rho_s(t)$	auto-convolution function for a signal; Eq. (18.16)
$\rho_f(t)$	auto-correlation function; Eq. (18.23)
$\rho_{fg}(t)$	cross-correlation function; Eq. (18.21)
σ	type of an entire function; Eq. (2.110)
$\sigma(H)$	symbol of H; Eq. (5.37)
$\sigma(0)$	conductivity at $\omega = 0$; §19.8, Eq. (19.176)
$\sigma(\omega)$	complex conductivity; Eq. (20.84)
$\sigma_t(\omega)$	total scattering cross-section; Eq. (19.316)
Σ	unit sphere; §15.1
τ_a	translation operator; Eqs. (4.64) and Eq. (15.63)
φ_n	eigenfunctions of the unperturbed Hamiltonian; §22.4, Eq. 22.57
ϕ	test function in a particular space; §2.15.1, §10.1
$\phi(\omega)$	optical rotatory dispersion; Eq. (21.192)
$\phi(t)$	instantaneous phase; Eq. (18.78)
$\phi_F(\omega)$	magneto-rotatory dispersion function; Eq. (21.94)
$\Phi(z, s, v)$	Lerch function; HT-01
$\Phi(\omega)$	complex optical rotation function; Eq. (21.197)
χ	(linear) electric susceptibility; §19.1
$\boldsymbol{\chi}^{(n)}$	nth-order electric susceptibility tensor; §22.1
χ_m	magnetic susceptibility; Eq. (20.112)
$\chi_S(x)$	characteristic function associated with the set S; Eq. (2.191)
$\chi_{[x_1,x_2]}$	characteristic function with the interval where the function is non-zero specified by a subscript; §2.11.1
$\{\psi_n\}$	sequence of step functions; §2.14
$\psi(x)$	step function; Eq. (2.190)
$\psi(z)$	psi (or digamma) function; Eq. (4.222), HT-01
$\psi^{(n)}(z)$	Polygamma function; HT-01
$\Psi(\omega)$	ellipticity function, Eq. (21.193)
ω	angular frequency
$\omega(t)$	instantaneous frequency; Eq. (18.79)
ω_z	complex angular frequency; §17.7, Eq. (17.53)
ω_r	real part of a complex angular frequency; §17.7, Eq. (17.53)

ω_{i}	imaginary part of a complex angular frequency; §17.7, Eq. (17.53)
ω_{p}	plasma frequency of the medium; Eq. (19.13)
ω_{c}	cyclotron frequency; Eq. (21.56)
ω_{mn}	energy separation (in frequency units) between the levels m and n; §22.4
Ω	part of the Calderón–Zygmund kernel; §15.1

Miscellaneous notations

$\sum_{k=-\infty}^{\infty}{}'$	summation with a particular value of k excluded (usually $k = 0$)
$\dfrac{\partial}{\partial x}$	partial derivative operator (with respect to x)
∂^m	shorthand for the mth derivative (with respect to the variable under discussion), for $m \geq 1$.
$(a)_k$	Pochhammer symbol; Eq. (5.31). The notation a_k is also employed
$\forall$	for all
∇^2	Laplacian (del-squared) operator; Eq. (7.3)
∇	gradient (del) operator
$\Rightarrow$	implies;
$\sim$	same order as; §2.2
$\sim$	correspondence; Eq. (3.269)
$\sim$	twiddle sign, employed to indicate asymptotic equivalence between functions in a particular limit; §8.1
$f[\,]$	functional notation, e.g. $f[g(x)]$; §2.15.2, Eq. (2.283)
$\mathcal{O}^+$	adjoint of a linear operator $\mathcal{O}$; §2.10
$[\alpha, \beta]$	closed interval, that is $\alpha \leq x \leq \beta$; §2.3
$[\mathcal{O}_1, \mathcal{O}_2]$	commutator of two operators; §2.10, Eq. (2.146)
$\{\mathcal{O}_1, \mathcal{O}_2\}$	anticommutator of two operators; Eq. (4.67)
(α, β)	open interval, that is $\alpha < x < \beta$; §2.3
(f, g)	scalar product for two functions $f(x)$ and $g(x)$; Eq. (14.52)
$\langle f, g \rangle$	inner product for two functions f and g in Dirac bra–ket notation; §2.10
$\binom{m}{n}$	binomial coefficient; Eq. (2.309)
$*$	convolution operator, e.g. $f * g$; Eq. (2.53)
$*$	complex conjugate of a function (e.g. z^*)
$\star$	pentagram symbol for the cross-correlation operation; Eq. (18.21)
$\int_a^b f(x)\mathrm{d}x$	Riemann or Lebegue integral (depending on context); §2.11
$\int_C$	integral along the specified contour C; §2.8.1. $\int_\Gamma$ is sometimes used when the contour is not closed

$\int_{\mathbb{R}} f(x)\mathrm{d}x$	integral over the real line; §2.11.1
$\int_{\mathbb{R}^2} f(x,y)dx\,\mathrm{d}y$	integral over the xy-plane; Eq. (2.213)
$\int_{\mathbb{T}} f(\theta)\mathrm{d}\theta$	integral over a 2π period
$\oint_C$	integral along the closed contour C taken in a specified orientation; §2.8.1
$\int_E f(x)\mathrm{d}x$	Lebesgue integral of f on E; Eq. (2.198)
$\int\limits_{\lvert x-t\rvert>\varepsilon} f(x,t)\mathrm{d}t$	integral for which a segment $(x-\varepsilon, x+\varepsilon)$ is excluded; Eq. (3.3)
$\exists$	there exists
$\in$	belongs to
$\notin$	does not belong to
$\mathscr{S}$	space of test functions that have rapid decay; §10.2
$\mathscr{S}_1$	space of test functions that belong to $\mathscr{S}$ and vanish on the interval function $(-a,\ a)$ for some $a>0$; §10.14
$\mathscr{S}'$	space of all tempered distributions; §10.2
$\langle\ \rvert$	Dirac bra; §2.10
$\lvert\ \rangle$	Dirac ket; §2.10
$\lvert f\rvert$	absolute value of the function f
$\lvert x\rvert$	norm of x in E^n; §15.1
$\lvert a\rvert$	sum of the components of the multi-index a; §15.5
$\lvert\theta\in[-\pi,\pi] : \lvert g(\theta)\rvert \geq \lambda\rvert$	distribution function of g, §7.2, Eq. (7.55)
$\lVert\phi\rVert$	norm of a vector ϕ; §2.10, Eq. (2.136)
$\lVert f\rVert_p$	pth-power norm of f; Eq. (2.202)
$\lVert f\rVert_\infty$	essential supremum of $\lvert f\rvert$; Eq. (2.203)
$\lVert f(\theta)\rVert_{\alpha,p}$	weighted norm; Eq. (7.186)
$\lVert f(\theta)\rVert_{\alpha,\infty}$	weighted norm; Eq. (7.187)
$\lVert f\rVert_{p,\mu}$	norm $(\int \lvert f\rvert^p\,\mathrm{d}\mu)^{1/p} < \infty$, $1<p<\infty$; Eq. (7.376)
$\lVert f\rVert_{W^{p,m}}$	Sobolev norm; §10.2, Eq. (10.40)
$\lVert p\rVert$	longest subinterval; Eq. (2.167)
$\{x_i : a\leq x\leq b\}$	set of points $\{x_i\}$ such that $a\leq x\leq b$; §2.10
$\varnothing$	empty set; §2.10
$\subset$	subset of, as in $A\subset B$, A is a (proper) subset of B; §2.10
$\subseteq$	included in, as in $A\subseteq B$, A is included in B; §2.10
$\cap$	intersection of sets, as in $A\cap B$; §2.10
$\cup$	union of sets, as in $A\cup B$; §2.10
$\bigcup\limits_k A_k$	union of the collection of sets A_k
$\backslash$	relative complement of a set, that is, the relative complement of B with respect to A (the difference of A and B) is denoted by $A\backslash B$; §2.10
$(m)!!$	double factorial; Eqs. (4.119) and (4.120)
$\lfloor m\rfloor$	floor function, the greatest integer less than or equal to m; Eq. (9.28)

$[m/2]$	value $m/2$ if m is an even integer or $(m-1)/2$ if m is an odd integer
$[\arg f(z)]_C$	change in $\arg f(z)$ as the contour is traversed; Eq. (11.174)
$\{x_n\}$	sequence of numbers
$\otimes$	tensor product (direct product); §10.6, Eq. (10.88)

Abbreviations

a.e.	almost everywhere: §2.11.1
CD	circular dichroism; §21.1
DFT	discrete Fourier transform; §13.2
DFHT	discrete fractional Hilbert transform; §18.14
DFRFT	discrete fractional Fourier transform; §18.13
EMD	empirical mode decomposition; §18.16
FFT	fast Fourier transform; §14.9, §14.10
FHT	fractional Hilbert transform; §18.9
FRFT	fractional Fourier transform; §18.9
FTNMR	Fourier transform nuclear magnetic resonance (spectroscopy); §1.1
FTIR	Fourier transform infrared (spectroscopy); §1.1
IDFT	inverse DFT; §13.4
IMF	intrinsic mode function; §18.16
MCD	magnetic circular dichroism; §21.1
MOR	magnetic optical rotation; §21.3
MRS	magnetic rotation spectra; §21.3
ORD	optical rotatory dispersion; §21.1
SHG	second-harmonic generation; §22.2
THG	third-harmonic generation; §22.2

15

Hilbert transforms in E^n

15.1 Definition of the Hilbert transform in E^n

In this chapter the elementary properties of the n-dimensional Hilbert transform are discussed. Basic aspects of the Calderón–Zygmund theory of singular integral operators in the n-dimensional Euclidean space, E^n, are also considered.

Applications of Hilbert transforms in E^n for $n \geq 2$ are significantly less numerous than the one-dimensional case; however, they do arise in important areas. These include problems in nonlinear optics that focus on deriving dispersion relations and sum rules for the nonlinear susceptibility. The publications of Smet and Smet (1974), Nieto-Vesperinas (1980), Peiponen, (1987b, 1988), Bassani and Scandolo (1991, 1992), and Peiponen, Vartiainen, and Asakura (1999), will give the reader a sense of the advances in this field. There are applications in signal processing (see Bose and Prabhu (1979), Zhu, Peyrin, and Goutte (1990), and Reddy *et al.* (1991a, 1991b)), and in spectroscopy (see Peiponen, Vartiainen, and Tsuboi (1990)). In scattering theory, the double dispersion relations, frequently referred to as the Mandelstam representation, express the scattering amplitude as a double iterated dispersion relation in the energy and momentum transfer variables; see Roman (1965) and Nussenzveig (1972). Some of these applications are touched upon in later chapters.

To proceed, some preliminary definitions are required. Let x denote the n-tuple $\{x_1, x_2, x_3, \ldots, x_n\}$, and let s denote the n-tuple $\{s_1, s_2, s_3, \ldots, s_n\}$. It is quite common in the literature to represent multi-dimensional integration factors by ds (or some similar variable), where the context is meant to signify an n-dimensional integration factor $\mathrm{d}s_1 \mathrm{d}s_2 \mathrm{d}s_3 \cdots \mathrm{d}s_n$. The quantity $|x|$ is defined by

$$|x| = \left(\sum_{k=1}^{n} x_k^2 \right)^{1/2}. \tag{15.1}$$

The sum $x + y$ is the n-tuple $\{x_1 + y_1, x_2 + y_2, \ldots, x_n + y_n\}$, and for a real constant α the quantity αx is $\{\alpha x_1, \alpha x_2, \ldots, \alpha x_n\}$. The scalar product is given by $a \cdot x = a_1 x_1 + a_2 x_2 + \cdots + a_n x_n$. Common notation for the scalar product in the physical sciences is to write $\boldsymbol{a} \cdot \boldsymbol{x}$, but the use of a bold font for vector notation is postponed to later chapters discussing applications.

By analogy with what was done for the Hilbert transform on $\mathbb{R}$, the Hilbert transform in n-dimensional Euclidean space is defined by the following convolution:

$$\mathcal{H}_n f(x) = (f * K)(x) = \int_{E^n} K(x-y) f(y) \mathrm{d}y. \tag{15.2}$$

The notation T in place of $\mathcal{H}_n$ is very commonly employed. Starting in the early 1950s, Alberto Calderón and Antoni Zygmund made seminal contributions to the study of this equation, and as a result $\mathcal{H}_n$ is also referred to as the Calderón–Zygmund singular operator associated with the kernel K (see the chapter end-notes for references). The kernels are taken to be of the following form:

$$K(x) = \frac{\Omega(x')}{|x|^n}, \tag{15.3}$$

where the function Ω is defined on the unit sphere, which is denoted by Σ, and $x' = x/|x|$. Recall that a homogeneous function $h(x)$ of degree α satisfies

$$h(tx) = t^\alpha \, h(x), \quad \text{for } t > 0. \tag{15.4}$$

In the sequel it is assumed that Ω is homogeneous of degree zero, that is

$$\Omega(tx) = \Omega(x), \quad \text{for } t > 0. \tag{15.5}$$

Then Eq. (15.2) can be rewritten as follows:

$$\mathcal{H}_n f(x) = \lim_{\varepsilon \to 0} \int_{|y|>\varepsilon} \frac{\Omega(y') f(x-y) \mathrm{d}y}{|y|^n}. \tag{15.6}$$

Up to this point, the terminology Hilbert transform of f has been used to signify a one-dimensional integral on $\mathbb{R}$ or $\mathbb{T}$, so, to avoid any possibility of confusion, Eq. (15.6) will be referred to as the generalized Hilbert transform of f on E^n. The names Calderón–Zygmund transform or Calderón–Zygmund singular integral would both be better, reflecting the contributions of these outstanding mathematicians. The reader might feel that a better name is the designation *n-dimensional Hilbert transform*; however, this choice will be reserved for a particular specialization of Eq. (15.6). Equations (15.2) and (15.3), or Eq. (15.6), will be regarded as defining the so-called classical Calderón–Zygmund operators.

A truncated version of Eq. (15.2) can be defined as follows:

$$\mathcal{H}_{n,\varepsilon} f(x) = (f * K_\varepsilon)(x) = \int_{E^n} K_\varepsilon(x-y) f(y) \mathrm{d}y, \tag{15.7}$$

with the truncated kernel defined for $\varepsilon > 0$ by

$$K_\varepsilon(x) = \begin{cases} K(x), & \text{for } |x| \geq \varepsilon \\ 0, & \text{otherwise}\,. \end{cases} \tag{15.8}$$

The operator appearing in Eq. (15.7) is sometimes referred to as the truncated Calderón–Zygmund operator, and Eq. (15.8) defines a truncated Calderón–Zygmund kernel. If the limit $\varepsilon \to 0$ exists, then

$$\lim_{\varepsilon \to 0} f * K_\varepsilon = \mathcal{H}_n f. \tag{15.9}$$

Two conditions are imposed on Ω:

$$\int_\Sigma \Omega(x')\mathrm{d}x' = 0 \tag{15.10}$$

and

$$\Omega \in L^1(\Sigma). \tag{15.11}$$

The first condition is a statement that the mean value of Ω on Σ is zero. This condition is used to advantage in the sequel. The second condition is sometimes replaced by the Lipschitz condition:

$$\Omega \in \text{Lip}\,\alpha, \quad \text{for } \alpha > 0. \tag{15.12}$$

Consider first the case $n = 1$; then, the sphere Σ reduces to the two points $x' = \pm 1$, and it follows from Eq. (15.10) that $\Omega(1) + \Omega(-1) = 0$. Writing $\Omega(x') = c \operatorname{sgn} x$, with c a constant, which is assigned the value π^{-1}, yields

$$K(x) = \frac{\Omega(x')}{|x|} = \frac{\operatorname{sgn} x}{\pi |x|} = \frac{1}{\pi x}, \tag{15.13}$$

and hence Eq. (15.6) becomes, on making the identification $\mathcal{H}_1 \equiv H$,

$$Hf(x) = \lim_{\varepsilon \to 0} \frac{1}{\pi} \int_{|y| > \varepsilon} \frac{f(x-y)\mathrm{d}y}{y}. \tag{15.14}$$

This is the standard definition of the Hilbert transform on the line. Because of this reduction process, Eq. (15.6) is sometimes termed simply the Hilbert transform of f, and for this reason the notation $\mathcal{H}_n f$ has been employed. If an alternative choice is made for the constant in Eq. (15.13), the reduction process just carried out leads to a scalar multiple of the Hilbert transform.

Two results concerning the existence of $\mathcal{H}_{n,\varepsilon} f$ are now examined. The following approach is based on Neri (1971, p. 83). If $f \in L^p(E^n)$, for $1 \leq p \leq \infty$, and Ω

is bounded by a constant C, then the existence of $\mathcal{H}_{n,\varepsilon}f$ can be established in the following manner:

$$
\begin{aligned}
|\mathcal{H}_{n,\varepsilon}f(x)| &\leq \int_{|y|>\varepsilon} |f(x-y)|\frac{|\Omega(y')|}{|y|^n}\,\mathrm{d}y \\
&\leq C\int_{|y|>\varepsilon} \frac{|f(x-y)|}{|y|^n}\,\mathrm{d}y \\
&\leq C\|f\|_p \left(\int_{|y|>\varepsilon} \frac{1}{|y|^{nq}}\,\mathrm{d}y\right)^{q^{-1}}, \qquad (15.15)
\end{aligned}
$$

where Hölder's inequality has been employed, and p and q are conjugate exponents. Both of the final integrals are finite, and hence the existence of $\mathcal{H}_{n,\varepsilon}f$ is established. The reader is requested to consider the outcome for the situations $p = 1$ and $p = \infty$. An alternative approach with weaker conditions on Ω is as follows. Let $f \in L^p(E^n)$, for $1 < p < \infty$, $\Omega \in L^1(\Sigma)$, in norm notation $\|\Omega\|_1 = \int_\Sigma |\Omega(y')|\mathrm{d}y'$, and define the integral $I(x)$ by

$$
I(x) = \int_{E^n} |f(y)||K_\varepsilon(x-y)|\mathrm{d}y = \int_{|y|>\varepsilon} |f(x-y)||K(y)|\mathrm{d}y. \qquad (15.16)
$$

If it can be shown that $I(x)$ is locally integrable, then it follows that $\mathcal{H}_{n,\varepsilon}f$ exists *a.e.*, since

$$
|\mathcal{H}_{n,\varepsilon}f(x)| \leq I(x). \qquad (15.17)
$$

Let B denote any bounded set in E^n, represent $\int_B \mathrm{d}x$ by $|B|$, and let C be a constant depending on p and ε, which is not necessarily the same at each occurrence. Let the conjugate exponent of p be q, set $r = |y|$, and $y' = r^{-1}y$; then,

$$
\begin{aligned}
\int_B I(x)\mathrm{d}x &= \int_B \mathrm{d}x \int_{|y|>\varepsilon} |f(x-y)|\frac{|\Omega(y')|}{|y|^n}\,\mathrm{d}y \\
&= \int_B \mathrm{d}x \int_\Sigma |\Omega(y')|\mathrm{d}y' \int_\varepsilon^\infty \frac{|f(x-ry')|}{r}\,\mathrm{d}r \\
&= \int_\Sigma |\Omega(y')|\mathrm{d}y' \int_B \mathrm{d}x \int_\varepsilon^\infty \frac{|f(x-ry')|}{r}\,\mathrm{d}r \\
&\leq C\int_\Sigma |\Omega(y')|\mathrm{d}y' \int_B \mathrm{d}x \left(\int_\varepsilon^\infty |f(x-ry')|^p\,\mathrm{d}r\right)^{p^{-1}} \\
&\leq C\int_\Sigma |\Omega(y')|\mathrm{d}y' \left(\int_B \mathrm{d}x\right)^{q^{-1}} \left(\int_B \mathrm{d}x \left(\int_\varepsilon^\infty |f(x-ry')|^p\,\mathrm{d}r\right)\right)^{p^{-1}} \\
&\leq C\|\Omega\|_1 |B|^{q^{-1}} \|f\|_p, \qquad (15.18)
\end{aligned}
$$

where Hölder's inequality has been applied twice. This is the required result. The case $p = 1$ is left as an exercise for the reader.

The connection between $\mathcal{H}_{n,\varepsilon} f$ and $\mathcal{H}_n f$ is now examined. Let Λ_α denote the space of Lipschitz continuous functions, with α denoting the order. Suppose $f \in L^p \cap \Lambda_\alpha$, with $1 \le p < \infty$ and $0 < \alpha \le 1$, and let Ω satisfy the conditions in Eqs. (15.10) and (15.11). Then

$$\mathcal{H}_n f = \lim_{\varepsilon \to 0} \mathcal{H}_{n,\varepsilon} f, \; a.e. \tag{15.19}$$

Select a δ such that $0 < \varepsilon < \delta$; then,

$$\mathcal{H}_{n,\varepsilon} f(x) = \int_{\delta \ge |y| \ge \varepsilon} f(x-y) \frac{\Omega(y)}{|y|^n} \, \mathrm{d}y + \int_{|y| > \delta} f(x-y) \frac{\Omega(y)}{|y|^n} \, \mathrm{d}y. \tag{15.20}$$

The first integral can be written as follows:

$$\int_{\delta \ge |y| \ge \varepsilon} |f(x-y)| \frac{\Omega(y)}{|y|^n} \, \mathrm{d}y = \int_{\delta \ge |y| \ge \varepsilon} [f(x-y) - f(x)] \frac{\Omega(y)}{|y|^n} \, \mathrm{d}y, \tag{15.21}$$

which is obtained on making use of the result

$$\int_{\delta \ge |y| \ge \varepsilon} \frac{\Omega(y)}{|y|^n} \, \mathrm{d}y = \int_\varepsilon^\delta \frac{\mathrm{d}r}{r} \int_\Sigma \Omega(y') \mathrm{d}y' = 0, \tag{15.22}$$

and Eq. (15.10) has been used. Since $f \in \operatorname{Lip} \alpha$, for $0 < \alpha$, then

$$|f(x-y) - f(x)| \le C|y|^\alpha, \quad \text{for } |y| \le \delta, \tag{15.23}$$

and the integral on the right-hand side of Eq. (15.21) is convergent in the limit $\varepsilon \to 0$, since

$$\left| [f(x-y) - f(x)] \frac{\Omega(y)}{|y|^n} \right| \le C\, |\Omega(y)| |y|^{\alpha - n}, \quad \text{for } |y| \le \delta, \tag{15.24}$$

and

$$\int_{\delta \ge |y|} |\Omega(y)| |y|^{\alpha-n} \, \mathrm{d}y = \int_0^\delta r^{\alpha-1} \, \mathrm{d}r \int_\Sigma |\Omega(y')| \mathrm{d}y' < \infty, \tag{15.25}$$

and Eq. (15.11) has been employed. Equation (15.19) follows from Eqs. (15.20) and (15.25). The importance of the constraint that Ω has a mean value of zero on Σ is made clear by the approach just employed.

15.2 Definition of the *n*-dimensional Hilbert transform

If the particular assignment of the constant in Eq. (15.7) is ignored, then there is only one singular integral that arises from Eq. (15.6) for the case $n = 1$. Beyond this case, there are an infinite number of choices, depending on the particular selection for Ω. One specific choice of singular integral, for $n \geq 2$, is as follows:

$$H_n f(x_1, x_2, x_3, \ldots, x_n) = \frac{1}{\pi^n} P \int_{-\infty}^{\infty} \int_{-\infty}^{\infty} \int_{-\infty}^{\infty} \times \ldots \int_{-\infty}^{\infty} \frac{f(s_1, s_2, s_3, \ldots, s_n)\, ds_1\, ds_2\, ds_3 \ldots ds_n}{\prod_{k=1}^{n} (x_k - s_k)}. \tag{15.26}$$

The case $n = 1$ is obviously the ordinary one-dimensional Hilbert transform, and the notation H_1 is reserved for the one-sided Hilbert transform defined in Eq. (8.18). The convention adopted in Eq. (15.26) is that the Cauchy principal value applies to each integral, and the P symbol is inserted in front of each integral only when there is some risk of confusion. Using the notation introduced at the start of Section 15.1, the preceding result can be written in the more compact form:

$$H_n f(x) = \frac{1}{\pi^n} P \int_{-\infty}^{\infty} f(s) \prod_{k=1}^{n} \frac{1}{(x_k - s_k)}\, ds. \tag{15.27}$$

Equation (15.27) is frequently expressed as follows:

$$H_n f(x) = \frac{1}{\pi^n} \left\{ \prod_{j=1}^{n} \lim_{\varepsilon_j \to 0} \right\} \int_{|x_j - s_j| > \varepsilon_j} f(s) \prod_{k=1}^{n} \frac{1}{(x_k - s_k)}\, ds, \tag{15.28}$$

where the product notation for the limits implies separate limits $\varepsilon_1 \to 0, \varepsilon_2 \to 0$, and so on. As for the one-dimensional case, the opposite sign convention to that given in Eq. (15.27) is sometimes employed, that is $(s_k - x_k)$ is employed in place of $(x_k - s_k)$. Some authors do not employ the subscript notation indicating the dimensionality of the transform. Symbols other than H are also used to denote this transform, of which the most frequently employed choice is T.

Three examples of the n-dimensional Hilbert transform are now examined. Consider first the case

$$f \equiv \sin(a \cdot x) = \sin(a_1 x_1 + a_2 x_2 + a_3 x_3 + \cdots + a_n x_n), \tag{15.29}$$

where a is a constant vector. Then it follows that

$$H_n\{\sin(a \cdot x)\} = \frac{1}{\pi^n} P \int_{-\infty}^{\infty} \frac{ds_1}{x_1 - s_1} \int_{-\infty}^{\infty} \frac{ds_2}{x_2 - s_2} \times \cdots \int_{-\infty}^{\infty} \frac{\sin(a_1 s_1 + a_2 s_2 + \cdots + a_n s_n) ds_n}{x_n - s_n}$$

$$= \frac{1}{\pi^n} P \int_{-\infty}^{\infty} \frac{\mathrm{d}s_1}{x_1 - s_1} \int_{-\infty}^{\infty} \frac{\mathrm{d}s_2}{x_2 - s_2}$$

$$\times \cdots \int_{-\infty}^{\infty} \frac{\sin(a_1 s_1 + a_2 s_2 + \cdots + a_n x_n - a_n y)\mathrm{d}y}{y}$$

$$= \frac{1}{\pi^n} P \int_{-\infty}^{\infty} \frac{\mathrm{d}s_1}{x_1 - s_1} \int_{-\infty}^{\infty} \frac{\mathrm{d}s_2}{x_2 - s_2} \cdots$$

$$\times \ldots \{-\cos(a_1 s_1 + a_2 s_2 + \cdots + a_n x_n)\} \int_{-\infty}^{\infty} \frac{\sin(a_n y)\mathrm{d}y}{y}$$

$$= \frac{1}{\pi^n} P \int_{-\infty}^{\infty} \frac{\mathrm{d}s_1}{x_1 - s_1} \int_{-\infty}^{\infty} \frac{\mathrm{d}s_2}{x_2 - s_2}$$

$$\times \cdots \{-\cos(a_1 s_1 + a_2 s_2 + \cdots + a_n x_n)\} \pi \operatorname{sgn} a_n$$

$$= -\frac{\operatorname{sgn} a_n}{\pi^{n-1}} P \int_{-\infty}^{\infty} \frac{\mathrm{d}s_1}{x_1 - s_1} \int_{-\infty}^{\infty} \frac{\mathrm{d}s_2}{x_2 - s_2} \cdots$$

$$\times \ldots \int_{-\infty}^{\infty} \frac{\cos(a_1 s_1 + a_2 s_2 + \cdots + a_{n-1} s_{n-1} + a_n x_n)\mathrm{d}s_{n-1}}{x_{n-1} - s_{n-1}}, \tag{15.30}$$

which simplifies, on continued integration, to yield

$$H_n \sin(a \cdot x) = \begin{cases} (-1)^{(n+1)/2} \cos(a \cdot x) \prod_{k=1}^{n} \operatorname{sgn} a_k, & \text{for } n \text{ odd} \\ (-1)^{n/2} \sin(a \cdot x) \prod_{k=1}^{n} \operatorname{sgn} a_k, & \text{for } n \text{ even.} \end{cases} \tag{15.31}$$

In a similar fashion,

$$H_n \cos(a \cdot x) = \begin{cases} (-1)^{(n-1)/2} \sin(a \cdot x) \prod_{k=1}^{n} \operatorname{sgn} a_k, & \text{for } n \text{ odd} \\ (-1)^{n/2} \cos(a \cdot x) \prod_{k=1}^{n} \operatorname{sgn} a_k, & \text{for } n \text{ even.} \end{cases} \tag{15.32}$$

As a third example, consider

$$f = \mathrm{e}^{-ax^2}, \tag{15.33}$$

where $a > 0$. Then $H_n\{e^{-ax^2}\}$ is evaluated as follows:

$$\begin{aligned}
H_n\{e^{-ax^2}\} &= \frac{1}{\pi^n} P \int_{-\infty}^{\infty} \frac{ds_1}{x_1 - s_1} \int_{-\infty}^{\infty} \frac{ds_2}{x_2 - s_2} \cdots \int_{-\infty}^{\infty} \frac{e^{-a(s_1^2+s_2^2+\cdots+s_n^2)}\, ds_n}{x_n - s_n} \\
&= \frac{1}{\pi^n} P \int_{-\infty}^{\infty} \frac{e^{-as_1^2}\, ds_1}{x_1 - s_1} \int_{-\infty}^{\infty} \frac{e^{-as_2^2}\, ds_2}{x_2 - s_2} \cdots \int_{-\infty}^{\infty} \frac{e^{-as_n^2}\, ds_n}{x_n - s_n} \\
&= \{-ie^{-ax_1^2} \operatorname{erf}(i\sqrt{(a)}\, x_1)\}\{-ie^{-ax_2^2} \operatorname{erf}(i\sqrt{(a)}\, x_2)\} \\
&\quad \times \cdots \{-ie^{-ax_n^2} \operatorname{erf}(i\sqrt{(a)}\, x_n)\} \\
&= (-i)^n e^{-ax^2} \prod_{k=1}^{n} \operatorname{erf}(i\sqrt{(a)}\, x_k), \qquad (15.34)
\end{aligned}$$

where Eq. (5.28) has been employed.

The n-dimensional Hilbert transform operator can be written in terms of a product of one-dimensional Hilbert transform operators. The variable on which the one-dimensional operator acts is specified by a subscript thus: $H_{(k)}$. So H_n can be written as

$$H_n = \prod_{k=1}^{n} H_{(k)}, \qquad (15.35)$$

where

$$\begin{aligned}
&H_{(k)} f(s_1, s_2, \ldots, s_{k-1}, x_k, s_{k+1}, \ldots, s_n) \\
&\quad = \frac{1}{\pi} P \int_{-\infty}^{\infty} \frac{f(s_1, s_2, \ldots, s_{k-1}, s_k, s_{k+1}, \ldots, s_n) ds_k}{x_k - s_k}. \qquad (15.36)
\end{aligned}$$

The operators $H_{(k)}$ satisfy the commutator condition

$$[H_{(k)}, H_{(j)}] = 0, \quad \text{for } j, k = 1, 2, \ldots, n. \qquad (15.37)$$

15.3 The double Hilbert transform

The double Hilbert transform can be used to indicate some of the issues that arise beyond the one-dimensional transform and can serve as a stepping point to higher dimensions, since the double Hilbert transform can be generalized in a rather straight-forward manner. This generalization is performed in subsequent sections for a number of topics. The double Hilbert transform is defined by the following equation:

$$\begin{aligned} H_2 f(x_1, x_2) &= \frac{1}{\pi^2} P \int_{-\infty}^{\infty} P \int_{-\infty}^{\infty} \frac{f(s_1, s_2) \mathrm{d}s_1 \, \mathrm{d}s_2}{(x_1 - s_1)(x_2 - s_2)} \\ &= \lim_{\varepsilon_1 \to 0} \lim_{\varepsilon_2 \to 0} \frac{1}{\pi^2} \int_{|x_1 - s_1| > \varepsilon_1} \int_{|x_2 - s_2| > \varepsilon_2} \frac{f(s_1, s_2) \mathrm{d}s_1 \, \mathrm{d}s_2}{(x_1 - s_1)(x_2 - s_2)} \\ &= \lim_{\varepsilon_1 \to 0} \lim_{\varepsilon_2 \to 0} \int_{|x_1 - s_1| > \varepsilon_1} \int_{|x_2 - s_2| > \varepsilon_2} K(x_1 - s_1) \\ &\quad \times K(x_2 - s_2) f(s_1, s_2) \mathrm{d}s_1 \, \mathrm{d}s_2 \end{aligned} \tag{15.38}$$

where the kernel function $K(x)$ given in Eq. (15.13) is employed. With the change of variables $t_j = x_j - s_j$, for $j = 1$ or 2, the integration region is the exterior of the cross shown in Figure 15.1.

The double Hilbert transform as just defined is unique. The two-dimensional version of Eq. (15.2), namely

$$\mathcal{H}_2 f(x) = (f * K)(x) = \int_{E^2} K(x - y) f(y) \mathrm{d}y, \tag{15.39}$$

leads to an infinite number of possibilities, depending on how the kernel function K is selected. As an example, the double Hilbert transform is constrasted with a different singular integral in E^2 (discussed by Calderón and Zygmund (1952)). This case arises in potential theory. Let $f \in L$, with E^2 denoting the plane (u, v). Let (x, y, z) designate a point in the half space $z > 0$, and let R signify the distance of (x, y, z) from (u, v), that is $R^2 = (x - u)^2 + (y - v)^2 + z^2$. The potential $U(x, y, z)$ is defined by

$$U(x, y, z) = \int_{E^2} \frac{f(u, v) \mathrm{d}u \, \mathrm{d}v}{R}. \tag{15.40}$$

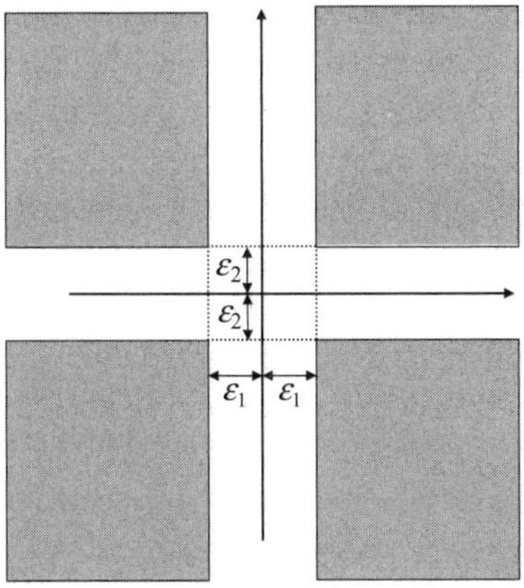

Figure 15.1. Integration domain for the double Hilbert transform.

The partial derivative of the potential, which is denoted by the appropriate subscript, is given by

$$U_x(x,y,z) = -\int_{E^2} \frac{f(u,v)(x-u)\mathrm{d}u\,\mathrm{d}v}{R^3}. \tag{15.41}$$

Set $z = 0$, then

$$U_x(x,y,0) = -\int_{E^2} f(u,v)K(x-u,y-v)\mathrm{d}u\,\mathrm{d}v, \tag{15.42}$$

where the singular kernel $K(x,y)$ is given by

$$K(x,y) = -\frac{x}{\surd[(x^2+y^2)^3]}. \tag{15.43}$$

In general, the integral in Eq. (15.42) does not converge, but it will exist *a.e.* as a principal value integral if f is a sufficiently smooth function, hence the stated requirement given for f. The principal value is interpreted here by carrying out the integral in Eq. (15.42) over the exterior of a circle of radius ε and center (x_0, y_0) and then letting $\varepsilon \to 0$.

It should be apparent to the reader from Section 15.2, and will become increasing evident in the following sections, that the factorization property of H_n allows many results about this operator to be derived relatively effortlessly from the corresponding properties for the case $n = 1$. However, there are problems where this is not the case. For example, let H_{M_n} denote the maximal Hilbert transform for the n-dimensional case. Recall Eq. (7.280) for the case $n = 1$; then, for $n = 2$,

$$H_{\mathrm{M}_2}f(x,y) = \sup_{\substack{\varepsilon_1>0,\\ \varepsilon_2>0}} \left| \frac{1}{\pi^2} \int_{|s|>\varepsilon_1} \int_{|t|>\varepsilon_2} \frac{f(x-s,y-t)}{st}\,\mathrm{d}s\,\mathrm{d}t \right|. \tag{15.44}$$

Let $f \in L\log^+ L(\mathbb{R}^2)$ and take $\operatorname{supp} f \leq [0,1] \times [0,1]$, then $(H_{\mathrm{M}_2}f)$ satisfies

$$|\{(x,y) \in [0,1]\times[0,1] : |H_{\mathrm{M}_2}f(x,y)| > \alpha\}| \leq \frac{C}{\alpha}\left\{ \left\| f \log^+ f \right\|_1 + C \right\}, \tag{15.45}$$

where the constants C and α are independent of f. Here, there is no simple factorization, and the proof of this result is relatively complicated. The interested reader can pursue this in Fefferman (1972).

15.4 Inversion property for the n-dimensional Hilbert transform

The results of Section 15.2 can be used to establish the inversion property for the n-dimensional Hilbert transform. Suppose

$$g(x) = H_n f(x), \tag{15.46}$$

then

$$H_n g(x) = (-1)^n f(x), \tag{15.47}$$

and hence

$$H_n^2 f(x) \equiv H_n(H_n f(x)) = (-1)^n f(x), \tag{15.48}$$

which represents a generalization of the corresponding one-dimensional inversion formula. This result can be written in operator form as

$$H_n^2 = (-1)^n I, \tag{15.49}$$

where I denotes the identity operator on $\mathbb{R}^n$. This result is derived in Section 15.6 after the action of the n-dimensional Fourier transform on H_n is discussed. The generalization of Eq. (15.49) is given by

$$H_n^m f(x) = \begin{cases} (-1)^{mn/2} f(x), & \text{for } m \geq 2 \text{ and even} \\ (-1)^{(m-1)n/2} g(x), & \text{for } m \geq 3 \text{ and odd.} \end{cases} \tag{15.50}$$

From Eq. (15.49), the inverse of the n-dimensional Hilbert transform operator is given by

$$H_n^{-1} = (-1)^n H_n, \tag{15.51}$$

and so

$$H_n^{-1} H_n f(x) = \frac{(-1)^n}{\pi^n} P \int_{-\infty}^{\infty} H_n f(s) \prod_{k=1}^{n} \frac{1}{(x_k - s_k)} \, \mathrm{d}s = f(x). \tag{15.52}$$

15.5 Derivative of the *n*-dimensional Hilbert transform

Let the multi-index $a = (a_1, a_2, \ldots, a_n)$, with $a_k \in \mathbb{Z}^+$, and denote $|a| = a_1 + a_2 + \cdots + a_n$. Vertical bars are being used, as is the common practice, with two different meanings: the first is for multi-indices as just defined, the second is for points in Euclidean space, as given in Section 15.1. The context will make the intended meaning obvious. For $f \in \mathbb{R}^n$, the partial derivative operator is defined by

$$D^a f \equiv \frac{\partial^{|a|} f}{\partial x_1^{a_1} \partial x_2^{a_2} \cdots \partial x_n^{a_n}}. \tag{15.53}$$

The effect of D^a on $H_n f$ is given by

$$D^a H_n f = D^a \prod_{k=1}^{n} H_{(k)} f(x_1, x_2, \ldots, x_n)$$
$$= \frac{\partial^{|a|}}{\partial x_1^{a_1} \partial x_2^{a_2} \ldots \partial x_n^{a_n}} \prod_{k=1}^{n} H_{(k)} f(x_1, x_2, \ldots, x_n), \qquad (15.54)$$

which simplifies, on using (see Eq. (4.138))

$$\frac{\partial^{a_k} H_{(k)} f(x_1, \ldots, x_k, \ldots, x_n)}{\partial x_k^{a_k}} = H_{(k)} \frac{\partial^{a_k} f(x_1, \ldots, x_k, \ldots, x_n)}{\partial x_k^{a_k}}, \qquad (15.55)$$

to yield

$$D^a H_n f = \prod_{k=1}^{n} H_{(k)} \frac{\partial^{|a|}}{\partial x_1^{a_1} \partial x_2^{a_2} \ldots \partial x_n^{a_n}} f(x_1, x_2, \ldots, x_n). \qquad (15.56)$$

Hence,

$$D^a H_n f = H_n D^a f. \qquad (15.57)$$

This is the n-dimensional analog of Eq. (4.138).

15.6 Fourier transform of the n-dimensional Hilbert transform

The n-dimensional generalization of the Fourier transform of the Hilbert transform is considered in this section. The Fourier transform of the Hilbert transform has some important practical applications, of which the principal one is providing a route to an effective evaluation strategy for the Hilbert transform. This was discussed in Section 5.2 for the one-dimensional case.

The n-dimensional Fourier transform, denoted by $\mathcal{F}_n f(x)$, is defined in the following manner:

$$\mathcal{F}_n f(x) = \int_{-\infty}^{\infty} \int_{-\infty}^{\infty} \int_{-\infty}^{\infty} \cdots \int_{-\infty}^{\infty} f(s) e^{-2\pi i x \cdot s} \, ds. \qquad (15.58)$$

Just as for the one-dimensional case, the 2π factor can be dealt with in more than one way. For the n-dimensional Fourier transform a common conventional treatment of this factor is as given in the preceding equation. When comparing formulas from different sources the reader will need to be cognizant of how this factor has been treated. Consider the action of $\mathcal{F}_n$ on the n-dimensional Hilbert transform. The expression

$\mathcal{F}_n H_n f(x)$ is calculated as follows:

$$\begin{aligned}
\mathcal{F}_n H_n f(x) &= \frac{1}{\pi^n} \mathcal{F}_n \left\{ P \int_{-\infty}^{\infty} f(t) \prod_{k=1}^{n} \frac{1}{s_k - t_k} \, \mathrm{d}t \right\} (x) \\
&= \frac{1}{\pi^n} \int_{-\infty}^{\infty} \mathrm{e}^{-2\pi \mathrm{i} x \cdot s} \mathrm{d}s P \int_{-\infty}^{\infty} f(t) \prod_{k=1}^{n} \frac{1}{s_k - t_k} \, \mathrm{d}t \\
&= \frac{1}{\pi^n} \int_{-\infty}^{\infty} f(t) \mathrm{d}t \prod_{k=1}^{n} P \int_{-\infty}^{\infty} \frac{e^{-2\pi \mathrm{i} x \cdot s} \, \mathrm{d}s}{s_k - t_k} \\
&= \frac{1}{\pi^n} \int_{-\infty}^{\infty} f(t) \mathrm{d}t \prod_{k=1}^{n} \mathrm{e}^{-2\pi \mathrm{i} x_k t_k} \int_{-\infty}^{\infty} \frac{\mathrm{e}^{-2\pi \mathrm{i} x_k y_k} \mathrm{d}y_k}{y_k} \\
&= \frac{1}{\pi^n} \int_{-\infty}^{\infty} f(t) (-\mathrm{i}\pi)^n \prod_{k=1}^{n} \mathrm{e}^{-2\pi \mathrm{i} x_k t_k} \operatorname{sgn} x_k \, \mathrm{d}t \\
&= \left\{ (-\mathrm{i})^n \prod_{k=1}^{n} \operatorname{sgn} x_k \right\} \int_{-\infty}^{\infty} f(t) \mathrm{e}^{-2\pi \mathrm{i} x \cdot t} \, \mathrm{d}t \\
&= \left\{ (-\mathrm{i})^n \prod_{k=1}^{n} \operatorname{sgn} x_k \right\} \mathcal{F}_n f(x). \qquad (15.59)
\end{aligned}$$

From this result it follows that

$$H_n f(x) = (-\mathrm{i})^n \mathcal{F}_n^{-1} \left\{ \prod_{k=1}^{n} \operatorname{sgn} y_k \, \mathcal{F}_n f \right\} (x). \qquad (15.60)$$

This formula can be used to evaluate n-dimensional Hilbert transforms. As an example, consider the determination of $H_n f(x)$, where $f(x) = \mathrm{e}^{\mathrm{i} a \cdot x}$. From Eq. (15.60) it follows that

$$\begin{aligned}
H_n\{\mathrm{e}^{\mathrm{i}a\cdot x}\} &= (-\mathrm{i})^n \mathcal{F}_n^{-1} \left\{ \left\{ \prod_{k=1}^{n} \operatorname{sgn} y_k \right\} \int_{-\infty}^{\infty} \mathrm{e}^{-\mathrm{i} y_1 t_1} \mathrm{e}^{\mathrm{i} a_1 t_1} \, \mathrm{d}t_1 \int_{-\infty}^{\infty} \mathrm{e}^{-\mathrm{i} y_2 t_2} \mathrm{e}^{-\mathrm{i} a_2 t_2} \, \mathrm{d}t_2 \right. \\
&\quad \left. \times \cdots \int_{-\infty}^{\infty} \mathrm{e}^{-\mathrm{i} y_n t_n} \mathrm{e}^{\mathrm{i} a_n t_n} \, \mathrm{d}t_n \right\} (x) \\
&= (-\mathrm{i})^n \mathcal{F}_n^{-1} \left\{ \left(\prod_{k=1}^{n} \operatorname{sgn} y_k \right) (2\pi)^n \left(\prod_{j=1}^{n} \delta(y_j - a_j) \right) \right\} (x)
\end{aligned}$$

$$
\begin{aligned}
&= (-\mathrm{i})^n \frac{1}{(2\pi)^n} \int_{-\infty}^{\infty} \mathrm{e}^{\mathrm{i}x_1y_1} \operatorname{sgn} y_1\{2\pi\delta(y_1 - a_1)\}\mathrm{d}y_1 \\
&\quad \times \int_{-\infty}^{\infty} \mathrm{e}^{\mathrm{i}x_2y_2} \operatorname{sgn} y_2\{2\pi\delta(y_2 - a_2)\}\mathrm{d}y_2 \\
&\quad \times \cdots \int_{-\infty}^{\infty} \mathrm{e}^{\mathrm{i}x_ny_n} \operatorname{sgn} y_n\{2\pi\delta(y_n - a_n)\}\mathrm{d}y_n \\
&= (-\mathrm{i})^n \prod_{k=1}^{n} \operatorname{sgn} a_k \mathrm{e}^{\mathrm{i}a_kx_k} \\
&= (-\mathrm{i})^n \mathrm{e}^{\mathrm{i}a\cdot x} \prod_{k=1}^{n} \operatorname{sgn} a_k.
\end{aligned} \tag{15.61}
$$

Equation (15.60) can be used to establish the inversion property. The notational shorthand $\operatorname{sgn} x \equiv \prod_{k=1}^{n} \operatorname{sgn} x_k$ is introduced, and let $H_n f(x) = g(x)$; then it follows that

$$
\begin{aligned}
H_n^2 f(x) &= H_n g(x) \\
&= (-\mathrm{i})^n \mathcal{F}_n^{-1}\{\operatorname{sgn} t\, \mathcal{F}_n g(t)\}(x) \\
&= (-\mathrm{i})^n \mathcal{F}_n^{-1}\{\operatorname{sgn} t\, \mathcal{F}_n H_n f(t)\}(x) \\
&= (-\mathrm{i})^n \mathcal{F}_n^{-1}\{\operatorname{sgn} t\, \mathcal{F}_n[(-\mathrm{i})^n \mathcal{F}_n^{-1}\{\operatorname{sgn} s\, \mathcal{F}_n f(s)\}](t)\}(x) \\
&= (-1)^n \mathcal{F}_n^{-1}\{\operatorname{sgn} t \operatorname{sgn} t\, \mathcal{F}_n f(t)\}(x) \\
&= (-1)^n \mathcal{F}_n^{-1} \mathcal{F}_n f(x) \\
&= (-1)^n f(x),
\end{aligned} \tag{15.62}
$$

which is the required result.

15.7 Relationship between the n-dimensional Hilbert transform and translation and dilation operators

The translation operator τ_a is defined in the following way. Let f be a function on $\mathbb{R}^n$ and let $a = (a_1, a_2, \ldots, a_n)$, with $a \in \mathbb{R}^n$, then

$$
\tau_a f(x) = f(x - a) = f(x_1 - a_1, x_2 - a_2, \ldots, x_n - a_n). \tag{15.63}
$$

The first issue is to determine whether the operator H_n is translation-invariant; that is, whether H_n commutes with τ_a. Using Eq. (15.35) and

$$
\tau_a = \prod_{k=1}^{n} \tau_{a_k}, \tag{15.64}
$$

where

$$\tau_{a_k} f(x) = f(x_1, x_2, \cdots, x_k - a_k, \cdots, x_n), \tag{15.65}$$

it follows, on using $\tau_{a_k} H_{(k)} f(x) = H_{(k)} \tau_{a_k} f(x)$ (see Eq. (4.65)), that

$$\prod_{k=1}^{n} \tau_{a_k} \prod_{j=1}^{n} H_{(j)} f(x) = \prod_{k=1}^{n} \prod_{j=1}^{n} H_{(j)} \tau_{a_k} f(x), \tag{15.66}$$

and so

$$[\tau_a, H_n] = 0, \tag{15.67}$$

which establishes that H_n is a translation-invariant operator.

The dilation operator is defined by

$$S_a f(x) = f(ax) = f(a_1 x_1, a_2 x_2, \ldots, a_n x_n), \quad \text{for } a_k > 0, \quad k = 1, 2, \ldots, n. \tag{15.68}$$

In an obvious manner, S_a can be written as

$$S_a = \prod_{k=1}^{n} S_{a_k}, \tag{15.69}$$

where

$$S_{a_k} f(x) = f(x_1, x_2, \ldots, a_k x_k, \ldots, x_n). \tag{15.70}$$

When all the a_k are equal, the dilation is isotropic, otherwise it is non-isotropic. Making use of the result

$$S_{a_k} H_{(k)} f(x) = H_{(k)} S_{a_k} f(x), \tag{15.71}$$

which is Eq. (4.71), yields

$$\prod_{k=1}^{n} S_{a_k} \prod_{j=1}^{n} H_{(j)} f(x) = \prod_{k=1}^{n} \prod_{j=1}^{n} H_{(j)} S_{a_k} f(x), \tag{15.72}$$

and hence

$$[S_a, H_n] = 0. \tag{15.73}$$

Let T be a bounded linear operator with $T \in L^p(\mathbb{R}^n)$ and suppose T commutes with translations and positive dilations. The discussion of Section 4.6 can be extended, and the analog of Eq. (4.87) can be written as follows:

$$\mathcal{F}_n T f(x) = m(x)\, \mathcal{F}_n f(x). \tag{15.74}$$

The n-dimensional version of Eq. (4.106) takes the following form:

$$m(x_1, x_2, \ldots, x_n) = \prod_{k=1}^{n} (a_k + b_k \ \mathrm{sgn}\, x_k), \tag{15.75}$$

where a_k and b_k are constants. The preceding result can be rewritten as

$$m(x_1, x_2, \ldots, x_n) = \alpha + \sum_{k=1}^{n} \beta'_k \ \mathrm{sgn}\ x_k + \sum_{j=1}^{n} \sum_{k>j}^{n} \gamma'_{jk} \ \mathrm{sgn}\ x_j \ \mathrm{sgn}\ x_k$$
$$+ \cdots + \omega' \ \mathrm{sgn}\, x, \tag{15.76}$$

where $\alpha, \beta'_k, \gamma'_{jk}$, and ω' are constants, and $\mathrm{sgn}\, x = \prod_{k=1}^{n} \mathrm{sgn}\ x_k$. Combining this result with Eq. (15.74), taking the inverse Fourier transform, using Eq. (15.59), and introducing the constants β_k, γ_{jk}, and ω, which are closely related to β'_k, γ'_{jk}, and ω', respectively, leads to

$$T = \alpha I + \sum_{k=1}^{n} \beta_k H_{(k)} + \sum_{j=1}^{n} \sum_{k>j}^{n} \gamma_{jk} H_{(j)} H_{(k)} + \cdots + \omega H_n. \tag{15.77}$$

Hence, the only bounded linear operator on $L^p(\mathbb{R}^n)$ that commutes with translations and positive dilations can be expressed as a constant multiple of the identity operator plus sums of products of the Hilbert transform operators $H_{(k)}$.

15.8 The Parseval-type formula

Let $f \in L^p(\mathbb{R}^n)$ and $g \in L^q(\mathbb{R}^n)$, with $p > 1$ and $p^{-1} + q^{-1} = 1$. The n-dimensional Hilbert transform satisfies the Parseval-type formula,

$$\int_{\mathbb{R}^n} H_n f(x)\, g(x) \mathrm{d}x = (-1)^n \int_{\mathbb{R}^n} f(x) H_n g(x) \mathrm{d}x. \tag{15.78}$$

To establish this result, proceed as follows:

$$\int_{-\infty}^{\infty} H_n f(x)\, g(x) \mathrm{d}x = \int_{-\infty}^{\infty} \prod_{k=1}^{n} H_{(k)} f(x)\, g(x) \mathrm{d}x$$
$$= \int_{-\infty}^{\infty} \mathrm{d}x_1 \int_{-\infty}^{\infty} \mathrm{d}x_2 \cdots \int_{-\infty}^{\infty} H_{(1)} H_{(2)} \cdots H_{(n)} f(x)\, g(x) \mathrm{d}x_n$$
$$= -\int_{-\infty}^{\infty} \mathrm{d}x_1 \int_{-\infty}^{\infty} \mathrm{d}x_2 \cdots \int_{-\infty}^{\infty} H_{(1)} H_{(2)} \cdots H_{(n-1)} f(x) H_{(n)} g(x) \mathrm{d}x_n$$

$$= (-1)^n \int_{-\infty}^{\infty} dx_1 \int_{-\infty}^{\infty} dx_2 \cdots \int_{-\infty}^{\infty} f(x) H_{(1)} H_{(2)} \cdots H_{(n)} g(x) dx_n$$

$$= (-1)^n \int_{-\infty}^{\infty} f(x) H_n g(x) dx, \tag{15.79}$$

and Eq. (4.176) has been employed.

Let $g(x) = H_n f(x)$, then, from Eqs. (15.79) and (15.62), it follows that

$$\int_{\mathbb{R}^n} \{H_n f(x)\}^2 \, dx = \int_{\mathbb{R}^n} \{f(x)\}^2 \, dx. \tag{15.80}$$

This is the n-dimensional analog of Eq. (4.172).

Let $f_1 = f$ and $g = H_n f_2$ in Eq. (15.78), then

$$\int_{\mathbb{R}^n} H_n f_1(x) \, H_n f_2(x) dx = \int_{\mathbb{R}^n} f_1(x) f_2(x) dx. \tag{15.81}$$

This is a generalization of Eq. (15.80) and represents the n-dimensional analog of Eq. (4.174).

Let $f \in L^2(\mathbb{R}^n)$ and set $g = f$ in Eq. (15.78), then

$$\int_{\mathbb{R}^n} H_n f(x) f(x) \, dx = (-1)^n \int_{\mathbb{R}^n} f(x) H_n f(x) \, dx; \tag{15.82}$$

that is,

$$\{1 - (-1)^n\} \int_{\mathbb{R}^n} f(x) \, H_n f(x) dx = 0. \tag{15.83}$$

If n is odd, then the orthogonality property follows:

$$\int_{\mathbb{R}^n} f(x) H_n f(x) dx = 0. \tag{15.84}$$

This is the n-dimensional analog of Eq. (4.198). For the case n is even, suppose f is odd in x_k, then $H_{(k)} f$ is even in x_k; if f is even in x_k, then $H_{(k)} f$ is odd in x_k. Hence, the product $f H_n f$ is odd, and the orthogonality property given in Eq. (15.84) follows.

15.9 Eigenvalues and eigenfunctions of the n-dimensional Hilbert transform

Let $f \in L^p(\mathbb{R}^n)$, for $1 < p < \infty$, and suppose

$$H_n f = \eta f. \tag{15.85}$$

Applying H_n to this equation and employing the inversion formula for the n-dimensional Hilbert transform yields $\eta^2 = (-1)^n$, so that the eigenvalues are given by

$$\eta = \begin{cases} \pm \mathrm{i}, & n \text{ odd} \\ \pm 1, & n \text{ even}. \end{cases} \tag{15.86}$$

A set of eigenfunctions of H_n in $L^2(\mathbb{R}^n)$ can be written based on Eq. (4.336), so that

$$f(x) = \prod_{k=1}^{n} \frac{(1 + \mathrm{i}x_k)^m}{(1 - \mathrm{i}x_k)^{m+1}}, \quad \text{for } m \in \mathbb{Z}^+. \tag{15.87}$$

The reader will note that the considerable ease with which the results in Sections 15.4 to 15.9 have been obtained is due to the obvious factorization of H_n.

15.10 Periodic functions

Some features of the multi-variable case for periodic functions are examined in this section. The focus is on the situation for two variables. Consider the power series given by

$$f(z_1, z_2) = \sum_{m=0}^{\infty} \sum_{n=0}^{\infty} (a_{mn} - \mathrm{i}b_{mn}) z_1^m z_2^n, \tag{15.88}$$

and take the real part, using $z_1 = \mathrm{e}^{\mathrm{i}x_1}$ and $z_2 = \mathrm{e}^{\mathrm{i}x_2}$; then,

$$g(x_1, x_2) = \operatorname{Re} f(z_1, z_2) = \sum_{m=0}^{\infty} \sum_{n=0}^{\infty} \{a_{mn} \cos(mx_1 + nx_2) + b_{mn} \sin(mx_1 + nx_2)\}, \tag{15.89}$$

and the imaginary part is given by

$$h(x_1, x_2) = \operatorname{Im} f(z_1, z_2) = \sum_{m=0}^{\infty} \sum_{n=0}^{\infty} \{a_{mn} \sin(mx_1 + nx_2) - b_{mn} \cos(mx_1 + nx_2)\}. \tag{15.90}$$

The immediate question is in what sense can Eq. (15.90) be regarded as the conjugate series of Eq. (15.89)? First, a recap of the single variable case is given. Recall for this situation that the conjugate series $\tilde{f}$ of

$$f(x) = \sum_{n=-\infty}^{\infty} c_n \mathrm{e}^{\mathrm{i}nx}, \quad \text{with } c_{-n} = \bar{c}_n, \tag{15.91}$$

takes the following form:

$$\tilde{f}(x) = \sum_{n=-\infty}^{\infty} (-\mathrm{i}\,\mathrm{sgn}\, n) c_n \mathrm{e}^{\mathrm{i}nx}. \tag{15.92}$$

The conjugate series of

$$f(x) = \frac{a_0}{2} + \sum_{k=1}^{\infty} \{a_k \cos kx + b_k \sin kx\} \tag{15.93}$$

is given by

$$\tilde{f}(x) = \sum_{k=1}^{\infty} \{a_k \sin kx - b_k \cos kx\}, \tag{15.94}$$

and these are the real and imaginary parts of the power series expansion

$$f(z) = \frac{a_0}{2} + \sum_{k=1}^{\infty} \{a_k - \mathrm{i} b_k\} z^k, \tag{15.95}$$

on the unit circle, with $z = \mathrm{e}^{\mathrm{i}x}$.

The two-variable series can be written as follows

$$f(x_1, x_2) = \sum_{m=-\infty}^{\infty} \sum_{n=-\infty}^{\infty} c_{mn} \mathrm{e}^{\mathrm{i}(mx_1 + nx_2)}, \tag{15.96}$$

with the assignment that $c_{-m-n} = \bar{c}_{mn}$. In the sequel it is assumed that $c_{mn} = 0$ if $mn = 0$. The following conjugate series can be defined:

$$\tilde{f}_{x_1}(x_1, x_2) = \sum_{m=-\infty}^{\infty} \sum_{n=-\infty}^{\infty} (-\mathrm{i}\,\mathrm{sgn}\, m) c_{mn} \mathrm{e}^{\mathrm{i}(mx_1 + nx_2)}, \tag{15.97}$$

$$\tilde{f}_{x_2}(x_1, x_2) = \sum_{m=-\infty}^{\infty} \sum_{n=-\infty}^{\infty} (-\mathrm{i}\,\mathrm{sgn}\, n) c_{mn} \mathrm{e}^{\mathrm{i}(mx_1 + nx_2)}, \tag{15.98}$$

and

$$\tilde{f}_{x_1, x_2}(x_1, x_2) = \sum_{m=-\infty}^{\infty} \sum_{n=-\infty}^{\infty} (-\mathrm{sgn}\, mn) c_{mn} \mathrm{e}^{\mathrm{i}(mx_1 + nx_2)}. \tag{15.99}$$

These conjugates are with respect to the variables x_1, x_2, and $\{x_1, x_2\}$, respectively, and the appropriate subscript is employed to signify which conjugate is being considered.

Equations (15.89) and (15.90) yield

$$h(x_1, x_2) \equiv \tilde{g}_{x_1}(x_1, x_2) \tag{15.100}$$

and

$$h(x_1, x_2) \equiv \tilde{g}_{x_2}(x_1, x_2). \tag{15.101}$$

That is, Eq. (15.90) can be regarded as the conjugate series of Eq. (15.89) in the variables x_1 and x_2 taken separately.

The type of relationship that exists between a trigonometric series and the corresponding conjugate series for the one-variable case does not carry over to the two-variable case, and, by extension, the general n-variable case. Zygmund (1949) formally identified the conjugate series of

$$g(x_1, x_2) = \sum_{m=0}^{\infty} \sum_{n=0}^{\infty} \{a_{mn} \cos(mx_1 + nx_2) + b_{mn} \sin(mx_1 + nx_2)\} \tag{15.102}$$

to be

$$\tilde{g}(x_1, x_2) = \sum_{m=0}^{\infty} \sum_{n=0}^{\infty} \{a_{mn} \sin(mx_1 + nx_2) - b_{mn} \cos(mx_1 + nx_2)\}. \tag{15.103}$$

The series formed from Eqs. (15.96) and (15.99) as

$$f(x_1, x_2) - \tilde{f}_{x_1,x_2}(x_1, x_2) = \sum_{m=-\infty}^{\infty} \sum_{n=-\infty}^{\infty} c_{mn}(1 + \operatorname{sgn}(mn))e^{i(mx_1+nx_2)}, \tag{15.104}$$

has $\tilde{f}_{x_1}(x_1, x_2) + \tilde{f}_{x_2}(x_1, x_2)$ as its conjugate series. On writing

$$c_{mn} = a_{mn} + ib_{mn}, \tag{15.105}$$

with a_{mn} and b_{mn} real, then

$$f(x_1, x_2) - \tilde{f}_{x_1,x_2}(x_1, x_2) = 4 \sum_{m=0}^{\infty} \sum_{n=0}^{\infty} \{a_{mn} \cos(mx_1 + nx_2) - b_{mn} \sin(mx_1 + nx_2)\} \tag{15.106}$$

and

$$\tilde{f}_{x_1}(x_1, x_2) + \tilde{f}_{x_2}(x_1, x_2) = 4 \sum_{m=0}^{\infty} \sum_{n=0}^{\infty} \{a_{mn} \sin(mx_1 + nx_2) + b_{mn} \cos(mx_1 + nx_2)\}. \tag{15.107}$$

By Zygmund's definition, the latter series is the conjugate of Eq. (15.106).

A number of the results given previously for the one-variable case have analogs for the case of several variables conjugated in *one* variable. For example, if $f(x,y) \in L\log L$, then $\tilde{f}_x(x,y) \in L^1$ and

$$\int_0^{2\pi}\int_0^{2\pi} \left|\tilde{f}_x(x,y)\right| \mathrm{d}x\,\mathrm{d}y \leq A\int_0^{2\pi}\int_0^{2\pi} |f(x,y)|\log^+|f(x,y)|\mathrm{d}x\,\mathrm{d}y + B, \tag{15.108}$$

where A and B are constants. This is an extension of Zygmund's inequality (see Eq. (7.76)).

The two-dimensional Hilbert transform on the domain $[-\pi,\pi]\times[-\pi,\pi]$ is defined by

$$\mathcal{H}_2 f(x,y) = \frac{1}{4\pi^2} P\int_{-\pi}^{\pi} P\int_{-\pi}^{\pi} f(x+s,y+t)\cot\frac{s}{2}\cot\frac{t}{2}\mathrm{d}s\,\mathrm{d}t. \tag{15.109}$$

The almost everywhere existence of $\mathcal{H}_2 f(x,y)$ for the case $f(x,y) \in L^p$ for $1 < p < \infty$, was first proved by Sokół–Sokółski (1947). He also established the following inequality:

$$\int_{-\pi}^{\pi}\int_{-\pi}^{\pi} |\mathcal{H}_2 f(x,y)|^p \,\mathrm{d}x\,\mathrm{d}y \leq C_p\int_{-\pi}^{\pi}\int_{-\pi}^{\pi} |f(x,y)|^p\,\mathrm{d}x\,\mathrm{d}y, \tag{15.110}$$

where C_p is a constant depending on p that is independent of f. The reader will recognize this as the extension of the M. Riesz theorem on $\mathbb{T}$ to two dimensions. If $f(x,y)\log^+|f(x,y)| \in L$, then $\mathcal{H}_2 f(x,y)$ exists *a.e.* (Zygmund, 1949), and

$$\left(\int_{-\pi}^{\pi}\int_{-\pi}^{\pi} |\mathcal{H}_2 f(x,y)|^p\,\mathrm{d}x\,\mathrm{d}y\right)^{p^{-1}} \leq A_p\int_{-\pi}^{\pi}\int_{-\pi}^{\pi} |f(x,y)|\log^+|f(x,y)|\mathrm{d}x\,\mathrm{d}y + A_p, \tag{15.111}$$

where A_p is a constant depending exclusively on p, and $0 < p < 1$. Zygmund also discusses an n-dimensional analog of this formula. The proofs of Eqs. (15.110) and (15.111) can be found in the cited references of Sokół–Sokółski and Zygmund.

15.11 A Calderón–Zygmund inequality

The generalization of the Riesz inequality to E^n is associated with Calderón and Zygmund and is referred to as the Calderón–Zygmund inequality. It will be instructive to consider first a specialized case. Let $f \in L^p(\mathbb{R}^2)$, with $1 < p < \infty$; then, the double Hilbert transform satisfies

$$\|H_2 f\|_p \leq C_p\|f\|_p. \tag{15.112}$$

Making use of the factorization indicated by Eq. (15.35),

$$H_2 f = H_{(1)} H_{(2)} f, \tag{15.113}$$

and so it follows that

$$\|H_2 f\|_p = \|H_{(1)} H_{(2)} f\|_p \le C'_p \|H_{(2)} f\|_p \le C_p \|f\|_p, \tag{15.114}$$

where the standard Riesz inequality for the Hilbert transform has been employed twice. The generalization of this argument to the n-dimensional Hilbert transform is immediate, and yields

$$\begin{aligned} \|H_n f\|_p &= \left\| \prod_{k=1}^{n} H_{(k)} f \right\|_p \\ &= \left\| H_{(1)} \prod_{k=2}^{n} H_{(k)} f \right\|_p \\ &\le C'_p \left\| \prod_{k=2}^{n} H_{(k)} f \right\|_p, \end{aligned} \tag{15.115}$$

using the Riesz inequality. Continuing the argument with repeated application of the Riesz inequality leads to

$$\|H_n f\|_p \le C_{p,n} \|f\|_p, \tag{15.116}$$

where C_p is a constant independent of f that depends on p and the dimension n. Equation (15.116) is a special case of the Calderon–Zygmund inequality.

The extension of Eq. (15.116) to cover the generalized n-dimensional Hilbert transform $\mathcal{H}_n f$, is now considered. For this case, the factorization argument is no longer available, so a different approach must be employed. Let $f \in L^p(E^n)$, with $1 < p < \infty$, and assume that Ω is an odd function and integrable on Σ. Then

$$\|\mathcal{H}_{n,\varepsilon} f\|_p \le C_{p,n} \|f\|_p, \tag{15.117}$$

$$\lim_{\varepsilon \to 0} \|\mathcal{H}_{n,\varepsilon} f - \mathcal{H}_n f\| \to 0, \tag{15.118}$$

and

$$\|\mathcal{H}_n f\|_p \le C_{p,n} \|f\|_p. \tag{15.119}$$

It is common to suppress the dependence of the constant $C_{p,n}$ on n, and henceforth this will be done. Equation (15.117) can be proved by reducing the problem to the corresponding one-dimensional case, and then taking advantage of previously developed results for the one-dimensional Hilbert transform. This is referred to as the *method of rotations*. Making use of the fact that Ω is odd and introducing polar coordinates leads to

$$\begin{aligned}
\mathcal{H}_{n,\varepsilon} f(x) &= \int_{|y|>\varepsilon} f(x-y) \frac{\Omega(y')}{|y|^n} \, \mathrm{d}y \\
&= -\int_{|y|>\varepsilon} f(x+y) \frac{\Omega(y')}{|y|^n} \, \mathrm{d}y \\
&= \frac{1}{2} \int_{|y|>\varepsilon} [f(x-y) - f(x+y)] \frac{\Omega(y')}{|y|^n} \, \mathrm{d}y \\
&= \frac{1}{2} \int_{\Sigma} \Omega(y') \mathrm{d}y' \int_{\varepsilon}^{\infty} \frac{[f(x-ry') - f(x+ry')]\mathrm{d}r}{r} \\
&= \frac{1}{2} \int_{\Sigma} \Omega(y') g_{\varepsilon}(x,y') \mathrm{d}y'.
\end{aligned} \tag{15.120}$$

Minkowski's integral inequality takes the following form. If $h \geq 0$ and $1 \leq p < \infty$, then

$$\left[\int_{E^n} \left(\int_{E^n} h(x,y) \mathrm{d}y \right)^p \mathrm{d}x \right]^{p^{-1}} \leq \int_{E^n} \left(\int_{E^n} h(x,y)^p \, \mathrm{d}x \right)^{p^{-1}} \mathrm{d}y, \tag{15.121}$$

assuming the integrals exist. This result is occasionally referred to as Minkowski's inequality. Related forms involving measures can also be written. Let $h(x,y) = \Omega(y) f(x,y)$; then,

$$\left[\int_{E^n} \left(\int_{E^n} \Omega(y) f(x,y) \mathrm{d}y \right)^p \mathrm{d}x \right]^{p^{-1}} \leq \int_{E^n} \Omega(y) \left(\int_{E^n} f(x,y)^p \, \mathrm{d}x \right)^{p^{-1}} \mathrm{d}y. \tag{15.122}$$

From Eq. (15.120),

$$\left[\int_{E^n} |\mathcal{H}_{n,\varepsilon} f(x)|^p \, \mathrm{d}x \right]^{p^{-1}} = \frac{1}{2} \left[\int_{E^n} \left| \int_{\Sigma} \Omega(y') g_{\varepsilon}(x,y') \mathrm{d}y' \right|^p \mathrm{d}x \right]^{p^{-1}}, \tag{15.123}$$

and applying Minkowski's integral inequality leads to

$$\|\mathcal{H}_{n,\varepsilon} f(x)\|_p \leq \frac{1}{2} \int_{\Sigma} |\Omega(y')| \left\| g_{\varepsilon}(x,y') \right\|_p \mathrm{d}y'. \tag{15.124}$$

To simplify the integration over E^n in $\|g_\varepsilon(x,y')\|_p$, let H denote the hyperplane orthogonal to the line $L = ty'$; then, for $w \in \mathrm{H}$, it follows that

$$\|g_\varepsilon(x,y')\|_p^p = \int_{E^n} \left| \int_\varepsilon^\infty \frac{[f(x-ry') - f(x+ry')]\mathrm{d}r}{r} \right|^p \mathrm{d}x$$

$$= \int_{\mathrm{H}} \mathrm{d}w \int_{-\infty}^{\infty} \mathrm{d}t \left| \int_\varepsilon^\infty \frac{[f(w+(t-r)y') - f(w+(t+r)y')]\mathrm{d}r}{r} \right|^p$$

$$= \int_{\mathrm{H}} \mathrm{d}w \int_{-\infty}^{\infty} \mathrm{d}t \left| \int_\varepsilon^\infty \frac{[h(t-r) - h(t+r)]\mathrm{d}r}{r} \right|^p, \tag{15.125}$$

where h, a function of a single variable, has been introduced using $h(s) = f(w + sy')$. The reader will recognize the integral over r as being related to the truncated Hilbert transform in one dimension (see Eq. (3.87)), and on using Eq. (7.324) it follows that

$$\|g_\varepsilon(x,y')\|_p^p = \pi^p \int_{\mathrm{H}} \mathrm{d}w \int_{-\infty}^{\infty} \mathrm{d}t |(H_\varepsilon h)(t)|^p$$

$$= \pi^p \int_{\mathrm{H}} \|H_\varepsilon h\|_p^p \,\mathrm{d}w$$

$$\le C_p \int_{\mathrm{H}} \|h\|_p^p \,\mathrm{d}w$$

$$= C_p \int_{\mathrm{H}} \mathrm{d}w \int_{-\infty}^{\infty} |f(w+sy')|^p \,\mathrm{d}s$$

$$= C_p \int_{E^n} |f(x)|^p \,\mathrm{d}x$$

$$= C_p \|f\|_p^p, \tag{15.126}$$

where as usual, C_p is a constant depending on p that is independent of f. Taking the p^{-1}th power of Eq. (15.126), and inserting the result into Eq. (15.124), leads to

$$\|\mathcal{H}_{n,\varepsilon} f\|_p \le C_p \|\Omega\|_1 \|f\|_p, \tag{15.127}$$

which gives the required result, Eq. (15.117).

The proof of Eq. (15.119) can be obtained directly from the results given in Eqs. (15.117) and (15.118). The proof of Eq. (15.119) can be found in a number of sources; see, for example, Neri (1971, p. 91). There is one obvious item left to address, and that is the situation where Ω is an even function. This case is treated at the end of the following section, using an approach based on the Riesz transform.

15.12 The Riesz transform

The Riesz transform is defined by

$$R_j f(x) = c_n P \int_{\mathbb{R}^n} \frac{y_j}{|y|^{n+1}} f(x-y)\mathrm{d}y, \quad \text{for } 1 \leq j \leq n, \tag{15.128}$$

where

$$c_n = \frac{\Gamma((n+1)/2)}{\pi^{(n+1)/2}}, \tag{15.129}$$

and Γ is the gamma function, which was defined in Chapter 4 by

$$\Gamma(z) = \int_0^\infty u^{z-1} \mathrm{e}^{-u}\, \mathrm{d}u, \quad \mathrm{Re}\, z > 0. \tag{15.130}$$

The reader is reminded that in Eq. (15.128), x and y denote, respectively, $(x_1, x_2, \ldots, x_n)$ and $(y_1, y_2, \ldots, y_n)$. The particular case $n = 1$ yields the correspondence $R_1 \equiv H$. Not surprisingly, the Riesz transform satisfies a number of results similar in structure to those satisfied by the Hilbert transform operator. A consideration of some of these results is now undertaken. The first result examined is the Fourier transform of the Riesz transform. This can be written as follows:

$$\mathcal{F}_n R_k f(x) = -\mathrm{i}\frac{x_k}{|x|} \mathcal{F}_n f(x). \tag{15.131}$$

Equation (15.131) can be obtained in a few different ways; a straightforward argument of Neri (1971, p. 101) is followed; see also Duoandikoetxea (2001, p. 76). First note that

$$\frac{\partial |x|^{1-n}}{\partial x_j} = (1-n)\, p.v. \frac{x_j}{|x|^{n+1}}. \tag{15.132}$$

Making use of this result for $n > 1$ leads to

$$\begin{aligned}
\mathcal{F}_n\, p.v. \frac{\xi_j}{|\xi|^{n+1}} &= \frac{1}{(1-n)} \mathcal{F}_n \frac{\partial\, |\xi|^{1-n}}{\partial \xi_j} \\
&= \frac{2\pi \mathrm{i}\xi_j}{(1-n)} \mathcal{F}_n |\xi|^{1-n} \\
&= \frac{2\pi \mathrm{i}\xi_j}{(1-n)} \frac{\pi^{(n-2)/2}\Gamma(1/2)}{\Gamma((n-1)/2)|\xi|} \\
&= -\mathrm{i} \frac{\pi^{(n+1)/2}}{\Gamma((n+1)/2)} \frac{\xi_j}{|\xi|} \\
&= -\frac{\mathrm{i}}{c_n} \frac{\xi_j}{|\xi|},
\end{aligned} \tag{15.133}$$

and the following result has been employed in the derivation:

$$\mathcal{F}_n\,|\xi|^{-a} = \frac{\pi^{(2a-n)/2}\Gamma((n-a)/2)\,|\xi|^{a-n}}{\Gamma(a/2)}. \tag{15.134}$$

Equation (15.134) can be proved by first evaluating the Fourier transform of the Gaussian function in one dimension. Let $f(t) = \mathrm{e}^{-\lambda t^2}$, where $\lambda > 0$, and set $F(x) = \mathcal{F}f(x)$, then observe that

$$f'(t) = -2\lambda t f(t). \tag{15.135}$$

Taking the Fourier transform yields

$$\mathcal{F}f'(w) = -2\lambda\,\mathcal{F}\{wf(w)\}. \tag{15.136}$$

Now,

$$\begin{aligned}\mathcal{F}f'(w) &= \int_{-\infty}^{\infty} \mathrm{e}^{-2\pi \mathrm{i}wt} f'(t)\mathrm{d}t \\ &= 2\pi\mathrm{i}w \int_{-\infty}^{\infty} \mathrm{e}^{-2\pi \mathrm{i}wt} f(t)\mathrm{d}t \\ &= 2\pi\mathrm{i}wF(w),\end{aligned} \tag{15.137}$$

and on taking the derivative with respective to w, leads to

$$\mathcal{F}wf(w) = \frac{\mathrm{i}}{2\pi}F'(w). \tag{15.138}$$

From Eqs. (15.137), and (15.138), the following differential equation is obtained:

$$F'(w) = -2\alpha wF(w), \tag{15.139}$$

where $\alpha = \lambda^{-1}\pi^2$. The differential equation can be solved in a straightforward fashion by multiplying by $\mathrm{e}^{\alpha w^2}$ to give

$$\mathrm{e}^{\alpha w^2}F'(w) + 2\alpha wF(w)\mathrm{e}^{\alpha w^2} = 0, \tag{15.140}$$

and hence

$$\frac{\mathrm{d}}{\mathrm{d}w}\{\mathrm{e}^{\alpha w^2}F(w)\} = 0, \tag{15.141}$$

so that

$$F(w) = c\mathrm{e}^{-\alpha w^2} \tag{15.142}$$

where the constant c can be found by evaluating the case $w = 0$; that is,

$$c = \int_{-\infty}^{\infty} \mathrm{e}^{-\lambda t^2}\,\mathrm{d}t = \surd\left(\frac{\pi}{\lambda}\right), \tag{15.143}$$

and hence

$$\mathcal{F}(\mathrm{e}^{-\lambda w^2}) = \surd\left(\frac{\pi}{\lambda}\right)\mathrm{e}^{-\pi^2\lambda^{-1}w^2}. \tag{15.144}$$

The integral in Eq. (15.143) is evaluated by the well known technique of squaring the integral and converting to polar coordinates; this is left as an exercise for the reader. The particular case for $\lambda = \pi$ yields the formula

$$\mathcal{F}(\mathrm{e}^{-\pi w^2}) = \mathrm{e}^{-\pi w^2}. \tag{15.145}$$

Equation (15.144) is now employed to evaluate $\mathcal{F}_n(\mathrm{e}^{-\pi\lambda|x|^2})$. The n-dimensional Fourier transform is separable into a product of one-dimensional Fourier transforms, so that

$$\begin{aligned}\mathcal{F}_n(\mathrm{e}^{-\pi\lambda|x|^2}) &= \prod_{k=1}^{n}\left\{\int_{-\infty}^{\infty}\mathrm{e}^{-2\pi\mathrm{i}x_k s_k}\mathrm{e}^{-\pi\lambda s_k^2}\,\mathrm{d}s_k\right\}\\ &= \prod_{k=1}^{n}\frac{\mathrm{e}^{-\pi\lambda^{-1}x_k^2}}{\surd\lambda}\\ &= \frac{\mathrm{e}^{-\pi\lambda^{-1}|x|^2}}{\lambda^{n/2}}.\end{aligned} \tag{15.146}$$

One more preliminary result is required. The definition of the gamma function given in Eq. (15.130) is employed, with the restriction that z is a real variable, and let $\alpha = z$. Using the change of variable $u = \pi r^2$ and the substitution $\beta = 2\alpha - 1$ leads to

$$\int_0^{\infty} r^{\beta}\mathrm{e}^{-\pi r^2}\,\mathrm{d}r = \frac{\Gamma((\beta+1)/2)}{2\pi^{(\beta+1)/2}}, \quad \text{for } \beta > -1. \tag{15.147}$$

From Eq. (15.130), and using the change of variable $u = ts$ with $t > 0$, it follows that

$$t^{-\alpha} = \frac{1}{\Gamma(\alpha)}\int_0^{\infty} s^{\alpha-1}\mathrm{e}^{-ts}\,\mathrm{d}s. \tag{15.148}$$

Introducing the substitution $t = \pi |x|^2$ into Eq. (15.148), and taking the Fourier transform, yields

$$\begin{aligned}\mathcal{F}_n\{\Gamma(\alpha)(\pi|x|^2)^{-\alpha}\}(w) &= \int_0^\infty s^{\alpha-1}\,\mathcal{F}_n\{e^{-\pi s|w|^2}\}ds \\ &= \int_0^\infty s^{\alpha-1} e^{-\pi s^{-1}|w|^2} s^{-n/2}\,ds \\ &= \int_0^\infty u^{(n/2)-\alpha-1} e^{-\pi|w|^2 u}\,du \\ &= \Gamma\left(\frac{n}{2}-\alpha\right)(\pi|w|^2)^{\alpha-n/2}, \end{aligned} \tag{15.149}$$

where Eq. (15.145) has been used, the change of variable $s = u^{-1}$ has been employed, and Eq. (15.148) has been utilized a second time. Equation (15.134) follows directly from Eq. (15.149). The Fourier transform of $f(x) = |x|^{-a}$ has an operational meaning, in the sense that

$$\int_{\mathbb{R}^n} \mathcal{F}_n f(x)\phi(x)^*\,dx = \int_{\mathbb{R}^n} f(x)\,\mathcal{F}_n\phi(x)^*\,dx, \tag{15.150}$$

where the test function ϕ is chosen such that $\phi \in \mathcal{S}(\mathbb{R}^n)$ (see Section 10.2). Equation (15.150) is the n-dimensional form of Eq. (10.62).

From the definition given in Eq. (15.128), it can be established that

$$R_j f(x) = c_n(f * K_j)(x), \tag{15.151}$$

where the kernel K_j is defined by $K_j = p.v.(x_j/|x|^{n+1})$. Taking the Fourier transform yields

$$\begin{aligned}\mathcal{F}_n R_j f(x) &= c_n \mathcal{F}_n f(x)\,\mathcal{F}_n K_j(x) \\ &= c_n \mathcal{F}_n f(x)\left(\frac{-i}{c_n}\frac{x_j}{|x|}\right),\end{aligned} \tag{15.152}$$

where Eq. (15.133) has been employed. Hence, Eq. (15.131) is obtained.

One result that follows immediately from Eq. (15.131) is a bound for $\|R_k f\|_2$. If $f \in L^2(\mathbb{R}^n)$, noting that $|x_k| \le |x|$, then

$$\|\mathcal{F}_n R_k f\|_2 = \left\|\frac{x_k}{|x|}\mathcal{F}_n f\right\|_2 \le \|\mathcal{F}_n f\|_2. \tag{15.153}$$

Applying the Parseval formula to each side of the last inequality leads to

$$\|R_k f\|_2 \le \|f\|_2. \tag{15.154}$$

The Riesz transform operators satisfy

$$\sum_{j=1}^{n} R_j^2 = -I. \tag{15.155}$$

This formula holds for functions $f \in L^p(E^n)$, with $1 < p < \infty$. The preceding formula also occurs in the literature without the minus sign, which is due to a slightly different definition of the Riesz transform: a $-\mathrm{i}$ factor is included in the definition of the constant c_n given in Eq. (15.129). The case $p = 2$ is considered first. Starting with

$$\mathcal{F}_n R_k g(x) = -\mathrm{i}\frac{x_k}{|x|}\mathcal{F}_n g(x), \tag{15.156}$$

and setting $g = R_k f$, yields

$$\begin{aligned} \mathcal{F}_n R_k^2 f(x) &= -\mathrm{i}\frac{x_k}{|x|}\mathcal{F}_n R_k f(x) \\ &= -\frac{x_k^2}{|x|^2}\mathcal{F}_n f(x). \end{aligned} \tag{15.157}$$

Summing over k leads to

$$\begin{aligned} \sum_{k=1}^{n} \mathcal{F}_n R_k^2 f(x) &= -\frac{1}{|x|^2}\sum_{k=1}^{n} x_k^2 \mathcal{F}_n f(x) \\ &= -\mathcal{F}_n f(x), \end{aligned} \tag{15.158}$$

and taking the inverse Fourier transform yields

$$\sum_{k=1}^{n} R_k^2 f(x) = -f(x), \tag{15.159}$$

and Eq. (15.155) follows. From the previous section (see Eq. (15.119)), the operators R_k are continuous in L^p, and Eq. (15.154) may be extended by continuity to $L^p(E^n)$, with $1 < p < \infty$, so that

$$\|R_k f\|_p \leq C_p \|f\|_p. \tag{15.160}$$

Riesz transforms find important applications in establishing bounds for various partial derivatives of a function. The Laplacian operator Δ is defined by

$$\Delta = \sum_{j=1}^{n} \frac{\partial^2}{\partial x_j^2}. \tag{15.161}$$

If f has compact support and satisfies $f \in C^2$, then the norm of $\partial^2 f/\partial x_i \partial x_j$ is bounded as follows:

$$\left\| \frac{\partial^2 f}{\partial x_i \partial x_j} \right\|_p \le C_p \|\Delta f\|_p, \quad \text{for } 1 < p < \infty, \tag{15.162}$$

where the constant C_p depends on p but is independent of f. To establish this result the following identity is proved first:

$$\frac{\partial^2 f}{\partial x_j \partial x_k} = -R_j R_k \Delta f. \tag{15.163}$$

Starting with the Fourier transform of $\partial f/\partial x_k$ and integrating by parts, yields

$$\begin{aligned} \mathcal{F}_n\left\{\frac{\partial f(x)}{\partial x_k}\right\} &= \int_{-\infty}^{\infty} \frac{\partial f(s)}{\partial s_k} \mathrm{e}^{-2\pi \mathrm{i} x\cdot s}\,\mathrm{d}s \\ &= 2\pi \mathrm{i} x_k \int_{-\infty}^{\infty} f(s) \mathrm{e}^{-2\pi \mathrm{i} x\cdot s}\,\mathrm{d}s \\ &= 2\pi \mathrm{i} x_k \mathcal{F}_n f(x). \end{aligned} \tag{15.164}$$

Repeating the same operation yields

$$\mathcal{F}_n\left\{\frac{\partial^2 f(x)}{\partial x_j \partial x_k}\right\} = -4\pi^2 x_j x_k \mathcal{F}_n f(x). \tag{15.165}$$

In a similar manner it follows that

$$\begin{aligned} \mathcal{F}_n \Delta f(x) &= \sum_{k=1}^{n} \mathcal{F}_n\left\{\frac{\partial^2 f(x)}{\partial x_k^2}\right\} \\ &= -4\pi^2 \sum_{k=1}^{n} x_k^2 \mathcal{F}_n f(x) \\ &= -4\pi^2 |x|^2 \mathcal{F}_n f(x). \end{aligned} \tag{15.166}$$

Employing Eqs. (15.131), (15.165), and (15.166), leads to

$$\begin{aligned} \mathcal{F}_n\left\{\frac{\partial^2 f(x)}{\partial x_j \partial x_k}\right\} &= \left(\frac{-\mathrm{i}x_j}{|x|}\right)\left(\frac{-\mathrm{i}x_k}{|x|}\right)(4\pi^2|x|^2)\mathcal{F}_n f(x) \\ &= -\left(\frac{-\mathrm{i}x_j}{|x|}\right)\left(\frac{-\mathrm{i}x_k}{|x|}\right)\mathcal{F}_n \Delta f(x) \\ &= -\left(\frac{-\mathrm{i}x_j}{|x|}\right)\mathcal{F}_n R_k \Delta f(x) \\ &= -\mathcal{F}_n R_j R_k \Delta f(x), \end{aligned} \tag{15.167}$$

from which Eq. (15.163) follows on taking the inverse Fourier transform. Noting that the Riesz transform is just a particular case of Eq. (15.6), Eqs. (15.119) and (15.163) may be used to obtain

$$\begin{aligned}\left\|\frac{\partial^2 f}{\partial x_j \partial x_k}\right\|_p &= \left\|R_j\{R_k \Delta f\}\right\|_p \\ &\le C_p'\|R_k \Delta f\|_p ,\end{aligned} \tag{15.168}$$

and hence

$$\left\|\frac{\partial^2 f}{\partial x_j \partial x_k}\right\|_p \le C_p\|\Delta f\|_p , \tag{15.169}$$

which is the required result.

An item of unfinished business from Section 15.11 is the case where Ω is an even function. Let $f \in L^p(E^n)$, with $1 < p < \infty$, assume that Ω is even and has mean value zero on Σ, and that $\Omega \in L^q(\Sigma)$ with $q > 1$; then,

$$\|\mathcal{H}_n f\|_p \le C_p\|f\|_p. \tag{15.170}$$

The main idea involved in the proof is now sketched. Consider the composition $R_j\mathcal{H}_n$ and denote this by the operator T_j. Suppose $T_j f$ satisfies

$$T_j f(x) = \int_{|y|>\varepsilon} f(x-y)\frac{\Omega(y')}{|y|^n}\,\mathrm{d}y. \tag{15.171}$$

Since the kernel function is odd, it is the composition of even and odd operators, and hence Eq. (15.119) can be employed to write

$$\left\|T_j f\right\|_p \le C_p\|f\|_p. \tag{15.172}$$

Making use of the property of the Riesz transforms given in Eq. (15.155),

$$\mathcal{H}_n f(x) = -\sum_{j=1}^{n} R_j^2\mathcal{H}_n f(x) = -\sum_{j=1}^{n} R_j T_j f(x). \tag{15.173}$$

From this last result it follows, on using Minkowski's inequality, that

$$\begin{aligned}\|\mathcal{H}_n f\|_p &= \left(\int\left|\sum_{j=1}^{n} R_j T_j f(x)\right|^p \mathrm{d}x\right)^{p^{-1}} \\ &\le \sum_{j=1}^{n}\left\|R_j T_j f\right\|_p .\end{aligned} \tag{15.174}$$

Equations (15.160) and (15.172) yield

$$\left\|R_j T_j f\right\|_p \leq C'_p \left\|T_j f\right\|_p \leq C_p \|f\|_p, \tag{15.175}$$

and therefore from Eqs. (15.172) and (15.174) it follows that

$$\|\mathcal{H}_n f\|_p \leq C_p \|f\|_p, \tag{15.176}$$

which is the required result. Discussion on the justification for the representation given in Eq. (15.171) can be found in Duoandikoetxea (2001, p. 77) or Stein and Weiss (1971, p. 224), to which the interested reader is directed for further details.

15.13 The *n*-dimensional Hilbert transform of distributions

A number of the results from Chapter 10 can be generalized to treat the n-dimensional Hilbert transform of distributions. There is more than one approach that can be employed, and the basic ideas that are involved in the different techniques are sketched.

Let X^m and Y^n denote, respectively, m- and n-dimensional Euclidean spaces. Points in the space X^m are denoted by $x = (x_1, x_2, \ldots, x_m)$ and those in the space Y^n by $y = (y_1, y_2, \ldots, y_n)$. The Cartesian product of these two spaces is denoted by $X^m \times Y^n$. Points in this product space are represented by $(x, y) = (x_1, x_2, \ldots, x_m, y_1, y_2, \ldots, y_n)$. The spaces of infinitely differentiable functions with compact support on X^m, Y^n, and $X^m \times Y^n$ are denoted by $\mathcal{D}_x$, $\mathcal{D}_y$, and $\mathcal{D}_{x,y}$, respectively, and the corresponding spaces of distributions are denoted by $\mathcal{D}'_x$, $\mathcal{D}'_y$, and $\mathcal{D}'_{x,y}$, respectively. Let $\phi(x, y)$ be a test function satisfying $\phi(x, y) \in \mathcal{D}_{x,y}$, and let T_x and S_y designate two distributions such that $T_x \in \mathcal{D}'_x$ and $S_y \in \mathcal{D}'_y$. The tensor product of these two distributions is denoted, as in the one-dimensional case, by $T \otimes S$, and represents a distribution belonging to $\mathcal{D}'_{x,y}$, which is defined by

$$\langle T_x \otimes S_y, \phi(x, y)\rangle = \langle T_x, \langle S_y, \phi(x, y)\rangle\rangle. \tag{15.177}$$

Suppose the test function $\phi(x, y)$ can be written in the form $\phi(x, y) = \varphi(x)\psi(y)$, where $\varphi(x)$ and $\psi(y)$ are test functions in $\mathcal{D}_x$ and $\mathcal{D}_y$, respectively. Then the preceding definition can be expressed as follows:

$$\langle T_x \otimes S_y, \phi(x, y)\rangle = \langle T_x, \varphi(x)\rangle\langle S_y, \psi(y)\rangle. \tag{15.178}$$

The commutative property for the tensor product can be then readily established. If R_z is a third distribution such that $R_z \in \mathcal{D}'_z$, and there is a test function $\phi(x, y, z)$ such that $\phi(x, y, z) \in \mathcal{D}_{x,y,z}$, where $\mathcal{D}_{x,y,z}$ denotes the space $X^m \times Y^n \times Z^p$, then the associative property,

$$T_x \otimes \{S_y \otimes R_z\} = \{T_x \otimes S_y\} \otimes R_z, \tag{15.179}$$

can be readily established.

The support of the tensor product satisfies

$$\operatorname{supp}\{T_x \otimes S_y\} = \operatorname{supp}\{T_x\} \times \operatorname{supp}\{S_y\}. \tag{15.180}$$

A proof of this result can be found in several sources; see, for example, Kanwal (1998, p. 177).

The convolution introduced in Eq. (10.102) can be generalized to higher dimensions as follows:

$$\langle T * S, \phi\rangle = \langle T_\eta \otimes S_\xi, \phi(\eta + \xi)\rangle, \tag{15.181}$$

where the distributions T and S belong to $\mathbb{R}^n$, and (η, ξ), denotes a point in the $2n$-dimensional Euclidean space. The support of the distribution $T_\eta \otimes S_\xi$ is the set of pairs (η, ξ), where $\eta \in \operatorname{supp}\{T\}$ and $\xi \in \operatorname{supp}\{S\}$. Let $K = \operatorname{supp}\{\phi\}$, then the support of $\phi(\eta + \xi)$ is the set of pairs (η, ξ) that satisfy $\eta + \xi \in K$, which leads to a strip in the $2n$-dimensional Euclidean space. Recall the discussion in Section 10.6 for the case $n = 1$. The support of $\phi(\eta + \xi)$ is not bounded. Just as for the case $n = 1$, Eq. (15.181) has a well defined meaning if the intersection of the supports of $T_\eta \otimes S_\xi$ and $\phi(\eta + \xi)$ is bounded. This condition is certainly satisfied if the support of T or S is bounded, or if both T and S have supports bounded from the left, or if both T and S have supports bounded from the right.

Equation (15.181) can be generalized to cover the convolution product of more than two distributions. If T, S, and R are distributions that belong to $\mathbb{R}^n$, then the convolution can be defined as follows:

$$\langle T * S * R, \phi\rangle = \langle T_\eta \otimes S_\xi \otimes R_\zeta, \phi(\eta + \xi + \zeta)\rangle, \tag{15.182}$$

and this exists if $(\eta + \xi + \zeta)$ is bounded, where η, ξ, and ζ belong to the sets $\operatorname{supp}\{T\}, \operatorname{supp}\{S\}$, and $\operatorname{supp}\{R\}$, respectively.

The convolution of distributions belonging to the space $\mathcal{D}'_{Lp}$ is now considered, and this leads directly to a definition of the n-dimensional Hilbert transform for distributions. The approach of Horváth (1956) is followed and his notation is employed. This notation is often used in the literature, but differs slightly from that employed in Section 15.1. If $x \in \mathbb{R}^n$, then the norm of x is given by Eq. (15.1). The unit sphere S_{n-1} is defined by the locus of points for which $|x| = 1$. The radial projection of x onto S_{n-1} is denoted by σ_x, and this satisfies $\sigma_x = x|x|^{-1}$. Each x can be written in the form $x = r\sigma_x$ and $\sigma_x \in S_{n-1}$. Let $k(\sigma)$ denote a function that is integrable on

S_{n-1} and satisfies

$$\int_{S_{n-1}} k(\sigma)\mathrm{d}\sigma = 0. \tag{15.183}$$

The distribution K is introduced according to the definition

$$K = p.v.\frac{k(\sigma)}{r^n}, \tag{15.184}$$

and

$$K(\phi) = \lim_{\varepsilon\to 0}\int_{r>\varepsilon} \frac{k(\sigma)}{r^n}\phi(x)\mathrm{d}x, \tag{15.185}$$

where the test function ϕ satisfies $\phi \in \mathcal{D}$. Horváth proved the following two propositions. The limit in the preceding equation exists and defines a distribution. To establish this, consider

$$K_\varepsilon(x) = \begin{cases} \dfrac{k(\sigma)}{r^n}, & r > \varepsilon \\ 0, & r \le \varepsilon. \end{cases} \tag{15.186}$$

Using ideas from the discussion of Section 15.1, the reader should try to establish that $K_\varepsilon(x)$ is a locally integrable function and that the right-hand side of Eq. (15.185) tends to a finite limit.

For the second proposition, suppose that

$$\int_{S_{n-1}} |k(\sigma)|^s\mathrm{d}\sigma < \infty, \quad \text{for } 1 < s < \infty, \tag{15.187}$$

then $K \in \mathcal{D}'_{L^s}$. This result can be established in the following manner. Let $\alpha(x)$ denote a test function with the following properties:

$$0 \le \alpha(x) \le 1 \tag{15.188}$$

and

$$\alpha(x) = 1, \quad \text{for } |x| \le 1, \tag{15.189}$$

and satisfying $\alpha(x) \in \mathcal{D}$. The distribution K can be expressed as follows:

$$K = \alpha K + (1-\alpha)K. \tag{15.190}$$

Since α has compact support, the term αK has compact support, and hence it follows, on making use of Eq. (10.35), that

$$\alpha K \in \mathcal{E}' \subset \mathcal{D}'_{L^s}. \tag{15.191}$$

The term $(1-\alpha)K$ belongs to L^s, which can be seen as follows:

$$\begin{aligned}\int |(1-\alpha)K|^s\,\mathrm{d}x &= \int_1^\infty \int_{S_{n-1}} \frac{|k(\sigma)|^s}{r^{ns}}\,\mathrm{d}x \\ &= \int_1^\infty \frac{\mathrm{d}r}{r^{ns-n+1}} \int_{S_{n-1}} |k(\sigma)|^s\,\mathrm{d}\sigma \\ &< \infty, \end{aligned} \tag{15.192}$$

where the last follows using Eq. (15.187). Hence, $(1-\alpha)K \in L^s$, and since $L^s \subset \mathcal{D}'_{L^s}$, the proposition is proved.

If Eq. (15.187) is satisfied for some $s > 1$, it follows that $K \in \mathcal{D}'_{L^p}$, for $1 < p < \infty$. This result is obtained by selecting a p such that $1 < p < s$, and using the proposition just given. If $s < p < \infty$, employ $\mathcal{D}'_{L^s} \subset \mathcal{D}'_{L^p}$.

Horváth defined a generalization of the Hilbert transform for a distribution T satisfying $T \in \mathcal{D}'_{L^p}$, by the following convolution formula:

$$\mathcal{H}_n T = (K * T), \tag{15.193}$$

and he proved the following theorem. Let Eq. (15.187) be satisfied for some $s > 1$; then, $T \to K * T$ is a continuous mapping of $\mathcal{D}'_{L^p}$ into $\mathcal{D}'_{L^p}$, for $1 < p < \infty$. The interested reader may pursue the proof of this result in Horváth (1956).

The ideas from Section 10.7 can be carried over to the n-dimensional space $\mathbb{R}^n$. If $T \in \mathcal{D}'_{L^p}(R^n)$ and $S \in \mathcal{D}'_{L^q}(\mathbb{R}^n)$, the convolution $S * T$ exists and $S * T \in \mathcal{D}'_{L^r}(\mathbb{R}^n)$. The n-dimensional Hilbert transform of the distribution $T \in \mathcal{D}'_{L^p}(\mathbb{R}^n)$ can be defined in terms of the convolution with the tensor product of the distributions $p.v.(1/x_k)$; that is, $p.v.(1/x_1) \otimes p.v.(1/x_2) \otimes \cdots \otimes p.v.(1/x_n)$, for $1 < p < \infty$, so that, in symbolic form,

$$H_n T = \frac{1}{\pi^n} T * p.v.\frac{1}{x_1} \otimes p.v.\frac{1}{x_2} \otimes \cdots \otimes p.v.\frac{1}{x_n}. \tag{15.194}$$

Pandey and his colleagues have discussed the Hilbert transform of distributions in n-dimensions in some different spaces, and a few of these ideas are concisely outlined in the sequel. For the two spaces, $\mathscr{S}_1$ and $\mathcal{Z}_1$, which were introduced in Section 10.14, Pandey (1982) defined the n-dimensional analogs in the following way. The space $\mathscr{S}_1(\mathbb{R}^n)$ consists of those functions belonging to $\mathscr{S}(\mathbb{R}^n)$ which vanish in an open interval containing the origin. The space $\mathcal{Z}_1(\mathbb{R}^n)$ consists of those functions belonging to $\mathscr{S}(\mathbb{R}^n)$ which are the Fourier transforms of functions belonging to $\mathscr{S}(\mathbb{R}^n)$. Pandey defined the Hilbert transform of a tempered distribution T in $\mathbb{R}^n$ by the following formula:

$$\langle H_n T, \mathcal{F}_n\varphi(x)\rangle = (-\mathrm{i})^n \langle T, \mathcal{F}_n(\operatorname{sgn} x\varphi(x))\rangle, \quad \forall\varphi \in \mathscr{S}_1(\mathbb{R}^n). \tag{15.195}$$

The inversion formula takes the following symbolic form:

$$H_n^2 T = (-1)^n T, \tag{15.196}$$

which can be derived in a manner similar to that displayed in Eq. (10.247).

Chaudhry and Pandey (1985) provide an alternative definition. Let $X(\mathbb{R}^n)$ denote the space of test functions $\phi \in \mathcal{D}(\mathbb{R}^n)$ such that, for finite k,

$$\phi = \sum_{m=1}^{k} \phi_{m,1}(x_1)\phi_{m,2}(x_2)\phi_{m,3}(x_3)\cdots\phi_{m,n}(x_n), \tag{15.197}$$

with $\phi_{m,1} \in \mathcal{D}(\mathbb{R})$, and suppose that $\partial^j\phi \in L^p(\mathbb{R}^n)$ for $j = 0, 1, \ldots$ and $1 < p < \infty$. The Hilbert transform of the test functions in $X(\mathbb{R}^n)$ is given by

$$H_n\phi(x) = \frac{1}{\pi^n}\sum_{m=1}^{k} P\int_{\mathbb{R}^n} \frac{\phi_{m,1}(t_1)\phi_{m,2}(t_2)\phi_{m,3}(t_3)\cdots\phi_{m,n}(t_n)}{\prod_{\mathrm{i}=1}^{n}(x_\mathrm{i} - t_\mathrm{i})}\,\mathrm{d}t. \tag{15.198}$$

With this definition it follows that

$$H_n^2\phi = (-1)^n\phi, \quad \forall\phi \in X(\mathbb{R}^n), \tag{15.199}$$

and the reader is left to ponder the proof.

Chaudhry and Pandey defined, by analogy to the classical Parseval-type formula, (see Eq. (10.81) and the extension for distributions in $\mathcal{D}'_{L^p}(\mathbb{R})$ given in Eq. (10.83)), the Hilbert transform for distributions $T \in \mathcal{D}'_{L^p}(\mathbb{R}^n)$, by the following formula:

$$\langle H_nT, \phi\rangle = \langle T, (-1)^n H_n\phi\rangle, \quad \forall\phi \in \mathcal{D}_{L^q}(\mathbb{R}^n), \tag{15.200}$$

where p and q denote conjugate exponents and $1 < p < \infty$. In symbolic form the inversion property

$$H_n^2 T = (-1)^n T, \quad \forall T \in \mathcal{D}'_{L^p}(\mathbb{R}^n), \tag{15.201}$$

can be derived from the definition in Eq. (15.200).

Singh and Pandey (1990a) introduced a space of test functions ϕ denoted by $\mathscr{H}(\mathbb{R}^n)$, defined in the following way. The test functions satisfy the following: (i) $\phi \in C^\infty$ on $\mathbb{R}^n$; (ii) there exists a $\varphi \in \mathcal{D}(\mathbb{R}^n)$, with $\phi(t) = H_n\varphi(t)$. This generalizes the start of the discussion in Section 10.14 to $\mathbb{R}^n$. The inclusion relation for this space is $\mathscr{H}(\mathbb{R}^n) \subset \mathcal{D}_{L^p}(\mathbb{R}^n)$. The space of continuous linear functionals on $\mathscr{H}(\mathbb{R}^n)$ is denoted by $\mathscr{H}'(\mathbb{R}^n)$. This space satisfies the inclusion condition $\mathcal{D}'_{L^p}(\mathbb{R}^n) \subset \mathscr{H}'(\mathbb{R}^n)$, with $1 < p < \infty$. Further details on this space can be found in the work of Singh and Pandey. The Hilbert transform of a distribution $T \in \mathcal{D}'(\mathbb{R}^n)$ is defined to be an ultradistribution $H_n T \in \mathscr{H}'(\mathbb{R}^n)$ such that

$$\langle H_n T, \varphi \rangle = \langle T, (-1)^n H_n \varphi \rangle, \quad \forall \varphi \in \mathscr{H}(\mathbb{R}^n). \tag{15.202}$$

If $T \in \mathscr{H}'(\mathbb{R}^n)$, then Singh and Pandey define the n-dimensional Hilbert transform $H_n T$ to be a Schwartz distribution given by the following formula:

$$\langle H_n T, \varphi \rangle = \langle T, (-1)^n H_n \varphi \rangle, \quad \forall \varphi \in \mathcal{D}(\mathbb{R}^n). \tag{15.203}$$

From this definition, the following result is obtained:

$$H_n^2 = (-1)^n I, \quad \text{on } \mathcal{D}'(\mathbb{R}^n), \tag{15.204}$$

which is left for the reader to prove.

The n-dimensional Hilbert transform is a homeomorphism from $\mathcal{D}_{L^p}(\mathbb{R}^n)$ to $\mathcal{D}_{L^p}(\mathbb{R}^n)$, for $1 < p < \infty$. This result can be established in a straightforward manner from the $n = 1$ case, and the proof is sketched following the approach of Pandey and Chaudhry (1983). Let $\phi \in \mathcal{D}_{L^p}(\mathbb{R}^n)$, for $1 < p < \infty$, and set $g = H\phi$; then,

$$\begin{aligned} g(x) &= \frac{1}{\pi} P \int_{-\infty}^{\infty} \frac{\phi(t)\mathrm{d}t}{x - t} \\ &= -\frac{1}{\pi} P \int_{-\infty}^{\infty} \frac{\phi(t + x)\mathrm{d}t}{t} \\ &= -\frac{1}{\pi} \int_{-\infty}^{-N} \frac{\phi(t + x)\mathrm{d}t}{t} - \frac{1}{\pi} P \int_{-N}^{N} \frac{\phi(t + x)\mathrm{d}t}{t} - \frac{1}{\pi} \int_{N}^{\infty} \frac{\phi(t + x)\mathrm{d}t}{t} \\ &= -\frac{1}{\pi} \int_{-\infty}^{-N} \frac{\phi(t + x)\mathrm{d}t}{t} - \frac{1}{\pi} \int_{-N}^{N} \frac{\{\phi(t + x) - \phi(x)\}\mathrm{d}t}{t} - \frac{1}{\pi} \int_{N}^{\infty} \frac{\phi(t + x)\mathrm{d}t}{t}. \end{aligned} \tag{15.205}$$

The function $\varphi(x, t)$ is introduced by

$$\varphi(x, t) = \begin{cases} \dfrac{\phi(t + x) - \phi(x)}{t}, & t \neq 0 \\ \phi'(x), & t = 0, \end{cases} \tag{15.206}$$

and then

$$g(x) = -\frac{1}{\pi} \int_{|t|>N} \frac{\phi(t + x)\mathrm{d}t}{t} - \frac{1}{\pi} \int_{N \geq |t|} \varphi(x, t)\mathrm{d}t. \tag{15.207}$$

Hence,

$$
\begin{aligned}
(H\phi(x))' &= -\frac{1}{\pi}\int_{|t|>N}\frac{\phi'(t+x)\mathrm{d}t}{t} - \frac{1}{\pi}\int_{N\geq|t|}\frac{\partial\varphi(x,t)}{\partial x}\,\mathrm{d}t \\
&= -\frac{1}{\pi}\int_{|t|>N}\frac{\phi'(t+x)\mathrm{d}t}{t} - \frac{1}{\pi}\lim_{\varepsilon\to0+}\int_{N>|t|>\varepsilon}\frac{\{\phi'(x+t)-\phi'(x)\}\mathrm{d}t}{t} \\
&= -\frac{1}{\pi}\int_{|t|>N}\frac{\phi'(t+x)\mathrm{d}t}{t} - \frac{1}{\pi}\lim_{\varepsilon\to0+}\int_{N>|t|>\varepsilon}\frac{\phi'(x+t)\mathrm{d}t}{t} \\
&= \frac{1}{\pi}\lim_{\varepsilon\to0+}\int_{|t|>\varepsilon}\frac{\phi'(x+t)\mathrm{d}t}{t} \\
&= H\phi'(x).
\end{aligned}
\tag{15.208}
$$

The argument can be extended by induction to show that

$$
\frac{\partial^k}{\partial x^k}H\phi(x) = H\frac{\partial^k}{\partial x^k}\phi(x), \quad \forall k\in\mathbb{N}. \tag{15.209}
$$

Hence, it follows that $H\phi \in \mathcal{D}_{L^p}(\mathbb{R})$, for $1 < p < \infty$. By the Riesz inequality, it follows that

$$
\left\|\frac{\partial^k}{\partial x^k}H\phi\right\|_p \leq C_p\left\|\frac{\partial^k\phi}{\partial x^k}\right\|_p, \quad \forall k\in\mathbb{Z}^+, \tag{15.210}
$$

and continuity of H follows. From the inversion formula $H^2 = -I$, it follows that $H^{-1} = -H$, and, because H is continuous, so is H^{-1}. Since H is invertible, it is bijective. Hence, the Hilbert transform is a homeomorphism from $\mathcal{D}_{L^p}(\mathbb{R})$ to $\mathcal{D}_{L^p}(\mathbb{R})$. Taking advantage of the factorization given by Eq. (15.35) allows the preceding result to be extended to the n-dimensional Hilbert transform. That is, H_n is a homeomorphism from $\mathcal{D}_{L^p}(\mathbb{R}^n)$ to $\mathcal{D}_{L^p}(\mathbb{R}^n)$, for $1 < p < \infty$. The details are left for the reader to complete.

From the discussion of this section it follows that the Hilbert transform operator is a linear isomorphism from $\mathcal{D}'_{L^p}(\mathbb{R})$ to $\mathcal{D}'_{L^p}(\mathbb{R})$ for $1 < p < \infty$ (Pandey and Chaudhry, 1983; Pandey, 1996, p. 97). This result can be generalized to the n-dimensional case, so that H_n is an isomorphism from $\mathcal{D}'_{L^p}(\mathbb{R}^n)$ to $\mathcal{D}'_{L^p}(\mathbb{R}^n)$, for $1 < p < \infty$.

15.14 Connection with analytic functions

The astute reader will have noticed the absence of any discussion so far on the connection of the Hilbert transforms in E^n with analytic functions. Historically, the important developments for the n-dimensional Hilbert transform theory did not evolve from the theory of several complex variables. This can be contrasted sharply with the situation for the one-dimensional case, where the theory of one complex variable played a central role. E. M. Stein, one of the modern masters of the subject, expressed this in the following way (Stein, 1998, p. 1132): "It is ironic that complex methods with their great power and success in the one-dimensional theory actually stood in the way of progress to higher dimensions and appeared to block further progress." In this section just a couple of points connected with the topic are mentioned very concisely; the end-notes give suggestions for further exploration.

The theory of functions of several complex variables is not a simple extension of the corresponding one-variable theory. There are results for the n-variable theory ($n > 1$) that do not have counterparts for the one-variable theory. Differences show up, for example, in the case of conformal mapping of an analytic function of one variable, versus attempts to conformally map an analytic function of more than one variable. An important area is to characterize the domains $D \subset \mathbb{C}^n$ ($n > 1$) that are or are not domains of holomorphy. This contrasts sharply with the corresponding situation for the case $n = 1$.

Let $f(z_1, z_2, \ldots, z_n)$ denote a function of n complex variables with the customary assignment

$$z_k = x_k + \mathrm{i}y_k, \quad k = 1, \ldots, n. \tag{15.211}$$

The function $f(z_1, z_2, \ldots, z_n)$ is analytic in a $2n$-dimensional domain $D \subset \mathbb{C}^n$ if, at every point in the domain, the derivative with respect to each complex variable is defined. This is often supplemented by the requirement that the function is continuous in D. The latter statement is a consequence of the former, but its adoption makes the proof of various theorems simpler. That the continuity statement is redundant was established at the turn of the twentieth century by F. Hartogs. An alternative approach is to define a holomorphic function as follows. The function $f(z_1, z_2, \ldots, z_n)$ is holomorphic in a domain $D \subset \mathbb{C}^n$ if a power series development of the form

$$f(z_1, z_2, \ldots, z_n) = \sum_{\mu_1}^{\infty} \sum_{\mu_2}^{\infty} \cdots \sum_{\mu_n}^{\infty} \alpha_{\mu_1 \mu_2 \cdots \mu_n} (z_1 - z_1^{(0)})^{\mu_1} (z_2 - z_2^{(0)})^{\mu_2} \cdots (z_n - z_n^{(0)})^{\mu_n} \tag{15.212}$$

can be written for each point $z^{(0)}$ in D. This definition of a holomorphic function implies the previous definition. Other statements for a function to be analytic in several variables can be given (see, for example, Krantz (1982, p. 3)).

Let u and v denote the real and imaginary parts of f. The Cauchy–Riemann equations take the form

$$\frac{\partial u}{\partial x_k} = \frac{\partial v}{\partial y_k}, \quad k = 1, 2, \ldots, n \tag{15.213}$$

and

$$\frac{\partial u}{\partial y_k} = -\frac{\partial v}{\partial x_k}, \quad k = 1, 2, \ldots, n. \tag{15.214}$$

From these results it follows that

$$\frac{\partial^2 u}{\partial x_k^2} + \frac{\partial^2 u}{\partial y_k^2} = 0, \quad k = 1, 2, \ldots, n, \tag{15.215}$$

$$\frac{\partial^2 u}{\partial x_k \partial x_l} + \frac{\partial^2 u}{\partial y_k \partial y_l} = 0, \quad k, l = 1, 2, \ldots, n, \tag{15.216}$$

and

$$\frac{\partial^2 u}{\partial x_k \partial y_l} - \frac{\partial^2 u}{\partial y_k \partial x_l} = 0, \quad k \neq l, \quad k, l = 1, 2, \ldots, n. \tag{15.217}$$

The reader is left to consider what results are obtained for the function v.

The Cauchy integral formula for a function of several variables can be cast in the following form. Let $f(z_1, z_2, \ldots, z_n)$ be holomorphic in a neighborhood of the domain D formed from the direct product $D_1 \times D_2 \times D_3 \times \cdots \times D_n$; then,

$$f(z_1, z_2, \ldots, z_n) = \frac{1}{(2\pi \mathrm{i})^n} \int_{\partial D_1 \times \partial D_2 \times \cdots \times \partial D_n} \frac{f(w_1, w_2, \ldots, w_n)\, \mathrm{d}w_1\, \mathrm{d}w_2 \cdots \mathrm{d}w_n}{(w_1 - z_1)(w_2 - z_2) \cdots (w_n - z_n)}, \tag{15.218}$$

where ∂D_k denotes the boundary of the domain D_k. There is a distinction between the one-variable formula and the preceding result. For the case $n = 1$, the integral is over the complete boundary of the domain, whereas in Eq. (15.218), applied to the case $n = 2$, the integral is over a two-dimensional manifold, but the complete topological boundary is of dimension $2n - 1$; that is, it would be a three-dimensional manifold.

If the function $f(z_1, z_2, \ldots, z_n)$ can be expressed as follows:

$$f(z_1, z_2, \ldots, z_n) = \frac{g(z_1, z_2, \ldots, z_n)}{h(z_1, z_2, \ldots, z_n)} \tag{15.219}$$

in the neighborhood of the point $(w_1, w_2, \ldots, w_n)$, where both the functions g and h are analytic at $(w_1, w_2, \ldots, w_n)$, and

$$g(w_1, w_2, \ldots, w_n) \neq 0 \tag{15.220}$$

and

$$h(w_1, w_2, \ldots, w_n) = 0, \tag{15.221}$$

then f has a pole at the point $(w_1, w_2, \ldots, w_n)$. The locus of points in the $2n-2$-dimensional manifold, where

$$h(z_1, z_2, \ldots, z_n) = 0, \tag{15.222}$$

will denote the other poles of the function f.

A result worth noting is the extension of the Sokhotsky–Plemelj relations to the case of more than one variable. Recall from Sections 3.7 and 3.8 that the Sokhotsky–Plemelj formulas play an important role in establishing the inversion formula for the Cauchy integral, and this can be applied directly to the Hilbert transform. For the case of more than one variable, the complexity increases somewhat. If $n = 2$, there are four relationships: two involving a single-variable Cauchy integral over the boundaries ∂D_1 and ∂D_2, respectively, a double integral over the product $\partial D_1 \times \partial D_2$, and a fourth relationship relating the kernel function to be determined in terms of the components of the function $f(z)$ that are holomorphic in various domains D_1, D_2, and their complements.

The ease with which Eq. (15.218) can be employed to derive useful connections between the real and imaginary components of a function f depends directly on establishing the domain of holomorphy for this function. There are a small number of applications that have employed this type of approach; namely a few problems in scattering theory, and some limited applications in nonlinear optics.

Notes

§15.1 For those wishing to explore the details of the Calderón–Zygmund methods, the papers of Calderón and Zygmund (1952, 1954, 1955, 1956a,b, 1957); the works by Zygmund (1956b, 1957, 1971); and the books by Neri (1971), Stein and Weiss (1971), Journé (1983), Torchinsky (1986), Christ (1990), Meyer and Coifman (1997), and Duoandikoetxea (2001), provide good starting points. Pioneering work on the topic was carried out by Tricomi (1926), Giraud (1934, 1936), and especially Mikhlin (see under the alternative spelling Mihlin (1950)). For further insights, see Horváth (1953b), Cotlar (1955), Fabes and Rivière (1966), and Ferrando, Jones, and Reinhold (1996). For a discussion of the Hilbert transform in the complex plane, see Reich (1967) and Iwaniec (1987). Parts of the discussion of this section are based on Neri (1971, chap. 4). For some connections with Clifford analysis, see Delanghe (2004), Brackx and De Schepper (2005a, 2005b), Brackx, De Knock, and De Schepper (2006a), Brackx, De Schepper, and Eelbode (2006b), and Brackx *et al.* (2006c).

§15.2 For additional reading, see Pandey (1996).

§15.3 Some further work on aspects of the double Hilbert transform can be found in Duffin (1957), Fefferman (1972), Basinger (1976), and Prestini (1985). For an application in power spectra measurements, see Stark, Bennett, and Arm (1969).

§15.4 For a discussion of the inversion of the multi-dimensional Hilbert transform, see Bitsadze (1987).

§15.7 See Pandey (1996, p. 186) for more discussion on dilation and translation operators in connection with the n-dimensional Hilbert transform. For some background reading on the one-dimensional case, see McLean and Elliott (1988).

§15.10 For further reading, see Sokół-Sokóskił (1947), Zygmund (1949), Calderón and Zygmund (1954), and Helson (1958, 1959).

§15.11 The method of rotations is due to Calderón and Zygmund (1956a).

§15.12 Some additional sources for information on the Riesz transform are the books by Stein (1970, 1993), and the papers by Horváth (1959), Berkson and Gillespie (1985), Bañuelos and Wang (1995), and Iwaniec and Martin (1996). For a discussion on hypersingular Hilbert–Riesz kernels, see Horváth *et al.* (1987). Further reading on the case of even kernels can be found in Neri (1971, p. 104), Stein and Weiss (1971, p. 224), and Duoandikoetxea (2001, p. 77). The Riesz transforms of the Gaussian are calculated in Kochneff (1995). For a discussion of a related operator, see Fefferman (1971).

§15.13 Additional reading on the n-dimensional Hilbert transform of distributions can be found in Horváth (1956), Vladimirov (1979), Pandey (1996), and a series of papers by Pandey and his collaborators: Pandey (1982), Chaudhry and Pandey (1985), and Singh and Pandey (1990a, 1990b). The latter authors have proved that the n-dimensional Hilbert transform is an automorphism on $\mathcal{D}'_{L^p}(\mathbb{R}^n)$ for $1 < p < \infty$ (Singh and Pandey, 1990a). For a proof of the statement connecting the invertible property of an operator with its bijective property, see, for example, Kantorovich and Akilov (1982, p. 152). An account of the many contibutions of John Horváth can be found in Ortner (2004).

§15.14 For some sources on the theory of several complex variables at the introductory level, see Cartan (1963), Bremermann (1965b), Osgood (1966), and Krantz (1987, 1990), and for a more advanced discussion, see Krantz (1982). The article by Wightman (1960) might be of interest to physics students. See Fuks (1963, p. 34) for a discussion of the Sokhotsky–Plemelj relations for $n > 1$.

Exercises

15.1 Suppose $f(s,t)$ has compact support and vanishes outside a circle of radius r_0, and that $f \in L$. Consider the logarithmic potential given by

$$U(x,y) = \int_{E^2} f(x-s, y-t) \log\left\{\frac{1}{\sqrt{(s^2+t^2)}}\right\} \mathrm{d}s\, \mathrm{d}t;$$

show that the integral converges absolutely.

15.2 Show that on almost every line parallel to either the x- or y-axis that $U(x,y)$ defined in Exercise 15.1 is continuously differentiable. Show that the integrals corresponding to $U_x(x,y)$ and $U_y(x,y)$ converge, and that these derivatives are absolutely continuous functions.

15.3 With $U(x,y)$ as defined in Exercise 15.1, evaluate $U_{xx}(x,y) + U_{yy}(x,y)$.

15.4 If $\Omega \in L^1(\Sigma)$, and $f \in L^1(E^n)$, show that $\mathcal{H}_{n,\varepsilon}f$ exists *a.e.*

15.5 Derive extensions to $\mathbb{R}^n$ for the key results of McLean and Elliott given in Section 4.24. Determine if an n-dimensional version of Eq. (4.550) exists.

15.6 Is there an analog of the Bedrosian theorem for the n-dimensional Hilbert transform?

15.7 If $f \in L^2(E^n)$, show that the Riesz transforms satisfy $\sum_{k=1}^{n} \|R_k f\|_2^2 = \|f\|_2^2$.

15.8 Is there an analog of the Tricomi theorem for the n-dimensional Hilbert transform?

15.9 If $f \in L^p(E^n)$, for $1 \leq p < \infty$ and $f \geq 0$, show that

$$\left[\int_{E^n}\left(\int_{E^n} f(x,y)\mathrm{d}y\right)^p \mathrm{d}x\right]^{p^{-1}} \leq \int_{E^n}\left(\int_{E^n} f(x,y)^p\,\mathrm{d}x\right)^{p^{-1}} \mathrm{d}y.$$

15.10 Determine whether the Riesz transforms commute with translations.

15.11 Do the Riesz transforms commute with dilations?

15.12 If the distribution T in Eq. (15.193) is replaced by a function $f \in L^p(\mathbb{R}^n)$, how is the statement after Eq. (15.193) modified?

15.13 Evaluate the n-dimensional Hilbert transform of the distribution $p.v.(1/x_1x_2x_3\cdots x_n)$ which belongs to $\mathcal{D}'_{L^p}(\mathbb{R}^n)$, for $1 < p < \infty$.

15.14 If $\delta(x)$ is shorthand for $\delta(x_1, x_2, \ldots, x_n)$, show that $\delta(x) \in \mathcal{D}'_{L^p}(\mathbb{R}^n)$.

15.15 Evaluate the n-dimensional Hilbert transform of the distribution $\delta(x_1, x_2, \ldots, x_n)$.

15.16 If $T \in \mathcal{D}'_{L^p}(\mathbb{R}^n)$ and $S \in \mathcal{D}'_{L^p}(\mathbb{R}^n)$, solve the symbolic equation $T = H_nT + S$. Consider both cases for even n and odd n.

15.17 Given a distribution $T \in \mathcal{D}'(\mathbb{R}^n)$, establish the derivative relationship $\partial^k H_n T = H_n \partial^k T$, for $k \in \mathbb{N}$.

15.18 Prove Eq. (15.203).

15.19 For distributions $T \in \mathcal{D}'(\mathbb{R}^n)$ and $S \in \mathcal{D}'(\mathbb{R}^n)$, determine if there is a solution for T of the equation $H_n\{\partial^2 T/\partial x_1 \partial x_2\} = S$.

16

Some further extensions of the classical Hilbert transform

16.1 Introduction

The Hilbert transform on $\mathbb{R}$ and on $\mathbb{T}$ are traditionally regarded as the classical Hilbert transforms. The truncated Hilbert transform is an extension of the classical transform to the interval (a, b). When a and b are both finite, this leads to the finite Hilbert transform, which was discussed in Chapter 11. When the interval is $(0, \infty)$, the one-sided Hilbert transform is obtained, which is also referred to as the reduced Hilbert transform, and this was dealt with in Sections 8.3 and 12.7. In this chapter some extensions of these forms of the Hilbert transform are discussed, including both one-dimensional and some specialized n-dimensional cases. The preceding chapter discussed the standard n-dimensional case. Part of the motivation for considering several of these extensions is that they represent generalizations of results presented earlier for the Hilbert transform and some of its most common variants. A number of the forms discussed have not shown up in applied problems. From a mathematical standpoint, it is occasionally the case that the underlying methods that are used to establish results about some of these Hilbert transform extensions take on considerably greater importance than the particular end result. Many of these more specialized variants of the standard Hilbert transform are less widely known among workers in the physical sciences and engineering.

16.2 An extension due to Redheffer

The following discussion is based on Redheffer (1968, 1971). The kernel of the classical Hilbert transform on $\mathbb{R}$ can be expanded formally in the following ways (with t the integration variable):

$$K(x - t) = \frac{1}{\pi x}\left\{1 + \frac{t}{x} + \frac{t^2}{x^2} + \cdots\right\}, \quad \text{near } t = 0, \tag{16.1}$$

and

$$K(x - t) = -\frac{1}{\pi x}\left\{\frac{x}{t} + \frac{x^2}{t^2} + \cdots\right\}, \quad \text{near } |t| = \infty. \tag{16.2}$$

The function $K_m(r)$ is introduced by

$$K_m(r) = 1 + r + r^2 + \cdots + r^{m-1}, \quad \text{for } |r| \leq 1, \tag{16.3}$$

and

$$K_m(r) + K_m(r^{-1}) = 1, \quad \text{for } |r| > 1. \tag{16.4}$$

Redheffer defined the modified Hilbert transform as follows:

$$H_{\mathrm{R}_m} f(x) = \frac{1}{\pi} P \int_{-\infty}^{\infty} \left\{ \frac{1}{x-t} - \frac{1}{x} K_m\left(\frac{t}{x}\right) \right\} f(t) \mathrm{d}t, \tag{16.5}$$

and the subscript on H is employed to distinguish the preceding integral from the classical Hilbert transform and to indicate the index m. Redheffer omits the factor of π, but it is inserted here to retain consistency with previous formulas. In the vicinity of $t \to 0$, the integral behaves like $\int_{|t|>\varepsilon} t^m f(t)\{1 + O(t)\}\mathrm{d}t$, and as $t \to \pm\infty$ it behaves like $\int_{|t|>\varepsilon^{-1}} t^{-m} f(t)\{1 + O(t^{-1})\}\mathrm{d}t$. If Hf exists, it follows that

$$H_{\mathrm{R}_1} f(x) = Hf(x) - \frac{1}{\pi x} \int_{-|x|}^{|x|} f(t)\mathrm{d}t; \tag{16.6}$$

and if f is an odd function, then

$$H_{\mathrm{R}_1} f(x) = Hf(x). \tag{16.7}$$

Redheffer gave a weighted norm inequality for $H_{\mathrm{R}_m} f$, which takes the form

$$\left\| x^a H_{\mathrm{R}_m} f \right\|_p \leq C \left(\mathfrak{R}_p + \frac{m}{m - |\alpha|} \right) \left\| x^a f \right\|_p, \tag{16.8}$$

where $\mathfrak{R}_p$ is the Riesz constant, C is a positive constant, $\alpha = a - (1 - p^{-1})$, and $|\alpha| < m$. The interested reader can pursue the proof of this result in Redheffer (1971). Several results discussed by Hardy and Littlewood (1936), Chen (1944), Babenko (1948), and Flett (1958) are obtained as special cases of Eq. (16.8), and some of these were discussed in Section 7.7. For example, consider the case $m = 1$, suppose f is an even function, and let

$$g(x) = \frac{2}{\pi x} \int_0^{|x|} f(t)\mathrm{d}t, \tag{16.9}$$

then from Eq. (16.6) it follows that

$$x^a Hf(x) = x^a H_{\mathrm{R}_1} f(x) + x^a g(x). \tag{16.10}$$

Applying Minkowski's inequality to this result yields

$$\left\{\int_0^\infty \left|x^a Hf(x)\right|^p \mathrm{d}x\right\}^{p^{-1}} \leq \left\{\int_0^\infty \left|x^a H_{\mathrm{R}_1} f(x)\right|^p \mathrm{d}x\right\}^{p^{-1}} + \left\{\int_0^\infty \left|x^a g(x)\right|^p \mathrm{d}x\right\}^{p^{-1}}. \tag{16.11}$$

The preceding result is simplified by applying an inequality of Hardy (Hardy *et al.*, 1952, p. 245): if $p > 1$, $r > 1$, and $K(x) = \int_0^x k(t)\mathrm{d}t$, then

$$\int_0^\infty x^{-r}|K(x)|^p \,\mathrm{d}x < \left(\frac{p}{|r-1|}\right)^p \int_0^\infty x^{p-r}|k(x)|^p \,\mathrm{d}x. \tag{16.12}$$

Let

$$K(x) = x^{a+rp^{-1}} \int_0^x k(t)\mathrm{d}t, \tag{16.13}$$

with

$$k(t) = \frac{1}{\pi x} f(t); \tag{16.14}$$

then

$$\int_0^\infty \left|x^a g(x)\right|^p \mathrm{d}x < C_{p,a} \int_0^\infty \left|x^a f(x)\right|^p \mathrm{d}x, \tag{16.15}$$

assuming that both integrals are well defined and that the constant $C_{p,a}$ depends on the indicated variables. In what follows, this constant is not necessarily the same at each occurrence. The preceding result can be rewritten as follows:

$$\left\|x^a g\right\|_p < C_{p,a} \left\|x^a f\right\|_p. \tag{16.16}$$

Taking advantage of Eqs. (16.15) and (16.8) allows Eq. (16.11) to be simplified to

$$\left\{\int_0^\infty \left|x^a Hf(x)\right|^p \mathrm{d}x\right\}^{p^{-1}} < C_{p,a} \left\{\int_0^\infty \left|x^a f(x)\right|^p \mathrm{d}x\right\}^{p^{-1}} + C'_{p,a} \left\{\int_0^\infty \left|x^a f(x)\right|^p \mathrm{d}x\right\}^{p^{-1}}, \tag{16.17}$$

and hence

$$\int_0^\infty \left|x^a Hf(x)\right|^p \mathrm{d}x < C_{p,a} \int_0^\infty \left|x^a f(x)\right|^p \mathrm{d}x. \tag{16.18}$$

This is the Hardy–Littlewood inequality discussed in Section 7.7. The reader might like to explore Eq. (16.8) for the case f is even and $m = 2$, for which $H_{\mathrm{R}_2} f = H_{\mathrm{R}_1} f$, and determine what inequality results for Hf.

Redheffer also gave inequalities of the following form:

$$\left\| \frac{H_{\mathrm{R}_m} f(x)}{(1+|x|)^a} \right\|_p \leq C_{p,m,a} \left\| \frac{f(x)}{(1+|x|)^a} \right\|_p \tag{16.19}$$

and

$$\int_{-\infty}^{\infty} \frac{\left|H_{\mathrm{R}_2} f(x)\right|^2}{1+x^2} \mathrm{d}x \leq C \int_{-\infty}^{\infty} \frac{|f(x)|^2}{1+x^2} \mathrm{d}x. \tag{16.20}$$

The interested reader can pursue further discussion of these results in Redheffer (1971).

16.3 Kober's definition for the L^∞ case

Kober (1943a) gave an extension of the definition of the Hilbert transform operator when $f \in L^\infty(\mathbb{R})$. He introduced the operator H_{K} via the definition

$$H_{\mathrm{K}} f(x) = \frac{1}{\pi} P \int_{-\infty}^{\infty} \left\{ \frac{t}{t^2+1} + \frac{1}{x-t} \right\} f(t) \mathrm{d}t, \tag{16.21}$$

where the sign has been reversed to be consistent with previous usage. This has been termed the *renormalized* Hilbert transform. Some writers use the same notation for both Kober's operator and for the Hilbert transform operator. If $f \in L^p(\mathbb{R})$ for $1 \leq p < \infty$, then $H_{\mathrm{K}} f$ and Hf differ only by a constant, which is equal to the integral

$$\frac{1}{\pi} \int_{-\infty}^{\infty} \frac{tf(t)}{t^2+1} \mathrm{d}t.$$

A key result is that the change of variables

$$t = \tan(\theta/2), \quad x = \tan(\phi/2), \tag{16.22}$$

with

$$g(\theta) = f\left(\tan \frac{\theta}{2}\right), \tag{16.23}$$

leads to (see Eqs. (3.128)–(3.133))

$$H_{\mathrm{K}} f(x) = \frac{1}{2\pi} P \int_{-\pi}^{\pi} g(\theta) \cot\left(\frac{\phi - \theta}{2}\right) \mathrm{d}\theta \equiv \mathcal{H} g(\phi). \tag{16.24}$$

Recall that if $F(z) = u(z) + \mathrm{i}v(z)$ is analytic in the unit disc and u and v are real-valued, then

$$F(z) = \frac{1}{2\pi}\int_{-\pi}^{\pi} \frac{\mathrm{e}^{\mathrm{i}\phi} + z}{\mathrm{e}^{\mathrm{i}\phi} - z} f(\mathrm{e}^{\mathrm{i}\phi})\mathrm{d}\phi, \quad |z| < 1. \tag{16.25}$$

Suppose that f is real-valued, then it follows that $v(0) = 0$, which is referred to as the *normalization* condition for the disc. For the upper-half complex plane, recall the following result:

$$v(z) = \int_{-\infty}^{\infty} Q(x - t, y) f(t)\mathrm{d}t = \frac{1}{\pi}\int_{-\infty}^{\infty} \frac{(x - t)}{(x - t)^2 + y^2} f(t)\mathrm{d}t, \tag{16.26}$$

where $Q(x, y)$ is the conjugate Poisson kernel. The normalization employed in this case is $\lim_{y\to\infty} v(x + \mathrm{i}y) = 0$. This result is often stated as $v(\mathrm{i}\infty) = 0$. For the case $f \in L^{\infty}(\mathbb{R})$, the normalization is set as $v(\mathrm{i}) = 0$, which follows directly from

$$v(z) = \int_{-\infty}^{\infty} \{Q(x - t, y) + \pi^{-1}(1 + t^2)^{-1}t\} f(t)\mathrm{d}t. \tag{16.27}$$

The preceding result leads directly to Kober's formula for the operator H_{K}. The normalization condition $v(\mathrm{i}) = 0$ ensures convergence of the integral in Eq. (16.21) as $t \to \infty$.

From Eqs. (16.24) and (6.34), it follows that

$$\begin{aligned} H_{\mathrm{K}}^2 f(x) &= \mathcal{H}^2 g(\phi) \\ &= -g(\phi) + \frac{1}{2\pi}\int_{-\pi}^{\pi} g(\theta)\mathrm{d}\theta, \end{aligned} \tag{16.28}$$

and reversing the change of variables introduced in Eq. (16.22) yields

$$H_{\mathrm{K}}^2 f(x) = -f(x) + \frac{1}{\pi}\int_{-\infty}^{\infty} \frac{f(t)\mathrm{d}t}{1 + t^2}. \tag{16.29}$$

In general, $H_{\mathrm{K}} f$ does not necessarily belong to $L^{\infty}(\mathbb{R})$.

A straightforward calculation shows that

$$(H_{\mathrm{K}} f(t^{-1}))(x^{-1}) = -H_{\mathrm{K}} f(x). \tag{16.30}$$

Kober gave the following bound:

$$\int_{-\infty}^{\infty} \frac{|H_{\mathrm{K}} f(t)|^p\,\mathrm{d}t}{(1 + t^2)^{\mu}} \le C_p^p \int_{-\infty}^{\infty} \frac{|f(t)|^p\,\mathrm{d}t}{1 + t^2}, \quad \text{for } 1 < p < \infty, \quad 1 \le \mu. \tag{16.31}$$

This result can be proved by changing the integration range from $(-\infty, \infty)$ to $(-\pi, \pi)$, taking note of Eq. (16.24), using a simple bound for $\cos(\phi/2)$, and applying the Riesz inequality Eq. (6.167). The reader is left to complete the details. Kober also gave the result that if $f \in L^{\infty}(\mathbb{R})$, and if

$$|f(t+h) - f(t)| \leq C|h|^{\alpha}, \quad 0 < \alpha < 1, \tag{16.32}$$

uniformly in t, then $H_{\mathrm{K}} f$ satisfies a Lipschitz condition with the same order α as f.

16.4 The Boas transform

Boas (1936) was interested in characterizing the class of functions having a Fourier transform which vanished on a finite interval. The Boas transform emerged from that investigation, and it is defined by the following relationship:

$$\begin{aligned} Bf(x) &= \frac{1}{\pi} P \int_0^{\infty} \frac{\{f(x+t) - f(x-t)\} \sin t \, dt}{t^2} \\ &= \frac{1}{\pi} P \int_{-\infty}^{\infty} \frac{f(x+t) \sin t \, dt}{t^2}. \end{aligned} \tag{16.33}$$

Note the close similarity to the Hilbert transform. Also note that the Hilbert transform has an intimate connection with Fourier transforms vanishing on a particular interval (see Section 4.15), which is one of the reasons for its central importance in numerous applications in engineering and the physical sciences.

16.4.1 Connection with the Hilbert transform

The first key result for the Boas transform is the following. If f is a trigonometric polynomial,

$$f(x) = \frac{a_0}{2} + \sum_{k=1}^{n} a_k \cos kx + b_k \sin kx, \tag{16.34}$$

then

$$Bf(x) = -Hf(x). \tag{16.35}$$

To prove the preceding result, the Boas transform of $\sin ax$ and $\cos ax$ are required. Let $f(x) = \sin ax$, where a is a constant; then,

$$\begin{aligned} B(\sin ax) &= \frac{1}{\pi} \int_0^{\infty} \frac{\{\sin a(x+t) - \sin a(x-t)\} \sin t \, dt}{t^2} \\ &= \frac{2 \cos ax}{\pi} \int_0^{\infty} \frac{\sin at \sin t \, dt}{t^2} \end{aligned}$$

$$= \frac{\cos ax}{\pi} \int_0^\infty \frac{\{(a+1)\sin(a+1)t + (1-a)\sin(a-1)t\}dt}{t}$$

$$= \frac{1}{2} \cos ax\{(a+1)\,\mathrm{sgn}(a+1) - (a-1)\,\mathrm{sgn}(a-1)\}, \qquad (16.36)$$

and an integration by parts has been used to simplify the integral. In a similar fashion, it follows for the case $f(x) = \cos ax$ that

$$B(\cos ax) = \frac{1}{\pi} \int_0^\infty \frac{\{\cos a(x+t) - \cos a(x-t)\}\sin t\, dt}{t^2}$$

$$= -\frac{2\sin ax}{\pi} \int_0^\infty \frac{\sin at \sin t\, dt}{t^2}$$

$$= -\frac{1}{2} \sin ax\{(a+1)\,\mathrm{sgn}(a+1) - (a-1)\,\mathrm{sgn}(a-1)\}. \qquad (16.37)$$

Since the Boas transform of a constant is zero, then, on using the last two results, Eq. (16.34) leads to

$$Bf(x) = \sum_{k=1}^{n} B\{a_k \cos kx + b_k \sin kx\}$$

$$= \frac{1}{2} \sum_{k=1}^{n} [-a_k \sin kx\{(k+1)\,\mathrm{sgn}(k+1) - (k-1)\,\mathrm{sgn}(k-1)\}$$

$$+ b_k \cos kx\{(k+1)\,\mathrm{sgn}(k+1) - (k-1)\,\mathrm{sgn}(k-1)\}]$$

$$= \sum_{k=1}^{n} [-a_k \sin kx + b_k \cos kx]. \qquad (16.38)$$

Since

$$Hf(x) = \sum_{k=1}^{n} a_k H(\cos kx) + b_k H(\sin kx)$$

$$= \sum_{k=1}^{n} a_k \sin kx - b_k \cos kx, \qquad (16.39)$$

Eq. (16.35) follows.

A number of the properties of the Boas transform, with the focus being the connections with the Hilbert transform, are examined in the remainder of this section. The first property considered is the relationship between the Boas and the Hilbert transform for a general function for which both transforms exist. Let

$$g(t) = \frac{1 - \cos t}{\pi t^2}; \qquad (16.40)$$

making use of $H[x^{-2}(1 - \cos x)] = x^{-2}(x - \sin x)$ yields

$$\begin{aligned}
(Hf * g)(x) &= \int_{-\infty}^{\infty} (Hf)(t)\, g(x-t)\mathrm{d}t \\
&= \frac{1}{\pi}\int_{-\infty}^{\infty} \frac{1}{\pi} P \int_{-\infty}^{\infty} \frac{f(s)\mathrm{d}s}{(x-t)-s} \frac{1-\cos t}{t^2}\mathrm{d}t \\
&= \frac{1}{\pi}\int_{-\infty}^{\infty} f(s)\mathrm{d}s \frac{1}{\pi} P \int_{-\infty}^{\infty} \frac{1-\cos t}{[(x-s)-t]t^2}\mathrm{d}t \\
&= \frac{1}{\pi}\int_{-\infty}^{\infty} f(s) \frac{x-s-\sin(x-s)}{(x-s)^2}\mathrm{d}s \\
&= \frac{1}{\pi} P \int_{-\infty}^{\infty} \frac{f(s)\mathrm{d}s}{x-s} - \frac{1}{\pi} P \int_{-\infty}^{\infty} f(x-t)\frac{\sin t}{t^2}\mathrm{d}t,
\end{aligned} \tag{16.41}$$

from which it follows that

$$Bf(x) = -Hf(x) + \{Hf * g\}(x). \tag{16.42}$$

16.4.2 Parseval-type formula for the Boas transform

A Parseval-type formula holds for the Boas transform:

$$\int_{-\infty}^{\infty} h(x)Bf(x)\mathrm{d}x = -\int_{-\infty}^{\infty} f(x)Bh(x)\mathrm{d}x, \tag{16.43}$$

where $f \in L^2(\mathbb{R})$ and $h \in L^2(\mathbb{R})$. Using Eq. (16.42) and utilizing the Parseval-type formula for the Hilbert transform, Eq. (4.176), leads to

$$\begin{aligned}
\int_{-\infty}^{\infty} h(x)Bf(x)\mathrm{d}x &= -\int_{-\infty}^{\infty} h(x)Hf(x)\mathrm{d}x + \int_{-\infty}^{\infty} h(x)\{Hf * g\}(x)\mathrm{d}x \\
&= \int_{-\infty}^{\infty} f(x)Hh(x)\mathrm{d}x + \int_{-\infty}^{\infty} Hf(x)\{h * g\}(x)\mathrm{d}x \\
&= \int_{-\infty}^{\infty} f(x)Hh(x)\mathrm{d}x - \int_{-\infty}^{\infty} f(x)H(h * g)(x)\mathrm{d}x \\
&= \int_{-\infty}^{\infty} f(x)[Hh(x) - (Hh * g)(x)]\mathrm{d}x \\
&= -\int_{-\infty}^{\infty} f(x)Bh(x)\mathrm{d}x.
\end{aligned} \tag{16.44}$$

Let $h(x) = f(x)$; then it follows that

$$\int_{-\infty}^{\infty} f(x)Bf(x)\mathrm{d}x = 0, \tag{16.45}$$

which is the analog of the orthogonality condition for the Hilbert transform (see Section 4.12). Let $h(x) = Bf(x)$; then Eq. (16.43) yields

$$\int_{-\infty}^{\infty} [Bf(x)]^2 \mathrm{d}x = -\int_{-\infty}^{\infty} f(x) B^2 f(x) \mathrm{d}x. \tag{16.46}$$

The Boas transform also satisfies

$$\int_{-\infty}^{\infty} Bf(x) \mathrm{d}x = 0. \tag{16.47}$$

The preceding formula is the analog of Eq. (4.204), and can be derived using the following result:

$$\begin{aligned}\int_{-\infty}^{\infty} Bf(x)\mathrm{d}x &= -\int_{-\infty}^{\infty} Hf(x)\mathrm{d}x + \int_{-\infty}^{\infty} \mathrm{d}x \int_{-\infty}^{\infty} Hf(t) g(x-t)\mathrm{d}t \\ &= \int_{-\infty}^{\infty} Hf(t)\mathrm{d}t \int_{-\infty}^{\infty} \frac{1-\cos(x-t)}{\pi(x-t)^2} \mathrm{d}x \\ &= \int_{-\infty}^{\infty} Hf(t)\mathrm{d}t \\ &= 0.\end{aligned} \tag{16.48}$$

16.4.3 Iteration formula for the Boas transform

The iteration property of the Boas transform is given by

$$B^2 f(x) = -f(x) + 2(f*g)(x) - (f*g*g)(x), \tag{16.49}$$

where g is defined in Eq. (16.40). This result can be established by taking advantage of the iteration property of the Hilbert transform:

$$\begin{aligned}B^2 f(x) &= B\{-Hf(x) + H(f*g)(x)\} \\ &= -H[-Hf + H(f*g)](x) + [H\{-Hf + H(f*g)\} * g](x) \\ &= -f(x) + (f*g)(x) + [\{f - (f*g)\} * g](x);\end{aligned} \tag{16.50}$$

hence, Eq. (16.49) follows.

16.4.4 Riesz-type bound for the Boas transform

There is an analog of the Riesz formula for the Boas transform. Consideration is restricted to the case $f \in L^2(\mathbb{R})$. Starting from Eq. (16.42), then employing Minkowski's inequality,

$$\|Bf\|_2 \le \|Hf\|_2 + \|H(f*g)\|_2, \tag{16.51}$$

and then using Eq. (4.172), leads to

$$\|Bf\|_2 \leq \|f\|_2 + \|f * g\|_2 . \tag{16.52}$$

The convolution term in the preceding result can be simplified by employing Young's inequality, which the reader will recall from Section 7.10: if $f \in L^p(\mathbb{R})$ and $h \in L^q(\mathbb{R})$, then $f * h \in L^r$, where then, $r^{-1} = p^{-1} + q^{-1} - 1$, and

$$\|f * h\|_r \leq \|f\|_p \, \|h\|_q . \tag{16.53}$$

Let $h = g$ and note that $g \in L^1(\mathbb{R})$; then, using Young's inequality with $q = 1$ allows Eq. (16.52) to be written as follows:

$$\|Bf\|_2 \leq (1 + \|g\|_1) \, \|f\|_2 , \tag{16.54}$$

that is

$$\|Bf\|_2 \leq 2 \, \|f\|_2 . \tag{16.55}$$

16.4.5 Fourier transform of the Boas transform

The Fourier transform of the Boas transform takes the form

$$\mathcal{F}Bf(x) = \mathrm{i} \operatorname{sgn} x \, \{1 - \mathcal{F}g(x)\}\mathcal{F}f(x), \tag{16.56}$$

where

$$\begin{aligned} \mathcal{F}g(x) &= \int_{-\infty}^{\infty} \frac{1 - \cos t}{\pi t^2} \mathrm{e}^{-\mathrm{i}xt} \, \mathrm{d}t \\ &= \begin{cases} 0, & |x| \geq 1 \\ (1 - |x|), & |x| < 1. \end{cases} \end{aligned} \tag{16.57}$$

To establish Eq. (16.56), the analogous formula for the Hilbert transform is employed (see Eq. (5.3)), so that

$$\begin{aligned} \mathcal{F}Bf(x) &= \mathcal{F}\{-Hf + H(f * g)\}(x) \\ &= \mathrm{i} \operatorname{sgn} x \, \mathcal{F}f(x) - \mathrm{i} \operatorname{sgn} x \, \mathcal{F}(f * g)(x) \\ &= \mathrm{i} \operatorname{sgn} x \, \{1 - \mathcal{F}g(x)\}\mathcal{F}f(x), \end{aligned} \tag{16.58}$$

which is the required result.

16.4.6 Two theorems due to Boas

If $f \in L^p$ for $1 < p \leq 2$, and $\mathcal{F}f(x)$ vanishes *a.e.* on the interval $(-1, 1)$, then $Bf(x) = -Hf(x)$. Taking the Fourier transform of Eq. (16.42) leads to

$$\begin{aligned}\mathcal{F}Bf(x) &= -\mathcal{F}Hf(x) + \mathcal{F}H\{f * g\}(x) \\ &= -\mathcal{F}Hf(x) - \mathrm{i}\,\mathrm{sgn}\,x\ \mathcal{F}f(x)\mathcal{F}g(x),\end{aligned} \tag{16.59}$$

and since $\mathcal{F}g(x) = 0$ for $|x| \geq 1$, and by the hypothesis on $\mathcal{F}f(x)$, it follows that

$$\mathcal{F}Bf(x) = -\mathcal{F}Hf(x). \tag{16.60}$$

Taking the inverse Fourier transform of the preceding equation completes the argument.

A key theorem of Boas, simplified to the particular case of Eq. (16.33), states that if $f \in L^2(\mathbb{R})$, then a necessary and sufficient condition that $\mathcal{F}f(x) = 0$ *a.e.* on $(-1, 1)$ is that $B^2f(x) = -f(x)$. To establish the necessary condition, proceed as follows. From Eq. (16.49), it follows, on taking the Fourier transform, that

$$\begin{aligned}\mathcal{F}B^2f(x) &= -\mathcal{F}f(x) + 2\mathcal{F}(f * g)(x) - \mathcal{F}(f * g * g)(x) \\ &= -\mathcal{F}f(x) + 2\mathcal{F}f(x)\mathcal{F}g(x) - \mathcal{F}f(x)\{\mathcal{F}g(x)\}^2.\end{aligned} \tag{16.61}$$

Since $\mathcal{F}g(x) = 0$ for $|x| \geq 1$, and if $\mathcal{F}f(x) = 0$ on $(-1, 1)$, it follows that

$$\mathcal{F}B^2f(x) = -\mathcal{F}f(x). \tag{16.62}$$

Taking the inverse Fourier transform leads to

$$B^2f(x) = -f(x), \tag{16.63}$$

which is the desired result. To handle the sufficient condition, assume $B^2f(x) = -f(x)$. It follows that

$$2(f * g)(x) - (f * g * g)(x) = 0, \tag{16.64}$$

and, on taking the Fourier transform,

$$\mathcal{F}f(x)[2 - \mathcal{F}g(x)]\mathcal{F}g(x) = 0. \tag{16.65}$$

On the interval $(-1, 1)$, $\mathcal{F}g(x) \neq 0$ and $[2 - \mathcal{F}g(x)] \neq 0$ (see Eq. (16.57)); it follows that on $(-1, 1)$, $\mathcal{F}f(x) = 0$.

16.4.7 Inversion of the Boas transform

Let $h(x) = Bf(x)$, then from Eq. (16.58) it follows that

$$\mathcal{F}f(x) = -\mathrm{i}\,\mathrm{sgn}\,x\,[1 - \mathcal{F}g(x)]^{-1}h(x). \tag{16.66}$$

Taking the inverse Fourier transform yields the formal relationship

$$f(x) = -\mathrm{i}\mathcal{F}^{-1}\{\mathrm{sgn}\,t\,[1 - \mathcal{F}g(t)]^{-1}h(t)\}(x). \tag{16.67}$$

An alternative and much more useful approach to the inversion of the Boas transform is as follows. Let

$$\phi(x) = (Bf * \psi)(x), \tag{16.68}$$

where

$$\psi(t) = \frac{1}{\pi}\left\{\frac{\cos t}{t} + \int_0^1 \frac{\sin ut\,\mathrm{d}u}{u}\right\}. \tag{16.69}$$

This particular choice for ψ is made so that the Fourier transform of ψ cancels part of the Fourier transform of Bf given in Eq. (16.58); that is, $\mathcal{F}\psi(x)$ multiplied by $\mathrm{i}\,\mathrm{sgn}\,x\;(1 - (\mathcal{F}g)(x))$ equals one. Taking the Fourier transform of Eq. (16.68) leads to

$$\mathcal{F}\phi(x) = \mathcal{F}\{Bf * \psi\}(x) = \mathcal{F}Bf(x)\,\mathcal{F}\psi(x). \tag{16.70}$$

Now $\mathcal{F}\psi(x)$ can be evaluated as follows:

$$\mathcal{F}\psi(x) = \frac{1}{\pi}\int_{-\infty}^{\infty} \mathrm{e}^{-\mathrm{i}xt}\left\{\frac{\cos t}{t} + \int_0^1 \frac{\sin ut\,\mathrm{d}u}{u}\right\}\mathrm{d}t, \tag{16.71}$$

and the first integral contribution is given by

$$\begin{aligned}\frac{1}{\pi}\int_{-\infty}^{\infty} \mathrm{e}^{-\mathrm{i}xt}\left\{\frac{\cos t}{t}\right\}\mathrm{d}t &= -\frac{\mathrm{i}}{2\pi}\int_{-\infty}^{\infty}\frac{\sin(x+1)t + \sin(x-1)t}{t}\,\mathrm{d}t\\ &= -\frac{\mathrm{i}}{2}\{\mathrm{sgn}(x+1) + \mathrm{sgn}(x-1)\}\\ &= \begin{cases} 0, & |x| < 1\\ -\mathrm{i}\,\mathrm{sgn}\,x, & |x| > 1.\end{cases}\end{aligned} \tag{16.72}$$

The second integral contribution is given by

$$\frac{1}{\pi}\int_{-\infty}^{\infty} e^{-ixt}\int_0^1 \frac{\sin ut\,du}{u}\,dt = \frac{1}{2i\pi}\int_0^1 \frac{du}{u}\int_{-\infty}^{\infty}\{e^{i(u-x)t} - e^{-i(u+x)t}\}dt$$

$$= -i\int_0^1 \frac{\{\delta(u-x) - \delta(u+x)\}}{u}\,du$$

$$= \begin{cases} -i\,p.v.\dfrac{1}{x}, & |x| < 1 \\ 0, & |x| > 1. \end{cases} \qquad (16.73)$$

The reader might like to investigate the starting double integral in the preceding equation without direct appeal to the Dirac delta function. From Eqs. (16.72) and (16.73), it follows that

$$\mathcal{F}\psi(x) = \begin{cases} -i\,p.v.\dfrac{1}{x}, & |x| < 1 \\ -i\,\mathrm{sgn}\,x, & |x| > 1. \end{cases} \qquad (16.74)$$

Using Eqs. (16.57), (16.58), (16.70), and (16.74) leads to

$$\mathcal{F}\phi(x) = \mathcal{F}Bf(x)\mathcal{F}\psi(x)$$

$$= \mathcal{F}f(x)\begin{cases} \{i\,\mathrm{sgn}\,x\,|x|\}\left\{-i\,p.v.\dfrac{1}{x}\right\} & |x| < 1 \\ \{i\,\mathrm{sgn}\,x\}\{-i\,\mathrm{sgn}\,x\}, & |x| > 1 \end{cases}$$

$$= \mathcal{F}f(x). \qquad (16.75)$$

Taking the inverse Fourier transform of the preceding equation leads to

$$f(x) = \frac{1}{\pi}\int_{-\infty}^{\infty} Bf(x-t)\left\{\frac{\cos t}{t} + \int_0^1 \frac{\sin ut}{u}\,du\right\}dt, \qquad (16.76)$$

which is the required inversion formula.

16.4.8 Generalization of the Boas transform

There is a generalization of the Boas transform which takes the form (Boas, 1936)

$$Bf(x) = \frac{1}{\pi}P\int_0^{\infty}\frac{\{f(x+t) - f(x-t)\}\mathsf{H}(t)dt}{t}. \qquad (16.77)$$

If $\mathsf{H}(t)$ is an even function, then Eq. (16.77) can be written as

$$Bf(x) = \frac{1}{\pi}P\int_{-\infty}^{\infty}\frac{f(x+t)\mathsf{H}(t)dt}{t}. \qquad (16.78)$$

The generalized transform is defined for any x for which the integral exists, taken in the sense of $\lim_{\varepsilon\to 0}\lim_{\eta\to\infty}\int_{\varepsilon}^{\eta}$. Let h denote a continuous function of bounded variation on the interval $(-1,1)$; then the function H is expressed in terms of the following integral:

$$\mathsf{H}(t) = \int_{-1}^{1} \cos st \, \mathrm{d}\mathsf{h}(s), \tag{16.79}$$

and $\mathsf{H}(x)$ has bounded variation on the interval (x_0, ∞) for some x_0, or $\mathsf{h}(s)$ is the integral of a function $k(s)$ of bounded variation on the interval $(-1,1)$. The choice $\mathsf{H}(t) = t^{-1} \sin t$, that is take $\mathsf{h}(s) = s/2$, recovers the first definition of the Boas transform given in Eq. (16.33). The interested reader can pursue further information on this generalization in the cited work of Boas.

The operator $\mathcal{B}$ is introduced by the following relationship:

$$\mathcal{B}f(x) = \frac{1}{\pi} P \int_{-\infty}^{\infty} f(x+t) \left\{ \frac{\cos t}{t} + \int_0^1 h(u) \sin ut \, \mathrm{d}u \right\} \mathrm{d}t. \tag{16.80}$$

A straightforward calculation shows that if $h(u) = 1$,

$$\mathcal{B}f(x) = -Hf(x), \tag{16.81}$$

and that if $h(u) = u$, then

$$\mathcal{B}f(x) = Bf(x). \tag{16.82}$$

So $\mathcal{B}$ can be regarded as incorporating both the Hilbert and Boas transform operators, and is therefore a generalization of both. Let $\mathsf{h}(s)$ be selected such that $\mathrm{d}\mathsf{h}(s) = (1/2)(\mathrm{d}h(s)/\mathrm{d}s)\,\mathrm{d}s$, with $h(s)$ odd and $h(1) = 1$; then, from Eq. (16.79), it follows that

$$\frac{\mathsf{H}(t)}{t} = \frac{\cos t}{t} + \int_0^1 h(u) \sin ut \, \mathrm{d}u, \tag{16.83}$$

which makes the connection between the generalized Boas transform given in Eq. (16.78) and the operator $\mathcal{B}$ defined by Eq. (16.80).

Heinig (1979) considered a generalization of the Boas transform of the form

$$B_n f(x) = \frac{1}{\pi} P \int_0^{\infty} \frac{\{f(x+t) - f(x-t)\}}{t} \left(\frac{\sin t}{t} \right)^n \mathrm{d}t. \tag{16.84}$$

This transform is connected to functions having a Fourier transform vanishing outside the interval $(-n, n)$. Not surprisingly, the Boas–Heinig transform $B_n f$ satisfies a number of properties closely related to Bf. Some of these are developed in the exercises.

16.5 The bilinear Hilbert transform

The bilinear singular integral operator is defined by

$$H_a(f(x), g(x)) = \pi^{-1} \lim_{\varepsilon \to 0} \int_{|t|>\varepsilon}^{\infty} \frac{f(x-t)g(x+at)\mathrm{d}t}{t}, \tag{16.85}$$

where a is a constant. To establish some of the key theorems for $H_a(f, g)$, it is common practice to take f and g belonging to the Schwartz class $\mathscr{S}(\mathbb{R})$ of smooth functions decaying rapidly at infinity on the real line. The bilinear singular integral operator is commonly defined without the factor of π^{-1}, but this factor is incorporated so that the special cases examined are related directly to the standard Hilbert transform, without the appearance of additional factors of π. The particular case $a = 1$ defines the bilinear Hilbert transform, that is,

$$H_1(f(x), g(x)) = \pi^{-1} \lim_{\varepsilon \to 0} \int_{|t|>\varepsilon}^{\infty} \frac{f(x-t)g(x+t)\mathrm{d}t}{t}. \tag{16.86}$$

There are three special cases for $H_a(f, g)$. The first is $a = 0$, which yields

$$H_0(f(x), g(x)) = g(x)Hf(x). \tag{16.87}$$

The second special case is $a = -1$, which leads to

$$H_{-1}(f(x), g(x)) = H\{f(x)g(x)\}. \tag{16.88}$$

The third case is $a = \infty$. Using the change of variable $w = at$ in Eq. (16.85), and taking the limit $a \to \infty$, leads to

$$H_\infty(f(x), g(x)) = \pi^{-1} \lim_{\varepsilon \to 0} \lim_{a \to \infty} \int_{|t|>\varepsilon}^{\infty} \frac{f(x - a^{-1}w)g(x+w)\mathrm{d}w}{w}, \tag{16.89}$$

and so

$$H_\infty(f(x), g(x)) = -f(x)Hg(x). \tag{16.90}$$

As a simple example, consider, for real constants α and β,

$$f(x) = \sin \alpha x, \quad g(x) = \cos \beta x; \tag{16.91}$$

then a straightforward calculation yields

$$\begin{aligned} H_a(f(x), g(x)) &= \frac{1}{\pi} P \int_{-\infty}^{\infty} \frac{\sin \alpha(x-t) \cos \beta(x+at)\mathrm{d}t}{t} \\ &= -\frac{1}{2}\{\mathrm{sgn}(\alpha + a\beta)\cos(\alpha - \beta)x + \mathrm{sgn}(\alpha - a\beta)\cos(\alpha + \beta)x\}, \end{aligned} \tag{16.92}$$

and hence the bilinear Hilbert transform evaluates to

$$H_1(f(x), g(x)) = -\frac{1}{2}\{\mathrm{sgn}(\alpha+\beta)\cos(\alpha-\beta)x + \mathrm{sgn}(\alpha-\beta)\cos(\alpha+\beta)x\}. \tag{16.93}$$

The other special cases $a=0,-1$, and ∞ of Eq. (16.92) can be quickly shown to correspond with the results obtained from Eqs. (16.87), (16.88), and (16.90), respectively.

The bilinear Hilbert transform has the obvious added complication relative to the normal Hilbert transform, in that its value is controlled by the behavior of two functions in the vicinity of the singularity. Let

$$f(x) = \mathrm{e}^{\mathrm{i}xw} f_1(x), \quad g(x) = \mathrm{e}^{\mathrm{i}xw} g_1(x), \tag{16.94}$$

and insert these into Eq. (16.86); then,

$$H_1(f(x), g(x)) = \pi^{-1}\mathrm{e}^{2\mathrm{i}wx} \lim_{\varepsilon\to 0}\int_{|t|>\varepsilon}^{\infty} \frac{f_1(x-t)g_1(x+t)\mathrm{d}t}{t}. \tag{16.95}$$

Taking the modulus leads to

$$|H_1(f,g)| = |H_1(f_1,g_1)|. \tag{16.96}$$

That is, the modulus of the bilinear Hilbert transform is invariant to a simultaneous introduction of identical phase components as in Eq. (16.94).

A key question concerning the operator $H_a(f,g)$ is whether or not there exist estimates of the form

$$\|H_a(f,g)\|_p \le C_{a,p_1,p_2}\|f\|_{p_1}\|g\|_{p_2}, \tag{16.97}$$

where the constant C_{a,p_1,p_2} depends on a, p_1, and p_2, but is independent of the functions f and g, and $p^{-1} = p_1^{-1} + p_2^{-1}$. Calderón (1965) was interested in the commutators of singular integral operators. A fundamental question that arises is whether or not $H_1(f,g)$ is a bounded operator from $L^2 \times L^2$ to L^1 (see Jones (1994)). This became known as Calderón's conjecture. From the boundedness of $H_1(f,g)$, certain bounds for commutators of singular integral operators follow. Equation (16.97) encompasses Calderón's conjecture. It took about thirty years to resolve this problem, which was finally accomplished by Lacey and Thiele (1997b, 1999). Their result is stated as follows. If $a \in \mathbb{R}\backslash\{0,-1\}$, $1 < p_1, p_2 \le \infty$, and $2/3 < p < \infty$, then Eq. (16.97) holds for $\forall f, g \in \mathscr{S}(\mathbb{R})$. The interested reader can pursue the proof, which involves some elaborate mathematical machinery, in the references just cited.

16.6 The vectorial Hilbert transform

Let h_j denote the vectorial distribution:

$$h_j = \frac{\Gamma((n+j)/2)}{\pi^{n/2}\Gamma(j/2)} p.v. \frac{x^j}{|x|^{n+j}}, \tag{16.98}$$

where $\Gamma(n)$ denotes a gamma function, $x = (x_1, x_2, \ldots, x_n)$ is a point in $\mathbb{R}^n$ with $n \geq 2$, $|x|$ is the norm of x, and the notation $r = |x|$ is employed. The components of h_j are given by

$$\frac{\Gamma((n+j)/2)}{\pi^{n/2}\Gamma(j/2)} p.v. \frac{Y_j(\sigma)}{r^n}, \ j = 1, 2, \ldots,$$

where $Y_j(\sigma)$ is a spherical harmonic, $x = r\sigma$, and $\sigma \in S^{n-1}$, where S^{n-1} denotes the unit sphere. The Fourier transform of h_j is given by (Horváth, 1956)

$$\mathcal{F}_n h_j(\xi) = (-i)^j \frac{\xi^j}{|\xi|^j}, \tag{16.99}$$

where $\xi = (\xi_1, \xi_2, \ldots, \xi_n)$ is a point in $\mathbb{R}^n$. Horváth (1956) defined the vectorial Hilbert transform of a distribution $T \in \mathcal{D}'_{L^p}$ as follows:

$$\mathbf{H}_j[T] = h_j * T, \quad j = 1, 2, \ldots \tag{16.100}$$

Note that

$$\begin{aligned}
\mathcal{F}_n[h_j * \{h_l * T\}] &= \{\mathcal{F}_n h_j\}\{\mathcal{F}_n[h_l * T]\} \\
&= \{\mathcal{F}_n h_j\}\{\mathcal{F}_n h_l\}\{\mathcal{F}_n T\} \\
&= (-\mathrm{i})^{j+l} \frac{\xi^{j+l}}{|\xi|^{j+l}} \{\mathcal{F}_n T\} \\
&= \{\mathcal{F}_n h_{j+l}\}\{\mathcal{F}_n T\} \\
&= \mathcal{F}_n\{h_{j+l} * T\}.
\end{aligned} \tag{16.101}$$

From this result it follows that

$$\mathbf{H}_j[\mathbf{H}_l[T]] = h_j * \{h_l * T\} = h_{j+l} * T = \mathbf{H}_{j+l}[T]. \tag{16.102}$$

Hence, it follows that the operators $\mathbf{H}_j$ form a semigroup under composition.

16.7 The directional Hilbert transform

In this section the form of certain directional operators in the plane is indicated. Let θ denote an angle, and $\mathrm{e}^{\mathrm{i}\theta}$ will be employed as shorthand notation to specify a point $(\cos\theta, \sin\theta)$. That is, $e^{i\theta}$ represents a direction in S^1. The truncated directional Hilbert transform operator is defined by

$$H_{\theta,\varepsilon} f(x) = \frac{1}{\pi} \int_{|t|>\varepsilon} \frac{f(x - te^{i\theta})dt}{t}. \tag{16.103}$$

Setting $\theta = 0$ recovers the standard truncated Hilbert transform; recall Eq. (3.3). The directional Hilbert transform operator is then defined by

$$H_\theta f(x) = \lim_{\varepsilon \to 0} \frac{1}{\pi} \int_{|t|>\varepsilon} \frac{f(x - te^{i\theta})dt}{t}. \tag{16.104}$$

The directional maximal Hilbert transform is defined by

$$H_{M_\theta} f(x) = \sup_{\varepsilon>0} \left| H_{\theta,\varepsilon} f(x) \right|. \tag{16.105}$$

In the literature, the notation most frequently employed to denote a maximal transform operator is to use a superscript *, so that the operator of Eq. (16.105) would be given as H_θ^*. In this work the symbol * has been reserved for the conjugation operation. A similar comment applies to other maximal operators introduced later in this chapter. The operators appearing in Eqs. (16.104) and 16.105) are bounded operators in $L^2(\mathbb{R}^2)$, for $1 < p < \infty$. Duoandikoetxea and Vargas (1995) gave the following lemma. Let f be a radial function; then,

$$\|H_\theta f - H_{\theta'} f\| \leq C_p \left|\theta - \theta'\right|^{1/p'} \|f\|_p, \quad \text{for} \quad 1 < p < 2, \tag{16.106}$$

where p' is the conjugate exponent of p and C_p is a constant independent of f. A related result for $2 \leq p < \infty$ can be given. The interested reader can pursue the proof in the given reference.

An alternative definition for the directional Hilbert transform is employed by Carro (1998). Adding a factor of π^{-1}, and noting the possible singular structure of the kernel, the directional Hilbert transform is defined by

$$Hf(x) = \frac{1}{\pi} \lim_{\varepsilon \to 0} \int_{|t|>\varepsilon} \frac{f(x - t)dt}{t}, \tag{16.107}$$

where $t \in \mathbb{R}$ and $x \in \mathbb{R}^n$, so that $(x - t) = (x_1 - t, x_2 - t, \ldots, x_n - t)$.

Garcia-Cuerva and Rubio De Francia (1985, p. 571) gave another definition for the directional Hilbert transform. Let $\{v_j\}_{j>1}$ be a set of unit vectors in $\mathbb{R}^2$. The directional Hilbert transform along v_j is denoted by H_j, and, for $f \in L^2 \cap L^p$ with $1 < p < \infty$, it is defined by the following formula:

$$\mathcal{F} H_j f(\xi) = -i \operatorname{sgn}(\xi \cdot v_j) \mathcal{F} f(\xi). \tag{16.108}$$

The directional Hilbert transform operator H_j is bounded in $L^p(\mathbb{R}^2)$, for $1 < p < \infty$.

16.8 Hilbert transforms along curves

The Hilbert transform along a curve is defined in the following manner:

$$H_\Gamma f(x) = \frac{1}{\pi} P \int_{-\infty}^{\infty} \frac{f(x - \Gamma(t))\mathrm{d}t}{t}, \quad x \in \mathbb{R}^n, \tag{16.109}$$

where $\Gamma(t)$ denotes a smooth curve in $\mathbb{R}^n$ and it is assumed in the sequel that $\Gamma(0) = 0$. It is common practice to define $H_\Gamma f$ with the factor of π^{-1} excluded, but it will be retained here so that, for the case $\Gamma(t) = t$, the standard definition of the Hilbert transform is obtained. The Hilbert transform along a curve specified by $(t, \Gamma(t))$ in $\mathbb{R}^2$ takes the following form:

$$H_\Gamma f(x,y) = \frac{1}{\pi} P \int_{-\infty}^{\infty} \frac{f(x - t, y - \Gamma(t))\mathrm{d}t}{t}. \tag{16.110}$$

A fundamental question is to determine for which curves $\Gamma(t)$ and for what values of p is H_Γ a bounded operator on $L^p(\mathbb{R}^n)$. This question, proposed about thirty years ago by E. M. Stein, has attracted considerable attention. There are known examples of curves belonging to C^∞ in $\mathbb{R}^n$ for which H_Γ is not bounded for all L^p.

It is a result of Stein and Wainger (1970) that H_Γ is bounded on $L^2(\mathbb{R}^n)$ if $\Gamma(t)$ is of the form $\Gamma(t) = (|t|^{\alpha_1} \operatorname{sgn} t, \ldots, |t|^{\alpha_n} \operatorname{sgn} t)$, for all $\alpha_i > 0$. Nagel, Rivière, and Wainger (1976) extended this result to the case $L^p(\mathbb{R}^n)$, for $1 < p < \infty$ and arbitrary n. That is,

$$\|H_\Gamma f\|_p \le C_p \|f\|_p, \tag{16.111}$$

where the constant C_p depends only on p and Γ. The reader will recognize this as the extension of the Riesz inequality to cover the case of the Hilbert transform on a particular class of curves.

Let a and b satisfy $0 < a < b < \infty$, then the expression for $H_\Gamma f$ can be decomposed in the following manner:

$$H_\Gamma f(x) = \frac{1}{\pi} P \int_{-a}^{a} \frac{f(x - \Gamma(t))\mathrm{d}t}{t} + \frac{1}{\pi} \int_{a<|t|<b} \frac{f(x - \Gamma(t))\mathrm{d}t}{t} + \frac{1}{\pi} P \int_{|t|>b}^{\infty} \frac{f(x - \Gamma(t))\mathrm{d}t}{t}. \tag{16.112}$$

The second integral on the right-hand side of this equation is straightforward to deal with for functions of the class $L^p(\mathbb{R}^n)$, for $1 \le p \le \infty$. The preceding dissection of the integral means that to determine the bounded properties of $H_\Gamma f$, it is necessary only to focus attention on the behavior of the curve near the origin and at infinity. Some authors define a modified Hilbert transform on a curve Γ in $\mathbb{R}^n$, with $n \ge 2$, by the following formula:

$$\overline{H}_\Gamma f(x) = \frac{1}{\pi} P \int_{-1}^{1} \frac{f(x - \Gamma(t))\mathrm{d}t}{t}, \quad \text{for } x \in \mathbb{R}^n, \tag{16.113}$$

in which, as previously mentioned, the π^{-1} factor is typically omitted. Studying the bounded properties of this integral is clearly an essential component of studying the bounded behavior of $H_\Gamma f$. Consider the plane curve $\Gamma \in \mathbb{R}^2$ having the following properties: $\Gamma(t) = (t, \gamma(t))$, with $\gamma(t)$ an odd function, and $\gamma(t) \in C^2$, such that $\gamma''(t) > 0$, for $t > 0$. Let $h(t) = t\gamma'(t) - \gamma(t)$. If h satisfies the condition

$$h(Ct) \geq 2h(t), \quad \text{for } 0 < t \leq \frac{1}{C}, \tag{16.114}$$

where C is a positive constant, then h is said to have a *bounded doubling "time,"* or, more simply, h is said to satisfy the *doubling property*. A key result of Nagel *et al.* (1986) is the following. If the plane curve $\Gamma(t)$ satisfies the preceding conditions, it follows that

$$\left\| \overline{H}_\Gamma f \right\|_2 \leq A \left\| f \right\|_2, \tag{16.115}$$

iff h satisfies the *doubling property*. The interested reader can pursue the proof in the cited reference. Related discussion for the operator H_Γ, including consideration of the case $p \neq 2$, can be found in Cordoba *et al.* (1986) and Carbery *et al.* (1989). The latter authors construct the following example. Let $\Gamma(t) = (t, \gamma(t))$ be a plane curve defined for $|t| \geq 1$, such that $\gamma(t)$ is an odd function and is linear on the intervals $2^j \leq t \leq 2^{j+1}$, for $j \in \mathbb{Z}^+$, and satisfies

$$\gamma(2^j) = j2^j, \quad \text{for } j \in \mathbb{Z}^+. \tag{16.116}$$

Then H_Γ is a bounded operator on $L^p(\mathbb{R}^2)$ only for $p = 2$.

A multiple Hilbert transform along a surface can be written (Nagel and Wainger, 1977) as a generalization of H_Γ. Let $\sigma(t)$ denote a surface and let $f \in C_0^\infty(\mathbb{R}^n)$; then, the multiple Hilbert transform on the surface $\sigma(t)$, denoted H_σ, is defined by

$$H_\sigma f(x,y) = \frac{1}{\pi^n} \lim_{\substack{\varepsilon \to 0 \\ N \to \infty}} \int_{\varepsilon \leq |t_1| \leq N} \cdots \int_{\varepsilon \leq |t_n| \leq N} f(x - \sigma(t)) \frac{\mathrm{d}t_1}{t_1} \cdots \frac{\mathrm{d}t_n}{t_n}. \tag{16.117}$$

With the appropriate choice of $\sigma(t)$, this form can be simplified to the n-dimensional Hilbert transform $H_n f$ given in Eq. (15.26).

16.9 The ergodic Hilbert transform

The ergodic Hilbert transform is introduced in this section. Some preliminary information on ergodic theory is considered first. Ergodic theory evolved from the ergodic hypothesis in statistical mechanics, advanced by Boltzmann in the latter part of the nineteenth century. The study of the relationship between time and ensemble averages for the properties of a system in classical statistical mechanics falls under the umbrella of ergodic theory. For example, suppose a property is observed over a long period of time, and further suppose that this long time interval can be subdivided into segments, so that the observations over the subintervals can be regarded as different systems comprising an ensemble. Observations on a single system at N arbitrary times can be made, or N arbitrary systems can be measured at the same time. The ergodic hypothesis is a statement about the equivalence of these two averages for properties in a statistical sense. When the equivalence of the two types of averages is postulated, it is often referred to as the quasi-ergodic hypothesis.

In mathematics, the focus of ergodic theory is the study of the properties of measurable transformations. Mean ergodic theory is concerned with the convergence of the sequence of averages of appropriate transformations. To introduce the basic ideas, some terminology is needed. Let X denote a non-empty set and denote by $\mathscr{R}$ any collection of subsets $\{A_1, A_2, \ldots\}$ of X which satisfy the following properties:

(i) if A_i and A_j belong to $\mathscr{R}$ then $A_i \cup A_j$ belongs to $\mathscr{R}$, and
(ii) if A_i and A_j belong to $\mathscr{R}$ then $A_i - A_j$ belongs to $\mathscr{R}$.

The collection $\mathscr{R}$ is called a ring of sets on X. It is called a σ-ring on X if it also contains the countable union of any collection of A_i. A ring with the property that X is also an element is called an algebra. A σ-ring with the property that X is also an element is referred to as a σ-algebra.

A non-negative countably additive set function μ defined on $\mathscr{R}$ is termed a measure on the σ-algebra $\mathscr{R}$. If $\mu(X) < \infty$, then μ is a finite measure. If X is the union of a countable collection of sets $A_j \in \mathscr{R}$, with the property $\mu(A_j) < \infty$, then μ is called a σ-finite measure. The combination $(X, \mathscr{R})$ is called a measurable space and $(X, \mathscr{R}, \mu)$ is termed a measure space. If the sets are Borel sets (recall Section 2.11) then the measure space is written as $(X, \mathscr{B}, \mu)$. If the Borel sets B_j satisfy $B_j \cap B_k = \varnothing$, then

$$\mu\left(\bigcup_{j=1}^{\infty} B_j\right) = \sum_{j=1}^{\infty} \mu(B_j), \tag{16.118}$$

which is termed the "countable additivity property" of the measure.

Consider a measure space $(X, \mathscr{R}, \mu)$ and a linear operator T such that $T \in L^1(X, \mathscr{R}, \mu)$. Suppose T satisfies the following properties:

$$f \geq 0,\ a.e. \Rightarrow Tf \geq 0, a.e., \tag{16.119}$$

$$\int_X |Tf| \mathrm{d}\mu \leq \int_X |f|\, \mathrm{d}\mu, \tag{16.120}$$

and, for $C > 0$,

$$|f| \leq C,\ a.e. \Rightarrow |Tf| \leq C, a.e. \tag{16.121}$$

For a non-negative n, let $R_n(f)$ denote the sum

$$R_n(f) = \frac{1}{n+1}\sum_{k=0}^{n} T^k f. \tag{16.122}$$

Just a couple of the key classical results concerning $R_n(f)$ are indicated. For a finite measure $\mu(\mathscr{R})$ and a measuring preserving transformation T of $\mathscr{R}$ into itself, then, for $f \in L^1$, the limit $\lim_{n\to\infty} \mathscr{R}_n(f)$ converges *a.e.* Suppose that the conditions in Eqs. (16.119)–(16.121) are satisfied and that $f \in L^p$ for $1 < p$, then there is an operator $\mathscr{P}$ such that

$$\lim_{n\to\infty}\left\| \frac{f + Tf + \cdots + T^n f}{n+1} - \mathscr{P}f \right\| = 0. \tag{16.123}$$

This is a statement of the mean ergodic theorem, and a proof can be found in Garsia (1970, p. 19).

The ergodic Hilbert transform associated with the invertible measure preserving transformation T is defined by

$$Hf = \lim_{n\to\infty} {\sum_{k=-n}^{n}}' \frac{1}{k} T^k f, \tag{16.124}$$

where the prime on the summation indicates that the term with $k = 0$ is excluded from the sum. The ergodic maximal Hilbert transform is defined by

$$H_{\mathrm{M}} f = \sup_{n\geq 1} \left| {\sum_{k=-n}^{n}}' \frac{1}{k} T^k f \right|. \tag{16.125}$$

By analogy with the definition of $R_n f$ in Eq. (16.122), let $R_n f$ be given by

$$R_n f = {\sum_{k=-n}^{n}}' \frac{1}{k} T^k f; \tag{16.126}$$

then

$$Hf = \lim_{n\to\infty} R_n(f), \tag{16.127}$$

taken in the pointwise sense. For $f \in L^p(\mathrm{d}\mu)$, with $1 < p < \infty$, Hf exists *a.e.*, and, in addition, $H_{\mathrm{M}} f < \infty$ *a.e.*

The ergodic Hilbert transform and the ergodic maximal Hilbert transform satisfy the following inequalities (Cotlar, 1955):

$$\int_X |Hf|^p \,\mathrm{d}\mu \leq C_p \int_X |f|^p \,\mathrm{d}\mu \tag{16.128}$$

and

$$\int_X |H_M f|^p \,\mathrm{d}\mu \le C_p \int_X |f|^p \,\mathrm{d}\mu, \tag{16.129}$$

where C_p denotes a constant independent of f, different in the two inequalities, $1 < p < \infty$, and $f \in L^p(\mathrm{d}\mu)$. There are several results that are equivalent to the statement immediately following Eq. (16.127), including Eqs. (16.128) and (16.129) (see, for example, Fernández-Cabrera, Martín-Reyes, and Torrea (1995)). In fact, both the statement and the preceding two inequalities can be generalized. The interested reader can pursue the details in Fernández-Cabrera *et al.* (1995).

16.10 The helical Hilbert transform

In this section $T : X \to X$ denotes an invertible measure-preserving transformation and $(X, \mathscr{R}, \mu)$ denotes a σ-finite measure space. The operator T is defined by

$$Tf(x) = f(Tx). \tag{16.130}$$

The helical Hilbert transform associated with the invertible measure-preserving transformation T is defined by

$$H_\theta f = \lim_{n\to\infty} \sum_{k=-n}^{n}{}' \frac{\mathrm{e}^{\mathrm{i}k\theta}}{k} T^k f, \tag{16.131}$$

where the prime on the summation indicates that the term with $k = 0$ is excluded from the sum. This is also sometimes termed the rotated ergodic Hilbert transform. The corresponding maximal helical Hilbert transform operator is defined by

$$H_{\mathrm{M}_\theta} f = \sup_{n\ge 0} \left| \sum_{k=-n}^{n}{}' \frac{\mathrm{e}^{\mathrm{i}k\theta}}{k} T^k f \right|. \tag{16.132}$$

There is a discrete analog of Eqs. (16.130) and (16.131). Let $a \in l^2(\mathbb{Z})$ and $\theta \in \mathbb{R}$; then the discrete helical Hilbert transform is defined by

$$H_\theta a(j) = \lim_{n\to\infty} \sum_{k=-n}^{n}{}' \frac{\mathrm{e}^{2\pi\mathrm{i}(j-k)\theta}}{k} a_{j-k} \tag{16.133}$$

and

$$H_{\mathrm{M}_\theta} a(j) = \sup_\theta |H_\theta a(j)|. \tag{16.134}$$

The operator H_{M_θ} is bounded, that is, for a constant C and $a \in l^2(\mathbb{Z})$,

$$\left\| H_{\mathrm{M}_\theta} a \right\|_{l^2} \le C \left\| a \right\|_{l^2}, \tag{16.135}$$

and C is independent of a.

A double maximal helical operator is defined (Campbell and Petersen, 1989; Assani and Petersen, 1992) as follows:

$$H_{\mathrm{M}_{n\theta}}f = \sup_{n\geq 1,\theta}\left|\sum_{k=-n}^{n}{}' \frac{\mathrm{e}^{2\pi ik\theta}}{k}T^k f\right|, \tag{16.136}$$

which is a bounded operator for all $f \in L^2(X, \mathcal{B}, \mu)$, so that

$$\left\|H_{\mathrm{M}_{n\theta}}f\right\|_2 \leq C\,\|f\|_2, \tag{16.137}$$

where the constant C is independent of f. For all $\lambda > 0$ and $f \in L^2$, there exists a constant $C > 0$ such that

$$\mu\{x : H_{\mathrm{M}_{n\theta}}f(x) > \lambda\} \leq \frac{C}{\lambda^2}\,\|f\|_2^2. \tag{16.138}$$

A reader interested in the proof of these results should consult the references just cited. Assani and Petersen (1992) also prove an analog of Eq. (16.134) for the discrete case as follows:

$$\left\|H_{\mathrm{M}_{n\theta}}a\right\|_{l^2} \leq C\,\|a\|_{l^2}, \quad \text{for } a \in l^2(\mathbb{Z}). \tag{16.139}$$

16.11 Some miscellaneous extensions of the Hilbert transform

Okikiolu (1967b, 1971, p. 432) gave the following form:

$$H_{(\nu)}f(x) = \frac{|x|^{-\nu}}{\pi}P\int_{-\infty}^{\infty}\frac{|t|^{\nu}f(t)\mathrm{d}t}{t-x}, \tag{16.140}$$

where ν is a real constant and Okikiolu's choice of sign convention has been retained. The notation $H_{(\nu)}$ in Eq. (16.140) is employed with a meaning different than in Chapter 15 (see Eq. (15.35)). The following results are due to Okikiolu (1967b, 1971, p. 432). If $p^{-1} - 1 < \nu < p^{-1}$, with $1 < p < \infty$, then $H_{(\nu)} : L^p(\mathbb{R}) \to L^p(\mathbb{R})$. For $f \in L^p(\mathbb{R})$, the iteration property takes the following form:

$$H_{(\nu)}^2 f(x) = -f(x), \quad a.e. \tag{16.141}$$

If $f \in L^p(\mathbb{R})$ and $g \in L^q(\mathbb{R})$, with q the conjugate exponent to p, then

$$\int_{-\infty}^{\infty}(H_{(\nu)}f)(x)\,g(x)\mathrm{d}x = -\int_{-\infty}^{\infty}f(x)(H_{(-\nu)}g)(x)\mathrm{d}x. \tag{16.142}$$

The preceding two results have the familiar form of the corresponding results for the operator H. It is left to the reader to provide the necessary proofs.

Okikiolu (1965a, 1967a) has investigated the following integrals:

$$G_\alpha f(x) = \frac{2}{\pi} P \int_{-\infty}^{\infty} \frac{\{\cos\alpha(t-x)[1-\cos\alpha(t-x)]\} f(t)\mathrm{d}t}{(t-x)^3} \tag{16.143}$$

and

$$I_\alpha f(x) = \frac{2}{\pi} P \int_{-\infty}^{\infty} \frac{\{1-\cos\alpha(t-x)\} f(t)\mathrm{d}t}{(t-x)^3}; \tag{16.144}$$

and Kober (1964) and Okikiolu (1965a, 1967a) have studied the related integrals,

$$B_\alpha f(x) = \frac{1}{\pi} P \int_{-\infty}^{\infty} \frac{\sin\alpha(t-x) f(t)\mathrm{d}t}{(t-x)^2}, \tag{16.145}$$

$$D_\alpha f(x) = \frac{1}{\pi} \int_{-\infty}^{\infty} \frac{\sin\alpha(t-x) f(t)\mathrm{d}t}{t-x}, \tag{16.146}$$

$$S_\alpha f(x) = \frac{1}{\pi} \int_{-\infty}^{\infty} \frac{\{1-\cos\alpha(t-x)\} f(t)\mathrm{d}t}{t-x}, \tag{16.147}$$

$$C_\alpha f(x) = \frac{1}{\pi} \int_{-\infty}^{\infty} \frac{\{1-\cos\alpha(t-x)\} f(t)\mathrm{d}t}{(t-x)^2}, \tag{16.148}$$

$$E_\alpha f(x) = \frac{2}{\pi} \int_{-\infty}^{\infty} \frac{\cos\alpha(t-x)\{1-\cos\alpha(t-x)\} f(t)\mathrm{d}t}{(t-x)^2}, \tag{16.149}$$

and

$$T_\alpha f(x) = \frac{2}{\pi} \int_{-\infty}^{\infty} \frac{\sin\alpha(t-x)\{1-\cos\alpha(t-x)\} f(t)\mathrm{d}t}{(t-x)^2}. \tag{16.150}$$

Assuming that f is well behaved, that is, f is continuous on $\mathbb{R}$, then the integrals in Eqs. (16.146)–(16.150) are not singular integrals. These integrals define auxilary functions which are useful in treating integrals like those occurring in Eqs. (16.143), (16.144), and the Boas transform, defined in a slightly more generalized way in Eq. (16.145) (compare with Eq. (16.33)). Each of the integrals in Eqs. (16.143)–(16.150) can be viewed as defining a particular transform operator. There are a quite a number of relationships between the operators just defined; just a couple of representative results are presented. The interested reader can explore the proofs of these and other related results in the Exercises.

If $f \in L^p(\mathbb{R})$, for $1 < p < \infty$, then $G_\alpha f \in L^p(\mathbb{R})$. For functions $f \in L^p(\mathbb{R})$, with $1 < p < \infty$, the operator G_α commutes with the operators $D_\beta, E_\beta, S_\beta$, and T_β. Under appropriate conditions, the operator G_α can be related to both the Hilbert transform operator and the Boas transform operator B_α. Further details can be pursued in Okikiolu (1965a).

Notes

§16.3 Additional discussion on Kober's form and the connection with the Hilbert transform for the disc can be found in Koizumi (1959b, 1960), Garnett (1981, pp. 109–110), Dyn'kin (1991, p. 210) and Koosis (1998, p. 108). For an extension of Koizumi's work, see Matsuoka (1991). In a recent work, Mashreghi (2001) considered the evaluation of $H_K f$ for $f(x) = \log|\sin x|$ and $f(x) = \log|p(x)|$, where $p(x)$ is a polynomial.

§16.4 For further reading on the Boas transform, see Goldberg (1960) and Zayed (1996, chap. 15).

§16.4.7 The inversion of the Boas transform is discussed in Heywood (1963) and Zaidi (1976, 1977). A consideration of transforms related to the Boas transform, as well as some additional relationships satisfied by this transform, can be found in Okikiolu (1965a, 1966, 1967a). A weighted inequality for the Boas transform is treated in Andersen (1995).

§16.5 Additional comments and investigations on the bilinear Hilbert transform can be found in Grafakos (1996/7), Lacey and Thiele (1997a, 1997b, 1998, 1999), Lacey (1998), Duoandikoetxea (2001), Buchkovska and Pilipović (2002), and Blasco and Villarroya (2003). Bounds for the bilinear form $P\int_{-\infty}^{\infty} f(x-\alpha t)g(x-\beta t)\mathrm{d}t/t$, with $\alpha, \beta \in \mathbb{R}$, are considered in Grafakos and Li (2004). A bilinear operator for the disc is discussed in Grafakos and Li (2006).

§16.8 For further reading on Hilbert transforms on curves and some extensions, see the following: Stein and Wainger (1970, 1978), Nagel, Rivière, and Wainger (1974), Nagel and Wainger (1976, 1977), Nagel *et al.* (1976, 1983, 1986), Nagel, Stein, and Wainger (1979), Weinberg (1981), Christ (1985), Cordoba *et al.* (1986), Carbery *et al.* (1989, 1995), Saal (1990), Saal and Urciuolo (1993), Ziesler (1994, 1995), Kim (1995), and Chandarana (1996). The latter author studied the hypersingular extension of the Hilbert transform along curves. For discussion of some results for the treatment of the maximal functions associated with H_Γ and $\overline{H}_\Gamma$, see Stein and Wainger (1978).

§16.9 A concise exposition on ergodic theory can be found in Garsia (1970). For a more in depth discussion, see Petersen (1983b). A proof of the existence of the ergodic Hilbert transform can be found in Petersen (1983a); see also Jajte (1987). Weighted versions of the inequalities of this section and extensions are discussed in Atencia and Martín-Reyes (1983, 1984), Gallardo (1989), Gallardo and Martín-Reyes (1989), Sato (1998 [2000]), and Panman (1999). Some further reading on the ergodic Hilbert transform can be found in Sato (1986, 1987a, 1987b), Ephremidze (1998, 2003), Derriennic and Lin (2001), and Cohen and Lin (2003).

§16.10 Further details on the helical Hilbert transform can be found in Assani and Petersen (1992).

§16.11 There is an extension of Okikiolu's Eq. (16.140) with $|x|^{-\nu}$ and $|t|^{\nu}$ replaced by $\rho(x)^{-1}$ and $\rho(t)$, respectively, where $\rho(x)$ is an even and *a.e.* positive function defined on $(-\infty, \infty)$; see Benedek and Panzone (1971). Kober (1967) has studied some further extensions of the Hilbert transform where the Hilbert kernel $(x-t)^{-1}$ is replaced by $(x-t)^{\alpha-1}$. Carbery, Ricci, and Wright (1998) have defined the "superhilbert transform" by

$$T_n f(x) = \sup_{p\in\mathcal{B}_n} \left|H_p f(x)\right| = \sup_{p\in\mathcal{B}_n} \left|\int_{-\infty}^{\infty} f(x - \surd(t))\frac{dt}{t}\right|,$$

where the class of polynomials of degree at most n in the variable t is denoted by B_n, and the condition $p(0) = 0$ is employed. The operator T_n is bounded on $L^q(\mathbb{R})$ for $q > n$. For some additional reading on miscellaneous forms of the Hilbert transform, see Artiaga (1964) and Heywood (1967). An extension of the Hilbert transform, called the Gaussian Hilbert transform, has been defined with a kernel function involving a Gaussian-type kernel function (Morán and Urbina, 1998).

Exercises

16.1 Evaluate $H_{\mathrm{R}_{m+1}}f(x) - H_{\mathrm{R}_m}f(x)$ for $m \geq 1$.

16.2 Given $g(x) = H_{\mathrm{R}_m}f(x)$, determine f. Indicate the conditions on f required to carry out this inversion.

16.3 If $(1+t^2)^{-1}f(t) \in L(\mathbb{R})$, determine which of the integrals in Eqs. (16.21), (16.26), and (16.27) are bounded.

16.4 If $(1+t^2)^{-1}f(t) \in L(\mathbb{R})$, what is the analogous result to Eq. (16.28)? Specify the range of any constant parameters that you introduce.

16.5 If the condition in Exercise 16.4 is changed to $(1+|t|)^{-1}f(t) \in L(\mathbb{R})$, what is the analogous result to Eq. (16.28)? Specify the range of any constant parameters that you introduce.

16.6 Evaluate the Boas transform of the functions cas ax and sinc ax, where a is a constant.

16.7 Is the Boas transform operator a linear operator?

16.8 Is the Boas transform a translation-invariant operator?

16.9 If

$$\int_0^{\infty} \frac{\sin|x|t}{t}\left(\frac{\sin t}{t}\right)^n dt = \frac{\pi}{2}[1 - \Delta_n(x)],\; n \in \mathbb{N},$$

determine $\Delta_n(x)$.

16.10 Determine the functions $g_n(x)$ such that $\mathcal{F}g_n(x) = \Delta_n(x)$, where Δ_n is defined in Exercise 16.9 and $n \in \mathbb{N}$.

16.11 If $f \in L^2(\mathbb{R})$, determine $B_n f$ in terms of Hf and $g_n(x)$, where the latter function is defined in Exercise 16.10.

16.12 If $f \in L^2(\mathbb{R})$, show that $\|B_n f\|_2 \leq \|f\|_2$.

16.13 If $f \in L^2(\mathbb{R})$, evaluate $B_n^2 f$ in terms of f and g_n.

16.14 Establish the conditions for $(f * g_n)(x) = 0$.

16.15 Determine a necessary and sufficient condition for $(\mathcal{F}f)(x) = 0$ *a.e.* on $(-n, n)$, with $n \in \mathbb{N}$.

16.16 If f and g belong to $L^2(\mathbb{R})$, determine, for $n \in \mathbb{N}$, if the following relationships are true:

(i) $\int_{-\infty}^{\infty} f(x)B_n f(x)\mathrm{d}x = 0$;

(ii) $\int_{-\infty}^{\infty} \{B_n f(x)\}^2\, \mathrm{d}x = -\int_{-\infty}^{\infty} f(x)B_n^2 f(x)\mathrm{d}x$;

(iii) $\int_{-\infty}^{\infty} f(x)B_n g(x)\mathrm{d}x = -\int_{-\infty}^{\infty} g(x)B_n f(x)\mathrm{d}x$;

(iv) $\int_{-\infty}^{\infty} B_n f(x)\, B_n g(x)\mathrm{d}x = -\int_{-\infty}^{\infty} f(x)\, B_n^2 g(x)\mathrm{d}x$.

16.17 Determine if the operator $H_a(f, g)$ commutes with the translation operator on $\mathbb{R}$.

16.18 Does the operator $H_a(f, g)$ commute with the dilation operator?

16.19 Let $\Gamma(t)$ be the curve t^α, for $0 < t < 1$ with $\alpha > 0$, and $at + b$ for large t with $b \neq 0$. Determine whether

$$H_\Gamma f(x, y) = \frac{1}{\pi} P \int_{-\infty}^{\infty} \frac{f(x - t, y - \Gamma(t))\mathrm{d}t}{t}$$

is bounded on $L^2(\mathbb{R}^2)$.

16.20 If $f \in L^p(\mathbb{R})$, for $1 < p < \infty$, and D_α is the operator defined in Eq. (16.146), evaluate $D_\alpha D_\beta f$.

16.21 If $f \in L^p(\mathbb{R})$, for $1 < p < \infty$, and S_α is the operator defined in Eq. (16.147), evaluate $S_\alpha S_\beta f$.

16.22 Evaluate $[D_\alpha, H]$ and $[S_\alpha, H]$.

16.23 If $f \in L^p(\mathbb{R})$, for $1 < p < \infty$, evaluate $HS_\alpha f$ and $HD_\alpha f$.

16.24 If $f \in L^p(\mathbb{R})$, for $1 < p < \infty$, and $g \in L^q(\mathbb{R})$ for suitable q, determine whether the following hold:

$$\int_{-\infty}^{\infty} f(x)D_\alpha g(x)\mathrm{d}x = \int_{-\infty}^{\infty} g(x)D_\alpha f(x)\mathrm{d}x,$$

and

$$\int_{-\infty}^{\infty} f(x)S_\alpha g(x)\mathrm{d}x = -\int_{-\infty}^{\infty} g(x)S_\alpha f(x)\mathrm{d}x.$$

16.25 If $f \in L^p(\mathbb{R})$, for $1 < p < \infty$, evaluate $HE_\alpha f$ and $HT_\alpha f$, where the operators E_α and T_α are defined in Eqs. (16.149) and (16.150), respectively.

16.26 If $f \in L^p(\mathbb{R})$, for $1 < p < \infty$, find expressions for $(\mathrm{d}/\mathrm{d}x)S_\alpha f$ and $(\mathrm{d}/\mathrm{d}x)D_\alpha f$.

16.27 Evaluate $[E_\alpha, T_\alpha]$.

16.28 If $f \in L^p(\mathbb{R})$, for $1 < p < \infty$, find expressions for $T_\alpha^2 f$ and $E_\alpha^2 f$.

16.29 Express the Boas transform operator B_α in terms of the operators H, D_α, and S_α.

16.30 Determine an inversion formula for $G_\alpha f$.

16.31 Okikiolu (1965b) investigated the following integral:

$$g_\alpha(x) = \frac{1}{2\Gamma(\alpha)\sin(\pi\alpha/2)} \int_0^\infty \frac{\{f(x+t) - f(x-t)\}t^\alpha \, dt}{t}, \quad \text{for } 0 < \alpha < 1.$$

Determine the inversion formula for this integral.

17

Linear systems and causality

17.1 Systems

This chapter is concerned with setting up the foundations that allow the connection between causality and analyticity to be established. The interplay between these two topics and the Hilbert transform is also treated. The material of this chapter lays the basis for many of the applications discussed in the following chapters.

Consider the arrangement in Figure 17.1, where the input to the system is denoted by i, and r designates the output response. The input and output could be of the same nature, for example a voltage, or very different variables, for example a voltage input, with the output being a physical displacement of a mass. As an example of Figure 17.1, consider the input to be the driving force acting on a simple oscillator arrangement with a mass hanging from a fixed point by a spring. The output is the displacement of the mass. The input has some dependence on time, and likewise the output response. Henceforth, the temporal dependence is made explicit. Often the input and output response are continuous functions of the time variable, but this is not a requirement. In later sections, the focus will include consideration of step function and impulse inputs.

The term "system" refers to a device capable of converting an input into some output response. The output response should be totally characterized by the system input and the characteristics of the system. The qualifier *active* is used to denote a system containing sources of energy. A *passive* system contains no such sources. In what follows, consideration is restricted to passive systems. For such systems, the internal workings of the system do not spontaneously generate an output response in the absence of an input.

A fundamental question to answer is the following. How is the response $r(t)$ related to $i(t)$? The second issue to resolve is what must be explicitly known about the system in order to understand the relationship between $r(t)$ and $i(t)$.

17.2 Linear systems

This section considers the simplest possible type of system – the so-called *linear system*. Discussion of the more complicated nonlinear case is postponed to

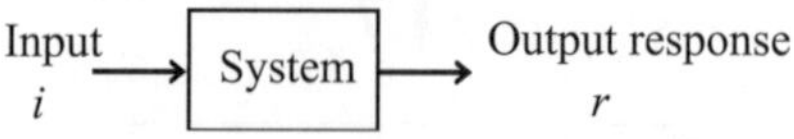

Figure 17.1. Arrangement of input, system, and output.

Section 22.3. For a linear system, the output response scales in a linear fashion with respect to the input. That is, if the input is $ci(t)$, then the output is $cr(t)$, where c is some constant. If S is used to denote an operator that represents the function of the system, then symbolically

$$r(t) = Si(t). \tag{17.1}$$

If $i_1(t)$ and $i_2(t)$ denote distinct inputs to a system, then a linear system is characterized by the following operation:

$$S\{i_1(t) + i_2(t)\} = Si_1(t) + Si_2(t). \tag{17.2}$$

This is called the *additivity property*. It follows from Eq. (17.2) that, for a constant c,

$$S\{ci(t)\} = cSi(t) = cr(t). \tag{17.3}$$

This is termed the *scaling* or *homogeneity property*. If the output response from $i_1(t)$ is $r_1(t)$, and that from $i_2(t)$ is $r_2(t)$, then the output response from $i_1(t) + i_2(t)$ is $r_1(t) + r_2(t)$, or, more generally, for constants c_1 and c_2,

$$\begin{aligned} S\{c_1 i_1(t) + c_2 i_2(t)\} &= c_1 S i_1(t) + c_2 S i_2(t) \\ &= c_1 r_1(t) + c_2 r_2(t). \end{aligned} \tag{17.4}$$

Equation (17.4) is a mathematical statement of the *principle of superposition*. Refinements to the preceding definition of a linear system and further discussion can be found in a number of sources, and a selection of references is given in the chapter end-notes.

A very simple example of a linear system operator is the differential operator. Consider a system that functions by taking a time derivative of the input. Readers familiar with circuit analysis can imagine the form that such a system would take. Equation (17.1) becomes

$$r(t) = \frac{\mathrm{d}i(t)}{\mathrm{d}t}, \tag{17.5}$$

and clearly

$$\frac{\mathrm{d}}{\mathrm{d}t}\{i_1(t) + i_2(t)\} = \frac{\mathrm{d}i_1(t)}{\mathrm{d}t} + \frac{\mathrm{d}i_2(t)}{\mathrm{d}t}. \tag{17.6}$$

Suppose that the function of the system is to take the square root of the input, that is

$$r(t) = \sqrt{i(t)}. \tag{17.7}$$

This is an example of a nonlinear system, since

$$\sqrt{\{i_1(t) + i_2(t)\}} \neq \sqrt{i_1(t)} + \sqrt{i_2(t)} \tag{17.8}$$

for general inputs $i_1(t)$ and $i_2(t)$.

An example is now investigated that is useful in later sections. Consider a damped harmonic oscillator, for which the displacement $x(t)$ satisfies the following equation of motion:

$$m\frac{\mathrm{d}^2x(t)}{\mathrm{d}t^2} + D\frac{\mathrm{d}x(t)}{\mathrm{d}t} + kx(t) = 0, \tag{17.9}$$

where m is the mass, D is the damping force per unit velocity, and k is a force constant. Trying $x(t) = c\mathrm{e}^{\lambda t}$ as a possible solution of the differential equation, where c and λ are constants, leads to

$$m\lambda^2 + D\lambda + k = 0, \tag{17.10}$$

and solutions of this quadratic equation are given by

$$\lambda = -\frac{D}{2m} \pm \sqrt{\left[\frac{D^2}{4m^2} - \frac{k}{m}\right]}, \tag{17.11}$$

and so the solution of Eq. (17.9) takes the form

$$x(t) = \left\{A\exp\left[t\sqrt{\left(\frac{D^2}{4m^2} - \frac{k}{m}\right)}\right] + B\exp\left[-t\sqrt{\left(\frac{D^2}{4m^2} - \frac{k}{m}\right)}\right]\right\}\mathrm{e}^{-(D/2m)t}, \tag{17.12}$$

where A and B are constants. The motion is described as overdamped if $D^2 > 4mk$, which corresponds to the situation where the damping force is larger than the restoring force, and hence the displaced mass comes slowly to the equilibrium position. The motion is called critically damped if $D^2 = 4mk$, and the displaced mass attains its permanent equilibrium position in the minimum time. When $D^2 < 4mk$, the damping force is smaller than the restoring force. Let

$$\omega_1 = \sqrt{\left[\frac{k}{m} - \frac{D^2}{4m^2}\right]} \tag{17.13}$$

and

$$\omega_2 = \frac{D}{2m}; \tag{17.14}$$

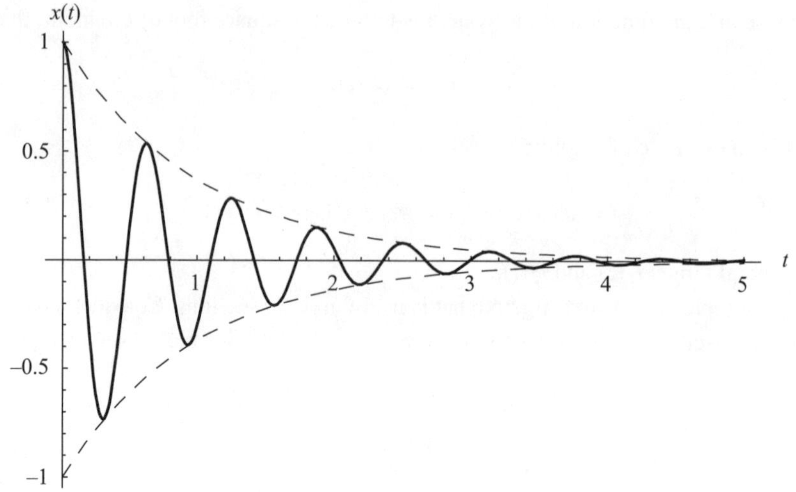

Figure 17.2. Temporal behavior of damped harmonic oscillator.

then

$$x(t) = \{A\mathrm{e}^{\mathrm{i}\omega_1 t} + B\mathrm{e}^{-\mathrm{i}\omega_1 t}\}\mathrm{e}^{-\omega_2 t}, \tag{17.15}$$

or

$$x(t) = \{\alpha \cos \omega_1 t + \beta \sin \omega_1 t\}\mathrm{e}^{-\omega_2 t}, \tag{17.16}$$

where α and β are constants. The behavior exhibited in Eq. (17.16) is referred to as "damped oscillatory motion." An example of this type of motion is depicted in Figure 17.2 for the case $\alpha = 1, \beta = 0, \omega_1 = 10$, and $\omega_2 = 1$. The dashed lines in the figure denote the boundary curves $\mathrm{e}^{-\omega_2 t}$ and $-\mathrm{e}^{-\omega_2 t}$.

Consider a spring subjected to a periodic force $f_0 \cos \omega t$, where f_0 denotes the maximum value of the applied force and ω is a circular frequency, which is equal to $2\pi\nu$, where ν is the frequency. The term "angular frequency" is widely used synonomously for circular frequency. The output response $x(t)$ in this linear problem is the sum of a transient solution, the solution of the homogeneous differential equation, Eq. (17.9), whose value decreases with time due to the presence of the damping factor, and a steady-state contribution, the solution of the following inhomogeneous differential equation:

$$m\frac{\mathrm{d}^2x(t)}{\mathrm{d}t^2} + D\frac{\mathrm{d}x(t)}{\mathrm{d}t} + k\,x(t) = f_0 \cos \omega t. \tag{17.17}$$

The steady-state solution of this differential equation can be found by trying

$$x(t) = A\cos(\omega t - \theta) \tag{17.18}$$

as a solution, where A and θ are time-independent constants. Inserting this into Eq. (17.17) leads to

$$x(t) = \frac{f_0 \cos(\omega t - \theta)}{\sqrt{\left[D^2\omega^2 + (k - m\omega^2)^2\right]}}, \tag{17.19}$$

where

$$\tan\theta = \frac{D\omega}{k - m\omega^2}. \tag{17.20}$$

This is the *particular solution* of Eq. (17.17). If $x_1(t)$ is the output response when the applied force f_0 is zero, and $x_2(t)$ is the output when the applied force is $f_0 \cos\omega t$, then it is a simple matter to prove that the total output response is given by

$$x(t) = x_1(t) + x_2(t), \tag{17.21}$$

which is a solution of the equation of motion, Eq. (17.17). In this example, and the linear systems considered previously, the requirement of *time invariance* of the system is implicitly assumed. That is, the characteristics of the system are assumed not to change with time. This is a highly idealized state, but an extremely convenient notion to introduce.

Consider the oscillator problem once again, but now think about the situation where the damping factor is not constant and has some complicated dependence on $x(t)$, or the restoring force has some complex dependence on $x(t)$. Both these situations would lead to nonlinear oscillation problems. Nonlinear problems are invariably much more difficult to solve than the linear counterparts. These cases are excluded from further discussion for the following reason. The superposition principle in the form that the output is a linear functional of the input no longer applies for nonlinear systems. This prevents the application of the causality principle in a simple form, and this in turn does not allow the application of the Titchmarsh theorem (recall the discussion of Section 4.22). The latter remarks will be become more transparent as the link is made in Sections 17.5–17.8 between the causality principle and analyticity.

If in place of the forcing term $f(t) = f_0 \cos\omega t, f(t) = f_0 \mathrm{e}^{-\mathrm{i}\omega t}$ is employed, then it follows that

$$x(t) = h(\omega)\mathrm{e}^{-\mathrm{i}\omega t}, \tag{17.22}$$

where

$$h(\omega) = -\frac{f_0}{m\omega^2 + \mathrm{i}\omega D - k} \tag{17.23}$$

is the particular solution of the inhomogeneous differential equation. A more general driving force has a Fourier transform representation, that is

$$f(t) = \int_{-\infty}^{\infty} F(\omega)\mathrm{e}^{-\mathrm{i}\omega t}\,\mathrm{d}\omega \tag{17.24}$$

and

$$F(\omega) = \frac{1}{2\pi}\int_{-\infty}^{\infty} f(t)\mathrm{e}^{\mathrm{i}\omega t}\,\mathrm{d}t. \tag{17.25}$$

It will be assumed that $f(t)$ is square integrable. This condition could be weakened, but for practical applications the preceding assumption covers the majority of situations likely to arise. The particular solution of the inhomogeneous equation of motion is then given by

$$x(t) = \int_{-\infty}^{\infty} X(\omega)\mathrm{e}^{-i\omega t}\,\mathrm{d}\omega, \tag{17.26}$$

where

$$X(\omega) = G(\omega)F(\omega) \tag{17.27}$$

and

$$G(\omega) = -\frac{1}{m\omega^2 + \mathrm{i}\omega D - k} = -\frac{1}{m(\omega + \mathrm{i}\,\omega_2 - \omega_1)(\omega + \mathrm{i}\,\omega_2 + \omega_1)}, \tag{17.28}$$

with ω_1 and ω_2 given by Eqs. (17.13) and (17.14), respectively. If the damping term is very small, $G(\omega)$ is sharply peaked when $\omega \approx \omega_1$, and this is referred to as a *resonance* condition. Equation (17.26) can be recast as follows. Employing Eq. (17.25), and denoting the Fourier transform of $G(\omega)$ by $g(t)$, yields

$$\begin{aligned} x(t) &= \int_{-\infty}^{\infty} G(\omega)F(\omega)\mathrm{e}^{-\mathrm{i}\omega t}\,\mathrm{d}\omega \\ &= \frac{1}{2\pi}\int_{-\infty}^{\infty} G(\omega)\mathrm{e}^{-\mathrm{i}\omega t}\left(\int_{-\infty}^{\infty} f(t')\mathrm{e}^{\mathrm{i}\omega t'}\,\mathrm{d}t'\right)\mathrm{d}\omega \\ &= \frac{1}{2\pi}\int_{-\infty}^{\infty} f(t')\mathrm{d}t' \int_{-\infty}^{\infty} G(\omega)\mathrm{e}^{-\mathrm{i}(t-t')\omega}\,\mathrm{d}\omega \\ &= \frac{1}{2\pi}\int_{-\infty}^{\infty} f(t')g(t-t')\mathrm{d}t', \end{aligned} \tag{17.29}$$

where Fubini's theorem has been employed to interchange the order of integration. The final result is of the form of a convolution, so that

$$x(t) = \frac{1}{2\pi}(f * g)(t). \tag{17.30}$$

Further use of this example is made in Section 17.7.

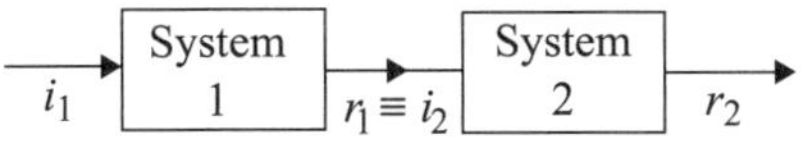

Figure 17.3. Arrangement of inputs and outputs for a sequential system array.

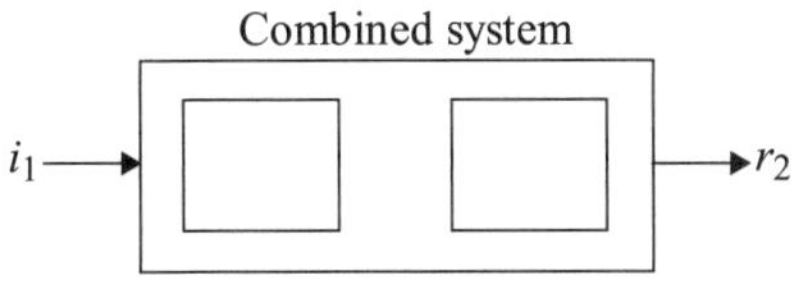

Figure 17.4. Configuration for a combined system.

17.3 Sequential systems

Suppose the output from one system is the input for a second system, as shown in Figure 17.3. Then

$$r_1 = S_1 i_1 \tag{17.31}$$

and

$$r_2 = S_2 i_2, \tag{17.32}$$

and hence, using $i_2 = r_1$,

$$r_2 = S_2 S_1 i_1. \tag{17.33}$$

The two systems in sequence can be effectively combined into a single system, as in Figure 17.4, with $S = S_1 S_2$. The order of operation of the individual systems can be important; that is, in general, $S_2 S_1 \neq S_1 S_2$. The system operations are not commutative under multiplication.

17.4 Stationary systems

An important group of systems comprises those that can be characterized as being stationary systems. Consider a linear system that can be analyzed by partitioning the input into components, and then summing the resulting output responses to each individual input. Suppose that an input is applied over a time interval δt, which is sufficiently small that the system is unchanged by the input. Systems satisfying this requirement are called *stationary systems*. The response to an input $i(t_0)$, applied for a short duration δt_0, is assumed to be

$$\delta r(t) \propto i(t_0)\delta t_0, \tag{17.34}$$

with $\delta t_0 = t - t_0$, and $i(t_0)$ is taken to be the average value of the input over the interval δt_0. The proportionality in Eq. (17.34) can be converted to an equality, thus

$$\delta r(t) = w(t, t_0)i(t_0)\delta t_0, \tag{17.35}$$

where $w(t, t_0)$ is referred to as the weighting function, the impulse response function, or the transient response function, of the system.

In order to simplify the discussion, the idea of *time-invariance* for the system is introduced. The input does not cause any time-dependent change in the operating behavior of the system. This means the system can be viewed as stationary with respect to time. This implies that $w(t, t_0)$ does not have a separate dependence on both t and t_0, but must be of the following form:

$$w(t, t_0) \equiv w(t - t_0). \tag{17.36}$$

The output depends only on the elapsed time from the application of the input, and not on the particular time at which the input was applied.

Consider the limit in Eq. (17.35), so that $\delta t_0 \to \mathrm{d}t_0$, then this equation can be integrated to obtain the total output response; hence,

$$r(t) = \int_{-\infty}^{\infty} w(t - t_0)i(t_0)\mathrm{d}t_0, \tag{17.37}$$

which follows directly from the principle of superposition. A basic requirement is that for a finite input, the output is finite. This condition is met by requiring

$$-\infty < \int_{-\infty}^{\infty} w(t - t_0)\mathrm{d}t_0 < \infty. \tag{17.38}$$

If the integral over the weight function is unbounded, a constant, but very small, input would lead to an unbounded response. Such systems are viewed as being unstable, and are excluded from the discussion. Note that the requirement of a bounded integral does not exclude the possibility of certain singularities for the weight function. Later developments will allow for the situation where the response, input, and weight functions are not normal functions, but behave as distributions.

It is possible to extend the discussion to systems described by variables other than the time, for example, a spatial variable might be most appropriate. For such systems the idea of time invariance is replaced by the notion of *shift invariance*. The shift applying to whatever variable is important for the description of the system.

17.5 Primitive statement of causality

There are several ways to approach the topic of causality in physical systems. The most primitive or basic definition of causality is stated in the following form: the cause must precede the outcome. That is, there is no output response from the system until *after* there has been an input to the system. A system satisfying this constraint is termed a *causal system*. Causality can be invoked by requiring in Eq. (17.35) that there is no response at time t, if $t < t_0$, and hence

$$w(t - t_0) = 0, \quad \text{if } t < t_0. \tag{17.39}$$

From Eq. (17.37) it follows that

$$\begin{aligned} r(t) &= \int_{-\infty}^{\infty} w(t - t_0) i(t_0) \mathrm{d}t_0 \\ &= \int_{-\infty}^{t} w(t - t_0) i(t_0) \mathrm{d}t_0 + \int_{t}^{\infty} w(t - t_0) i(t_0) \mathrm{d}t_0 \\ &= \int_{-\infty}^{t} w(t - t_0) i(t_0) \mathrm{d}t_0, \end{aligned} \tag{17.40}$$

and hence

$$r(t) = \int_0^{\infty} w(\tau) i(t - \tau) \mathrm{d}\tau. \tag{17.41}$$

At this juncture, the conditions that are assumed to be satisfied by the system can be summarized as follows:

(1) linear;
(2) stationary;
(3) integral of the system weight function is bounded;
(4) the principle of causality is satisfied.

Further issues connected with causality are explored in Sections 17.7 and 17.8, but the discussion is now shifted to an examination of the implications of the preceding definition of causality.

17.6 The frequency domain

The developments so far have focused attention on the time domain; that is, time is the central variable characterizing the input–output behavior of the system. The emphasis is now switched and the discussion concentrates on the frequency domain. There are two key reasons for this change in focus. The obvious one is that many properties that characterize a system (think of the refractive index, or the absorption of electromagnetic radiation by a system, as examples), depend on the frequency of the input electromagnetic radiation. A second, and less obvious, reason is that a connection can be made with analytic function theory, treating the angular frequency as a complex variable, and arriving at a large number of important relationships.

Let $R(\omega), I(\omega)$, and $W(\omega)$ denote the Fourier transforms of the output response, the input, and the system weight function, respectively:

$$R(\omega) = \int_{-\infty}^{\infty} r(t)\, \mathrm{e}^{\mathrm{i}\omega t}\, \mathrm{d}t, \tag{17.42}$$

$$I(\omega) = \int_{-\infty}^{\infty} i(t)\, \mathrm{e}^{\mathrm{i}\omega t}\, \mathrm{d}t, \tag{17.43}$$

and

$$W(\omega) = \int_{-\infty}^{\infty} w(t)\, \mathrm{e}^{\mathrm{i}\omega t}\, \mathrm{d}t. \tag{17.44}$$

The function $W(\omega)$ is termed the *system transfer function*. The reason for the choice of the positive sign in the exponent of the Fourier transform will become apparent in Section 17.8. In Eq. (17.43) no confusion should arise between the input, i, which is a function of time, and the complex number i. It is assumed that all three integrals in Eqs. (17.42)–(17.44) exist. It suffices to work with square integrable functions $r(t), i(t)$, and $w(t)$, though weaker conditions can be imposed. Inspection of Eq. (17.37) shows it to be of the form of a convolution (see Section 2.6.2), and hence

$$r(t) = \{w * i\}(t), \tag{17.45}$$

from which it follows, on taking the Fourier transform, that

$$\begin{aligned} \mathcal{F}r(t) &= \mathcal{F}\{w * i\}(t) \\ &= \mathcal{F}w(t)\mathcal{F}i(t), \end{aligned} \tag{17.46}$$

and the preceding line follows on using the convolution theorem for Fourier transforms (see Eq. (2.54)). Using Eqs. (17.42)–(17.44), Eq. (17.46) can be written as follows:

$$R(\omega) = W(\omega)I(\omega). \tag{17.47}$$

The simplicity of the structure of Eq. (17.47) is a large part of the driving force behind the switch from the time domain to the frequency domain.

From Eq. (17.44) and the causality condition, it follows that

$$\begin{aligned} W(\omega) &= \int_{-\infty}^{\infty} w(t)\, \mathrm{e}^{\mathrm{i}\omega t}\, \mathrm{d}t \\ &= \int_{0}^{\infty} w(t)\, \mathrm{e}^{\mathrm{i}\omega t}\, \mathrm{d}t, \end{aligned} \tag{17.48}$$

because

$$w(t) = 0, \quad \text{for } t < 0. \tag{17.49}$$

The preceding developments allow the connection with analytic function theory to be established.

Before closing this section, two theorems are indicated that are useful in making the connection between the time and frequency domains. The results are due to Bochner (Bochner and Chandrasekharan, 1949, pp. 142–144). Let $f(t)$ and $g(t)$ belong to $L^2(\mathbb{R})$ and denote the corresponding Fourier transforms as $F(\omega)$ and $G(\omega)$, respectively. If T is a bounded linear transformation given by

$$Tf(t + t_0) = g(t + t_0), \tag{17.50}$$

where t_0 is a real-time translation, then there exists a bounded function $W(\omega)$ measurable for almost all ω, such that

$$G(\omega) = TF(\omega) = W(\omega)F(\omega). \tag{17.51}$$

There is a theorem that is the converse of the preceding result. If Eq. (17.51) holds, then

$$\tau_{-t_0}Tf(t) = \tau_{-t_0}g(t) = g(t + t_0) = Tf(t + t_0) = T\tau_{-t_0}f(t), \tag{17.52}$$

where τ_a is the normal translation operator. The consequence is that T commutes with the translation operator. These ideas have immediate implications for the jump from Eq. (17.41) to Eq. (17.47).

17.7 Connection to analyticity

Allowing the frequency to be a complex variable, an extension of Eq. (17.44) can be made, which provides a connection to the upper half complex angular frequency plane. From the practical standpoint of an experimental measurement, the only quantities of interest are real. To link up with experimental results, ultimately requires a restriction back to the real angular frequency axis, and in particular, for frequencies, which cannot be negative, the positive real angular frequency axis. It is not difficult to find literature sources where the angular frequency (the observable) and the complex angular frequency are both denoted ω, with the reader left to sort out the intended meaning. This is usually not hard in most contexts, but in others, it can lead to possible confusion. To avoid this type of difficulty when complex frequencies are under discussion, the following notational convention is employed. Complex frequencies are denoted by ω_z, with the subscript intended to be suggestive of the complex plane, and the variable ω_z can be written in terms of its real and imaginary parts as

$$\omega_z = \omega_r + i\omega_i. \tag{17.53}$$

The real part ω_r, for $\omega_r \geq 0$, denotes the observable frequency. When only the real frequency enters into the discussion, the subscript r is dropped and the notation ω employed to denote the angular frequency.

In the discussion that follows, it is assumed that the systems under consideration are described by a weight function, $w(t)$, that is a continuous and bounded function. The weight function is assumed to vanish sufficiently quickly as $|t| \to \infty$, so that its Fourier transform is bounded. Generally, the weight function is regarded as belonging to the class of square integrable functions.

The function $W(\omega_r)$ can be analytically continued into the upper half of the complex angular frequency plane. That is, using the form of Eq. (17.44) where causality is assumed, Eq. (17.48), leads to

$$\begin{aligned} W(\omega_z) &= \int_0^\infty w(t)\, e^{i\omega_z t}\, dt \\ &= \int_0^\infty w(t)\, e^{-\omega_i t} e^{i\omega_r t}\, dt. \end{aligned} \tag{17.54}$$

If the integral in Eq. (17.48) is bounded, then so is the integral in Eq. (17.54), since the factor $e^{-\omega_i t}$, for $\omega_i > 0$, can only improve the convergence of the integral. In the lower half of the complex angular frequency plane, writing $\omega_z = \omega_r - i\omega_i$, with $\omega_i > 0$, yields

$$W(\omega_z) = \int_0^\infty w(t)\, e^{\omega_i t} e^{i\omega_r t}\, dt, \tag{17.55}$$

and there is no guarantee, for a sufficiently wide class of $w(t)$, that the integral in Eq. (17.55) is bounded. For example, if $w(t)$ behaves like $(c + t^2)^{-1}$, with c some time-independent constant, then it is straightforward to demonstrate that the integral in Eq. (17.55) diverges.

Note that causality has played a central role. Without this constraint, it would not have been possible to continue $W(\omega_r)$ analytically into the upper half of the complex angular frequency plane, except possibly for some rather restrictive choices of $w(t)$. So, at this point in the development, the conditions on the system weight function $w(t)$ are rather general. Also, it should be noted that the development so far is *model-independent* in terms of the nature of the material under consideration.

The alert reader may have wondered about one loose end left in the previous discussion. The domain of analyticity of $W(\omega_r)$ was extended to the upper half complex angular frequency plane, without making any reference to the significance of W evaluated at negative values of ω_r. Equation (17.48) can be used to extend the definition of $W(\omega_r)$ to include formally negative values of ω_r. Recalling that the system weight function in the time domain is a real function, and by writing Eq. (17.48) as

$$W(\omega_{\rm r}) = \int_0^\infty w(t)\cos\omega_{\rm r}t\,{\rm d}t + {\rm i}\int_0^\infty w(t)\sin\omega_{\rm r}t\,{\rm d}t, \qquad (17.56)$$

then it is apparent that the real part of $W(\omega_{\rm r})$ is invariant to a change in sign for $\omega_{\rm r}$, and the imaginary part of $W(\omega_{\rm r})$ is an odd function of $\omega_{\rm r}$, and so

$$\mathrm{Re}\{W(-\omega_{\rm r})\} = \mathrm{Re}\{W(\omega_{\rm r})\} \qquad (17.57)$$

and

$$\mathrm{Im}\{W(-\omega_{\rm r})\} = -\mathrm{Im}\{W(\omega_{\rm r})\}, \qquad (17.58)$$

where Re and Im stand for the real and imaginary parts, respectively. Equations (17.57) and (17.58) provide the purely formal connections for the weight function in the frequency domain at negative angular frequencies, with the experimentally accessible weight function at positive frequencies. These two results are special cases of a more general relationship that can be obtained from Eq. (17.54) by taking the complex conjugate, which is denoted by an asterisk, hence

$$W(\omega_z)^* = \int_0^\infty w(t)\mathrm{e}^{-\mathrm{i}\omega_z^* t}\,{\rm d}t; \qquad (17.59)$$

that is,

$$W(\omega_z)^* = \int_0^\infty w(t)\mathrm{e}^{-\omega_{\rm i}t}\cos\omega_{\rm r}t\,{\rm d}t - {\rm i}\int_0^\infty w(t)\mathrm{e}^{-\omega_{\rm i}t}\sin\omega_{\rm r}t\,{\rm d}t, \qquad (17.60)$$

and, by considering Eq. (17.54) as a function of ω_z^*, this leads to

$$W(-\omega_z^*) = \int_0^\infty w(t)\mathrm{e}^{-\mathrm{i}\omega_z^* t}\,{\rm d}t. \qquad (17.61)$$

Comparing Eqs. (17.59) and (17.61) yields

$$W(\omega_z)^* = W(-\omega_z^*). \qquad (17.62)$$

Equation (17.62) is called a *crossing symmetry relation*, and allows the weight function in the frequency domain at negative complex frequencies to be determined from $W(\omega_z)$. When $\omega_{\rm i} = 0$, Eq. (17.62) becomes

$$W(\omega_{\rm r})^* = W(-\omega_{\rm r}), \qquad (17.63)$$

from which it follows, without a detailed knowledge of the structure of the real and imaginary parts of $W(\omega_{\rm r})$, that the real part of $W(\omega_{\rm r})$ is an even function of $\omega_{\rm r}$, that is Eq. (17.57) holds, and the imaginary part of $W(\omega_{\rm r})$ is an odd function of $\omega_{\rm r}$, and hence Eq. (17.58).

The example discussed in Section 17.2 is now revisited. Consider from Eq. (17.28) the integral

$$g(t) = \int_{-\infty}^{\infty} G(\omega)\mathrm{e}^{-\mathrm{i}\omega t}\,\mathrm{d}\omega, \tag{17.64}$$

and examine the case $t < 0$. The function $G(\omega_z)$ is analytic in the upper half of the complex angular frequency plane. From Eq. (17.28), the poles of $G(\omega_z)$ are located in the lower half of the complex plane. Consider the integral $\oint_C G(\omega_z)\mathrm{e}^{-\mathrm{i}\omega_z t}\,\mathrm{d}\omega_z$, where C is a semicircular contour in the upper half of the complex angular frequency plane of radius R, diameter along the real axis, and center the origin. From the Cauchy integral theorem, it follows that

$$\int_{-R}^{R} G(\omega_\mathrm{r})\mathrm{e}^{-\mathrm{i}\omega_\mathrm{r} t}\,\mathrm{d}\omega_\mathrm{r} + \int_{C_R} G(\omega_z)\mathrm{e}^{-\mathrm{i}\omega_z t}\,\mathrm{d}\omega_z = 0, \tag{17.65}$$

where the second of the two integrals, which is over the semicircular section of the contour, denoted C_R, can be simplified with the substitution $\omega_z = R\mathrm{e}^{\mathrm{i}\theta}$. In the limit $R \to \infty$ in Eq. (17.65), application of Jordan's lemma (see Section 2.8.4) leads to

$$\int_{-\infty}^{\infty} G(\omega_\mathrm{r})\mathrm{e}^{-\mathrm{i}\omega_\mathrm{r} t}\,\mathrm{d}\omega_\mathrm{r} = 0, \quad \text{for } t < 0. \tag{17.66}$$

Hence, from Eq. (17.64),

$$g(t) = 0, \quad \text{for } t < 0, \tag{17.67}$$

and so g is a causal function (refer to Section 4.22). Since real frequencies are in use, the notation is simplified by replacing ω_r by ω; then, applying Titchmarsh's theorem (Section 4.22), leads to

$$\operatorname{Im} G(\omega) = \frac{1}{\pi} P \int_{-\infty}^{\infty} \frac{\operatorname{Re} G(\omega')\mathrm{d}\omega'}{\omega - \omega'} \tag{17.68}$$

and

$$\operatorname{Re} G(\omega) = -\frac{1}{\pi} P \int_{-\infty}^{\infty} \frac{\operatorname{Im} G(\omega')\mathrm{d}\omega'}{\omega - \omega'}. \tag{17.69}$$

Equation (17.27) is of the form of Eq. (17.47) with the obvious identifications $I(\omega) \equiv F(\omega), R(\omega) \equiv X(\omega)$, and $W(\omega) \equiv G(\omega)$. Using Eqs. (17.67) and (17.29), it follows for the damped harmonic oscillator model that

$$\begin{aligned} x(t) &= \frac{1}{2\pi}\int_{-\infty}^{\infty} f(t-t')g(t')\mathrm{d}t' \\ &= \frac{1}{2\pi}\int_{0}^{\infty} f(t-t')g(t')\mathrm{d}t' \\ &= \frac{1}{2\pi}\int_{-\infty}^{t} f(t')g(t-t')\mathrm{d}t', \end{aligned} \tag{17.70}$$

and hence the displacement at time t is determined by the force applied up to time t. This model is therefore causal.

A key feature of the model is the fact that the damping constant satisfies $D > 0$, so from Eq. (17.14) $\omega_2 > 0$ and hence both poles of $G(\omega)$ occur in the lower half of the complex angular frequency plane. Consequently, the system is causal. The damping factor $D\dot{x}(t)$ leads to energy dissipation, and accordingly, the model describes a linear passive system.

17.7.1 A generalized response function

In the preceding development a key requirement employed for the function $w(t)$ was that it satisfied both a causal condition and was real. In this subsection, the consequence of working with a complex causal $w(t)$ is examined. Let

$$W(\omega) = \int_{-\infty}^{\infty} w(t)\mathrm{e}^{\mathrm{i}\omega t}\,\mathrm{d}t, \tag{17.71}$$

with $w(t)$ given by

$$w(t) = \alpha(t) + \mathrm{i}\beta(t), \tag{17.72}$$

and the functions $\alpha(t)$ and $\beta(t)$ are assumed to be both real and causal. Let

$$a(\omega) = \int_{0}^{\infty} \alpha(t)\mathrm{e}^{\mathrm{i}\omega t}\,\mathrm{d}t \tag{17.73}$$

and

$$b(\omega) = \int_{0}^{\infty} \beta(t)\mathrm{e}^{\mathrm{i}\omega t}\,\mathrm{d}t. \tag{17.74}$$

The first observation is that

$$W(\omega_z)^* \neq W(-\omega_z^*), \tag{17.75}$$

which follows directly from Eq. (17.71) and on noting the fact that $w(t)^* \neq w(t)$. The immediate consequence is that the crossing symmetry relation no longer applies in

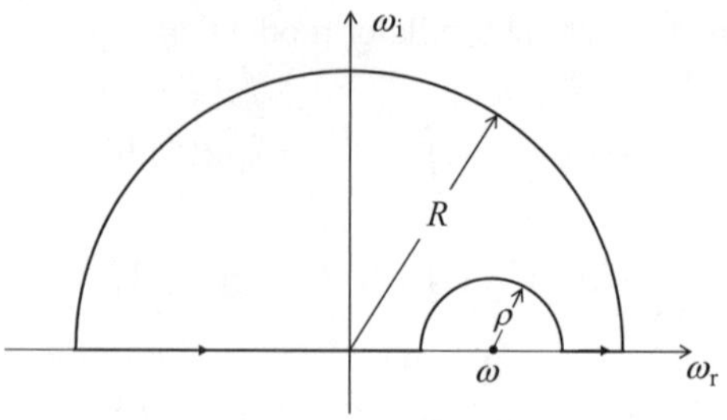

Figure 17.5. Contour for the evaluation of dispersion relations based on a complex $w(t)$.

the simple form given by Eq. (17.63). Later, in Chapters 19–22, it will become evident that the crossing symmetry condition is a central component in the simplification of the Hilbert transform relations, which connect the real and imaginary components of a complex optical property. The crossing symmetry condition allows the Hilbert transform on the unphysical spectral interval $(-\infty, \infty)$ to be converted to the physical angular frequency range $[0, \infty)$. This has an important implication for the analysis of experimental optical data, which is confined to the interval $[0, \infty)$.

Suppose the functions $a(\omega)$ and $b(\omega)$ satisfy $a(\omega) \in L^2(\mathbb{R})$ and $b(\omega) \in L^2(\mathbb{R})$, and that $W(\omega_z) = O(\omega_z^{-1-\delta})$ as $\omega_z \to \infty$, with $\delta > 0$. Consider the evaluation of the contour integral

$$\oint_C \frac{W(\omega_z)\mathrm{d}\omega_z}{\omega - \omega_z},$$

where C denotes the contour shown in Figure 17.5.

From the Cauchy integral theorem, it follows that

$$\int_{-R}^{\omega-\rho} \frac{W(\omega_{\mathrm{r}})\mathrm{d}\omega_{\mathrm{r}}}{\omega - \omega_{\mathrm{r}}} + \int_{\omega+\rho}^{R} \frac{W(\omega_{\mathrm{r}})\mathrm{d}\omega_{\mathrm{r}}}{\omega - \omega_{\mathrm{r}}} - \int_{\pi}^{0} \frac{W(\omega + \rho\mathrm{e}^{\mathrm{i}\theta})\mathrm{i}\rho\mathrm{e}^{\mathrm{i}\theta}\mathrm{d}\theta}{\rho\mathrm{e}^{\mathrm{i}\theta}} + \int_{0}^{\pi} \frac{W(R\mathrm{e}^{\mathrm{i}\theta})\mathrm{i}R\mathrm{e}^{\mathrm{i}\theta}\mathrm{d}\theta}{\omega - R\mathrm{e}^{\mathrm{i}\theta}} = 0. \tag{17.76}$$

In the limit $R \to \infty$, the preceding equation reduces to

$$\int_{-\infty}^{\omega-\rho} \frac{W(\omega_{\mathrm{r}})\mathrm{d}\omega_{\mathrm{r}}}{\omega - \omega_{\mathrm{r}}} + \int_{\omega+\rho}^{\infty} \frac{W(\omega_{\mathrm{r}})\mathrm{d}\omega_{\mathrm{r}}}{\omega - \omega_{\mathrm{r}}} = -\mathrm{i}\int_{0}^{\pi} W(\omega + \rho\mathrm{e}^{\mathrm{i}\theta})\mathrm{d}\theta, \tag{17.77}$$

which simplifies, on taking the limit $\rho \to 0$, to yield

$$\frac{1}{\pi} P \int_{-\infty}^{\infty} \frac{W(\omega')\mathrm{d}\omega'}{\omega - \omega'} = -\mathrm{i}W(\omega). \tag{17.78}$$

Expressing W in terms of its real and imaginary parts, $W_{\rm r}$ and $W_{\rm i}$, allows the real and imaginary parts of the preceding result to be written as follows:

$$\frac{1}{\pi}P\int_{-\infty}^{\infty}\frac{W_{\rm r}(\omega')\mathrm{d}\omega'}{\omega-\omega'}=W_{\rm i}(\omega) \tag{17.79}$$

and

$$\frac{1}{\pi}P\int_{-\infty}^{\infty}\frac{W_{\rm i}(\omega')\mathrm{d}\omega'}{\omega-\omega'}=-W_{\rm r}(\omega). \tag{17.80}$$

The preceding two equations represent the Hilbert transform pair arising from a complex $w(t)$.

From Eqs. (17.71), (17.73), and (17.74), it follows that

$$W_{\rm r}(\omega)=a_{\rm r}(\omega)-b_{\rm i}(\omega), \tag{17.81}$$

$$W_{\rm r}(-\omega)=a_{\rm r}(\omega)+b_{\rm i}(\omega), \tag{17.82}$$

$$W_{\rm i}(\omega)=a_{\rm i}(\omega)+b_{\rm r}(\omega), \tag{17.83}$$

and

$$W_{\rm i}(-\omega)=-a_{\rm i}(\omega)+b_{\rm r}(\omega), \tag{17.84}$$

where the real parts of $a(\omega)$ and $b(\omega)$ are denoted by a subscript r, and the imaginary parts are indicated by a subscript i. The functions $a_{\rm r}(\omega)$ and $b_{\rm r}(\omega)$ are even functions and $a_{\rm i}(\omega)$ and $b_{\rm i}(\omega)$ are odd functions. Inserting Eqs. (17.81)–(17.84) into Eqs. (17.79) and (17.80) leads to

$$\frac{2}{\pi}P\int_{0}^{\infty}\frac{\{\omega a_{\rm r}(\omega')-\omega' b_{\rm i}(\omega')\}\mathrm{d}\omega'}{\omega^2-\omega'^2}=a_{\rm i}(\omega)+b_{\rm r}(\omega) \tag{17.85}$$

and

$$\frac{2}{\pi}P\int_{0}^{\infty}\frac{\{\omega' a_{\rm i}(\omega')+\omega b_{\rm r}(\omega')\}\mathrm{d}\omega'}{\omega^2-\omega'^2}=-a_{\rm r}(\omega)+b_{\rm i}(\omega). \tag{17.86}$$

These results represent the generalization of the Hilbert transform connections for a complex $w(t)$. An obvious drawback of these formulas is the coupling that arises; components of both $a(\omega)$ and $b(\omega)$ occur as part of the integrand and outside the integral. One area of application of the preceding development is the inversion of Raman cross-section data, where the causal time correlation function is treated as a complex function (Remacle and Levine, 1993).

17.8 Application of a theorem due to Titchmarsh

Based on the developments in the preceding two sections, the link between causality and the theory of Hilbert transforms can now be established. This can be directly accomplished by application of the principal result given in Section 4.22, Eqs. (4.450)–(4.454), most often referred to as Titchmarsh's theorem. To simplify the notation, let $\omega = \omega_r$ since there is no risk of confusing ω as a complex variable, and let

$$W_r(\omega) = \operatorname{Re} W(\omega_r) \tag{17.87}$$

and

$$W_i(\omega) = \operatorname{Im} W(\omega_r), \tag{17.88}$$

so that, on the real axis,

$$W(\omega) = W_r(\omega) + iW_i(\omega). \tag{17.89}$$

The notation (ω_r, ω_i) is retained for a point in the complex angular frequency plane.

By imposing the causality condition Eq. (17.49), and by the requirement placed on the weight function to be square integrable, then Eq. (4.452) is satisfied. It follows immediately that the real and imaginary components are related by a Hilbert transform pair (see Eqs. (4.450) and (4.451)), so that

$$W_i(\omega) = \frac{1}{\pi} P \int_{-\infty}^{\infty} \frac{W_r(\omega')d\omega'}{\omega - \omega'} \tag{17.90}$$

and

$$W_r(\omega) = -\frac{1}{\pi} P \int_{-\infty}^{\infty} \frac{W_i(\omega')d\omega'}{\omega - \omega'}. \tag{17.91}$$

It also follows that, for $\omega_i > 0$, the function $W(\omega_r + i\omega_i)$ is analytic in the upper half of the complex angular frequency plane. Furthermore, $W(\omega_r + i\omega_i)$ is square integrable over any line parallel to the real axis, for all $\omega_i > 0$, so that

$$\int_{-\infty}^{\infty} \left|W(\omega_r + i\,\omega_i)\right|^2 d\omega_r < C, \tag{17.92}$$

where C is a positive constant.

Some examples are now examined, both cases that satisfy and those that do not satisfy the conditions of Titchmarsh's theorem. No particular concern will be paid to forcing the system response function to be real. Suppose that

$$W(\omega) = \frac{e^{-i\lambda\omega}}{\omega - \alpha + i\beta}, \tag{17.93}$$

with the constants α, β, and λ all real and greater than zero. This function is analytic in the upper half of the complex angular frequency plane. However,

$$\int_{-\infty}^{\infty} \left|W(\omega_\mathrm{r} + \mathrm{i}\omega_\mathrm{i})\right|^2 \mathrm{d}\omega_\mathrm{r} = \mathrm{e}^{2\lambda\omega_\mathrm{i}} \int_{-\infty}^{\infty} \frac{\mathrm{d}\omega_\mathrm{r}}{(\omega_\mathrm{r} - \alpha)^2 + (\omega_\mathrm{i} + \beta)^2}, \tag{17.94}$$

and the term on the right-hand side of the preceding equation diverges as $\omega_\mathrm{i} \to \infty$. Evaluating the inverse Fourier transform of $W(\omega)$ leads to

$$\begin{aligned} w(t) &= \frac{1}{2\pi} \int_{-\infty}^{\infty} W(\omega)\mathrm{e}^{-\mathrm{i}t\omega}\,\mathrm{d}\omega \\ &= \frac{1}{2\pi} \int_{-\infty}^{\infty} \frac{\mathrm{e}^{-\mathrm{i}(t+\lambda)\omega}\,\mathrm{d}\omega}{\omega - \alpha + \mathrm{i}\beta} \\ &= -\frac{\mathrm{i}\mathrm{e}^{-\mathrm{i}\alpha(\lambda+t)}}{2\pi} \int_{-\infty}^{\infty} \frac{\{x \sin ax + \cos ax\}\mathrm{d}x}{x^2 + 1} \\ &= \begin{cases} 0, & \text{for } t < -\lambda \\ -\mathrm{i}\mathrm{e}^{-\mathrm{i}(\alpha - \mathrm{i}\beta)(\lambda + t)}, & \text{for } t > -\lambda, \end{cases} \end{aligned} \tag{17.95}$$

where the change of variable $\omega - \alpha = \beta x$ has been used and the substitution $a = \beta(\lambda + t)$ employed. It follows from the preceding calculation that $w(t)$ is not causal in the sense of Titchmarsh's theorem; that is, the function only vanishes for $t < -\lambda$ rather than the required $t < 0$. This example makes it clear that a square integrable function that is the boundary value of a function analytic in the upper half of the complex plane is not necessarily the inverse Fourier transform of a causal function.

Suppose the sign in the exponent in Eq. (17.93) is reversed, so that

$$W(\omega) = \frac{\mathrm{e}^{\mathrm{i}\lambda\omega}}{\omega - \alpha + \mathrm{i}\beta}, \tag{17.96}$$

with α, β, and λ all real and greater than zero; then, repeating the calculation in Eq. (17.94), which simply amounts to reversing the sign of λ, yields

$$\int_{-\infty}^{\infty} \left|W(\omega_\mathrm{r} + \mathrm{i}\,\omega_\mathrm{i})\right|^2 \mathrm{d}\omega_\mathrm{r} = \mathrm{e}^{-2\lambda\omega_\mathrm{i}} \int_{-\infty}^{\infty} \frac{\mathrm{d}\omega_\mathrm{r}}{(\omega_\mathrm{r} - \alpha)^2 + (\omega_\mathrm{i} + \beta)^2}. \tag{17.97}$$

The right-hand side of Eq. (17.97) is clearly bounded for all $\omega_\mathrm{i} > 0$. By Titchmarsh's theorem it follows that the inverse Fourier transform of $W(\omega)$ is a causal function. That this is indeed the case follows directly from Eq. (17.95), and so

$$w(t) = \begin{cases} 0, & \text{for } t < \lambda \\ -\mathrm{i}\mathrm{e}^{-\mathrm{i}(\alpha - \mathrm{i}\beta)(t - \lambda)}, & \text{for } t > \lambda. \end{cases} \tag{17.98}$$

Since $\lambda > 0$, it follows that $w(t) < 0$ for $t < 0$, which is the expectation based on Titchmarsh's theorem. From this pair of examples the following points can be inferred.

On the real axis, the factor $e^{\pm i\lambda\omega}$ can be treated as a phase factor, and is of no consequence in establishing that $\int_{-\infty}^{\infty} |W(\omega)|^2 \, d\omega$ is bounded on the real line. However, Titchmarsh's theorem places a stronger constraint on $W(\omega)$ that must be satisfied: the requirement that $W(\omega_r + i\omega_i)$ should be square integrable along every line in the upper half of the complex angular frequency plane that is parallel to the real axis. In this case, the sign of the phase factor is critical, as is apparent from the two examples given. The second case discussed is actually an example of a more general result (Titchmarsh, 1948, pp. 129–130). A necessary and sufficient condition that $F(z)$ should be the limit as $y \to 0$ of an analytic function $F(z)$, with

$$\int_{-\infty}^{\infty} |F(x+iy)|^2 \, dx = O(e^{2ky}), \tag{17.99}$$

with k a real constant, is that the inverse Fourier transform of $F(x)$, denoted $f(x)$, satisfies

$$f(x) = 0, \quad \text{for } x < -k. \tag{17.100}$$

The interested reader can pursue the proof of this result in Titchmarsh (1948).

The limit $\lambda \to 0$ in the previous two examples leads to

$$W(\omega) = \frac{1}{\omega - \alpha + i\beta} = \frac{\omega - \alpha}{(\omega - \alpha)^2 + \beta^2} - i\frac{\beta}{(\omega - \alpha)^2 + \beta^2}, \tag{17.101}$$

a function that satisfies, for a positive constant C, the condition

$$\int_{-\infty}^{\infty} \left|W(\omega_r + i\omega_i)\right|^2 d\omega_r \le C, \tag{17.102}$$

and the inverse Fourier transform of W is a causal function, by Titchmarsh's theorem, which can be determined directly from Eq. (17.95) or Eq. (17.98). The Hilbert transform pair that follows as a consequence of Eq. (17.102) is the following familiar set of equations:

$$H\left[\frac{\beta}{(\omega - \alpha)^2 + \beta^2}\right] = \frac{\omega - \alpha}{(\omega - \alpha)^2 + \beta^2} \tag{17.103}$$

and

$$H\left[\frac{\omega - \alpha}{(\omega - \alpha)^2 + \beta^2}\right] = -\frac{\beta}{(\omega - \alpha)^2 + \beta^2}. \tag{17.104}$$

As a final example, consider the case

$$W(\omega) = e^{-\lambda\omega^2}, \tag{17.105}$$

with $\lambda > 0$. This function is analytic in the complex angular frequency plane, and it is square integrable on the real axis. However,

$$\int_{-\infty}^{\infty} \left|W(\omega_r + i\,\omega_i)\right|^2 d\omega_r = e^{2\lambda\omega_i^2} \int_{-\infty}^{\infty} e^{-2\lambda\omega_r^2}\, d\omega_r, \tag{17.106}$$

and so the integral on the left-hand side is unbounded as $\omega_i \to \infty$. The reader is requested to demonstrate that the inverse Fourier transform of W is not a causal function.

17.9 An acausal example

In this section, a well known problem is examined where the outcome is acausal. Interest will focus on the underlying analytic behavior of the model. The classical motion of a harmonically bound charged particle of mass m in an electric field is governed by the following equation of motion:

$$m\ddot{\boldsymbol{x}}(t) + m\omega_0^2\boldsymbol{x}(t) = \boldsymbol{F}_{\text{ext}} + \boldsymbol{F}_{\text{rad}}, \tag{17.107}$$

where $\boldsymbol{F}_{\text{ext}}$ and $\boldsymbol{F}_{\text{rad}}$ denote the external force and the radiative reaction force, respectively. Bold type denotes vector quantities and derivatives with respect to time are signified by an over dot. If the charged particle is assumed to be an electron, the forces involved are given by

$$\boldsymbol{F}_{\text{ext}} = -e[\mathbf{E}(t) + \mathbf{v}(t) \times \mathbf{B}(t)] \tag{17.108}$$

and

$$\boldsymbol{F}_{\text{rad}} = m\tau\dddot{\boldsymbol{x}}(t), \tag{17.109}$$

where $\mathbf{E}(t)$ and $\mathbf{B}(t)$ denote the electric field and magnetic induction, respectively, $-e$ is the charge on the electron, and τ is a parameter characterizing the strength of the radiative reaction force. The justification of the form of $\boldsymbol{F}_{\text{rad}}$ can be found in standard texts on electrodynamics (see, for example, Jackson, 1999, p. 748). The electron is assumed to be acted upon by a spherically symmetric harmonic restoring force of strength $m\omega_0^2 x(t)$, and the particle speed is assumed to be small compared with the speed of light, so that the magnetic term in Eq. (17.108) can be dropped. Equation (17.107) with the forces just defined is termed the Abraham–Lorentz equation.

The solution of

$$\ddot{\boldsymbol{x}}(t) + \omega_0^2\boldsymbol{x}(t) - \tau\dddot{\boldsymbol{x}}(t) = -\frac{e}{m}\mathbf{E}(t), \tag{17.110}$$

with the assumption $\mathbf{E}(t) = \mathbf{E}_0 \mathrm{e}^{-\mathrm{i}\omega t}$, can be obtained using the same approach sketched in Section 17.2. On suppressing the ω-dependence of $x(t)$, let

$$\boldsymbol{x}(t) = \boldsymbol{h}(\omega)\mathrm{e}^{-\mathrm{i}\omega t} \tag{17.111}$$

then,

$$\boldsymbol{h}(\omega) = \frac{-e\mathbf{E}_0}{m(\omega_0^2 - \omega^2 - \mathrm{i}\tau\omega^3)}. \tag{17.112}$$

A more general external force has a Fourier transform representation, that is

$$\mathbf{E}(t) = \int_{-\infty}^{\infty} \mathbf{E}(\omega)\mathrm{e}^{-\mathrm{i}\omega t}\,\mathrm{d}\omega \tag{17.113}$$

and

$$\mathbf{E}(\omega) = \frac{1}{2\pi}\int_{-\infty}^{\infty} \mathbf{E}(t)\mathrm{e}^{\mathrm{i}\omega t}\,\mathrm{d}t. \tag{17.114}$$

The reader is alerted to the convention where the same symbol is employed for a function and its Fourier transform, and the argument signifies whether it is the field in the time domain or the frequency domain. The particular solution of the inhomogeneous equation of motion is then given by

$$\boldsymbol{x}(t) = \int_{-\infty}^{\infty} \boldsymbol{X}(\omega)\mathrm{e}^{-\mathrm{i}\omega t}\,\mathrm{d}\omega, \tag{17.115}$$

where

$$\boldsymbol{X}(\omega) = G(\omega)\mathbf{E}(\omega) \tag{17.116}$$

and

$$G(\omega) = \frac{1}{m(\omega_0^2 - \omega^2 - \mathrm{i}\tau\omega^3)}. \tag{17.117}$$

Consider Eq. (17.117) as a function of the complex variable ω_z. If τ is treated as a very small parameter, then the roots of Eq. (17.117) can be written as follows:

$$\omega_z \approx \omega_0 - \frac{\mathrm{i}\omega_0^2\tau}{2}, \tag{17.118}$$

$$\omega_z \approx -\omega_0 - \frac{\mathrm{i}\omega_0^2\tau}{2}, \tag{17.119}$$

and

$$\omega_z \approx \mathrm{i}(\tau^{-1} + \omega_0^2\tau). \tag{17.120}$$

These solutions ignore terms of order τ^2 and higher, which can be quickly checked by showing that

$$\omega_0^2 - \omega_z^2 - \mathrm{i}\tau\omega_z^3 = O(\tau^2), \tag{17.121}$$

using the preceding approximations for the roots. If

$$g(t) = \int_{-\infty}^{\infty} G(\omega)\mathrm{e}^{-\mathrm{i}\omega t}\,\mathrm{d}\omega, \tag{17.122}$$

is the condition

$$g(t) = 0, \quad \text{for } t < 0, \tag{17.123}$$

satisfied? This question is resolved by evaluating the preceding integral for $t < 0$. The integral can be most readily handled by using a contour integration approach, and working to terms of $O(\tau^2)$ is sufficient. This is left as an exercise for the reader. It will come as no surprise to the reader to find that Eq. (17.123) is not satisfied, because of the pole of G in the upper half of the complex plane. The displacement in terms of the external electric field is given by

$$\boldsymbol{x}(t) = \frac{1}{2\pi}\int_{-\infty}^{\infty} \mathbf{E}(t')g(t-t')\mathrm{d}t', \tag{17.124}$$

and since $g(t)$ does not satisfy Eq. (17.123), $\boldsymbol{x}(t)$ is determined by the applied field for times $t' > t$. Hence, the system described by Eq. (17.107) is acausal. The reader is left to explore how the pole of G in the upper half of the complex plane changes the form of the Hilbert transform connections between Re $G(\omega)$ and Im $G(\omega)$.

17.10 The Paley–Wiener log-integral theorem

In this section the relationship between Fourier transforms and causal functions is explored further. The main result considered is a theorem of Paley and Wiener (1934, pp. 16–17). Let $f \in L^2(\mathbb{R})$ and let it satisfy $f(t) = 0$ for $t \leq t_0$, and let $F(\omega)$ denote the Fourier transform of f. Then, a necessary and sufficient condition that there exists a function $A(\omega)$, such that $A(\omega) = |F(\omega)|$, is the requirement

$$\int_{-\infty}^{\infty} \frac{|\log A(\omega)|\mathrm{d}\omega}{1+\omega^2} < \infty. \tag{17.125}$$

An immediate consequence of requiring the preceding integral to be convergent is that the support of $A(\omega)$ is unbounded. If this were not the case, $A(\omega)$ would be zero over finite intervals, and hence $|\log A(\omega)|$ would be infinite over the same intervals, and the integral in Eq. (17.125) would be divergent. The second consequence is that $A(\omega)$ should approach zero asymptotically as $\omega \to \pm\infty$ slower than any exponential function. For example, suppose $A(\omega) = A_0 \mathrm{e}^{-\alpha|\omega|}$, where α and A_0 are constants with $\alpha > 0$; then,

$$\int_{-\infty}^{\infty} \frac{|\log A(\omega)|\mathrm{d}\omega}{1+\omega^2} = 2\int_0^{\infty} \frac{|\log A_0 - \alpha\omega|\mathrm{d}\omega}{1+\omega^2} = \infty. \tag{17.126}$$

To establish the necessity and sufficiency of Eq. (17.125), a few preliminary results are required. Let $f(z)$ be analytic and non-zero for $|z| \leq R$; then, from the Cauchy integral formula, restricted to the case where the closed contour is a circle of radius R centered at the point z,

$$f(z) = \frac{1}{2\pi}\int_0^{2\pi} f(z + R\mathrm{e}^{\mathrm{i}\theta})\mathrm{d}\theta, \tag{17.127}$$

and this is sometimes referred to as the *mean-value theorem* for analytic functions. From this result it follows that

$$\log|f(0)| = \frac{1}{2\pi}\int_0^{2\pi} \log\left|f(R\mathrm{e}^{\mathrm{i}\theta})\right|\mathrm{d}\theta. \tag{17.128}$$

The generalization of this result to cover the case where $f(z)$ has zeros inside the circle $|z| = R$ goes as follows. Let $f(0) \neq 0$, and let the zeros of f inside the circle be denoted by $z_1, z_2, z_3, \ldots, z_n$, arranged so that the moduli of these points, denoted $a_1, a_2, a_3, \ldots, a_n$, are in increasing order. Multiple zeros are represented by identical z_k, in the sequence and higher-order zeros being dealt with in a similar fashion. The function $g(z)$, defined by

$$g(z) = f(z)\prod_{k=1}^{n} \frac{1}{(z - z_k)}, \tag{17.129}$$

is analytic and non-zero for $|z| \leq R$. The special case where $f(0) = 0$ can be dealt with by considering the Taylor expansion of f about $z = 0$ and working with a modified choice of function. This is left as an exercise for the reader to consider. Since $g(z)$ is analytic and non-zero for $|z| \leq R$, Eq. (17.127) applied to $\log|g(z)|$ leads to

$$\log|f(0)| - \sum_{k=1}^{n}\log a_k = \frac{1}{2\pi}\int_0^{2\pi}\left\{\log\left|f(R\mathrm{e}^{\mathrm{i}\theta})\right| - \sum_{k=1}^{n}\log\left|R\mathrm{e}^{\mathrm{i}\theta} - z_k\right|\right\}\mathrm{d}\theta. \tag{17.130}$$

Using

$$\frac{1}{2\pi}\int_0^{2\pi}\log\left|R\mathrm{e}^{\mathrm{i}\theta} - z_k\right|\mathrm{d}\theta = \log R, \tag{17.131}$$

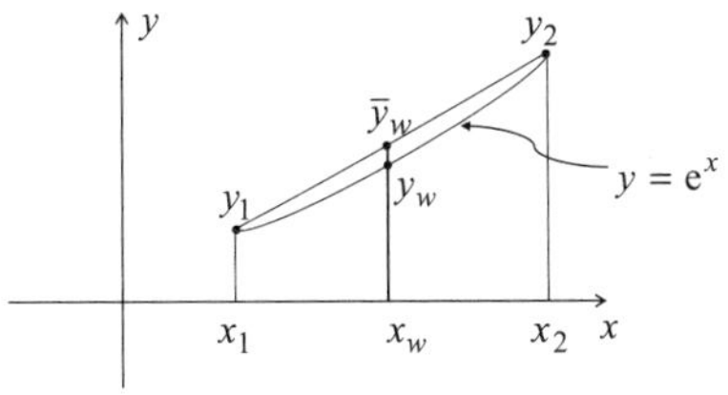

Figure 17.6. Convex behavior for the function e^x.

then

$$\frac{1}{2\pi}\int_0^{2\pi} \log\left|f(Re^{i\theta})\right| d\theta = \log|f(0)| + \log\left\{\frac{R^n}{a_1 a_2 \ldots a_n}\right\}, \tag{17.132}$$

which is called Jensen's formula.

A second result that is needed in the sequel is as follows:

$$\exp\left\{\int_{-\infty}^{\infty} w(\omega)\,x(\omega) d\omega\right\} \leq \int_{-\infty}^{\infty} w(\omega) e^{x(\omega)}\, d\omega, \tag{17.133}$$

where $w(\omega) \geq 0$ and

$$\int_{-\infty}^{\infty} w(\omega) d\omega = 1. \tag{17.134}$$

To obtain the preceding formula, the convex property of the function e^x is employed, and so, for $x_1 < x_2$,

$$e^{wx_1+(1-w)x_2} \leq we^{x_1} + (1-w)e^{x_2}, \quad \text{for } w \in [0, 1]. \tag{17.135}$$

This can be readily seen from the situation depicted in Figure 17.6.

Let x_w denote a point on the interval $[x_1, x_2]$; then,

$$x_w = wx_1 + (1-w)x_2, \tag{17.136}$$

and the corresponding point on the curve is

$$y_w \equiv y(x_w) = e^{wx_1+(1-w)x_2}. \tag{17.137}$$

Using the notation $y_j = e^{x_j}, j = 1, 2,$ or w, and $\bar{y}_w = \bar{y}(x_w)$, where $\bar{y}$ denotes the straight line connecting the points (x_1, y_1) and (x_2, y_2) shown in Figure 17.6, and is given by

$$\bar{y}(x) = \frac{1}{x_2 - x_1}[(y_2 - y_1)x + x_2 y_1 - x_1 y_2], \tag{17.138}$$

and hence the distance $\bar{y}_w$ is given by

$$\bar{y}_w = wy_1 + (1 - w)y_2. \tag{17.139}$$

Inspection of the figure makes it obvious that $y_w \leq \bar{y}_w$ because of the convexity of the function e^x, and hence Eq. (17.135) follows. The inequality in Eq. (17.135) can be generalized to read

$$\exp\left\{\sum_{k=1}^{n} w_k e^{x_k}\right\} \leq \sum_{k=1}^{n} w_k e^{x_k}, \tag{17.140}$$

with

$$\sum_{k=1}^{n} w_k = 1. \tag{17.141}$$

On changing from the discrete w_k to a continuous $w(x)$, by taking the appropriate limit, Eqs. (17.140) and (17.141) lead to Eqs. (17.133) and (17.134), respectively.

With the results of the preceding two paragraphs, the Paley–Wiener theorem can now be established. The function $F(z)$ is analytic in the upper half of the complex plane. A conformal mapping from the upper half plane to the interior of the unit circle centered at the origin is made using the transformation

$$\varsigma = \frac{z - i}{z + i}. \tag{17.142}$$

On setting $\Phi(\varsigma) = F(-i(\varsigma + 1)(\varsigma - 1)^{-1})$ and using $\varsigma = e^{i\theta}$, yields

$$\int_{-\infty}^{\infty} \frac{|F(x)|^2 \, dx}{1 + x^2} = \frac{1}{2} \int_{-\pi}^{\pi} \left|\Phi(e^{i\theta})\right|^2 d\theta. \tag{17.143}$$

The integral on the right-hand side of Eq. (17.143) satisfies

$$\int_{-\pi}^{\pi} \left|\Phi(e^{i\theta})\right|^2 d\theta < \infty, \tag{17.144}$$

which can be established using the following argument. By assumption, $f \in L^2(\mathbb{R})$, from which it follows that $F \in L^2(\mathbb{R})$; therefore,

$$\int_{-\infty}^{\infty} \frac{|F(x)|^2 \, dx}{1 + x^2} < \int_{-\infty}^{\infty} |F(x)|^2 \, dx < \infty, \tag{17.145}$$

and hence Eq. (17.144) follows. In a similar fashion to writing Eq. (17.143), it follows that

$$2\int_{-\infty}^{\infty}\frac{|\log A(x)|\mathrm{d}x}{1+x^2}=\int_{-\pi}^{\pi}|\log|\Phi(\mathrm{e}^{\mathrm{i}\theta})||\mathrm{d}\theta. \tag{17.146}$$

It is necessary to show that this integral is bounded. To that end, set

$$g_+(\theta)=\max\left\{0,\log\left|\Phi(\mathrm{e}^{\mathrm{i}\theta})\right|\right\}, \tag{17.147}$$

and

$$g_-(\theta)=-\min\left\{0,\log\left|\Phi(\mathrm{e}^{\mathrm{i}\theta})\right|\right\}, \tag{17.148}$$

so that

$$\log\left|\Phi(\mathrm{e}^{\mathrm{i}\theta})\right|=g_+(\theta)-g_-(\theta). \tag{17.149}$$

Hence,

$$\int_{-\pi}^{\pi}\log|\Phi(\mathrm{e}^{\mathrm{i}\theta})|\mathrm{d}\theta=\int_{-\pi}^{\pi}g_+(\theta)\mathrm{d}\theta-\int_{-\pi}^{\pi}g_-(\theta)\mathrm{d}\theta \tag{17.150}$$

and

$$\int_{-\pi}^{\pi}\left|\log\left|\Phi(\mathrm{e}^{\mathrm{i}\theta})\right|\right|\mathrm{d}\theta=\int_{-\pi}^{\pi}g_+(\theta)\mathrm{d}\theta+\int_{-\pi}^{\pi}g_-(\theta)\mathrm{d}\theta. \tag{17.151}$$

From the elementary inequality $\log x < x^2$, for $x > 0$, it follows that

$$\int_{-\pi}^{\pi}g_+(\theta)\mathrm{d}\theta<\int_{-\pi}^{\pi}\left|\Phi(\mathrm{e}^{\mathrm{i}\theta})\right|^2\mathrm{d}\theta. \tag{17.152}$$

Assume that $\Phi(0)\neq 0$; then, making use of Jensen's formula, Eq. (17.132), it follows from Eqs. (17.151) and (17.152) that

$$\begin{aligned}\frac{1}{2\pi}\int_{-\pi}^{\pi}\left|\log\left|\Phi(\mathrm{e}^{\mathrm{i}\theta})\right|\right|\mathrm{d}\theta&=\frac{2}{2\pi}\int_{-\pi}^{\pi}g_+(\theta)\mathrm{d}\theta-\frac{1}{2\pi}\int_{-\pi}^{\pi}\{g_+(\theta)-g_-(\theta)\}\mathrm{d}\theta\\&=\frac{1}{\pi}\int_{-\pi}^{\pi}g_+(\theta)\mathrm{d}\theta-\frac{1}{2\pi}\int_{-\pi}^{\pi}\log\left|\Phi(\mathrm{e}^{\mathrm{i}\theta})\right|\mathrm{d}\theta\\&<\frac{1}{\pi}\int_{-\pi}^{\pi}\left|\Phi(\mathrm{e}^{\mathrm{i}\theta})\right|^2\mathrm{d}\theta-\log|\Phi(0)|-\log\left\{\frac{1}{a_1a_2\cdots a_n}\right\},\end{aligned} \tag{17.153}$$

which yields, on making use of Eq. (17.144), the following formula:

$$\frac{1}{2\pi}\int_{-\pi}^{\pi}\left|\log\left|\Phi(\mathrm{e}^{\mathrm{i}\theta})\right|\right|\mathrm{d}\theta < \infty. \tag{17.154}$$

On using Eq. (17.146), Eq. (17.125) has been established, which is the desired result. If $\Phi(0) = 0$, then $\Phi(z)$ can be expanded in a Taylor series about $z = 0$, so that

$$\Phi(z) = \alpha_n z^n + \alpha_{n-1} z^{n-1} + \cdots, \tag{17.155}$$

and an appropriate modification of Jensen's formula can be obtained.

Consider the integral

$$h(x+\mathrm{i}y) = \frac{1}{\pi}\int_{-\infty}^{\infty}\frac{y\log A(x')}{(x-x')^2+y^2}\,\mathrm{d}x', \tag{17.156}$$

and suppose Eq. (17.125) is satisfied, then the preceding integral converges. The limit $y \to 0$ yields

$$\lim_{y\to 0} h(x+\mathrm{i}y) = \lim_{y\to 0}\frac{1}{\pi}\int_{-\infty}^{\infty}\frac{y\log A(x')}{(x-x')^2+y^2}\,\mathrm{d}x' = \log A(x), \tag{17.157}$$

where the following result has been employed:

$$\lim_{y\to 0}\frac{1}{\pi}\frac{y}{(x-x')^2+y^2} = \delta(x'-x). \tag{17.158}$$

Noting that

$$\lim_{y\to 0}\mathrm{e}^{h(x+\mathrm{i}y)} = A(x) \tag{17.159}$$

and

$$\int_{-\infty}^{\infty}\frac{1}{\pi}\frac{y\,\mathrm{d}x'}{(x-x')^2+y^2} = 1, \tag{17.160}$$

using Eqs. (17.156) and (17.133) yields

$$\mathrm{e}^{h(x+\mathrm{i}y)} = \exp\left\{\frac{1}{\pi}\int_{-\infty}^{\infty}\frac{y\log A(x')}{(x-x')^2+y^2}\,\mathrm{d}x'\right\} \le \frac{1}{\pi}\int_{-\infty}^{\infty}\frac{yA(x')}{(x-x')^2+y^2}\,\mathrm{d}x'. \tag{17.161}$$

Applying the Cauchy–Schwarz–Buniakowski inequality leads to

$$
\begin{aligned}
e^{h(x+iy)} &\le \frac{y}{\pi}\int_{-\infty}^{\infty} \frac{A(x')}{\sqrt{[(x-x')^2+y^2]}}\frac{1}{\sqrt{[(x-x')^2+y^2]}}\,dx' \\
&\le \frac{y}{\pi}\sqrt{\left(\int_{-\infty}^{\infty}\frac{A^2(x')dx'}{(x-x')^2+y^2}\right)}\sqrt{\left(\int_{-\infty}^{\infty}\frac{dx''}{(x-x'')^2+y^2}\right)}. \qquad (17.162)
\end{aligned}
$$

Squaring this last inequality, integrating over x and making use of the results

$$\int_{-\infty}^{\infty}\frac{dx''}{(x'-x'')^2+4y^2}=\frac{\pi}{2y} \qquad (17.163)$$

and

$$\int_{-\infty}^{\infty}\frac{dx}{[(x-x')^2+y^2][(x-x'')^2+y^2]}=\frac{2\pi}{y\{(x'-x'')^2+4y^2\}}, \qquad (17.164)$$

leads to

$$
\begin{aligned}
\int_{-\infty}^{\infty} e^{2h(x+iy)}\,dx &\le \frac{y^2}{\pi^2}\int_{-\infty}^{\infty}dx\int_{-\infty}^{\infty}\frac{A^2(x')dx'}{(x-x')^2+y^2}\int_{-\infty}^{\infty}\frac{dx''}{(x-x'')^2+y^2} \\
&= \frac{y^2}{\pi^2}\int_{-\infty}^{\infty}A^2(x')dx'\int_{-\infty}^{\infty}dx''\int_{-\infty}^{\infty}\frac{dx}{[(x-x')^2+y^2][(x-x'')^2+y^2]} \\
&= \frac{2y}{\pi}\int_{-\infty}^{\infty}A^2(x')dx'\int_{-\infty}^{\infty}\frac{dx''}{\{(x'-x'')^2+4y^2\}} \\
&= \int_{-\infty}^{\infty}A^2(x')dx'. \qquad (17.165)
\end{aligned}
$$

The reader should justify the change of integration order carried out. So, for all $y>0$ and the square integrable assumption on f, which implies that A is square integrable, it follows that

$$\int_{-\infty}^{\infty} e^{2h(x+iy)}\,dx \le \int_{-\infty}^{\infty}A^2(x)dx < \infty. \qquad (17.166)$$

Let the harmonic conjugate function of $h(x+iy)$ be denoted by $k(x+iy)$, then the following function can be defined:

$$\theta(z)=h(z)+ik(z), \qquad (17.167)$$

and $e^{\theta(z)}$ is analytic for all $y>0$. Making use of

$$\left|e^{\theta(z)}\right|^2=e^{2h(z)}, \qquad (17.168)$$

and employing Eq. (17.166), leads to

$$\int_{-\infty}^{\infty} \left| e^{\theta(z)} \right|^2 dx < \infty, \quad \text{for } y > 0. \tag{17.169}$$

It then follows immediately from Titchmarsh's theorem (Section 4.22) that there exists a function $f(t)$ such that $f(t) = 0$ for $t < 0$, and

$$e^{\theta(x)} = \int_{-\infty}^{\infty} f(t) e^{-ixt}\, dt. \tag{17.170}$$

Hence, if $A(\omega)$ is square integrable and Eq. (17.125) is satisfied, there exists a causal function whose Fourier transform is directly related to $A(\omega)$.

17.11 Extensions of the causality concept

In Section 17.8 the intuitive or primitive concept of causality was employed, namely, that the output cannot precede the input. Some extensions of the basic ideas connected with causality are amplified upon in this section. These extensions arise in the context of scattering theory. Dispersion relations in scattering theory are the subject of an extensive literature, and, for this reason, the present and following section should be viewed as providing a very concise introduction to some of the most elementary aspects of the topic.

Classical scattering from a spherically symmetric scatterer of radius r_0 is the focus problem of this section. The equation governing scattering is as follows:

$$\nabla^2 \psi(r,t) = \frac{1}{v^2} \frac{\partial \psi(r,t)}{\partial t^2}, \tag{17.171}$$

where v is the velocity of propagation of the wave front, and the solution is sought for $r > r_0$. The general solution of Eq. (17.171) can be written in the following form:

$$\psi(k,r,t) = \psi_{\text{in}}(k,r,t) + \psi_{\text{out}}(k,r,t), \tag{17.172}$$

where $\psi_{\text{in}}(k,r,t)$ denotes the incident ingoing wave, $\psi_{\text{out}}(k,r,t)$ designates the outgoing wave, and k is the wave number. These functions take the following form, for $r > r_0$:

$$\psi_{\text{in}}(k,r,t) = -a(k) \frac{e^{-ik(r+vt)}}{r} \tag{17.173}$$

and

$$\psi_{\text{out}}(k,r,t) = b(k) \frac{e^{ik(r-vt)}}{r}, \tag{17.174}$$

where the coefficients $a(k)$ and $b(k)$ are independent of the spatial and temporal variables. The functions ψ_{in} and ψ_{out} are spherically symmetric and describe *s-wave* scattering. The case of higher multipole waves is treated in the references cited in the chapter end-notes. If consideration is restricted to wave packets of finite energy, then it can be deduced that the $a(k)$ coefficient is square integrable (Van Kampen, 1953a). Conservation of energy then requires that the $b(k)$ coefficient is also square integrable.

It is convenient to introduce an S-function as the ratio of the coefficient of the outgoing wave to the coefficient of the incoming wave, that is

$$S(k) = \frac{b(k)}{a(k)}. \tag{17.175}$$

The scattered wave is given by

$$\psi_{\mathrm{sc}}(k, r, t) = a(k)\{S(k) - 1\}\frac{\mathrm{e}^{\mathrm{i}k(r-vt)}}{r}. \tag{17.176}$$

The general solution of Eq. (17.171) can be written as follows:

$$\psi(r, t) = \int_{-\infty}^{\infty} \psi(k, r, t)\mathrm{d}k, \tag{17.177}$$

with similar results for ψ_{in} and ψ_{out}:

$$\psi_{\mathrm{in}}(r, t) = \int_{-\infty}^{\infty} \psi_{\mathrm{in}}(k, r, t)\mathrm{d}k \tag{17.178}$$

and

$$\psi_{\mathrm{out}}(r, t) = \int_{-\infty}^{\infty} \psi_{\mathrm{out}}(k, r, t)\mathrm{d}k. \tag{17.179}$$

The preceding three results assume explicitly that the superposition principle applies, which, in turn, requires system linearity. In Eqs. (17.177)–(17.178), a common literature convention is employed, where the same symbol is employed for the fields depending on k, r, and t, and for the corresponding fields depending on only r and t. These are of course different functions: the reader should have no problem discerning which is intended, since the variable dependence makes it clear which field is under discussion. The fields $\psi_{\mathrm{in}}(r, t)$ and $\psi_{\mathrm{out}}(r, t)$ are real. From this result and Eqs. (17.175), (17.178), and (17.179), it follows that

$$S(-k) = S(k)^*, \tag{17.180}$$

which is the crossing symmetry relation for the S-function. Similar results hold for the coefficients $a(k)$ and $b(k)$.

The causality condition is employed in the following form. The incoming wave must appear at the scatterer before there can be an outgoing wave. Suppose the incoming wave has a sharp front, which arrives at the scatterer at time t_0; then,

$$\psi_{\text{in}}(r_0, t) = -\frac{1}{r_0}\int_{-\infty}^{\infty} a(k)\mathrm{e}^{-\mathrm{i}kv(t+t_0)}\,\mathrm{d}k \tag{17.181}$$

and

$$\psi_{\text{in}}(r_0, t) = 0, \quad \text{if } t < t_0. \tag{17.182}$$

The outgoing wave can be written as follows:

$$\psi_{\text{out}}(r_0, t) = \frac{1}{r_0}\int_{-\infty}^{\infty} b(k)\mathrm{e}^{-\mathrm{i}kv(t-t_0)}\,\mathrm{d}k \tag{17.183}$$

and

$$\psi_{\text{out}}(r_0, t) = 0, \quad \text{if } t < t_0. \tag{17.184}$$

Equations (17.182) and (17.184) become the mathematical representations of the causality statement just given.

Taking the inverse Fourier transform of Eq. (17.181) leads to

$$a(k) = -\frac{vr_0}{2\pi}\int_{-\infty}^{\infty} \psi_{\text{in}}(r_0, t)\mathrm{e}^{\mathrm{i}kv(t+t_0)}\,\mathrm{d}t, \tag{17.185}$$

which, on using Eq. (17.182), reduces to

$$a(k) = -\frac{vr_0\mathrm{e}^{\mathrm{i}kr_0}}{2\pi}\int_{t_0}^{\infty} \psi_{\text{in}}(r_0, t)\mathrm{e}^{\mathrm{i}kvt}\,\mathrm{d}t. \tag{17.186}$$

If for convenience the scatterer location is fixed so that t_0 is positive, then the last result can be written as follows:

$$a(k) = -\frac{vr_0\mathrm{e}^{\mathrm{i}kr_0}}{2\pi}\int_{0}^{\infty} \psi_{\text{in}}(r_0, t)\mathrm{e}^{\mathrm{i}kvt}\,\mathrm{d}t. \tag{17.187}$$

The coefficient $a(k)\mathrm{e}^{-\mathrm{i}kr_0}$ is therefore a causal transform and $\psi_{\text{in}}(r_0, t)$ is a causal function. Let k be a complex variable. Application of Titchmarsh's theorem yields the conclusion that $a(k)$ is an analytic function in the upper half complex k-plane. Hilbert transform connections can therefore be written between the real and imaginary parts of $a(k)$. The interested reader is left to explore this as an exercise.

It follows in a similar fashion that

$$b(k) = \frac{vr_0\mathrm{e}^{-\mathrm{i}kr_0}}{2\pi}\int_{0}^{\infty} \psi_{\text{out}}(r_0, t)\mathrm{e}^{\mathrm{i}kvt}\,\mathrm{d}t. \tag{17.188}$$

From this result, it is deduced that $b(k)\mathrm{e}^{\mathrm{i}kr_0}$ is a causal transform and that $\psi_{\mathrm{out}}(r_0, t)$ a causal function. Employing Eq. (17.175) leads to

$$S_0(k) \equiv \frac{b(k)\mathrm{e}^{\mathrm{i}kr_0}}{a(k)\mathrm{e}^{-\mathrm{i}kr_0}} = \mathrm{e}^{2\mathrm{i}kr_0} S(k). \tag{17.189}$$

Since $a(k)\mathrm{e}^{-\mathrm{i}kr_0}$ and $b(k)\mathrm{e}^{\mathrm{i}kr_0}$ are both analytic in the upper half complex k-plane, it follows that $S_0(k)$ is analytic in the same region, provided that $S_0(k)$ has no poles at the zeros of $a(k)\mathrm{e}^{-\mathrm{i}kr_0}$ in the upper half complex k-plane. The connection $b(k) = S(k)a(k)$ must hold for any complex k, and at a zero of $a(k)$ the function $S(k)$ must behave in such a manner as to offset this vanishing behavior of $a(k)$, since $b(k)$ does not vanish at a zero of $a(k)$ for almost all k. Hence, it is concluded that the scattering function $S(k)$ is analytic in the upper half complex k-plane. Dispersion relations can therefore be obtained for the scattering function. If the scattering function is written in terms of the phase shift $\eta(k)$ via

$$S(k) = \mathrm{e}^{2\mathrm{i}\eta(k)}, \tag{17.190}$$

then the cross-section for s-wave scattering, $\sigma_0(k)$, can be written as follows:

$$\sigma_0(k) = \frac{\pi}{k^2}\,|1 - S(k)|^2 = \frac{4\pi}{k^2}\sin^2\eta(k). \tag{17.191}$$

It is left for the reader to explore in the exercises the dispersion relations that can be obtained for the scattering function, cross-section, and phase shift.

17.12 Basic quantum scattering: causality conditions

Elementary aspects of non-relativistic quantum scattering with an emphasis on the connection with causality are considered in this section. The same problem discussed in the preceding section is reconsidered, but from a quantum perspective. The discussion follows the accounts of Schützer and Tiomno (1951), Van Kampen (1953a, 1953b), and particularly Nussenzveig (1972). The problem is governed by the time-dependent Schrödinger equation,

$$-\frac{\hbar^2}{2m}\nabla^2\psi + V\psi = \mathrm{i}\hbar\frac{\partial\psi}{\partial t}, \tag{17.192}$$

where m is the particle mass, $\hbar$ is Planck's constant divided by 2π, and V is the interaction potential. The scattering potential is taken to be zero outside $r > r_0$. The solutions of Eq. (17.192) can be expressed as

$$\psi(r,t) = r^{-1}\varphi_{\mathrm{in}}(r,t) + r^{-1}\varphi_{\mathrm{out}}(r,t), \tag{17.193}$$

where $\varphi_{in}(r,t)$ denotes the incident incoming wave packet and $\varphi_{\mathrm{out}}(r,t)$ designates the outgoing wave packet. Assuming the superposition principle to hold, and consequently the implied linearity of the system, it follows that

$$\psi(r,t) = \int_0^\infty \psi(E,r,t)\mathrm{d}E, \tag{17.194}$$

where E is the energy and $\psi(E,r,t)$ is given by

$$\psi(E,r,t) = r^{-1}\varphi_{\mathrm{in}}(E,r,t) + r^{-1}\varphi_{\mathrm{out}}(E,r,t). \tag{17.195}$$

The particle energy is given in terms of the wave number k by

$$E = \frac{\hbar^2 k^2}{2m}, \tag{17.196}$$

and $\varphi_{\mathrm{in}}(r,t)$ and $\varphi_{\mathrm{out}}(r,t)$ can be expressed as follows:

$$\varphi_{\mathrm{in}}(r,t) = -\int_0^\infty a(E)\mathrm{e}^{-\mathrm{i}kr - \mathrm{i}Et/\hbar}\,\mathrm{d}E \tag{17.197}$$

and

$$\varphi_{\mathrm{out}}(r,t) = \int_0^\infty b(E)\mathrm{e}^{\mathrm{i}kr - \mathrm{i}Et/\hbar}\,\mathrm{d}E. \tag{17.198}$$

From Eq. (17.196), it follows for real k that E is positive, and hence the integration intervals for the integrals in Eqs. (17.194), (17.197), and (17.198) are $[0,\infty)$. The scattering S-function is defined in the same manner employed in Section 17.11:

$$S(E) = \frac{b(E)}{a(E)}. \tag{17.199}$$

The terminology for the more general scattering problem is *S-matrix*, but, in the context of the present problem, it is a function. The scattered wave packet is given by

$$\varphi_{\mathrm{sc}}(r,t) = \int_0^\infty \{1 - S(E)\}a(E)\mathrm{e}^{\mathrm{i}kr - \mathrm{i}Et/\hbar}\,\mathrm{d}E. \tag{17.200}$$

The conservation of the norm of the wave function can be cast in terms of

$$\nabla\cdot\boldsymbol{J} + \frac{\partial\rho}{\partial t} = 0, \tag{17.201}$$

where ρ is the probability density given by

$$\rho = \psi^*\psi \tag{17.202}$$

and $\boldsymbol{J}$ is the probability current density defined by

$$\boldsymbol{J}(r,t) = -\frac{\mathrm{i}\hbar}{m}\{\psi^*\nabla\psi\}, \tag{17.203}$$

and the real part is understood in this expression.

Let $\Phi(r,t)$ denote the probability flux through a sphere of radius greater than r_0; being mindful of the limitations of this interpretation arising from uncertainty principle considerations,

$$\Phi(r,t) = \frac{2\pi \mathrm{i}h}{m}\left\{\varphi^* \frac{\partial\varphi}{\partial r} - \varphi\frac{\partial\varphi^*}{\partial r}\right\}, \tag{17.204}$$

where h is Planck's constant and $\varphi(r,t) = r\psi(r,t)$. Employing Eqs. (17.193), (17.197), and (17.198) leads to

$$\begin{aligned}\Phi(r,t) = \frac{2\pi h}{m}\int_0^\infty \mathrm{d}E\int_0^\infty k'\Big\{&\mathrm{e}^{\mathrm{i}(E-E')t}\Big[a(E)^*a(E')\mathrm{e}^{\mathrm{i}(k-k')r} - b(E)^*b(E')\mathrm{e}^{\mathrm{i}(k'-k)r}\\ &+a(E)^*b(E')\mathrm{e}^{\mathrm{i}(k+k')r} - b(E)^*a(E')\mathrm{e}^{-\mathrm{i}(k'+k)r}\Big]\\ &+\mathrm{e}^{-\mathrm{i}(E-E')t}\Big[a(E)a(E')^*\mathrm{e}^{\mathrm{i}(k'-k)r} - b(E)b(E')^*\mathrm{e}^{\mathrm{i}(k-k')r}\\ &+a(E)b(E')^*\mathrm{e}^{-\mathrm{i}(k+k')r} - b(E)a(E')^*\mathrm{e}^{\mathrm{i}(k'+k)r}\Big]\Big\}\mathrm{d}E'.\end{aligned} \tag{17.205}$$

The total integrated flux over all time is given by

$$\begin{aligned}\int_{-\infty}^{\infty}\Phi(r,t)\mathrm{d}t &= \frac{2\pi h}{m}\int_{-\infty}^{\infty}\mathrm{d}t\int_0^\infty \mathrm{d}E\int_0^\infty k'\left\{\mathrm{e}^{\mathrm{i}(E-E')t}[\cdots]+\mathrm{e}^{-\mathrm{i}(E-E')t}[\cdots]\right\}\mathrm{d}E'\\ &= \frac{4\pi^2 h}{m}\int_0^\infty \mathrm{d}E\int_0^\infty k'\{[\cdots]+[\cdots]\}\delta(E-E')\mathrm{d}E',\end{aligned} \tag{17.206}$$

where the terms $[\cdots]$ should be obvious by comparison with Eq. (17.205) for $\Phi(r,t)$, and Eq. (2.269) has been employed. Evaluating the integral over E', Eq. (17.206) simplifies to

$$\int_{-\infty}^{\infty}\Phi(r,t)\mathrm{d}t = \frac{8\pi^2 h}{m}\int_0^\infty k\{|a(E)|^2 - |b(E)|^2\}\mathrm{d}E. \tag{17.207}$$

Assuming conservation of probability, it can be shown that

$$\int_{-\infty}^{\infty}\Phi(r,t)\mathrm{d}t = 0, \tag{17.208}$$

and hence

$$\int_0^\infty k\{|a(E)|^2 - |b(E)|^2\}\mathrm{d}E = 0. \tag{17.209}$$

It is assumed that the functions $E^{1/4}a(E)$ and $E^{1/4}b(E)$ are square integrable. Equation (17.209) is consistent with the unitary condition for the S-function:

$$|S(E)|^2 = 1. \tag{17.210}$$

In non-relativistic theory there is no limiting velocity. It might appear reasonable to the reader that the causality statement of Section 17.11 could be transferred to the present problem, that is, if $\psi_{\text{in}}(r,t)$ vanishes at the boundary of the scatterer for all times $t < t_0$, then $\psi_{\text{out}}(r,t)$ vanishes for all points outside the scatterer for $t < t_0$. The difficulty with this is that there are no ingoing or outgoing wave packets that are exactly zero up to a certain time. A Schrödinger wave packet will instantaneously spread out spatially (see, for example, Merzbacher (1970, p. 221). A feasible approach, to circumvent this difficulty, is to adopt the idea that, at a distance r and time t, the scattered wave packet does not depend on the incident wave packet at the same position for times greater than t (Schützer and Tiomno, 1951). If the following connection between the scattered wave packet and the incoming wave packet for $r > r_0$ is written as

$$\varphi_{\text{sc}}(r,t) = \int_{-\infty}^{\infty} g(r,t-t')\varphi_{\text{in}}(r,t')\text{d}t', \tag{17.211}$$

then the Schützer–Tiomno requirement leads to

$$g(r,t-t') = 0, \quad \text{for } t - t' < 0. \tag{17.212}$$

Taking the inverse Fourier transform of Eq. (17.211), using the Fourier convolution theorem in the form

$$\begin{aligned} \mathcal{F}^{-1}\{\varphi_{\text{sc}}(r,t)\}(E) &= \mathcal{F}^{-1}\{g * \varphi_{\text{in}}\}(E) \\ &= 2\pi\mathcal{F}^{-1}g(E)\mathcal{F}^{-1}\varphi_{\text{in}}(E), \end{aligned} \tag{17.213}$$

and using Eqs. (17.197) and (17.200) to evaluate the inverse Fourier transforms of φ_{in} and φ_{sc}, leads to the following result:

$$G(r,E) = \frac{1}{2\pi}\{S(E) - 1\}\text{e}^{2ikr}, \tag{17.214}$$

where $G(r,E)$ denotes the inverse Fourier transform of $g(r,t)$,

$$G(r,E) = \frac{1}{2\pi}\int_{-\infty}^{\infty} g(r,t)\text{e}^{\text{i}Et/\hbar}\,\text{d}t. \tag{17.215}$$

Note that the inverse Fourier transforms of $\varphi_{\text{sc}}(r,t)$ and $\varphi_{\text{in}}(r,t)$ each involve the Heaviside function. To write Eq. (17.214), it has been implicitly assumed that the requisite inverse Fourier transforms exist, which in turn requires that the functions φ_{in}, φ_{sc}, and g are square integrable. Weaker conditions could be invoked, but it is also necessary to have the appropriate conditions in place so that Titchmarsh's theorem can be applied. It might happen in cases of interest that the Fourier transform of g behaves like a tempered distribution. In this situation, it would be necessary to invoke some of the mathematical machinery discussed in Chapter 10. The condition given in Eq. (17.212) allows Eq. (17.215) to be written as follows:

$$G(r,E) = \frac{1}{2\pi}\int_0^\infty g(r,t)e^{iEt/\hbar}\,dt. \tag{17.216}$$

From the preceding result it follows that $G(r, E)$ is analytic in the upper half complex E-plane. The connection arrived at in Eq. (17.214) then leads to the conclusion that $S(E)$ is an analytic function in the same region. To obtain the standard form of the dispersion relations for $S(E)$ requires information for $E < 0$. This information is not forthcoming from the causality principle.

To discuss the scattering function in the complex k-plane, from Eq. (17.196) employ $k = \hbar^{-1}\surd(2mE)$, and adopting the positive root for $E > 0$ leads to the conclusion that $S(k)$ is an analytic function in the first quadrant of the complex k-plane. A cut can be made along the positive E-axis, and the upper half complex E-plane of the first Riemann sheet then corresponds to the first quadrant of the complex k-plane. Setting up dispersion relations for $S(k)$ still requires information for $S(-k)$. With certain restrictions on the form of the interaction potential, it can be shown for real k (Wigner, 1964) that

$$S(-k) = S(k)^*, \tag{17.217}$$

which is of the same form as the result given for the classical case, Eq. (17.180). The function $S(k)$ can be analytically continued into the second quadrant of the complex k-plane. There is a complication from bound states of the system. Bound states can be incorporated by making the replacement $k \to ik_b$ with k_b real, and setting $a(E_b) = 0$, then the definition of $S(E)$ makes it clear that the scattering function has a pole at each bound state. These poles appear on the positive imaginary axis in the complex k-plane.

Van Kampen (1953b) introduced the following causality condition. The probability of finding an outgoing particle at some position r_1 prior to the time t_1 cannot exceed the probability of finding an ingoing particle at r_1 before t_1. For finite r_1 this statement has limitations, as Van Kampen points out. For a finite r_1 there is no unique decomposition of the total probability into a probability for the ingoing particle and a probability for the outgoing particle. This is due to the interference term that arises between ingoing and outgoing wave packets. To circumvent this problem, Van Kampen proposed the following alternative definition for the causality condition. The outgoing probability current integrated over the interval $(-\infty, t_1]$ cannot exceed the integrated current by an amount that is larger than the absolute value of the integral of the interference term. This statement can be cast in the following form:

$$4\pi \int_{r_0}^{\infty} r^2 \left|\varphi_{\text{in}}(r,0) + \varphi_{\text{out}}(r,0)\right|^2 \mathrm{d}r \leq 1. \tag{17.218}$$

Van Kampen gave a detailed account of the analytic properties of $S(k)$ and considered the crossing symmetry constraint given in Eq. (17.217). He also derived dispersion relations involving the function $S(k)$. The interested reader can pursue the details in the cited reference.

Because scattering theory plays such a central role in many areas of modern physics, the reader should not be surprised to learn that significant work has been carried out on the derivation of dispersion relations for various types of scattering processes. Some of this work has been performed at a fairly sophisticated mathematical level. The treatment of this and the preceding section have of necessity been rather condensed. Readers with the requisite background in scattering theory can pursue further developments, including the generalization to cover the situations where more than one scattering variable is considered a complex variable, by consulting the references in the chapter end-notes.

17.13 Extension of Titchmarsh's theorem for distributions

In this section, the analog of Titchmarsh's theorem for distributions is considered. (Refer to Section 4.22 for a discussion of the case of ordinary functions and to Section 10.2 for the notation employed for distributions.) Symbolic notation is used for most occurrences of distributions in this section, which is a common custom in the literature. The equations can all be recast in the more formal notation for distributions employed in Chapter 10. The principal objective is to make a connection between analytic properties and Hilbert transform relations when distributions are permitted. This allows a discussion of causal behavior to be made when the input is a distribution.

In this section the system is assumed to be linear. Let i_t and r_t be the input and output response of a system, respectively. Subscript notation is typically used to indicate the variable dependence of the distribution, and helps remind the reader that distributions are under consideration. Suppose that the input is a distribution such that $i_t \in \mathcal{D}'_+$, and that the output can be obtained by the action of the operator $\mathcal{G}$ on the input, thus

$$r_t = \mathcal{G} i_t, \tag{17.219}$$

where $\mathcal{D}'_+$ denotes the space of distributions whose support is restricted to $[0, \infty)$ and the subscript t for these distributions indicates a temporal dependence. The operator $\mathcal{G}$ is taken to be linear. It is also assumed that it commutes with translations in time. That is, for a time shift by τ,

$$r_{t+\tau} = \mathcal{G} i_{t+\tau}. \tag{17.220}$$

The operator $\mathscr{G}$ can be written as a convolution as follows:

$$\mathscr{G}\mathrm{i}_t = g_t * i_t, \tag{17.221}$$

so that

$$r_t = g_t * i_t, \tag{17.222}$$

where g_t is a system response function, which in this case is a distribution, and $g_t \in \mathcal{D}'_+$. Let the input i_t be a Dirac delta distribution $\delta(t)$, and suppose the system response is $r_t = g_t$. If the input is changed to a time-shifted Dirac delta distribution $\delta(t - \tau)$, where τ is treated as a continuous variable, then

$$i_t = \int_0^\infty i_\tau \delta(t - \tau) \mathrm{d}\tau. \tag{17.223}$$

It follows from the linearity condition that the response to a time-shifted Dirac delta distribution $\delta(t - \tau)$ can be written as a convolution of the form of Eq. (17.222). To cover other types of inputs, and to improve the rigor of the preceding argument, it is necessary to add a continuity condition.

The continuity assumption takes the following form. If a sequence of inputs $\{i_{t,j}\}$ converges to the input i_t in the limit $j \to \infty$, and to the input $i_{t,j}$ there is an output response $r_{t,j}$, then the sequence of output responses $\{r_{t,j}\}$ converges to the output response r_t in the limit $j \to \infty$. With this postulate in place, a more rigorous sketch of the derivation of Eq. (17.222) can be given. This formula is true for $i_t = \delta(t)$, using the definition of g_t of the preceding paragraph, and, by the time-translational invariance requirement, the equation is also true for $i_t = \delta(t - \tau)$. From the linearity condition it follows that

$$i_t = \sum_k c_k i_{t,k} \delta(t - \tau_k), \tag{17.224}$$

where c_k are constants and also

$$r_t = g_t * \sum_k c_k i_{t,k} \delta(t - \tau_k). \tag{17.225}$$

In the limit that τ_k is taken as a continuous variable, the convolution formula results. Since any distribution in $\mathcal{D}'_+$ can be expressed as series expansion of Dirac delta distributions, Eq. (17.222) is obtained for the case of general distributions belonging to $\mathcal{D}'_+$.

The following result, due to Schwartz (1966a), is useful in the sequel. If T and S are distributions belonging to $\mathcal{D}'_+$, then the convolution of T and S is defined for all test functions $\phi \in \mathcal{D}$ as follows:

$$\langle T * S, \phi \rangle = \langle T_x, \langle S_y, \phi(x + y) \rangle \rangle = \langle S_x, \langle T_y, \phi(x + y) \rangle \rangle. \tag{17.226}$$

Since $\text{supp}(T) = [0,\infty)$ and $\text{supp}(S) = [0,\infty)$, the new distribution $T * S$ has $\text{supp}(T * S) = [0,\infty)$ and $T * S \in \mathcal{D}'_+$. A sketch of the essential idea behind this result is as follows. If the support of ϕ is selected so that $\text{supp}(\phi) \subset (-\infty, 0)$, and if the function φ is defined by

$$\varphi(x) = \langle T_y, \phi(x+y)\rangle, \tag{17.227}$$

then the support of φ is in $(-\infty, 0)$; therefore,

$$\text{supp}(\varphi) \cap \text{supp}(S) = 0. \tag{17.228}$$

Hence, it follows that

$$\langle T * S, \phi\rangle = 0, \tag{17.229}$$

and so

$$\text{supp}(T * S) = [0,\infty), \tag{17.230}$$

and thus $T * S \in \mathcal{D}'_+$.

The intuitive notion of strict causality is that there is no output before the application of the input. The preceding convolution result can be applied directly to Eq. (17.222) so that $r_t \in \mathcal{D}'_+$, and hence r_t is a casual distribution. To summarize, the output response of a causal system to a causal distributional input is also a causal distribution, and hence strict causality applies.

Taylor (1958) established the counterpart of Titchmarsh's theorem for distributions. To proceed, it is necessary to convert to the frequency domain, which is achieved by taking the appropriate Fourier transforms. The distributions i_t, r_t, and g_t are restricted to belong to $\mathcal{D}'_+ \cap \mathcal{S}'$, so that

$$I_\omega = \mathcal{F} i_t, \tag{17.231}$$

$$R_\omega = \mathcal{F} r_t, \tag{17.232}$$

and

$$G_\omega = \mathcal{F} g_t. \tag{17.233}$$

The subscript ω denotes an angular frequency. The notation $\mathcal{S}'_+$ is often used to denote the space $\mathcal{D}'_+ \cap \mathcal{S}'$. Application of the convolution theorem of Fourier transforms to Eq. (17.222) leads to

$$R_\omega = G_\omega I_\omega. \tag{17.234}$$

The Fourier transform of the convolution of distributions in Eq. (17.222) can be taken under different conditions. For cxamplc, if both i_t and g_t have bounded supports, that is, if i_t and $g_t \in \mathcal{E}'$, or if i_t or $g_t \in \mathcal{D}'$ and the other distribution has bounded support (Zemanian, 1965, pp. 191, 206), or if one of i_t or g_t belongs to $\mathcal{S}'$ and the other distribution belongs to $\mathcal{E}'$.

The extension of Titchmarsh's theorem can be cast for distributions as the following fundamental theorem. Let $G_\omega = \mathcal{F}g_t$, and let $G(\omega + \mathrm{i}\omega')$ denote an analytic function in the upper half complex angular frequency plane. The necessary and sufficient conditions that the distribution G_ω is the boundary value of $G(\omega + \mathrm{i}\omega')$ as $\omega' \to 0+$, is that one of the following conditions is satisfied:

(i) $g_t \in \mathcal{D}'_+ \cap \mathcal{S}'$, or
(ii) $G = -\dfrac{1}{\pi \mathrm{i}} G * p.v.\dfrac{1}{\omega}$.

The condition in (ii) can be expressed in a more familiar form. If

$$\lim_{\omega' \to 0+} G(\omega + \mathrm{i}\omega') \equiv G_\omega = A_\omega + \mathrm{i}B_\omega, \tag{17.235}$$

where G_ω, A_ω, and B_ω are distributions, then (ii) becomes

$$A_\omega = -\frac{1}{\pi} P \int_{-\infty}^{\infty} \frac{B_{\omega'}\, \mathrm{d}\omega'}{\omega - \omega'} \tag{17.236}$$

and

$$B_\omega = \frac{1}{\pi} P \int_{-\infty}^{\infty} \frac{A_{\omega'}\, \mathrm{d}\omega'}{\omega - \omega'}. \tag{17.237}$$

The reader should recognize these last two results as a Hilbert transform pair. Note that they are written in symbolic form, since the Hilbert transform of a distribution is evaluated on a suitable test function via $\langle HA_\omega, \phi \rangle$. The proof of the preceding theorem can be found in Beltrami and Wohlers (1966b). This theorem also admits some further generalizations, which the interested reader can seek out in the references indicated in the end-notes. The fundamental idea is that the necessary and sufficient condition for a tempered distribution to be a causal tempered distribution is that the Fourier transform of this distribution satisfies a Hilbert transform relationship. This result was developed and extended by Taylor, Schwartz, Beltrami, and Wohlers. It is the key to the connectivity between causality and Fourier transforms of distributions.

A fundamental result, due to Schwartz (1962), is the following. Let $G_\omega = \mathcal{F}g_t \in \mathcal{S}'$, then a necessary and sufficient condition that $g_t \in \mathcal{D}'_+$ is that $G_\omega = -(1/\pi \mathrm{i})\, G_\omega * p.v.(1/\omega)$.

Other alternative necessary and sufficient conditions can be given (see Beltrami and Wohlers (1967)).

The second condition of the fundamental theorem involving Hilbert transforms can be arrived at in two ways. Taylor (1958) showed that the necessary and sufficient condition for a distribution $g_t \in \mathscr{S}'_+$, is that the distribution G_ω, defined by

$$G_\omega = (1+\omega^2)^{-M}\,\mathcal{F}g_t, \tag{17.238}$$

where $M \in \mathbb{Z}^+$, belongs to the space $\mathcal{B}'$, the set of continuous linear functionals on $\mathcal{D}_{L^1}$ and satisfies a *subtracted* dispersion relationship of the following form:

$$G_\omega = \sum_{m=0}^{2M+1} \alpha_m \omega^m (1+\omega^2)^{-M} + \frac{\mathrm{i}(1+\omega^2)}{\pi} P\int_{-\infty}^{\infty} \frac{G_{\omega'}\,\mathrm{d}\omega'}{(\omega-\omega')(1+\omega'^2)}, \tag{17.239}$$

where the α_m are expansion coefficients. The proof of Taylor's theorem requires the analytic extension of $\mathcal{F}g_t$ to the upper half complex angular frequency plane, and then complex variable methods are used to derive the dispersion formula.

An alternative way to proceed to establish dispersion relations is to employ ideas developed by Schwartz. The strategy is to write a convolution of G_ω with the distribution $p.v.(\omega^{-1})$ and ascertain for which distributions this convolution is defined. This objective can be achieved by introducing a new space of distributions denoted by $\mathcal{D}'_{\mathcal{L}}$. A distribution G_ω satisfies

$$G_\omega \in \mathcal{D}'_{\mathcal{L}} \tag{17.240}$$

if

$$\frac{1}{\sqrt{(1+\omega^2)}} G_\omega \in \mathcal{D}'_{\mathcal{L}}. \tag{17.241}$$

The factor $1/\sqrt{(1+\omega^2)}$ is inserted to lead to an improved asymptotic behavior as $\omega \to \infty$ in a convolution integral with $p.v.(\omega^{-1})$, and the particular form is selected to avoid introducing any additional complications at the origin. Many such factors can of course be chosen. Any distribution $G_\omega \in \mathcal{E}'$ or $G_\omega \in \mathcal{O}'_C$ belongs to $\mathcal{D}'_{\mathcal{L}}$. The Dirac delta distribution belongs to $\mathcal{D}'_{\mathcal{L}}$ and $p.v.(\omega^{-1}) \in \mathcal{D}'_{\mathcal{L}}$. The convolution $G_\omega * p.v.(\omega^{-1})$ is well defined for any $G_\omega \in \mathcal{E}'$, and in such cases it is unnecessary to introduce the additional factor $1/\sqrt{(1+\omega^2)}$.

Any distribution g_t satisfies $g_t \in \mathcal{D}'_+$ if it has a decomposition of the form

$$g_t = g(t)\mathrm{H}(t), \tag{17.242}$$

where $g(t)$ is some locally integrable function and $\mathrm{H}(t)$ denotes the Heaviside distribution (step function). It is assumed that this product is well defined. A key result, which is obtained without recourse to analytic function theory ideas, is the following. If the distribution G_ω satisfies $G_\omega = \mathcal{F}g_t \in \mathcal{D}'_{\mathcal{L}}$, then $g_t \in \mathcal{D}'_+$, iff

$$G_\omega = G_\omega * \delta^+, \tag{17.243}$$

where δ^+ is a Heisenberg delta function (recall Eq. (10.9)). Since

$$G_\omega = G_\omega * \delta = G_\omega * (\delta^+ + \delta^-), \tag{17.244}$$

an alternative condition is that

$$G_\omega * \delta^- = 0. \tag{17.245}$$

Using the definition of δ^+ leads to the following reformulation of Eq. (17.243):

$$G_\omega = \frac{1}{2} G_\omega * \delta - \frac{1}{2\pi \mathrm{i}} G_\omega * p.v.\frac{1}{\omega} = -\frac{1}{\pi \mathrm{i}} G_\omega * p.v.\frac{1}{\omega}. \tag{17.246}$$

Expressing G_ω in terms of its real and imaginary parts, $G_\omega = A_\omega + \mathrm{i}B_\omega$, allows Eq. (17.244) to be written in the following manner:

$$B_\omega = \frac{1}{\pi} A_\omega * p.v.\frac{1}{\omega} \tag{17.247}$$

and

$$A_\omega = -\frac{1}{\pi} B_\omega * p.v.\frac{1}{\omega}, \tag{17.248}$$

which are the Hilbert transform pair in symbolic notation.

A sketch of the derivation of the requirement in Eq. (17.243) is as follows. First note that the inverse Fourier transforms of the Heisenberg delta distributions can be written in terms of the Heaviside function as follows:

$$\mathcal{F}^{-1}\delta^{\pm}(x) = \frac{1}{2\pi}\mathrm{H}(\pm x). \tag{17.249}$$

There is an issue of phase choice for the Fourier transform that plays a role here. If the same phase choice as in Section 10.4 is employed, then it follows that $\mathcal{F}^{-1}\delta^{\pm}(x) = (1/2\pi)\mathrm{H}(\mp x)$. Recall the comment after Eq. (10.75). In discussing causal functions and distributions, the conventional choice for the Fourier transform exponent is a positive sign, and the inverse Fourier transform consequently has a negative sign. This is opposite to the sign choice in Section 10.4. Taking the inverse Fourier transform of Eq. (17.243) leads to

$$\begin{aligned}
\mathcal{F}^{-1}G_\omega(t) &= 2\pi \mathcal{F}^{-1}G_\omega(t)\mathcal{F}^{-1}\delta^+(t) \\
&= \{2\pi g(t)\}\left\{\frac{1}{2\pi}\mathrm{H}(t)\right\} \\
&= g(t)\mathrm{H}(t) \\
&= g_t,
\end{aligned} \tag{17.250}$$

and hence g_t belongs to the space $g_t \in \mathcal{D}'_+$.

Notes

§17.1 For some further reading on passive systems, see Page (1955, chap. 12).

§17.2 For further reading on introductory aspects of linear systems see for example, Gross (1956), Brown (1961), Guillemin (1963), Zadeh and Desoer (1963), Fratila (1982), Bendat (1990), Oppenheim, Willsky, and Nawab (1997). Readers interested in examining the complications that start to emerge for nonlinear systems could consult Nayfeh and Mook (1979).

§17.4 For further discussion on weighting and response functions, see Brown (1961) or Pippard (1986).

§17.5 The most authoritative source for discussions on causality, particularly in the context of the connection to Hilbert transform relations, is Nussenzveig (1972). The books by Hilgevoord (1962) and Roos (1969) are also recommended, as is the paper by Toll (1956). A general introductory discussion can be found in Holbrow and Davidon (1964).

§17.6 Conversion of a time-dependent signal to the frequency domain is usually made via the Fourier transform, but the Hartley transform can also be employed; see, for example, Agneni (1990) and Pei and Jaw (1990). For further discussion on Bochner's theorems for linear passive systems, see Youla, Castriota, and Carlin (1959).

§17.7 It is possible to replace the Fourier transform in the argument of this section with the Hartley transform; see Millane (1994). For more advanced mathematical treatments of the issues involving causality and analyticity, see Fourès and Segal (1955) and Freedman, Falb, and Anton (1969).

§17.7.1 For additional reading on the subject of this subsection, see Remacle and Levine (1993), Kircheva and Hadjichristov (1994), and Lee (1995).

§17.8 The books by Roos (1969) and Nussenzveig (1972) represent good places for further reading on the link between analytic behavior and the application of Titchmarsh's theorem. To derive Hilbert transform connections for a linear, time-translational invariant, passive, causal system, it may be necessary to add further requirements. For some additional reading in this direction, see Wang (2002).

§17.9 For further discussion on the acausal situation, see Bennett (1987a,b).

§17.10 The key paper on this topic is Paley and Wiener (1933). For a generalization treating tempered distributions, see Pfaffelhuber (1971a).

§17.11 For an introduction to more advanced ideas on causality and the connection with scattering, Nussenzveig (1972) is the definitive source, and this reference has served as the basis for the discussion given in this section. For additional reading, see Van Kampen (1958a,b, 1961). Goldberger and Watson (1964) provide a detailed account of dispersion relations from a scattering perspective. For additional background reading on the basic aspects of scattering theory, see for example, Schiff (1955), Wu and Ohmura (1962), or Merzbacher (1970).

§17.12 Key early papers are Schützer and Tiomno (1951), Van Kampen (1953a,b, 1957), and Giambiagi and Saavedra (1963). For further reading on causality and analyticity in scattering theory see Minerbo (1971), and Nussenzveig (1972) gives a detailed discussion. The original suggestion to employ the causality condition as a supplementary constraint on the scattering matrix (S-matrix) is due to Kronig (1946). For some additional reading on dispersion relations in scattering theory, see Bogolyubov, Medvedev, and Polivanov (1958), Corinaldesi (1959), Thirring (1959), Goldberger (1960), Hamilton (1960), Sugawara and Kanazawa (1961), Hilgevoord (1962), Wong (1964), Barton (1965), Burkhardt (1969), and Nishijima (1974). For a discussion on some issues associated with the determination of the S-matrix and its relation to causality, the papers by Van Kampen (1957) and Wigner (1964) are worth a read. For one-dimensional scattering dispersion relations, Jin and Martin (1964) make a connection between the behavior of the scattering amplitude across the cut and its asymptotic behavior.

§17.13 For further reading on the topics of this section, see Taylor (1958), Beltrami and Wohlers (1965, p. 72, 1966a,b, 1967), Beltrami (1967), Roos (1969, pp. 349–360), and Nussenzveig (1972, pp. 33–43). The underlying ideas for much of the discussion are due to Schwartz (1962, 1966a,b). The previously cited works of Beltrami and Wohlers contain extensions of a number of the results mentioned in this section.

Exercises

17.1 If a particular property represented by $f(t)$ satisfies $f(t) = 0$, for $t < 0$, does it follow that the real and imaginary parts of the Fourier transform of $f(t)$ satisfy a pair of Hilbert transform relations?

17.2 Evaluate $\int_{-\infty}^{\infty} G(\omega)\mathrm{e}^{-\mathrm{i}\omega t}\,\mathrm{d}\omega$ for the damped harmonic model for $t > 0$.

17.3 Determine the displacement $x(t)$ for the damped harmonic oscillator model when the driving force is given by

$$f(t) = \begin{cases} 0, & t < 0 \\ \mathrm{e}^{-t}, & t \geq 0. \end{cases}$$

17.4 If the damping constant D in the damped harmonic oscillator model satisfied $D < 0$, would the model be causal?

17.5 If the damping constant D in the damped harmonic oscillator model was negative, would the system be passive?

17.6 Evaluate $g(t) = \int_{-\infty}^{\infty} G(\omega)\mathrm{e}^{-\mathrm{i}\omega t}\,\mathrm{d}\omega$ for $t < 0$ using $G(\omega) = 1/[m(\omega_0^2 - \omega^2 - \mathrm{i}\tau\omega^3)]$ and ignoring terms of $O(\tau^2)$ and higher.

17.7 If every bounded input $i(t)$ produces a bounded output response $r(t)$, is the condition $\int_{-\infty}^{\infty} |w(\tau)|\mathrm{d}\tau < \infty$, where $w(\tau)$ is the weight function, both necessary and sufficient for a system to be stable?

17.8 Determine the form of the Hilbert transform connections between Re $G(\omega)$ and Im $G(\omega)$ for the problem represented by Eq. (17.110).

17.9 Assume a harmonic oscillator, initially at rest at time $t = -\infty$, is acted upon by an external force $f(t)$ and is subject to damping characterized by Γ, with $\Gamma > 0$. Show for this passive system that $\int_{-\infty}^{t} f(t')\dot{x}(t')\mathrm{d}t' \geq 0$.

17.10 Discuss the solution of the equation of motion

$$\ddot{x}(t) + \Gamma\dot{x}(t) + \omega_0^2 x(t) - \tau\dddot{x}(t) = -\frac{e}{m}\mathbf{E}(t),$$

where Γ and τ are independent of time.

17.11 For the model given in Exercise 17.10, is the system causal? Is the system passive?

17.12 Prove Eq. (17.131).

17.13 Derive the dispersion relations for the $a(k)$ coefficient in Eq. (17.173).

17.14 Derive the dispersion relations for the scattering function $S(k)$ for classical s-wave scattering from a spherical scatterer.

17.15 From the result of Exercise 17.14, what dispersion relation can be written for the phase shift?

17.16 What dispersion relations can be written for the scattering cross-section for classical s-wave scattering from a spherical scatterer?

17.17 Derive Eq. (17.204).

17.18 Prove Eq. (17.208).

17.19 Determine the form of the dispersion relations for the scattering function $S(k)$ for non-relativistic quantum s-wave scattering from a spherical scatterer.

17.20 Decide which of the following belong to $\mathcal{D}'_{\mathcal{L}}$: (i) $p.v.(\omega^{-1})$, (ii) a constant function, and (iii) any L^2 function.

17.21 Show that the Heisenberg delta distribution δ^+ belongs to $\mathcal{D}'_{\mathcal{L}}$. Hence, or otherwise, prove $\delta^+ * \delta^- = 0$. By writing $p.v.(\omega^{-1}) = i\pi\{\delta_\omega^- - \delta_\omega^+\}$, show that for a suitable distribution T, the following result (written in symbolic form) holds:

$$H^2T = -\{\delta^- - \delta^+\} * [\{\delta^- - \delta^+\} * T] = -T.$$

What restrictions must be placed on T?

18

The Hilbert transform of waveforms and signal processing

18.1 Introductory ideas on signal processing

Signal processing plays a central role in a multitude of modern technologies. Think about those industries that are dependent on data communication, or on radar, to give just two examples, and the importance of signal processing becomes self-evident. The Hilbert transform plays a central role in a number of signal processing applications. Pioneering work on the application of Hilbert transforms to signal theory was carried out by Gabor (1946).

A notational alert to the reader is appropriate at the start of this chapter. In the following sections the standard Hilbert transform operator H, the Heaviside step function $\mathrm{H}(x)$, the Hermite polynomials $H_n(x)$, the Hilbert transfer function $\mathsf{H}(\omega)$, the fractional Hilbert transform H_α, and the fractional Hilbert transform filter $\mathsf{H}_p(\omega)$ all appear, sometimes in close proximity, so the reader should pay careful attention to the particular symbols in use.

Broadly defined, a signal provides a means for transmission of information about a system. For the signals of interest in this chapter, it is assumed that a mathematical representation of the signal is known. There are two important types of signals. The first are the continuous or analog signals – sometimes referred to as continuous-time signals. Unless something explicitly to the contrary is indicated, it is assumed throughout this chapter that all signals of this group belong to the class $L^2(\mathbb{R})$. In a number of places, this requirement can be generalized. From a practical standpoint, the class $L^2(\mathbb{R})$ will cover many signals of interest.

The second group contains the discrete or digital signals – also referred to as discrete-time signals. Chapter 13 dealt with aspects of the discrete signal. In this chapter, the focus is on continuous signals, which can be represented by a function of time, $s(t)$, and these signals may be represented as real or complex functions. The system input and output are both continuous functions. If the input is measured at particular discrete points, a suitable interpolation scheme can be performed to render the data continuous.

In order to extract information from a signal, some type of processing of the signal must be performed. Processing a signal corresponds mathematically to performing some operation that changes its properties, or in some way extracts information from it. A key variable associated with a signal is the energy, E, defined by

$$E = \int_{-\infty}^{\infty} |s(t)|^2 \, \mathrm{d}t. \tag{18.1}$$

A signal with finite energy obviously requires this integral to be bounded.

A shift from the time domain to the frequency domain can be made by taking the Fourier transform of the signal. That is, the signal can be characterized by $s(t)$, or by $S(\omega)$, where

$$S(\omega) \equiv \mathcal{F}s(\omega) = \int_{-\infty}^{\infty} s(t)\mathrm{e}^{-\mathrm{i}\omega t} \, \mathrm{d}t. \tag{18.2}$$

Parseval's theorem for Fourier transforms (see Section 2.6.3) allows the signal energy to be expressed in terms of the function $S(\omega)$ via

$$E = \int_{-\infty}^{\infty} |S(\omega)|^2 \, \mathrm{d}\omega. \tag{18.3}$$

The term *filter* is employed to describe a mechanism for selectively clipping or screening out certain frequency components from a signal, so that the support for the signal is some restricted frequency interval. There are a number of specific filters, but some common categories are defined in terms of the system transfer function $W(\omega)$ (see Section 17.6) as follows. Low-pass filters satisfy

$$W(\omega) = \begin{cases} 0, & \text{for } |\omega| > \omega_\mathrm{c} \\ h(\omega), & \text{for } |\omega| \leq \omega_\mathrm{c}, \end{cases} \tag{18.4}$$

where $h(\omega)$ is some (non-zero) function of frequency and ω_c denotes the cut-off frequency beyond which the system transfer function is zero. A high-pass filter has a system transfer function whose value is zero for $|\omega|$ below some cut-off frequency ω_c:

$$W(\omega) = \begin{cases} 0, & \text{for } |\omega| < \omega_\mathrm{c} \\ h(\omega), & \text{for } |\omega| \geq \omega_\mathrm{c}. \end{cases} \tag{18.5}$$

A band-pass filter has a system transfer function whose support is $[\omega_{\mathrm{min}}, \omega_{\mathrm{max}}]$, that is

$$W(\omega) = \begin{cases} 0, & \text{for } \omega < \omega_{\mathrm{min}} \text{ or } \omega > \omega_{\mathrm{max}} \\ h(\omega), & \text{for } \omega_{\mathrm{min}} \leq \omega \leq \omega_{\mathrm{max}}. \end{cases} \tag{18.6}$$

For a band-stop filter, as the name implies, the system transfer function has the support $(-\infty, \omega_{\mathrm{min}}) \bigcup (\omega_{\mathrm{max}}, \infty)$, that is

$$W(\omega) = \begin{cases} 0, & \text{for } \omega_{\mathrm{min}} \leq \omega \leq \omega_{\mathrm{max}} \\ h(\omega), & \text{for } \omega < \omega_{\mathrm{min}} \text{ or } \omega > \omega_{\mathrm{max}}. \end{cases} \tag{18.7}$$

18.2 The Hilbert filter

The Hilbert filter takes an input signal and returns the Hilbert transform of the signal as an output signal. This is also referred to as a Hilbert transformer, a quadrature filter, or a 90° phase shifter. Suppose a signal $s(t)$ has the form

$$s(t) = \sum_{k=1}^{\infty} a_k \sin(k\omega t + \varphi_k), \tag{18.8}$$

where ω is a positive constant and the a_k are expansion coefficients; then,

$$Hs(t) = \sum_{k=1}^{\infty} a_k \sin(k\omega t + \varphi_k - 90^\circ), \tag{18.9}$$

and for a signal with the form

$$s(t) = \sum_{k=1}^{\infty} a_k \cos(k\omega t + \varphi_k), \tag{18.10}$$

it follows that

$$Hs(t) = \sum_{k=1}^{\infty} a_k \cos(k\omega t + \varphi_k - 90^\circ). \tag{18.11}$$

From Eqs. (18.9) and (18.11), it is clear that there is a shift of phase of 90° on taking the Hilbert transform, hence the alternative name, 90° phase shifter.

The Hilbert transformer functions by performing a convolution operation of $1/\pi t$ with the signal, or in the frequency domain by employing a Fourier transform procedure. Let the input signal be denoted by $s(t)$ and its corresponding Fourier transform in the frequency domain be $S(\omega)$, and let $h(t)$ denote $1/\pi t$. On the basis of the discussion in Section 17.6, it follows that the output response signal $R(\omega)$ in the frequency domain, after carrying out the convolution $h * s$, is given by

$$\begin{aligned} \mathcal{F}r(\omega) &= \mathcal{F}[h(t) * s(t)](\omega) \\ &= \mathcal{F}h(\omega)\mathcal{F}s(\omega), \end{aligned} \tag{18.12}$$

so that

$$R(\omega) = \mathsf{H}(\omega)S(\omega), \tag{18.13}$$

where the Hilbert transfer function, written in standard engineering format, is given by

$$\mathsf{H}(\mathrm{i}\omega) = -\mathrm{i}\,\mathrm{sgn}\,\omega. \tag{18.14}$$

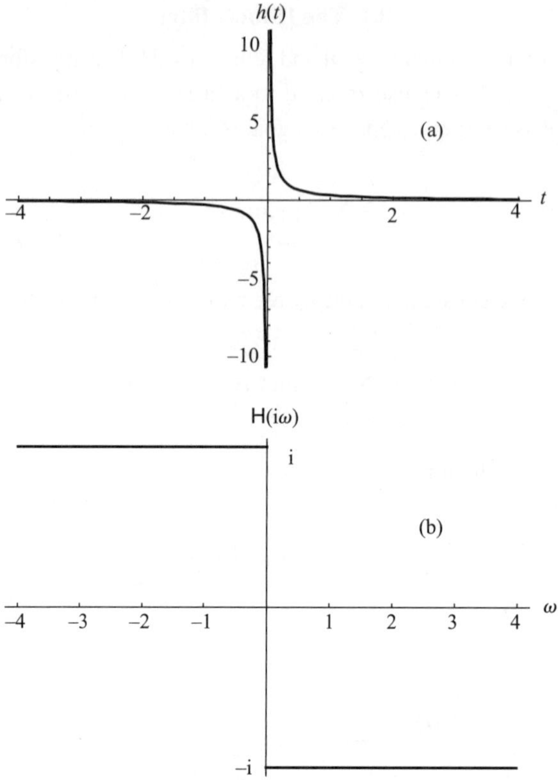

Figure 18.1. (a) The function $h(t)$; (b) the Hilbert transfer function $\mathsf{H}(\mathrm{i}\omega)$.

The function $h(t)$ and the transfer function $\mathsf{H}(\mathrm{i}\omega)$ are shown in Figure 18.1.

The magnitude of $\mathsf{H}(\mathrm{i}\omega)$ is a constant, except at $\omega = 0$. If the standard definition of the sgn function is employed (see Eq. (1.14)), then $\mathsf{H}(\mathrm{i}\omega)$ has a discontinuity at the frequency origin, where $\mathsf{H}(\mathrm{i}\omega) = 0$. The phase angle, denoted by $\angle\mathsf{H}(\mathrm{i}\omega)$, switches sign at $\omega = 0$. These two functions are illustrated in Figure 18.2.

The phase function $\varphi(\omega)$ is defined by

$$\varphi(\omega) = \arg\{\mathsf{H}(\mathrm{i}\omega)\} = -\frac{\pi}{2}\,\mathrm{sgn}\,\omega. \tag{18.15}$$

The idealized Hilbert filter discussed so far is not physically realizable. At best, one can only obtain close approximations to it.

The Hilbert filter for the treatment of discrete time signals was discussed in Section 13.8 (see in particular Eq. (13.71)). The interested reader could revisit that section to see the role of the Hilbert filter in the determination of the connection between the real and imaginary parts of $X(\omega)$, the Fourier transform of the time sequence $x[n]$.

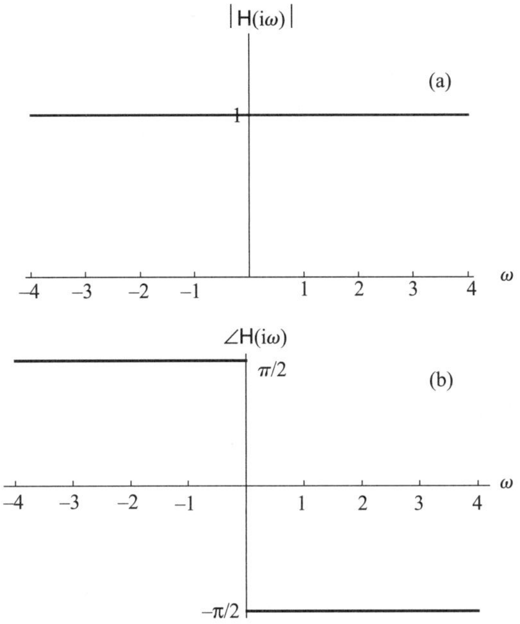

Figure 18.2. (a) The function $|\mathsf{H}(i\omega)|$; (b) the phase angle of $\mathsf{H}(i\omega)$.

18.3 The auto-convolution, cross-correlation, and auto-correlation functions

The auto-correlation, cross-correlation, and auto-convolution functions find a number of applications in signal analysis. For example, the energy associated with a signal can be identified with the auto-correlation function. This section explores these three functions and considers some connections with the Hilbert transform.

Let $s(t)$ denote a signal and let $s_H(t)$ be employed as shorthand to denote the Hilbert transform of $s(t)$. The auto-convolution function is defined by

$$\rho_s(t) = \{s * s\}(t) = \int_{-\infty}^{\infty} s(t')s(t-t')\mathrm{d}t', \tag{18.16}$$

and there is an analogous definition for $\rho_{s_H}(t)$. The auto-convolutions $\rho_s(t)$ and $\rho_{s_H}(t)$ are connected by the following formula:

$$\rho_s(t) = -\rho_{s_H}(t). \tag{18.17}$$

To prove this result, take the Fourier transform of $\rho_s(t)$, and let the Fourier transform of $s(t)$ be denoted $S(\omega)$, then using Eq. (2.54) yields

$$\mathcal{F}\rho_s(\omega) = \mathcal{F}\{s * s\}(\omega) = \mathcal{F}s(\omega)\mathcal{F}s(\omega) = S^2(\omega), \tag{18.18}$$

and on using Eq. (5.3) it follows that

$$\begin{aligned}\mathcal{F}\rho_{s_H}(t) &= \mathcal{F}\{s_H * s_H\}\\ &= \mathcal{F}Hs(\omega)\mathcal{F}Hs(\omega)\\ &= \{-\mathrm{i}\,\mathrm{sgn}\,\omega\,\mathcal{F}s(\omega)\}^2\\ &= -S^2(\omega),\end{aligned} \tag{18.19}$$

so that

$$\mathcal{F}\rho_s(t) = -\mathcal{F}\rho_{s_H}(t), \tag{18.20}$$

which leads, on taking the inverse Fourier transform of this result, to Eq. (18.17).

The cross-correlation function (also called the correlation function) is defined by

$$\rho_{fg}(t) = \{f \star g\}(t) = \int_{-\infty}^{\infty} f(t')g(t'+t)\mathrm{d}t', \tag{18.21}$$

where f and g are two functions defined on $\mathbb{R}$ and the integral is assumed to exist. This result can be written, on making the change of variable $t' + t = \tau$, in the following alternative form:

$$\rho_{fg}(t) = \{f \star g\}(t) = \int_{-\infty}^{\infty} f(\tau - t)g(\tau)\mathrm{d}\tau. \tag{18.22}$$

In the preceding two equations the pentagram symbol $\star$ plays a role similar to the convolution symbol $*$. If $g = f$ then the auto-correlation function is defined by

$$\rho_f(t) = \{f \star f\}(t) = \int_{-\infty}^{\infty} f(\tau - t)f(\tau)\mathrm{d}\tau = \int_{-\infty}^{\infty} f(\tau + t)f(\tau)\mathrm{d}\tau. \tag{18.23}$$

The reader needs to be alert to the similar symbolism employed for the auto-convolution and the auto-correlation functions. The cross-correlation and auto-correlation operations satisfy a number of correlation identities, and the reader interested in pursuing further discussion on this topic could consult Howell (2001, p. 396).

The following identity holds for the cross-correlation function:

$$\rho_{ss_H}(t) = -\rho_{s_Hs}(t). \tag{18.24}$$

Let $f(t)$ denote the signal $s(t)$ and let $g(t)$ designate $s_H(t)$. The proof of Eq. (18.24) takes advantage of the following correlation identity:

$$\mathcal{F}\{f \star g\}(\omega) = F^*(\omega)G(\omega), \tag{18.25}$$

where F and G are the Fourier transforms of f and g, respectively, and the superscript * denotes complex conjugation. The proof of Eq. (18.25) is left as an exercise for the reader. From Eq. (18.22) it follows, on taking the Fourier transform and employing Eq. (18.25), that

$$\begin{aligned} \mathcal{F}\{s \star Hs\}(\omega) &= S^*(\omega)\mathcal{F}Hs(\omega) \\ &= -\mathrm{i}\,\mathrm{sgn}\,\omega\,|S(\omega)|^2, \end{aligned} \tag{18.26}$$

and similarly

$$\begin{aligned} \mathcal{F}\{Hs \star s\}(\omega) &= [\mathcal{F}Hs(\omega)]^* S(\omega) \\ &= \mathrm{i}\,\mathrm{sgn}\,\omega\,|S(\omega)|^2, \end{aligned} \tag{18.27}$$

and hence

$$\mathcal{F}\rho_{ss_H}(\omega) = -\mathcal{F}\rho_{s_H s}(\omega). \tag{18.28}$$

Taking the inverse Fourier transform leads to Eq. (18.24). If the auto-correlation function is evaluated at $t = 0$, it follows from the definition in Eq. (18.22) that

$$\rho_{ss_H}(0) = \rho_{s_H s}(0), \tag{18.29}$$

and hence from Eq. (18.24) that

$$\rho_{ss_H}(0) = 0. \tag{18.30}$$

This means that, at $t = 0$, $s(t)$ and $s_H(t)$ are uncorrelated.

The application of the auto-correlation function to $s(t)$ and $s_H(t)$ leads to the following identity:

$$\rho_s(t) = \rho_{s_H}(t). \tag{18.31}$$

This can be proved in a similar fashion to that outlined for Eq. (18.24). The correlation identity required can be obtained directly from Eq. (18.25), that is

$$\mathcal{F}\{f \star f\}(\omega) = |F(\omega)|^2. \tag{18.32}$$

The energy associated with the real signal $s(t)$ is defined by

$$E = \rho_s(0) = \int_{-\infty}^{\infty} s^2(t)\mathrm{d}t. \tag{18.33}$$

Employing Eq. (4.172) allows the energy to be expressed as follows:

$$E = \int_{-\infty}^{\infty} s^2(t)\mathrm{d}t = \int_{-\infty}^{\infty} [Hs(t)]^2\,\mathrm{d}t. \tag{18.34}$$

This result means that energy is conserved after taking the Hilbert transform of a signal.

In some cases it is useful to introduce a normalized auto-correlation function, defined for a complex function f by

$$\gamma(t) = \frac{\int_{-\infty}^{\infty} f(\tau)^* f(\tau + t)\mathrm{d}\tau}{\int_{-\infty}^{\infty} f(\tau)^* f(\tau)\mathrm{d}\tau}. \tag{18.35}$$

In situations where neither integral exists separately, it may be possible to evaluate

$$\gamma_T(t) = \frac{\int_{-T}^{T} f(\tau)^* f(\tau + t)\mathrm{d}\tau}{\int_{-T}^{T} f(\tau)^* f(\tau)\mathrm{d}\tau}, \tag{18.36}$$

as a finite quantity, and then examine the limit $T \to \infty$, so that

$$\gamma_\infty(t) = \lim_{T\to\infty} \frac{\int_{-T}^{T} f(\tau)^* f(\tau + t)\mathrm{d}\tau}{\int_{-T}^{T} f(\tau)^* f(\tau)\mathrm{d}\tau}. \tag{18.37}$$

In some situations it may be known that $\int_{-T}^{T} f(\tau)^* f(\tau)\mathrm{d}\tau$ behaves linearly with T, so that an alternative convergent form for the auto-correlation function can be defined as follows:

$$\bar{\gamma}_\infty(t) = \lim_{T\to\infty} \frac{1}{2T} \int_{-T}^{T} f(\tau)^* f(\tau + t)\mathrm{d}\tau. \tag{18.38}$$

There are counterparts for the results of this section for discrete-time signals. The reader with an interest in this direction can consult Oppenheim *et al.* (1999, p. 65).

18.4 The analytic signal

The analytic signal arising from the function g in the time domain is defined in the following way:

$$f(t) = g(t) + \mathrm{i}Hg(t). \tag{18.39}$$

Some authors call f the *pre-envelope* of the signal $g(t)$. As a reminder to the reader, the complex unity is represented throughout this chapter by i, and not the more conventional j that is employed in the engineering literature. The origin of the definition in Eq. (18.39) is clear. The imaginary part of an analytic signal is connected to its real part by a Hilbert transform. As remarked in Chapter 1, the Hilbert transform is sometimes referred to as the *quadrature* function, and so the second term of the analytic signal in Eq. (18.39) is occasionally called the quadrature term. In the literature, the analytic signal is also seen with a minus sign in front of the quadrature term, which reflects the opposite sign convention employed for the definition of the Hilbert transform, relative to the one used in this book. A simple illustrative example is the function $e^{i\omega t}$. Recalling that the Hilbert transform of $\cos\omega t$ is $\sin\omega t$, for a constant $\omega > 0$, then

$$e^{i\omega t} = \cos\omega t + iH[\cos\omega t]. \tag{18.40}$$

Therefore, the analytic signal corresponding to $\cos\omega t$ is $e^{i\omega t}$. If the signal is $\sin\omega t$ for $\omega > 0$, the corresponding analytic signal is $-ie^{i\omega t} = e^{i(\omega t - \pi/2)}$.

The key historical contribution is that of Gabor (1946). He started with a real signal of the form

$$s(t) = a\cos\omega t + b\sin\omega t, \tag{18.41}$$

where a and b are constants, and then formed the complex signal

$$\psi(t) = s(t) + i\sigma(t) = (a - ib)e^{i\omega t}, \tag{18.42}$$

where $\sigma(t)$ is formed from $s(t)$ by replacing $\cos\omega t$ by $\sin\omega t$ and replacing $\sin\omega t$ by $-\cos\omega t$. The significance of $\sigma(t)$ is that it is the quadrature of the signal $s(t)$. The reader will immediately note the relationship between Gabor's analytic signal and the conjugate series discussed in Section 3.13. Gabor went on to make the connection with analytic function theory, and he noted the Hilbert transform connection between $\sigma(t)$ and $s(t)$.

Recalling the definition of the projection operator P_+ from Eq. (4.352), the analytic operator $\mathcal{A}$ can be introduced by the following definition:

$$\mathcal{A} = 2P_+ = (I + iH), \tag{18.43}$$

so that, if $z(t)$ denotes the analytic signal corresponding to the real signal $s(t)$,

$$z(t) \equiv (\mathcal{A}s)(t). \tag{18.44}$$

Here are some examples. If $s(t) = e^{i\omega t}$, then the corresponding analytic symbol is given by

$$(\mathcal{A}s)(t) = (1 + \operatorname{sgn}\omega)e^{i\omega t}. \tag{18.45}$$

If $s(t) = \cos\omega t$, then

$$\mathcal{A}\{\cos\omega t\} = \cos\omega t + i\operatorname{sgn}\omega\sin\omega t \tag{18.46}$$

and

$$\mathcal{A}\{\sin\omega t\} = \sin\omega t - i\operatorname{sgn}\omega\cos\omega t. \tag{18.47}$$

For the case $s(t) = \cos(\omega_1 t + \varphi_1)\cos(\omega_2 t + \varphi_2)$, $(\mathcal{A}s)(t)$ can be evaluated either by employing a trigonometric expansion for the product of cosine functions, or by Euler's formula. Using the latter approach it follows that

$$\begin{aligned}(\mathcal{A}s)(t) &= \frac{1}{4}\mathcal{A}\{[e^{i(\omega_1 t+\varphi_1)} + e^{-i(\omega_1 t+\varphi_1)}][e^{i(\omega_2 t+\varphi_2)} + e^{-i(\omega_2 t+\varphi_2)}]\}\\ &= \frac{1}{4}\{[1+\operatorname{sgn}(\omega_1+\omega_2)]e^{i(\omega_1 t+\omega_2 t+\varphi_1+\varphi_2)}\\ &\quad + [1+\operatorname{sgn}(\omega_1-\omega_2)]e^{i(\omega_1 t-\omega_2 t+\varphi_1-\varphi_2)}\\ &\quad + [1+\operatorname{sgn}(\omega_2-\omega_1)]e^{i(\omega_2 t-\omega_1 t-\varphi_1+\varphi_2)}\\ &\quad + [1-\operatorname{sgn}(\omega_1+\omega_2)]e^{-i(\omega_1 t+\omega_2 t+\varphi_1+\varphi_2)}\}. \end{aligned} \tag{18.48}$$

Restricting to the case $0 \le \omega_1 < \omega_2$, the preceding result simplifies to

$$\begin{aligned}\mathcal{A}\{\cos(\omega_1 t+\varphi_1)\cos(\omega_2 t+\varphi_2)\} &= \frac{1}{2}\{e^{i(\omega_1 t+\omega_2 t+\varphi_1+\varphi_2)} + e^{i(\omega_2 t-\omega_1 t-\varphi_1+\varphi_2)}\}\\ &= \cos(\omega_1 t+\varphi_1)e^{i(\omega_2 t+\varphi_2)}. \end{aligned} \tag{18.49}$$

An important feature of an analytic signal is that, on making the switch from the time domain to the frequency domain, the signal contains only positive frequencies, which is often termed the one-sided character of the analytic signal. This can be demonstrated in a straightforward fashion by taking the Fourier transform of the analytic signal. From Eq. (18.39) it follows that

$$\mathcal{F}f(t) = \mathcal{F}g(t) + i\mathcal{F}Hg(t). \tag{18.50}$$

Suppose $F(\omega)$ and $G(\omega)$ denote the Fourier transforms of $f(t)$ and $g(t)$, respectively. Then Eq. (18.50) simplifies, on using Eq. (5.3), to

$$\begin{aligned}F(\omega) &= G(\omega) + \operatorname{sgn}\omega\,\mathcal{F}g(\omega)\\ &= G(\omega)\{1+\operatorname{sgn}\omega\}. \end{aligned} \tag{18.51}$$

It follows that

$$F(\omega) = \begin{cases} 0, & \omega < 0 \\ 2G(\omega), & \omega > 0. \end{cases} \tag{18.52}$$

The preceding result establishes the principal property of the analytic signal: that its Fourier transform contains no contributions at negative frequencies. As an example, consider the analytic signal

$$f(t) = \frac{\sin at}{\pi t} + i\frac{(1-\cos at)}{\pi t}, \quad \text{for } a > 0. \tag{18.53}$$

The Fourier transform of the real part of $f(t)$ is given by

$$\mathcal{F}\left[\frac{\sin a\omega}{\pi\omega}\right] = \frac{1}{2}\{\operatorname{sgn}(\omega + a) + \operatorname{sgn}(a - \omega)\}, \tag{18.54}$$

and the Fourier transform of the imaginary part of $f(t)$ is given by

$$\mathcal{F}\left[\frac{\mathrm{i}(1 - \cos a\omega)}{\pi\omega}\right] = \operatorname{sgn}\omega - \frac{1}{2}\{\operatorname{sgn}(\omega - a) + \operatorname{sgn}(\omega + a)\}. \tag{18.55}$$

Hence, the Fourier transform of the analytic signal is given by

$$\mathcal{F}f(\omega) = \operatorname{sgn}\omega - \operatorname{sgn}(\omega - a). \tag{18.56}$$

The latter result is clearly one-sided, being zero for frequencies $\omega < 0$. The behavior of $(\pi t)^{-1}\sin at$, $(\pi t)^{-1}(1 - \cos at)$, and the corresponding Fourier transforms are shown in Figure 18.3 for the value $a = \pi$. The one-sided character of the analytic signal in the frequency domain is clearly evident in this figure.

It is not uncommon to see in the literature the convention that $S(t)$ denotes a signal in the time domain, and the same symbol is employed for the frequency domain, that is, $S(\omega)$ is the corresponding signal in the frequency domain. This approach is avoided; with $s(t)$ and $S(\omega)$ employed as distinctive symbols to denote, respectively, the signal in the time domain and its corresponding representation in the frequency domain. Recalling Eq. (5.3),

$$\mathcal{F}HS(\omega) = -\mathrm{i}\operatorname{sgn}\omega\,\mathcal{F}S(\omega), \tag{18.57}$$

and the multiplier for the Hilbert transform operator based on the preceding formula (see Eq. (4.87)) is given by

$$m_H(\omega) = -\mathrm{i}\operatorname{sgn}\omega, \tag{18.58}$$

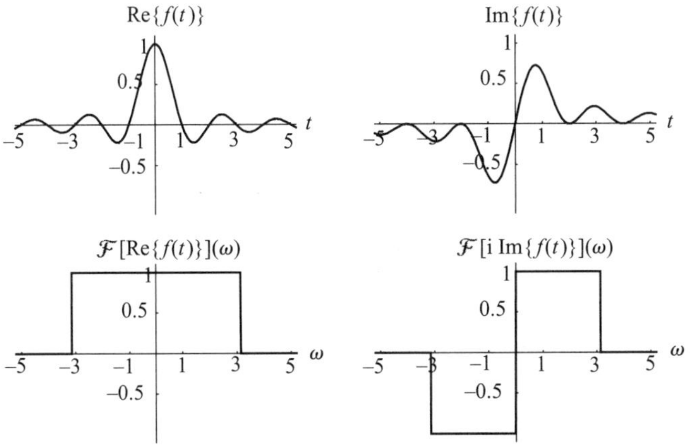

Figure 18.3. The components of the analytic signal and their Fourier transforms.

so that

$$\mathcal{F}HS(\omega) = m_H(\omega)\mathcal{F}S(\omega). \tag{18.59}$$

The reader will note the correspondence with $\sigma(H)$, the symbol of H, discussed in Section 5.2 (see Eq. (5.37)). A relative multiplier $m_{X,Y}(\omega)$ connecting $X(\omega)$ with $Y(\omega)$ can be defined in the following manner:

$$\mathcal{F}X(\omega) = m_{X,Y}(\omega)\,\mathcal{F}Y(\omega). \tag{18.60}$$

If the analytic signal arising from the real signal $s(t)$ is written as

$$z(t) = s(t) + \mathrm{i}Hs(t), \tag{18.61}$$

then, on taking the Fourier transform, it follows that

$$\begin{aligned}\mathcal{F}z(\omega) &= \mathcal{F}s(\omega) + \operatorname{sgn}\omega\,\mathcal{F}s(\omega)\\ &= m_{z,s}(\omega)\mathcal{F}s(\omega),\end{aligned} \tag{18.62}$$

where

$$m_{z,s}(\omega) = 1 + \operatorname{sgn}\omega \tag{18.63}$$

and

$$m_{z,s}(\omega) = 1 + \mathrm{i}\,m_H(\omega). \tag{18.64}$$

Several properties of the Hilbert transform that were discussed in Chapter 4 can be used to derive the following results. From Eq. (4.198) it follows, using inner product notation, that

$$(\operatorname{Re} z(t), \operatorname{Im} z(t)) = 0. \tag{18.65}$$

From Eq. (4.175), for two analytic signals $z_1(t)$ and $z_2(t)$,

$$(\operatorname{Re} z_1(t), \operatorname{Re} z_2(t)) = (\operatorname{Im} z_1(t), \operatorname{Im} z_2(t)), \tag{18.66}$$

which yields, in the special case $z_1(t) = z_2(t) \equiv z(t)$, the following result:

$$\int_{-\infty}^{\infty} \{\operatorname{Re} z(t)\}^2\,\mathrm{d}t = \int_{-\infty}^{\infty} \{\operatorname{Im} z(t)\}^2\,\mathrm{d}t. \tag{18.67}$$

From Eq. (4.204), it follows that

$$\int_{-\infty}^{\infty} \operatorname{Im} z(t)\mathrm{d}t = 0. \tag{18.68}$$

The analytic signal $z(t)$ is an eigenfunction of the Hilbert transform operator with eigenvalue $-\mathrm{i}$. That is, applying the Hilbert transform operator to $z(t)$ leads to

$$Hz(t) = -\mathrm{i}z(t). \tag{18.69}$$

Let $\boldsymbol{A}$ operate on a complex signal, and in particular consider its action on an analytic signal $z(t)$; then, from the definition Eq. (18.43) and the eigenfunction property just given, it follows that

$$\boldsymbol{A}z(t) = 2z(t). \tag{18.70}$$

Recalling the derivative property of the Hilbert transform, Eq. (4.137), it follows immediately that the operators $\boldsymbol{A}$ and $\mathrm{d}/\mathrm{d}t$ commute, so that

$$\boldsymbol{A}\frac{\mathrm{d}s(t)}{\mathrm{d}t} = \frac{\mathrm{d}}{\mathrm{d}t}\boldsymbol{A}s(t), \tag{18.71}$$

or, more generally for $n \in \mathbb{N}$,

$$\boldsymbol{A}\frac{\mathrm{d}^n s(t)}{\mathrm{d}t^n} = \frac{\mathrm{d}^n}{\mathrm{d}t^n}\boldsymbol{A}s(t). \tag{18.72}$$

If $s_1(t)$ and $s_2(t)$ are two real signals, the action of the analytic operator on a linear combination of the two signals can be evaluated as follows:

$$\boldsymbol{A}\{\alpha s_1(t) + \beta s_2(t)\} = \alpha\boldsymbol{A}s_1(t) + \beta\boldsymbol{A}s_2(t), \tag{18.73}$$

where α and β are constants. This follows directly from the fact that $\boldsymbol{A}$ is a linear operator. A more complicated situation arises for the evaluation of $\boldsymbol{A}\{s_1(t)s_2(t)\}$. In general there is no simple resolution of this quantity; however, in a particular special case there is a major simplification. Bedrosian's theorem (recall Section 4.15) can be applied to treat one special case. Let $S_1(\omega)$ denote the Fourier spectrum of $s_1(t)$, which is defined by

$$S_1(\omega) = \mathcal{F}s_1(\omega). \tag{18.74}$$

The descriptor "Fourier" is often omitted in the term Fourier spectrum, and it is common to simply refer to the spectrum of the signal. Suppose that the spectrum of $s_1(t)$ vanishes for $|\omega| > \omega_1$, with $\omega_1 > 0$, and that the spectrum of $s_2(t)$ vanishes for $|\omega| < \omega_1$. Then, by making use of Eq. (4.257), it follows that

$$\boldsymbol{A}\{s_1(t)s_2(t)\} = s_1(t)\boldsymbol{A}s_2(t), \tag{18.75}$$

which amounts to a major simplification.

A quantity of interest is the envelope function, defined by reference to Eq. (18.39) as follows:

$$\mathcal{E}(t) = |f(t)| = \sqrt{\{g^2(t) + [Hg(t)]^2\}}. \tag{18.76}$$

If Eq. (18.39) is written the form

$$f(t) = \mathcal{E}(t)e^{i\phi(t)}, \tag{18.77}$$

then $\mathcal{E}(t)$ is defined by Eq. (18.76) and $\phi(t)$ is given by

$$\phi(t) = \tan^{-1}\left(\frac{Hg(t)}{g(t)}\right), \tag{18.78}$$

which is called the *instantaneous phase*. In connection with the analytic signal, the *instantaneous frequency* can be defined by the following expression:

$$\omega(t) = \frac{d\phi(t)}{dt}. \tag{18.79}$$

The envelope function is often referred to as simply the envelope. The quantity $\mathcal{E}(t)$ is often referred to as the *instantaneous amplitude* of the analytical signal. There are other definitions that have been employed for the envelope function. One alternative definition involves a time derivative of the signal in place of the Hilbert transform of the signal. For various classes of signals, this alternative definition coincides with that given in Eq. (18.76). As an example of an envelope function, consider the case $g(t) = \operatorname{sinc} t$; then, using entry (5.42) from Table 1.5 in Appendix 1, the upper and lower envelopes are given by

$$\mathcal{E}(t) = \pm\left|\operatorname{sinc}\frac{t}{2}\right|. \tag{18.80}$$

This example is illustrated in Figure 18.4. At the points t_i, where the function $g(t)$ intersects its envelope $\mathcal{E}(t)$, $Hg(t_i) = 0$.

If the construction of a complex signal is considered in a more abstract way, a principal problem is to determine an operator $\mathcal{O}$ such that a signal written in the form

$$w(t) = u(t) + iv(t) \tag{18.81}$$

is constructed using

$$v(t) = \mathcal{O}u(t). \tag{18.82}$$

What requirements must be placed on $\mathcal{O}$ such that the resulting complex signal has certain desirable properties? Vakman (1996) has given three conditions that should be satisfied by the operator $\mathcal{O}$. If the initial signal $u(t)$ is slightly perturbed by the

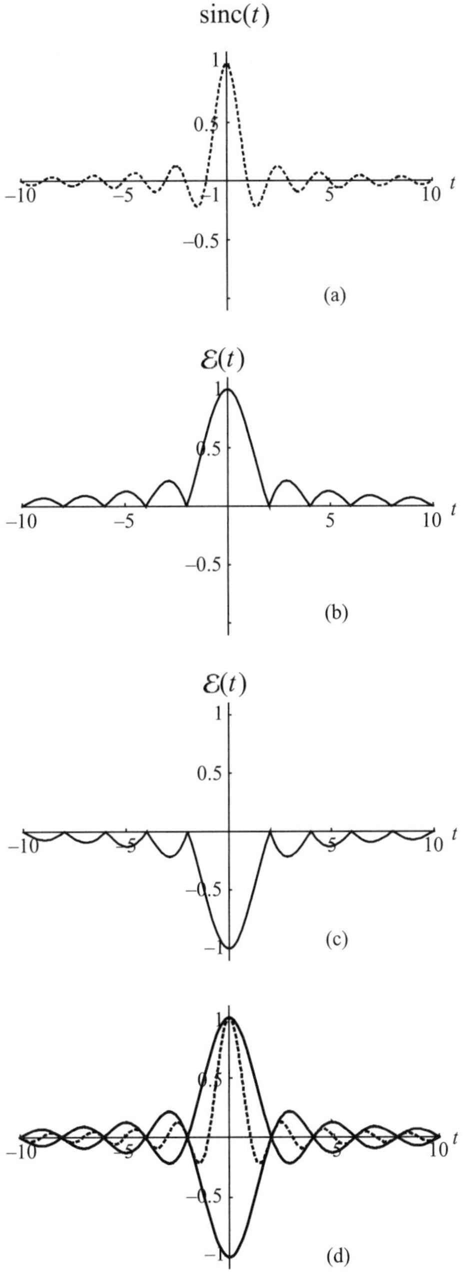

Figure 18.4. (a) Plot of the function $g(t) = \text{sinc}\, t$; (b) upper envelope of function; (c) lower envelope of function; (d) combination of graphs from (a)–(c).

addition of a very small signal $\delta u(t)$, then, in general, it is reasonable to expect that the signal $v(t)$ is perturbed by a very small amount, say $\delta v(t)$, that is

$$\mathcal{O}\{u(t) + \delta u(t)\} = v(t) + \delta v(t), \tag{18.83}$$

and it is desirable that

$$\mathcal{O}\{u(t) + \delta u(t)\} \to v(t), \quad \text{as } \|\delta u\| \to 0. \tag{18.84}$$

It is highly desirable that $\mathcal{O}$ is a continuous operator.

A second condition that should be imposed is that the phase and frequency ought to remain unaltered when the signal is multiplied by a scalar. Let the signal $u(t)$ be multiplied by the scalar c, with $c > 0$. If the phase is defined by

$$\phi(t) = \tan^{-1}\left(\frac{\mathcal{O}cu(t)}{cu(t)}\right), \tag{18.85}$$

then it is required that

$$\frac{\mathcal{O}cu(t)}{cu(t)} = \frac{c\mathcal{O}u(t)}{cu(t)} = \frac{\mathcal{O}u(t)}{u(t)}, \tag{18.86}$$

in order for the phase to be invariant under scalar multiplication by a constant. Hence,

$$\mathcal{O}cu(t) = c\mathcal{O}u(t), \tag{18.87}$$

which is a homogeneity condition (recall Eq. (7.166)). This can be treated as a phase invariance to scalar multiplication condition.

The third condition invoked by Vakman, is that the amplitude and frequency of a sinusoidal signal should be preserved under the action of the operator $\mathcal{O}$. That is, for a a constant,

$$\mathcal{O}a\cos(\omega t + \phi) = a\sin(\omega t + \phi). \tag{18.88}$$

This has been called a harmonic correspondence condition. It is not difficult to demonstrate that the Hilbert transform satisfies the three conditions that have been given. It is left as an exercise for the reader to verify that this is indeed the case.

Under the action of a translation of the signal $u(t)$, it is reasonable to expect that the signal satisfies

$$u(t) \to u(t-a), \quad \text{then } w(t) \to w(t-a) = u(t-a) + \mathrm{i}v(t-a). \tag{18.89}$$

Since $v(t)$ is determined from $u(t)$, it is required that the following sequence apply:

$$\mathcal{O}u(t-a) = \mathcal{O}\tau_a u(t) = \tau_a \mathcal{O}u(t) = \tau_a v(t) = v(t-a), \tag{18.90}$$

where τ_a is the translation operator defined in Eq. (4.64). In a similar fashion for a constant $b > 0$, it is reasonable to expect that the signal satisfies

$$u(t) \to u(bt), \quad \text{then } w(t) \to w(bt) = u(bt) + \mathrm{i}v(bt). \tag{18.91}$$

It is desirable that the following sequence applies:

$$\mathcal{O}u(bt) = \mathcal{O}S_b u(t) = S_b \mathcal{O}u(t) = S_b v(t) = v(bt), \tag{18.92}$$

where S_b is the dilation operator defined in Eq. (4.70). The reader will recognize that the two key requirements imposed in Eqs. (18.90) and (18.92) are the commutator conditions

$$[\mathcal{O}, \tau_a] = 0 \tag{18.93}$$

and

$$[\mathcal{O}, S_b] = 0. \tag{18.94}$$

If $\mathcal{O}$ is a bounded linear operator acting on functions of the class $L^p(\mathbb{R})$, for $1 < p < \infty$, and if $\mathcal{O}$ is required to commute with the translation operator and the dilation operator for positive dilations, then $\mathcal{O}$ takes the following form:

$$\mathcal{O} = \alpha I + \beta H, \tag{18.95}$$

where α and β are constants. This result was proved in Section 4.6. With the appropriate choice of constants in Eq. (18.95), the very general constraints of having a translation-invariant operator (recall Eq. (4.69)) and the commutator condition $[\mathcal{O}, S_b] = 0$, for $b > 0$, lead to the identification of $\mathcal{O}$ in the equation $v(t) = \mathcal{O}u(t)$ with the Hilbert transform operator. The choice $\alpha = 1$ and $\beta = \mathrm{i}$ in Eq. (18.95) leads to the definition of the analytic operator $\boldsymbol{A}$.

18.5 Amplitude modulation

Let $s(t)$ denote a signal dependent on time and let $S(\omega)$ denote its Fourier spectrum, $S(\omega) = \mathcal{F}s$, and assume that this is band-limited to the some frequency interval $[-\omega_0, \omega_0]$. That is, $S(\omega)$ has its support restricted to $[-\omega_0, \omega_0]$. If $S \in L^2$, then the time-dependent signal can be obtained by Fourier inversion as follows:

$$s(t) = \frac{1}{2\pi} \int_{-\omega_0}^{\omega_0} S(\omega) \mathrm{e}^{\mathrm{i}\omega t} \, \mathrm{d}\omega. \tag{18.96}$$

Suppose that $s(t)$ is a radio transmission. It is obviously advantageous that different radio stations in a given geographic setting broadcast signals with distinct frequency ranges. This can be achieved by carrying out a frequency modulation (FM) or an amplitude modulation (AM) of the signal. Electromagnetic radiation of the appropriate frequency serves as the carrier, and the process whereby the message modifies the carrier is termed "modulation." Recovery of the message at the receiver is called demodulation. In what follows, the focus is on amplitude modulation, which is achieved by constructing the new signal $s_{\mathrm{m}}(t)$ as

$$s_{\mathrm{m}}(t) = s(t)\cos(\omega_{\mathrm{c}} t), \tag{18.97}$$

where the carrier frequency is denoted by ω_{c}. The carrier frequency is much higher than the message frequencies. In the United States, the AM radio carrier frequencies are in the range 540 to 1700 kHz, and modulation frequencies for voice and radio are typically in the range 50 Hz to 10 kHz (Lindsey and Doelitzsch, 1993). The Fourier transform of $s_{\mathrm{m}}(t)$ leads to

$$\begin{aligned} S_{\mathrm{m}}(\omega) = \mathcal{F}s_{\mathrm{m}}(\omega) &= \int_{-\infty}^{\infty} s(t)\cos(\omega_{\mathrm{c}} t)\mathrm{e}^{-\mathrm{i}\omega t}\,\mathrm{d}t \\ &= \frac{1}{2}S(\omega-\omega_{\mathrm{c}}) + \frac{1}{2}S(\omega+\omega_{\mathrm{c}}). \end{aligned} \tag{18.98}$$

On the positive frequency scale, S_{m} spans the range $[\omega_{\mathrm{c}} - \omega_0, \omega_{\mathrm{c}} + \omega_0]$, with a corresponding range $[-\omega_{\mathrm{c}} - \omega_0, -\omega_{\mathrm{c}} + \omega_0]$ on the negative frequency scale. The signal that arrives at the receiver must be demodulated, or detected, which is accomplished by multiplying $s_{\mathrm{m}}(t)$ by the factor $\cos(\omega_{\mathrm{c}} t)$, that is

$$s_{\mathrm{d}}(t) = s_{\mathrm{m}}(t)\cos(\omega_{\mathrm{c}} t). \tag{18.99}$$

The Fourier spectrum of $s_{\mathrm{d}}(t)$ is given by

$$\begin{aligned} S_{\mathrm{d}}(\omega) = \mathcal{F}s_{\mathrm{d}}(\omega) &= \int_{-\infty}^{\infty} s_{\mathrm{m}}(t)\cos(\omega_{\mathrm{c}} t)\mathrm{e}^{-\mathrm{i}\omega t}\,\mathrm{d}t \\ &= \frac{1}{2}\int_{-\infty}^{\infty} s(t)\{1+\cos(2\omega_{\mathrm{c}} t)\}\mathrm{e}^{-\mathrm{i}\omega t}\,\mathrm{d}t \\ &= \frac{1}{2}S(\omega) + \frac{1}{4}S(\omega - 2\omega_{\mathrm{c}}) + \frac{1}{4}S(\omega+2\omega_{\mathrm{c}}). \end{aligned} \tag{18.100}$$

The components with frequencies shifted from ω are filtered out, leaving the ear to detect the frequency component corresponding to the signal $s(t)$. The contributions on each side of the frequency ω are referred to as the sidebands. The approach just outlined is called *double-sideband modulation*.

A common form of modulation for radio broadcasting is amplitude modulation. Here, the carrier is combined with the signal to be transmitted, so that

$$s_{\mathrm{am}}(t) = A\{1 + s_{\mathrm{m}}(t)\}\cos(\omega_{\mathrm{c}} t), \tag{18.101}$$

with the condition $|s_{\mathrm{m}}(t)| \leq 1$ so that $A\{1 + s_{\mathrm{m}}(t)\}$ is non-negative. It is left as an exercise for the reader to determine the spectrum corresponding to $s_{\mathrm{am}}(t)$.

In the double sideband modulation scheme the frequency range of the initial unmodulated signal is $2\omega_0$, while the total range of the modulated signal is $4\omega_0$. It is desirable to reduce this doubling of the frequency range, and this can be accomplished by an approach called *single-sideband modulation*. In place of the modulation employed in Eq. (18.97), the following modulation is employed:

$$s_{\mathrm{ssm}}(t) = s(t)\cos(\omega_{\mathrm{c}}t) - Hs(t)\sin(\omega_{\mathrm{c}}t). \tag{18.102}$$

This signal can be viewed as being formed by taking the real part of the analytical signal formed from $s(t)$ and multiplied by $\mathrm{e}^{\mathrm{i}\omega_{\mathrm{c}}t}$, that is,

$$s_{\mathrm{ssm}}(t) = \mathrm{Re}[\{s(t) + \mathrm{i}Hs(t)\}\mathrm{e}^{\mathrm{i}\omega_{\mathrm{c}}t}]. \tag{18.103}$$

Demodulation of this signal at the receiver end is carried out by multiplying $s_{\mathrm{ssm}}(t)$ by $\cos(\omega_{\mathrm{c}}t)$ and taking the Fourier transform, so that

$$\begin{aligned} S_{\mathrm{ssm}}(\omega) &= \int_{-\infty}^{\infty} \{s(t)\cos(\omega_{\mathrm{c}}t) - Hs(t)\sin(\omega_{\mathrm{c}}t)\}\cos(\omega_{\mathrm{c}}t)\mathrm{e}^{-\mathrm{i}\omega t}\,\mathrm{d}t \\ &= \frac{1}{2}\int_{-\infty}^{\infty} s(t)\{1 + \cos(2\omega_{\mathrm{c}}t)\}\mathrm{e}^{-\mathrm{i}\omega t}\,\mathrm{d}t \\ &\quad - \frac{1}{2}\int_{-\infty}^{\infty} \sin(2\omega_{\mathrm{c}}t)Hs(t)\,\mathrm{e}^{-\mathrm{i}\omega t}\,\mathrm{d}t. \end{aligned} \tag{18.104}$$

The preceding integral is evaluated as follows:

$$\begin{aligned} \int_{-\infty}^{\infty} \sin(2\omega_{\mathrm{c}}t)Hs(t)\,\mathrm{e}^{-\mathrm{i}\omega t}\,\mathrm{d}t &= \frac{1}{2\mathrm{i}}\int_{-\infty}^{\infty} Hs(t)\{\mathrm{e}^{-\mathrm{i}(\omega-2\omega_{\mathrm{c}})t} - \mathrm{e}^{-\mathrm{i}(\omega+2\omega_{\mathrm{c}})t}\}\mathrm{d}t \\ &= \frac{1}{2\mathrm{i}}\{\mathcal{F}Hs(\omega - 2\omega_{\mathrm{c}}) - \mathcal{F}Hs(\omega + 2\omega_{\mathrm{c}})\} \\ &= \frac{1}{2\mathrm{i}}\{-\mathrm{i}\,\mathrm{sgn}\,(\omega - 2\omega_{\mathrm{c}})\,\mathcal{F}s(\omega - 2\omega_{\mathrm{c}}) \\ &\quad + \mathrm{i}\,\mathrm{sgn}\,(\omega + 2\omega_{\mathrm{c}})\,\mathcal{F}s(\omega + 2\omega_{\mathrm{c}})\} \\ &= \frac{1}{2}\{\mathrm{sgn}(\omega + 2\omega_{\mathrm{c}})\,S(\omega + 2\omega_{\mathrm{c}})\} \\ &\quad - \mathrm{sgn}(\omega - 2\omega_{\mathrm{c}})\,S(\omega - 2\omega_{\mathrm{c}})\}, \end{aligned} \tag{18.105}$$

and Eq. (5.3) has been employed. Equation (18.104) reduces to the following:

$$\begin{aligned} S_{\mathrm{ssm}}(\omega) &= \frac{1}{2}S(\omega) + \frac{1}{4}S(\omega - 2\omega_{\mathrm{c}}) + \frac{1}{4}S(\omega + 2\omega_{\mathrm{c}}) \\ &\quad - \frac{1}{4}\,\mathrm{sgn}(\omega + 2\omega_{\mathrm{c}})\,S(\omega + 2\omega_{\mathrm{c}}) + \frac{1}{4}\,\mathrm{sgn}(\omega - 2\omega_{\mathrm{c}})\,S(\omega - 2\omega_{\mathrm{c}}), \end{aligned} \tag{18.106}$$

which simplifies to yield

$$S_{\mathrm{ssm}}(\omega) = \frac{1}{2}S(\omega) + \frac{1}{4}\{1 + \mathrm{sgn}(\omega - 2\omega_{\mathrm{c}})\}S(\omega - 2\omega_{\mathrm{c}}) + \frac{1}{4}\{1 - \mathrm{sgn}(\omega + 2\omega_{\mathrm{c}})\}S(\omega + 2\omega_{\mathrm{c}}). \quad (18.107)$$

For a signal $s(t)$ and its Hilbert transform $Hs(t)$, the corresponding moduli of the amplitudes of the Fourier transforms of these two signals are identical. The two Fourier signals differ only in phase. The ear is fairly insensitive to the detection of phase changes for the Fourier components of a signal, with the result that if $s(t)$ denotes a speech signal; taking the Hilbert transform of the signal does not make it indecipherable.

18.6 The frequency domain

Consider a function with the analogous structure of the analytic signal, but defined in the frequency domain:

$$F(\omega) = G(\omega) + \mathrm{i}HG(\omega). \quad (18.108)$$

Take the Fourier transform of this expression, and denote the corresponding Fourier transforms of $F(\omega)$ and $G(\omega)$ by $f(t)$ and $g(t)$, respectively; then

$$f(t) = g(t)\{1 + \mathrm{sgn}\, t\}. \quad (18.109)$$

The outcome is that $f(t)$ is a causal function. As an example, consider the Lorentzian function defined by

$$I(\omega) = \frac{1}{\pi}\frac{a}{a^2 + (\omega - \omega_0)^2}, \quad (18.110)$$

where a and ω_0 are constants. The Hilbert transform of the Lorentzian is given by

$$HI(\omega) = \frac{1}{\pi}\frac{(\omega - \omega_0)}{a^2 + (\omega - \omega_0)^2}, \quad (18.111)$$

and $F(\omega)$ is therefore given by

$$F(\omega) = \frac{1}{\pi}\frac{a}{a^2 + (\omega - \omega_0)^2} + \mathrm{i}\frac{1}{\pi}\frac{(\omega - \omega_0)}{a^2 + (\omega - \omega_0)^2}. \quad (18.112)$$

The Fourier transforms of the real and imaginary parts of $F(\omega)$ are, for $a > 0$,

$$g(t) = \mathcal{F}[\mathrm{Re}\, F(t)] = \mathrm{e}^{-\mathrm{i}\omega_0 t - a|t|}, \quad (18.113)$$

and

$$iHg(t) = i\mathcal{F}[\operatorname{Im} F(t)] = \operatorname{sgn} t\, e^{-i\omega_0 t - a|t|}, \tag{18.114}$$

and hence

$$f(t) = \mathcal{F}[F(t)] = e^{-i\omega_0 t - a|t|}\{1 + \operatorname{sgn} t\}. \tag{18.115}$$

This final expression clearly displays the causal character of $f(t)$. For this example, the separate behavior for $g(t), iHg(t)$, and $f(t)$ are illustrated in Figure 18.5 for the choice $\omega_0 = 0$ and $a = 1$.

18.7 Some useful step and pulse functions

Some functions that have considerable value in determining the shape of a signal are examined in this section. These step functions also find considerable use in many areas of applied mathematics. Section 18.8 examines which of these step functions have a non-divergent Hilbert transform.

18.7.1 The Heaviside function

The first step function considered is the Heaviside function, which is arguably one of the most important of the step functions in practical applications. Heaviside's unit step function is defined by

$$\mathrm{H}(x) = \begin{cases} 0, & x < 0 \\ 1/2, & x = 0 \\ 1, & x > 0. \end{cases} \tag{18.116}$$

Two definitions are in common use. The first definition leaves H(0) undefined, and $\mathrm{H}(x)$ for $|x| > 0$ involves the assignment given in the preceding equation. The second definition assigns a constant c to the value of $\mathrm{H}(x)$ at $x = 0$, with the choice of constant most commonly selected as the value 1/2. The Heaviside function has a jump discontinuity at $x = 0$. It is represented graphically as shown in Figure 18.6.

The Heaviside step function is particularly useful for representing physical processes that are switched on at some particular time. For example, suppose a constant voltage is applied to a system, starting at time zero, then the shape of the voltage–time curve would be that of a Heaviside function.

The step function shown in Figure 18.7(a) can be represented by a Heaviside function with a shift of the argument, that is $\mathrm{H}(x - a)$. With a change in sign for the argument of the latter Heaviside function, that is, $\mathrm{H}(a - x)$, the graphical

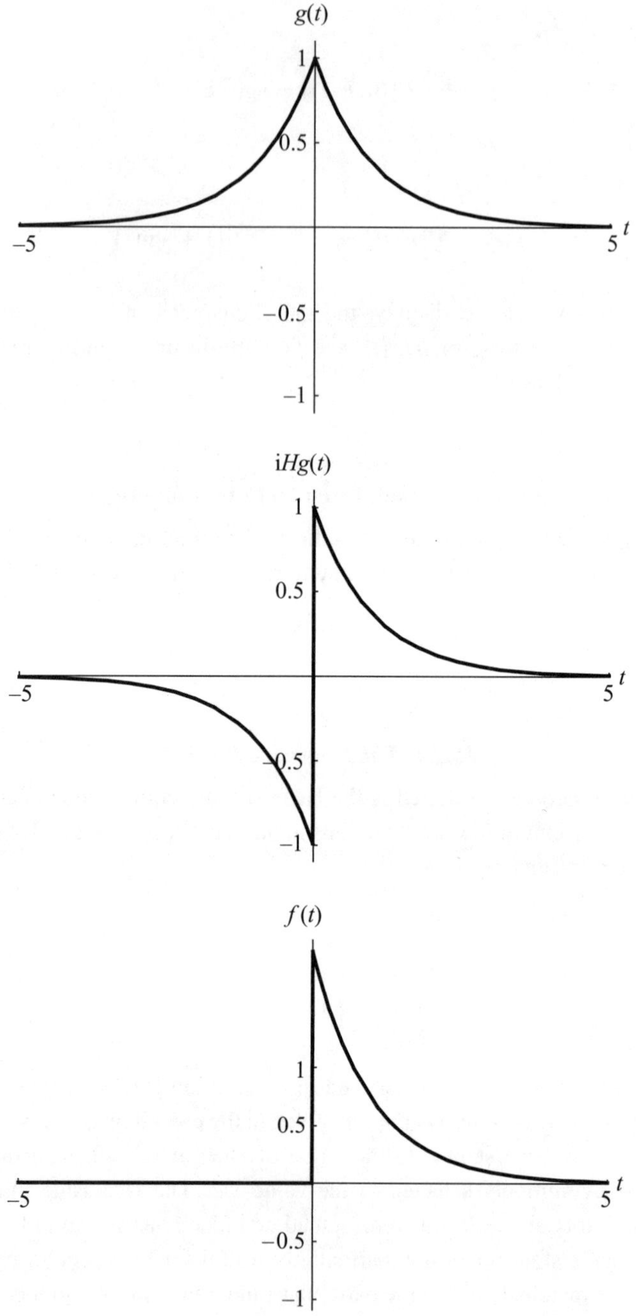

Figure 18.5. The functions $g(t)$, $\mathrm{i}Hg(t)$, and $f(t)$ given in Eqs. (18.113)–(18.115).

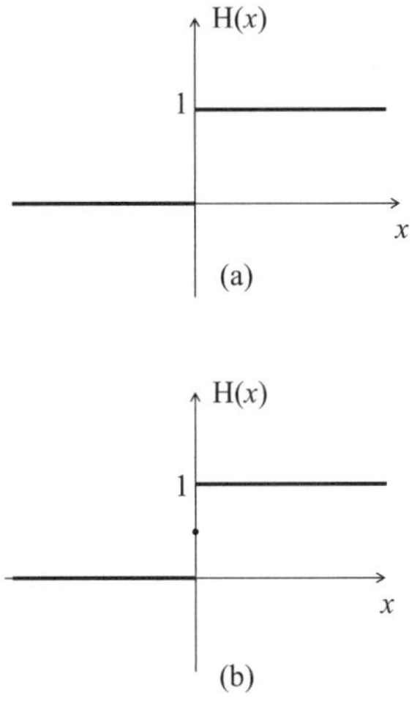

Figure 18.6. The Heaviside unit step function. (a) The form when $\mathrm{H}(x)$ is not defined at $x = 0$. (b) The form when the function is assigned a value 1/2 at $x = 0$.

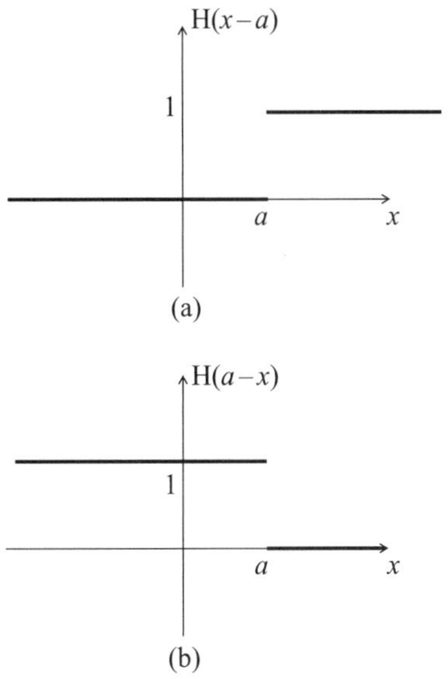

Figure 18.7. The Heaviside step function $\mathrm{H}(x - a)$ with a displaced origin. (b) The Heaviside function $\mathrm{H}(a - x)$.

representation is as shown in Figure 18.7(b). This form of the Heaviside function could be used to model a process where a constant signal is turned off at some particular time.

The Heaviside function can be used as an effective means for representing functions with one or more finite jump discontinuities. The following example illustrates this usage. Let

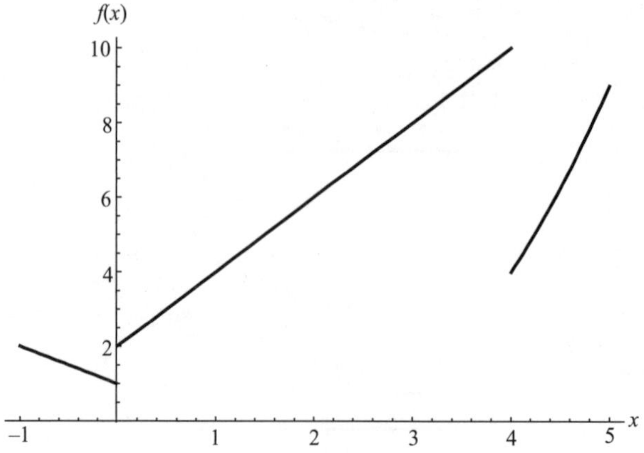

Figure 18.8. The function $f(x)$ defined by Eqs. (18.117)–(18.118).

$$
\begin{aligned}
g_1(x) &= -x+1, \quad -1 < x < 0,\\
g_2(x) &= 2x+2, \quad 0 < x < 4,\\
g_3(x) &= (x-2)^2, \quad 4 < x,
\end{aligned} \tag{18.117}
$$

and define the function f by

$$f(x) = g_1(x) + g_2(x) + g_3(x). \tag{18.118}$$

This function is shown in Figure 18.8. The function has finite jump discontinuities at $x = 0$ and $x = 4$. Equation (18.118) can be written as follows:

$$f(x) = (1-x)\mathrm{H}(x+1) + (3x+1)\mathrm{H}(x) + (x^2 - 6x + 2)\mathrm{H}(x-4). \tag{18.119}$$

18.7.2 The signum function

The signum (or sign) function has arisen in several places in the book so far, and was defined in Eq. (1.14). An alternative definition that is commonly employed is as follows:

$$\operatorname{sgn} x = \begin{cases} 1, & x > 0 \\ -1, & x < 0, \end{cases} \tag{18.120}$$

where no assignment is made for the value at $x = 0$. The signum function can be written in terms of the Heaviside function:

$$\operatorname{sgn} x = 2\mathrm{H}(x) - 1. \tag{18.121}$$

18.7.3 The rectangular pulse function

The rectangular pulse function (sometimes termed the "boxcar") of unit height was defined in Eq. (9.19) and illustrated in Figure 9.1. The alternative definition for $a > 0$,

$$\Pi_{2a}(x) = \begin{cases} 0, & |x| > a \\ 1/2, & |x| = a \\ 1, & |x| < a, \end{cases} \tag{18.122}$$

is also in use. The rectangular pulse function for the case $a = 1/2$ is often written as $\mathrm{rect}(x)$. The function $\Pi_{2a}(x)$ can be written as a combination of two Heaviside step functions:

$$\Pi_{2a}(x) = \mathrm{H}(x + a) - \mathrm{H}(x - a). \tag{18.123}$$

The rectangular waveform shown in Figure 18.9 can be represented mathematically in different ways. Consider the cosine function $\cos(\pi x/2a)$ depicted in Figure 18.10. The function is negative for x in the interval $(4k + 1)a < x < (4k + 3)a$, for integer k, and positive for $(4k - 1)a < x < (4k + 1)a$. Hence, the Heaviside function can provide a convenient representation of the rectangular waveform:

$$\mathrm{H}\Big(\cos\Big(\frac{\pi x}{2a}\Big)\Big) = \begin{cases} 0, & (4k + 1)a < x < (4k + 3)a \\ 1, & (4k - 1)a < x < (4k + 1)a. \end{cases} \tag{18.124}$$

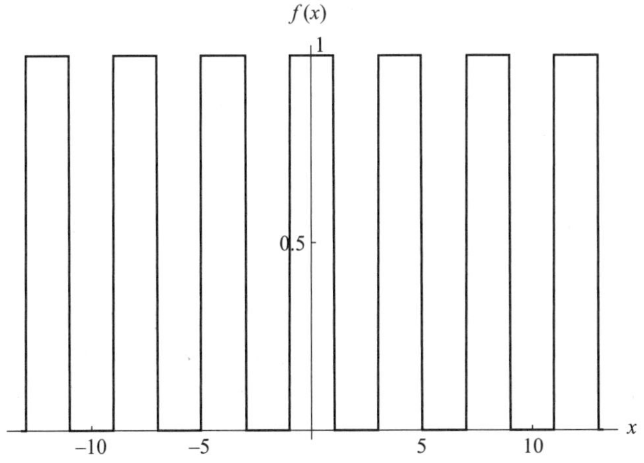

Figure 18.9. Rectangular waveform.

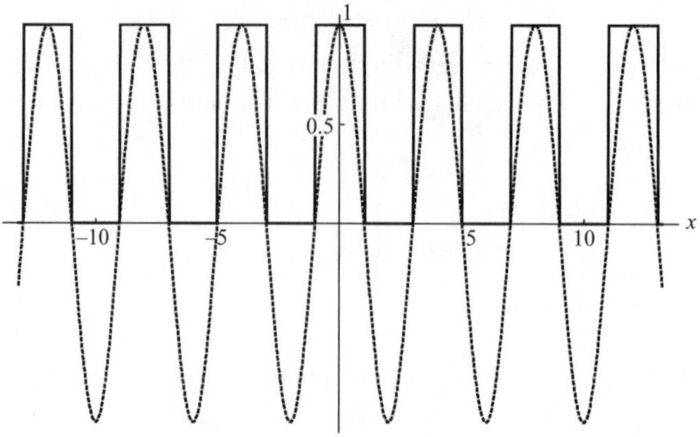

Figure 18.10. Graph of the cosine function for the case $a = 1$ and the rectangular waveform resulting from this function.

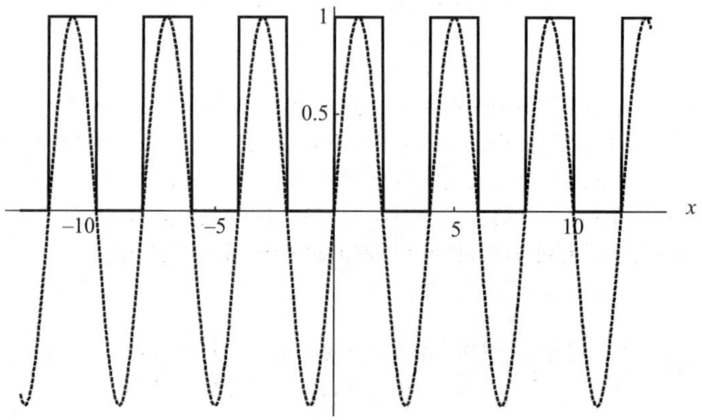

Figure 18.11. Graph of the sine function for the case $a = 1$ and the rectangular waveform resulting from this function.

Using the same type of subtraction procedure indicated in Eq. (18.123), it is possible to write $\mathrm{H}(\cos(\pi x/2a))$ in the following form:

$$\mathrm{H}\left(\cos\left(\frac{\pi x}{2a}\right)\right) = \sum_{k=-\infty}^{\infty} \{\mathrm{H}(x - (4k - 1)a) - \mathrm{H}(x - (4k + 1)a)\}. \tag{18.125}$$

A rectangular waveform shifted by one unit can be obtained by employing the sine function $\sin(\pi x/2a)$. The situation is illustrated in Figure 18.11. This form of the rectangular wave can be written in terms of $\mathrm{H}(\sin(\pi x/2a))$ by employing

$$\mathrm{H}\left(\sin\left(\frac{\pi x}{2a}\right)\right) = \begin{cases} 0, & (4k + 2)a < x < (4k + 4)a \\ 1, & 4ka < x < (4k + 2)a. \end{cases} \tag{18.126}$$

Hence,

$$\mathrm{H}\left(\sin\left(\frac{\pi x}{2a}\right)\right) = \sum_{k=-\infty}^{\infty} \{\mathrm{H}(x - 4ka) - \mathrm{H}(x - (4k + 2)a)\}. \tag{18.127}$$

18.7.4 The triangular pulse function

Recall from Eq. (4.265) that the unit triangular function is defined by

$$\Lambda(x) = \begin{cases} 0, & -\infty < x < -1 \\ 1 + x, & -1 < x < 0 \\ 1 - x, & 0 < x < 1 \\ 0, & 1 < x < \infty. \end{cases} \tag{18.128}$$

The graphical representation of $\Lambda(x)$ is displayed in Figure 18.12. The triangle function $\Lambda(x)$ can be cast in terms of the Heaviside step function as follows:

$$\Lambda(x) = (x + 1)\mathrm{H}(x + 1) - 2x\mathrm{H}(x) + (x - 1)\mathrm{H}(x - 1), \tag{18.129}$$

and it can also be written in terms of the signum function.

18.7.5 The sinc pulse function

The sinc function was defined in Eq. (4.260) as $\operatorname{sinc} x = \sin \pi x/(\pi x)$. Note that some authors define the sinc function without the factor of pi included. The sinc function encloses unit area on the interval $(-\infty, \infty)$, that is

$$\int_{-\infty}^{\infty} \operatorname{sinc} x \, \mathrm{d}x = 1. \tag{18.130}$$

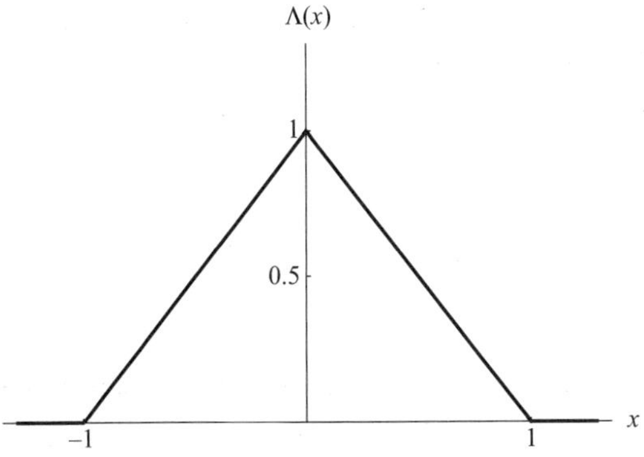

Figure 18.12. The unit triangular function $\Lambda(x)$.

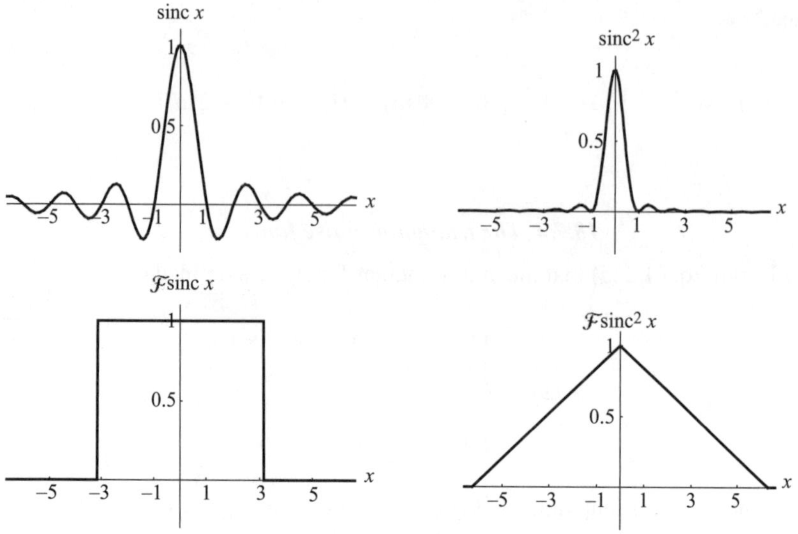

Figure 18.13. The functions $\operatorname{sinc} x$ and $\operatorname{sinc}^2 x$, and the Fourier transforms of these functions.

Of related interest is the $\operatorname{sinc}^2 x$ function. This function also encloses unit area on the interval $(-\infty, \infty)$. The sinc function was illustrated in Figure 4.3. In Figure 18.13 the sinc function and the function $\operatorname{sinc}^2 x$ are compared, together with the corresponding Fourier transforms of these functions.

The sinc function has the interesting feature that its Fourier transform is the rectangular function $\Pi_{2\pi}(x)$, and the Fourier transform of $\operatorname{sinc}^2 x$ is $\Lambda(x/2\pi)$.

18.8 The Hilbert transform of step functions and pulse forms

The Hilbert transform of both the Heaviside and sgn functions diverge. The problem lies not with the singular behavior on the real axis, but with the behavior at infinity. This can be contrasted with the Hilbert transform of a constant function, which is well behaved and has the value zero.

The rectangular pulse $\Pi_{2a}(x)$ is defined in Eqs. (9.19) and (18.122) and has a Hilbert transform given by

$$H\Pi_{2a}(x) = \frac{1}{\pi} \log \left| \frac{a+x}{a-x} \right|. \tag{18.131}$$

The result is well defined except at $x = \pm a$, which correspond to the points of jump discontinuity for $\Pi_{2a}(x)$. For the related pulse functions given by $\mathrm{H}(\cos(\pi x/2a))$ and $\mathrm{H}(\sin(\pi x/2a))$, the reader is left to decide if the Hilbert transforms of these functions are well behaved.

The Hilbert transform of the unit triangular pulse function is given by

$$H\Lambda(x) = \frac{1}{\pi}\left\{\log\left|\frac{x+1}{x-1}\right| + x\log\left|\frac{x^2-1}{x^2}\right|\right\}, \tag{18.132}$$

which is well behaved except at the discontinuous points of $\Lambda(x)$ at $x = \pm 1$. The Hilbert transform of the more general triangular pulse of the form

$$\Lambda_a(x) = \begin{cases} 0, & |x| > a \\ 1 - \dfrac{|x|}{a}, & |x| < a \end{cases} \tag{18.133}$$

is left as a case for the reader to consider.

The sinc function arises widely in applications. The Hilbert transform of this pulse function is straightforward to evaluate:

$$\begin{aligned} H\,\mathrm{sinc}(ax) &= \frac{1}{\pi}P\int_{-\infty}^{\infty}\frac{\sin(a\pi t)\mathrm{d}t}{a\pi t(x-t)} \\ &= \frac{1}{xa\pi^2}\left\{P\int_{-\infty}^{\infty}\frac{\sin(a\pi t)\mathrm{d}t}{x-t} + \int_{-\infty}^{\infty}\frac{\sin(a\pi t)\mathrm{d}t}{t}\right\} \\ &= \frac{-\mathrm{sgn}\,a\cos(a\pi x)}{a\pi x} + \frac{\mathrm{sgn}\,a}{xa\pi} \\ &= \frac{\mathrm{sgn}\,a}{a\pi x}\{1-\cos(a\pi x)\}, \end{aligned} \tag{18.134}$$

and $H(\sin ax) = -\mathrm{sgn}\,a\cos ax$ has been employed. As an exercise, the reader is left to try the example $H\,\mathrm{sinc}^2 ax$.

18.9 The fractional Hilbert transform: the Lohmann–Mendlovic–Zalevsky definition

The idea of a fractional Hilbert transform (FHT) operation as it is applied in engineering applications is examined in this section. Three distinct definitions that have been introduced in the literature are described. The standard Hilbert transform is applied in image processing because it can be used to enhance selectively particular features of an image. A motivation behind introducing a fractional Hilbert transform is that its application can affect the type of edge enhancement obtained for an image. This behavior is illustrated shortly.

An example of Lohmann, Mendlovic, and Zalevsky (1998) is considered first. Let $g_i, i = 0,\ldots,3$ denote, respectively, the images $u(x,y)$, $u(y,-x), u(-x,-y)$, and $u(-y,x)$, which are constructed by successive rotation by 90°. The set g_i can be replaced by the following function:

$$g(x,y,p) = u(x\cos\phi + y\sin\phi, -x\sin\phi + y\cos\phi), \tag{18.135}$$

where

$$\phi = \frac{\pi}{2}p. \tag{18.136}$$

Each of the g_i can be recovered with the choices $p = 0, 1, 2,$ and 3, respectively. In a sense it follows that

$$g_i \rightarrow g(i) \rightarrow g(p), \tag{18.137}$$

which allows the four images to be recovered for the four integer values of p. It is possible to interpolate between images using fractional values of p.

Lohmann, Mendlovic, and Zalevsky (1996a) introduced two different definitions of a modified Hilbert transform that they termed the "fractional Hilbert transform." The first definition of the FHT introduced was based on a spatial filter with a fractional parameter, introduced in the same way as illustrated in Eqs. (18.135)–(18.136). The second definition was based on the fractional Fourier transform. The first of these two approaches is now considered. The transfer function of the Hilbert transform takes the following form:

$$\mathsf{H}_1(\omega) = \begin{cases} \mathrm{i}, & \omega > 0 \\ 0, & \omega = 0 \\ -\mathrm{i}, & \omega < 0. \end{cases} \tag{18.138}$$

The appearance of the extra subscript on H is explained shortly. Equation (18.138) can be rewritten in terms of the Heaviside step function as follows:

$$\mathsf{H}_1(\omega) = \mathrm{i}\mathrm{H}(\omega) - \mathrm{i}\mathrm{H}(-\omega). \tag{18.139}$$

Equation (18.139) can be put in the following form:

$$\mathsf{H}_1(\omega) = \exp(\mathrm{i}\pi/2)\mathrm{H}(\omega) + \exp(-\mathrm{i}\pi/2)\mathrm{H}(-\omega). \tag{18.140}$$

A fractional generalization of this result can be written as follows:

$$\mathsf{H}_p(\omega) = \exp(\mathrm{i}p\pi/2)\mathrm{H}(\omega) + \exp(-\mathrm{i}p\pi/2)\mathrm{H}(-\omega). \tag{18.141}$$

Employing Eqs. (18.121) and (18.136) allows the preceding result to be rewritten as follows:

$$\mathsf{H}_p(\omega) = \cos\phi + \mathrm{i}\sin\phi \,\mathrm{sgn}\,\omega. \tag{18.142}$$

The parameter p is called the order.

Using Eq. (18.141) as the definition of the fractional Hilbert filter function introduced by Lohmann *et al.* (1996a), several properties follow.

(1) The conventional Hilbert filter is recovered when $p = 1$.
(2) For the case $p = 0, \mathsf{H}_0(\omega) = 1$, and a signal is unaltered by the action of $\mathsf{H}_0(\omega)$.
(3) There is order additivity; that is

$$\mathsf{H}_{p1+p2}(\omega) = \mathsf{H}_{p1}(\omega)\mathsf{H}_{p2}(\omega), \tag{18.143}$$

which follows directly from Eq. (18.141) on using $\mathrm{H}(\omega)\mathrm{H}(-\omega) = 0$. A signal subject to $\mathsf{H}_{p_1}(\omega)$ and $\mathsf{H}_{p_2}(\omega)$ sequentially is equivalent to the single action of $\mathsf{H}_{p1+p2}(\omega)$. From Eq. (18.143), it follows that the fractional Hilbert filter operation forms a semigroup.
(4) The function $\mathsf{H}_p(\omega)$ is periodic in the variable p and satisfies

$$\mathsf{H}_p(\omega) = \mathsf{H}_{p+4}(\omega); \tag{18.144}$$

that is, the period in the variable p is four.

An alternative approach to defining the fractional Hilbert transform is via the corresponding fractional Fourier transform (FRFT). The FRFT is discussed in detail in Section 18.10. A transfer function V_Q is defined by (Lohmann *et al.*, 1996a)

$$V_Q = \mathcal{F}_{-Q}\{\mathsf{H}_1\mathcal{F}_Q\}, \tag{18.145}$$

where $\mathcal{F}_Q$ denotes the fractional Fourier transform. This transfer function corresponds to a fractional Hilbert transform. The similarity with the corresponding result for the Hilbert transform, Eq. (5.3), should be obvious to the reader. To bring this definition in line with Eq. (5.3) requires the choice $p = 3$ in place of $p = 1$ in Eq. (18.142), to switch the sign. The parameter Q serves to specify the fractional order. A generalization of Eq. (18.145) has been given as follows (Lohmann *et al.*, 1998):

$$V_{PQ} = \mathcal{F}_{-Q}\{\mathsf{H}_P\mathcal{F}_Q\}. \tag{18.146}$$

To deal with both Eqs. (18.145) and (18.146), a definition of the fractional Fourier transform is required, a topic which is now addressed.

18.10 The fractional Fourier transform

The fractional Fourier transform has been employed in mathematics for many years, but its use in signal processing started around the early 1990s. The fractional Fourier transform is defined by

$$S_\alpha(u) = \mathcal{F}_\alpha s(u) = \int_{-\infty}^{\infty} s(t)K_\alpha(t,u)\mathrm{d}t, \tag{18.147}$$

where the kernel function $K_\alpha(t,u)$ is defined by

$$K_\alpha(t,u) = \begin{cases} \sqrt{\left(\dfrac{1-\mathrm{i}\cot\alpha}{2\pi}\right)} \exp\left[\mathrm{i}\dfrac{t^2+u^2}{2}\cot\alpha - \mathrm{i}ut\csc\alpha\right], & \alpha \neq n\pi, \text{for integer } n \\ \delta(t-u), & \alpha = 2n\pi, \text{for integer } n \\ \delta(t+u), & \alpha = n\pi, \text{for odd integer } n. \end{cases} \tag{18.148}$$

An alternative definition in use employs $\alpha = a\pi/2$ and utilizes the kernel function $K_a(t,u)$ in place of $K_\alpha(t,u)$, which simplifies the notation in some cases, such as $K_1(t,\omega)$ in place of $K_{\pi/2}(t,\omega)$. In the literature the fractional Fourier transform operator is also often defined with a superscript in place of a subscript designation. Why the fractional Fourier transform is defined using the kernel function of Eq. (18.148) will become apparent later.

Some particular properties of the kernel function are now examined. The case $\alpha = \pi/2$ leads to

$$K_{\pi/2}(t,\omega) = \sqrt{\left(\frac{1}{2\pi}\right)}\,\mathrm{e}^{-\mathrm{i}\omega t}, \tag{18.149}$$

which is the kernel function for the normal Fourier transform. To prove various relationships for the kernel function $K_\alpha(t,u)$, it is advantageous to introduce the following simplifying notation:

$$a(\alpha) = \sqrt{\left(\frac{1-\mathrm{i}\cot\alpha}{2\pi}\right)}, \quad b(\alpha) = \frac{1}{2}\cot\alpha, \quad c(\alpha) = \csc\alpha. \tag{18.150}$$

The principal properties satisfied by the kernel $K_\alpha(t,u)$ are as follows:

$$K_\alpha(t,u) = K_\alpha(u,t), \tag{18.151}$$

$$K_{-\alpha}(t,u) = K_\alpha^*(t,u), \tag{18.152}$$

and

$$K_\alpha(-t,u) = K_\alpha(t,-u), \tag{18.153}$$

which are all obvious by inspection of the definition of $K_\alpha(t,u)$, and the key integral relationships are given by

$$\int_{-\infty}^{\infty} K_\alpha(t,\lambda)K_\beta(\lambda,u)\mathrm{d}\lambda = K_{\alpha+\beta}(t,u) \tag{18.154}$$

and

$$\int_{-\infty}^{\infty} K_\alpha(t,u)K_\alpha^*(t,u_0)\mathrm{d}t = \delta(u-u_0). \tag{18.155}$$

The kernel functions $K_\alpha(t, u)$ form an orthonormal set in the variable t, which is apparent from the last result. The proof of Eq. (18.155) follows directly from Eq. (18.152) and the definition given in Eq. (18.148). The proof of Eq. (18.154) requires some dexterity with trigonometric identities. It follows that

$$\int_{-\infty}^{\infty} K_\alpha(t,\lambda)K_\beta(\lambda,u)\mathrm{d}\lambda = a(\alpha)a(\beta)\mathrm{e}^{\mathrm{i}t^2 b(\alpha)+\mathrm{i}u^2 b(\beta)} \times \int_{-\infty}^{\infty} \exp\{\mathrm{i}\lambda^2[b(\alpha)+b(\beta)] - \mathrm{i}\lambda[tc(\alpha)+uc(\beta)]\}\mathrm{d}\lambda. \tag{18.156}$$

Employ the result for $B \in \mathbb{R}$, that

$$\int_{-\infty}^{\infty} \mathrm{e}^{\mathrm{i}C\lambda^2 - \mathrm{i}B\lambda}\mathrm{d}\lambda = \sqrt{\pi}\,\mathrm{e}^{-\mathrm{i}B^2/4C} \begin{cases} \dfrac{\mathrm{e}^{\mathrm{i}\pi/4}}{\sqrt{C}}, & C > 0 \\ \dfrac{\mathrm{e}^{-\mathrm{i}\pi/4}}{\sqrt{(-C)}}, & C < 0, \end{cases} \tag{18.157}$$

which allows Eq. (18.156) to be written as follows:

$$\int_{-\infty}^{\infty} K_\alpha(t,\lambda)K_\beta(\lambda,u)\mathrm{d}\lambda = a(\alpha)a(\beta)\mathrm{e}^{\mathrm{i}t^2 b(\alpha)+\mathrm{i}u^2 b(\beta)}\sqrt{\pi}\exp\left\{-\mathrm{i}\frac{[tc(\alpha)+uc(\beta)]^2}{4[b(\alpha)+b(\beta)]}\right\} \times \begin{cases} \dfrac{\mathrm{e}^{\mathrm{i}\pi/4}}{\sqrt{[b(\alpha)+b(\beta)]}}, & [b(\alpha)+b(\beta)] > 0 \\ \dfrac{\mathrm{e}^{-\mathrm{i}\pi/4}}{\sqrt{-[b(\alpha)+b(\beta)]}}, & [b(\alpha)+b(\beta)] < 0. \end{cases} \tag{18.158}$$

The exponent of the lead exponential term simplifies as follows:

$$\mathrm{i}t^2 b(\alpha) + \mathrm{i}u^2 b(\beta) - \mathrm{i}\frac{[tc(\alpha)+uc(\beta)]^2}{4[b(\alpha)+b(\beta)]} = \mathrm{i}t^2\left\{b(\alpha) - \frac{c^2(\alpha)}{4[b(\alpha)+b(\beta)]}\right\} + \mathrm{i}u^2\left\{b(\beta) - \frac{c^2(\beta)}{4[b(\alpha)+b(\beta)]}\right\} - \mathrm{i}tu\frac{c(\alpha)c(\beta)}{2[b(\alpha)+b(\beta)]}. \tag{18.159}$$

On making use of the trigonometric identity

$$\cot(\alpha+\beta) = \frac{\cot\alpha\cot\beta - 1}{\cot\alpha + \cot\beta}, \tag{18.160}$$

a short calculation shows that

$$b(\alpha) - \frac{c^2(\alpha)}{4[b(\alpha)+b(\beta)]} = \frac{1}{2}\cot(\alpha+\beta) = b(\alpha+\beta), \tag{18.161}$$

and also that

$$\frac{c(\alpha)c(\beta)}{2[b(\alpha)+b(\beta)]} = c(\alpha+\beta), \tag{18.162}$$

and hence

$$\mathrm{i}t^2 b(\alpha) + \mathrm{i}u^2 b(\beta) - \mathrm{i}\frac{[tc(\alpha)+uc(\beta)]^2}{4[b(\alpha)+b(\beta)]} = \mathrm{i}(t^2+u^2)b(\alpha+\beta) - \mathrm{i}tuc(\alpha+\beta). \tag{18.163}$$

Considering for the moment the choice $[b(\alpha)+b(\beta)] > 0$, the pre-exponential factor of Eq. (18.158) simplifies as follows:

$$\begin{aligned} a(\alpha)a(\beta)\frac{\sqrt{\pi}}{\sqrt{[b(\alpha)+b(\beta)]}} &= \sqrt{\left[\frac{1}{2\pi}\left(\frac{1-\cot\alpha\cot\beta}{\cot\alpha+\cot\beta} - \mathrm{i}\right)\right]} \\ &= \sqrt{\left[\frac{1}{2\pi}\left(-\cot(\alpha+\beta) - \mathrm{i}\right)\right]} \\ &= \sqrt{(-\mathrm{i})}\, a(\alpha+\beta), \end{aligned} \tag{18.164}$$

so that

$$a(\alpha)a(\beta)\frac{\sqrt{\pi}\,\mathrm{e}^{\mathrm{i}\pi/4}}{\sqrt{[b(\alpha)+b(\beta)]}} = a(\alpha+\beta). \tag{18.165}$$

For the case $[b(\alpha)+b(\beta)] < 0$,

$$a(\alpha)a(\beta)\frac{\sqrt{\pi}\,\mathrm{e}^{-\mathrm{i}\pi/4}}{\sqrt{-[b(\alpha)+b(\beta)]}} = \sqrt{(\mathrm{i})}\,\mathrm{e}^{-\mathrm{i}\pi/4}a(\alpha+\beta) = a(\alpha+\beta). \tag{18.166}$$

Inserting the results from Eqs. (18.163), (18.165), and (18.166) into Eq. (18.158) leads to

$$\begin{aligned} \int_{-\infty}^{\infty} K_\alpha(t,\lambda)K_\beta(\lambda,u)\mathrm{d}\lambda &= a(\alpha+\beta)\exp\Big[\mathrm{i}(t^2+u^2)b(\alpha+\beta) - \mathrm{i}tuc(\alpha+\beta)\Big] \\ &= K_{\alpha+\beta}(t,u), \end{aligned} \tag{18.167}$$

which is the desired result.

A geometric motivation for the introduction of the fractional Fourier transform can be given. Consider the temporal and frequency behavior of a signal to be displayed in the time–frequency plane as shown in Figure 18.14.

The construction of the signal in the frequency direction can be obtained by the operation $(\mathcal{F}s)(\omega)$. That is, the Fourier transform operation is equated with a $\pi/2$ counter-clockwise rotation in the time–frequency plane. Applying the Fourier

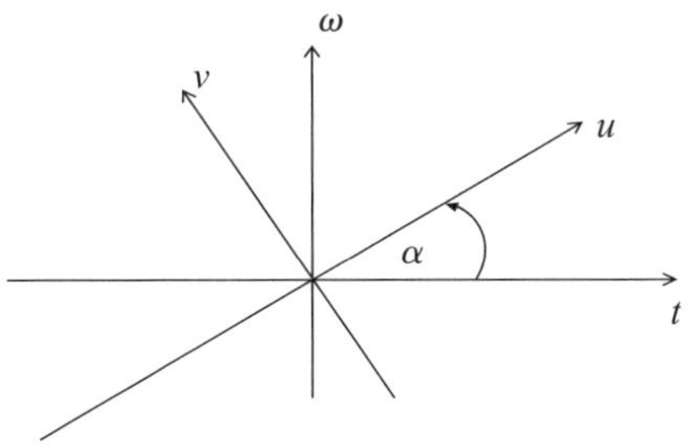

Figure 18.14. Time–frequency plane and the angle α characterizing the fractional Fourier transform.

operation a second time yields

$$\mathcal{F}\mathcal{F}s(t) = s(-t), \tag{18.168}$$

which can be viewed as two successive counter-clockwise rotations by $\pi/2$ to yield the signal $s(-t)$ on the negative portion of the time axis. The obvious question is as follows: starting from the time axis, what is the outcome of making a general rotation by α radians in the time–frequency plane? Let R_α denote a rotation operator whose action is given by

$$R_\alpha s = S_\alpha, \tag{18.169}$$

where S_α denotes the signal evaluated along the direction u in the time–frequency plane. The rotation operator is required to satisfy the following properties:

$$R_0 = I, \tag{18.170}$$

$$R_{2\pi} = I, \tag{18.171}$$

$$R_{\pi/2} = \mathcal{F}, \tag{18.172}$$

and

$$R_\alpha R_\beta = R_{\alpha+\beta}. \tag{18.173}$$

The first two of these requirements involving the identity operator and the third are obvious; the fourth condition is reasonable based on the expectation that a rotation by α radians followed by a rotation by β radians should be equivalent to a rotation by $\alpha + \beta$ radians. The function $S_\alpha(u)$ can be expressed as follows:

$$S_\alpha(u) = \int_{-\infty}^{\infty} s(t) K_\alpha(t, u) \mathrm{d}t. \tag{18.174}$$

The property given in Eq. (18.170) follows directly from the definition of $K_\alpha(t, u)$, as do the properties given in Eqs. (18.171) and (18.172). Equation (18.173) can be obtained as follows:

$$\begin{aligned} R_\alpha R_\beta s(u) &= \int_{-\infty}^{\infty} K_\alpha(t,u)\mathrm{d}t \int_{-\infty}^{\infty} s(v)K_\beta(v,t)\mathrm{d}v \\ &= \int_{-\infty}^{\infty} s(v)\mathrm{d}v \int_{-\infty}^{\infty} K_\alpha(t,u)K_\beta(v,t)\mathrm{d}t \\ &= \int_{-\infty}^{\infty} s(v)K_{\alpha+\beta}(v,u)\mathrm{d}v \\ &= R_{\alpha+\beta}s(u), \end{aligned} \tag{18.175}$$

where it has been assumed that the integration orders can be switched and Eq. (18.154) has been employed. It also follows that

$$\mathcal{F}_\alpha \mathcal{F}_\beta = \mathcal{F}_{\alpha+\beta}, \tag{18.176}$$

which is called the index additivity property of the FRFT. The definition given in Eq. (18.147) is therefore consistent with the desirable properties required for the FRFT. There is an alternative definition that makes more transparent the particular functional form for the kernel function.

The inverse operation, that is the recovery of $s(t)$ from $S_\alpha(u)$, is given by

$$s(t) = \int_{-\infty}^{\infty} S_\alpha(u)K_{-\alpha}(u,t)\mathrm{d}u. \tag{18.177}$$

Inserting Eq. (18.174) into the preceding result and making use of Eq. (18.154) leads to

$$\begin{aligned} \int_{-\infty}^{\infty} S_\alpha(u)K_{-\alpha}(u,t)\mathrm{d}u &= \int_{-\infty}^{\infty} K_{-\alpha}(u,t)\mathrm{d}u \int_{-\infty}^{\infty} s(t')K_\alpha(t',u)\mathrm{d}t' \\ &= \int_{-\infty}^{\infty} s(t')\mathrm{d}t' \int_{-\infty}^{\infty} K_{-\alpha}(t,u)K_\alpha(u,t')\mathrm{d}u \\ &= \int_{-\infty}^{\infty} s(t')K_0(t,t')\mathrm{d}t' \\ &= \int_{-\infty}^{\infty} s(t')\delta(t-t')\mathrm{d}t' \\ &= s(t), \end{aligned} \tag{18.178}$$

where it has been assumed that the integration order can be inverted.

An alternative approach to the FRFT is the following. The Hermite–Gaussian functions are defined by

$$h_n(x) = A_n H_n(\surd(2\pi)\, x)\mathrm{e}^{-\pi x^2}, \tag{18.179}$$

where

$$A_n = \frac{1}{\sqrt{(2^{(2n-1)/2}n!)}}, \tag{18.180}$$

and $H_n(x)$ are the Hermite polynomials. These functions form an orthonormal basis on the interval $(-\infty, \infty)$ as follows:

$$\int_{-\infty}^{\infty} h_n(x)h_m(x)\mathrm{d}x = \delta_{nm}, \tag{18.181}$$

and are eigenfunctions of the Fourier transform operator:

$$\mathcal{F}h_n(x) = \mathrm{e}^{-\mathrm{i}n\pi/2}h_n(x). \tag{18.182}$$

The reader is requested to verify both of these results. A second definition of the FRFT is introduced in the following way. The FRFT is defined to be a linear operator that satisfies the eigenvalue equation

$$\mathcal{F}_\alpha h_n(x) = \left(\mathrm{e}^{-\mathrm{i}n}\right)^\alpha h_n(x) = \mathrm{e}^{-\mathrm{i}n\alpha}h_n(x). \tag{18.183}$$

Let the signal $s(t)$ be expanded in the form

$$s(t) = \sum_{n=0}^{\infty} c_n h_n(t), \tag{18.184}$$

where the coefficients can be found in the obvious fashion:

$$c_n = \int_{-\infty}^{\infty} h_n(t)s(t)\mathrm{d}t. \tag{18.185}$$

The evaluation of $\mathcal{F}_\alpha s(t)$ can be carried out as follows:

$$\begin{aligned}
\mathcal{F}_\alpha s(t) &= \mathcal{F}_\alpha \sum_{n=0}^{\infty} c_n h_n(t) \\
&= \sum_{n=0}^{\infty} c_n \mathcal{F}_\alpha h_n(t) \\
&= \sum_{n=0}^{\infty} c_n \mathrm{e}^{-\mathrm{i}n\alpha} h_n(t) \\
&= \sum_{n=0}^{\infty} \mathrm{e}^{-\mathrm{i}n\alpha} h_n(t) \int_{-\infty}^{\infty} h_n(t')s(t')\mathrm{d}t' \\
&= \int_{-\infty}^{\infty} s(t')\mathrm{d}t' \sum_{n=0}^{\infty} \mathrm{e}^{-\mathrm{i}n\alpha} h_n(t)h_n(t'),
\end{aligned} \tag{18.186}$$

which simplifies on introducing

$$K_\alpha(t,t') = \sum_{n=0}^{\infty} \mathrm{e}^{-in\alpha} h_n(t) h_n(t'), \tag{18.187}$$

to yield

$$\mathcal{F}_\alpha s(t) = \int_{-\infty}^{\infty} s(t') K_\alpha(t,t') \mathrm{d}t'. \tag{18.188}$$

Equation (18.187) is the spectral expansion for the kernel function. A standard property of the Hermite–Gaussian functions is given by

$$\sum_{n=0}^{\infty} \mathrm{e}^{-in\alpha} h_n(t) h_n(t') = \sqrt{(1 - \mathrm{i}\cot\alpha)} \exp\left\{\mathrm{i}\pi\left[(t^2 + t'^2)\cot\alpha - 2tt'\csc\alpha\right]\right\}. \tag{18.189}$$

With a slight change of variables, $t \to t/\sqrt{(2\pi)}$, $t' \to t'/\sqrt{(2\pi)}$, and the introduction of a factor $1/\sqrt{(2\pi)}$, this is the definition of the kernel function given in Eq. (18.148). This second definition makes the reason for the particular choice of kernel function much more transparent.

The proof of Eq. (18.189) starts first with a proof of Mehler's formula:

$$\sum_{n=0}^{\infty} \frac{H_n(x) H_n(y) r^n}{2^n n!} = (1/\sqrt{(1-r^2)}) \exp\left\{\frac{2rxy - (x^2 + y^2) r^2}{1 - r^2}\right\}. \tag{18.190}$$

From the result

$$\mathrm{e}^{-x^2} = \frac{1}{\sqrt{\pi}} \int_{-\infty}^{\infty} \mathrm{e}^{-t^2 + 2\mathrm{i}xt}\, \mathrm{d}t, \tag{18.191}$$

it follows that

$$\frac{\mathrm{d}^n \mathrm{e}^{-x^2}}{\mathrm{d}x^n} = \frac{(2\mathrm{i})^n}{\sqrt{\pi}} \int_{-\infty}^{\infty} t^n \mathrm{e}^{-t^2 + 2\mathrm{i}xt}\, \mathrm{d}t, \tag{18.192}$$

and by employing Rodrigues' formula,

$$H_n(x) = (-1)^n \mathrm{e}^{x^2} \frac{\mathrm{d}^n \mathrm{e}^{-x^2}}{\mathrm{d}x^n}, \tag{18.193}$$

the following integral representation for the Hermite polynomials is obtained:

$$H_n(x) = \frac{(-2\mathrm{i})^n}{\sqrt{\pi}} \mathrm{e}^{x^2} \int_{-\infty}^{\infty} t^n \mathrm{e}^{-t^2 + 2\mathrm{i}xt}\, \mathrm{d}t. \tag{18.194}$$

Hence,

$$\sum_{n=0}^{\infty}\frac{H_n(x)H_n(y)r^n}{2^n n!} = \frac{e^{x^2+y^2}}{\pi}\sum_{n=0}^{\infty}\frac{(-2r)^n}{n!}\int_{-\infty}^{\infty} t^n e^{-t^2+2ixt}\,dt\int_{-\infty}^{\infty} s^n e^{-s^2+2iys}\,ds$$

$$= \frac{e^{x^2+y^2}}{\pi}\int_{-\infty}^{\infty} e^{-t^2+2ixt}\,dt\int_{-\infty}^{\infty} e^{-s^2+2iys}\,ds\sum_{n=0}^{\infty}\frac{(-2rst)^n}{n!}$$

$$= \frac{e^{x^2+y^2}}{\pi}\int_{-\infty}^{\infty} e^{-t^2+2ixt}\,dt\int_{-\infty}^{\infty} e^{-s^2-2(rt-iy)s}\,ds. \qquad (18.195)$$

For $\alpha > 0$, note that

$$\int_{-\infty}^{\infty} e^{-\alpha^2 s^2-2\beta s}\,ds = \frac{\sqrt{\pi}}{\alpha}e^{\beta^2/\alpha^2}, \qquad (18.196)$$

and therefore that Eq. (18.195) simplifies as follows:

$$\sum_{n=0}^{\infty}\frac{H_n(x)H_n(y)r^n}{2^n n!} = \frac{e^{x^2+y^2}}{\sqrt{\pi}}\int_{-\infty}^{\infty} e^{-t^2+2ixt}e^{(rt-iy)^2}\,dt$$

$$= \frac{e^{x^2}}{\sqrt{\pi}}\int_{-\infty}^{\infty}\exp\left\{-(1-r^2)t^2-2ti(ry-x)\right\}dt$$

$$= \frac{e^{x^2}}{\sqrt{(1-r^2)}}\exp\left\{\frac{-(ry-x)^2}{1-r^2}\right\}$$

$$= \frac{1}{\sqrt{(1-r^2)}}\exp\left\{\frac{2rxy-r^2(x^2+y^2)}{1-r^2}\right\}, \qquad (18.197)$$

which completes the proof of Eq. (18.190). Using the definition of the Hermite–Gaussian functions given in Eq. (18.179), employing Eq. (18.190), setting $r = e^{-i\alpha}$, replacing x by $\sqrt{(2\pi)}\,x$, y by $\sqrt{(2\pi)}\,y$, and then multiplying both sides of the equation by $\sqrt{2}\,e^{-\pi x^2-\pi y^2}$, it follows that

$$\sum_{n=0}^{\infty} e^{-in\alpha}h_n(x)h_n(y) = \sqrt{[2/(1-e^{-2i\alpha})]}\,e^{-\pi x^2-\pi y^2}$$

$$\times\exp\left[2\pi\left\{\frac{2e^{-i\alpha}xy-(x^2+y^2)e^{-2i\alpha}}{1-e^{-2i\alpha}}\right\}\right]. \qquad (18.198)$$

On noting

$$\frac{2}{1-e^{-2i\alpha}} = 1-i\cot\alpha, \qquad (18.199)$$

$$e^{-i\alpha}(1-i\cot\alpha) = -i\csc\alpha, \qquad (18.200)$$

and

$$1 + (1 - \mathrm{i}\cot\alpha)\mathrm{e}^{-2\mathrm{i}\alpha} = -\mathrm{i}\cot\alpha, \tag{18.201}$$

this leads to

$$\begin{aligned}\sum_{n=0}^{\infty} \mathrm{e}^{-\mathrm{i}n\alpha} h_n(x)h_n(y) &= \sqrt{(1 - \mathrm{i}\cot\alpha)} \exp\pi\{(x^2 + y^2)[-1 - (1 - \mathrm{i}\cot\alpha)\mathrm{e}^{-2\mathrm{i}\alpha}] \\ &\quad + 2xy\mathrm{e}^{-\mathrm{i}\alpha}(1 - \mathrm{i}\cot\alpha)\} \\ &= \sqrt{(1 - \mathrm{i}\cot\alpha)} \exp[\mathrm{i}\pi\{(x^2 + y^2)\cot\alpha - 2xy\csc\alpha\}],\end{aligned} \tag{18.202}$$

which completes the proof of Eq. (18.189).

A simpler proof of Eq. (18.154) can be given in the following manner. Making use of Eq. (18.187), it follows that

$$\begin{aligned}\int_{-\infty}^{\infty} K_\alpha(t,\lambda)K_\beta(\lambda,u)\mathrm{d}\lambda &= \int_{-\infty}^{\infty} \sum_{n=0}^{\infty} \mathrm{e}^{-\mathrm{i}n\alpha} h_n(t)h_n(\lambda) \sum_{m=0}^{\infty} \mathrm{e}^{-\mathrm{i}m\beta} h_{\mathrm{m}}(\lambda)h_{\mathrm{m}}(u)\mathrm{d}\lambda \\ &= \sum_{n=0}^{\infty} \mathrm{e}^{-\mathrm{i}n\alpha} h_n(t) \sum_{m=0}^{\infty} \mathrm{e}^{-\mathrm{i}m\beta} h_{\mathrm{m}}(u) \int_{-\infty}^{\infty} h_n(\lambda)h_{\mathrm{m}}(\lambda)\mathrm{d}\lambda \\ &= \sum_{n=0}^{\infty} \mathrm{e}^{-\mathrm{i}n\alpha} h_n(t) \sum_{m=0}^{\infty} \mathrm{e}^{-\mathrm{i}m\beta} h_{\mathrm{m}}(u)\delta_{nm} \\ &= \sum_{n=0}^{\infty} \mathrm{e}^{-\mathrm{i}n(\alpha+\beta)} h_n(t)h_n(u) \\ &= K_{\alpha+\beta}(t,u).\end{aligned} \tag{18.203}$$

A couple of key properties of the fractional Fourier transform are as follows. Let $X_\alpha(u)$ and $Y_\alpha(u)$ denote the fractional Fourier transforms of $x(t)$ and $y(t)$, respectively; then, the Parseval relation holds for the FRFT and takes the following form:

$$\int_{-\infty}^{\infty} x(t)y(t)^*\mathrm{d}t = \int_{-\infty}^{\infty} X_\alpha(u)Y_\alpha(u)^*\,\mathrm{d}u. \tag{18.204}$$

The particular case $y(t) = x(t)$ yields

$$\int_{-\infty}^{\infty} |x(t)|^2\,\mathrm{d}t = \int_{-\infty}^{\infty} |X_\alpha(u)|^2\,\mathrm{d}u. \tag{18.205}$$

If $x(t)$ is real, then

$$X_{-\alpha}(u) = X_\alpha^*(u). \tag{18.206}$$

In addition, the following two properties hold:

$$\mathcal{F}_\alpha x'(u) = \cos\alpha\, X'_\alpha(u) + \mathrm{i}\sin\alpha\, uX_\alpha(u) \tag{18.207}$$

and

$$\mathcal{F}_\alpha\{ux(u)\} = \cos\alpha\, uX_\alpha(u) + \mathrm{i}\sin\alpha\, X'_\alpha(u). \tag{18.208}$$

The reader is asked to provide the proofs of the preceding five relationships.

18.11 The fractional Hilbert transform: Zayed's definition

Zayed (1998) introduced a generalization of the Hilbert transform which might rightly be called a fractional Hilbert transform. His definition is as follows:

$$H_\alpha f(t) = \frac{\mathrm{e}^{-\mathrm{i}\cot\alpha\, t^2/2}}{\pi} P\int_{-\infty}^{\infty} \frac{f(x)\mathrm{e}^{\mathrm{i}\cot\alpha\, x^2/2}\,\mathrm{d}x}{t-x}. \tag{18.209}$$

The case $\alpha = (2n+1)\pi/2$, with $n \in \mathbb{Z}$, reduces to the standard Hilbert transform. A key result that is obtained from this definition is the following:

$$\mathcal{F}_\alpha H_\alpha s(u) = -\mathrm{i}\,\mathrm{sgn}\, u\, \mathcal{F}_\alpha s(u), \quad \text{for } 0 < \alpha < \pi. \tag{18.210}$$

The same result holds for $\pi < \alpha < 2\pi$, with the sign of the right-hand side of the equation reversed. The reader will note immediately that Eq. (18.210) has the same form as the important result:

$$\mathcal{F}Hf(x) = -\mathrm{i}\,\mathrm{sgn}\, x\, \mathcal{F}f(x). \tag{18.211}$$

The proof of Eq. (18.210) is straightforward and proceeds in the following manner. Let

$$g(t) = \frac{1}{\pi} P\int_{-\infty}^{\infty} \frac{f(x)\mathrm{e}^{\mathrm{i}\cot\alpha\, x^2/2}\,\mathrm{d}x}{t-x}, \tag{18.212}$$

so that

$$H_\alpha f(t) = \mathrm{e}^{-\mathrm{i}\cot\alpha\, t^2/2} g(t), \tag{18.213}$$

and hence

$$\begin{aligned}
\mathcal{F}_\alpha H_\alpha s(u) &= \int_{-\infty}^{\infty} K_\alpha(t,u)\mathrm{e}^{-\mathrm{i}\cot\alpha\, t^2/2} g(t)\mathrm{d}t \\
&= \int_{-\infty}^{\infty} K_\alpha(t,u)\mathrm{e}^{-\mathrm{i}\cot\alpha\, t^2/2}\,\mathrm{d}t \frac{1}{\pi} P\int_{-\infty}^{\infty} \frac{f(x)\mathrm{e}^{\mathrm{i}\cot\alpha\, x^2/2}\,\mathrm{d}x}{t-x} \\
&= -\int_{-\infty}^{\infty} f(x)\mathrm{e}^{\mathrm{i}\cot\alpha\, x^2/2}\,\mathrm{d}x \frac{1}{\pi} P\int_{-\infty}^{\infty} \frac{K_\alpha(t,u)\mathrm{e}^{-\mathrm{i}\cot\alpha\, t^2/2}\,\mathrm{d}t}{x-t} \\
&= -a(\alpha)\mathrm{e}^{ib(\alpha)u^2}\int_{-\infty}^{\infty} f(x)\mathrm{e}^{ib(\alpha)x^2}\,\mathrm{d}x \frac{1}{\pi} P\int_{-\infty}^{\infty} \frac{\mathrm{e}^{-\mathrm{i}\csc\alpha\, ut}}{x-t}\,\mathrm{d}t \\
&= -a(\alpha)\mathrm{e}^{ib(\alpha)u^2}\int_{-\infty}^{\infty} f(x)\mathrm{e}^{ib(\alpha)x^2}\,\mathrm{i}\,\mathrm{sgn}\,[u\csc\alpha]\mathrm{e}^{-\mathrm{i}\csc\alpha\, ux}\,\mathrm{d}x \\
&= -\mathrm{i}\,\mathrm{sgn}[u\csc\alpha]\int_{-\infty}^{\infty} f(x)a(\alpha)\mathrm{e}^{ib(\alpha)(x^2+u^2)-\mathrm{i}\csc\alpha\, ux}\,\mathrm{d}x \\
&= -\mathrm{i}\,\mathrm{sgn}\,[u\csc\alpha]\int_{-\infty}^{\infty} f(x)K_\alpha(x,u)\mathrm{d}x \\
&= -\mathrm{i}\,\mathrm{sgn}\,u\,\mathrm{sgn}(\csc\alpha)\mathcal{F}_\alpha f(u),
\end{aligned} \tag{18.214}$$

and Eq. (18.210) follows for $0 < \alpha < \pi$; for $\pi < \alpha < 2\pi$,

$$\mathcal{F}_\alpha H_\alpha f(u) = \mathrm{i}\,\mathrm{sgn}\,u\,\mathcal{F}_\alpha f(u). \tag{18.215}$$

The result $H(\mathrm{e}^{\mathrm{i}\beta x}) = -\mathrm{i}\,\mathrm{sgn}\,\beta\, H(\mathrm{e}^{\mathrm{i}\beta x})$ has been employed and the reader is asked to justify the interchange of integration order that has been made.

18.12 The fractional Hilbert transform: the Cusmariu definition

An alternative definition of the fractional Hilbert transform has been given by Cusmariu (2002), who defined the fractional Hilbert transform operator directly in terms of the normal Hilbert transform operator by the following relation:

$$H_\alpha = \cos\alpha + \sin\alpha\, H. \tag{18.216}$$

An advantage of this definition is that the forms of several of the standard properties of the Hilbert transform are preserved.

From Cusmariu's definition, the following particular cases are obtained directly from Eq. (18.216):

$$H_0 = I, \tag{18.217}$$

with I the unit operator and

$$H_{\pi/2} = H. \tag{18.218}$$

Equation (18.59) has a counterpart for Cusmariu's definition of the fractional Hilbert transform:

$$\mathcal{F} H_\alpha S(\omega) = m_{H_\alpha}(\omega) \mathcal{F} S(\omega), \tag{18.219}$$

where

$$m_{H_\alpha}(\omega) = \cos\alpha - \mathrm{i}\,\mathrm{sgn}\,\omega \sin\alpha. \tag{18.220}$$

The proof of Eq. (18.219) can be performed as follows:

$$\begin{aligned} \mathcal{F} H_\alpha S(\omega) &= \mathcal{F}\{\cos\alpha + \sin\alpha\, H\} S(\omega) \\ &= \cos\alpha\ \mathcal{F} S(\omega) + \sin\alpha\ \mathcal{F} H S(\omega) \\ &= \cos\alpha\ \mathcal{F} S(\omega) + \sin\alpha(-\mathrm{i}\,\mathrm{sgn}\,\omega) \mathcal{F} S(\omega) \\ &= m_{H_\alpha}(\omega) \mathcal{F} S(\omega). \end{aligned} \tag{18.221}$$

Cusmariu's definition leads to the following result:

$$H_\alpha H_\beta = H_\beta H_\alpha, \tag{18.222}$$

that is, a commutative condition holds. Also

$$H_\alpha H_\beta = H_{\alpha+\beta}, \tag{18.223}$$

so that the operator H_α satisfies a semigroup condition under composition. Equation (18.223) can be readily established on recalling $H^2 = -I$, so that

$$\begin{aligned} H_\alpha H_\beta &= (\cos\alpha + \sin\alpha\, H)(\cos\beta + \sin\beta\, H) \\ &= \cos\alpha\cos\beta - \sin\alpha\sin\beta + \{\sin\alpha\cos\beta + \cos\alpha\sin\beta\} H \\ &= \cos(\alpha+\beta) + \sin(\alpha+\beta) H \\ &= H_{\alpha+\beta}. \end{aligned} \tag{18.224}$$

A number of additional properties for H_α can be given. From Eq. (18.223) it follows, on setting $\beta = -\alpha$, that

$$H_\alpha H_{-\alpha} = I, \tag{18.225}$$

and applying H_α^{-1} to both sides of this result yields

$$H_{-\alpha} = H_\alpha^{-1}. \tag{18.226}$$

Using inner product notation, for two signals $s_1(t)$ and $s_2(t)$,

$$(H_\alpha s_1, s_2) = (s_1, H_{-\alpha} s_2), \tag{18.227}$$

which can be derived as follows:

$$\begin{aligned}(H_\alpha s_1, s_2) &= \cos\alpha (s_1, s_2) + \sin\alpha (Hs_1, s_2)\\ &= \cos\alpha (s_1, s_2) - \sin\alpha (s_1, Hs_2)\\ &= (s_1, H_{-\alpha} s_2),\end{aligned} \tag{18.228}$$

and Eq. (4.176) has been employed. An immediate consequence of Eqs. (18.227) and the semigroup property is

$$(H_{\alpha-\beta} s_1, s_2) = (H_\alpha s_1, H_\beta s_2). \tag{18.229}$$

From this result the special case for $\beta = \alpha$ is given by

$$(s_1, s_2) = (H_\alpha s_1, H_\alpha s_2), \tag{18.230}$$

which is the analog of Eq. (4.175). For $s_1(t) = s_2(t)$,

$$\int_{-\infty}^{\infty} s^2(t)\mathrm{d}t = \int_{-\infty}^{\infty} [H_\alpha s(t)]^2 \,\mathrm{d}t. \tag{18.231}$$

This result indicates that the action of the fractional Hilbert transform on the signal preserves the energy of the signal. This is a generalization of Eq. (18.34).

Some additional relationships that are straightforward to prove are as follows:

$$(H_\alpha s_1, s_2) + (s_1, H_\alpha s_2) = 2\cos\alpha (s_1, s_2), \tag{18.232}$$

$$(H_\alpha s, H_\beta s) = \cos(\alpha - \beta)(s, s), \tag{18.233}$$

and using Eq. (4.204) yields

$$\int_{-\infty}^{\infty} H_\alpha s(t)\mathrm{d}t = \cos\alpha \int_{-\infty}^{\infty} s(t)\mathrm{d}t. \tag{18.234}$$

The analytic signal $z(t)$ corresponding to $s(t)$ is an eigenfunction of H_α:

$$H_\alpha z(t) = \mathrm{e}^{-\mathrm{i}\alpha} z(t). \tag{18.235}$$

Let $s_\alpha(t) \equiv H_\alpha s(t)$, then the action of the derivative operator $\partial/\partial\alpha$ on $s_\alpha(t)$ is given by

$$\frac{\partial s_\alpha(t)}{\partial\alpha} = H_{(2\alpha+\pi)/2} s(t) = s_{(2\alpha+\pi)/2}(t), \tag{18.236}$$

so that

$$\frac{\partial H_\alpha}{\partial \alpha} = H_{(2\alpha+\pi)/2}. \tag{18.237}$$

A related result is

$$H\frac{\partial H_\alpha}{\partial \alpha} = -H_\alpha. \tag{18.238}$$

Hence it follows on using Eq. (4.198) that

$$\left(H_\alpha s, \frac{\partial}{\partial \alpha} H_\alpha s\right) = 0. \tag{18.239}$$

18.13 The discrete fractional Fourier transform

As a precursor to discussing the discrete fractional Hilbert transform, it is necessary to introduce first the discrete fractional Fourier transform (DFRFT). The discrete fractional Fourier transform can be defined by reference to Eqs. (18.187) and (18.188) in the following fashion. Let $\mathcal{F}_\alpha$ denote the DFRFT matrix whose elements are discussed by the following (Pei and Yeh, 1997; Pei, Yeh, and Tseng, 1999; Ozaktas, Zalevsky, and Kutay, 2001, p. 210):

$$(\mathcal{F}_\alpha)_{lm} = \begin{cases} \displaystyle\sum_{k=0}^{N-1} \mathrm{e}^{-ik\alpha\pi/2} \mathrm{h}_{kl}\mathrm{h}_{km}, & \text{for } N \text{ odd} \\ \mathrm{e}^{-\mathrm{i}N\alpha\pi/2}\mathrm{h}_{(N-1)l}\mathrm{h}_{(N-1)m} + \displaystyle\sum_{k=0}^{N-2} \mathrm{e}^{-ik\alpha\pi/2}\mathrm{h}_{kl}\mathrm{h}_{km}, & \text{for } N \text{ even,} \end{cases} \tag{18.240}$$

where h_{kl} is the lth element of the vector $\mathbf{h}_k$, and these functions play the role of the discrete analogs of the Hermite–Gaussian functions that are defined in Eq. (18.179).

A digression is made to discuss the basics concerning the discrete Hermite–Gaussian functions, following the approach of Candan, Kutay, and Ozaktas (2000). The idea is to set up a finite difference analog of the differential equation satisfied by the Hermite–Gaussian functions, which is then solved in matrix form. Given a constant λ, the Hermite–Gaussian functions $h_n(x)$ satisfy the differential equation

$$\frac{\mathrm{d}^2 f(t)}{\mathrm{d}t^2} - 4\pi^2 t^2 f(t) = \lambda f(t), \tag{18.241}$$

where $\lambda = -2\pi(2n+1)$. With the shorthand $\mathrm{D} = \mathrm{d}/\mathrm{d}t$ and the usual notation for the Fourier transform, the preceding equation can be cast in the form

$$\mathcal{S} f(t) = \lambda f(t), \tag{18.242}$$

where the operator S is introduced as follows:

$$S = \mathrm{D}^2 + \mathcal{F}\mathrm{D}^2\mathcal{F}^{-1}, \tag{18.243}$$

and the Fourier transform operator employed has a 2π factor in the exponent. The operator D^2 is approximated by a difference operator $h^{-2}\mathcal{D}^2$, defined by

$$h^{-2}\mathcal{D}^2 f(t) = \frac{f(t+h) - 2f(t) + f(t-h)}{h^2}, \tag{18.244}$$

where h represents a small increment in the evaluation point. From the Taylor series expansion of $f(t+h)$, the following formal connection can be written:

$$f(t+h) = \mathrm{e}^{h\mathrm{D}} f(t), \tag{18.245}$$

so that the operator in Eq. (18.244) can be written as

$$h^{-2}\mathcal{D}^2 = \frac{\mathrm{e}^{h\mathrm{D}} - 2 + \mathrm{e}^{-h\mathrm{D}}}{h^2} = \mathrm{D}^2 + O(h^2), \tag{18.246}$$

and hence $h^{-2}\mathcal{D}^2$ is an approximation to D^2. An operator $\mathcal{S}$ is defined by the replacement of D^2 by $h^{-2}\mathcal{D}^2$ in Eq. (18.243), so that

$$\begin{aligned} \mathcal{S} &= h^{-2}\mathcal{D}^2 + \mathcal{F}h^{-2}\mathcal{D}^2\mathcal{F}^{-1} \\ &= h^{-2}\mathcal{D}^2 + \mathcal{F}h^{-2}(\mathrm{e}^{h\mathrm{D}} - 2 + \mathrm{e}^{-h\mathrm{D}})\mathcal{F}^{-1} \\ &= h^{-2}\mathcal{D}^2 + 2h^{-2}\{\cos(2\pi ht) - 1\}, \end{aligned} \tag{18.247}$$

where $\mathcal{F}\mathrm{e}^{h\mathrm{D}}\mathcal{F}^{-1} = \mathrm{e}^{2\pi iht}$ has been employed. Using the preceding result, Eq. (18.243) can be expressed in the following form:

$$f(t+h) + f(t-h) + 2\{\cos(2\pi ht) - 2\}f(t) = h^2\lambda f(t). \tag{18.248}$$

Discretization of the time variable by introducing $t = nh$, with $h = 1/\sqrt{N}$, where N is the number of discrete points, and setting $f[n] = f(nh)/h^2$, leads to

$$f[n+1] + f[n-1] + c[n]f[n] = \lambda N^{-1} f[n], \tag{18.249}$$

where

$$c[n] = 2\cos(2\pi n/N) - 4. \tag{18.250}$$

Equation (18.249) can be written in the matrix form

$$\mathbf{Sf} = \lambda\mathbf{f}, \tag{18.251}$$

where the factor N^{-1} has been incorporated into λ. If the DFT matrix is denoted by $\mathbf{F}$, then it can be shown that the matrices $\mathbf{S}$ and $\mathbf{F}$ commute, and it follows that $\mathbf{S}$ and $\mathbf{F}$ share a common set of eigenvectors.

The matrix $\mathbf{S}$ can be put into the following form:

$$\mathbf{PSP}^{-1} = \begin{pmatrix} \mathbf{E} & \mathbf{0} \\ \mathbf{0} & \mathbf{O} \end{pmatrix}, \tag{18.252}$$

where $\mathbf{E}$ is a tridiagonal matrix of dimension $\lfloor (N/2)+1 \rfloor$ and $\mathbf{O}$ is a tridiagonal matrix of dimension $\lfloor (N-1)/2 \rfloor$. The matrix $\mathbf{P}$ is constructed such that the vector $\mathbf{g}$ calculated from $\mathbf{g} = \mathbf{Pf}$ has the first $\lfloor (N/2)+1 \rfloor$ components constructed from the even combinations $f[k] + f[-k]$, and the following $\lfloor (N/2)+1 \rfloor$ components constructed from the odd combinations $f[k] - f[-k]$, with the first component of $\mathbf{g}$ being taken as $\sqrt{(2)}\, f[0]$. The periodic condition

$$f[k+N] = f[k] \tag{18.253}$$

is employed to convert $f[-k]$ to $f[N-k]$. The eigenvectors of $\mathbf{E}$ are right-padded with zeros to form the vectors $\mathbf{e}_k$ with $k = 0$ to $\lfloor (N/2) \rfloor$, and the eigenvectors of $\mathbf{O}$ are left-padded with zeros to form the vectors $\mathbf{o}_k$ with $k = 1$ to $\lfloor (N-1)/2 \rfloor$. The vectors $\mathbf{h}_k$ are formed in the following manner:

$$\mathbf{h}_{2k} = \mathbf{Pe}_k, \tag{18.254}$$

$$\mathbf{h}_{2k+1} = \mathbf{Po}_k, \tag{18.255}$$

and are referred to as the discrete Hermite–Gaussian functions.

As an example of the preceding scheme, set $N = 9$; then the matrix $\mathbf{S}$ takes the following form:

$$\mathbf{S} = \begin{pmatrix}
-2 & 1 & 0 & 0 & 0 & 0 & 0 & 0 & 1 \\
1 & 2\cos\left(\frac{2\pi}{9}\right) - 4 & 1 & 0 & 0 & 0 & 0 & 0 & 0 \\
0 & 1 & 2\cos\left(\frac{4\pi}{9}\right) - 4 & 1 & 0 & 0 & 0 & 0 & 0 \\
0 & 0 & 1 & -5 & 1 & 0 & 0 & 0 & 0 \\
0 & 0 & 0 & 1 & 2\cos\left(\frac{8\pi}{9}\right) - 4 & 1 & 0 & 0 & 0 \\
0 & 0 & 0 & 0 & 1 & 2\cos\left(\frac{10\pi}{9}\right) - 4 & 1 & 0 & 0 \\
0 & 0 & 0 & 0 & 0 & 1 & -5 & 1 & 0 \\
0 & 0 & 0 & 0 & 0 & 0 & 1 & 2\cos\left(\frac{14\pi}{9}\right) - 4 & 1 \\
1 & 0 & 0 & 0 & 0 & 0 & 0 & 1 & 2\cos\left(\frac{16\pi}{9}\right) - 4
\end{pmatrix}, \tag{18.256}$$

and the corresponding **P** matrix is given by

$$\mathbf{P} = \frac{1}{\sqrt{2}} \begin{pmatrix} \sqrt{2} & 0 & 0 & 0 & 0 & 0 & 0 & 0 & 0 \\ 0 & 1 & 0 & 0 & 0 & 0 & 0 & 0 & 1 \\ 0 & 0 & 1 & 0 & 0 & 0 & 0 & 1 & 0 \\ 0 & 0 & 0 & 1 & 0 & 0 & 1 & 0 & 0 \\ 0 & 0 & 0 & 0 & 1 & 1 & 0 & 0 & 0 \\ 0 & 0 & 0 & 0 & 1 & -1 & 0 & 0 & 0 \\ 0 & 0 & 0 & 1 & 0 & 0 & -1 & 0 & 0 \\ 0 & 0 & 1 & 0 & 0 & 0 & 0 & -1 & 0 \\ 0 & 1 & 0 & 0 & 0 & 0 & 0 & 0 & -1 \end{pmatrix}. \tag{18.257}$$

The matrix $\mathbf{PSP}^{-1}$ takes the following form:

$$\mathbf{PSP}^{-1} = \begin{pmatrix} -2 & \sqrt{2} & 0 & 0 & 0 & 0 & 0 & 0 & 0 \\ \sqrt{2} & 2\cos\left(\frac{2\pi}{9}\right) - 4 & 1 & 0 & 0 & 0 & 0 & 0 & 0 \\ 0 & 1 & 2\cos\left(\frac{4\pi}{9}\right) - 4 & 1 & 0 & 0 & 0 & 0 & 0 \\ 0 & 0 & 1 & -5 & 1 & 0 & 0 & 0 & 0 \\ 0 & 0 & 0 & 1 & 2\cos\left(\frac{8\pi}{9}\right) - 3 & 0 & 0 & 0 & 0 \\ 0 & 0 & 0 & 0 & 0 & 2\cos\left(\frac{8\pi}{9}\right) - 5 & 1 & 0 & 0 \\ 0 & 0 & 0 & 0 & 0 & 1 & -5 & 1 & 0 \\ 0 & 0 & 0 & 0 & 0 & 0 & 1 & 2\cos\left(\frac{4\pi}{9}\right) - 4 & 1 \\ 0 & 0 & 0 & 0 & 0 & 0 & 0 & 1 & 2\cos\left(\frac{2\pi}{9}\right) - 4 \end{pmatrix}, \tag{18.258}$$

from which the matrices and **E** and **O** follow:

$$\mathbf{E} = \begin{pmatrix} -2 & \sqrt{2} & 0 & 0 & 0 \\ \sqrt{2} & 2\cos\left(\frac{2\pi}{9}\right) - 4 & 1 & 0 & 0 \\ 0 & 1 & 2\cos\left(\frac{4\pi}{9}\right) - 4 & 1 & 0 \\ 0 & 0 & 1 & -5 & 1 \\ 0 & 0 & 0 & 1 & 2\cos\left(\frac{8\pi}{9}\right) - 3 \end{pmatrix} \tag{18.259}$$

and

$$\mathbf{O} = \begin{pmatrix} 2\cos\left(\frac{8\pi}{9}\right) - 5 & 1 & 0 & 0 \\ 1 & -5 & 1 & 0 \\ 0 & 1 & 2\cos\left(\frac{4\pi}{9}\right) - 4 & 1 \\ 0 & 0 & 1 & 2\cos\left(\frac{2\pi}{9}\right) - 4 \end{pmatrix}. \tag{18.260}$$

The eigenvectors of $\mathbf{E}$ are in zero-padded form:

$$\left.\begin{aligned} \mathbf{e}_0 &= \{0.697\,551, 0.571\,624, 0.242\,29, 0.058\,724\,8, 0.013\,846\,8, 0, 0, 0, 0\} \\ \mathbf{e}_1 &= \{-0.532\,553, 0.268\,717, 0.687\,125, 0.376\,572, 0.173\,872, 0, 0, 0, 0\} \\ \mathbf{e}_2 &= \{-0.422\,972, 0.542\,044, -0.130\,564, -0.521\,202, -0.488\,451, 0, 0, 0, 0\} \\ \mathbf{e}_3 &= \{0.222\,982, -0.415\,262, 0.584\,029, -0.157\,672, -0.641\,783, 0, 0, 0, 0\} \\ \mathbf{e}_4 &= \{0.0344\,177, -0.102\,265, 0.333\,193, -0.747\,15, 0.564\,895, 0, 0, 0, 0\} \end{aligned}\right\}, \tag{18.261}$$

and the eigenvectors of $\mathbf{O}$ in zero-padded form are given by

$$\left.\begin{aligned} \mathbf{o}_1 &= \{0, 0, 0, 0, 0, 0.035\,667\,5, 0.180\,932, 0.542\,108, 0.819\,823\} \\ \mathbf{o}_2 &= \{0, 0, 0, 0, 0, 0.171\,478, 0.560\,888, 0.609\,01, -0.533\,954\} \\ \mathbf{o}_3 &= \{0, 0, 0, 0, 0, 0.412\,866, 0.683\,567, -0.565\,796, 0.205\,31\} \\ \mathbf{o}_4 &= \{0, 0, 0, 0, 0, -0.893\,792, 0.430\,587, -0.122\,882, 0.025\,112\,6\} \end{aligned}\right\}. \tag{18.262}$$

The eigenvectors $\mathbf{h}_k$ calculated from Eqs. (18.254) and (18.255) are in ordered form:

$$\left.\begin{aligned} \mathbf{h}_0 &= \{0.697\,551, 0.474\,91, 0.171\,325, 0.041\,524\,7, 0.009\,791\,18, 0.009\,791\,18, \\ &\quad 0.041\,524\,7, 0.171\,325, 0.474\,91\} \\ \mathbf{h}_1 &= \{0, -0.579\,703, -0.383\,328, -0.127\,938, -0.025\,220\,8, 0.025\,220\,8, \\ &\quad 0.127\,938, 0.383\,328, 0.579\,703\} \\ \mathbf{h}_2 &= \{-0.532\,553, 0.190\,012, 0.485\,871, 0.266\,276, 0.122\,946, 0.122\,946, \\ &\quad 0.266\,276, 0.485\,871, 0.190\,012\} \\ \mathbf{h}_3 &= \{0, -0.377\,562, 0.430\,635, 0.396\,608, 0.121\,253, -0.121\,253, -0.396\,608, \\ &\quad -0.430\,635, 0.377\,562\} \\ \mathbf{h}_4 &= \{-0.422\,972, 0.383\,283, -0.092\,322\,5, -0.368\,546, -0.345\,387, \\ &\quad -0.345\,387, -0.368\,546, -0.092\,322\,5, 0.383\,283\} \\ \mathbf{h}_5 &= \{0.222\,982, -0.293\,635, 0.412\,971, -0.111\,491, -0.453\,809, -0.453\,809, \\ &\quad -0.111\,491, 0.412\,971, -0.293\,635\} \\ \mathbf{h}_6 &= \{0, 0.145\,176, -0.400\,078, 0.483\,355, 0.291\,941, -0.291\,941, -0.483\,355, \\ &\quad 0.400\,078, -0.145\,176\} \\ \mathbf{h}_7 &= \{0.034\,4177, -0.072\,312, 0.235\,603, -0.528\,315, 0.399\,441, 0.399\,441, \\ &\quad -0.528\,315, 0.235\,603, -0.072\,312\} \\ \mathbf{h}_8 &= \{0, -0.017\,757\,3, 0.086\,890\,6, -0.304\,471, 0.632\,007, -0.632\,007, \\ &\quad 0.304\,471, -0.086\,890\,6, 0.017\,757\,3\} \end{aligned}\right\}. \tag{18.263}$$

The eigenvectors are normalized to unity and are orthogonal.

18.14 The discrete fractional Hilbert transform

The work of the previous section is now exploited to discuss the discrete fractional Hilbert transform (DFHT). Recall that the transfer function for the discrete Hilbert transform is defined by

$$\mathsf{H}(\omega) = \begin{cases} -\mathrm{i}, & 0 < \omega < \pi \\ 0, & \omega = 0, \omega = \pi \\ \mathrm{i}, & -\pi < \omega < 0. \end{cases} \tag{18.264}$$

To review, the discrete Hilbert transform can be evaluated in the following manner. For a discrete signal $x[k]$, let $X[n]$ denote the corresponding discrete Fourier transform. Now multiply $X[n]$ by the mask M_1, where M_1 is defined by

$$M_1 = \begin{cases} [0, \underbrace{-\mathrm{i}, -\mathrm{i}, \ldots, -\mathrm{i}}_{(N/2)-1}, 0, \underbrace{\mathrm{i}, \mathrm{i}, \ldots, \mathrm{i}}_{(N/2)-1}], & N \text{ is even} \\ [0, \underbrace{-\mathrm{i}, -\mathrm{i}, \ldots, -\mathrm{i}}_{(N-1)/2}, \underbrace{\mathrm{i}, \mathrm{i}, \ldots, \mathrm{i}}_{(N-1)/2)}], & N \text{ is odd.} \end{cases} \tag{18.265}$$

The discrete Hilbert transform is computed from $X[n]M_1[n]$ by taking the inverse discrete Fourier transform, so that

$$\mathcal{H}_{\mathrm{D}}x[n] = \mathrm{IDFT}[\mathrm{X}[n]M_1[n]]. \tag{18.266}$$

The opposite sign convention can also be found in the literature for Eqs. (18.264) and (18.265).

Pei and Yeh (1998, 2000) have given the following scheme for determining the discrete fractional Hilbert transform. First compute the DFRFT of the signal $x[k]$ as

$$X_q[n] = \mathcal{F}_q x[k]. \tag{18.267}$$

Let the mask function M_p be defined by

$$M_p = \begin{cases} [\cos\alpha, \underbrace{\mathrm{e}^{-\mathrm{i}\alpha}, \mathrm{e}^{-\mathrm{i}\alpha}, \ldots, \mathrm{e}^{-\mathrm{i}\alpha}}_{(N/2)-1}, \cos\alpha, \underbrace{\mathrm{e}^{\mathrm{i}\alpha}, \mathrm{e}^{\mathrm{i}\alpha}, \ldots, \mathrm{e}^{\mathrm{i}\alpha}}_{(N/2)-1}], & N \text{ is even} \\ [\cos\alpha, \underbrace{\mathrm{e}^{-\mathrm{i}\alpha}, \mathrm{e}^{-\mathrm{i}\alpha}, \ldots, \mathrm{e}^{-\mathrm{i}\alpha}}_{(N-1)/2}, \underbrace{\mathrm{e}^{\mathrm{i}\alpha}, \mathrm{e}^{\mathrm{i}\alpha}, \ldots, \mathrm{e}^{\mathrm{i}\alpha}}_{(N-1)/2)}], & N \text{ is odd,} \end{cases} \tag{18.268}$$

where $\alpha = p\pi/2$. Now multiply $X_q[n]$ by the mask M_p and then take the inverse fractional Fourier transform to form

$$\mathcal{H}_{\mathrm{D}_{pq}} x[n] = \mathcal{F}_{-q}\{X_q[n]M_p[n]\}, \tag{18.269}$$

and $\mathcal{H}_{\mathrm{D}_{pq}} x[n]$ is referred to as the discrete fractional Hilbert transform. When $p = 1$, Eq. (18.268) reduces to Eq. (18.265). The choice $p = 0$ gives the mask function $M_0[1, 1, \ldots, 1]$, and the input signal is unaltered by this mask. When $p = 2$, the mask function is $M_2[-1, -1, \ldots, -1]$, and the output signal is unaltered except for a change in sign. For the case $q = 1$, the DFRFT $\mathcal{F}_q$ reduces to the DFT. So the scheme just outlined for the DFRHT reduces to the conventional discrete Hilbert transform for the appropriate choice of p and q.

18.15 The fractional analytic signal

Four different suggestions for the definition of the fractional analytic signal are examined, and each is labeled with an appropriate subscript. Zayed (1998) has defined the fractional analytical signal as follows:

$$z_{1_\alpha}(t) = s(t) + \mathrm{i}H_\alpha s(t), \tag{18.270}$$

where H_α is defined in Eq. (18.209). If $0 < \alpha < \pi$, it follows from this definition that $Z_{1_\alpha}(\omega)$ vanishes for $\omega < 0$. To see this, take the FRFT of order α of Eq. (18.270) and employ Eq. (18.210), then

$$\begin{aligned} Z_{1_\alpha}(\omega) &= S_\alpha(\omega) + \mathrm{i}\mathcal{F}_\alpha H_\alpha s(\omega) \\ &= S_\alpha(\omega) - \mathrm{i}^2 \operatorname{sgn} \omega \; \mathcal{F}_\alpha s(\omega) \\ &= S_\alpha(\omega)\{1 + \operatorname{sgn} \omega\}, \end{aligned} \tag{18.271}$$

which is the analog of the result for the non-fractional case.

The fractional Hilbert transform

$$H_\alpha s(t) = s(t)\cos\alpha + \sin\alpha \, Hs(t), \tag{18.272}$$

can serve as a basis for formulating a definition of a fractional analytic signal. Cusmariu (2002) has suggested three different possible definitions. The first is also given by Eq. (18.270), with Eq. (18.272) being employed in place of Zayed's definition for H_α, so that

$$z_{2_\alpha}(t) = s(t) + \mathrm{i}H_\alpha s(t). \tag{18.273}$$

It is left as an exercise for the reader to determine whether $Z_{2_\alpha}(\omega)$ generated from Eq. (18.273) satisfies a result similar to Eq. (18.271).

Cusmariu's second suggestion was to define the fractional analytic signal by the following result:

$$z_{3_\alpha}(t) = H_\alpha z(t). \tag{18.274}$$

This choice has some geometric appeal. By employing the eigenfunction property given in Eq. (18.235), $(H_\alpha z)(t)$ can be quickly cast in terms of $z(t)$. It follows immediately that

$$\left|z_{3_\alpha}(t)\right|^2 = |z(t)|^2 . \tag{18.275}$$

It also follows, by taking the Fourier transform of Eq. (18.274), that

$$\left|Z_{3_\alpha}(\omega)\right| = |Z(\omega)| . \tag{18.276}$$

The third suggestion of Cusmariu involves the definition

$$z_{4_\alpha}(t) = \cos\alpha\, s(t) + \mathrm{i}\sin\,\alpha\, Hs(t). \tag{18.277}$$

Which of these definitions of the fractional analytic signal will turn out to be the most useful in practical applications is yet to be determined.

18.16 Empirical mode decomposition: the Hilbert–Huang transform

In this section the focus is time series. A time series is a collection of data for a fluctuating variable sampled sequentially at differing times. The notation $X(t)$ is used to denote the value of a time series at a particular time t. Interest will center on continuous time series, where the sampling takes place continuously in time.

A new and promising technique that has been developed to deal with the analysis of time series data is explored. The method has been recently labeled the Hilbert–Huang transform, named for Norbert Huang, the principal architect of the approach. One of the traditional techniques employed to analyze time series data is the Fourier transform. This approach has limitations when the data is nonlinear and non-stationary. A stationary time series has statistical properties such as a mean and variance that are independent of time. An individual random variable x_t is characterized by a probability distribution function,

$$P_{x_{t_n}}(x_{t_0}) = P[x_{t_n} \leq x_{t_0}], \tag{18.278}$$

where $P[x_{t_n} \leq x_{t_0}]$ is the probability associated with the set of all outcomes such that $x_{t_n} \leq x_{t_0}$, where x_{t_0} is a particular value of x_{t_n}. For two random variables a joint distribution function is defined by

$$P_{x_{tm},x_{t_n}}(x_{t_1},x_{t_2}) = P[x_{t_m} \leq x_{t_1}, x_{t_n} \leq x_{t_2}]. \tag{18.279}$$

Formally, a stationary time series is defined by the requirement that the joint probability distribution of $\{X_{t_1}, X_{t_2}, \ldots, X_{t_n}\}$, is equivalent to the joint probability distribution of $\{X_{t_1+\tau}, X_{t_2+\tau}, \ldots, X_{t_n+\tau}\}$. That is, the joint probability distribution is independent of a shift of the time origin by τ.

The starting point in the Hilbert–Huang approach is called empirical mode decomposition (EMD). The essential idea in EMD is to decompose any time-dependent data series into a series of *intrinsic mode functions* (IMFs). Each IMF represents a particular intrinsic oscillation of the system. N. E. Huang *et al.* (1998) defined an IMF by the following two requirements: (i) the number of extrema and zero crossings must be equal, or differ by at most one; (ii) the mean value of the envelopes formed from the extrema should be zero at any time.

Here is the algorithm used to carry out the EMD procedure.

(1) Locate each maximum and construct the envelope curve connecting the maxima using a cubic spline interpolation.
(2) Locate each minimum and construct the envelope curve connecting the minima using a cubic spline interpolation.
(3) Construct the mean of the two envelopes – call it $m_{11}(t)$. The first subscript index refers to the particular IMF under construction and the second index indicates the particular cycle under consideration.
(4) Construct the difference between the time series and $m_{11}(t)$, so that

$$h_{11}(t) = h_{10}(t) - m_{11}(t), \tag{18.280}$$

where $h_{10}(t) \equiv X(t)$. The function $h_{11}(t)$ should be close to an IMF, but, due to fitting imperfections, this function requires additional refinement.
(5) Treat $h_{11}(t)$ as the new input data and repeat steps (1), (2), and (3), calling the new mean $m_{12}(t)$ and constructing the new difference function

$$h_{12}(t) = h_{11}(t) - m_{12}(t). \tag{18.281}$$

(6) Repeat the procedure k times so that

$$h_{1k}(t) = h_{1(k-1)}(t) - m_{1k}(t), \tag{18.282}$$

with appropriate stopping criteria applied.
(7) The final $h_{1k}(t)$ is subtracted from the original $X(t)$ to yield $h_{20}(t)$, and the process is repeated starting with the data set $h_{20}(t)$ and the IMF $h_{2k'}(t)$ determined in a similar fashion.
(8) The following assignment is made:

$$c_i(t) = h_{ik}(t), \tag{18.283}$$

where the k subscript counts the number of steps required for the stopping criteria to terminate the cycling process. In general, the value of k employed will be different for each $c_i(t)$. The outcome is

$$X(t) = \sum_{i=1}^{n} c_i(t) + r_n(t), \tag{18.284}$$

where $r_n(t)$ represents a residue term.

The first part of the preceding cycle is illustrated in Figure 18.15 for the discrete time sequence constructed from the function

$$X(t) = \text{floor}[2 - \sin \pi t] \sin 2\pi t, \tag{18.285}$$

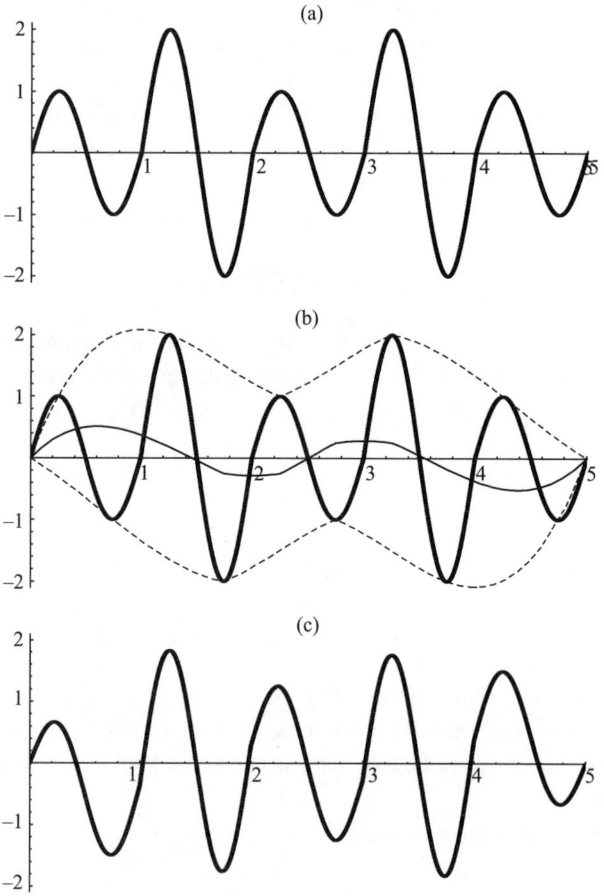

Figure 18.15. (a) Initial time series $X(t)$. Part (b) Envelopes of the maxima and minima in place and the mean $m_{11}(t)$ of the envelopes. (c) First estimation of the intrinsic mode function given by $h_{11}(t) = X(t) - m_{11}(t)$.

where floor[x] is defined to be the greatest integer $\leq x$. The function is sampled at time increments of 10^{-3} s. The second stage of the cycle is shown in Figure 18.16. Continuing the process leads to the output shown in Figure 18.17. For this simple example the $r_n(t)$ in Eq. (18.284) is effectively zero and can be ignored. The deconvolution of the original $X(t)$ into a sum of three intrinsic mode functions is shown to be excellent based on the difference plot in Figure 18.17. This example is relatively free of some of the complicating issues that can arise.

The process of isolating the different functions $c_i(t)$ is called *sifting*. An advantage of the method is that it is adaptive, being tied directly to the form of the data and the changes induced by the sifting process. A couple of different ideas have been advanced (N. E. Huang *et al.*, 1998, 1999, 2003a) to terminate a cycle. The process

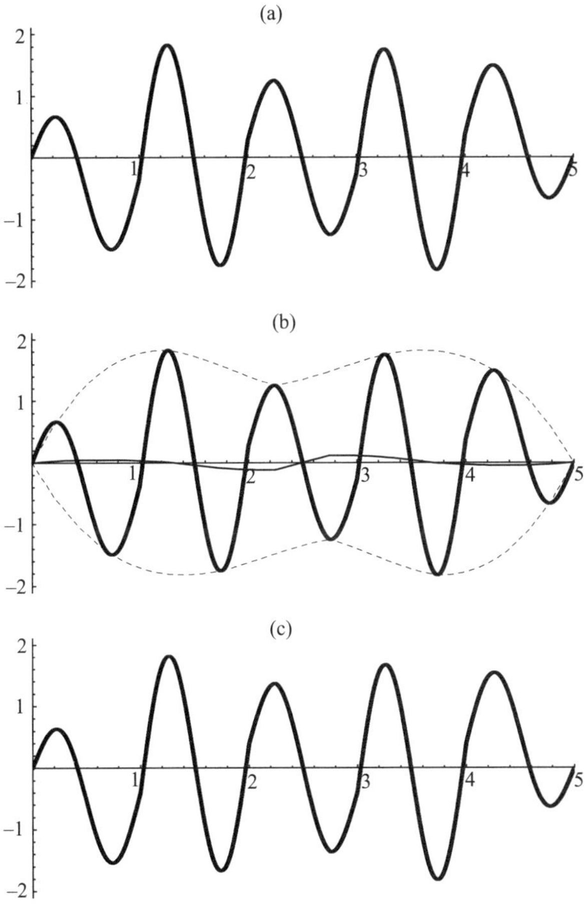

Figure 18.16. (a) Time series $h_{11}(t)$ from the previous cycle. (b) Envelopes of the maxima and minima in place and the mean $m_{12}(t)$ of the envelopes. (c) Second estimation of the intrinsic mode function given by $h_{12}(t) = h_{11}(t) - m_{12}(t)$.

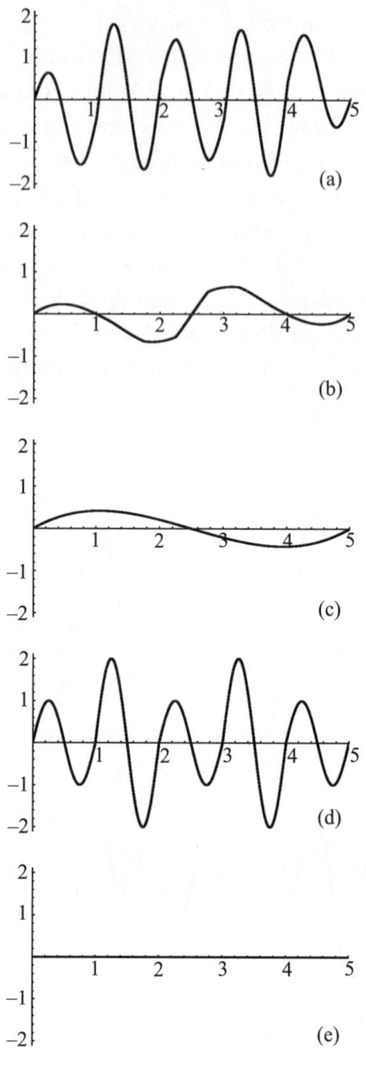

Figure 18.17. (a)–(c) Intrinsic mode functions $c_k(t)$, $k = 1, 2, 3$, constructed for the original data series. (d) $\sum_{k=1}^{3} c_k(t)$. (e) The difference $X(t) - \sum_{k=1}^{3} c_k(t)$.

is terminated when the function $h_{jk}(t)$ satisfies the two conditions stated for an IMF. A second approach is to define an S-value such that the sifting process gives the same number of extrema and zero crossings for S successive sifting cycles. With this approach, different IMF sets can be generated according to the choices of S_i used for each $c_i(t)$ determination. Values of S_i in the range of about 3 to 5 appear to be effective. A third idea is to employ a curvature criterion (N. E. Huang *et al.*, 1999, 2003a). Another way in which the sifting process can be terminated is by assigning some predetermined cut-off value to the weighted sum-of-squared deviations, denoted

D_w, and given by

$$D_w = \sum_{t=0}^{T} \frac{[h_{i(k-1)}(t) - h_{ik}(t)]^2}{h_{i(k-1)}^2(t)}, \tag{18.286}$$

where T is the total time interval under consideration and zero crossing times are excluded from the sum.

The overall process is terminated when the residue $r_n(t)$ is either a constant, a monotonic function, or a function with a single extremum. If the initial data set has a particular mean trend, then the residue term should reflect that trend.

A second discrete time example is constructed based on the choice

$$X(t) = \cos\left(\frac{3}{4}\pi t\right) + \cos\left(\frac{5}{4}\pi t\right) + \cos\left(\frac{7}{4}\pi t\right) + \cos\left(\frac{9}{4}\pi t\right) - 3, \tag{18.287}$$

with the function displayed in Figure 18.18. The function is sampled at time increments of 10^{-3} s. The remaining two IMFs are displayed, along with the residual contribution, in Figure 18.19.

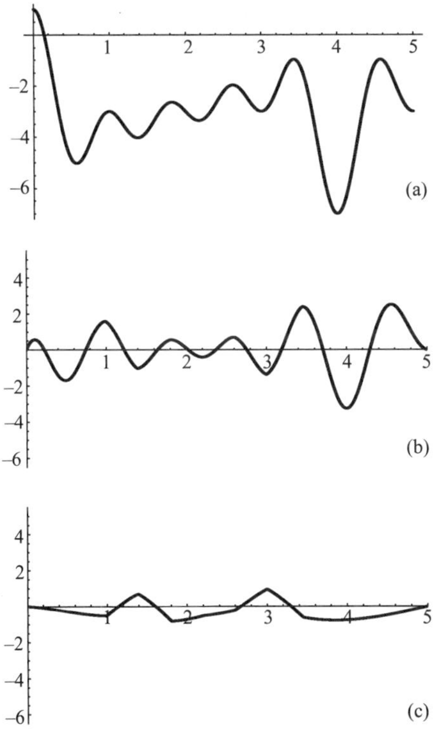

Figure 18.18. (a) Initial time series $X(t)$ based on Eq. (18.287). The first two intrinsic mode functions are shown in parts (b) and (c), respectively.

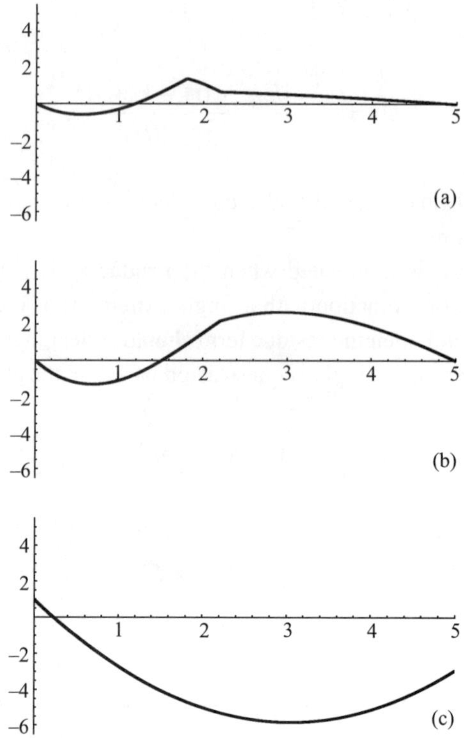

Figure 18.19. (a), (b) Third and fourth IMFs; (c) resulting residual term $r_4(t)$.

For readers having access to *Mathematica* or related software, a simple program can be readily constructed to carry out the type of calculations just discussed. This is left as an exercise for the interested reader.

Equation (18.284) gives the temporal behavior of the time series. To convert to a frequency–time description, the residue term is ignored and the Hilbert transform of the IMF set is taken, so that

$$\begin{aligned} Y(t) \equiv HX(t) &= \frac{1}{\pi} P \int_{-\infty}^{\infty} \frac{X(t')\mathrm{d}t'}{t-t'} \\ &= \frac{1}{\pi} \sum_{i=1}^{n} P \int_{-\infty}^{\infty} \frac{c_i(t')\mathrm{d}t'}{t-t'}. \end{aligned} \tag{18.288}$$

In the Fourier transform approach, taking the Fourier transform of a time series converts from the temporal domain to the frequency domain. The Hilbert transform of a time signal is another time-dependent signal. The frequency connection is made via the corresponding analytic signal, which is given by

$$Z(t) = X(t) + \mathrm{i}Y(t) = a(t)\mathrm{e}^{\mathrm{i}\theta(t)}. \tag{18.289}$$

The factors $a(t)$ and $\theta(t)$ are given by

$$a(t) = \surd[X^2(t) + Y^2(t)], \tag{18.290}$$

and

$$\theta(t) = \tan^{-1}\left(\frac{Y(t)}{X(t)}\right), \tag{18.291}$$

where $a(t)$ is the instantaneous amplitude of the analytic signal and $\theta(t)$ is the instantaneous phase. The instantaneous frequency is introduced by the following expression:

$$\omega(t) = \frac{\mathrm{d}\theta(t)}{\mathrm{d}t}. \tag{18.292}$$

The notion of instantaneous frequency has its origins in the work of Ville (1948). The instantaneous phase can be derived from Eq. (18.292) as follows:

$$\theta(t) = \theta(0) + \int_0^t \omega(t')\mathrm{d}t'. \tag{18.293}$$

The concept and definition of the instantaneous frequency has proved to be a rather contentious one.

The Hilbert transformed IMFs serve as a basis for the original time series, so that the original time series is given by

$$X(t) = \mathrm{Re}\left\{\sum_{k=1}^{n} a_k(t) \exp\left(\mathrm{i}\int \omega_k(t')\,\mathrm{d}t'\right)\right\}, \tag{18.294}$$

with Re denoting the real part. By comparison, a Fourier analysis would lead to a result of the form

$$X(t) = \mathrm{Re}\left\{\sum_{k=1}^{\infty} \alpha_k \mathrm{e}^{\mathrm{i}\omega_k t}\right\} = \frac{a_0}{2} + \sum_{k=1}^{\infty}\{a_k \cos\omega_k t + b_k \sin\omega_k t\}, \tag{18.295}$$

where $\omega_k = k\pi/t_0$ and the Fourier coefficients a_k and b_k are determined in the standard manner for the expansion of $X(t)$ on the interval $(0, 2t_0)$ by the following formulas:

$$a_k = \frac{1}{t_0}\int_0^{2t_0} X(t)\cos\omega_k t\,\mathrm{d}t, \tag{18.296}$$

and

$$b_k = \frac{1}{t_0} \int_0^{2t_0} X(t) \sin \omega_k t \, \mathrm{d}t. \tag{18.297}$$

The Fourier coefficients and circular frequencies in Eq. (18.295) are time-independent. By comparison of the two expansions for $X(t)$, it is clear that Eq. (18.294) is the more flexible form; in addition, it is more likely to provide a better physical interpretation in many instances where the Fourier technique is limited.

Using Eq. (18.294) it is possible to construct contour diagrams, with the amplitude and frequency represented as functions of time. This is a most commonly employed mode of data presentation. The frequency–time representation of the amplitude is called the Hilbert amplitude spectrum, or, more concisely, the Hilbert spectrum, and is denoted by $H(\omega, t)$. Various statistics can be defined for the Hilbert spectrum. The marginal spectrum is introduced by

$$h(\omega) = \int_0^T H(\omega, t) \mathrm{d}t, \tag{18.298}$$

where T denotes the total time span for the data set. The marginal spectrum provides a measure of the total amplitude. The mean marginal spectrum is defined by

$$n(\omega) = \frac{1}{T} \int_0^T H(\omega, t) \mathrm{d}t. \tag{18.299}$$

The instantaneous energy is given by

$$IE(t) = \frac{1}{T} \int_0^{\omega_N} H^2(\omega, t) \mathrm{d}\omega, \tag{18.300}$$

where ω_N represents the uppermost frequency for the data set. Measures for the degree of stationarity of the data can also be defined. The interested reader can pursue further details in N.E. Huang *et al.* (1998).

Notes

§18.1 For a short survey of key ideas, see Bogner (2001); for more detailed discussions, see Hahn (1996a,b) and Oppenheim *et al.* (1999). One way to carry out data interpolation is via the sinc approach; see, for example, Olkkonen (1990).

§18.2 For some general background, see Jackson (1989). Discussion on the practical implementation of Hilbert filters can be found in Hahn (1996a,b). For some further reading on the topic of Hilbert filters, see Cain (1972), Eu and Lohmann (1973), Read and Treitel (1973), Sabri and Steenaart (1974, 1975, 1976, 1977), Dutta Roy and Agrawal (1978), Ansari (1987), Wang (1990), Reddy *et al.* (1991b), Damera-Venkata, Evans, and McCaslin (2000), and Zhechev (2005). Additional background on filters, and some of the other topics of this chapter, can be found in Dorf (1993).

For some applications dealing with error-distored and noisy data, see Simpson and Blackwell (1966), Hinich and Weber (1984, 1994), Nakano and Tagami (1988), and Perry and Brazil (1998). For a mathematical analysis, see Parker and Anderson (1990). Anderson and Green (1988) discuss the situation for error propagation for signals that belong to L^{∞}.

§18.3 For further reading on the auto-convolution, cross-correlation, and autocorrelation functions involving the Hilbert transform of a signal, see Dugundji (1958), Spanos and Miller (1994), and Hahn (1996a). For an application to the estimation of time delay differences, see Grennberg and Sandell (1994), and for discussion in the treatment of two-dimensional correlation spectra, see Noda (2000). The technical report by Bendat (1985) contains a number of useful formulas, and much of the report can be found in Bendat and Piersol (2000, Chap. 13).

§18.4 A discussion of two of the principal definitions of the envelope function can be found in Dugundji (1958). A good account of the conditions required in formulating the analytic signal, as well as discussion of some alternative conditions, is given by Vakman (1996). For an important alternative consideration, see Loughlin (1998). The analytic signal can be developed in terms of the Hartley transform; see Pei and Jaw (1990). A two-dimensional application is discussed by Barnes (1996), and a more general analysis is provided by Bülow and Sommer (2001) and Felsberg and Sommer (2001). For some additional reading on the analytic signal, see Oswald (1956), Wolf (1958), Urkowitz (1964), Voelcker (1966a), Brown (1974), Vakman and Vaĭnshteĭn (1977), Langley (1986), Vakman (1994, 1997), Carcaterra and Sestieri (1997), Feldman (2001), and Qian (2006). For a different approach to the analytic signal, see Craig (1996). See also Rao and Kumaresan (1998). For some key comments on the instantaneous frequency, see Loughlin and Tacer (1997) and Nho and Loughlin (1999). For signals impacted by Doppler shifts, see Lerner (1960). A discussion of analytic signals on the circle can be found in Qian, Chen, and Li (2005).

§18.5 For a recent work describing an implementation scheme for demodulation, see Skwarek and Hans (2001). A comparative study of different approaches to demodulation is carried out by Potamianos and Maragos (1994). For additional reading, see Voelcker (1966b), Leuthold (1974), Logan (1978), and Nahin (2006).

§18.7.5 The sinc function occurs widely in signal processing; for a practical application, see Boche and Protzmann (1997).

§18.8 For the evaluation of an extended Hilbert transform of $\operatorname{sgn} x$, see Duoandikoetxea (2001, p. 120). Wavelet bases that form a Hilbert transform pair find application in signal processing. For a discussion of the design issues, see Selesnick (2001, 2002).

§18.9 Lohmann *et al.* (1996a) provide further discussion, Lohmann *et al.* (1998) review applications of fractional transforms in optics, and Alieva, Bastiaans, and Calvo (2005) do the same for optical signal processing. Applications in image processing can be found in Lohmann, Ojeda-Castañeda, and Diaz-Santana (1996b), Lohmann, Tepichín, and Ramírez (1997), Davis *et al.* (1998, 2000, 2001), and Davis and Nowak (2002).

§18.10 For some historical references, see Wiener (1929) and Condon (1937b). Some later developments can be found in the following works: Namias (1980), McBride and Kerr (1987), Kerr (1988), Almeida (1994), Ozaktas *et al.* (1996, 2001), Santhanam and McClellan (1996), Zayed and García (1999), and Pei and Ding (2001).

§18.13 The discrete fractional Fourier transform is discussed in Pei and Yeh (1997), Pei *et al.* (1998, 1999), Candan *et al.* (2000), and Ozaktas *et al.* (2001, chap. 6). For a discussion of the eigenvalues and eigenvectors of the discrete Fourier transform, see McClellan and Parks (1972) and Dickinson and Steiglitz (1982).

§18.14 For further reading, see Pei and Yeh (1998, 2000), Tseng and Pei (2000), and Pei and Wang (2001).

§18.15 See Zayed (1998) and Cusmariu (2002) for further discussion on the fractional analytic signal.

§18.16 The key papers on the topics of this section are by N. E. Huang *et al.* (1996, 1998, 1999, 2003a). For applications in nonlinear water waves, see Schlurmann (2002) and Veltcheva (2002); in system identification of linear structures, see Yang *et al.* (2003a, 2003b); in seismic surface waves, see Chen, Li, and Teng (2002); in molecular dynamics, see Phillips *et al.* (2003); in the treatment of satellite data for plasma structures, see Chen *et al.* (2001); for some medical applications, see Echeverría *et al.* (2001) and W. Huang *et al.* (1998, 1999); and for applications to financial time series, see Huang *et al.* (2003b). The volume by Huang and Shen (2005) is highly recommended for its discussion of basic ideas and applications. For further reading on the mathematical analysis of intrinsic mode functions, see Sharpley and Vatchev (2006) and Chen *et al.* (2006). For an analytical approach to the sifting process, see Deléchelle, Lemoine, and Niang (2005). A wavelet based determination of the Hilbert spectrum for non-stationary signals is treated by Olhede and Walden (2004). For discussion on alternative approaches to nonlinear analysis, see, for example, Simon and Tomlinson (1984), Tomlinson (1987), Braun and Feldman (1997), Feldman (1997), and Gottlieb and Feldman (1997). Further discussion on time series can be found in Wiener (1960), Papoulis (1965), Priestley (1988), and Chatfield (1989).

Exercises

18.1 Suggest a form for an idealized transfer function whose purpose would be to act as a differentiator and a Hilbert transformer.

18.2 Determine the envelope function for the signal $s(t)\cos(\omega_0 t + \varphi)$, where $s(t)$ is a non-negative band-limited function. What role does the value of ω_0 play in relation to the bandwidth of $s(t)$?

18.3 If $\star$ denotes a cross-correlation, is $f \star g = g \star f$ true for general functions on $\mathbb{R}$ such that $f \star g$ exists?

18.4 Can $f \star g$ be expressed as a convolution of functions simply related to f and g?

18.5 Show that the auto-correlation function $\rho_f(t)$ satisfies the inequality

$$\rho_f(t) \leq \int_{-\infty}^{\infty} f^2(\tau)\mathrm{d}\tau.$$

18.6 Let $s_\mathrm{a}(t)$ denote the analytic signal formed from the real signal $s(t)$ whose Fourier transform is $S(\omega)$. Express $s_\mathrm{a}(t)$ in terms of $S(\omega)$.

18.7 Does a bounded signal have a bounded instantaneous amplitude? Discuss.

18.8 Evaluate the limits (i) $\lim_{a\to 0+} a^{-1}\,\mathrm{rect}(x/a)$ and (ii) $\lim_{a\to 0+} a^{-1}\,\mathrm{sinc}(x/a)$.

18.9 Evaluate the Hilbert transform of the following: (i) $\mathrm{H}(x)\mathrm{e}^{-x}$, (ii) $\mathrm{sgn}\,x\,\mathrm{e}^{-|x|}$, (iii) $\Pi_2(x)\,\mathrm{e}^{-|x|}$, and (iv) $\mathrm{H}(x)\Pi_2(x)\mathrm{e}^{-x}$.

18.10 Determine the Hilbert transforms of the following: (i) $\mathrm{H}(x)\cos\pi x$, (ii) $\mathrm{H}(x)\sin\pi x$, (iii) $\Pi_{2\pi}(x)\cos x$, and (iv) $\Pi_{2\pi}(x)\sin x$.

18.11 Evaluate the Hilbert transform of the following function:

$$f(x,\alpha) = \frac{1}{\pi}\left[\tan^{-1}\left(\frac{\alpha}{x-1}\right) + \tan^{-1}\left(\frac{\alpha}{x+1}\right) - 2\tan^{-1}\left(\frac{\alpha}{x}\right)\right],$$

where α is a constant, and plot the functions $f(x,\alpha)$ and Hf for the values $\alpha = 0, 0.5$, and 1.

18.12 Prove Eqs. (18.181) and (18.182).

18.13 Evaluate the fractional Fourier transforms of (i) rect (x), (ii) sinc x, and (iii) chirp(x), where the chirp function is defined by $\mathrm{chirp}(x) = \mathrm{e}^{-\mathrm{i}\pi/4}\mathrm{e}^{\mathrm{i}\pi x^2}$.

18.14 Determine the fractional Fourier transform of the following functions: (i) $\mathrm{e}^{-\pi x^2}$, (ii) $h_\mathrm{m}(x)$, and (iii) $\mathrm{e}^{\mathrm{i}\pi x}$.

18.15 Using Zayed's definition of the fractional Hilbert transform, Eq. (18.209), determine if the semigroup property under composition, $H_\alpha H_\beta = H_{\alpha+\beta}$, holds.

18.16 Evaluate $|\mathcal{F}H_\alpha S(\omega)|$ using Cusmariu's definition of the fractional Hilbert transform operator, where $S(\omega)$ is the Fourier transform of the signal $s(t)$.

18.17 Let the matrix **S** have a distinct set of eigenvalues. If the matrix **F** commutes with **S**, show that the eigenvectors of **S** are also eigenvectors of **F**.

18.18 For readers with some basic *Mathematica* skills, or equivalent knowledge with a different software package, write a short program to deal with the following problem. For the choice of function given by Eq. (18.285), explore the construction of the IMFs on a larger time scale, for example from [0, 15]. What problems, if any, emerge? How can any problems that emerge be overcome?

18.19 For the same example as in Exercise 18.18, set up a stopping procedure in the construction of the IMFs based on a weighted sum-of-squared deviations criterion.

18.20 Determine the IMFs that result from a discrete time series (with a sampling interval of 10^{-3} s) for the function $X(t) = \cos(a\pi t) + \cos(b\pi t) + \cos(c\pi t) + \cos(d\pi t)$ on the time interval from 0 to 5 s, using some different choices for the set of constants $\{a, b, c, d\}$.

19

Kramers–Kronig relations

19.1 Some background from classical electrodynamics

The principal intent of this chapter is to arrive at the classical Hilbert transform connections that apply between the real and imaginary components of the generalized (complex) refractive index, and for the complex dielectric constant. Connections of this type are frequently termed dispersion relations in the physics literature. But for the two functions just mentioned, and for many associated results, they are most often referred to as the Kramers–Kronig relations. Historically, these were the first applications of the Hilbert transform concept in the physical sciences, and were discovered by Kronig (1926) and independently by Kramers (1927). These authors were interested in issues connected with the dispersion of light, and from this emerged the term dispersion relation to describe the Hilbert transform relations found by Kramers and Kronig. The reader will recall that dispersion refers to the frequency variation of the refractive index (or some other optical property), and dispersion formulas provide a connection between the refractive index and the frequency. Functions such as the dielectric constant, refractive index, and permeability, which will be defined shortly, are referred to as *optical constants*. These functions characterize the interaction of electromagnetic radiation with matter. Though in widespread use, this terminology is somewhat of a misnomer, since the optical constants actually depend on the frequency of the incident electromagnetic radiation interacting with the material, and are hence not true constants.

The electric polarization $\mathbf{P}(\mathbf{x})$ of a medium, that is, the dipole moment per unit volume for a collection of molecules, is given by

$$\mathbf{P}(\mathbf{x}) = \sum_i \mathcal{N}_i \langle \mathbf{p}_i \rangle, \tag{19.1}$$

where $\mathcal{N}_i$ and $\mathbf{p}_i$ are, respectively, the average number per unit volume and the associated dipole moment of molecules of type i in a small volume element centered at $\mathbf{x}$, and the angular brackets denote an average taken over this volume element. (Vector quantities are shown in bold type in this chapter.) The electric displacement **D** is defined in terms of the electric field **E** and electric polarization as follows:

$$\mathbf{D} = \varepsilon_0 \mathbf{E} + \mathbf{P}, \tag{19.2}$$

where ε_0 denotes the vacuum permittivity. The variable dependence of the various fields and associated electrodynamic quantities will often be suppressed, unless there is a specific purpose to make this apparent, for example making the jump between the time and frequency domains. The reader needs to be cognizant of the units employed when comparing the formulas of the present work with other sources. The SI system of units (Système International d'Unités) is employed in this chapter. In other unit systems, such as the esu and emu systems, various changes occur. The factors to be alert for are $4\pi, \varepsilon_0, c$ (the speed of light in a vacuum), and μ_0 (the permeability of the vacuum). In particular, to match up with many results in the literature, a factor of $(4\pi\varepsilon_0)^{-1}$ in equations based on SI units is replaced by unity, which converts to the esu (electrostatic) and emu (electromagnetic) unit systems. For the remainder of this chapter, the applied electric field is assumed to be of low strength, and consequently the system response can be treated as linear in the field strength. The simplifying assumption that the medium is isotropic is adopted, which means that the vectors **E** and **P** are oriented in the same direction. In this case,

$$\mathbf{P} = \varepsilon_0 \chi \mathbf{E}, \tag{19.3}$$

where χ denotes the electric susceptibility. The latter quantity is also called the dielectric susceptibility by some authors. Employing this result allows the electric displacement to be written as follows:

$$\mathbf{D} = \varepsilon \mathbf{E}, \tag{19.4}$$

where ε, the permittivity of the medium, is given by

$$\varepsilon = \varepsilon_0(1 + \chi). \tag{19.5}$$

Some authors refer to the permittivity as the electric permittivity, or, less commonly, the dielectric permittivity. The dimensionless ratio $\varepsilon/\varepsilon_0$ is called the relative permittivity or the dielectric constant. The electric polarization has units $\mathrm{C\,m^{-2}}$, where C denotes the SI unit of charge, the coulomb, m is the SI unit of length, the meter, and the electric field strength is given in units of $\mathrm{V\,m^{-1}}$, where V is the SI unit of potential, the volt. The vacuum permittivity has the units $\mathrm{J^{-1}\,C^2\,m^{-1}}$, where J is the SI unit of energy, the joule. The electric susceptibility is dimensionless.

When an electric field is applied to a collection of identical molecules, there are two principal consequences. If the molecule has a permanent electric dipole moment, then the electric field will cause an alignment of the individual molecular moments.

The electric field can also induce a temporary dipole moment in each of the molecules, the strength of which is proportional to the applied electric field, that is

$$\mathbf{p} = \alpha \mathbf{E}, \tag{19.6}$$

where α is the polarizability.

The analog of Eq. (19.4) for magnetic interactions, adopting the same assumptions previously stated for the medium, is

$$\mathbf{B} = \mu \mathbf{H}, \tag{19.7}$$

where **H** denotes the magnetic field, **B** is the magnetic induction, and μ designates the permeability. The latter term is also referred to as the magnetic permeability.

19.2 Kramers–Kronig relations: a simple derivation

The Hilbert transform connections between the real and imaginary parts of the dielectric constant are derived in this section. The asymptotic behavior of the dielectric constant for large frequencies is the first topic considered. The sample is subjected to a beam of electromagnetic radiation of given frequency; that is, the beam is treated as monochromatic. If the frequency of the external field is significantly larger than the binding energy of the electrons, then a simple model to describe the motion of the electrons can be adopted. Writing the charge on the electron as $-e$, the equation of motion is given by

$$m\ddot{\mathbf{x}} = -e\mathbf{E}, \tag{19.8}$$

where $\ddot{\mathbf{x}}$ denotes the second derivative of the particle displacement with respect to time and m is the mass of the electron. The spatial variation of the electric field can be assumed negligible, since the displacements of the electrons from their equilibrium positions are small compared with the wavelength of the incident electromagnetic radiation. The electric field is assumed to vary harmonically in time as

$$\mathbf{E} = \mathbf{E}_0 \mathrm{e}^{-i\omega t}, \tag{19.9}$$

where $\mathbf{E}_0$ incorporates the amplitude and polarization vector of the field. Experimental electric fields are real quantities, and this fact is handled in the equations by commonly using two different approaches. One procedure is to assume that the real part of equations like Eq. (19.9) is taken. In the literature, this is sometimes made explicit by the appearance of the symbol Re, and often it is implicitly implied, simply by stating that the real part is to be taken, and the reader needs to keep this fact in mind. The latter approach is taken in this book. The alternative procedure used is to add a complex conjugate term, so that Eq. (19.9) would be written as $\mathbf{E} = \mathbf{E}_0 \mathrm{e}^{-i\omega t} + \mathbf{E}_0^* \mathrm{e}^{i\omega t}$, or the commonly seen form: $\mathbf{E} = \mathbf{E}_0 \mathrm{e}^{-i\omega t} + c.c.$, where $c.c.$ stands for the complex conjugate of the preceding term. In this chapter the discussion is restricted to the case of linear polarization (plane polarized light), which has the electric field vector always oriented in one direction. Other types of polarization are dealt with in Chapter 21. The solution of Eq. (19.8) is given by

$$\mathbf{x} = \frac{e\mathbf{E}}{m\omega^2}, \tag{19.10}$$

and the resulting electric polarization is as follows:

$$\mathbf{P} = -\mathcal{N} Ze\,\mathbf{x} = -\frac{\mathcal{N} Ze^2\mathbf{E}}{m\omega^2}, \tag{19.11}$$

where $\mathcal{N}$ is the number of molecules per unit volume and Z is the number of electrons per molecule. Using Eqs. (19.3), (19.5), and (19.11), it follows that

$$\frac{\varepsilon(\omega)}{\varepsilon_0} = 1 - \frac{\omega_\mathrm{p}^2}{\omega^2}, \quad \text{for } \omega \to \infty, \tag{19.12}$$

and

$$\omega_\mathrm{p}^2 = \frac{\mathcal{N} Ze^2}{\varepsilon_0 m}. \tag{19.13}$$

The quantity ω_p is termed the plasma frequency of the medium. Equation (19.12) shows that the dielectric constant is bounded asymptotically as $\omega \to \infty$.

The next step is to develop a simple model for the frequency dependence of $\varepsilon(\omega)$, and from this derive dispersion relations for the permittivity. Following this, a generalization is made to a model-independent discussion of the dispersion relations for $\varepsilon(\omega)$. Suppose the forces that hold the electrons near their equilibrium positions are analogous to the restoring forces in a spring. If a light beam is incident on the dielectric, the electrons will undergo oscillation about their equilibrium positions due to the electric and magnetic fields associated with the light. If the model is restricted to the non-relativistic regime, the magnetic effects can be ignored. The equation of motion for an electron (with charge $-e$) bound by a harmonic restoring force, subject to a damping force $-m\gamma\dot{\mathbf{x}}$, and acted upon by an external field $\mathbf{E}(\mathbf{x}, t)$, is (suppressing the variable dependence of $\mathbf{x}$ for the field) as follows:

$$m\ddot{\mathbf{x}} + m\gamma\dot{\mathbf{x}} + m\omega_0^2\mathbf{x} = -e\mathbf{E}. \tag{19.14}$$

Assuming $\mathbf{E}(\mathbf{x}, t) = \mathbf{E}_0 \mathrm{e}^{-\mathrm{i}\omega t}$, then the solution of the preceding equation is given by

$$\mathbf{x} = \frac{-e\mathbf{E}}{m(\omega_0^2 - \omega^2 - \mathrm{i}\gamma\omega)}. \tag{19.15}$$

The electric polarization is given by

$$\mathbf{P} = -\mathcal{N}Ze\mathbf{x} = \frac{\mathcal{N}Ze^2\mathbf{E}}{m(\omega_0^2 - \omega^2 - \mathrm{i}\gamma\omega)}. \tag{19.16}$$

A generalization can be made by assuming that the binding frequency is not the same for all electrons. Let f_j denote the number of electrons with binding frequency ω_j and damping constant γ_j; then,

$$\mathbf{P} = \frac{\mathcal{N}e^2\mathbf{E}}{m}\sum_j \frac{f_j}{\omega_j^2 - \omega^2 - \mathrm{i}\gamma_j\omega}. \tag{19.17}$$

Employing Eqs. (19.3), (19.5), and (19.17) leads to

$$\frac{\varepsilon(\omega)}{\varepsilon_0} = 1 + \frac{\mathcal{N}e^2}{\varepsilon_0 m}\sum_j \frac{f_j}{\omega_j^2 - \omega^2 - \mathrm{i}\gamma_j\omega}. \tag{19.18}$$

This is the Lorentz model for a dielectric material. The factors f_j are referred to as oscillator strengths, and they satisfy the obvious sum rule condition

$$\sum_j f_j = Z. \tag{19.19}$$

In a more advanced treatment using quantum theory, the oscillator strengths are defined in terms of the intensities of the transitions from the ground electronic state to the excited states. Equation (19.19) still applies, but the sum is interpreted as a summation over the discrete states together with an integral over the continuum states involved.

In the vicinity of an absorption band, the dielectric constant has the behavior shown in Figure 19.1. The typical observation is that $\varepsilon(\omega)$ becomes larger with increasing angular frequency, this is the so-called normal dispersion behavior. In the vicinity of a region with strong absorption, $\varepsilon(\omega)$ shows abrupt changes as a function of frequency, and decreases as the angular frequency increases for a certain range of frequencies. This is the region of anomalous dispersion, and it is clearly displayed in Figure 19.1. The refractive index shows a frequency behavior similar to the dielectric constant.

In the limit $\omega \to \infty$, Eq. (19.18) yields

$$\lim_{\omega\to\infty} \frac{\varepsilon(\omega)}{\varepsilon_0} \approx 1 - \frac{\mathcal{N}e^2}{\varepsilon_0 m\omega^2}\sum_j f_j, \tag{19.20}$$

and, on using Eqs. (19.19) and (19.13), Eq. (19.12) follows. So the model is asymptotically consistent with the free-electron model described earlier.

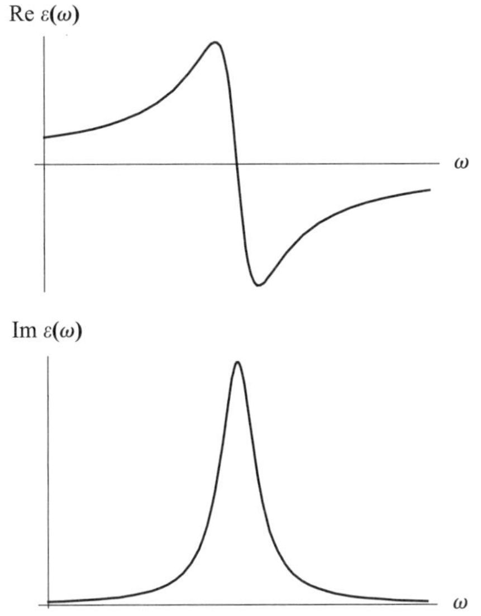

Figure 19.1. Behavior of the real and imaginary parts of the dielectric constant in a region where absorption occurs.

From Eqs. (19.18) and (19.5), the electric susceptibility is given by

$$\chi(\omega) = \frac{\mathcal{N}e^2}{\varepsilon_0 m} \sum_j \frac{f_j}{\omega_j^2 - \omega^2 - \mathrm{i}\gamma_j \omega}. \tag{19.21}$$

The generalization to complex frequencies is now examined. To avoid confusion, the convention of Chapter 17 (Section 17.7) is employed, where a complex angular frequency is written as $\omega_z = \omega_r + \mathrm{i}\omega_i$, and the assignment $\omega = \omega_r$ is used when there is no risk of confusion. The first observation to note from Eq. (19.21) is that

$$\chi(-\omega_z^*) = \chi(\omega_z)^*. \tag{19.22}$$

This is an example of a *crossing symmetry* relation of the type discussed in Section 17.7. The crossing symmetry relation provides a means to extend the range of definition of the electric susceptibility to negative frequencies. Negative frequencies are of course unphysical, but the generalization to such frequencies is a useful, though not essential, feature employed to derive the Kramers–Kronig relations. The function $\chi(\omega_z)$ is analytic in the upper half complex angular frequency plane. Let $\beta_j = \sqrt{\left(\omega_j^2 - \gamma_j^2/4\right)}$; then, for $\omega_j > \gamma_j/2$ and $\gamma_j > 0$, the poles of $\chi(\omega_z)$ lie in the

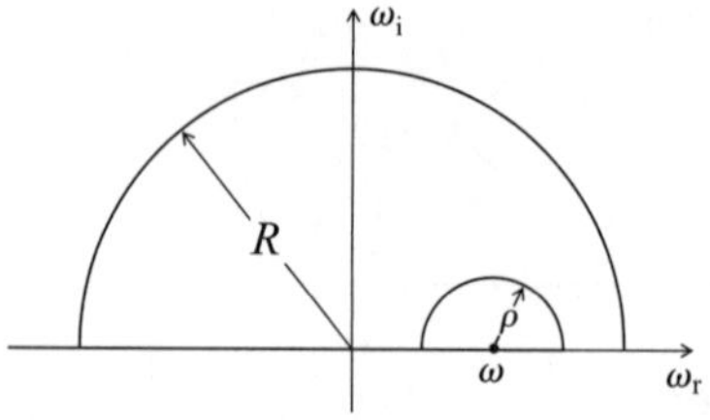

Figure 19.2. Contour for the evaluation of dispersion relations for the electric susceptibility.

lower half of the complex angular frequency plane at $\omega_z = \pm\beta_j - (1/2)\mathrm{i}\gamma_j$. Consider the contour integral

$$\oint_C \frac{\chi(\omega_z)\mathrm{d}\omega_z}{\omega - \omega_z},$$

where C denotes the contour shown in Figure 19.2. This contour will find multiple applications in both this and the following chapters. It is the most frequently used contour employed to derive dispersion relations. From the Cauchy integral theorem, it follows that

$$\int_{-R}^{\omega-\rho} \frac{\chi(\omega_r)\mathrm{d}\omega_r}{\omega - \omega_r} + \int_{\omega+\rho}^{R} \frac{\chi(\omega_r)\mathrm{d}\omega_r}{\omega - \omega_r} - \int_{\pi}^{0} \frac{\chi(\omega + \rho\mathrm{e}^{\mathrm{i}\theta})\mathrm{i}\rho\mathrm{e}^{\mathrm{i}\theta}\mathrm{d}\theta}{\rho\mathrm{e}^{\mathrm{i}\theta}}$$
$$+ \int_0^{\pi} \frac{\chi(R\mathrm{e}^{\mathrm{i}\theta})\mathrm{i}R\mathrm{e}^{\mathrm{i}\theta}\mathrm{d}\theta}{\omega - R\mathrm{e}^{\mathrm{i}\theta}} = 0. \tag{19.23}$$

In the limit $R \to \infty$, Eq. (19.23) simplifies to

$$\int_{-\infty}^{\omega-\rho} \frac{\chi(\omega_r)\mathrm{d}\omega_r}{\omega - \omega_r} + \int_{\omega+\rho}^{\infty} \frac{\chi(\omega_r)\mathrm{d}\omega_r}{\omega - \omega_r} = -\mathrm{i}\int_0^{\pi} \chi(\omega + \rho\mathrm{e}^{\mathrm{i}\theta})\mathrm{d}\theta, \tag{19.24}$$

where the following asymptotic behavior has been employed:

$$\chi(\omega) = -\frac{\omega_p^2}{\omega^2}, \quad \text{for } \omega \to \infty. \tag{19.25}$$

Taking the limit $\rho \to 0$ in Eq. (19.24) yields

$$\frac{1}{\pi} P \int_{-\infty}^{\infty} \frac{\chi(\omega_r)\mathrm{d}\omega_r}{\omega - \omega_r} = -\mathrm{i}\chi(\omega). \tag{19.26}$$

If the electric susceptibility is written in terms of its real and imaginary components, thus:

$$\chi(\omega) = \chi_r(\omega) + \mathrm{i}\chi_i(\omega), \tag{19.27}$$

it follows from Eq. (19.26) that

$$\chi_{\mathrm{i}}(\omega) = \frac{1}{\pi} P \int_{-\infty}^{\infty} \frac{\chi_{\mathrm{r}}(\omega_{\mathrm{r}})\mathrm{d}\omega_{\mathrm{r}}}{\omega - \omega_{\mathrm{r}}} \tag{19.28}$$

and

$$\chi_{\mathrm{r}}(\omega) = -\frac{1}{\pi} P \int_{-\infty}^{\infty} \frac{\chi_{\mathrm{i}}(\omega_{\mathrm{r}})\mathrm{d}\omega_{\mathrm{r}}}{\omega - \omega_{\mathrm{r}}}. \tag{19.29}$$

These are the Hilbert transform connections between the real and imaginary parts of the electric susceptibility. From the crossing symmetry relation Eq. (19.22), it follows that

$$\chi_{\mathrm{r}}(-\omega) = \chi_{\mathrm{r}}(\omega) \tag{19.30}$$

and

$$\chi_{\mathrm{i}}(-\omega) = -\chi_{\mathrm{i}}(\omega). \tag{19.31}$$

Using this last pair of results allows Eqs. (19.28) and (19.29) to be written as follows:

$$\chi_{\mathrm{i}}(\omega) = \frac{2\omega}{\pi} P \int_{0}^{\infty} \frac{\chi_{\mathrm{r}}(\omega_{\mathrm{r}})\mathrm{d}\omega_{\mathrm{r}}}{\omega^2 - \omega_{\mathrm{r}}^2} \tag{19.32}$$

and

$$\chi_{\mathrm{r}}(\omega) = -\frac{2}{\pi} P \int_{0}^{\infty} \frac{\omega_{\mathrm{r}}\chi_{\mathrm{i}}(\omega_{\mathrm{r}})\mathrm{d}\omega_{\mathrm{r}}}{\omega^2 - \omega_{\mathrm{r}}^2}. \tag{19.33}$$

These are the Kramers–Kronig relations for the electric susceptibility. They are also referred to as the dispersion relations for the electric susceptibility. The corresponding results for the dielectric constant are

$$\varepsilon_{\mathrm{i}}(\omega) = \frac{2\omega}{\pi} P \int_{0}^{\infty} \frac{[\varepsilon_{\mathrm{r}}(\omega_{\mathrm{r}}) - \varepsilon_0]\mathrm{d}\omega_{\mathrm{r}}}{\omega^2 - \omega_{\mathrm{r}}^2} \tag{19.34}$$

and

$$\varepsilon_{\mathrm{r}}(\omega) - \varepsilon_0 = -\frac{2}{\pi} P \int_{0}^{\infty} \frac{\omega_{\mathrm{r}}\varepsilon_{\mathrm{i}}(\omega_{\mathrm{r}})\mathrm{d}\omega_{\mathrm{r}}}{\omega^2 - \omega_{\mathrm{r}}^2}. \tag{19.35}$$

In Eq. (19.34) the reader should note the significance of the contribution $-\varepsilon_0$ in the integrand. This point will be amplified in the following section.

The underlying assumptions that have been utilized directly or implicitly to arrive at the Kramers–Kronig relationships are summarized as follows.

(i) The electronic motion is sufficiently slow so that magnetic effects can be ignored; that is, the non-relativistic regime is employed.

(ii) The strength of the external electric field is assumed to be sufficiently small, so that nonlinear effects can be ignored.
(iii) The medium has been assumed to be isotropic, which allows the tensor character of the electric susceptibility and the permittivity to be simplified to a single term. For an anisotropic medium this simplification no longer holds.
(iv) The density of the medium has been assumed to be low.
(v) The spatial variation of the electric field has been ignored.
(vi) The difference between the applied external field and the local field has been ignored.
(vii) No fraction of the electronic population is assumed to be freely mobile.
(viii) No radiative reaction forces are included.
(ix) Binding forces for the electrons are assumed to be harmonic.
(x) The application of classical ideas has been assumed in place of a quantum mechanical description.

Some of these issues are revisited in this and the following chapter. At this juncture it is noted in particular that no explicit reference to causality has been invoked. In summary, the derivation of the Kramers–Kronig relations given in this section for the electric susceptibility and the permittivity is to be regarded as being highly model-dependent. For the latter reason, it might be supposed that these relations would be of restricted validity. This is not the case, as the following section explains.

19.3 Kramers–Kronig relations: a more rigorous derivation

In this section the Kramers–Kronig relations are derived under much more general conditions than were employed in the previous section. Some of the simplifying assumptions given in the list of the preceding section are retained, but the classical damped harmonically bound electron model can be dispensed with. This means the Kramers–Kronig relations can be applied much more widely in applications.

The electric displacement $\mathbf{D}(\mathbf{x}, t)$ and the electric field $\mathbf{E}(\mathbf{x}, t)$ are related at a particular frequency by

$$\mathbf{D}(\mathbf{x}, \omega) = \varepsilon(\omega)\mathbf{E}(\mathbf{x}, \omega). \tag{19.36}$$

In this chapter, a common convention of using the same symbol for the field and its Fourier transform is adopted. The field and its corresponding Fourier transform are, of course, two different functions in general, so this convention does force upon the reader the requirement for added diligence in deciding which field is under discussion. To aid the reader, the arguments employed for the field will convey the required information.

The time and frequency domains are connected by assuming the general Fourier transform connections:

$$\mathbf{D}(\mathbf{x}, t) = \int_{-\infty}^{\infty} \mathbf{D}(\mathbf{x}, \omega)\mathrm{e}^{-\mathrm{i}\omega t}\, \mathrm{d}\omega \tag{19.37}$$

and

$$\mathbf{E}(\mathbf{x},t) = \int_{-\infty}^{\infty} \mathbf{E}(\mathbf{x},\omega)\mathrm{e}^{-\mathrm{i}\omega t}\,\mathrm{d}\omega, \tag{19.38}$$

and the inverse Fourier transform relationships are given by

$$\mathbf{D}(\mathbf{x},\omega) = \frac{1}{2\pi}\int_{-\infty}^{\infty} \mathbf{D}(\mathbf{x},t)\mathrm{e}^{\mathrm{i}\omega t}\,\mathrm{d}t \tag{19.39}$$

and

$$\mathbf{E}(\mathbf{x},\omega) = \frac{1}{2\pi}\int_{-\infty}^{\infty} \mathbf{E}(\mathbf{x},t)\mathrm{e}^{\mathrm{i}\omega t}\,\mathrm{d}t. \tag{19.40}$$

Employing Eqs. (19.36), (19.37), and (19.40) gives

$$\begin{aligned}
\mathbf{D}(\mathbf{x},t) &= \int_{-\infty}^{\infty} \varepsilon(\omega)\mathbf{E}(\mathbf{x},\omega)\mathrm{e}^{-\mathrm{i}\omega t}\,\mathrm{d}\omega \\
&= \frac{1}{2\pi}\int_{-\infty}^{\infty} \varepsilon(\omega)\mathrm{e}^{-\mathrm{i}\omega t}\,\mathrm{d}\omega \int_{-\infty}^{\infty} \mathbf{E}(\mathbf{x},t')\mathrm{e}^{\mathrm{i}\omega t'}\,\mathrm{d}t' \\
&= \frac{\varepsilon_0}{2\pi}\int_{-\infty}^{\infty} \mathrm{e}^{-\mathrm{i}\omega t}\,\mathrm{d}\omega \int_{-\infty}^{\infty} \mathbf{E}(\mathbf{x},t')\mathrm{e}^{\mathrm{i}\omega t'}\,\mathrm{d}t' \\
&\quad + \frac{\varepsilon_0}{2\pi}\int_{-\infty}^{\infty} \left[\frac{\varepsilon(\omega)}{\varepsilon_0} - 1\right]\mathrm{e}^{-\mathrm{i}\omega t}\,\mathrm{d}\omega \int_{-\infty}^{\infty} \mathbf{E}(\mathbf{x},t')\mathrm{e}^{\mathrm{i}\omega t'}\,\mathrm{d}t' \\
&= \frac{\varepsilon_0}{2\pi}\int_{-\infty}^{\infty} \mathbf{E}(\mathbf{x},t')\mathrm{d}t' \int_{-\infty}^{\infty} \mathrm{e}^{-\mathrm{i}\omega(t-t')}\,\mathrm{d}\omega \\
&\quad + \frac{\varepsilon_0}{2\pi}\int_{-\infty}^{\infty} \mathbf{E}(\mathbf{x},t')\mathrm{d}t' \int_{-\infty}^{\infty} \left[\frac{\varepsilon(\omega)}{\varepsilon_0} - 1\right]\mathrm{e}^{-\mathrm{i}\omega(t-t')}\,\mathrm{d}\omega \\
&= \varepsilon_0\int_{-\infty}^{\infty} \mathbf{E}(\mathbf{x},t')\delta(t-t')\mathrm{d}t' \quad + \varepsilon_0\int_{-\infty}^{\infty} \mathbf{E}(\mathbf{x},t')G(t-t')\mathrm{d}t'.
\end{aligned} \tag{19.41}$$

The functions $\varepsilon(\omega)$ and $\mathbf{E}(\mathbf{x},t')$ are assumed to be sufficiently well behaved to allow the interchange of the order of integration, and the following definition has been employed:

$$G(t) = \frac{1}{2\pi}\int_{-\infty}^{\infty} \left[\frac{\varepsilon(\omega)}{\varepsilon_0} - 1\right]\mathrm{e}^{-\mathrm{i}\omega t}\,\mathrm{d}\omega. \tag{19.42}$$

The function $G(t)$ plays the role of a response function. For this definition to be useful, it is assumed that $\{\varepsilon(\omega)\varepsilon_0^{-1} - 1\} \in L^2(\mathbb{R})$, and this will allow the inverse Fourier transform relationship to be obtained. Actually, this requirement could be modified somewhat to include functions of other classes, but the given condition is likely to the most useful in a practical setting. The stated condition implies an appropriate asymptotic behavior for $\{\varepsilon(\omega)\varepsilon_0^{-1} - 1\}$ as $\omega \to \pm\infty$. This issue is revisited shortly. Systems with conducting properties are excluded from the present discussion. Conductors require an important modification, and this is addressed later. Equation (19.41) simplifies to give

$$\mathbf{D}(\mathbf{x}, t) = \varepsilon_0 \mathbf{E}(\mathbf{x}, t) + \varepsilon_0 \int_{-\infty}^{\infty} G(t')\mathbf{E}(\mathbf{x}, t - t')\mathrm{d}t'. \tag{19.43}$$

Since $\mathbf{D}(\mathbf{x}, t)$ and $\mathbf{E}(\mathbf{x}, t)$ are real quantities, the function $G(t)$ is also real. Equation (19.43) can be rewritten as follows:

$$\mathbf{D}(\mathbf{x}, t) = \varepsilon_0 \left\{ \mathbf{E}(\mathbf{x}, t) + \int_0^{\infty} G(t')\mathbf{E}(\mathbf{x}, t - t')\mathrm{d}t' + \int_{-\infty}^{0} G(t')\mathbf{E}(\mathbf{x}, t - t')\mathrm{d}t' \right\}. \tag{19.44}$$

This result makes it clear that there is a non-local temporal relationship between the electric displacement and the electric field. If the electric field is switched on in the infinite past, then the first integral on the right-hand side of Eq. (19.44) shows, depending on the exact structure of $G(t)$, that the displacement $\mathbf{D}(\mathbf{x}, t)$ depends non-locally on the electric field up to time t. The second integral, however, indicates that $\mathbf{D}(\mathbf{x}, t)$ depends on the electric field for times in advance of the time t. This would be a breakdown of causality. As a fundamental assumption, the system is assumed to be causal, which requires that

$$\int_{-\infty}^{0} G(t')\mathbf{E}(\mathbf{x}, t - t')\mathrm{d}t' = 0; \tag{19.45}$$

that is,

$$G(t) = 0, \quad \text{for } t < 0. \tag{19.46}$$

This means that the fundamental connection between the displacement and the electric field takes the following form:

$$\mathbf{D}(\mathbf{x}, t) = \varepsilon_0 \left\{ \mathbf{E}(\mathbf{x}, t) + \int_0^{\infty} G(t')\mathbf{E}(\mathbf{x}, t - t')\mathrm{d}t' \right\}. \tag{19.47}$$

Making use of the definition of G in Eq. (19.42), it follows, on taking the inverse Fourier transform, that

$$\frac{\varepsilon(\omega)}{\varepsilon_0} - 1 = \int_{-\infty}^{\infty} G(t)\mathrm{e}^{\mathrm{i}\omega t}\,\mathrm{d}t, \tag{19.48}$$

and from the causality condition, Eq. (19.46), it follows that

$$\frac{\varepsilon(\omega)}{\varepsilon_0} - 1 = \int_0^{\infty} G(t)\mathrm{e}^{\mathrm{i}\omega t}\,\mathrm{d}t. \tag{19.49}$$

Taking the real and imaginary parts of this formula yields

$$\frac{\varepsilon_{\mathrm{r}}(\omega)}{\varepsilon_0} - 1 = \int_0^{\infty} G(t)\cos\omega t\,\mathrm{d}t \tag{19.50}$$

and

$$\frac{\varepsilon_{\mathrm{i}}(\omega)}{\varepsilon_0} = \int_0^{\infty} G(t)\sin\omega t\,\mathrm{d}t. \tag{19.51}$$

Because the real and imaginary parts of the dielectric constant can be derived from knowledge of G, it is anticipated that these two functions are interrelated.

Since $G(t)$ is real, it follows from Eq. (19.49) and on setting, for complex frequencies, $\omega_z = \omega_{\mathrm{r}} + \mathrm{i}\omega_{\mathrm{i}}$, that

$$\chi(-\omega_z^*) = \chi(\omega_z)^* \tag{19.52}$$

and

$$\varepsilon(-\omega_z^*) = \varepsilon(\omega_z)^*, \tag{19.53}$$

which are the crossing symmetry relations for the electric susceptibility and dielectric constant, respectively. From Eq. (19.53), it follows, for real angular frequencies, that

$$\varepsilon_{\mathrm{r}}(-\omega) = \varepsilon_{\mathrm{r}}(\omega) \tag{19.54}$$

and

$$\varepsilon_{\mathrm{i}}(-\omega) = -\varepsilon_{\mathrm{i}}(\omega). \tag{19.55}$$

That is, the real part of $\varepsilon(\omega)$ is an even function of the angular frequency and the imaginary part of $\varepsilon(\omega)$ is an odd function of the angular frequency. The crossing symmetry results were derived in Section 19.2 from a model-dependent approach; see the jump from the model given in Eq. (19.21) to Eq. (19.22). What has just been done is clearly a much more general approach relative to the treatment of the preceding section.

The behavior of $G(t)$ for $t \to \infty$ can be obtained from Eq. (19.42). Since

$$G(t) = \frac{1}{2\pi}\left\{\int_{-\infty}^{\infty}\left[\frac{\varepsilon(\omega)}{\varepsilon_0} - 1\right]\cos\omega t\,\mathrm{d}\omega - \mathrm{i}\int_{-\infty}^{\infty}\left[\frac{\varepsilon(\omega)}{\varepsilon_0} - 1\right]\sin\omega t\,\mathrm{d}\omega\right\}, \tag{19.56}$$

it follows on using the Riemann–Lebesgue lemma (see Section 2.14) and the previous restriction on $\{\varepsilon(\omega)\varepsilon_0^{-1} - 1\}$, that

$$\lim_{t\to\infty} G(t) = 0. \tag{19.57}$$

The discussion so far has assumed that distribution functions are not under consideration. Consequently, the response function $G(t)$ does not have a jump discontinuity. Since the causality condition specifies the behavior of $G(t)$ for $t < 0$, it follows that

$$\lim_{t\to 0+} G(t) = 0, \tag{19.58}$$

so that the behavior of $G(t)$ for $t > 0$ links up with the behavior for $t < 0$. The interpretation of this result is that there is no *instantaneous* dielectric response to an applied electromagnetic field.

The question of the asymptotic behavior of the dielectric constant is now considered. Applying integration by parts to Eq. (19.49) leads to

$$\frac{\varepsilon(\omega)}{\varepsilon_0} - 1 = \mathrm{i}\frac{G(0)}{\omega} - \frac{G'(0)}{\omega^2} - \mathrm{i}\frac{G''(0)}{\omega^3} - \frac{\mathrm{i}}{\omega^3}\int_0^\infty G'''(t)\mathrm{e}^{\mathrm{i}\omega t}\,\mathrm{d}\omega. \tag{19.59}$$

Employing the result for $G(0)$ given in Eq. (19.58) yields the following asymptotic conditions for the real and imaginary parts of the dielectric constant:

$$\frac{\varepsilon_\mathrm{r}(\omega)}{\varepsilon_0} - 1 \approx O\left(\frac{1}{\omega^2}\right), \quad \text{as } \omega \to \infty, \tag{19.60}$$

and

$$\frac{\varepsilon_\mathrm{i}(\omega)}{\varepsilon_0} \approx O\left(\frac{1}{\omega^3}\right), \quad \text{as } \omega \to \infty. \tag{19.61}$$

In several later applications it will be sufficient to assume $\varepsilon_\mathrm{i}(\omega) \approx O(1/\omega^{2+\delta})$ as $\omega \to \infty$ with $\delta > 0$. In these same applications the alternative assumption that $\varepsilon_\mathrm{i}(\omega) \approx O(1/\omega^2 \log^\delta \omega)$, for $\delta > 1$, could be employed, without changing the outcome. Equation (19.60) coincides with the result obtained from the free-electron model discussed in the previous section; see Eq. (19.12). The Phragmén–Lindelöf theorem (Section 3.4.2) is employed to make the link between the asymptotic behavior on the real axis, with the asymptotic behavior for complex angular frequencies. Equations (19.60) and (19.61) hold for complex angular frequencies.

Equation (19.49) can be used to extend the dielectric function or the electric susceptibility as a function of a real angular frequency to a function of the complex angular frequency ω_z in the upper half complex angular frequency plane, so that

$$\frac{\varepsilon(\omega_z)}{\varepsilon_0} - 1 = \int_0^\infty G(t)\mathrm{e}^{\mathrm{i}\omega_\mathrm{r} t - \omega_\mathrm{i} t}\,\mathrm{d}\omega. \tag{19.62}$$

It is clear from Eq. (19.62) that, for $\omega_\mathrm{i} > 0$, the additional factor $\mathrm{e}^{-\omega_\mathrm{i} t}$ in the integrand will improve the convergence of the integral. If the discussion of this section had commenced with a harmonic time-dependent factor $\mathrm{e}^{\mathrm{i}\omega t}$, then $G(t)$ would also have the opposite phase choice in Eq. (19.42). In this case, the dielectric constant would be continued analytically as a function of a complex angular frequency into the lower half complex angular frequency plane.

Consider the contour integral

$$\oint_C \frac{\{\varepsilon(\omega_z)\varepsilon_0^{-1} - 1\}\mathrm{d}\omega_z}{\omega - \omega_z},$$

where C denotes the contour shown in Figure 19.2. From the Cauchy integral theorem, it follows that

$$\int_{-R}^{\omega-\rho} \frac{\{\varepsilon(\omega_\mathrm{r})\varepsilon_0^{-1} - 1\}\mathrm{d}\omega_\mathrm{r}}{\omega - \omega_\mathrm{r}} + \int_{\omega+\rho}^{R} \frac{\{\varepsilon(\omega_\mathrm{r})\varepsilon_0^{-1} - 1\}\mathrm{d}\omega_\mathrm{r}}{\omega - \omega_\mathrm{r}}$$
$$- \int_{\pi}^{0} \frac{\{\varepsilon(\omega + \rho \mathrm{e}^{\mathrm{i}\theta})\varepsilon_0^{-1} - 1\}\mathrm{i}\rho \mathrm{e}^{\mathrm{i}\theta}\,\mathrm{d}\theta}{\rho \mathrm{e}^{\mathrm{i}\theta}} + \int_{0}^{\pi} \frac{\{\varepsilon(R\mathrm{e}^{\mathrm{i}\theta})\varepsilon_0^{-1} - 1\}\mathrm{i}R\mathrm{e}^{\mathrm{i}\theta}\,\mathrm{d}\theta}{\omega - R\mathrm{e}^{\mathrm{i}\theta}} = 0. \tag{19.63}$$

Taking the limits $\rho \to 0$ and $R \to \infty$ leads to

$$\frac{\mathrm{i}}{\pi} P\int_{-\infty}^{\infty} \frac{\{\varepsilon(\omega_\mathrm{r})\varepsilon_0^{-1} - 1\}\mathrm{d}\omega_\mathrm{r}}{\omega - \omega_\mathrm{r}} - \pi^{-1} \lim_{R\to\infty} R \int_{0}^{\pi} \frac{\{\varepsilon(R\mathrm{e}^{\mathrm{i}\theta})\varepsilon_0^{-1} - 1\}\mathrm{e}^{\mathrm{i}\theta}\,\mathrm{d}\theta}{\omega - R\mathrm{e}^{\mathrm{i}\theta}} = \frac{\varepsilon(\omega)}{\varepsilon_0} - 1. \tag{19.64}$$

The integral on the large semicircular section of the contour in Figure 19.2 was previously dealt with by employing the known algebraic decay of the electric susceptibility, which was derived from the harmonically bound electron model discussed in Section 19.2. Here, the more general asymptotic expansion given in Eq. (19.59) is employed for $\varepsilon(\omega)$. Hence, it follows that

$$\frac{\mathrm{i}}{\pi} P\int_{-\infty}^{\infty} \frac{\{\varepsilon(\omega_\mathrm{r})\varepsilon_0^{-1} - 1\}\mathrm{d}\omega_\mathrm{r}}{\omega - \omega_\mathrm{r}} = \frac{\varepsilon(\omega)}{\varepsilon_0} - 1. \tag{19.65}$$

Equating the real and imaginary parts, and using the even character of $\varepsilon_\mathrm{r}(\omega)$ and the odd behavior of $\varepsilon_\mathrm{i}(\omega)$, leads to the Kramers–Kronig relations given in Eqs. (19.34) and (19.35).

A few more comments on the asymptotic behavior are appropriate. Suppose that $\lim_{\omega\to\infty} \varepsilon_0^{-1}\varepsilon(\omega) = \varepsilon_\infty$, where ε_∞ is some constant other than one; then, the preceding derivation could be carried through unaltered, except for the replacement of $\varepsilon_0^{-1}\varepsilon(\omega) - 1$ by $\varepsilon_0^{-1}\varepsilon(\omega) - \varepsilon_\infty$. The resulting dispersion relations are modified appropriately. Relativistic considerations do play a role, and this issue is discussed in Section 19.6. A second point concerns the following question. What is the effect of working directly with $\varepsilon_0^{-1}\varepsilon(\omega)$ in place of $\varepsilon_0^{-1}\varepsilon(\omega) - 1$ in the preceding analysis? If an angular frequency ω_L is selected much larger than ω in Eq. (19.63), then the integral along the real frequency axis can be written as follows:

$$\lim_{R\to\infty}\int_{-R}^{-\omega_L}\frac{\varepsilon(\omega_r)\varepsilon_0^{-1}\,d\omega_r}{\omega-\omega_r}+P\int_{-\omega_L}^{\omega_L}\frac{\varepsilon(\omega_r)\varepsilon_0^{-1}\,d\omega_r}{\omega-\omega_r}+\int_{\omega_L}^{R}\frac{\varepsilon(\omega_r)\varepsilon_0^{-1}\,d\omega_r}{\omega-\omega_r}.$$

Suppose $\lim_{\omega\to\infty}\varepsilon_0^{-1}\varepsilon(\omega)=\varepsilon_\infty$ and ω_L is selected sufficiently large so that $\varepsilon_0^{-1}\varepsilon(\omega)$ can be replaced by ε_∞, then the first and third integrals in the preceding expression can be written as follows:

$$\begin{aligned}\varepsilon_0^{-1}\lim_{R\to\infty}\left\{\int_{-R}^{-\omega_L}\frac{\varepsilon(\omega_r)d\omega_r}{\omega-\omega_r}+\int_{\omega_L}^{R}\frac{\varepsilon(\omega_r)d\omega_r}{\omega-\omega_r}\right\}&=\varepsilon_\infty\lim_{R\to\infty}\int_{\omega_L}^{R}\left\{\frac{1}{\omega_r+\omega}-\frac{1}{\omega_r-\omega}\right\}d\omega_r\\&=\varepsilon_\infty\log\left\{\frac{\omega_L-\omega}{\omega_L+\omega}\right\}\\&\to 0,\quad \text{as}\,\omega_L\to\infty.\end{aligned}\tag{19.66}$$

This is the cancellation feature of the Hilbert transform in the asymptotic regime, a feature first explored in Chapter 3. The Hilbert transform of a constant is zero because of the same cancellation effect. The integral along the large semicircular arc in Eq. (19.63) is given by

$$\varepsilon_0^{-1}\lim_{R\to\infty}\int_0^{\pi}\frac{\varepsilon(Re^{i\theta})iRe^{i\theta}\,d\theta}{\omega-Re^{i\theta}}=-i\varepsilon_0^{-1}\int_0^{\pi}\lim_{R\to\infty}\varepsilon(Re^{i\theta})d\theta.\tag{19.67}$$

If $\lim_{\omega_z\to\infty}\varepsilon_0^{-1}\varepsilon(\omega_z)=\varepsilon_\infty$, that is, suppose $\varepsilon(\omega_z)$ behaves asymptotically like a constant for complex frequencies, then

$$\varepsilon_0^{-1}\lim_{R\to\infty}\int_0^{\pi}\frac{\varepsilon(Re^{i\theta})iRe^{i\theta}\,d\theta}{\omega-Re^{i\theta}}=-i\pi\varepsilon_\infty.\tag{19.68}$$

Since a constant may be inserted in the numerator of the Hilbert transform without changing the value of the integral, the dispersion relations obtained are identical in form to those given previously. The essential modification is that there is a non-zero contribution from the integral on the semicircular arc, which gives the required constant term in the dispersion relation expression. Working directly with the function $\varepsilon_0^{-1}\varepsilon(\omega_z)-\varepsilon_\infty$ avoids the need for a detailed evaluation of the integral along the semicircular arc, since this contribution can be shown to vanish as $R\to\infty$. Note also that working directly with $\varepsilon_0^{-1}\varepsilon(\omega_z)-\varepsilon_\infty$ helps with the convergence properties of the Cauchy principal value integral, and hence has implications for the numerical approximation of the integral.

If $\varepsilon(\omega_z)$ became unbounded, then the derivation leading to Eq. (19.65) breaks down. In this case it would be necessary either to insert some type of weighting function, or to subtract from $\varepsilon(\omega_z)$ some appropriate factor such that the resulting function remained bounded in the limits $\omega_z\to\pm\infty$.

Here is a summary of the principal ideas that have been employed. The causality principle has played a central role. The assumption of system linearity has been imposed. A bounded condition on the dielectric constant or electric susceptibility is also required, with the link between the asymptotic behavior on the real frequency axis and in a general direction in the complex angular frequency plane being made by way of the Phragmén–Lindelöf theorem. All three of these constraints are required to derive the Kramers–Kronig relations in the form that has been given. Extension of the basic form of the Kramers–Kronig relations is developed later in this chapter.

19.4 An alternative approach to the Kramers–Kronig relations

In this section an alternative approach to obtain the Kramers–Kronig relations is considered which bypasses the use of complex variable theory. The key idea is to examine the Fourier transform of the optical constant. The reader should recall that the Hilbert transform and the Fourier transform are closely linked; see Section 5.2. Taking the Fourier transform of Eq. (19.50) leads to

$$\begin{aligned}\int_{-\infty}^{\infty}\left\{\frac{\varepsilon_r(\omega)}{\varepsilon_0}-1\right\}\mathrm{e}^{-\mathrm{i}\omega t'}\,\mathrm{d}\omega &= \int_{-\infty}^{\infty}\mathrm{e}^{-\mathrm{i}\omega t'}\,\mathrm{d}\omega\int_0^{\infty}G(t)\cos\omega t\,\mathrm{d}t\\ &= \lim_{\lambda\to\infty}\int_{-\lambda}^{\lambda}\cos\omega t'\,\mathrm{d}\omega\int_0^{\infty}G(t)\cos\omega t\,\mathrm{d}t\\ &= \lim_{\lambda\to\infty}\int_0^{\infty}G(t)\mathrm{d}t\int_{-\lambda}^{\lambda}\cos\omega t'\cos\omega t\,\mathrm{d}\omega\\ &= \lim_{\lambda\to\infty}\int_0^{\infty}G(t)\left\{\frac{\sin(t'-t)\lambda}{t'-t}+\frac{\sin(t'+t)\lambda}{t'+t}\right\}\mathrm{d}t\\ &= \pi\int_0^{\infty}G(t)\{\delta(t'-t)+\delta(t'+t)\}\mathrm{d}t, \qquad (19.69)\end{aligned}$$

where the last line follows on using Eq. (2.254). Employing the even character of $\varepsilon_r(\omega)$ leads to

$$G(t)=\frac{2}{\pi}\int_0^{\infty}\left\{\frac{\varepsilon_r(\omega)}{\varepsilon_0}-1\right\}\cos\omega t\,\mathrm{d}\omega. \qquad (19.70)$$

In a similar fashion, it follows from Eq. (19.51) that

$$\begin{aligned}\int_{-\infty}^{\infty}\frac{\varepsilon_i(\omega)}{\varepsilon_0}\mathrm{e}^{-\mathrm{i}\omega t'}\,\mathrm{d}\omega &= \int_{-\infty}^{\infty}\mathrm{e}^{-\mathrm{i}\omega t'}\,\mathrm{d}\omega\int_0^{\infty}G(t)\sin\omega t\,\mathrm{d}t\\ &= \lim_{\lambda\to\infty}\int_{-\lambda}^{\lambda}\sin\omega t'\,\mathrm{d}\omega\int_0^{\infty}G(t)\sin\omega t\,\mathrm{d}t\end{aligned}$$

$$= \lim_{\lambda \to \infty} \int_0^\infty G(t) \left\{ \frac{\sin(t'-t)\lambda}{t'-t} - \frac{\sin(t'+t)\lambda}{t'+t} \right\} dt$$

$$= \pi \int_0^\infty G(t)\{\delta(t'-t) - \delta(t'+t)\}dt. \tag{19.71}$$

Employing the odd character of $\varepsilon_i(\omega)$ leads to

$$G(t) = \frac{2}{\pi} \int_0^\infty \frac{\varepsilon_i(\omega)}{\varepsilon_0} \sin \omega t \, d\omega. \tag{19.72}$$

It follows directly from Eqs. (19.70) and (19.72) that

$$\int_0^\infty \left\{ \frac{\varepsilon_r(\omega)}{\varepsilon_0} - 1 \right\} \cos \omega t \, d\omega = \int_0^\infty \frac{\varepsilon_i(\omega)}{\varepsilon_0} \sin \omega t \, d\omega. \tag{19.73}$$

This is an example of a *sum rule*. The first sum rule encountered, Eq. (19.19), involved an ordinary sum. The terminology sum rule is also used to embrace extensions from the discrete to the continuous case, that is, integral constraints. The reader should note that this sum rule applies under rather general conditions, and is not tied to a particular model of a dielectric material. It can be used as a consistency check on the quality of measured values of $\varepsilon_r(\omega)$ and $\varepsilon_i(\omega)$. There are some issues associated with the use of sum rules, and that discussion is postponed to the latter part of this chapter, and additional discussion is given in the following two chapters.

Starting with Eq. (19.50) and using Eq. (19.72) leads to

$$\frac{\varepsilon_r(\omega)}{\varepsilon_0} - 1 = \frac{2}{\pi} \int_0^\infty \cos \omega t \, dt \int_0^\infty \frac{\varepsilon_i(\omega')}{\varepsilon_0} \sin \omega' t \, d\omega'$$

$$= \frac{2}{\pi} \lim_{\lambda \to \infty} \int_0^\infty \frac{\varepsilon_i(\omega')}{\varepsilon_0} d\omega' \int_0^\lambda \sin \omega' t \cos \omega t \, dt$$

$$= \frac{1}{\pi} \lim_{\lambda \to \infty} \int_0^\infty \frac{\varepsilon_i(\omega')}{\varepsilon_0} \left\{ \frac{1 - \cos(\omega'-\omega)\lambda}{\omega'-\omega} + \frac{1-\cos(\omega'+\omega)\lambda}{\omega'+\omega} \right\} d\omega'$$

$$= \frac{1}{\pi \varepsilon_0} \lim_{\eta \to 0} \left\{ \int_0^{\omega-\eta} \left[\varepsilon_i(\omega') \left\{ \frac{1}{\omega'-\omega} + \frac{1}{\omega'+\omega} \right\} \right] d\omega' \right.$$

$$\left. + \int_{\omega+\eta}^0 \left[\varepsilon_i(\omega') \left\{ \frac{1}{\omega'-\omega} + \frac{1}{\omega'+\omega} \right\} \right] d\omega' \right\}$$

$$- \lim_{\eta \to 0} \lim_{\lambda \to \infty} \left\{ \int_0^{\omega-\eta} \left[\varepsilon_i(\omega') \left\{ \frac{\cos(\omega'-\omega)\lambda}{\omega'-\omega} + \frac{\cos(\omega'+\omega)\lambda}{\omega'+\omega} \right\} \right] d\omega' \right.$$

$$\left. + \int_{\omega+\eta}^\infty \left[\varepsilon_i(\omega') \left\{ \frac{\cos(\omega'-\omega)\lambda}{\omega'-\omega} + \frac{\cos(\omega'+\omega)\lambda}{\omega'+\omega} \right\} \right] d\omega' \right\}. \tag{19.74}$$

Provided $\varepsilon_i(\omega)$ is a well behaved function on the interval $[0, \infty)$, which is expected to be the case since the experimental dielectric constant is assumed to be an integrable function, then the second pair of integrals Eq. (19.74) make a zero contribution, by the Riemann–Lebesgue lemma. Hence,

$$\varepsilon_{\mathrm{r}}(\omega) - \varepsilon_0 = -\frac{2}{\pi} P \int_0^\infty \frac{\omega' \varepsilon_{\mathrm{i}}(\omega') \mathrm{d}\omega'}{\omega^2 - \omega'^2}. \tag{19.75}$$

In a similar fashion, it follows from Eqs. (19.51) and (19.70) that

$$\begin{aligned}
\varepsilon_{\mathrm{i}}(\omega) &= \frac{2}{\pi} \int_0^\infty \sin \omega t \, \mathrm{d}t \int_0^\infty \left\{\varepsilon_{\mathrm{r}}(\omega') - \varepsilon_0\right\} \cos \omega' t \, \mathrm{d}\omega' \\
&= \frac{2}{\pi} \lim_{\lambda \to \infty} \int_0^\infty \left\{\varepsilon_{\mathrm{r}}(\omega') - \varepsilon_0\right\} \mathrm{d}\omega' \int_0^\lambda \sin \omega t \cos \omega' t \, \mathrm{d}t \\
&= \frac{1}{\pi} \lim_{\eta \to 0} \left\{ \int_0^{\omega - \eta} [\varepsilon_{\mathrm{r}}(\omega') - \varepsilon_0] \left\{ \frac{1}{\omega - \omega'} + \frac{1}{\omega + \omega'} \right\} \mathrm{d}\omega' \right. \\
&\quad \left. + \int_{\omega + \eta}^\infty \left[\varepsilon_{\mathrm{r}}(\omega') - \varepsilon_0\right] \left\{ \frac{1}{\omega - \omega'} + \frac{1}{\omega + \omega'} \right\} \mathrm{d}\omega' \right\} \\
&\quad - \lim_{\eta \to 0} \lim_{\lambda \to \infty} \left\{ \int_0^{\omega - \eta} \left[\varepsilon_{\mathrm{i}}(\omega') - \varepsilon_0\right] \left\{ \frac{\cos(\omega + \omega')\lambda}{\omega + \omega'} + \frac{\cos(\omega - \omega')\lambda}{\omega - \omega'} \right\} \mathrm{d}\omega' \right. \\
&\quad \left. + \int_{\omega + \eta}^\infty \left[\varepsilon_{\mathrm{i}}(\omega') - \varepsilon_0\right] \left\{ \frac{\cos(\omega + \omega')\lambda}{\omega + \omega'} + \frac{\cos(\omega - \omega')\lambda}{\omega - \omega'} \right\} \mathrm{d}\omega' \right\},
\end{aligned} \tag{19.76}$$

and hence

$$\varepsilon_{\mathrm{i}}(\omega) = \frac{2\omega}{\pi} P \int_0^\infty \frac{\{\varepsilon_{\mathrm{r}}(\omega') - \varepsilon_0\} \mathrm{d}\omega'}{\omega^2 - \omega'^2}. \tag{19.77}$$

Equations (19.75) and (19.77) are the Kramers–Kronig relations for the dielectric constant. The preceding derivation did not involve explicit reference to the theory of functions of a complex variable.

19.5 Direct derivation of the Kramers–Kronig relations on the interval $[0, \infty)$

In this section the derivation of the Kramers–Kronig relations is considered once again, but this time directly on the interval $[0, \infty)$, thereby avoiding the need to invoke the crossing symmetry relations that were used in Sections 19.2 and 19.3. First note from Eq. (19.49) that

$$\frac{\varepsilon(\mathrm{i}\omega)}{\varepsilon_0} - 1 = \int_0^\infty G(t) \mathrm{e}^{-\omega t} \, \mathrm{d}t, \tag{19.78}$$

and, since $G(t)$ is real, then, along the imaginary axis in the upper half complex angular frequency plane, $\varepsilon(\mathrm{i}\omega)$ is a real quantity. This fact is employed in the sequel.

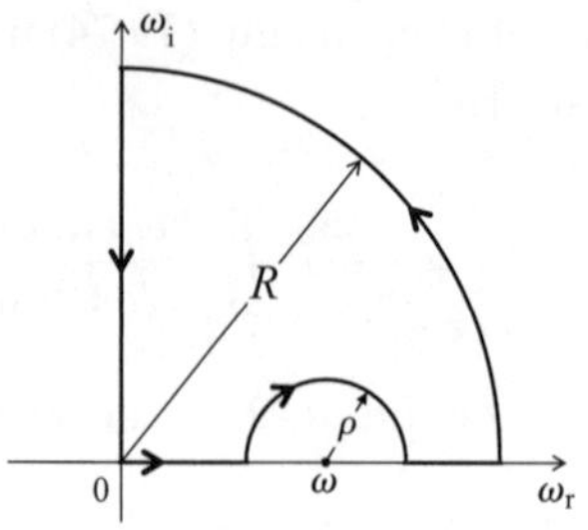

Figure 19.3. Contour for the derivation of the Kramers–Kronig relations on the interval $[0, \infty)$.

Consider the integral

$$\oint_C \frac{\{\varepsilon(\omega_z)\varepsilon_0^{-1} - 1\}\, \mathrm{d}\omega_z}{\omega^2 - \omega_z^2},$$

where C is the contour indicated in Figure 19.3. Application of the Cauchy integral theorem leads to

$$\begin{aligned}
&\int_0^{\omega-\rho} \frac{\left\{\varepsilon(\omega_r)\varepsilon_0^{-1} - 1\right\}\mathrm{d}\omega_r}{\omega^2 - \omega_r^2} + \int_{\omega+\rho}^{R} \frac{\left\{\varepsilon(\omega_r)\varepsilon_0^{-1} - 1\right\}\mathrm{d}\omega_r}{\omega^2 - \omega_r^2} \\
&\quad - \int_\pi^0 \frac{\left\{\varepsilon\left(\omega + \rho e^{i\theta}\right)\varepsilon_0^{-1} - 1\right\} i\rho e^{i\theta}\, \mathrm{d}\theta}{\rho e^{i\theta}\left(2\omega + \rho e^{i\theta}\right)} + \int_0^{\pi/2} \frac{\left\{\varepsilon\left(R e^{i\theta}\right)\varepsilon_0^{-1} - 1\right\} iR e^{i\theta}\, \mathrm{d}\theta}{\omega^2 - R^2 e^{2i\theta}} \\
&\quad + \int_R^0 \frac{\left\{\varepsilon(i\omega_i)\varepsilon_0^{-1} - 1\right\} i\, \mathrm{d}\omega_i}{\omega^2 + \omega_i^2} = 0.
\end{aligned} \tag{19.79}$$

Taking the limits $\rho \to 0$ and $R \to \infty$ yields

$$P\int_0^\infty \frac{\left\{\varepsilon(\omega_r)\varepsilon_0^{-1} - 1\right\}\mathrm{d}\omega_r}{\omega^2 - \omega_r^2} + \frac{i\pi\left\{\varepsilon(\omega)\varepsilon_0^{-1} - 1\right\}}{2\omega} = i\int_0^\infty \frac{\left\{\varepsilon(i\omega_i)\varepsilon_0^{-1} - 1\right\}\mathrm{d}\omega_i}{\omega^2 + \omega_i^2}. \tag{19.80}$$

Noting that the integrand of the integral on the right-hand side of the preceding result is real, taking the real part of this equation gives

$$\varepsilon_i(\omega) = \frac{2\omega}{\pi} P\int_0^\infty \frac{\{\varepsilon_r(\omega') - \varepsilon_0\}\mathrm{d}\omega'}{\omega^2 - \omega'^2}. \tag{19.81}$$

Starting instead with the integral

$$\oint_C \frac{\omega_z\{\varepsilon(\omega_z)\varepsilon_0^{-1} - 1\}\mathrm{d}\omega_z}{\omega^2 - \omega_z^2},$$

using the same contour shown in Figure 19.3, and applying the Cauchy integral theorem, gives

$$\begin{aligned}
&\int_0^{\omega-\rho} \frac{\omega_\mathrm{r}\{\varepsilon(\omega_\mathrm{r})\varepsilon_0^{-1} - 1\}\mathrm{d}\omega_\mathrm{r}}{\omega^2 - \omega_\mathrm{r}^2} + \int_{\omega+\rho}^{R} \frac{\omega_\mathrm{r}\{\varepsilon(\omega_\mathrm{r})\varepsilon_0^{-1} - 1\}\mathrm{d}\omega_\mathrm{r}}{\omega^2 - \omega_\mathrm{r}^2} \\
&\quad - \int_\pi^0 \frac{\{\omega + \rho\mathrm{e}^{\mathrm{i}\theta}\}\{\varepsilon(\omega + \rho\mathrm{e}^{\mathrm{i}\theta})\varepsilon_0^{-1} - 1\}\mathrm{i}\rho\mathrm{e}^{\mathrm{i}\theta}\,\mathrm{d}\theta}{\rho\mathrm{e}^{\mathrm{i}\theta}(2\omega + \rho\mathrm{e}^{\mathrm{i}\theta})} \\
&\quad + \int_0^{\pi/2} \frac{\{\varepsilon(R\mathrm{e}^{\mathrm{i}\theta})\varepsilon_0^{-1} - 1\}\mathrm{i}R^2\mathrm{e}^{2\mathrm{i}\theta}\,\mathrm{d}\theta}{\omega^2 - R^2\mathrm{e}^{2\mathrm{i}\theta}} - \int_R^0 \frac{\omega_\mathrm{i}\{\varepsilon(\mathrm{i}\omega_\mathrm{i})\varepsilon_0^{-1} - 1\}\mathrm{d}\omega_\mathrm{i}}{\omega^2 + \omega_\mathrm{i}^2} = 0.
\end{aligned} \tag{19.82}$$

Evaluating the limits $\rho \to 0$ and $R \to \infty$ leads to

$$P\int_0^\infty \frac{\omega'\{\varepsilon(\omega')\varepsilon_0^{-1} - 1\}\mathrm{d}\omega'}{\omega^2 - \omega'^2} + \frac{\mathrm{i}\pi\{\varepsilon(\omega)\varepsilon_0^{-1} - 1\}}{2} = -\int_0^\infty \frac{\omega_\mathrm{i}\{\varepsilon(\mathrm{i}\omega_\mathrm{i})\varepsilon_0^{-1} - 1\}\mathrm{d}\omega_\mathrm{i}}{\omega^2 + \omega_\mathrm{i}^2}. \tag{19.83}$$

The imaginary part of the last result yields

$$\varepsilon_\mathrm{r}(\omega) - \varepsilon_0 = -\frac{2}{\pi}P\int_0^\infty \frac{\omega'\varepsilon_\mathrm{i}(\omega')\mathrm{d}\omega'}{\omega^2 - \omega'^2}. \tag{19.84}$$

Equations (19.81) and (19.84) are the standard form of the Kramers–Kronig relations. This approach clearly establishes that the crossing symmetry conditions are not a required ingredient in the derivation of the Kramers–Kronig relations. The preceding approach also avoids dealing with the optical constant evaluated at unphysical negative angular frequencies.

19.6 The refractive index: Kramers–Kronig relations

In this section the focus is on a non-conducting medium, and the discussion is restricted to the homogenous isotropic linear case. With the assumption of a harmonic time dependence of the standard form $\mathrm{e}^{-\mathrm{i}\omega t}$, the Maxwell equations applied to an infinite medium lead to the Helmholtz equation

$$(\nabla^2 + \mu\varepsilon\omega^2)\mathbf{E} = 0, \tag{19.85}$$

with an analogous result for the magnetic induction. Suppose the propagation direction is along the x-axis and the solution is of the form of a plane wave, then, from

Eq. (19.85), it follows that

$$\mathbf{E}(x,t) = \mathbf{E}_0(\alpha e^{ikx - i\omega t} + \beta e^{-ikx - i\omega t}), \tag{19.86}$$

where k denotes the wave number and α and β are constants. Since experimental fields are real quantities, the standard convention of taking the real part of an expression for $\mathbf{E}(x,t)$ in terms of complex quantities is adopted. From Eqs. (19.85) and (19.86), the wave number is obtained from

$$k^2 = \mu\varepsilon\omega^2. \tag{19.87}$$

The phase velocity of the wave is defined by

$$v = \frac{c}{n}, \tag{19.88}$$

where c is the speed of light in a vacuum, taken by definition to be 299 792 458 m s^{-1}, and n is the index of refraction. The term refractive index is used synonymously with index of refraction. The phase velocity can also be written in terms of the wave number as

$$v = \frac{\omega}{k}. \tag{19.89}$$

Equations (19.87)–(19.89) can be combined with the result $c = 1/\sqrt{(\mu_0\varepsilon_0)}$ (using $n = 1$ for a vacuum), where μ_0 denotes the permeability of the vacuum, to give

$$N^2 = \mu\varepsilon c^2 = \frac{\mu\varepsilon}{\mu_0\varepsilon_0}. \tag{19.90}$$

The symbol change $n \to N$ has been made. The quantity N is called the complex refractive index, and is written as follows:

$$N(\omega) = n(\omega) + i\kappa(\omega), \tag{19.91}$$

where, as before, $n(\omega)$ denotes the ordinary refractive index and $\kappa(\omega)$ is a direct measure of the absorption of the wave as the beam propagates through the medium. On writing

$$\kappa(\omega) = \frac{c\alpha(\omega)}{2\omega}, \tag{19.92}$$

where $\alpha(\omega)$ denotes the absorption coefficient of the medium, the propagating beam intensity in the x-direction is attenuated like $e^{-\alpha(\omega)x}$. The reader should be alert to the almost identical notational representation commonly employed for the polarizability and the absorption coefficient; the context should always make it clear which quantity is being discussed.

The permeability often satisfies $\mu \approx \mu_0$ to a very good approximation, and accordingly the factor $\mu\mu_0^{-1}$ in Eq. (19.90) is dropped. Using Eq. (19.91) allows $n(\omega)$ and $\kappa(\omega)$ to be written in terms of $\varepsilon_{\rm r}(\omega)/\varepsilon_0$ and $\varepsilon_{\rm i}(\omega)/\varepsilon_0$, so that

$$n(\omega) = \sqrt{\left\{\frac{1}{2\varepsilon_0}\left[\varepsilon_{\rm r}(\omega) + \sqrt{\left(\varepsilon_{\rm r}(\omega)^2 + \varepsilon_{\rm i}(\omega)^2\right)}\right]\right\}}, \tag{19.93}$$

and

$$\kappa(\omega) = \sqrt{\left\{\frac{1}{2\varepsilon_0}\left[-\varepsilon_{\rm r}(\omega) + \sqrt{\left(\varepsilon_{\rm r}(\omega)^2 + \varepsilon_{\rm i}(\omega)^2\right)}\right]\right\}}. \tag{19.94}$$

Note that $\varepsilon_{\rm i}(\omega) > 0$ for $\omega > 0$, which requires that the medium in the absence of a variable electric field is in a state of thermodynamic equilibrium, a fact that has been implicitly assumed. The particular choice of signs in Eqs. (19.93) and (19.94) is chosen so that wave propagation takes place in the positive x-direction.

Neglecting the permeability, it follows from the connection given in Eq. (19.90) that $N(\omega_{\rm r} + {\rm i}\omega_{\rm i})^2 - 1$ is an analytic function in the upper half of the complex angular frequency plane, and hence, employing the discussion in Section 19.3, the following results are obtained:

$$n(\omega)\kappa(\omega) = \frac{\omega}{\pi} P \int_0^\infty \frac{[n(\omega')^2 - \kappa(\omega')^2 - 1]{\rm d}\omega'}{\omega^2 - \omega'^2} \tag{19.95}$$

and

$$n(\omega)^2 - \kappa(\omega)^2 - 1 = -\frac{4}{\pi} P \int_0^\infty \frac{\omega' n(\omega')\kappa(\omega'){\rm d}\omega'}{\omega^2 - \omega'^2}. \tag{19.96}$$

From a practical standpoint, these dispersion relations are not the most useful forms, since $n(\omega)$ and $\kappa(\omega)$ are both present in the integrands. These formulas could be used to provide consistency checks on experimental data, but would not allow the construction of one optical function from the other. Sum rules can be derived directly from these dispersion relations. More important dispersion relations are obtained by considering the function $N(\omega_{\rm r} + {\rm i}\omega_{\rm i}) - 1$. The latter function is analytic in the upper half of the complex angular frequency plane. The justification for this assertion is given momentarily. In the lower half plane this function has branch-cut features, and the degree of complexity of these is significant if a realistic model of the medium is employed.

There is no linear relationship between physically measurable quantities for which the proportionality factor is the refractive index. Contrast this situation with the relationship between **D** and **E**, where the proportionality term is the permittivity (see Eq. (19.4)). In order to obtain the dispersion relations for $N(\omega_{\rm r} + {\rm i}\omega_{\rm i}) - 1$, it is necessary to proceed in a different manner.

The asymptotic behavior of the complex refractive index can be evaluated from Eqs. (19.93) and (19.94), and hence

$$n(\omega) \approx \surd\left(\frac{\varepsilon_{\mathrm{r}}(\omega)}{\varepsilon_0}\right), \quad \text{as } \omega \to \infty; \tag{19.97}$$

that is,

$$n(\omega)^2 - 1 \approx \frac{\varepsilon_{\mathrm{r}}(\omega)}{\varepsilon_0} - 1 = O(\omega^{-2}), \quad \text{as } \omega \to \infty, \tag{19.98}$$

and for the imaginary part

$$\kappa(\omega) \approx \frac{\varepsilon_{\mathrm{i}}(\omega)}{2\varepsilon_0} = O(\omega^{-3}), \quad \text{as } \omega \to \infty, \tag{19.99}$$

where Eqs. (19.60) and (19.61) have been employed.

Since the function $\varepsilon(\omega_z)$ is analytic in the upper half complex angular frequency plane, it follows that $N(\omega_z) - 1$ is also analytic in the same region. For a system in thermodynamic equilibrium, the heat dissipation of an electromagnetic field, which is always greater than zero, is proportional to the imaginary component of the permittivity. In the upper half of the complex angular frequency plane, the permittivity has a non-zero imaginary component everywhere except along the imaginary frequency axis. The latter fact is proved later (see Eq. (19.235)). Along the imaginary axis, the permittivity decreases monotonically, approaching ε_0 as $\omega \to \infty$. Thus, there are no zeros for the permittivity in the upper half of the complex frequency plane, or along the real axis. The preceding arguments assume explicitly that spatial dispersion effects can be ignored. The issues associated with spatial dispersion are considered later, in Section 20.9. It is therefore concluded that the complex refractive index has no zeros in the upper half of the complex angular frequency plane and on the real axis. The only possible singularity in the upper half plane for the complex refractive index resides at $\omega = 0$ for the case of conductors, and when this occurs either the contour to be employed is suitably modified, or the integrand function involving the complex refractive index must be tailored so as to eliminate this singularity.

Consider the function $f(\omega_z) = c\{N(\omega_z) - 1\}$, with c a suitably chosen constant; then, from Eqs. (19.97) and (19.99), it follows that

$$|f(\omega_{\mathrm{r}})| \leq 1. \tag{19.100}$$

Let

$$M(R) = \max_{0 \leq \theta \leq \pi} \left| f(R\mathrm{e}^{\mathrm{i}\theta}) \right|, \tag{19.101}$$

over a semicircle in the upper half complex plane, of radius R and centered at the origin. From the Phragmén–Lindelöf theorem (recall Section 3.4.2) it follows that either $f(\omega_z)$ is bounded in the upper half plane,

$$|f(\omega_z)| \leq 1, \tag{19.102}$$

or

$$\lim_{R\to\infty} \inf \left\{ \frac{\log M(R)}{R} \right\} = \alpha > 0, \tag{19.103}$$

where α is a constant. If $M(R)$ exhibits an asymptotic behavior for large R of the form $M(R) \approx e^{\alpha R}$, with $\alpha > 0$, then clearly Eq. (19.103) results. If $M(R)$ behaves for large R like $M(R) \approx R^{-m}$, with $m > 0$, then Eq. (19.102) applies and the left-hand side of Eq. (19.103) evaluates to zero. The function $\{\varepsilon(\omega_z) - \varepsilon_0\}$ leads to an $M(R)$ of the form $M(R) \approx R^{-m}$, with $m > 0$, a result following from the bounded condition for $\{\varepsilon(\omega_r) - \varepsilon_0\}$ and Titchmarsh's theorem (see Section 4.22). From the definition Eq. (19.90), and assuming the permeability factor can be ignored, it is concluded that $N(\omega_z) - 1$ also has an $M(R)$ satisfying $M(R) \approx R^{-m}$, for some $m > 0$. Consequently, the condition given by Eq. (19.102) holds, and $N(\omega_z) - 1$ is a bounded function in the upper half complex angular frequency plane. It follows that $N(\omega_z) + 1$ is also a bounded function in the same region.

From the definition Eq. (19.90), and retaining the assumption that the permeability factor can be ignored, it follows by Titchmarsh's theorem and from the results of Section 19.3 that

$$\int_{-\infty}^{\infty} \left| \frac{\varepsilon(\omega_r + i\omega_i)}{\varepsilon_0} - 1 \right|^2 d\omega_r < C, \quad \text{for } \omega_i > 0, \tag{19.104}$$

where C is a positive constant, and hence

$$\int_{-\infty}^{\infty} \left| N(\omega_r + i\omega_i)^2 - 1 \right|^2 d\omega_r < C, \quad \text{for } \omega_i > 0. \tag{19.105}$$

In view of the fact that $N(\omega_z) + 1$ is a bounded function for $\omega_i > 0$,

$$\int_{-\infty}^{\infty} |N(\omega_r + i\omega_i) - 1|^2 \, d\omega_r < C', \tag{19.106}$$

where C' is a positive constant. Applying Titchmarsh's theorem allows the Hilbert transform pair to be written, and so

$$\kappa(\omega) = \frac{1}{\pi} P \int_{-\infty}^{\infty} \frac{\{n(\omega_r) - 1\} d\omega_r}{\omega - \omega_r} \tag{19.107}$$

and

$$n(\omega) - 1 = -\frac{1}{\pi} P \int_{-\infty}^{\infty} \frac{\kappa(\omega_r) d\omega_r}{\omega - \omega_r}. \tag{19.108}$$

From Titchmarsh's theorem it follows that

$$N(\omega) - 1 = \int_0^\infty R(t)\mathrm{e}^{\mathrm{i}\omega t}\,\mathrm{d}t, \tag{19.109}$$

where $R(t)$, which plays the role of a response function, does not have the same physical realization as does the response function $G(t)$ given in Eq. (19.49), which links, via Eq. (19.43), the electric displacement with the electric field.

The crossing symmetry relation for the refractive index is given by

$$N(-\omega_z^*) = N(\omega_z)^*. \tag{19.110}$$

This result follows directly from Eq. (19.109). From this result it follows that

$$n(-\omega) = n(\omega) \tag{19.111}$$

and

$$\kappa(-\omega) = -\kappa(\omega). \tag{19.112}$$

Using these two conditions, the Hilbert transform pair can be written as follows:

$$\kappa(\omega) = \frac{2\omega}{\pi} P \int_0^\infty \frac{\{n(\omega') - 1\}\mathrm{d}\omega'}{\omega^2 - \omega'^2} \tag{19.113}$$

and

$$n(\omega) - 1 = -\frac{2}{\pi} P \int_0^\infty \frac{\omega'\kappa(\omega')\mathrm{d}\omega'}{\omega^2 - \omega'^2}. \tag{19.114}$$

These are the Kramers–Kronig relations for the refractive index. They are also commonly referred to as the dispersion relations for the refractive index.

In contrast to the dispersion relations given in Eqs. (19.95) and (19.96), the preceding pair of dispersion relations allows the function $\kappa(\omega)$ to be determined from knowledge of $n(\omega) - 1$ and vice versa. A principal difficulty in the determination of one optical constant from another is the limitation that the experimentally determined optical constant is never measured over an infinite spectral interval. There are two principal ways to proceed. The first is to assume that, to a very good approximation, the integration interval may be restricted, for example,

$$n(\omega) - 1 \approx -\frac{2}{\pi} P \int_{\omega_1}^{\omega_2} \frac{\omega'\kappa(\omega')\mathrm{d}\omega'}{\omega^2 - \omega'^2}, \tag{19.115}$$

where the angular frequencies ω_1 and ω_2 are selected so that outside the interval $[\omega_1, \omega_2]$, the function $\kappa(\omega)$ is sufficiently small so that the integrals

$$P\int_0^{\omega_1} \frac{\omega'\kappa(\omega')\mathrm{d}\omega'}{\omega^2 - \omega'^2}$$

and

$$P\int_{\omega_2}^{\infty} \frac{\omega'\kappa(\omega')\mathrm{d}\omega'}{\omega^2 - \omega'^2}$$

make a very small contribution relative to the value obtained for the integral over the interval $[\omega_1, \omega_2]$. An alternative strategy is to attempt an extrapolation of the measured experimental data to the asymptotic regime $\omega \to \infty$, and to the low-frequency domain. This is achieved by adopting a model of the medium. For example, the high-frequency limit could be modeled along the lines discussed in Section 19.2.

The other issue that needs to be considered is the requirement for data smoothing. Since the measured data are typically discrete, or comprise a combination of discrete and continuous data over some different frequency intervals, good data fitting routines with reliable error estimations are essential. High quality data fitting is particularly important in the region around the singularity in the integrand. Errors in data fitting close to the singular frequency have the potential to be significantly magnified as the contributions to the integral are evaluated as $\omega' \to \omega$. In optical data analysis the position of the singularity is usually varied over a wide spectral interval, and hence excellent data fitting over the same range is essential if an accurate Kramers–Kronig transform of the data set is to be obtained.

Employing the definition of the absorption coefficient given in Eq. (19.92), and making use of Eq. (19.114), leads to

$$n(\omega) = 1 - \frac{c}{\pi} P\int_0^{\infty} \frac{\alpha(\omega')\mathrm{d}\omega'}{\omega^2 - \omega'^2}. \tag{19.116}$$

Historically, this is the first form of the dispersion relation obtained by Kronig (1926).

If in the preceding developments the asymptotic behavior of $n(\omega)$ was $n(\omega) \to n(\infty)$, as $\omega \to \infty$, and $n(\infty)$ was some constant other than unity, then the derivation of the dispersion relations could be repeated with $n(\omega) - 1$ simply replaced by $n(\omega) - n(\infty)$. However, if the restriction is imposed that the *front velocity* of the wave does not exceed the speed of light as $\omega \to \infty$, then the requirement $n(\omega) \to 1$ as $\omega \to \infty$ follows. This can be viewed as a relativistic constraint on the system. The front is the leading edge of the electromagnetic pulse. The issue involved here is commonly referred to as the principle of relativistic causality: causal influences, that is, information, cannot propagate at speeds in excess of the speed of light. This is also commonly stated in the form that no signal can transmit information outside of the source light-cone or, as it is commonly termed, the forward light-cone. The light-cone is the space-time surface on which the electromagnetic disturbance travels from its source. Relativistic causality is also often stated in more than one way. The idea that is invoked in the preceding discussion is often referred to as Einstein causality. The reader is directed to Nussenzveig (1972) for further information on causality.

19.7 Application of Herglotz functions

In this section a somewhat lesser known approach to the derivation of the Kramers–Kronig relations is considered. The derivation is carried forward with a fairly general optical property X, with the principal restriction being that the imaginary part of X satisfies the following condition:

$$\mathrm{Im}\, X(\omega) \geq 0. \tag{19.117}$$

Since the imaginary component of X typically represents the dissipative behavior of the system, or can be selected in a manner so that it is directly proportional to this contribution, Eq. (19.117) applies. Herglotz functions play an essential role in the sequel.

The following theorem due to Herglotz is central to the following development. Let $g(w)$ be analytic in $|w| < 1$, and suppose $\mathrm{Im}\, g(w) \geq 0$ in the same domain; then the function $g(w)$ admits the integral representation

$$g(w) = \mathrm{i} \int_0^{2\pi} \frac{\mathrm{e}^{\mathrm{i}\theta} + w}{\mathrm{e}^{\mathrm{i}\theta} - w} \,\mathrm{d}\beta(\theta) + C. \tag{19.118}$$

In Eq. (19.118), $\beta(\theta)$ is a non-decreasing bounded real function and C is a real constant. A function satisfying these two conditions is called a Herglotz function. Equation (19.118) can be converted to a result for the upper half plane in the following way. Employ the conformal mapping

$$z = \mathrm{i}\frac{1 + w}{1 - w}, \tag{19.119}$$

which converts the interior of the disc to the upper half complex plane. Introducing the change of variable

$$t = -\mathrm{i}\frac{\mathrm{e}^{\mathrm{i}\theta} + 1}{\mathrm{e}^{\mathrm{i}\theta} - 1}, \tag{19.120}$$

that is

$$t = -\cot\frac{\theta}{2}, \tag{19.121}$$

and making use of the substitutions $f(z) = g(w)$ and $\alpha(t) = \beta(\theta)$, yields

$$f(z) = Az + \int_{-\infty}^{\infty} \frac{1 + tz}{t - z} \,\mathrm{d}\alpha(t) + C. \tag{19.122}$$

In Eq. (19.122), A is a positive real constant, and the Az term arises from the possible jump of $\beta(\theta)$ at $\theta=0$ and $\theta=2\pi$. Cauer (1932) and Wall (1948, p. 277) can be consulted for some additional comments on the derivation. Equation (19.122) is the key formula needed to derive the Kramers–Kronig relations between the real and imaginary parts of a general optical property.

Suppose $X(\omega_z)$ satisfies the condition given in Eq. (19.117) and is analytic in the upper half complex frequency plane. Then it is a Herglotz function, so that

$$\begin{aligned} X(\omega_z) &= A\omega_z + \int_{-\infty}^{\infty} \frac{1+\omega'\omega_z}{\omega'-\omega_z}\,\mathrm{d}\alpha(\omega') + C \\ &= A(\omega_\mathrm{r} + \mathrm{i}\omega_\mathrm{i}) + C \\ &\quad + \int_{-\infty}^{\infty} \frac{[(1+\omega'\omega_\mathrm{r})(\omega'-\omega_\mathrm{r}) - \omega'\omega_\mathrm{i}^2] + \mathrm{i}\omega_\mathrm{i}(\omega'^2+1)}{(\omega'-\omega_\mathrm{r})^2+\omega_\mathrm{i}^2}\,\mathrm{d}\alpha(\omega'). \end{aligned} \quad (19.123)$$

In Eq. (19.123) and the remainder of this section, frequency terms are rendered dimensionless by dividing each by 1 Hz, when this is necessary for dimensional considerations. At this juncture it is assumed that $\alpha(t)$ is differentiable everywhere; then the Stieltjes integral can be rewritten using $\mathrm{d}\alpha(\omega) = \alpha'(\omega)\mathrm{d}\omega$. This assumption is considered further in the following development. Taking the limit $\omega_\mathrm{i} \to 0$, using Eq. (10.1), and denoting ω_r by ω, then Eq. (19.123) can be recast as follows:

$$\begin{aligned} X(\omega) &= A\omega + C + \lim_{\omega_\mathrm{i}\to 0} \int_{-\infty}^{\infty} \frac{[(1+\omega'\omega)(\omega'-\omega) - \omega'\omega_\mathrm{i}^2]}{(\omega'-\omega)^2+\omega_\mathrm{i}^2}\alpha'(\omega')\mathrm{d}\omega' \\ &\quad + \lim_{\omega_\mathrm{i}\to 0} \int_{-\infty}^{\infty} \frac{\mathrm{i}\omega_\mathrm{i}(\omega'^2+1)}{(\omega'-\omega)^2+\omega_\mathrm{i}^2}\alpha'(\omega')\mathrm{d}\omega', \end{aligned} \quad (19.124)$$

which simplifies to yield

$$\begin{aligned} X(\omega) &= A\omega + C + P\int_{-\infty}^{\infty} \frac{1+\omega'\omega}{\omega'-\omega}\alpha'(\omega')\mathrm{d}\omega' \\ &\quad + \mathrm{i}\pi \int_{-\infty}^{\infty} \delta(\omega'-\omega)[\omega'^2+1]\alpha'(\omega')\mathrm{d}\omega' \\ &= A\omega + C + P\int_{-\infty}^{\infty} \frac{1+\omega'\omega}{\omega'-\omega}\alpha'(\omega')\mathrm{d}\omega' + \mathrm{i}\pi(1+\omega^2)\alpha'(\omega). \end{aligned} \quad (19.125)$$

The imaginary part of Eq. (19.125) leads to

$$\alpha'(\omega) = \frac{\operatorname{Im} X(\omega)}{\pi(1+\omega^2)}. \quad (19.126)$$

For a homogeneous medium, the right-hand side of Eq. (19.126) is expected to be a continuous function for insulators. For conducting materials, the same continuity condition applies, except at $\omega = 0$. For inhomogeneous materials, the situation is less obvious, and it is not at all clear that $\operatorname{Im} X(\omega)$ is necessarily a continuous function. Inserting Eq. (19.126) into Eq. (19.125), and taking the real part, yields

$$\operatorname{Re} X(\omega) = A\omega + C - \frac{1}{\pi} P \int_{-\infty}^{\infty} \frac{(1+\omega'\omega) \operatorname{Im} X(\omega')\mathrm{d}\omega'}{(\omega-\omega')(1+\omega'^2)}. \tag{19.127}$$

One possibility is that $X(\omega)$ satisfies a crossing symmetry constraint of the following form:

$$X(-\omega) = -X^*(\omega), \tag{19.128}$$

which implies that the imaginary component is an even function and the real part is an odd function. If Eq. (19.128) holds, then Eq. (19.127) can be rewritten as follows:

$$\operatorname{Re} X(\omega) = A\omega + C - \frac{2\omega}{\pi} P \int_0^{\infty} \frac{\operatorname{Im} X(\omega')\mathrm{d}\omega'}{\omega^2-\omega'^2}. \tag{19.129}$$

An alternative possibility is that $X(\omega)$ satisfies the crossing symmetry constraint

$$X(-\omega) = X^*(\omega). \tag{19.130}$$

In this case, Eq. (19.127) can be recast as follows:

$$\operatorname{Re} X(\omega) = A\omega + C - \frac{2(1+\omega^2)}{\pi} P \int_0^{\infty} \frac{\omega' \operatorname{Im} X(\omega')\mathrm{d}\omega'}{(\omega^2-\omega'^2)(1+\omega'^2)}. \tag{19.131}$$

Employing a partial fraction simplification of the integrand yields

$$\operatorname{Re} X(\omega) = A\omega + C - \frac{2}{\pi} \int_0^{\infty} \frac{\omega' \operatorname{Im} X(\omega')\mathrm{d}\omega'}{1+\omega'^2} - \frac{2}{\pi} P \int_0^{\infty} \frac{\omega' \operatorname{Im} X(\omega')\mathrm{d}\omega'}{\omega^2-\omega'^2}. \tag{19.132}$$

The constants A and C in Eq. (19.129) can be determined in the following manner. The limit $\omega \to 0$ yields

$$C = \operatorname{Re} X(0). \tag{19.133}$$

The behavior of the integral in Eq. (19.129), as $\omega \to 0$, can be best examined by writing $2\omega(\omega^2 - \omega'^2)^{-1} = (\omega - \omega')^{-1} + (\omega + \omega')^{-1}$. Dividing both sides of Eq. (19.129) by ω and examining the limit $\omega \to \infty$ leads to

$$A = \left.\frac{\operatorname{Re} X(\omega)}{\omega}\right|_{\omega\to\infty}. \tag{19.134}$$

Substituting these results into Eq. (19.129) yields

$$\operatorname{Re}\big\{X(\omega) - X(0) - \omega\{\omega^{-1}X(\omega)\}_{\omega\to\infty}\big\} = -\frac{2\omega}{\pi} P \int_0^{\infty} \frac{\operatorname{Im} X(\omega')\mathrm{d}\omega'}{\omega^2-\omega'^2}. \tag{19.135}$$

In a similar fashion, Eq. (19.132) can be rewritten as

$$\begin{aligned}
&\operatorname{Re}\big\{X(\omega) - X(0) - \omega\{\omega^{-1}X(\omega)\}_{\omega\to\infty}\big\} \\
&= -\frac{2}{\pi}P\int_0^\infty \frac{\omega' \operatorname{Im} X(\omega')\mathrm{d}\omega'}{\omega^2 - \omega'^2} - \frac{2}{\pi}P\int_0^\infty \frac{\operatorname{Im} X(\omega')\mathrm{d}\omega'}{\omega'} \\
&= -\frac{2\omega^2}{\pi}P\int_0^\infty \frac{\operatorname{Im} X(\omega')\mathrm{d}\omega'}{\omega'(\omega^2 - \omega'^2)}.
\end{aligned} \tag{19.136}$$

In Eq. (19.136) it is assumed that the integral $P\int_0^\infty \omega'^{-1} \operatorname{Im} X(\omega')\mathrm{d}\omega'$ is convergent.

To see how the preceding development applies, some applications involving the complex refractive index $N(\omega)$ are considered. Let

$$X(\omega) = N(\omega) - 1. \tag{19.137}$$

This particular choice is analytic in the upper half complex frequency plane, and $\operatorname{Im} X(\omega) = \kappa(\omega)$, which satisfies Eq. (19.117). From the result that $\operatorname{Re} X(\omega) = n(\omega) - 1$, and using the asymptotic behavior $n(\omega) - 1 = O(\omega^{-2})$, as $\omega \to \infty$, which follows from Eq. (19.98), leads to

$$\omega^{-1} \operatorname{Re} X(\omega)\Big|_{\omega\to\infty} = 0. \tag{19.138}$$

The crossing symmetry condition given by Eq. (19.130) applies for the choice employed in Eq. (19.137), and hence from Eq. (19.136) it follows that

$$n(\omega) - n(0) = -\frac{2}{\pi}P\int_0^\infty \frac{\omega'\kappa(\omega')\mathrm{d}\omega'}{\omega^2 - \omega'^2} - \frac{2}{\pi}\int_0^\infty \frac{\kappa(\omega')\mathrm{d}\omega'}{\omega'}. \tag{19.139}$$

In this form, the result applies to insulators. The case of conductors is excluded because of the behavior of $\kappa(\omega)$ as $\omega \to 0$, $\kappa(\omega) \approx 1/\sqrt{\omega}$, which can be obtained from Eq. (19.178). The refractive index $n(\omega)$ for a conductor displays a similar behavior as $\omega \to 0$, that is $n(\omega) \approx 1/\sqrt{\omega}$. Equation (19.139) can be recast as follows:

$$n(\omega) - n(0) = -\frac{2\omega^2}{\pi}P\int_0^\infty \frac{\kappa(\omega')\mathrm{d}\omega'}{\omega'(\omega^2 - \omega'^2)}, \tag{19.140}$$

which is a single subtracted form of the dispersion relation for the refractive index. A more general single subtractive dispersion relation can be derived by subtracting the dispersion relation for $n(\omega'')$ from the dispersion relation for $n(\omega)$.

As an alternative selection for $X(\omega)$, consider

$$X(\omega) = \omega N(\omega). \tag{19.141}$$

For this case, Eq. (19.117) is satisfied, $\mathrm{Re}\,X(0) = 0$, and

$$\omega^{-1}\,\mathrm{Re}\,X(\omega)\Big|_{\omega\to\infty} = \lim_{\omega\to\infty} n(\omega) = 1. \tag{19.142}$$

The crossing symmetry condition given in Eq. (19.128) applies in this case, and hence Eq. (19.135) yields

$$n(\omega) - 1 = -\frac{2}{\pi} P \int_0^\infty \frac{\omega'\kappa(\omega')\mathrm{d}\omega'}{\omega^2 - \omega'^2}. \tag{19.143}$$

This is the standard form of the Kramers–Kronig connection relating $n(\omega) - 1$ to $\kappa(\omega)$ (recall Eq. (19.114)).

Another choice for $X(\omega)$ is

$$X(\omega) = \omega\{N(\omega) - 1\}, \tag{19.144}$$

which satisfies Eq. (19.117), $\mathrm{Re}\,X(0) = 0$, and

$$\omega^{-1}\,\mathrm{Re}\,X(\omega)\Big|_{\omega\to\infty} = 0, \tag{19.145}$$

with the result that Eq. (19.143) is obtained from Eq. (19.135).

The Kramers–Kronig relations come in pairs. Some basic properties of the Hilbert transform can be employed to obtain the second Kramers–Kronig formula. Equation (19.127) can be written as

$$\mathrm{Re}\,X(\omega) = A\omega + C - Hf(\omega) - \omega Hg(\omega), \tag{19.146}$$

where the functions f and g are introduced as follows:

$$f(\omega) = \frac{\mathrm{Im}\,X(\omega)}{1+\omega^2} \tag{19.147}$$

and

$$g(\omega) = \omega f(\omega). \tag{19.148}$$

To proceed further, the following restrictions are imposed: (i) $A = 0$, which assumes that $\mathrm{Re}\,X(\omega)$ has an appropriate asymptotic behavior as $\omega \to \infty$; (ii) $X(-\omega) = X^*(\omega)$, which is a common crossing symmetry condition satisfied by many optical properties; (iii) $g(t) \in L^p(\mathbb{R})$, for $1 < p < \infty$, and (iv) $\mathrm{Re}\,X(\omega) \in L^q(\mathbb{R})$, for $1 \le q < \infty$. The latter requirement is sufficient to ensure that the Hilbert transform of $\mathrm{Re}\,X(\omega)$ exists; in order to ensure that it is bounded, the condition on q must be changed to $1 < q < \infty$. Applying the Hilbert transform operator to Eq. (19.146) leads to

$$H\,\mathrm{Re}\,X(\omega) = H[C] - H^2 f(\omega) - H[\omega Hg(\omega)]. \tag{19.149}$$

Let $h(\omega) = Hg(\omega)$, then the moment formula of the Hilbert transform (recall Eq. (4.111)) can be applied to yield

$$H[\omega h(\omega)] = \omega Hh(\omega) - \frac{1}{\pi}\int_{-\infty}^{\infty} h(x)\mathrm{d}x. \tag{19.150}$$

If condition (ii) applies, then $\mathrm{Im}\, X(t)$ is an odd function and hence $g(t)$ is an even function. Recalling that the Hilbert transform of an even function is an odd function (see Section 4.2), then $h(x)$ is odd, and so Eq. (19.150) simplifies to give

$$H[\omega h(\omega)] = \omega H^2 g(\omega). \tag{19.151}$$

Condition (iii) allows the inversion formula of the Hilbert transform to be applied as follows:

$$H^2 f(\omega) = -f(\omega), \quad H^2 g(\omega) = -g(\omega), \tag{19.152}$$

and, recalling that the Hilbert transform of a constant is zero, Eq. (19.149) can be recast as

$$H\,\mathrm{Re}\, X(\omega) = f(\omega) + \omega g(\omega). \tag{19.153}$$

This result simplifies on inserting the definitions of f and g to yield

$$\mathrm{Im}\, X(\omega) = H\,\mathrm{Re}\, X(\omega). \tag{19.154}$$

Employing condition (ii) leads to

$$\mathrm{Im}\, X(\omega) = \frac{2\omega}{\pi} P\int_0^{\infty} \frac{\mathrm{Re}\, X(\omega')\mathrm{d}\omega'}{\omega^2 - \omega'^2}. \tag{19.155}$$

This is the standard form of one of the Kramers–Kronig relations.

As an application, consider the choice given in Eq. (19.137), and note that

$$\omega^{-1}\,\mathrm{Re}\, X(\omega)\Big|_{\omega\to\infty} = \omega^{-1}\{n(\omega) - 1\}\Big|_{\omega\to\infty} = 0. \tag{19.156}$$

Conditions (ii), (iii), and (iv) are satisfied, so that Eq. (19.155) yields

$$\kappa(\omega) = \frac{2\omega}{\pi} P\int_0^{\infty} \frac{\{n(\omega') - 1\}\mathrm{d}\omega'}{\omega^2 - \omega'^2}, \tag{19.157}$$

which is the well known Kramers–Kronig connection between $\kappa(\omega)$ and $n(\omega) - 1$.

An alternative approach can be given to obtain the second of the Kramers–Kronig pair. The real part of Eq. (19.125) yields

$$\mathrm{Re}\, X(\omega) = A\omega + C + P\int_{-\infty}^{\infty} \frac{1 + \omega^2 + (\omega' - \omega)\omega}{\omega' - \omega} \alpha'(\omega')\mathrm{d}\omega'. \tag{19.158}$$

This result can be rewritten as follows:

$$\begin{aligned}\operatorname{Re} X(\omega) &= A\omega + C - \pi(1+\omega^2)H[\alpha'(\omega)] + \omega\{\alpha(\infty) - \alpha(-\infty)\}\\ &= A\omega + C - \pi(1+\omega^2)H[\alpha'(\omega)],\end{aligned} \tag{19.159}$$

where $\alpha(\omega)$ is taken to vanish as $|\omega| \to \infty$. The latter result can be justified by reference to Eq. (19.126). If both sides of the preceding result are multiplied by $(1+\omega^2)^{-1}$ and the Hilbert transform operator is applied, then application of the inversion property of the Hilbert transform yields

$$\pi\alpha'(\omega) = \frac{A}{1+\omega^2} - \frac{C\omega}{1+\omega^2} + H\left\{\frac{\operatorname{Re} X(\omega)}{1+\omega^2}\right\}. \tag{19.160}$$

Inserting this result into Eq. (19.126) yields

$$\operatorname{Im} X(\omega) = A - C\omega + (1+\omega^2)H\left\{\frac{\operatorname{Re} X(\omega)}{1+\omega^2}\right\}. \tag{19.161}$$

The Hilbert transform of $(1+\omega^2)^{-1}\operatorname{Re} X(\omega)$ can be simplified by writing

$$\frac{1}{(\omega-\omega')(1+\omega'^2)} = \frac{1}{(1+\omega^2)}\left\{\frac{\omega}{1+\omega'^2} + \frac{1}{\omega-\omega'} + \frac{\omega'}{1+\omega'^2}\right\}, \tag{19.162}$$

and hence

$$\begin{aligned}\operatorname{Im} X(\omega) = A - C\omega + H[\operatorname{Re} X(\omega)] + \frac{\omega}{\pi}\int_{-\infty}^{\infty}\frac{\operatorname{Re} X(\omega')\mathrm{d}\omega'}{1+\omega'^2}\\ + \frac{1}{\pi}\int_{-\infty}^{\infty}\frac{\omega'\operatorname{Re} X(\omega')\mathrm{d}\omega'}{1+\omega'^2}.\end{aligned} \tag{19.163}$$

The two constants A' and C' are introduced by the following definitions:

$$A' = \frac{1}{\pi}\int_{-\infty}^{\infty}\frac{\omega'\operatorname{Re} X(\omega')\mathrm{d}\omega'}{1+\omega'^2} \tag{19.164}$$

and

$$C' = \frac{1}{\pi}\int_{-\infty}^{\infty}\frac{\operatorname{Re} X(\omega')\mathrm{d}\omega'}{1+\omega'^2}, \tag{19.165}$$

with both integrals assumed to be convergent. On assigning the values $c_1 = A + A'$ and $c_2 = C' - C$, Eq. (19.163) yields

$$\operatorname{Im} X(\omega) = c_1 + c_2\omega + H[\operatorname{Re} X(\omega)]. \tag{19.166}$$

If $X(\omega)$ satisfies the crossing symmetry condition given in Eq. (19.128), then

$$\operatorname{Im} X(\omega) = c_1 + c_2\omega + \frac{2}{\pi} P \int_0^\infty \frac{\omega' \operatorname{Re} X(\omega')\mathrm{d}\omega'}{\omega^2 - \omega'^2}. \tag{19.167}$$

If $X(\omega)$ satisfies the alternative crossing symmetry condition of Eq. (19.130), then Eq. (19.166) leads to

$$\operatorname{Im} X(\omega) = c_1 + c_2\omega + \frac{2\omega}{\pi} P \int_0^\infty \frac{\operatorname{Re} X(\omega')\mathrm{d}\omega'}{\omega^2 - \omega'^2}. \tag{19.168}$$

The constants Eqs. (19.167 and 19.168) can be fixed in the same manner as previously discussed. The results are as follows:

$$\begin{aligned}\operatorname{Im} X(\omega) &= \operatorname{Im} X(0) + \omega\left\{\omega^{-1} \operatorname{Im} X(\omega)\right\}_{\omega\to\infty} \\ &\quad + \frac{2}{\pi} P \int_0^\infty \omega' \operatorname{Re} X(\omega') \left\{\frac{1}{\omega^2 - \omega'^2} + \frac{1}{\omega'^2}\right\} \mathrm{d}\omega'\end{aligned} \tag{19.169}$$

and

$$\operatorname{Im} X(\omega) = \operatorname{Im} X(0) + \omega\{\omega^{-1} \operatorname{Im} X(\omega)\}_{\omega\to\infty} + \frac{2\omega}{\pi} P \int_0^\infty \frac{\operatorname{Re} X(\omega')\mathrm{d}\omega'}{\omega^2 - \omega'^2}. \tag{19.170}$$

To arrive at Eq. (19.169) requires that the integral $\int_0^\infty \omega'^{-1} \operatorname{Re} X(\omega')\mathrm{d}\omega'$ be convergent, and to obtain Eq. (19.170) assumes that $\lim_{\omega\to 0}\left\{\omega P \int_0^\infty (\omega^2 - \omega'^2)^{-1} \operatorname{Re} X(\omega')\,\mathrm{d}\omega'\right\}$ is zero.

For the choice of $X(\omega)$ given in Eq. (19.137), Im $X(0) = 0$ for insulators, $\left\{\omega^{-1} \operatorname{Im} X(\omega)\right\}_{\omega\to\infty} = 0$, and Eq. (19.130) applies, so Eq. (19.170) leads to

$$\kappa(\omega) = \frac{2\omega}{\pi} P \int_0^\infty \frac{\{n(\omega') - 1\}\mathrm{d}\omega'}{\omega^2 - \omega'^2}, \tag{19.171}$$

which is the standard form of one of the Kramers–Kronig pair.

For the choice of $X(\omega)$ indicated in Eq. (19.141), the crossing symmetry constraint given in Eq. (19.128) applies; however,

$$\int_0^\infty \frac{\operatorname{Re}\{\omega N(\omega)\}}{\omega} \mathrm{d}\omega = \int_0^\infty n(\omega)\mathrm{d}\omega = \infty, \tag{19.172}$$

so that Eq. (19.169) cannot be applied for this choice of $X(\omega)$.

A small modification of the previous choice gives the case shown in Eq. (19.144), and this fixes the previous divergent integral problem. For this choice, $\operatorname{Im} X(0) = 0$ and $\left\{\omega^{-1} \operatorname{Im} X(\omega)\right\}_{\omega\to\infty} = 0$, and the crossing symmetry condition of Eq. (19.128) applies. Hence, Eq. (19.169) leads to

$$\omega\kappa(\omega) = \frac{2}{\pi}P\int_0^\infty \omega'^2\{n(\omega') - 1\}\left\{\frac{1}{\omega^2 - \omega'^2} + \frac{1}{\omega'^2}\right\}\mathrm{d}\omega'. \tag{19.173}$$

This result simplifies to give

$$\begin{aligned}\omega\kappa(\omega) &= \frac{2}{\pi}P\int_0^\infty \frac{(\omega'^2 - \omega^2 + \omega^2)\{n(\omega') - 1\}\mathrm{d}\omega'}{\omega^2 - \omega'^2} + \frac{2}{\pi}\int_0^\infty \{n(\omega) - 1\}\mathrm{d}\omega \\ &= \frac{2\omega^2}{\pi}P\int_0^\infty \frac{\{n(\omega') - 1\}\mathrm{d}\omega'}{\omega^2 - \omega'^2}\end{aligned} \tag{19.174}$$

and hence Eq. (19.171) is recovered.

To summarize: an alternative to the standard approach of using contour integration and the Cauchy integral formula for deriving the Kramers–Kronig relations has been demonstrated. The derivation still utilizes the analytic behavior of the property under consideration. Only elementary properties of the Hilbert transform are required to obtain the second of the Kramers–Kronig relations from the Herglotz representation of the optical property as a Herglotz function.

19.8 Conducting materials

The previous sections have dealt with the situation for non-conducting materials. In this section the case of a conducting medium is investigated, with the focus on the changes that arise in the derivation of the Kramers–Kronig relations for this case. The key starting point is the Maxwell equation

$$\nabla \times \mathbf{H} = \mathbf{J} + \frac{\partial \mathbf{D}}{\partial t}, \tag{19.175}$$

where $\mathbf{J}$ is the current density, which is given by $\mathbf{J} = \sigma\mathbf{E}$ and σ denotes the conductivity. If the electric field is assumed to have a time dependence of the form $\mathrm{e}^{-\mathrm{i}\omega t}$, then $\partial\mathbf{D}/\partial t = -\mathrm{i}\omega\varepsilon_\mathrm{b}(\omega)\mathbf{E}$, where the subscript b denotes the component of the permittivity arising from bound electrons, and hence

$$\nabla \times \mathbf{H} = -\mathrm{i}\omega\varepsilon_\mathrm{b}(\omega)\mathbf{E} + \sigma(0)\mathbf{E}, \tag{19.176}$$

where $\sigma(0)$ denotes the conductivity at $\omega = 0$. On writing the last result as

$$\nabla \times \mathbf{H} = -\mathrm{i}\omega\varepsilon(\omega)\mathbf{E}, \tag{19.177}$$

the permittivity is identified as

$$\varepsilon(\omega) = \varepsilon_\mathrm{r}(\omega) + \mathrm{i}\varepsilon_\mathrm{i}(\omega) = \varepsilon_\mathrm{b}(\omega) + \frac{\mathrm{i}\sigma(0)}{\omega}, \tag{19.178}$$

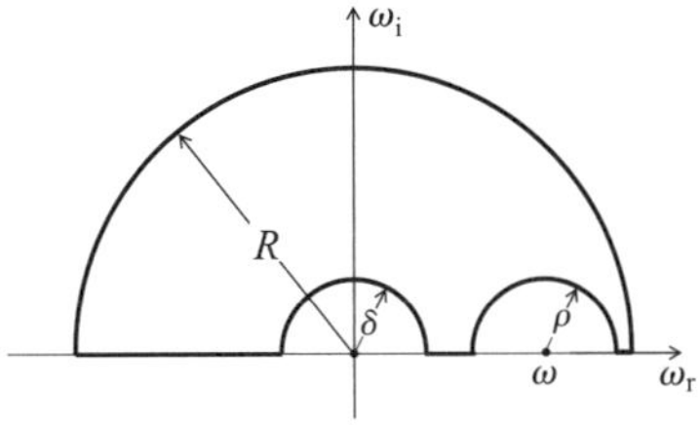

Figure 19.4. Contour for the derivation of dispersion relations for the permittivity of a conductor.

and the bound-electron contribution $\varepsilon_b(\omega)$ can be split into real and imaginary components. The form of the conductivity term in the preceding result can be modified to incorporate free-electron effects away from $\omega = 0$; however, that issue is not pursued at this point, since the focus of interest here is the structure of $\varepsilon(\omega)$ at zero angular frequency. From Eq. (19.178) it is clear that, in the limit $\omega \to 0$, the permittivity of a conductor exhibits a singularity. Since the dc conductivity $\sigma(0)$ is real, the singularity of $\varepsilon(\omega)$ resides in the imaginary component of the permittivity.

To derive Kramers–Kronig relations for the permittivity of a conductor, consider the integral

$$\oint_C \frac{\{\varepsilon(\omega_z)\varepsilon_0^{-1} - 1\}\mathrm{d}\omega_z}{\omega - \omega_z},$$

where C denotes the contour shown in Figure 19.4. The permittivity of a conductor is an analytic function in the upper half of the complex angular frequency plane, but there is the added complication of a pole at zero angular frequency, and hence the necessity to modify the contour used previously in Figure 19.2. From the Cauchy integral theorem it follows that

$$\int_{-R}^{-\delta} \frac{\{\varepsilon(\omega_r)\varepsilon_0^{-1} - 1\}\mathrm{d}\omega_r}{\omega - \omega_r} + \int_{\pi}^{0} \frac{\{\varepsilon(\delta e^{i\theta})\varepsilon_0^{-1} - 1\}i\delta e^{i\theta}\,\mathrm{d}\theta}{\omega - \delta e^{i\theta}} + \int_{\delta}^{\omega-\rho} \frac{\{\varepsilon(\omega_r)\varepsilon_0^{-1} - 1\}\mathrm{d}\omega_r}{\omega - \omega_r}$$
$$- \int_{\pi}^{0} \frac{\{\varepsilon(\omega + \rho e^{i\theta})\varepsilon_0^{-1} - 1\}i\rho e^{i\theta}\,\mathrm{d}\theta}{\rho e^{i\theta}} + \int_{\omega+\rho}^{R} \frac{\{\varepsilon(\omega_r)\varepsilon_0^{-1} - 1\}\mathrm{d}\omega_r}{\omega - \omega_r}$$
$$+ \int_{0}^{\pi} \frac{\{\varepsilon(Re^{i\theta})\varepsilon_0^{-1} - 1\}i\,Re^{i\theta}\,\mathrm{d}\theta}{\omega - Re^{i\theta}} = 0. \qquad (19.179)$$

Taking the limits $\rho \to 0$, $\delta \to 0$, and $R \to \infty$ leads to

$$P\int_{-\infty}^{\infty} \frac{\{\varepsilon(\omega_r)\varepsilon_0^{-1} - 1\}\mathrm{d}\omega_r}{\omega - \omega_r} + i\pi\{\varepsilon(\omega)\varepsilon_0^{-1} - 1\} + \frac{\pi\sigma(0)}{\omega\varepsilon_0} = 0, \qquad (19.180)$$

where the following limit has been employed:

$$\lim_{\delta\to 0} \delta\varepsilon(\delta e^{i\theta}) = i\sigma(0)e^{-i\theta}. \tag{19.181}$$

The Cauchy principal value in Eq. (19.180) includes both the obvious singular point at $\omega_r = \omega$ and the singularity at $\omega_r = 0$. Taking the real and imaginary parts of Eq. (19.180) leads to

$$\varepsilon_i(\omega) - \frac{\sigma(0)}{\omega} = \frac{1}{\pi} P \int_{-\infty}^{\infty} \frac{\{\varepsilon_r(\omega') - \varepsilon_0\}d\omega'}{\omega - \omega'} \tag{19.182}$$

and

$$\varepsilon_r(\omega) - \varepsilon_0 = -\frac{1}{\pi} P \int_{-\infty}^{\infty} \frac{\varepsilon_i(\omega')d\omega'}{\omega - \omega'}. \tag{19.183}$$

Using the crossing symmetry relations in Eqs. (19.54) and (19.55) yields

$$\varepsilon_i(\omega) - \frac{\sigma(0)}{\omega} = \frac{2\omega}{\pi} P \int_{0}^{\infty} \frac{\{\varepsilon_r(\omega') - \varepsilon_0\}d\omega'}{\omega^2 - \omega'^2} \tag{19.184}$$

and

$$\varepsilon_r(\omega) - \varepsilon_0 = -\frac{2}{\pi} P \int_{0}^{\infty} \frac{\omega'\varepsilon_i(\omega')d\omega'}{\omega^2 - \omega'^2}. \tag{19.185}$$

The reader should note that as $\omega' \to 0$ in the integrand in the last result, $\omega'\varepsilon_i(\omega')$ is well behaved. From the preceding derivation it is clear that the dispersion relation expressing $\varepsilon_r(\omega) - \varepsilon_0$ as the Hilbert transform of $\varepsilon_i(\omega)$ applies to both non-conductors and conductors. However, the dispersion relation representing $\varepsilon_i(\omega)$ in terms of the Hilbert transform of $\varepsilon_r(\omega) - \varepsilon_0$ contains, for conductors, an additional contribution, depending on the conductivity of the medium, which is not present for the corresponding dispersion relation for an insulator.

For the case of the refractive index of a conducting system, the principal point that needs additional discussion is the limit $\omega \to 0$. Assuming the permeability contribution can be ignored, that is, considering the restriction to non-magnetic materials, it follows that

$$\lim_{\omega\to 0} N(\omega) \to \sqrt{\left[\frac{i\sigma(0)}{\omega}\right]} = (1+i)\sqrt{\left[\frac{\sigma(0)}{2\omega}\right]}. \tag{19.186}$$

Hence, the complex refractive index does not have a simple pole at $\omega = 0$ but a singularity of the form $N(\omega) \approx 1/\sqrt{\omega}$ as $\omega \to 0$, which is integrable. As a result, Eqs. (19.113) and (19.114) can also be applied to conductors.

The conductivity could be treated as an independent optical property distinct from the dielectric constant. However, since these two optical properties are directly related, it is more a matter of convenience to use one function versus the other. There are two definitions commonly employed to relate the conductivity and the permittivity. These connections, along with the derivation of dispersion relations, are discussed further in Section 20.4.

19.9 Asymptotic behavior of the dispersion relations

In this section, the asymptotic behavior of the dispersion relations is investigated for some particular choices of the asymptotic behavior of the integrand. This builds on and extends ideas developed in Section 8.3. The approach considered is based closely on a treatment of Frye and Warnock (1963). The results obtained find application in the development of sum rules for optical constants. Examples of this type are developed in the following two sections.

Let $f(x) = O(x^{-1}\log^{-\lambda} x)$ as $x \to \infty$, with $\lambda > 0$, and suppose that $f'(x)$ is continuous, then the Cauchy principal value integral $tP\int_{x_0}^{\infty} f(x)\mathrm{d}x/(t-x)$, where $x_0 > 0$, satisfies

$$tP\int_{x_0}^{\infty}\frac{f(x)\mathrm{d}x}{t-x} = \int_{x_0}^{t} f(x)\mathrm{d}x + O(\log^{-\lambda} t), \quad \text{as } t \to \infty. \tag{19.187}$$

To obtain Eq. (19.187), the integral $tP\int_{x_0}^{\infty} f(x)\mathrm{d}x/(t-x)$ is partitioned in the following manner:

$$\begin{aligned} tP\int_{x_0}^{\infty}\frac{f(x)\mathrm{d}x}{t-x} &= \int_{x_0}^{t(1+\rho)} f(x)\mathrm{d}x + \int_{x_0}^{t(1-\rho)}\frac{xf(x)\mathrm{d}x}{t-x} \\ &\quad + P\int_{t(1-\rho)}^{t(1+\rho)}\frac{xf(x)\mathrm{d}x}{t-x} + t\int_{t(1+\rho)}^{\infty}\frac{f(x)\mathrm{d}x}{t-x} \\ &= I_1 + I_2 + I_3 + I_4. \end{aligned} \tag{19.188}$$

The asymptotic behavior of each of these integrals is now examined separately. Let C denote a positive constant that is not necessarily the same at each occurrence. The I_1 integral can be written for $t \to \infty$ as follows:

$$\begin{aligned} I_1 &= \int_{x_0}^{t} f(x)\mathrm{d}x + \int_{t}^{t(1+\rho)} f(x)\mathrm{d}x \\ &= \int_{x_0}^{t} f(x)\mathrm{d}x + \int_{t}^{t(1+\rho)} x^{-1}\log^{-\lambda} x\,\mathrm{d}x \\ &= \int_{x_0}^{t} f(x)\mathrm{d}x + \frac{1}{1-\lambda}\int_{t}^{t(1+\rho)}\frac{\mathrm{d}}{\mathrm{d}x}\{\log^{1-\lambda} x\}\mathrm{d}x \\ &= \int_{x_0}^{t} f(x)\mathrm{d}x + O(\log^{-\lambda} t), \quad \text{as } t \to \infty. \end{aligned} \tag{19.189}$$

The particular case of $\lambda = 1$ requires a different treatment of the second integral on the right-hand side of Eq. (19.189); however, the outcome is unchanged. To handle I_2, employ the fact that $x \leq t - t\rho$, that is, $(t-x)^{-1} \leq (t\rho)^{-1}$, then it follows that

$$|I_2| \leq (t\rho)^{-1} \int_{x_0}^{t(1-\rho)} \log^{-\lambda} x \, dx, \quad \text{as } t \to \infty, \tag{19.190}$$

and, on introducing the substitution $x = e^y$, this leads to

$$|I_2| \leq (t\rho)^{-1} \int_{\log x_0}^{\log t(1-\rho)} \frac{e^y \, dy}{y^\lambda}, \quad \text{as } t \to \infty. \tag{19.191}$$

Integration by parts yields

$$|I_2| \leq (t\rho)^{-1} \left\{ O\left(\frac{t}{\log^\lambda t}\right) + \lambda \int_{\log x_0}^{\log t(1-\rho)} \frac{e^y \, dy}{y^{\lambda+1}} \right\}; \tag{19.192}$$

hence

$$|I_2| = O\left(\frac{1}{\log^\lambda t}\right), \quad \text{as } t \to \infty. \tag{19.193}$$

To deal with the I_3 integral, first note that

$$P \int_{t(1-\rho)}^{t(1+\rho)} \frac{dx}{t-x} = 0, \tag{19.194}$$

then

$$I_3 = -\int_{t(1-\rho)}^{t(1+\rho)} \frac{\{xf(x) - tf(t)\}dx}{x-t}, \tag{19.195}$$

which simplifies on using the mean value theorem to give

$$I_3 = -\int_{t(1-\rho)}^{t(1+\rho)} \{xf(x)\}'\big|_{x=\xi} \, dx, \tag{19.196}$$

where $x < \xi < t$. The I_3 integral can be evaluated asymptotically, for a non-zero constant C, as

$$I_3 = \frac{2\lambda t\rho}{\xi \log^{\lambda+1} \xi} = C \frac{2\lambda\rho}{\log^{\lambda+1} t}, \tag{19.197}$$

and hence

$$I_3 = O\left(\frac{1}{\log^{\lambda+1} t}\right), \quad \text{as } t \to \infty. \tag{19.198}$$

The I_4 integral can be written as follows:

$$|I_4| \le \frac{C}{\log^\lambda t}\int_{t(1+\rho)}^{\infty}\left\{\frac{1}{x}+\frac{1}{t-x}\right\}\mathrm{d}x$$
$$= \frac{C}{\log^\lambda t}\log\left\{\frac{\rho}{1+\rho}\right\}, \tag{19.199}$$

and therefore

$$I_4 = O(\log^{-\lambda} t), \quad \text{as } t \to \infty. \tag{19.200}$$

Putting Eqs. (19.189), (19.193), (19.198), and (19.200) together leads to Eq. (19.187).

Using a similar procedure, it can be shown that if $f(x) = O(x^{-\lambda})$ as $x \to \infty$, with $1 < \lambda < 2$, or $\lambda > 2$, and $f \in C^1$ for some $x \ge x_0$ with $x_0 > 0$, then, as $t \to \infty$,

$$tP\int_{x_0}^{\infty}\frac{f(x)\mathrm{d}x}{t-x} = \int_{x_0}^{t} f(x)\mathrm{d}x + O\left(\frac{1}{t^{\lambda-1}}\right), \quad \text{as } t \to \infty. \tag{19.201}$$

For the case $f(x) = O(x^{-2})$ as $x \to \infty$, with $f \in C^1$ for some $x \ge x_0$ and $x_0 > 0$,

$$I_2 = \frac{1}{t}\int_{x_0}^{t(1-\rho)}\left\{\frac{1}{x}+\frac{1}{t-x}\right\}\mathrm{d}x, \tag{19.202}$$

and hence

$$I_2 = O\left(\frac{\log t}{t}\right), \quad \text{as } t \to \infty. \tag{19.203}$$

It is left as an exercise for the reader to examine the behavior of I_1, I_3, and I_4 for this case. The outcome is as follows:

$$tP\int_{x_0}^{\infty}\frac{f(x)\mathrm{d}x}{t-x} = \int_{x_0}^{t} f(x)\mathrm{d}x + O\left(\frac{\log t}{t}\right), \quad \text{as } t \to \infty. \tag{19.204}$$

Let $g(y)$ denote the integral

$$g(y) = P\int_0^{\infty}\frac{f(x)\mathrm{d}x}{y-x}, \tag{19.205}$$

and write, for $y \gg x_0$,

$$yg(y) = y\int_0^{x_0}\frac{f(x)\mathrm{d}x}{y-x} + yP\int_{x_0}^{\infty}\frac{f(x)\mathrm{d}x}{y-x}. \tag{19.206}$$

Now,

$$y\int_0^{x_0}\frac{f(x)\mathrm{d}x}{y-x} = \int_0^{x_0} f(x)\mathrm{d}x + O(y^{-1}), \quad \text{as } y \to \infty, \tag{19.207}$$

and

$$yP\int_{x_0}^{\infty}\frac{f(x)\mathrm{d}x}{y-x}=\int_{x_0}^{\infty}f(x)\mathrm{d}x+O(k(y)),\quad \text{as } y\to\infty, \tag{19.208}$$

where the function $k(y)$ is determined from the asymptotic behavior of $f(x)$ as $x\to\infty$. Combining these last two results allows the asymptotic behavior of $g(y)$ as $y\to\infty$ to be determined.

The principal contents of this section can be summarized by the following theorem. Let

$$g(y)=P\int_0^{\infty}\frac{f(x)\mathrm{d}x}{y-x}, \tag{19.209}$$

and assume $f\in C^1$ for some $x\geq x_0$. If

$$f(x)=O(x^{-\lambda}),\quad 1<\lambda,\quad \text{as } x\to\infty, \tag{19.210}$$

then

$$g(y)=y^{-1}\int_0^{\infty}f(x)\mathrm{d}x+O(y^{-\lambda}),\quad 1<\lambda<2,\quad \text{or } \lambda>2,\quad \text{as } y\to\infty, \tag{19.211}$$

and

$$g(y)=y^{-1}\int_0^{\infty}f(x)\mathrm{d}x+O(y^{-2}\log y),\quad \lambda=2,\quad \text{as } y\to\infty. \tag{19.212}$$

If

$$f(x)=O\left(\frac{1}{x\log^{\lambda}x}\right),\quad \lambda>0,\quad \text{as } x\to\infty, \tag{19.213}$$

then

$$g(y)=y^{-1}\int_0^{\infty}f(x)\mathrm{d}x+O\left(\frac{1}{y\log^{\lambda}y}\right),\quad \lambda>0,\quad \text{as } y\to\infty. \tag{19.214}$$

This collection of results, or parts thereof, is sometimes referred to in the physics literature as the *superconvergence theorem*. Applications of this theorem are given in the following developments.

19.10 Sum rules for the dielectric constant

Examination of the limit $\omega\to 0$ for Eq. (19.185) yields

$$\varepsilon_{\mathrm{r}}(0)-\varepsilon_0=\frac{2}{\pi}\int_0^{\infty}\frac{\varepsilon_{\mathrm{i}}(\omega)\mathrm{d}\omega}{\omega}. \tag{19.215}$$

This sum rule applies for non-conductors. Dielectric losses in a material are governed by the quantity $\varepsilon_{\mathrm{i}}(\omega)$. It follows from Eq. (19.215) that insulators, for which $\varepsilon_{\mathrm{r}}(0)-\varepsilon_0$ is fairly small, do not show significant dielectric losses. Making use of Eq. (19.178), for the case of conductors it is possible to write

$$\varepsilon_{\mathrm{r}}(\omega)-\varepsilon_0=-\frac{2}{\pi}P\int_0^{\infty}\frac{\omega'\left\{\varepsilon_{\mathrm{i}}(\omega')-\sigma(0)/\omega'\right\}\mathrm{d}\omega'}{\omega^2-\omega'^2}, \tag{19.216}$$

and this is an example of a dispersion relation with a subtraction. Since

$$P\int_0^{\infty}\frac{\mathrm{d}\omega'}{\omega^2-\omega'^2}=0, \quad \omega\neq 0, \tag{19.217}$$

which the reader will recall from the fact that the Hilbert transform of a constant is zero, the subtraction in Eq. (19.216) does not alter the value of the integral. Examination of the limit $\omega\to 0$ in Eq. (19.216) leads to

$$\varepsilon_{\mathrm{r}}(0)-\varepsilon_0=\frac{2}{\pi}\int_0^{\infty}\frac{\left\{\varepsilon_{\mathrm{i}}(\omega')-\sigma(0)/\omega'\right\}\mathrm{d}\omega'}{\omega'}. \tag{19.218}$$

The integrand of this sum rule is well behaved in the vicinity of $\omega'\to 0$. This can be demonstrated by using Eq. (19.18) for the permittivity of the bound electrons. If the fraction of electrons in free motion is denoted by f_0 and the assignment $\omega_j=0$ is used in Eq. (19.18), then the dielectric constant for the free charge carriers is given by

$$\begin{aligned}\frac{\varepsilon(\omega)}{\varepsilon_0}-1&=-\frac{\mathcal{N}e^2}{\varepsilon_0 m}\frac{f_0}{\omega^2+\mathrm{i}\gamma_0\omega}\\&=\frac{\mathcal{N}e^2f_0}{\varepsilon_0 m\gamma_0^2}\left(\frac{\mathrm{i}\gamma_0}{\omega}-1\right)\left(1+\frac{\omega^2}{\gamma_0^2}\right)^{-1}.\end{aligned} \tag{19.219}$$

The restriction of the Lorentz model to free charge carriers is often referred to as the Drude model. The latter model allows for a theoretical treatment of the susceptibility of metals to be made within the framework of classical theory. From Eq. (19.219), it follows that

$$\frac{\varepsilon_{\mathrm{r}}(\omega)}{\varepsilon_0}-1=-\frac{\mathcal{N}e^2f_0}{\varepsilon_0 m\gamma_0^2}(1+O(\omega^2)), \quad \text{as } \omega\to 0, \tag{19.220}$$

and

$$\frac{\varepsilon_{\mathrm{i}}(\omega)}{\varepsilon_0}=\frac{\mathcal{N}e^2f_0}{\varepsilon_0 m\gamma_0}\left\{\frac{1}{\omega}+O(\omega)\right\}, \quad \text{as } \omega\to 0. \tag{19.221}$$

The limit $\omega\to 0$ for Eq. (19.184) does not yield a sum rule.

Equation (19.34) can be rewritten using the substitutions $y = \omega^2, x = \omega_r^2$, and $f(x) = \{\varepsilon_r(\sqrt{x}) - \varepsilon_0\}/\sqrt{x}$ as

$$\frac{\pi\varepsilon_i(\omega)}{\omega} = P\int_0^\infty \frac{f(x)\mathrm{d}x}{y-x}. \tag{19.222}$$

The change of variables for the integral has been made so the central theorem in Section 19.8 can be directly employed. Making use of the result in Eq. (19.60) yields $f(x) = O(x^{-3/2})$, as $x \to \infty$, and using Eq. (19.211) leads to

$$\frac{\pi\varepsilon_i(\omega)}{\omega} = \frac{1}{y}\int_0^\infty f(x)\mathrm{d}x + O(y^{-3/2}), \quad \text{as } y \to \infty, \tag{19.223}$$

and hence

$$\varepsilon_i(\omega) = \frac{2}{\pi\omega}\int_0^\infty \{\varepsilon_r(\omega') - \varepsilon_0\}\mathrm{d}\omega' + O(\omega^{-3}), \quad \text{as } \omega \to \infty. \tag{19.224}$$

Comparison of Eqs. (19.224) and (19.61) leads to

$$\int_0^\infty [\varepsilon_r(\omega) - \varepsilon_0]\mathrm{d}\omega = 0. \tag{19.225}$$

This result is directly tied to the short-time behavior of the response function $G(t)$. The preceding sum rule can also be obtained from Eq. (19.42) by examining the $t \to 0+$ limit and using Eq. (19.58). Altarelli and Smith (1974) point out that at very short times after the incident electromagnetic field arrives, the electronic displacements from the equilibrium positions caused by the applied field are very small, and the restoring and damping forces can be treated as negligible. The system is, in a sense, "free-electron-like." The key idea is that the detailed interatomic and intermolecular interactions that would characterize the response of each different material are not yet in play in restoring the system to a new equilibrium state. Hence, a very general sum rule emerges that is independent of the specific nature of the material. The only broad division is into conductors and insulators, for the reason that the generalized dielectric constant for a conductor has a singularity at zero frequency, but insulators do not. The result given in Eq. (19.225) applies for non-conductors, and be obtained under the weaker asymptotic condition that $\varepsilon_i(\omega) \approx O(\omega^{-1-\alpha})$, as $\omega \to \infty$, with $\alpha > 0$. For the case of a conductor, it follows, on taking the limit $\omega \to \infty$ for Eq. (19.184), that

$$\int_0^\infty \{\varepsilon_r(\omega) - \varepsilon_0\}\mathrm{d}\omega = -\frac{\pi\sigma(0)}{2}. \tag{19.226}$$

Introducing the change of variables $y = \omega^2$, $x = \omega_r^2$, and $f(x) = \varepsilon_i(\sqrt{x})$ in Eq. (19.35) gives

$$\pi\{\varepsilon_r(\omega) - \varepsilon_0\} = -P\int_0^\infty \frac{f(x)\mathrm{d}x}{y - x}. \tag{19.227}$$

Making use of Eq. (19.61), so that $f(x) = O(x^{-3/2})$ as $\omega \to \infty$, and employing Eq. (19.211), leads to

$$\pi\{\varepsilon_r(\omega) - \varepsilon_0\} = -\frac{1}{y}\int_0^\infty f(x)\mathrm{d}x + O(y^{-3/2}), \quad \text{as } y \to \infty, \tag{19.228}$$

and hence

$$\varepsilon_r(\omega) - \varepsilon_0 = -\frac{2}{\pi\omega^2}\int_0^\infty \omega'\varepsilon_i(\omega')\mathrm{d}\omega' + O(\omega^{-3}), \quad \text{as } \omega \to \infty. \tag{19.229}$$

Comparison of Eqs. (19.229) and (19.12) leads to

$$\int_0^\infty \omega\varepsilon_i(\omega)\mathrm{d}\omega = \frac{\pi\varepsilon_0\omega_p^2}{2}. \tag{19.230}$$

The preceding formula is often called the f sum rule, and it applies to both conductors and non-conductors. This is a particular case of the Thomas–Reiche–Kuhn sum rule, which involves a summation over all the oscillator strengths. The idea for a dispersion theoretic derivation of Eq. (19.230) appears first in the work of Kronig (1926). The reader should note that the derivation of the f sum rule rests on a specific model of the large-frequency behavior of the medium. This contrasts with the more general assumptions employed to obtain the sum rules in Eqs. (19.215), (19.218), (19.225), and (19.226). Although, of course, it is necessary to have some knowledge of the asymptotic behavior of the optical functions to obtain these sum rules.

The problem of obtaining general sum rules for powers of the dielectric constant is now considered. Consider the integral

$$\oint_C \omega_z^m\{\varepsilon(\omega_z) - \varepsilon_0\}^p\,\mathrm{d}\omega_z,$$

where C denotes a closed semicircular contour in the upper half complex angular frequency plane, center the origin and radius R. The exponents m and p take on selected integer values satisfying $m \geq 0$ and $p \geq 1$. In order that the integral does not diverge in the limit as $R \to \infty$ requires $m + 1 < 2p$, which assumes that Eqs. (19.60) and (19.61) hold. For insulators there is no constraint imposed on m and p as $\omega \to 0$; however, to treat conductors requires $p < m + 1$, and hence the values of p are fixed by the following conditions:

$$m + 1 < 2p, \quad \text{for insulators}, \tag{19.231}$$

and

$$m + 1 < 2p < 2m + 2, \quad \text{for conductors.} \tag{19.232}$$

Evaluation of the contour integral subject to the preceding two constraints leads to

$$\int_{-\infty}^{\infty} \omega^m \{\varepsilon(\omega) - \varepsilon_0\}^p \, \mathrm{d}\omega = 0. \tag{19.233}$$

For insulators, Eq. (19.225) is obtained using $m = 0$ and $p = 1$. Several additional cases arising for $m = 0, 1, 2,$ and 3 are summarized in Section 19.14. The reader is invited to derive these sum rules and identify whether they are appropriate for insulators, or for both insulators and conductors.

The behavior of the dielectric constant at complex angular frequencies is now examined. Consider the integral

$$\oint_C \frac{\omega_z \{\varepsilon(\omega_z) - \varepsilon_0\} \mathrm{d}\omega_z}{\omega^2 + \omega_z^2},$$

where C denotes the contour shown in Figure 19.5. From the Cauchy residue theorem and taking the limit $R \to \infty$, this leads to

$$\int_{-\infty}^{\infty} \frac{\omega_\mathrm{r} \{\varepsilon(\omega_\mathrm{r}) - \varepsilon_0\} \mathrm{d}\omega_\mathrm{r}}{\omega^2 + \omega_\mathrm{r}^2} = \pi \mathrm{i} \{\varepsilon(\mathrm{i}\omega) - \varepsilon_0\}. \tag{19.234}$$

Recalling that $\varepsilon_\mathrm{r}(\omega)$ and $\varepsilon_\mathrm{i}(\omega)$ are even and odd functions of the angular frequency, respectively, it follows that

$$\varepsilon(\mathrm{i}\omega) - \varepsilon_0 = \frac{2}{\pi} \int_0^{\infty} \frac{\omega' \varepsilon_\mathrm{i}(\omega') \mathrm{d}\omega'}{\omega^2 + \omega'^2}. \tag{19.235}$$

It is clear from this expression that the dielectric constant on the imaginary angular frequency axis is a real quantity. If the preceding result is integrated over the interval

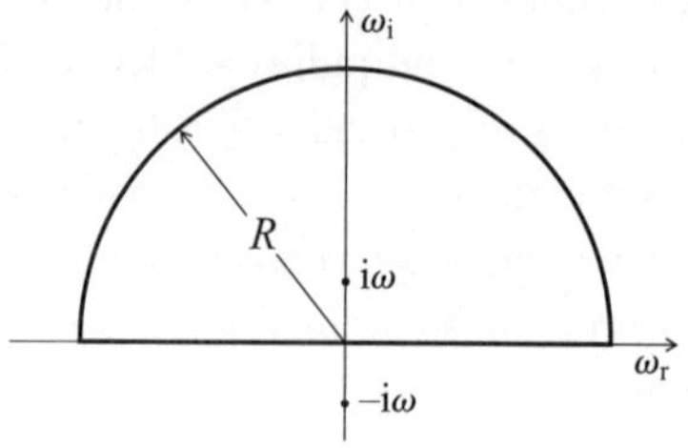

Figure 19.5. Contour employed to derive a sum rule along the complex angular frequency axis.

$[0, \omega_0]$, then the following sum rule the following sum rule is obtained:

$$\int_0^{\omega_0} \{\varepsilon(\mathrm{i}\omega) - \varepsilon_0\} \mathrm{d}\omega = \frac{2}{\pi} \int_0^\infty \varepsilon_\mathrm{i}(\omega) \tan^{-1}\left\{\frac{\omega_0}{\omega}\right\} \mathrm{d}\omega. \tag{19.236}$$

The limit $\omega_0 \to \infty$ yields the sum rule

$$\int_0^\infty \{\varepsilon(\mathrm{i}\omega) - \varepsilon_0\} \mathrm{d}\omega = \int_0^\infty \varepsilon_\mathrm{i}(\omega) \mathrm{d}\omega. \tag{19.237}$$

Since experimental data are not obtained at complex frequencies, a sum rule of the preceding type would be most useful in checking the validity of proposed theoretical models for the frequency dependence of the dielectric constant.

19.11 Sum rules for the refractive index

From Eq. (19.12) it follows, on ignoring contributions of the $O(\omega^{-m})$, with $m > 2$, that

$$n(\omega) - 1 \approx -\frac{\omega_\mathrm{p}^2}{2\omega^2}, \quad \text{as } \omega \to \infty. \tag{19.238}$$

Equation (19.114) can be recast using the substitutions $y = \omega^2, x = \omega'^2$, and $f(x) = \kappa(\sqrt{x})$ as follows:

$$n(\omega) - 1 = -\frac{1}{\pi} P \int_0^\infty \frac{f(x)\mathrm{d}x}{y - x}. \tag{19.239}$$

Suppose $\kappa(\omega)$ is assumed to satisfy the asymptotic condition

$$\kappa(\omega) = O(\omega^{-2} \log^{-\lambda} \omega), \quad \text{with } \lambda > 0, \quad \text{as } \omega \to \infty. \tag{19.240}$$

This is actually a conservative estimate for the asymptotic falloff of $\kappa(\omega)$. The reader should note that the crossing symmetry result for $\kappa(\omega)$ must also dictate how this function behaves asymptotically for large values of the angular frequency. Making use of the discussion from Section 19.9 leads to the following:

$$\begin{aligned} n(\omega) - 1 &= -\frac{1}{y\pi} \int_0^\infty f(x)\mathrm{d}x + O\left(\frac{1}{y \log^\lambda y}\right) \\ &= -\frac{2}{\omega^2 \pi} \int_0^\infty \omega' \kappa(\omega')\mathrm{d}\omega' + O\left(\frac{1}{\omega^2 \log^\lambda \omega}\right), \quad \text{as } \omega \to \infty. \end{aligned} \tag{19.241}$$

A comparison of Eqs. (19.238) and (19.241) yields

$$\int_0^\infty \omega \kappa(\omega)\mathrm{d}\omega = \frac{\pi}{4} \omega_\mathrm{p}^2. \tag{19.242}$$

This is the f sum rule for $\kappa(\omega)$ and it applies for both insulators and conductors. A consequence of this preceding result and the analogous formula for $\varepsilon_{\mathrm{i}}(\omega)$ Eq. (19.230), is the sum rule

$$\int_0^\infty \omega\kappa(\omega)\{n(\omega) - 1\}\mathrm{d}\omega = 0. \tag{19.243}$$

This formula was noted by Stern (1963, p. 341).

Equation (19.113) can be rewritten using the substitutions $y = \omega^2, x = \omega'^2$, and $f(x) = \{n(\sqrt{x}) - 1\}/\sqrt{x}$, to give

$$\kappa(\omega) = \frac{\omega}{\pi} P \int_0^\infty \frac{f(x)\mathrm{d}x}{y - x}. \tag{19.244}$$

Taking the limit $\omega \to \infty$ and employing Eq. (19.238) leads to

$$\kappa(\omega) = \frac{\omega}{\pi}\left\{\frac{1}{y}\int_0^\infty f(x)\mathrm{d}x + O(y^{-3/2})\right\}, \quad \text{as } \omega \to \infty. \tag{19.245}$$

Comparing Eqs. (19.245) and (19.240) (or using Eq. (19.99) in place of Eq. (19.240) yields

$$\int_0^\infty f(x)\mathrm{d}x = 0; \tag{19.246}$$

that is,

$$\int_0^\infty \{n(\omega) - 1\}\mathrm{d}\omega = 0. \tag{19.247}$$

The sum rule in Eq. (19.247) applies for both conductors and insulators. It is clear from either Eq. (19.247) or Eq. (19.243) that $n(\omega) < 1$ for some range of angular frequencies, otherwise the integrals could not equal zero. Using the definition of the phase velocity given in Eq. (19.88), does the fact that $n(\omega) < 1$ lead to a violation of the principle of special relativity, that the limiting velocity is the speed of light? The group velocity v_{g} is given by

$$v_{\mathrm{g}} = \frac{\mathrm{d}\omega}{\mathrm{d}k}. \tag{19.248}$$

Normally, energy transmission is associated with the group velocity. Using Eqs. (19.88) and (19.89) yields

$$v_{\mathrm{g}} = \frac{\mathrm{d}}{\mathrm{d}k}\left(\frac{ck}{n}\right) = \frac{c}{n} - \frac{ck}{n^2}\frac{\mathrm{d}n}{\mathrm{d}k} = \frac{c}{n} - \frac{ck}{n^2}\frac{\mathrm{d}n}{\mathrm{d}\omega}\frac{\mathrm{d}\omega}{\mathrm{d}k}, \tag{19.249}$$

and hence

$$v_g = \frac{c/n}{1 + (ck/n^2)(dn/d\omega)} = \frac{c}{n + \omega\,(dn/d\omega)}. \tag{19.250}$$

Typically, $n(\omega) > 1$ and $dn/d\omega > 0$, so that $v_g < c$. From Eq. (19.250),

$$n + \omega\frac{dn}{d\omega} = \frac{c}{v_g}, \tag{19.251}$$

and it follows that, in a region of anomalous dispersion where $dn/d\omega < 0$, and for $n(\omega) < 1$, that $c/v_g < 1$, so that $v_g > c$. In the presence of absorption, the group velocity is devoid of meaning in so far as signal propagation is concerned. The concept of associating energy transmission with the group velocity also breaks down. The signal front never moves with a velocity exceeding the speed of light, and information transmission also does not occur at speeds which exceed c.

For an insulator, Eq. (19.114) can be written, on taking the limit $\omega \to 0$, as follows:

$$n(0) - 1 = \frac{2}{\pi}\int_0^\infty \frac{\kappa(\omega)d\omega}{\omega}. \tag{19.252}$$

For the case of a conductor, both the integral and the left-hand side of the preceding equation diverge. From Eq. (19.113) the limit $\omega \to 0$ yields no general sum rule.

The problem of obtaining sum rules for powers of the refractive index is now examined. To do this, consider the integral

$$\oint_C \omega_z^m \{N(\omega_z) - 1\}^p \, d\omega_z,$$

where C denotes a closed semicircular contour in the upper half complex angular frequency plane, center the origin and radius R. The integer exponents m and p are selected so that the integral does not diverge in the limit as $R \to \infty$ and that there is no pole at $\omega = 0$. The choice $m \geq 0$ and $p \geq 1$ is made. For the integral not to diverge in the limit as $R \to \infty$ requires $m + 1 < 2p$, which assumes that Eqs. (19.98) and (19.99) hold. For insulators there is no constraint imposed on m and p as $\omega \to 0$; however, to treat conductors requires $p < 2m + 2$, and hence the values of p are fixed by the conditions

$$m + 1 < 2p, \quad \text{for insulators,} \tag{19.253}$$

and

$$m + 1 < 2p < 4m + 4, \quad \text{for conductors.} \tag{19.254}$$

Evaluation of the contour integral subject to the preceding two constraints leads to

$$\int_{-\infty}^{\infty} \omega^m \{N(\omega) - 1\}^p \, d\omega = 0. \tag{19.255}$$

For $m = 0$ and $p = 1$, Eq. (19.247) is obtained. This approach to Eq. (19.247) is achieved without the need to examine the asymptotic behavior of a principal value integral. A number of additional sum rules arising for $m = 0, 1, 2$, and 3 are summarized in Section 19.14. The reader is invited to derive these sum rules and to determine which are appropriate for insulators, and which apply to both insulators and conductors.

The function $n(\omega) - 1$ clearly must have at least one zero on the spectral interval $[0, \infty)$. Let such frequencies, if there are more than one, be designated as $\omega_0, \omega'_0, \omega''_0$, and so on. Then it follows directly from the dispersion relation for the refractive index that

$$P\int_0^{\infty} \frac{\omega\kappa(\omega)d\omega}{\omega_0^2 - \omega^2} = P\int_0^{\infty} \frac{\omega\kappa(\omega)d\omega}{\omega'^2_0 - \omega^2} = P\int_0^{\infty} \frac{\omega\kappa(\omega)d\omega}{\omega''^2_0 - \omega^2} = \cdots = 0. \tag{19.256}$$

Combining any one of these results with the f sum rule gives

$$P\int_0^{\infty} \frac{\omega^3\kappa(\omega)d\omega}{\omega_0^2 - \omega^2} = -\frac{\pi\omega_p^2}{4}, \tag{19.257}$$

and it is assumed that Eq. (19.99) holds. The principal difficulty in applying sum rules of this type is that the zeros of $n(\omega) - 1$ depend on the particular medium under study, and must be found experimentally. Alternatively, sum rules of this type might be used to estimate the locations of the zeros of $n(\omega) - 1$.

A different approach to deriving sum rules is as follows. Integrating Eq. (19.114) leads to

$$\int_0^{\infty} \{n(\omega) - 1\}d\omega = -\frac{2}{\pi}\int_0^{\infty} d\omega \, P\int_0^{\infty} \frac{\omega'\kappa(\omega')d\omega'}{\omega^2 - \omega'^2}. \tag{19.258}$$

If the interchange of integration order can be justified,

$$\int_0^{\infty} \{n(\omega) - 1\}d\omega = \frac{2}{\pi}\int_0^{\infty} \omega'\kappa(\omega')d\omega' \, P\int_0^{\infty} \frac{d\omega}{\omega'^2 - \omega^2}, \tag{19.259}$$

and employing the result

$$P\int_0^{\infty} \frac{d\omega}{\omega'^2 - \omega^2} = -\frac{\pi^2}{4}\delta(\omega') \tag{19.260}$$

leads to

$$\int_0^\infty \{n(\omega) - 1\}d\omega = -\frac{\pi}{2}\int_0^\infty \omega\kappa(\omega)\delta(\omega)d\omega = 0, \tag{19.261}$$

which is Eq. (19.247). From Eq. (19.113), it follows on squaring that

$$\kappa^2(\omega) = \frac{2\omega}{\pi}P\int_0^\infty \frac{\{n(\omega') - 1\}d\omega'}{\omega^2 - \omega'^2}\frac{2\omega}{\pi}P\int_0^\infty \frac{\{n(\omega'') - 1\}d\omega''}{\omega^2 - \omega''^2}, \tag{19.262}$$

and hence

$$\int_0^\infty \kappa^2(\omega)d\omega = \frac{4}{\pi^2}\int_0^\infty \omega^2\, d\omega\, P\int_0^\infty \frac{\{n(\omega') - 1\}d\omega'}{\omega^2 - \omega'^2}P\int_0^\infty \frac{\{n(\omega'') - 1\}d\omega''}{\omega^2 - \omega''^2}. \tag{19.263}$$

If the interchange of integration order can be justified, then

$$\begin{aligned}\int_0^\infty \kappa^2(\omega)d\omega = \frac{4}{\pi^2}\int_0^\infty \{n(\omega') - 1\}d\omega' P\int_0^\infty \{n(\omega'') - 1\}d\omega''\\ \times P\int_0^\infty \frac{\omega^2\, d\omega}{(\omega^2 - \omega'^2)(\omega^2 - \omega''^2)},\end{aligned} \tag{19.264}$$

and employing

$$P\int_0^\infty \frac{\omega^2\, d\omega}{(\omega^2 - \omega'^2)(\omega^2 - \omega''^2)} = \frac{\pi^2}{4}\{\delta(\omega' - \omega'') + \delta(\omega' + \omega'')\} \tag{19.265}$$

leads to

$$\int_0^\infty \kappa^2(\omega)d\omega = \int_0^\infty \{n(\omega') - 1\}d\omega')\int_0^\infty \{n(\omega'') - 1\}d\omega''\{\delta(\omega' - \omega'') + \delta(\omega' + \omega'')\}, \tag{19.266}$$

and hence

$$\int_0^\infty \kappa^2(\omega)d\omega = \int_0^\infty \{n(\omega) - 1\}^2\, d\omega. \tag{19.267}$$

19.12 Application of some properties of the Hilbert transform

A number of sum rules for the optical constants can be obtained by direct application of some of the properties satisfied by the Hilbert transform. In this section some applications to the complex dielectric constant and the complex refractive index are considered.

If the Parseval-type identity for the Hilbert transform (see Eq. (4.172)) is applied to Eq. (19.107) or Eq. (19.108), along with the crossing symmetry conditions in Eqs. (19.111) and (19.112), then the following sum rule is obtained:

$$\int_0^\infty \kappa(\omega)^2\,\mathrm{d}\omega = \int_0^\infty \{n(\omega)-1\}^2\,\mathrm{d}\omega. \tag{19.268}$$

The same Parseval-type identity applied to Eq. (19.34) or Eq. (19.35), along with Eqs. (19.54) and (19.55), leads to

$$\int_0^\infty \varepsilon_\mathrm{i}(\omega)^2\,\mathrm{d}\omega = \int_0^\infty \{\varepsilon_\mathrm{r}(\omega)-\varepsilon_0\}^2\,\mathrm{d}\omega. \tag{19.269}$$

The preceding two sum rules apply for insulators. Application of the Parseval-type identity given in Eq.(4.174), with $f_1(\omega) = \omega\kappa(\omega)$ and $f_2(\omega) = n(\omega) - 1$, and noting from Eq. (4.111) that

$$H[\omega\kappa(\omega)] = \omega H\kappa(\omega) - \frac{1}{\pi}\int_{-\infty}^\infty \kappa(\omega)\mathrm{d}\omega = \omega H\kappa(\omega), \tag{19.270}$$

leads to the sum rule

$$\int_0^\infty \omega\kappa(\omega)\{n(\omega)-1\}\mathrm{d}\omega = 0. \tag{19.271}$$

The analogous result for the dielectric constant becomes

$$\int_0^\infty \omega\varepsilon_\mathrm{i}(\omega)\{\varepsilon_\mathrm{r}(\omega)-\varepsilon_0\}\mathrm{d}\omega = 0. \tag{19.272}$$

The orthogonality condition for the Hilbert transform, Eq.(4.198), applied to the function $f(\omega) = \kappa(\omega) + n(\omega) - 1$ leads to Eq. (19.268) and applied to the function $f(\omega) = \varepsilon_\mathrm{i}(\omega) + \varepsilon_\mathrm{r}(\omega) - \varepsilon_0$ gives Eq. (19.269). The property given in Eq. (4.204) applied to $f(\omega) = \kappa(\omega)$ leads directly to

$$\int_0^\infty \{n(\omega)-1\}\mathrm{d}\omega = 0. \tag{19.273}$$

Applying the same result to $f(\omega) = \varepsilon_\mathrm{i}(\omega)$ leads to

$$\int_0^\infty \{\varepsilon_\mathrm{r}(\omega)-\varepsilon_0\}\mathrm{d}\omega = 0, \tag{19.274}$$

which applies for both insulators and conductors.

The property given in Eq. (4.205) can be recast for the interval $[0, \infty)$, and with $g(y) = f(y)Hf(y)$, then

$$\int_{-\infty}^{\infty} \mathrm{d}x \int_{-\infty}^{x} g(y)\mathrm{d}y = \int_{0}^{\infty} \mathrm{d}x \left\{ \int_{0}^{x} [g(y) + g(-y)]\mathrm{d}y + 2\int_{x}^{\infty} g(-y)\mathrm{d}y \right\} = 0. \tag{19.275}$$

If f is an even or an odd function, then g is an odd function, so that the integral $\int_0^x [g(y) + g(-y)]\mathrm{d}y$ vanishes. Let $f(\omega) = \kappa(\omega)$, then $g(\omega) = -\kappa(\omega)\{n(\omega) - 1\}$ and hence

$$\int_{0}^{\infty} \mathrm{d}\omega \int_{\omega}^{\infty} \kappa(\omega')\{n(\omega') - 1\}\mathrm{d}\omega' = 0. \tag{19.276}$$

The choice $f(\omega) = \{n(\omega) - 1\}$ also leads to Eq. (19.276). The analogous result for the dielectric constant can be written with the choice $f(\omega) = \varepsilon_{\mathrm{i}}(\omega)$, leading to

$$\int_{0}^{\infty} \mathrm{d}\omega \int_{\omega}^{\infty} \varepsilon_{\mathrm{i}}(\omega')\{\varepsilon_{\mathrm{r}}(\omega') - \varepsilon_0\}\mathrm{d}\omega' = 0. \tag{19.277}$$

If $f(\omega) = \omega\kappa(\omega)$, then $g(\omega) = -\omega^2\kappa(\omega)\{n(\omega) - 1\}$, and hence

$$\int_{0}^{\infty} \mathrm{d}\omega \int_{\omega}^{\infty} \omega'^2\kappa(\omega')\{n(\omega') - 1\}\mathrm{d}\omega' = 0. \tag{19.278}$$

Equations (4.181) and (4.183) can be written for the interval $[0, \infty)$ as follows:

$$\int_{0}^{x} Hf(t)\mathrm{d}t = \frac{1}{\pi}\int_{0}^{\infty} \left\{ f(t)\log\left|1 - \frac{x}{t}\right| + f(-t)\log\left|1 + \frac{x}{t}\right| \right\}\mathrm{d}t \tag{19.279}$$

and

$$\int_{0}^{x} f(t)\mathrm{d}t = -\frac{1}{\pi}\int_{0}^{\infty} \left\{ Hf(t)\log\left|1 - \frac{x}{t}\right| + Hf(-t)\log\left|1 + \frac{x}{t}\right| \right\}\mathrm{d}t. \tag{19.280}$$

Application of Eq. (19.279) with the choice $f(\omega) = \kappa(\omega)$ yields

$$\int_{0}^{\omega} \{n(\omega') - 1\}\mathrm{d}\omega' = \frac{1}{\pi}\int_{0}^{\infty} \kappa(\omega')\log\left|\frac{\omega' + \omega}{\omega' - \omega}\right|\mathrm{d}\omega', \tag{19.281}$$

and using Eq. (19.280) with $f(\omega) = \kappa(\omega)$ yields

$$\int_{0}^{\omega} \kappa(\omega')\mathrm{d}\omega' = \frac{1}{\pi}\int_{0}^{\infty} \{n(\omega') - 1\}\log\left|\frac{\omega'^2 - \omega^2}{\omega'^2}\right|\mathrm{d}\omega'. \tag{19.282}$$

It might look tempting to the reader to examine the limit $\omega \to \infty$ in Eq. (19.281), thereby obtaining Eq. (19.247). Taking this limit is not a valid operation, since the derivation of Eq. (19.279) rests on the assumption that $\chi_{(0,x)}(t)$ belongs to $L^p(\mathbb{R})$ for $p > 1$, which is clearly false in the limit $x \to \infty$, and hence the derivation of

Eq. (4.181) does not apply. In a similar manner, the limit $x \to \infty$ cannot be taken in Eq. (19.280), because the inversion formula for the Hilbert transform cannot be applied to obtain Eq. (4.183). Hence, the limit $\omega \to \infty$ cannot be taken in Eq. (19.282).

The analytic signal concept employed to obtain Eq. (4.245) can be used to derive dispersion relations involving powers of the optical constants. For example, let $f_1(\omega) = f_2(\omega) = N(\omega) - 1$, then Eq. (4.245) yields, on taking the real and imaginary parts,

$$H[\{n(\omega) - 1\}^2 - \kappa(\omega)^2] = 2\kappa(\omega)\{n(\omega) - 1\} \tag{19.283}$$

and

$$H[2\kappa(\omega)\{n(\omega) - 1\}] = -[\{n(\omega) - 1\}^2 - \kappa(\omega)^2]. \tag{19.284}$$

Equation (19.283) also follows directly from the Tricomi identity, Eq. (4.270), on using $f(\omega) = n(\omega) - 1$ and $g(\omega) = \kappa(\omega)$. The choice $f(\omega) = g(\omega) = \kappa(\omega)$ and Tricomi's formula yields Eq. (19.284).

Employing Eq. (4.196) for the choice $f(\omega) = \kappa(\omega)$ yields Eq. (19.268), and for $f(\omega) = \varepsilon_{\mathrm{i}}(\omega)$ Eq. (19.269) is obtained. The choice $f(\omega) = \omega\kappa(\omega)$ leads to

$$\int_0^\infty \omega^2\kappa(\omega)^2\,\mathrm{d}\omega = \int_0^\infty \omega^2\{n(\omega) - 1\}^2\,\mathrm{d}\omega, \tag{19.285}$$

and $f(\omega) = \omega\varepsilon_{\mathrm{i}}(\omega)$ gives

$$\int_0^\infty \omega^2\varepsilon_{\mathrm{i}}(\omega)^2\,\mathrm{d}\omega = \int_0^\infty \omega^2\{\varepsilon_{\mathrm{r}}(\omega) - \varepsilon_0\}^2\,\mathrm{d}\omega. \tag{19.286}$$

Both equations apply for conductors and insulators. The unitary property is actually a particular case of the Riesz inequality, Eq. (4.382). If Eq. (4.427) is employed, and f is restricted to be an even or an odd function, as is done in the sequel, then the integration interval can be restricted to $[0, \infty)$, so for the case $p > 1$, it follows that

$$\{\Re_p\}^{-p}\int_0^\infty |f(x)|^p\,\mathrm{d}x \le \int_0^\infty |Hf(x)|^p\,\mathrm{d}x \le \{\Re_p\}^p\int_0^\infty |f(x)|^p\,\mathrm{d}x, \tag{19.287}$$

where $\Re_p$ is the Riesz constant given by

$$\Re_p = \begin{cases}\tan(\pi/2p), & 1 < p \le 2\\ \cot(\pi/2p), & 2 \le p < \infty.\end{cases} \tag{19.288}$$

Selecting $f(\omega) = \kappa(\omega)$ leads to the inequality

$$\{\Re_p\}^{-p}\int_0^\infty \kappa(\omega)^p\,\mathrm{d}\omega \le \int_0^\infty |n(\omega) - 1|^p\,\mathrm{d}\omega \le \{\Re_p\}^p\int_0^\infty \kappa(\omega)^p\,\mathrm{d}\omega, \tag{19.289}$$

which applies for insulators for $p > 1$ and for conductors for $1 < p < 2$. Setting $f(\omega) = \omega\kappa(\omega)$, then

$$\{\mathfrak{R}_p\}^{-p} \int_0^\infty \omega^p \kappa(\omega)^p \, d\omega \leq \int_0^\infty |\omega\{n(\omega) - 1\}|^p \, d\omega \leq \{\mathfrak{R}_p\}^p \int_0^\infty \omega^p \kappa(\omega)^p \, d\omega, \tag{19.290}$$

which applies for both insulators and conductors for $p > 1$. Using $f(\omega) = n(\omega) - 1$ yields

$$\{\mathfrak{R}_p\}^{-p} \int_0^\infty |n(\omega) - 1|^p \, d\omega \leq \int_0^\infty \kappa(\omega)^p \, d\omega \leq \{\mathfrak{R}_p\}^p \int_0^\infty |n(\omega) - 1|^p \, d\omega, \tag{19.291}$$

which applies for insulators for $p > 1$ and for conductors for $1 < p < 2$. The case $f(\omega) = \omega\{n(\omega) - 1\}$ leads to

$$\begin{aligned} \{\mathfrak{R}_p\}^{-p} \int_0^\infty |\omega\{n(\omega) - 1\}|^p \, d\omega &\leq \int_0^\infty \omega^p \kappa(\omega)^p \, d\omega \\ &\leq \{\mathfrak{R}_p\}^p \int_0^\infty |\omega\{n(\omega) - 1\}|^p \, d\omega, \end{aligned} \tag{19.292}$$

which applies for both insulators and conductors for $p > 1$. For the choices $f(\omega) = \varepsilon_i(\omega)$, and $f(\omega) = \varepsilon_r(\omega) - \varepsilon_0$, it follows that

$$\{\mathfrak{R}_p\}^{-p} \int_0^\infty \varepsilon_i(\omega)^p \, d\omega \leq \int_0^\infty |\varepsilon_r(\omega) - \varepsilon_0|^p \, d\omega \leq \{\mathfrak{R}_p\}^p \int_0^\infty \varepsilon_i(\omega)^p \, d\omega \tag{19.293}$$

and

$$\{\mathfrak{R}_p\}^{-p} \int_0^\infty |\varepsilon_r(\omega) - \varepsilon_0|^p \, d\omega \leq \int_0^\infty \varepsilon_i(\omega)^p \, d\omega \leq \{\mathfrak{R}_p\}^p \int_0^\infty |\varepsilon_r(\omega) - \varepsilon_0|^p \, d\omega, \tag{19.294}$$

which apply for insulators for $p > 1$, but do not apply for conductors because of the singular structure at $\omega = 0$. The choices $f(\omega) = \omega\varepsilon_i(\omega)$ and $f(\omega) = \omega\{\varepsilon_r(\omega) - \varepsilon_0\}$ lead to

$$\{\mathfrak{R}_p\}^{-p} \int_0^\infty \omega^p \varepsilon_i(\omega)^p \, d\omega \leq \int_0^\infty |\omega\{\varepsilon_r(\omega) - \varepsilon_0\}|^p \, d\omega \leq \{\mathfrak{R}_p\}^p \int_0^\infty \omega^p \varepsilon_i(\omega)^p \, d\omega \tag{19.295}$$

and

$$\{\mathfrak{R}_p\}^{-p} \int_0^\infty \omega^p |\varepsilon_r(\omega) - \varepsilon_0|^p \, d\omega \leq \int_0^\infty \omega^p \varepsilon_i(\omega)^p \, d\omega$$
$$\leq \{\mathfrak{R}_p\}^p \int_0^\infty \omega^p |\varepsilon_r(\omega) - \varepsilon_0|^p \, d\omega, \qquad (19.296)$$

which apply for both insulators and conductors for $p > 1$. Note that for $p = 2$, the Riesz inequality is an equality. So using $p = 2$ yields Eq. (19.268) from Eq. (19.289), Eq. (19.269) from Eq. (19.294), Eq. (19.285) from Eq. (19.290), and Eq. (19.286) from (19.295).

19.13 Sum rules involving weight functions

In this section the question of deriving weighted sum rules for optical constants is considered. The basic idea in play here is that, from a practical standpoint, experimental data are not available for a particular optical constant over a wide spectral range. It has already been indicated that this circumstance might be dealt with by looking for appropriate extrapolations of the collected data beyond the measured spectral interval. An alternative idea is to find dispersion relations for which the integrand vanishes sufficiently rapidly as $\omega \to \infty$, so that the high-frequency data for the optical constant in question makes only a very small contribution to the overall value of the integral.

An interesting approach, introduced by Liu and Okubo (1967, 1968), is discussed first. Their technique was developed to deal with the scattering amplitude in a branch of elementary particle scattering. The Liu–Okubo idea will be applied to the refractive index of a non-conducting material. This application is due to Villani and Zimerman (1973b). Consider the integral

$$\oint_C \frac{\{N(\omega_z) - 1\}e^{i\pi\beta} \, d\omega_z}{(\omega_z - \omega_a)^\beta (\omega_z + \omega_a)^\beta},$$

where the parameter β satisfies

$$-\frac{1}{2} < \beta < 1, \qquad (19.297)$$

the frequency $\omega_a > 0$, and C denotes the contour shown in Figure 19.6. The contour sections along $(-\infty, -\omega_a)$ and $(\omega_a, \infty)$are placed a distance k above the branch cuts, and the limit $k \to 0+$ is examined. From the Cauchy integral theorem, it follows that

$$\int_{-\infty}^{-\omega_a} \frac{\{N(\omega_r + ik) - 1\}e^{i\pi\beta} \, d\omega_r}{(\omega_r + ik - \omega_a)^\beta (\omega_r + ik + \omega_a)^\beta} + \int_{ik}^{0} \frac{\{N(-\omega_a + i\omega_i) - 1\}e^{i\pi\beta} i \, d\omega_i}{(i\omega_i - 2\omega_a)^\beta (i\omega_i)^\beta}$$

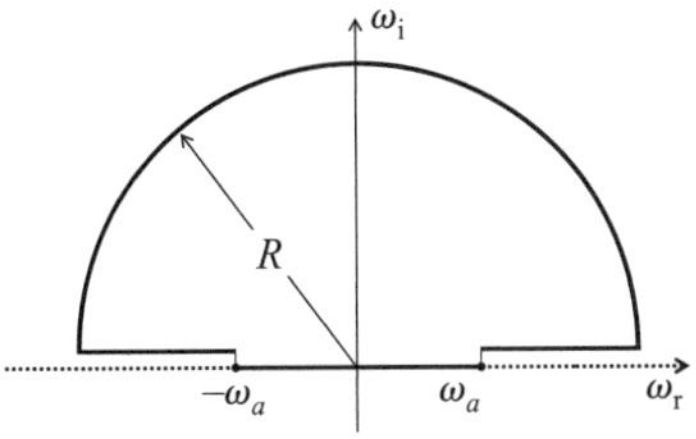

Figure 19.6. Contour employed to derive weighted sum rules for the complex refractive index.

$$+\int_{-\omega_a}^{\omega_a}\frac{\{N(\omega_r)-1\}e^{i\pi\beta}\,d\omega_r}{(\omega_r-\omega_a)^\beta(\omega_r+\omega_a)^\beta}+\int_0^{ik}\frac{\{N(\omega_a+i\omega_i)-1\}e^{i\pi\beta}i\,d\omega_i}{(i\omega_i)^\beta(2\omega_a+i\omega_i)^\beta}$$
$$+\int_{\omega_a}^{\infty}\frac{\{N(\omega_r+ik)-1\}e^{i\pi\beta}\,d\omega_r}{(\omega_r+ik-\omega_a)^\beta(\omega_r+ik+\omega_a)^\beta}+\int_0^{\pi}\frac{\{N(Re^{i\theta})-1\}e^{i\pi\beta}iRe^{i\theta}\,d\theta}{(Re^{i\theta}-\omega_a)^\beta(Re^{i\theta}+\omega_a)^\beta}=0. \tag{19.298}$$

Taking the limits $R\to\infty$ and $k\to 0+$ in Eq. (19.298) leads to

$$\int_{-\infty}^{-\omega_a}\frac{\{N(\omega_r)-1\}e^{i\pi\beta}\,d\omega_r}{(\omega_r-\omega_a)^\beta(\omega_r+\omega_a)^\beta}+\int_{-\omega_a}^{\omega_a}\frac{\{N(\omega_r)-1\}d\omega_r}{(\omega_a^2-\omega_r^2)^\beta}$$
$$+\int_{\omega_a}^{\infty}\frac{\{N(\omega_r)-1\}e^{i\pi\beta}\,d\omega_r}{(\omega_r-\omega_a)^\beta(\omega_r+\omega_a)^\beta}=0. \tag{19.299}$$

Taking advantage of the symmetry conditions given in Eqs. (19.111) and (19.112) allows the preceding result to be simplified as follows:

$$\int_{\omega_a}^{\infty}\frac{\{n(\omega)-1-i\kappa(\omega)\}e^{i\pi\beta}\,d\omega}{(-\omega-\omega_a)^\beta(-\omega+\omega_a)^\beta}+2\int_0^{\omega_a}\frac{\{n(\omega)-1\}d\omega}{(\omega_a^2-\omega^2)^\beta}$$
$$+\int_{\omega_a}^{\infty}\frac{\{n(\omega)-1+i\kappa(\omega)\}e^{i\pi\beta}\,d\omega}{(\omega-\omega_a)^\beta(\omega+\omega_a)^\beta}=0. \tag{19.300}$$

That is,

$$\int_{\omega_a}^{\infty}\frac{[\{n(\omega)-1\}(e^{i\pi\beta}+e^{-i\pi\beta})+i\kappa(\omega)(e^{i\pi\beta}-e^{-i\pi\beta})]d\omega}{(\omega-\omega_a)^\beta(\omega+\omega_a)^\beta}$$
$$+2\int_0^{\omega_a}\frac{\{n(\omega)-1\}d\omega}{(\omega_a^2-\omega^2)^\beta}=0, \tag{19.301}$$

and hence the following sum rule is obtained:

$$\cos\pi\beta\int_{\omega_a}^{\infty}\frac{\{n(\omega)-1\}d\omega}{(\omega^2-\omega_a^2)^\beta}-\sin\pi\beta\int_{\omega_a}^{\infty}\frac{\kappa(\omega)d\omega}{(\omega^2-\omega_a^2)^\beta}+\int_0^{\omega_a}\frac{\{n(\omega)-1\}d\omega}{(\omega_a^2-\omega^2)^\beta}=0. \tag{19.302}$$

The particular choice of $\beta = 0$ leads to

$$\int_0^{\infty} \{n(\omega) - 1\} d\omega = 0, \tag{19.303}$$

which was derived previously in Section 19.11 by a different approach. The case $\beta = 1/2$ yields the following sum rule:

$$\int_0^{\omega_a} \frac{\{n(\omega) - 1\} d\omega}{\surd(\omega_a^2 - \omega^2)} = \int_{\omega_a}^{\infty} \frac{\kappa(\omega) d\omega}{\surd(\omega^2 - \omega_a^2)}. \tag{19.304}$$

Because $\kappa(\omega) > 0$, for $\omega > 0$, the following inequality is obtained:

$$\int_0^{\omega_a} \frac{\{n(\omega) - 1\} d\omega}{\surd(\omega_a^2 - \omega^2)} > 0. \tag{19.305}$$

Since ω_a is an arbitrary frequency, the preceding inequality has the potential to be useful as a self-consistency check for refractive index data measured over the spectral interval $[0, \omega_a)$.

Another approach that has been employed (King, 1976a) consists of considering the contour integral

$$\oint_C \frac{\{N(\omega_z) - 1\} d\omega_z}{\omega_z + i\omega},$$

where C denotes the contour shown in Figure 19.7 and ω is an arbitrary angular frequency satisfying $\omega > 0$.

Applying the Cauchy integral theorem leads to

$$\int_{-\infty}^{\infty} \frac{\{N(\omega') - 1\} d\omega'}{\omega' + i\omega} = 0. \tag{19.306}$$

Employing the crossing symmetry condition for $N(\omega) - 1$ leads to

$$\int_0^{\infty} \frac{\{\omega'[N(\omega') - N(\omega')^*] - i\omega[N(\omega') + N(\omega')^* - 2]\} d\omega'}{\omega'^2 + \omega^2} = 0, \tag{19.307}$$

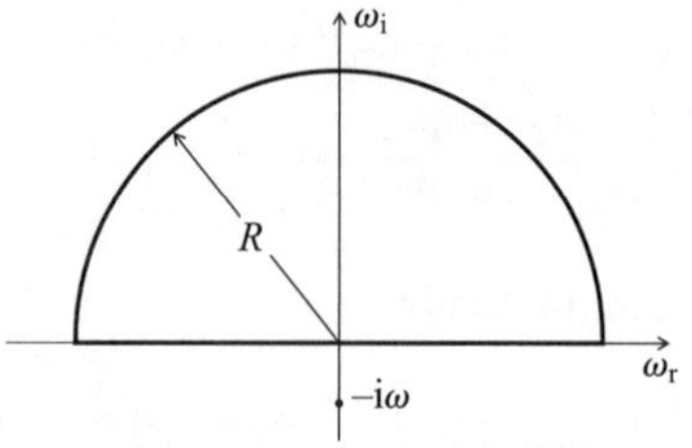

Figure 19.7. Contour for the evaluation of contour integrals of the optical functions with no poles in the upper half plane or on the real axis.

from which it follows that

$$\int_0^\infty \frac{\omega'\kappa(\omega')\mathrm{d}\omega'}{\omega'^2+\omega^2} = \int_0^\infty \frac{\omega\{n(\omega')-1\}\mathrm{d}\omega'}{\omega'^2+\omega^2}. \tag{19.308}$$

Relations such as Eq. (19.308) provide necessary conditions that must be satisfied by theoretical models or experimental data.

19.14 Summary of sum rules for the dielectric constant and refractive index

In this section a number of the sum rules that have been considered so far for the dielectric constant and refractive index, plus a number of additional rules that can be derived using assumptions about the analytic behavior of various powers and related functions of these two optical constants, are summarized. The principal assumptions in play are that the materials are isotropic, spatial dispersion effects are ignored, and the materials are non-magnetic, so the permeability does not arise in any of the given formulas. Unless specifically indicated, the sum rules apply to both conductors and insulators. Some of the exercises at the end of this chapter provide hints as to their derivation. The references given indicate where the results were first derived or discussed in relation to other sum rules. Some references have been supplied where these results are discussed in practical applications. The chapter end-notes provide further sources.

19.15 Light scattering: the forward scattering amplitude

In this section a connection between the developments of Section 19.6 and the scattering of light is made. The classical theory of light scattering leads to the following result (due to Lorentz):

$$N(\omega) - 1 = \frac{2\pi c^2}{\omega^2}\mathcal{N} f(\omega, 0), \tag{19.309}$$

where $\mathcal{N}$ is the number of scattering centers per unit volume and $f(\omega, 0)$ is the scattering amplitude at the angle $\theta = 0$, that is the forward direction. Henceforth in this section, the notation $f(\omega) \equiv f(\omega, 0)$ is adopted. Equation (19.309) applies to the situation where the density of scattering centers is sufficiently small that the incident field and the field at the scattering center can be treated as identical. This would apply to a gas at low-density or a very thin slice of a dielectric. The low-density assumption implies that intermolecular interactions can be regarded as small, and, as a consequence, can be ignored in a first treatment. The derivation and the assumptions underlying the result can be pursued in standard sources (Goldberger and Watson, 1964; Jackson, 1999). The reader needs to be alert to the fact that any dispersion relations derived using Eq. (19.309) also depend on the assumptions just stated. The

Table 19.1. *Summary of sum rules for the dielectric constant*

Number	Sum rule	Reference
(1)	$\int_0^\infty \frac{\varepsilon_i(\omega)d\omega}{\omega} = \frac{\pi}{2}\{\varepsilon_r(0) - \varepsilon_0\}$ (insulators)	Gorter and Kronig (1936)
(2)	$\int_0^\infty \frac{\{\varepsilon_i(\omega) - \sigma(0)/\omega\}d\omega}{\omega} = \frac{\pi}{2}\{\varepsilon_r(0) - \varepsilon_0\}$	
(3)	$\int_0^\infty [\varepsilon_r(\omega) - \varepsilon_0]d\omega = 0$ (insulators)	Saslow (1970); Scaife (1972)
(4)	$\int_0^\infty \{\varepsilon_r(\omega) - \varepsilon_0\}d\omega = -\frac{\pi\sigma(0)}{2}$	Saslow (1970)
(5)	$\int_0^\infty \omega\varepsilon_i(\omega)d\omega = \frac{\pi\varepsilon_0\omega_p^2}{2}$	Landau and Lifshitz (1960); Stern (1963)
(6)	$\int_0^\infty \{\varepsilon_r(\omega) - \varepsilon_0\}\cos\omega t\, d\omega = \int_0^\infty \varepsilon_i(\omega)\sin\omega t\, d\omega,\ t > 0$	Cole and Cole (1942); Scaife (1972); King (1978a)
(7)	$\int_0^\infty \{\varepsilon_r(\omega) - \varepsilon_0\}^2\, d\omega = \int_0^\infty \varepsilon_i(\omega)^2\, d\omega$ (insulators)	
(8)	$\int_0^\infty \{\varepsilon_r(\omega) - \varepsilon_0\}[\{\varepsilon_r(\omega) - \varepsilon_0\}^2 - 3\varepsilon_i(\omega)^2]d\omega = 0$ (insulators)	
(9)	$\int_0^\infty \omega\varepsilon_i(\omega)\{\varepsilon_r(\omega) - \varepsilon_0\}d\omega = 0$ (insulators)	Villani and Zimerman (1973b)

(10)	$\int_0^\infty \omega^2\{\varepsilon_r(\omega)-\varepsilon_0\}^2\,d\omega = \int_0^\infty \omega^2\varepsilon_i(\omega)^2\,d\omega$	
(11)	$\int_0^\infty \omega^2\{\varepsilon_r(\omega)-\varepsilon_0\}[\{\varepsilon_r(\omega)-\varepsilon_0\}^2-3\varepsilon_i(\omega)^2]d\omega = 0$	
(12)	$\int_0^\infty \omega^3\varepsilon_i(\omega)[3\{\varepsilon_r(\omega)-\varepsilon_0\}^2-\varepsilon_i(\omega)^2]d\omega = 0$	
(13)	$\int_0^\infty \omega^3\varepsilon_i(\omega)\{\varepsilon_r(\omega)-\varepsilon_0\}d\omega = -\frac{\pi\varepsilon_0^2\omega_p^4}{4}$	Altarelli and Smith (1974)
(14)	$\int_0^\infty \omega^5\varepsilon_i(\omega)[3\{\varepsilon_r(\omega)-\varepsilon_0\}^2-\varepsilon_i(\omega)^2]d\omega = \frac{\pi\varepsilon_0^3\omega_p^6}{2}$	Villani and Zimerman (1973a)
(15)	$\int_0^\infty \frac{\omega\varepsilon_i(\omega)d\omega}{\omega^2+\omega_0^2} = \int_0^\infty \frac{\omega_0\{\varepsilon_r(\omega)-\varepsilon_0\}d\omega}{\omega^2+\omega_0^2}, \quad \omega_0 > 0$	King (1976a)

Table 19.2. *Summary of sum rules for the refractive index*

Number	Sum rule	Reference
(1)	$\int_0^\infty \{n(\omega) - 1\}\mathrm{d}\omega = 0$	Saslow (1970); Altarelli *et al.* (1972); Smith (1985)
(2)	$\int_0^\infty \omega\kappa(\omega)\mathrm{d}\omega = \frac{\pi}{4}\omega_\mathrm{p}^2$	Kronig (1926)
(3)	$\int_0^\infty \frac{\kappa(\omega)\mathrm{d}\omega}{\omega} = \frac{\pi}{2}\{n(0) - 1\}$ (insulators)	Moss (1961)
(4)	$\int_0^\infty \omega\kappa(\omega)n(\omega)\mathrm{d}\omega = \frac{\pi}{4}\omega_\mathrm{p}^2$	Villani and Zimerman (1973a)
(5)	$\int_0^\infty \{n(\omega) - 1\}\cos\omega t\,\mathrm{d}\omega = \int_0^\infty \kappa(\omega)\sin\omega t\,\mathrm{d}\omega, \quad t > 0$	
(6)	$\int_0^\infty \omega\kappa(\omega)[3n(\omega)^2 - \kappa(\omega)^2]\mathrm{d}\omega = \frac{3\pi}{4}\omega_\mathrm{p}^2$	
(7)	$\int_0^\infty \omega\kappa(\omega)\{n(\omega) - 1\}\mathrm{d}\omega = 0$	Stern (1963); Altarelli *et al.* (1972)
(8)	$\int_0^\infty \omega^m\kappa(\omega)[3\{n(\omega) - 1\}^2 - \kappa(\omega)^2]\mathrm{d}\omega = 0, \quad m = 1, 3$	Villani and Zimerman (1973b)
(9)	$\int_0^\infty \omega^m\{n(\omega) - 1\}[\{n(\omega) - 1\}^2 - 3\kappa(\omega)^2]\mathrm{d}\omega = 0,\ m = 2, 4$	Villani and Zimerman (1973b)

(10)	$\int_0^\infty \{n(\omega)-1\}[\{n(\omega)-1\}^2-3\kappa(\omega)^2]\mathrm{d}\omega = 0$ (insulators)	Villani and Zimerman (1973b)
(11)	$\int_0^\infty \omega^2[\{n(\omega)-1\}^2-\kappa(\omega)^2]\mathrm{d}\omega = 0$	Villani and Zimerman (1973b)
(12)	$\int_0^\infty [\{n(\omega)-1\}^2-\kappa(\omega)^2]\mathrm{d}\omega = 0$ (insulators)	Altarelli and Smith (1974)
(13)	$\int_0^\infty [\{n(\omega)-1\}^2-\kappa(\omega)^2]\mathrm{d}\omega = -\frac{\pi\sigma(0)}{2\varepsilon_0}$	Smith (1985)
(14)	$\int_0^\infty \omega^3\kappa(\omega)\{n(\omega)-1\}\mathrm{d}\omega = -\frac{\pi}{16}\omega_{\mathrm{p}}^4$	Villani and Zimerman (1973a); Altarelli and Smith (1974)
(15)	$\int_0^\infty \omega^5\kappa(\omega)[3\{n(\omega)-1\}^2-\kappa(\omega)^2]\mathrm{d}\omega = \frac{\pi}{16}\omega_{\mathrm{p}}^6$	Villani and Zimerman (1973a)
(16)	$\int_0^\infty \frac{\omega\kappa(\omega)\mathrm{d}\omega}{\omega^2+\omega_0^2} = \int_0^\infty \frac{\omega_0\{n(\omega)-1\}\mathrm{d}\omega}{\omega^2+\omega_0^2}, \quad \omega_0 > 0$	King (1976a)
(17)	$\int_0^{\omega_0} \frac{\{n(\omega)-1\}\mathrm{d}\omega}{\sqrt{(\omega_0^2-\omega^2)}} = \int_{\omega_0}^\infty \frac{\kappa(\omega)\mathrm{d}\omega}{\sqrt{(\omega^2-\omega_0^2)}}, \quad \omega_0 > 0$	Villani and Zimerman (1973b)
(18)	$\cos\pi\beta\int_{\omega_0}^\infty \frac{\{n(\omega)-1\}\mathrm{d}\omega}{(\omega^2-\omega_0^2)^\beta} - \sin\pi\beta\int_{\omega_0}^\infty \frac{\kappa(\omega)\mathrm{d}\omega}{(\omega^2-\omega_0^2)^\beta}$ $+\int_0^{\omega_0} \frac{\{n(\omega)-1\}\mathrm{d}\omega}{(\omega_0^2-\omega^2)^\beta} = 0, \omega_0 > 0, -1/2 < \beta < 1$	Villani and Zimerman (1973b)

classical model given in Eq. (19.18) leads to the following result (due to Lorentz):

$$N(\omega) - 1 = \frac{\mathcal{N}e^2}{2\varepsilon_0 m}\sum_j \frac{f_j}{\omega_j^2 - \omega^2 - \mathrm{i}\gamma_j\omega}. \tag{19.310}$$

This is called a *dispersion formula*. The reader should note the terminology that is employed. A dispersion formula gives the frequency variation of $N(\omega)$, whereas a dispersion relation states a Hilbert transform connection between functions. Coupling Eqs. (19.309) and (19.310) leads to an expression for the forward scattering amplitude in terms of the classical Lorentz model.

Let $f_\mathrm{r}(\omega)$ and $f_\mathrm{i}(\omega)$ denote the real and imaginary parts of the forward scattering amplitude, respectively. The dispersion relations for $f(\omega)$ can be obtained from the corresponding results for $\kappa(\omega)$ and $n(\omega) - 1$. The function $f(\omega)$ is analytic in the upper half of the complex angular frequency plane. The asymptotic behavior for $f(\omega)$ as $\omega \to \infty$ can be worked out from Eqs. (19.309), (19.98), and (19.99). As an alternative, readers familiar with the Kramers–Heisenberg formula can use this result to deduce the asymptotic behavior for $f(\omega)$. The crossing symmetry relations for the forward scattering amplitude take the form

$$f_\mathrm{r}(-\omega) = f_\mathrm{r}(\omega) \tag{19.311}$$

and

$$f_\mathrm{i}(-\omega) = -f_\mathrm{i}(\omega). \tag{19.312}$$

From Eq. (19.113) it follows that

$$f_\mathrm{i}(\omega) = \frac{2\omega^3}{\pi} P\int_0^\infty \frac{f_\mathrm{r}(\omega')\mathrm{d}\omega'}{\omega'^2(\omega^2 - \omega'^2)}, \tag{19.313}$$

and from Eq. (19.114) the following dispersion relation is obtained:

$$f_\mathrm{r}(\omega) = -\frac{2\omega^2}{\pi} P\int_0^\infty \frac{f_\mathrm{i}(\omega')\mathrm{d}\omega'}{\omega'(\omega^2 - \omega'^2)}. \tag{19.314}$$

As an alternative, these results can be obtained by considering the integral

$$\oint_C \frac{\{f(\omega_z)\omega_z^{-2}\}\mathrm{d}\omega_z}{\omega - \omega_z},$$

where C denotes the contour shown in Figure 19.2. Application of the Phragmén–Lindelöf theorem is used to establish the asymptotic behavior of $f(\omega_z)\omega_z^{-2}$ in the upper half of the complex angular frequency plane.

As an electromagnetic beam traverses a distance x through a thin slice of a medium, the intensity is diminished by the factor

$$\left|\mathrm{e}^{\mathrm{i}N(\omega)\omega x/c}\right|^2 = \mathrm{e}^{-2\kappa(\omega)\omega x/c}. \tag{19.315}$$

The diminution of the beam intensity can also be written in terms of the total scattering cross-section $\sigma_{\mathrm{t}}(\omega)$, and is given by $\mathrm{e}^{-N\sigma_{\mathrm{t}}(\omega)x}$. From the latter result and Eqs. (19.309) and (19.315), it follows that the total scattering cross-section is related to the imaginary part of the forward scattering amplitude:

$$\sigma_{\mathrm{t}}(\omega) = \frac{4\pi c}{\omega} f_{\mathrm{i}}(\omega). \tag{19.316}$$

This is the termed the *optical theorem*. Employing this result, Eq. (19.314) can be rewritten as follows:

$$f_{\mathrm{r}}(\omega) = -\frac{\omega^2}{2\pi^2 c} P \int_0^\infty \frac{\sigma_{\mathrm{t}}(\omega')\mathrm{d}\omega'}{\omega^2 - \omega'^2}. \tag{19.317}$$

It therefore follows that knowledge of the total cross-section over the complete spectral domain determines the complex forward scattering amplitude at all frequencies.

On writing Eq. (19.313) using the substitutions $y = \omega^2$, $x = \omega'^2$, and $g(x) = f_{\mathrm{r}}(\sqrt{x})x^{-3/2}$, leads to

$$\frac{\pi f_{\mathrm{i}}(\omega)}{\omega^3} = P \int_0^\infty \frac{g(x)\mathrm{d}x}{y - x}. \tag{19.318}$$

Taking the limit $\omega \to \infty$ and employing Eq. (19.211) yields

$$\frac{\pi f_{\mathrm{i}}(\omega)}{\omega^3} = y^{-1} \int_0^\infty g(x)\mathrm{d}x + O(y^{-3/2}), \quad \text{as } y \to \infty. \tag{19.319}$$

Hence,

$$f_{\mathrm{i}}(\omega) = \frac{2\omega}{\pi} \int_0^\infty \frac{f_{\mathrm{r}}(\omega')\mathrm{d}\omega'}{\omega'^2} + O(1), \quad \text{as } \omega \to \infty. \tag{19.320}$$

From Eqs. (19.99) and (19.309), it follows that

$$f_{\mathrm{i}}(\omega) = O(\omega^{-1}), \quad \text{as } \omega \to \infty, \tag{19.321}$$

hence,

$$\int_0^\infty \frac{f_{\mathrm{r}}(\omega)\mathrm{d}\omega}{\omega^2} = 0. \tag{19.322}$$

A similar argument applied to Eq. (19.314) yields

$$-\frac{\pi}{2\omega^2} f_{\rm r}(\omega) = \frac{1}{\omega^2}\int_0^\infty \frac{f_{\rm i}(\omega')\mathrm{d}\omega'}{\omega'} + O(\omega^{-3}), \quad \text{as } \omega \to \infty, \tag{19.323}$$

and hence

$$\int_0^\infty \frac{f_{\rm i}(\omega)\mathrm{d}\omega}{\omega} = \frac{\omega_{\rm p}^2}{8c^2\mathcal{N}}. \tag{19.324}$$

This is the analog of the f sum rule for the scattering amplitude in the forward direction. Equation (19.324) can also be obtained directly from Eq. (19.309) and the f sum rule for $\kappa(\omega)$, Eq. (19.242). Making use of Eq. (19.316), the preceeding result can be written as follows:

$$\int_0^\infty \sigma_{\rm t}(\omega)\mathrm{d}\omega = \frac{\pi\omega_{\rm p}^2}{2c\mathcal{N}}. \tag{19.325}$$

Consider the integral

$$\oint_C \frac{\{f(\omega_z)^2\omega_z^{-2}\}\mathrm{d}\omega_z}{\omega - \omega_z},$$

where C denotes the contour shown in Figure 19.2. The following dispersion relations are obtained:

$$f_{\rm r}(\omega)f_{\rm i}(\omega) = \frac{\omega^3}{\pi}P\int_0^\infty \frac{\{f_{\rm r}(\omega')^2 - f_{\rm i}(\omega')^2\}\mathrm{d}\omega'}{\omega'^2(\omega^2 - \omega'^2)} \tag{19.326}$$

and

$$f_{\rm r}(\omega)^2 - f_{\rm i}(\omega)^2 = -\frac{4\omega^2}{\pi}P\int_0^\infty \frac{f_{\rm r}(\omega')f_{\rm i}(\omega')\mathrm{d}\omega'}{\omega'(\omega^2 - \omega'^2)}. \tag{19.327}$$

Using the familiar substitutions $y = \omega^2$ and $x = \omega'^2$, and setting $g(x) = 2^{-1}\{f_{\rm r}(\sqrt{x})^2 - f_{\rm i}(\sqrt{x})^2\}x^{-3/2}$, leads to

$$\frac{\pi}{\omega^3} f_{\rm r}(\omega)f_{\rm i}(\omega) = P\int_0^\infty \frac{g(x)\mathrm{d}x}{y - x}, \tag{19.328}$$

and hence

$$\frac{\pi}{\omega^3} f_{\rm r}(\omega)f_{\rm i}(\omega) = \frac{1}{y}\int_0^\infty g(x)\mathrm{d}x + O(y^{-3/2}), \quad \text{as } y \to \infty. \tag{19.329}$$

This last result can be rewritten as follows:

$$f_{\rm r}(\omega)f_{\rm i}(\omega) = \frac{\omega}{\pi}\int_0^\infty \frac{\{f_{\rm r}(\omega')^2 - f_{\rm r}(\omega')^2\}}{\omega'^2}\mathrm{d}\omega + O(1), \quad \text{as } \omega \to \infty. \tag{19.330}$$

Since $f_r(\omega)f_i(\omega) = O(\omega^{-\lambda})$ for $\lambda > 0$, as $\omega \to \infty$, then it follows from Eq. (19.330) that

$$\int_0^\infty \frac{\{f_r(\omega)^2 - f_i(\omega)^2\}\mathrm{d}\omega}{\omega^2} = 0. \tag{19.331}$$

The differential elastic scattering cross-section is denoted by $\mathrm{d}\sigma_e(\omega)/\mathrm{d}\Omega$, and can be expressed in terms of the forward scattering amplitude by the following expression:

$$\frac{\mathrm{d}\sigma_e(\omega)}{\mathrm{d}\Omega} = |f(\omega)|^2 = f_r(\omega)^2 + f_i(\omega)^2. \tag{19.332}$$

Equation (19.331) can be recast in terms of the differential elastic scattering cross-section and the total cross-section, thus:

$$\int_0^\infty \left[\frac{c^2}{\omega^2}\frac{\mathrm{d}\sigma_e(\omega)}{\mathrm{d}\Omega} - \frac{\sigma_t(\omega)^2}{8\pi^2}\right]\mathrm{d}\omega = 0. \tag{19.333}$$

Notes

§19.1 An excellent exposition and standard text on classical electrodynamics is Jackson (1999), and Landau and Lifshitz (1960) is also highly recommended reading. Also worthy of consideration for further reading on the subject are the accounts by Lorentz (1952) and Rosenfeld (1965). On the topic of dielectrics, see Fröhlich (1958), and on electrodynamics, see Podolsky and Kunz (1969). Kramers' work in the area of absorption and refraction of X-rays is first referred to in an abstract of a presentation to the Royal Danish Academy of Sciences and Letters on November 27, 1925 (Kramers, 1926). Ter Haar (1998) reports that the account of this work published later (Kramers, 1929) is based on an excerpt of a presentation at a meeting in Zürich in July 1929, and was written by F. Bloch. Another paper appearing about the same time was Kronig (1929). Kramers' work can also be found in the publication of his collected papers (Kramers, 1956). Slightly later papers include the works of Gorter and Kronig (1936) and Kronig (1938). In parallel with the work of Kramers and Kronig, related advances using Fourier transform methods were made around the same time in the engineering sciences. Carson (1926, chap. 11) points out that if either the real or imaginary component of the complex steady state admittance is specified over the entire frequency range, then the behavior of a network is completely determined. Dispersion relations for the polarizability are discussed by Scaife (1971), Mukhtarov (1979), and Keefe (2001). Refer to Chapter 14 for references on the numerical evaluation of the Kramers–Kronig transforms. For a discussion on the treatment of the Kramers–Kronig transforms as a filtering problem, see Shtrauss (2006), and for a constrained variational analysis of optical spectra utilizing the Kramers–Kronig transforms, see Kuzmenko (2005).

§19.2 For supplementary reading, see Silva and Gross (1941), Van Vleck (1945), Brachman and MacDonald (1956), MacDonald and Brachman (1956),

Van Wijngaarden (1963), Sharnoff (1964), Bolton (1969a,b), Nussenzveig (1972), Troup and Bambini (1973), Gross (1975), Peiponen and Vartiainen (1991), Lee and Sindoni (1992), Castro and Nabet (1999), and Reick (2002). For some applications, see MacDonald (1952), Parke (1965), Lovell (1974, 1975), Castaño González, de Dios Leyva, and Pérez Alvarez (1975), Peiponen and Hämäläinen (1986), Tomiyama, Clough, and Kneizys (1987), Smith, Inokuti, and Karstens (2001), Debiais (2002), and Nitsche and Fritz (2004). The first effort to derive the Kramers–Kronig type relations in the context of quantum field theory was carried out by Gell-Mann, Goldberger, and Thirring (1954). A more advanced discussion of the permittivity can be found in Nozières and Pines (1999). If $\varepsilon(\omega)/\varepsilon_0$ is replaced by $n^2(\omega)$ and the terms γ_j set to zero in Eq. (19.18), the resulting expression is often termed Sellmeier's dispersion formula.

§19.3 The essential source for additional study on the derivation of dispersion relations for optical constants is Nussenzveig (1972). The book by Peiponen *et al.* (1999) and the papers by Toll (1956) and Altarelli *et al.* (1972) are also recommended for further reading. See also Davydov (1963), Ginzburg and Meĭman (1964), Cardona (1969), and Rhodes (1977). For discussion on issues associated with the physical interpretation of the response function, see Altarelli and Smith (1974). For a recent book devoted to the application of the Kramers–Kronig relations to optical properties, see Lucarini *et al.* (2005b). For the application of the Kramers–Kronig relations to optical data analysis, some type of data extrapolation is invariably required. An alternative approach is to consider the possibility of reconstructing the required optical data from more limited information. The following three papers have a bearing on this topic, and are highly recommended reading: Hulthén (1982), Milton, Eyre, and Mantese (1997), and Dienstfrey and Greengard (2001). For an introduction to optical properties, see Wooten (1972).

§19.4 The papers by Peterson and Knight (1973) and King (1978a) contain further discussion on the topic of this section.

§19.6 An early treatment on the behavior of the generalized complex refractive index can be found in Brillouin (1914), and an English translation of this work is given in Brillouin (1960). Nussenzveig (1972) is a key source for parts of the discussion of this section. The literature terminology for $\kappa(\omega)$ in Eq. (19.91) is rather variable. For example, Stern (1963) calls $\kappa(\omega)$ the extinction coefficient, and Born and Wolf (1999, p. 737) call $n(\omega)\kappa(\omega)$ the attenuation index and indicate that this is also referred to as the extinction coefficient. The problem is compounded by the fact that the absorption coefficient, $\alpha(\omega)$, defined by Eq. (19.92), is also referred to as the extinction coefficient. For solution studies, it is common to work with the molar absorption coefficient, $\alpha_X(\omega)$, where the beam intensity is attenuated like $e^{-\alpha_X(\omega)[X]x}$, and $[X]$ denotes the molar concentration of the absorbing species X. The molar absorption coefficient is also widely referred to as the extinction coefficient! For an application in seismology, see Futterman (1962), and for one with geophysical implications, see Lamb (1962). For an application of the subtractive Kramers–Kronig approach to the treatment of the refractive index of ice, see Wallis and Wickramasinghe (1999).

Additional discussion on the statement following Eq. (19.94) can be found in Landau and Lifshitz (1969, p. 384–391). An authoritative discussion on wave propagation can be found in the account by Brillouin (1960). For some further reading, see Mojahedi *et al.* (2003). A concise discussion of the determination of optical properties using ellipsometry, with comments on extrapolation issues, can be found in Aspnes (1985).

§19.7 The original paper is Herglotz (1911). Additional accounts on Herglotz functions and some general issues connected to these functions can be found in Nevanlinna (1922), Cauer (1932), Wall (1948, p. 275), Aheizer and Krein (1962), Akhiezer (1965), Shohat and Tamarkin (1970, p. 23), and Nussenzveig (1972, p. 393). The papers by Symanzik (1960, app. B), Wu (1962), Jin and Martin (1964), and the key work of Weaver and Pao (1981), discuss applications of Herglotz functions. The approach of this section is based on King (2006).

§19.8 Smith (1985) and Altarelli *et al.* (1972) can be consulted for some additional information.

§19.9 For an application of the idea that asymptotic limits can yield sum rules for high-energy scattering, see De Alfaro *et al.* (1966).

§19.10 Sum rules obtained from dispersion relations, or derived from the analytic behavior of particular choices of functions, were first discussed at length for problems arising in electric circuit analysis. A number of sum rules for optical constants can be obtained in a similar manner to the approaches employed for electric circuit analysis. Some references for these developments are Bode (1945), Page (1955, chap. 12), and Hamilton (1960). An authoritative discussion of the practical application of sum rules can be found in the review by Smith (1985), and the account of Bassani and Altarelli (1983) is also highly recommended reading. The sum rule given in Eq. (19.225) was first reported in Saslow (1970); later, it was given implicitly in the work of Scaife (1972), who gave the sum rule for the real part of the linear electric susceptibility, and in the work of Altarelli *et al.* (1972). Some useful references for further information on sum rules for the dielectric constant and the refractive index are Altarelli *et al.* (1972), Altarelli and Smith (1974), King (1976a), Smith and Shiles (1978), Smith and Graham (1980), Gründler (1983), and Lévêque (1986), along with the references cited in Tables 19.1 and 19.2 in Section 19.14. See also the work by Hopfield (1970), and for some quadrupole sum rules, consult Wu, Mahler, and Birman (1978). For the situations where relativistic effects cannot be ignored, see Scandolo, Bassani, and Lucarini (2001). A moment expansion approach is discussed by Kubo and Ichimura (1972). Sum rules of the type given in Eq. (19.225) are sometimes referred to as *superconvergence relations*. The name comes from elementary particle scattering, where the quantity of interest, usually a scattering amplitude, may decay too slowly at high energies to give rise to convergent sum rules; in contrast, there are situations where the scattering amplitude has a sufficiently rapid decay at high energies so that the resulting sum rule converges. For the latter case the terminology superconvergence relation is applied to the sum rule. The terminology of calling functions with rapid decay superconvergent has, fortunately, found limited application outside of sum rule development. For further reading, see the review by Ferro Fontán, Queen, and Violini (1972) and the paper by Fischer *et al.* (1969).

§19.11 The first two key papers that point out sum rules for the refractive index are Saslow (1970) and Altarelli *et al.* (1972). For additional reading, see Altarelli and Smith (1974), Peiponen (1985), and Smith (1985). For a detailed discussion on issues associated with phase velocity, group velocity and signal transmission, see Brillouin (1914) and Sommerfeld (1914). An English translation of this Sommerfeld paper can be found in Brillouin (1960). For a treatment of sum rules and Kramers–Kronig relations for media with a negative index of refraction, see Peiponen, *et al.* (2004b).

§19.12 The most common property of the Hilbert transform that has been employed in the literature to derive sum rules for optical properties is the Parseval-type formula. Very few of the other properties have been exploited for this purpose.

§19.13 For further reading, see the papers by Villani and Zimerman (1973b), King (1976a), and Kimel (1982a).

§19.14 The review by Smith (1985) provides a good starting point for further reading together with a survey of the published literature.

§19.15 For further discussion along the lines of this section, see Maximon and O'Connell (1974) and King (1976b).

Exercises

19.1 Use Eq. (19.18) with the restriction to two terms in the sum to obtain a description of the branch cut structure for $N(\omega_z)$.

19.2 Can the dispersion relations derived in Eqs. (19.28) and (19.29) be obtained by replacing the contour of Figure 19.2 with a rectangular contour with vertices at $(\pm R, 0)$ and $(\pm R, \mathrm{i}\omega_0)$, together with a semicircular indentation at the frequency ω?

19.3 Making use of the asymptotic behavior derived from the Lorentz model, and the causality condition, determine $\lim_{t\to 0+} G'(t)$ and $\lim_{t\to 0-} G'(t)$. Recall Eq. (19.42) for the definition of $G(t)$. Hence show that the f sum rule follows from Eq. (19.56). How would the argument be modified to handle conductors? What physical interpretation can you assign to the f sum rule on the basis of this calculation?

19.4 Suppose a medium had both a negative permittivity and a negative permeability (see Veselago (1967, 1968) for an exploration of this assumption). Is the derivation of dispersion relations for the permittivity altered in any way?

19.5 Show that any non-dissipative medium is also non-dispersive and that any non-dispersive medium is non-dissipative.

19.6 Prove that a dissipative medium is dispersive and that a dispersive medium is dissipative.

19.7 Explore the consequences if a non-conducting material has a non-zero value for $\kappa(0)$. What would happen to the standard formulation of the Kramers–Kronig relations?

19.8 Let $g(\omega)=\omega^{-1}\{(2\omega^2[N(\omega)-1])^m-(-1)^m\omega_\mathrm{p}^{2m}\}$ for integer $m\geq 1$. By selecting a suitable contour integral, derive the first few sum rules that emerge. Indicate whether the rules apply for both conductors and insulators, or to insulators only. The idea for this choice of function is due to Villani and Zimerman (1973a).

19.9 Let $g(\omega)=\omega^{-1}\{(\omega^2[\varepsilon(\omega)\varepsilon_0^{-1}-1])^m-(-1)^m\omega_\mathrm{p}^{2m}\}$ for integer $m\geq 1$. Using an appropriate contour integration, determine the first few sum rules that emerge. Which of the rules you derive are restricted to insulators and which apply to both insulators and conductors? The idea for this choice of function is due to Villani and Zimerman (1973a).

19.10 How are the first three sum rules in Table 19.2 altered if the permeability is not ignored?

19.11 In Eq. (19.113), can the term -1 in the numerator of the integrand be dropped? If it can, what might be the consequences?

19.12 Sketch a plot of the behavior of $n(\omega)$ in the anomalous dispersion region. Indicate how the phase velocity and group velocity vary in the same spectral region. Discuss the implications for special relativity.

19.13 Consider the evaluation of the integral $\oint_C \omega_z^m\{N(\omega_z)-1\}^p\,\mathrm{d}\omega_z$, where C is a suitably selected contour and the integers m and p satisfy $m\geq 0$ and $p\geq 1$. Derive the sum rule entries in Table 19.2 that arise from this contour integral. Explain which rules apply to conductors and insulators and which rules are restricted to insulators.

19.14 Evaluate $\oint_C \omega_z^m\{\varepsilon(\omega_z)\varepsilon_0^{-1}-1\}^p\,\mathrm{d}\omega_z$, where C is a suitably chosen contour and the integers m and p satisfy $m\geq 0$ and $p\geq 1$. Obtain the sum rule entries in Table 19.1 that arise from this contour integral. Explain which rules apply to conductors and insulators and which rules are restricted to insulators.

19.15 Derive the analogs of the sum rules number (16) and number (17) in Table 19.2 for the dielectric constant.

19.16 Prove $\int_0^\infty\{\varepsilon_\mathrm{r}(\omega)-\varepsilon_0\}\cos\omega t\,\mathrm{d}\omega=\int_0^\infty\varepsilon_\mathrm{i}(\omega)\sin\omega t\,\mathrm{d}\omega$ for $t>0$. Explain whether this sum rule applies to conductors and insulators or only to insulators.

19.17 Let the angular frequency ω_0 be selected sufficiently large so that Eq. (19.238) applies. Show that $\int_0^{\omega_0}\{n(\omega)-1\}\mathrm{d}\omega\approx\omega_\mathrm{p}^2/2\omega_0$, a result given by Altarelli *et al.* (1972).

19.18 Apply Tricomi's formula to the case $f_1(\omega)=\kappa(\omega)$ and $f_2(\omega)=\phi_n(\omega)$, where $\phi_n(\omega)$ are eigenfunctions of the Hilbert transform operator (see Eq. (4.336)). Examine the particular case $n=0$ and determine the form of the dispersion relations that arise.

19.19 Employ the Liu–Okubo technique to obtain sum rules for the forward elastic low-energy scattering of light.

20

Dispersion relations for some linear optical properties

20.1 Introduction

The developments of the previous chapter are continued in this chapter. The principal focus is on the reflectance and the energy loss function. Knowledge of the reflectance and phase allows the real and imaginary parts of the complex refractive index to be evaluated, and hence the real and imaginary parts of the complex dielectric constant can be determined. Measurement of the reflectance and calculation of the phase provides a convenient route to a number of optical properties.

The procedure for dealing with the formulation of dispersion relations for the reflectance and phase has some features that are different from the cases of the dielectric constant and the refractive index. This is tied directly to the fact that the real and imaginary parts of the complex reflectivity depend upon both the reflectance and the phase. To uncouple these quantities requires consideration of the logarithm of the complex reflectivity, which introduces some additional issues into the discussion. This also has a direct bearing on the types of sum rules that can be derived.

20.2 Dispersion relations for the normal-incident reflectance and phase

The topic of obtaining dispersion relations for the normal-incident reflectivity is examined in this section. Consideration is restricted to the normal-incidence case, and it will be assumed that light is impinging on the material surface from a vacuum. The reflectivity, $R(\omega)$, is the ratio of the reflected to the incident light intensities.

There has been considerable attention focused on the study of dispersion relations for the reflectance, due mainly to the fact that the determination of the generalized complex reflectivity provides a route to other optical properties. The generalized or complex reflectivity, denoted $\tilde{r}(\omega)$, is defined by

$$\tilde{r}(\omega) = r(\omega)e^{i\theta(\omega)}, \tag{20.1}$$

where $\theta(\omega)$ is the phase, and the reflectivity amplitude $r(\omega)$ is related to the reflectivity by

$$R(\omega) = r(\omega)^2. \tag{20.2}$$

The quantity $R(\omega)$ is often referred to as the intensity reflectivity, and is accessible experimentally. If $\mathbf{E_i}(\omega)$ and $\mathbf{E_r}(\omega)$ are used to denote, respectively, the incident and reflected electric fields associated with an electromagnetic beam impinging on a material, then

$$\mathbf{E_r}(\omega) = \tilde{r}(\omega)\mathbf{E_i}(\omega). \tag{20.3}$$

The fields $\mathbf{E_i}(\omega)$ and $\mathbf{E_r}(\omega)$ can be written as follows:

$$\mathbf{E_r}(\omega) = \int_{-\infty}^{\infty} \mathbf{E_r}(t)\mathrm{e}^{\mathrm{i}\omega t}\,\mathrm{d}t \tag{20.4}$$

and

$$\mathbf{E_i}(\omega) = \int_{-\infty}^{\infty} \mathbf{E_i}(t)\mathrm{e}^{\mathrm{i}\omega t}\,\mathrm{d}t. \tag{20.5}$$

The reader should note that the notational convention indicated in Section 19.3, about the use of the same symbol for the field and its Fourier transform, is again employed in this chapter. From Eqs. (20.4) and (20.5) it follows, on taking the inverse Fourier transform, that

$$\begin{aligned}
\mathbf{E_r}(t) &= \frac{1}{2\pi}\int_{-\infty}^{\infty} \mathbf{E_r}(\omega)\mathrm{e}^{-\mathrm{i}\omega t}\,\mathrm{d}\omega \\
&= \frac{1}{2\pi}\int_{-\infty}^{\infty} \tilde{r}(\omega)\mathbf{E_i}(\omega)\mathrm{e}^{-\mathrm{i}\omega t}\,\mathrm{d}\omega \\
&= \frac{1}{2\pi}\int_{-\infty}^{\infty} \tilde{r}(\omega)\mathrm{e}^{-\mathrm{i}\omega t}\mathrm{d}\omega \int_{-\infty}^{\infty} \mathbf{E_i}(t')\mathrm{e}^{\mathrm{i}\omega t'}\,\mathrm{d}t' \\
&= \frac{1}{2\pi}\int_{-\infty}^{\infty} \mathbf{E_i}(t')\mathrm{d}t' \int_{-\infty}^{\infty} \tilde{r}(\omega)\mathrm{e}^{-\mathrm{i}(t-t')\omega}\,\mathrm{d}\omega,
\end{aligned} \tag{20.6}$$

and hence

$$\mathbf{E_r}(t) = \int_{-\infty}^{\infty} \mathbf{E_i}(t')G(t-t')\mathrm{d}t', \tag{20.7}$$

where

$$G(\tau) = \frac{1}{2\pi}\int_{-\infty}^{\infty} \tilde{r}(\omega)\mathrm{e}^{-\mathrm{i}\tau\omega}\,\mathrm{d}\omega. \tag{20.8}$$

From Eq. (20.7) it follows that $G(\tau)$ is a real function since the experimentally observable electric fields are real quantities. The quantity $G(\tau)$ plays the role of a response function. Taking the inverse Fourier transform of Eq. (20.8) yields

$$\widetilde{r}(\omega) = \int_{-\infty}^{\infty} G(\tau)\mathrm{e}^{\mathrm{i}\tau\omega}\,\mathrm{d}\tau. \tag{20.9}$$

The causality principle requires that

$$G(\tau) = 0, \quad \text{for } \tau < 0. \tag{20.10}$$

Repeating the argument of Section 17.7 leads to the conclusion that $\widetilde{r}(\omega)$ is an analytic function in the upper half of the complex angular frequency plane.

The complex reflectivity can be expressed in terms of the complex refractive index as follows:

$$\widetilde{r}(\omega) = \frac{N(\omega) - 1}{N(\omega) + 1}. \tag{20.11}$$

From this result it follows that

$$R(\omega) = \frac{(n(\omega) - 1)^2 + \kappa(\omega)^2}{(n(\omega) + 1)^2 + \kappa(\omega)^2}. \tag{20.12}$$

Equations (20.1) and (20.11) yield

$$\cos\theta(\omega) = \frac{n(\omega)^2 + \kappa(\omega)^2 - 1}{\surd\{[(n(\omega) + 1)^2 + \kappa(\omega)^2][(n(\omega) - 1)^2 + \kappa(\omega)^2]\}} \tag{20.13}$$

and

$$\sin\theta(\omega) = \frac{2\kappa(\omega)}{\surd\{[(n(\omega) + 1)^2 + \kappa(\omega)^2][(n(\omega) - 1)^2 + \kappa(\omega)^2]\}}. \tag{20.14}$$

The real and imaginary parts of the complex refractive index can be expressed in terms of $r(\omega)$ and $\theta(\omega)$ as follows:

$$n(\omega) = \frac{1 - r(\omega)^2}{1 - 2r(\omega)\cos\theta(\omega) + r(\omega)^2} \tag{20.15}$$

and

$$\kappa(\omega) = \frac{2r(\omega)\sin\theta(\omega)}{1 - 2r(\omega)\cos\theta(\omega) + r(\omega)^2}. \tag{20.16}$$

Thus, the determination of $r(\omega)$ and $\theta(\omega)$ allows the optical constants $n(\omega)$ and $\kappa(\omega)$ to be evaluated, and, from the results given in Section 19.6, $\varepsilon_{\mathrm{r}}(\omega)$ and $\varepsilon_{\mathrm{i}}(\omega)$ can also be obtained.

The first observation of importance to be noted from Eq. (20.1) is that the real and imaginary parts of $\tilde{r}(\omega)$ involve both the reflectivity and the phase. The reader can contrast this situation with the form of the real and imaginary parts of the complex refractive index or the complex dielectric constant. The obvious way to uncouple $r(\omega)$ and $\theta(\omega)$ is to take the log of Eq. (20.1), so that

$$\log \tilde{r}(\omega) = \log r(\omega) + \mathrm{i}\theta(\omega). \tag{20.17}$$

Although this last result provides the necessary separability, it introduces some extra issues to resolve. The first point to consider is the asymptotic behavior of $\log \tilde{r}(\omega)$ for $\omega \to \infty$. Making use of Eqs. (19.98) and (19.99), and employing Eq. (20.12), leads to

$$R(\omega) = O(\omega^{-4}), \quad \text{as } \omega \to \infty. \tag{20.18}$$

Hence, $\log r(\omega)$ exhibits a logarithmic divergence as $\omega \to \infty$. This fact eliminates some of the simple and useful sum rules that might have been obtained by analogy with the results derived in Sections 19.10 and 19.11 for the dielectric constant and refractive index. This particular asymptotic behavior also prevents the direct application of the asymptotic technique discussed in Section 19.9. The other item that needs attention is the more subtle question of the existence of zeros for the function $\tilde{r}(\omega)$. To address this question, the idea of a Blaschke factor is introduced.

On mathematical grounds, the function $\tilde{r}(\omega)$ could equally well be replaced by $\tilde{r}(\omega)B(\omega)$, where $B(\omega)$ denotes the Blaschke product given by

$$B(\omega) = \prod_{n=1}^{N} \frac{\omega - \beta_n}{\beta_n^* - \omega}, \tag{20.19}$$

and $\beta_n = a_n + \mathrm{i}b_n$, where a_n and b_n are real constants, and N denotes the number of factors multiplied. For real angular frequencies it follows that

$$|\tilde{r}(\omega)B(\omega)|^2 = |\tilde{r}(\omega)|^2 = r(\omega)^2. \tag{20.20}$$

Since $|B(\omega)|^2 = 1$, the reflectivity is unchanged. The Blaschke factor must satisfy two key requirements: (i) the constants b_n must be non-negative in order to ensure that $\tilde{r}(\omega)$ remains analytic in the upper half complex angular frequency plane; (ii) the crossing symmetry relation for the complex reflectivity must be satisfied, and this takes the form

$$\tilde{r}(-\omega_z^*) = \tilde{r}(\omega_z)^*. \tag{20.21}$$

Equation (20.21) is proved in the following paragraph. To satisfy this result means that the Blaschke factor must contain terms that, in general, are paired together, with an appropriate sign change for the a_n term in order to ensure that the crossing symmetry relation holds for $\widetilde{r}(\omega_z)B(\omega_z)$. That is, if the Blaschke factor contains a term with the factor $\beta_n = a_n + ib_n$, there must also be a term with the factor $\beta_n' = -a_n + ib_n$.

The crossing symmetry relation for the complex reflectivity follows directly from Eqs. (20.11) and (19.110). Alternatively, this result can be obtained without recourse to the crossing symmetry result for the complex refractive index. The requirement that the observable incident and reflected electric fields are real can be expressed as follows:

$$\mathbf{E}_\mathrm{i}(t)^* = \mathbf{E}_\mathrm{i}(t) \tag{20.22}$$

and

$$\mathbf{E}_\mathrm{r}(t)^* = \mathbf{E}_\mathrm{r}(t). \tag{20.23}$$

From this pair of results and Eqs. (20.4) and (20.5) it follows that

$$\mathbf{E}_\mathrm{i}(-\omega_z^*) = \mathbf{E}_\mathrm{i}(\omega_z)^* \tag{20.24}$$

and

$$\mathbf{E}_\mathrm{r}(-\omega_z^*) = \mathbf{E}_\mathrm{r}(\omega_z)^*. \tag{20.25}$$

Using Eq. (20.3), the crossing symmetry relationship is given by

$$\widetilde{r}(-\omega_z^*) = \frac{\mathbf{E}_\mathrm{r}(-\omega_z^*)}{\mathbf{E}_\mathrm{i}(-\omega_z^*)} = \frac{\mathbf{E}_\mathrm{r}(\omega_z)^*}{\mathbf{E}_\mathrm{i}(\omega_z)^*} = \widetilde{r}(\omega_z)^*, \tag{20.26}$$

which is the desired result.

From Eqs. (19.90) and (20.11) it follows, for a non-magnetic material, that

$$\widetilde{r}(\omega_z) = \frac{\sqrt{[\varepsilon(\omega_z)]} - \sqrt{\varepsilon_0}}{\sqrt{[\varepsilon(\omega_z)]} + \sqrt{\varepsilon_0}}. \tag{20.27}$$

The quantity $\widetilde{r}(\omega_z)B(\omega_z)$ has zeros in the upper half of the complex angular frequency plane when $\omega_z = \beta_n$ for $b_n > 0$. For this to arise, it is necessary to find the solutions of

$$\varepsilon(\omega_z) = \varepsilon_0. \tag{20.28}$$

From Eq. (19.49) it is clear that $\varepsilon(\omega_z)$ is a complex function everywhere in the upper half of the complex angular frequency plane, except on the imaginary frequency axis, where it is real. It takes the value

$$\varepsilon(0) = \varepsilon_0 + \frac{2}{\pi}\int_0^\infty \frac{\varepsilon_\mathrm{i}(\omega)\mathrm{d}\omega}{\omega}, \tag{20.29}$$

at $\omega = 0$, and decreases as ω increases according to Eq. (19.235), with $\varepsilon(\mathrm{i}\infty) = \varepsilon_0$. From this it is concluded that there are no solutions of Eq. (20.28) in the finite upper half of the complex angular frequency plane. Hence, it is deduced that, for the case of the normal reflectivity, there is no Blaschke factor present and no zeros of $\tilde{r}(\omega_z)$ in the domain under discussion.

The dispersion relations for the reflectance and phase take the following form:

$$\theta(\omega) = \frac{\omega}{\pi} P \int_0^{\infty} \frac{\log R(\omega')\mathrm{d}\omega'}{\omega^2 - \omega'^2} \tag{20.30}$$

and

$$\log R(\omega) - \log R(\omega_0) = \frac{4}{\pi} P \int_0^{\infty} \omega'\theta(\omega') \left(\frac{1}{\omega_0^2 - \omega'^2} - \frac{1}{\omega^2 - \omega'^2} \right) \mathrm{d}\omega'. \tag{20.31}$$

The derivation of these two formulas is now presented. The first thing to note is that these two results do not form a Hilbert transform pair, allowing for the even–odd character of the functions involved. The reason why a skew-symmetric pair of integral transforms does not arise will become evident in the derivation presented.

The crossing symmetry relation for the complex reflectivity was given in Eq. (20.21), from which it follows that

$$R(-\omega) = R(\omega) \tag{20.32}$$

and

$$\theta(-\omega) = -\theta(\omega). \tag{20.33}$$

These results are useful for what follows.

The function $\log \tilde{r}(\omega_z)$ is analytic in the upper half of the complex angular frequency plane. This follows directly from Eq. (20.11) and the fact that $\log\{N(\omega_z) - 1\}$ and $\log\{N(\omega_z) + 1\}$ are both analytic functions in the upper half of the complex frequency plane. Consider the integral

$$\oint_C \frac{\log \tilde{r}(\omega_z)\mathrm{d}\omega_z}{\omega - \omega_z},$$

where C denotes the contour shown in Figure 19.2 (see p. 188). The notation for the radius of the semicircle is changed from R to $\mathcal{R}$, since the reflectance is being denoted by the symbol R in this chapter. From the Cauchy integral theorem it follows that

$$\int_{-\mathcal{R}}^{\omega-\rho} \frac{\log \tilde{r}(\omega_\mathrm{r})\mathrm{d}\omega_\mathrm{r}}{\omega - \omega_\mathrm{r}} + \int_{\omega+\rho}^{\mathcal{R}} \frac{\log \tilde{r}(\omega_\mathrm{r})\mathrm{d}\omega_\mathrm{r}}{\omega - \omega_\mathrm{r}} + \mathrm{i}\int_0^{\pi} \log \tilde{r}(\omega + \rho\mathrm{e}^{\mathrm{i}\phi})\mathrm{d}\phi$$
$$+ \mathrm{i}\mathcal{R}\int_0^{\pi} \frac{\log \tilde{r}(\mathcal{R}\mathrm{e}^{\mathrm{i}\vartheta})\mathrm{e}^{\mathrm{i}\vartheta}\mathrm{d}\vartheta}{\omega - \mathcal{R}\mathrm{e}^{\mathrm{i}\vartheta}} = 0. \tag{20.34}$$

Taking the limit $\rho \to 0$ in Eq. (20.34) leads to

$$P\int_{-\mathcal{R}}^{\mathcal{R}} \frac{\log \widetilde{r}(\omega_r)\mathrm{d}\omega_r}{\omega - \omega_r} + \mathrm{i}\pi \log \widetilde{r}(\omega) + \mathrm{i}\int_0^{\pi} \frac{\log \widetilde{r}(\mathcal{R}e^{\mathrm{i}\vartheta})\mathrm{d}\vartheta}{\omega\mathcal{R}^{-1}e^{-\mathrm{i}\vartheta} - 1} = 0. \tag{20.35}$$

On taking the real part of this result, and employing Eq. (20.32),

$$\omega P\int_0^{\mathcal{R}} \frac{\log R(\omega_r)\mathrm{d}\omega_r}{\omega^2 - \omega_r^2} - \pi\theta(\omega) - \mathrm{Im}\int_0^{\pi} \frac{\log \widetilde{r}(\mathcal{R}e^{\mathrm{i}\vartheta})\mathrm{d}\vartheta}{\omega\mathcal{R}^{-1}e^{-\mathrm{i}\vartheta} - 1} = 0. \tag{20.36}$$

The asymptotic behavior of $\widetilde{r}(\omega)$ can be written using Eq. (19.12) as follows:

$$\widetilde{r}(\omega) = \frac{\sqrt{[\varepsilon(\omega)/\varepsilon_0]} - 1}{\sqrt{[\varepsilon(\omega)/\varepsilon_0]} + 1} \approx -\frac{\omega_p^2}{4\omega^2}, \quad \text{as } \omega \to \infty. \tag{20.37}$$

In the limit $\mathcal{R} \to \infty$, the last term of Eq. (20.36) becomes

$$\begin{aligned}
-\mathrm{Im}\int_0^{\pi} \frac{\log \widetilde{r}(\mathcal{R}e^{\mathrm{i}\vartheta})\mathrm{d}\vartheta}{\omega\mathcal{R}^{-1}e^{-\mathrm{i}\vartheta} - 1} &\to \mathrm{Im}\int_0^{\pi} \log \widetilde{r}(\mathcal{R}e^{\mathrm{i}\vartheta})\mathrm{d}\vartheta \\
&= \mathrm{Im}\int_0^{\pi} \log(-\omega_p^2 2^{-2}\mathcal{R}^{-2}e^{-2\mathrm{i}\vartheta})\mathrm{d}\vartheta \\
&= \mathrm{Im}\int_0^{\pi} \left[\log(-1) + 2\log\left(\frac{\omega_p}{2\mathcal{R}}\right) - 2\mathrm{i}\vartheta\right]\mathrm{d}\vartheta \\
&= \mathrm{Im}\left[\mathrm{i}\pi^2 + 2\pi\log\left(\frac{\omega_p}{2\mathcal{R}}\right) - 2\mathrm{i}\int_0^{\pi} \vartheta\,\mathrm{d}\vartheta\right] \\
&= 0.
\end{aligned} \tag{20.38}$$

Making use of Eq. (20.38) and taking the limit $\mathcal{R} \to \infty$ in Eq. (20.36) gives Eq. (20.30), which is the required result.

Taking the imaginary part of Eq. (20.35) yields

$$2P\int_0^{\mathcal{R}} \frac{\omega_r\theta(\omega_r)\mathrm{d}\omega_r}{\omega^2 - \omega_r^2} + \pi\log r(\omega) + \mathrm{Re}\int_0^{\pi} \frac{\log \widetilde{r}(\mathcal{R}e^{\mathrm{i}\vartheta})\mathrm{d}\vartheta}{\omega\mathcal{R}^{-1}e^{-\mathrm{i}\vartheta} - 1} = 0, \tag{20.39}$$

and it is immediately evident that the contribution from the semicircular section of the contour integral diverges in the limit $\mathcal{R} \to \infty$.

A simpler approach is to consider the integral

$$\oint_C \log \widetilde{r}(\omega_z)\left(\frac{1}{\omega - \omega_z} - \frac{1}{\omega_0 - \omega_z}\right)\mathrm{d}\omega_z,$$

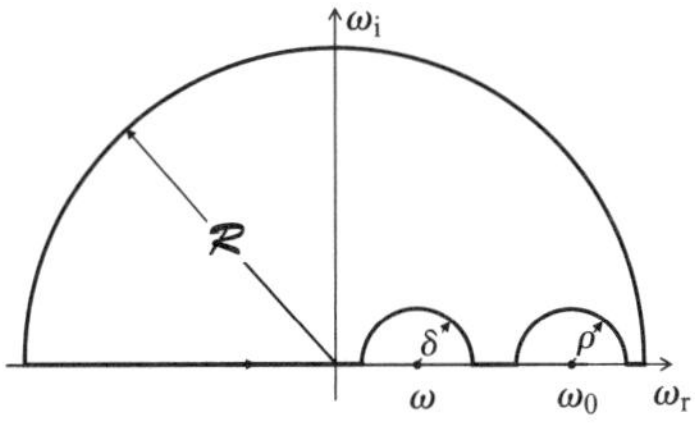

Figure 20.1. Contour for the determination of dispersion relations for the reflectance and phase.

where C denotes the contour shown in Figure 20.1. From the Cauchy integral theorem it follows that

$$\begin{aligned}
&\int_{-\mathcal{R}}^{\omega-\delta} \log \widetilde{r}(\omega_r)\left(\frac{1}{\omega-\omega_r}-\frac{1}{\omega_0-\omega_r}\right)\mathrm{d}\omega_r \\
&\quad+\int_{\omega+\delta}^{\omega_0-\rho} \log \widetilde{r}(\omega_r)\left(\frac{1}{\omega-\omega_r}-\frac{1}{\omega_0-\omega_r}\right)\mathrm{d}\omega_r \\
&\quad+\int_{\omega_0+\rho}^{\mathcal{R}} \log \widetilde{r}(\omega_r)\left(\frac{1}{\omega-\omega_r}-\frac{1}{\omega_0-\omega_r}\right)\mathrm{d}\omega_r \\
&\quad+\int_0^{\pi} \log \widetilde{r}(\omega+\delta e^{i\phi})\left(\frac{1}{\delta e^{i\phi}}+\frac{1}{\omega_0-\omega-\delta e^{i\phi}}\right) i\delta e^{i\phi}\,\mathrm{d}\phi \\
&\quad-\int_0^{\pi} \log \widetilde{r}(\omega_0+\rho e^{i\phi})\left(\frac{1}{\omega-\omega_0-\rho e^{i\phi}}+\frac{1}{\rho e^{i\phi}}\right) i\rho e^{i\phi}\,\mathrm{d}\phi \\
&\quad+\int_0^{\pi} \log \widetilde{r}(\mathcal{R}e^{i\vartheta})\left(\frac{1}{\omega-\mathcal{R}e^{i\vartheta}}-\frac{1}{\omega_0-\mathcal{R}e^{i\vartheta}}\right) i\mathcal{R}e^{i\vartheta}\,\mathrm{d}\vartheta = 0. \qquad (20.40)
\end{aligned}$$

Taking the limits $\delta \to 0$ and $\rho \to 0$ in Eq. (20.40) leads to

$$\begin{aligned}
&P\int_{-\mathcal{R}}^{\mathcal{R}} \log \widetilde{r}(\omega_r)\left(\frac{1}{\omega-\omega_r}-\frac{1}{\omega_0-\omega_r}\right)\mathrm{d}\omega_r + i\pi\{\log \widetilde{r}(\omega) - \log \widetilde{r}(\omega_0)\} \\
&\quad+\int_0^{\pi} \log \widetilde{r}(\mathcal{R}e^{i\vartheta})\left(\frac{1}{\omega-\mathcal{R}e^{i\vartheta}}-\frac{1}{\omega_0-\mathcal{R}e^{i\vartheta}}\right) i\mathcal{R}e^{i\vartheta}\,\mathrm{d}\vartheta = 0. \qquad (20.41)
\end{aligned}$$

On taking the limit $\mathcal{R} \to \infty$, the integral along the large semicircular arc vanishes, and hence

$$i\pi\{\log \widetilde{r}(\omega) - \log \widetilde{r}(\omega_0)\} = -P\int_{-\infty}^{\infty} \log \widetilde{r}(\omega')\left(\frac{1}{\omega-\omega'}-\frac{1}{\omega_0-\omega'}\right)\mathrm{d}\omega'. \qquad (20.42)$$

From the real and imaginary parts of this result, it follows that

$$\theta(\omega) - \theta(\omega_0) = \frac{1}{\pi}P\int_{-\infty}^{\infty} \log r(\omega')\left(\frac{1}{\omega-\omega'}-\frac{1}{\omega_0-\omega'}\right)\mathrm{d}\omega' \qquad (20.43)$$

and

$$\log r(\omega) - \log r(\omega_0) = -\frac{1}{\pi} P \int_{-\infty}^{\infty} \theta(\omega') \left(\frac{1}{\omega - \omega'} - \frac{1}{\omega_0 - \omega'} \right) \mathrm{d}\omega'. \tag{20.44}$$

Using Eqs. (20.32) and (20.33) leads to

$$\theta(\omega) - \theta(\omega_0) = \frac{1}{\pi} P \int_0^{\infty} \log R(\omega') \left(\frac{\omega}{\omega^2 - \omega'^2} - \frac{\omega_0}{\omega_0^2 - \omega'^2} \right) \mathrm{d}\omega' \tag{20.45}$$

and

$$\log R(\omega) - \log R(\omega_0) = \frac{4}{\pi} P \int_0^{\infty} \omega' \theta(\omega') \left(\frac{1}{\omega_0^2 - \omega'^2} - \frac{1}{\omega^2 - \omega'^2} \right) \mathrm{d}\omega'. \tag{20.46}$$

From Eq. (20.11), $\tilde{r}(0)$ is given by

$$\tilde{r}(0) = \frac{n(0) + \mathrm{i}\kappa(0) - 1}{n(0) + \mathrm{i}\kappa(0) + 1} = \frac{n(0) - 1}{n(0) + 1}, \tag{20.47}$$

and hence

$$\theta(0) = 0. \tag{20.48}$$

This result applies for both conductors and insulators. The limit $\omega_0 \to 0$ in Eq. (20.45) yields Eq. (20.30). From Eq. (20.11) it follows that

$$\tilde{r}(\omega) = \frac{n(\omega)^2 - 1 + \kappa(\omega)^2 + 2\mathrm{i}\kappa(\omega)}{(n(\omega) + 1)^2 + \kappa(\omega)^2}, \tag{20.49}$$

and hence

$$r(\omega) \cos\theta(\omega) \approx -\frac{\omega_\mathrm{p}^2}{4\omega^2}, \quad \text{as } \omega \to \infty, \tag{20.50}$$

and

$$r(\omega) \sin\theta(\omega) \approx O(\omega^{-2-\beta}), \quad \beta > 0, \quad \text{as } \omega \to \infty. \tag{20.51}$$

From Eq. (20.2), using the positive root, and Eq. (20.12), it follows that

$$r(\omega) \approx \frac{\omega_\mathrm{p}^2}{4\omega^2}, \quad \text{as } \omega \to \infty. \tag{20.52}$$

For a discussion on aspects of the issue associated with the possible flexibility in the form for $r(\omega)$, the work of Smith and Manogue (1981) is recommended reading.

Comparison of Eqs. (20.50) and (20.52) yields

$$\cos\theta(\omega) = -1, \quad \text{as } \omega \to \infty, \tag{20.53}$$

that is,

$$\theta(\omega) = \pi, \quad \text{as } \omega \to \infty. \tag{20.54}$$

On the basis of Eqs. (20.48) and (20.54), the phase is restricted to the interval $0 \le \theta(\omega) \le \pi$. Making use of Eq. (20.54), it is clear why the second point ω_0 must be selected in Eq. (20.46). The integrand in this equation cannot be broken into the two obvious component integrals, since each is separately divergent.

A related approach starts by considering the integral

$$\oint_C \frac{(1+\omega\omega_z)\log\tilde{r}(\omega_z)\mathrm{d}\omega_z}{(1+\omega_z^2)(\omega-\omega_z)},$$

where C denotes the contour shown in Figure 19.2, and for dimensional reasons each angular frequency has been divided by 1 Hz to give dimensionless variables. The contribution $(1+\omega\omega_z)(1+\omega_z^2)^{-1}$ serves as a weighting factor, which improves the convergence behavior of the integrand as $\omega_z \to \infty$. There is a pole located at $\omega_z = \mathrm{i}$ in the upper half of the complex angular frequency plane. The Cauchy residue theorem yields

$$P\int_{-\infty}^{\infty} \frac{(1+\omega\omega_{\mathrm{r}})\log\tilde{r}(\omega_{\mathrm{r}})\mathrm{d}\omega_{\mathrm{r}}}{(1+\omega_{\mathrm{r}}^2)(\omega-\omega_{\mathrm{r}})} + \mathrm{i}\pi\log\tilde{r}(\omega) = \mathrm{i}\pi\log\tilde{r}(\mathrm{i}). \tag{20.55}$$

Using the symmetry properties given in Eqs. (20.32) and (20.33) leads to

$$\mathrm{i}\{\log\tilde{r}(\mathrm{i}) - \log r(\omega)\} + \theta(\omega) = \frac{2}{\pi}P\int_0^{\infty}\left\{\omega\log r(\omega') + \frac{\mathrm{i}(1+\omega^2)\omega'\theta(\omega')}{(1+\omega'^2)}\right\}\frac{\mathrm{d}\omega'}{\omega^2-\omega'^2}. \tag{20.56}$$

From Eq. (20.9) it follows that $\tilde{r}(\mathrm{i})$ is real. Taking the real and imaginary parts of Eq. (20.56) leads to Eq. (20.30) and

$$\log\tilde{r}(\mathrm{i}) - \log r(\omega) = \frac{2(1+\omega^2)}{\pi}P\int_0^{\infty}\frac{\omega'\theta(\omega')\mathrm{d}\omega'}{(1+\omega'^2)(\omega^2-\omega'^2)}. \tag{20.57}$$

An alternative derivation of the dispersion relations for the reflectance and phase can be carried out by considering the integral

$$\oint_C \frac{\log\tilde{r}(\omega_z)\mathrm{d}\omega_z}{\omega^2-\omega_z^2},$$

where C denotes the contour shown in Figure 19.3 (with the change of notation that the radius of the large quarter circle R is replaced by $\mathcal{R}$). Application of the Cauchy integral theorem yields

$$\int_0^{\omega-\rho} \frac{\log \tilde{r}(\omega_r)d\omega_r}{\omega^2 - \omega_r^2} + \int_{\omega+\rho}^{\mathcal{R}} \frac{\log \tilde{r}(\omega_r)d\omega_r}{\omega^2 - \omega_r^2} + i\int_0^{\pi} \frac{\log \tilde{r}(\omega + \rho e^{i\phi})d\phi}{2\omega + \rho e^{i\phi}}$$
$$+ i\mathcal{R}\int_0^{\pi} \frac{\log \tilde{r}(\mathcal{R}e^{i\vartheta})e^{i\vartheta}d\vartheta}{\omega^2 - \mathcal{R}^2 e^{2i\vartheta}} + \int_{\mathcal{R}}^{0} \frac{\log \tilde{r}(i\omega_i)i\, d\omega_i}{\omega^2 + \omega_i^2} = 0. \tag{20.58}$$

Taking the limits $\rho \to 0$ and $\mathcal{R} \to \infty$ in Eq. (20.58) leads to

$$P\int_0^{\infty} \frac{\log \tilde{r}(\omega')d\omega'}{\omega^2 - \omega'^2} = \frac{-i\pi \log \tilde{r}(\omega)}{2\omega} + i\int_0^{\infty} \frac{\log \tilde{r}(i\omega_i)d\omega_i}{\omega^2 + \omega_i^2}. \tag{20.59}$$

Since $\tilde{r}(i\omega_i)$ is real on the imaginary axis, on taking the real part of Eq. (20.59), Eq. (20.30) is obtained. It might be expected that the other dispersion relation relating the reflectivity to the phase could be obtained by considering the integral

$$\oint_C \frac{\omega_z \log \tilde{r}(\omega_z)d\omega_z}{\omega^2 - \omega_z^2},$$

with the same contour just employed; however, this does not work. Why? Instead, consider

$$\oint_C \omega_z \log \tilde{r}(\omega_z)\left(\frac{1}{\omega_1^2 - \omega_z^2} - \frac{1}{\omega_2^2 - \omega_z^2}\right)d\omega_z,$$

where the contour is shown in Figure 20.2. From the Cauchy integral theorem it follows that

$$\int_0^{\omega_1-\delta} \omega_r \log \tilde{r}(\omega_r)\left(\frac{1}{\omega_1^2 - \omega_r^2} - \frac{1}{\omega_2^2 - \omega_r^2}\right)d\omega_r$$
$$+ \int_{\omega_1+\delta}^{\omega_2-\rho} \omega_r \log \tilde{r}(\omega_r)\left(\frac{1}{\omega_1^2 - \omega_r^2} - \frac{1}{\omega_2^2 - \omega_r^2}\right)d\omega_r$$
$$+ \int_{\omega_2+\rho}^{\mathcal{R}} \omega_r \log \tilde{r}(\omega_r)\left(\frac{1}{\omega_1^2 - \omega_r^2} - \frac{1}{\omega_2^2 - \omega_r^2}\right)d\omega_r$$
$$+ \int_0^{\pi} \{\omega_1 + \delta e^{i\phi}\} \log \tilde{r}(\omega_1 + \delta e^{i\phi})$$
$$\times \left(\frac{1}{\delta e^{i\phi}(2\omega_1 + \delta e^{i\phi})} + \frac{1}{\omega_2^2 - (\omega_1 + \delta e^{i\phi})^2}\right) i\delta e^{i\phi}\, d\phi$$

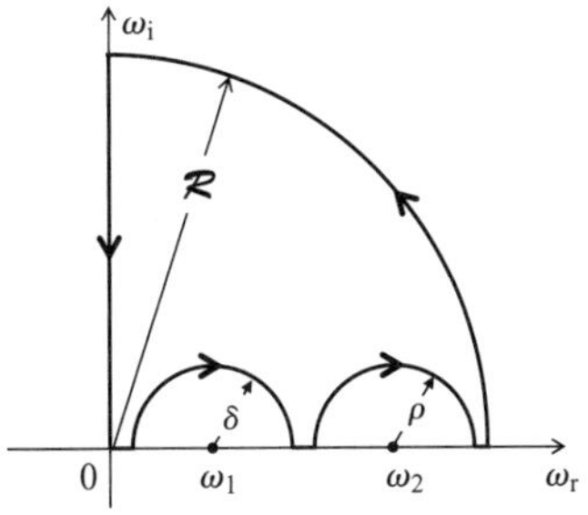

Figure 20.2. Contour for the determination of the reflectance in terms of the phase.

$$
\begin{aligned}
&- \int_0^{\pi} \{\omega_2 + \rho e^{i\phi}\} \log \tilde{r}(\omega_2 + \rho e^{i\phi}) \\
&\times \left(\frac{1}{\omega_1^2 - (\omega_2 + \rho e^{i\phi})^2} + \frac{1}{\rho e^{i\phi}(2\omega_2 + \rho e^{i\phi})} \right) i\rho e^{i\phi}\, d\phi \\
&+ \int_0^{\pi} \mathcal{R} e^{i\phi} \log \tilde{r}(\mathcal{R} e^{i\phi}) \left(\frac{1}{\omega_1^2 - \mathcal{R}^2 e^{2i\phi}} - \frac{1}{\omega_2^2 - \mathcal{R}^2 e^{2i\phi}} \right) i\mathcal{R} e^{i\phi}\, d\phi \\
&+ \int_0^{\mathcal{R}} \omega_i \log \tilde{r}(i\omega_i) \left(\frac{1}{\omega_1^2 + \omega_i^2} - \frac{1}{\omega_2^2 + \omega_i^2} \right) d\omega_i = 0. \qquad (20.60)
\end{aligned}
$$

Taking the limits $\delta \to 0$, $\rho \to 0$, and $\mathcal{R} \to \infty$ in Eq. (20.60) leads to

$$
\begin{aligned}
&P \int_0^{\infty} \omega_r \log \tilde{r}(\omega_r) \left(\frac{1}{\omega_1^2 - \omega_r^2} - \frac{1}{\omega_2^2 - \omega_r^2} \right) d\omega_r + \frac{i\pi \log \tilde{r}(\omega_1)}{2} - \frac{i\pi \log \tilde{r}(\omega_2)}{2} \\
&= - \int_0^{\infty} \omega_i \log \tilde{r}(i\omega_i) \left(\frac{1}{\omega_1^2 + \omega_i^2} - \frac{1}{\omega_2^2 + \omega_i^2} \right) d\omega_i. \qquad (20.61)
\end{aligned}
$$

Taking the imaginary part of this last result leads to

$$
\log \frac{R(\omega_1)}{R(\omega_2)} = -\frac{4}{\pi} P \int_0^{\infty} \omega' \theta(\omega') \left(\frac{1}{\omega_1^2 - \omega'^2} - \frac{1}{\omega_2^2 - \omega'^2} \right) d\omega', \qquad (20.62)
$$

which is the desired result.

20.3 Sum rules for the reflectance and phase

Consider the integral

$$
\oint_C \frac{\tilde{r}(\omega_z) d\omega_z}{\omega - \omega_z},
$$

where C denotes the contour shown in Figure 19.2. Application of the Cauchy integral theorem gives

$$\int_{-\mathcal{R}}^{\omega-\rho} \frac{\tilde{r}(\omega_r)\mathrm{d}\omega_r}{\omega - \omega_r} + \int_{\omega+\rho}^{\mathcal{R}} \frac{\tilde{r}(\omega_r)\mathrm{d}\omega_r}{\omega - \omega_r} + \mathrm{i}\int_0^{\pi} \tilde{r}(\omega + \rho \mathrm{e}^{\mathrm{i}\phi})\mathrm{d}\phi + \mathrm{i}\mathcal{R}\int_0^{\pi} \frac{\tilde{r}(\mathcal{R}\mathrm{e}^{\mathrm{i}\phi})\mathrm{e}^{\mathrm{i}\phi}\,\mathrm{d}\phi}{\omega - \mathcal{R}\mathrm{e}^{\mathrm{i}\phi}} = 0. \tag{20.63}$$

Taking the limits $\rho \to 0$ and $\mathcal{R} \to \infty$ in Eq. (20.63) leads to

$$P\int_{-\infty}^{\infty} \frac{\tilde{r}(\omega_r)\mathrm{d}\omega_r}{\omega - \omega_r} = -\mathrm{i}\pi\tilde{r}(\omega). \tag{20.64}$$

The real and imaginary parts of this result yield

$$P\int_{-\infty}^{\infty} \frac{r(\omega')\cos\theta(\omega')\mathrm{d}\omega'}{\omega - \omega'} = \pi r(\omega)\sin\theta(\omega) \tag{20.65}$$

and

$$P\int_{-\infty}^{\infty} \frac{r(\omega')\sin\theta(\omega')\mathrm{d}\omega'}{\omega - \omega'} = -\pi r(\omega)\cos\theta(\omega). \tag{20.66}$$

Applying the crossing symmetry conditions in Eqs. (20.32) and (20.33) leads to

$$r(\omega)\sin\theta(\omega) = \frac{2\omega}{\pi}P\int_0^{\infty} \frac{r(\omega')\cos\theta(\omega')\mathrm{d}\omega'}{\omega^2 - \omega'^2} \tag{20.67}$$

and

$$r(\omega)\cos\theta(\omega) = -\frac{2}{\pi}P\int_0^{\infty} \frac{\omega' r(\omega')\sin\theta(\omega')\mathrm{d}\omega'}{\omega^2 - \omega'^2}. \tag{20.68}$$

The reader will quickly spot the limitation of this pair of dispersion relations. The phase and reflectivity are coupled together on both sides of each equation. So these expressions are not of direct practical value for determining $\theta(\omega)$ from $r(\omega)$ or vice versa. However, Eqs. (20.67) and (20.68) can be used to derive sum rules. Application of the Parseval-type property of the Hilbert transform (recall Eq. (4.176)) leads immediately to the sum rule

$$\int_0^{\infty} R(\omega)\cos 2\theta(\omega)\mathrm{d}\omega = 0. \tag{20.69}$$

Examination of the limit $\omega \to 0$ in Eq. (20.68) yields

$$\int_0^{\infty} \frac{\sqrt{R(\omega)}\sin\theta(\omega)\mathrm{d}\omega}{\omega} = \frac{\pi}{2}\sqrt{R(0)}, \tag{20.70}$$

where the result $\theta(0) = 0$ has been employed. From Eq. (20.49) it follows that, for insulators,

$$\lim_{\omega\to 0} r(\omega)\sin\theta(\omega) = \lim_{\omega\to 0} \frac{2\kappa(\omega)}{(n(\omega)+1)^2+\kappa(\omega)^2} = O(\omega^\alpha), \quad \text{for } \alpha > 0, \tag{20.71}$$

and, for conductors,

$$\lim_{\omega\to 0} r(\omega)\sin\theta(\omega) = \lim_{\omega\to 0} \frac{2\kappa(\omega)}{(n(\omega)+1)^2+\kappa(\omega)^2} = O(\sqrt{\omega}), \tag{20.72}$$

which implies that the integral in Eq. (20.70) does not diverge as $\omega \to 0$. Hence, this sum rule holds for both conductors and insulators. For conductors $R(0) = 1$ and for insulators $R(0)$ must be determined experimentally.

Using the substitutions $y = \omega^2, x = \omega'^2$, and $g(x) = r(\sqrt{x})\sin\theta(\sqrt{x})$, Eq. (20.68) can be written as follows:

$$\pi r(\omega)\cos\theta(\omega) = -P\int_0^\infty \frac{g(x)\mathrm{d}x}{y-x}. \tag{20.73}$$

If Eq. (20.51) is employed together with the key theorem of Section 19.9, then for $0 < \beta < 2$, or $\beta > 2$, it follows that

$$\pi r(\omega)\cos\theta(\omega) = -\frac{1}{y}\int_0^\infty g(x)\mathrm{d}x + O(y^{-(2+\beta)/2}), \quad \text{as } y \to \infty. \tag{20.74}$$

The case $\beta = 2$ is handled using Eq. (19.212). Hence,

$$r(\omega)\cos\theta(\omega) = -\frac{2}{\pi\omega^2}\int_0^\infty \omega' r(\omega')\sin\theta(\omega')\mathrm{d}\omega' + O(\omega^{-(2+\beta)}), \quad \text{as } \omega \to \infty. \tag{20.75}$$

Making use of Eq. (20.50) leads to the following sum rule:

$$\int_0^\infty \omega\sqrt{R(\omega)}\sin\theta(\omega)\mathrm{d}\omega = \frac{\pi}{8}\omega_\mathrm{p}^2. \tag{20.76}$$

Employing the change of variables $y = \omega^2$, $x = \omega'^2$, and $g(x) = \{r(\sqrt{x})\cos\theta(\sqrt{x})\}(1/\sqrt{x})$ in Eq. (20.67), it follows that

$$\frac{\pi r(\omega)\sin\theta(\omega)}{\omega} = P\int_0^\infty \frac{g(x)\mathrm{d}x}{y-x}, \tag{20.77}$$

and from Eq. (20.50) it follows that $g(x) = O(x^{-3/2})$ as $x \to \infty$, so

$$\frac{\pi r(\omega)\sin\theta(\omega)}{\omega} = \frac{1}{y}\int_0^\infty g(x)\mathrm{d}x + O(y^{-3/2}), \quad \text{as } y \to \infty; \tag{20.78}$$

that is,

$$r(\omega)\sin\theta(\omega) = \frac{2}{\pi\omega}\int_0^\infty r(\omega')\cos\theta(\omega')\mathrm{d}\omega' + O(\omega^{-2}), \quad \text{as } \omega \to \infty. \tag{20.79}$$

Making use of Eq. (20.51) leads to the following sum rule:

$$\int_0^\infty \sqrt{R(\omega)}\cos\theta(\omega)\mathrm{d}\omega = 0. \tag{20.80}$$

A different pair of dispersion relations can be derived by starting with the integral

$$\oint_C \frac{\tilde{r}(\omega_z)^2 \mathrm{d}\omega_z}{\omega - \omega_z},$$

where C denotes the contour shown in Figure 19.2. From the resulting dispersion relations, additional sum rules can be obtained. This is left as an exercise for the reader to pursue. In order to obtain new sum rules, an alternative strategy is to consider the integral $\oint_C \omega_z^m \tilde{r}(\omega_z)^p \,\mathrm{d}\omega_z$, where C denotes a closed semicircular contour in the upper half of the complex angular frequency plane, center the origin and radius $\mathcal{R}$. The integer exponents m and p are selected so that the integral does not diverge in the limit as $\mathcal{R} \to \infty$ and that there is no pole at $\omega = 0$. The choice $m \geq 0$ and $p \geq 1$ is made. In order that the integral not diverge in the limit as $\mathcal{R} \to \infty$ requires

$$m + 1 < 2p, \tag{20.81}$$

which assumes that Eqs. (20.50) and (20.51) hold. For both conductors and insulators there is no additional constraint imposed on m and p as $\omega \to 0$. Evaluation of the contour integral subject to the preceding constraint leads to

$$\int_{-\infty}^\infty \omega^m r(\omega)^p \mathrm{e}^{\mathrm{i}p\theta(\omega)}\,\mathrm{d}\omega = 0. \tag{20.82}$$

For $m = 0$ and $p = 1$, Eq. (20.80) is obtained; for $m = 0$ and $p = 2$, Eq. (20.69) results, and in general, for $m = 0$,

$$\int_0^\infty R(\omega)^{p/2}\cos p\theta(\omega)\mathrm{d}\omega = 0, \quad \text{for } p \geq 1. \tag{20.83}$$

A number of additional sum rules arising for $m = 1, 2$, and 3 are summarized in Table 20.1, along with some of the other well known sum rules satisfied by the normal reflectivity. The sum rules apply to both conductors and insulators. The references provided serve as sources for further reading, and indicate where the results were first derived or where other discussion on these sum rules is given. Additional sum rules satisfied by the reflectivity can be found in the references given in Table 20.1.

Table 20.1. *Summary of sum rules for the normal reflectance and phase*

Number	Sum rule	Reference
(1)	$\int_0^\infty \frac{\sqrt{R(\omega)}\sin\theta(\omega)\mathrm{d}\omega}{\omega} = \frac{\pi}{2}\sqrt{R(0)}$	King (1979)
(2)	$\int_0^\infty \sqrt{R(\omega)}\cos\theta(\omega)\mathrm{d}\omega = 0$	Inagaki, Ueda, and Kawata (1978); King (1979)
(3)	$\int_0^\infty \omega^2 R(\omega)\cos 2\theta(\omega)\mathrm{d}\omega = 0$	King (1979)
(4)	$\int_0^\infty R(\omega)\cos 2\theta(\omega)\mathrm{d}\omega = 0$	Inagaki *et al.* (1978); King (1979)
(5)	$\int_0^\infty \omega R(\omega)\sin 2\theta(\omega)\mathrm{d}\omega = 0$	Inagaki *et al.* (1978); King (1979)
(6)	$\int_0^\infty \omega\sqrt{R(\omega)}\sin\theta(\omega)\mathrm{d}\omega = \frac{\pi}{8}\omega_{\mathrm{p}}^2$	Inagaki *et al.* (1978); King (1979)
(7)	$\int_0^\infty R(\omega)^{p/2}\cos p\theta(\omega)\mathrm{d}\omega = 0, \quad \text{for } p \geq 1$	Smith and Manogue (1981)
(8)	$\int_0^\infty \omega R(\omega)^{p/2}\sin p\theta(\omega)\mathrm{d}\omega = 0, \quad \text{for } p \geq 2$	Smith and Manogue (1981)
(9)	$\int_0^\infty \omega^2 R(\omega)^{p/2}\cos p\theta(\omega)\mathrm{d}\omega = 0, \quad \text{for } p \geq 2$	

20.4 The conductance: dispersion relations

The complex conductivity $\sigma(\omega)$ can be defined in terms of the complex permittivity of the medium using the following formula (in SI units):

$$\sigma(\omega) = -\mathrm{i}\omega\varepsilon(\omega). \tag{20.84}$$

A factor of 4π or $4\pi\varepsilon_0$ can be found in the denominator of this definition when different unit systems are employed. Some authors also define the complex conductivity by

$$\sigma(\omega)\varepsilon_0^{-1} = -\mathrm{i}\omega[\varepsilon(\omega)\varepsilon_0^{-1} - 1]. \tag{20.85}$$

This latter definition leads to the asymptotic condition $\sigma(\omega) = O(\omega^{-1})$, as $\omega \to \infty$, in contrast with the result $\sigma(\omega) = O(\omega)$, as $\omega \to \infty$, obtained from Eq. (20.84).

These alternative definitions will obviously lead to different dispersion relations for the conductivity. As far as optical properties are concerned, Eq. (20.84) allows a unified discussion of insulators and conductors to be made; the particular domain that requires careful deliberation is $\omega \to 0$. The conductivity is written in terms of its real and imaginary parts as follows:

$$\sigma(\omega) = \sigma_{\mathrm{r}}(\omega) + \mathrm{i}\sigma_{\mathrm{i}}(\omega). \tag{20.86}$$

The obvious connections follow:

$$\sigma_{\mathrm{r}}(\omega) = \omega\varepsilon_{\mathrm{i}}(\omega) \tag{20.87}$$

and

$$\sigma_{\mathrm{i}}(\omega) = -\omega\varepsilon_{\mathrm{r}}(\omega). \tag{20.88}$$

The dispersion relations for the conductivity based on the definition given in Eq. (20.84) can be obtained by considering the integral

$$\oint_C \frac{\{\sigma(\omega_z)\varepsilon_0^{-1} + \mathrm{i}\omega_z\}\mathrm{d}\omega_z}{\omega - \omega_z},$$

where C denotes the contour shown in Figure 19.2. The additional factor in the numerator is selected so that $\sigma(\omega)\varepsilon_0^{-1} + \mathrm{i}\omega = O(\omega^{-1})$ as $\omega \to \infty$, and it is assumed that $\varepsilon(\omega)\varepsilon_0^{-1} = 1$ as $\omega \to \infty$. If this limit yields some other dimensionless constant, say ε_∞, then the numerator of the integrand would be replaced by $\sigma(\omega)\varepsilon_0^{-1} + \mathrm{i}\omega\varepsilon_\infty$. Evaluation of the contour integral leads to the dispersion relations

$$\sigma_{\mathrm{i}}(\omega)\varepsilon_0^{-1} + \omega = \frac{2\omega}{\pi} P \int_0^\infty \frac{\sigma_{\mathrm{r}}(\omega')\varepsilon_0^{-1}\,\mathrm{d}\omega'}{\omega^2 - \omega'^2} \tag{20.89}$$

and

$$\sigma_{\mathrm{r}}(\omega)\varepsilon_0^{-1} = -\frac{2}{\pi} P \int_0^\infty \frac{\omega'\{\sigma_{\mathrm{i}}(\omega')\varepsilon_0^{-1} + \omega'\}\mathrm{d}\omega'}{\omega^2 - \omega'^2}. \tag{20.90}$$

Examination of the limit $\omega \to \infty$ in Eq. (20.89), and noting that

$$\{\sigma_{\mathrm{i}}(\omega)\varepsilon_0^{-1} + \omega\} = \frac{\omega_{\mathrm{p}}^2}{\omega}, \quad \text{as } \omega \to \infty, \tag{20.91}$$

leads, on application of the method of Section 19.9, to the following sum rule:

$$\int_0^\infty \sigma_{\mathrm{r}}(\omega)\mathrm{d}\omega = \frac{\pi\varepsilon_0\omega_{\mathrm{p}}^2}{2}. \tag{20.92}$$

This is the f sum rule for the conductivity.

With the definition given in Eq. (20.85), the corresponding dispersion relations are given by

$$\sigma_{\mathrm{i}}(\omega) = \frac{2\omega}{\pi} P \int_0^\infty \frac{\sigma_{\mathrm{r}}(\omega')\mathrm{d}\omega'}{\omega^2 - \omega'^2} \tag{20.93}$$

and

$$\sigma_{\mathrm{r}}(\omega) = -\frac{2}{\pi} P \int_0^\infty \frac{\omega'\sigma_{\mathrm{i}}(\omega')\mathrm{d}\omega'}{\omega^2 - \omega'^2}. \tag{20.94}$$

20.5 The energy loss function: dispersion relations

The power dissipation of an electromagnetic beam in a material due to dielectric losses is proportional to $\varepsilon_{\mathrm{i}}(\omega)\varepsilon_0^{-1}$. An alternative way to probe the structure and properties of materials is via the use of electron beams. The power dissipation of a beam of fast electrons due to dielectric losses is proportional to $\varepsilon_{\mathrm{i}}(\omega)\varepsilon_0^{-1} \Big/ \left|\varepsilon(\omega)\varepsilon_0^{-1}\right|^2$. This latter factor can be written as follows:

$$\frac{\varepsilon_{\mathrm{i}}(\omega)\varepsilon_0^{-1}}{\left|\varepsilon(\omega)\varepsilon_0^{-1}\right|^2} = -\mathrm{Im}\{\varepsilon(\omega)^{-1}\varepsilon_0\}, \tag{20.95}$$

and is called the *energy loss function*. An experimental determination of this function allows $\varepsilon_{\mathrm{r}}(\omega)$ to be evaluated, and hence $n(\omega)$, $\kappa(\omega)$, and $R(\omega)$ can be calculated. The asymptotic form for the energy loss function as $\omega \to \infty$ can be determined from the corresponding asymptotes for $\varepsilon_{\mathrm{r}}(\omega)$ and $\varepsilon_{\mathrm{i}}(\omega)$ by using the following formulas:

$$\mathrm{Im}\{\varepsilon(\omega)^{-1}\varepsilon_0\} = -\frac{\varepsilon_{\mathrm{i}}(\omega)\varepsilon_0}{\varepsilon_{\mathrm{r}}(\omega)^2 + \varepsilon_{\mathrm{i}}(\omega)^2} \tag{20.96}$$

and

$$\mathrm{Re}\{\varepsilon(\omega)^{-1}\varepsilon_0\} = \frac{\varepsilon_{\mathrm{r}}(\omega)\varepsilon_0}{\varepsilon_{\mathrm{r}}(\omega)^2 + \varepsilon_{\mathrm{i}}(\omega)^2}. \tag{20.97}$$

The function $\varepsilon_0\varepsilon(\omega_z)^{-1}$ is analytic in the upper half of the complex angular frequency plane. Note that the function $\varepsilon(\omega_z)$ has no zeros in the upper half plane or on the real frequency axis. The function $\varepsilon(\omega_z)$ is real only on the imaginary frequency axis, and from Eq. (19.235) it follows that $\varepsilon(\omega_z)$ is not zero on this axis. The dispersion relations for the energy loss function can be determined by considering the integral

$$\oint_C \frac{\{\varepsilon_0\varepsilon(\omega_z)^{-1} - 1\}\mathrm{d}\omega_z}{\omega - \omega_z},$$

where C denotes the contour shown in Figure 19.2. The function $\{\varepsilon_0\varepsilon(\omega)^{-1} - 1\}$ behaves asymptotically like

$$\{\varepsilon_0\varepsilon(\omega)^{-1} - 1\} = \frac{\omega_p^2}{\omega^2}, \quad \text{as } \omega \to \infty. \tag{20.98}$$

Evaluation of the contour integral leads to

$$P\int_{-\infty}^{\infty} \frac{\{\varepsilon_0\varepsilon(\omega_r)^{-1} - 1\}\mathrm{d}\omega_r}{\omega - \omega_r} = -\mathrm{i}\pi\{\varepsilon_0\varepsilon(\omega)^{-1} - 1\}, \tag{20.99}$$

which, on taking the real and imaginary parts and making use of the crossing symmetry relations for $\varepsilon_r(\omega)$ and $\varepsilon_i(\omega)$, Eqs. (19.54) and (19.55), yields the following dispersion relations:

$$\mathrm{Im}[\varepsilon_0\varepsilon(\omega)^{-1}] = \frac{2\omega}{\pi}P\int_0^{\infty} \frac{\mathrm{Re}[\varepsilon_0\varepsilon(\omega')^{-1} - 1]\mathrm{d}\omega'}{\omega^2 - \omega'^2} \tag{20.100}$$

and

$$\mathrm{Re}[\varepsilon_0\varepsilon(\omega)^{-1} - 1] = -\frac{2}{\pi}P\int_0^{\infty} \frac{\omega'\,\mathrm{Im}[\varepsilon_0\,\varepsilon(\omega')^{-1}]\mathrm{d}\omega'}{\omega^2 - \omega'^2}. \tag{20.101}$$

Both of these dispersion relations have a practical value. The former allows a useful sum rule constraint to be obtained, the latter is of value in data analysis. From experimental data for the energy loss function, the quantity $\mathrm{Re}[\varepsilon_0\varepsilon(\omega)^{-1}]$ can be obtained from Eq. (20.101), and hence $\varepsilon_r(\omega)$ and $\varepsilon_i(\omega)$ may be determined. This opens up a pathway to other optical constants, as previously indicated.

Sum rules for the energy loss function can be derived in a similar fashion to the approaches discussed in Section 19.10. Applying the asymptotic technique of Section 19.9 to Eq. (20.101) leads to the following result:

$$\int_0^{\infty} \omega\,\mathrm{Im}[\varepsilon(\omega)^{-1}]\mathrm{d}\omega = -\frac{\pi\omega_p^2}{2\varepsilon_0}, \tag{20.102}$$

which is sometimes referred to as the f sum rule for the energy loss function. Application of the asymptotic technique to Eq. (20.100) yields the sum rule

$$\int_0^{\infty} \mathrm{Re}[\varepsilon_0\varepsilon(\omega)^{-1} - 1]\mathrm{d}\omega = 0. \tag{20.103}$$

For an insulator, the limit $\omega \to 0$ in Eq. (20.101) yields

$$\int_0^{\infty} \frac{\mathrm{Im}[\varepsilon_0\varepsilon(\omega)^{-1}]\mathrm{d}\omega}{\omega} = \frac{\pi}{2}\{\varepsilon_0\varepsilon_r(0)^{-1} - 1\}; \tag{20.104}$$

for a conductor,

$$\int_0^\infty \frac{\mathrm{Im}[\varepsilon_0\varepsilon(\omega)^{-1}]\mathrm{d}\omega}{\omega} = -\frac{\pi}{2}. \tag{20.105}$$

In order to obtain additional sum rules, consider the integral

$$\oint_C \omega_z^m\{\varepsilon_0\varepsilon(\omega_z)^{-1} - 1\}^p\,\mathrm{d}\omega_z,$$

where C denotes a closed semicircular contour in the upper half of the complex angular frequency plane with center the origin and radius $\mathcal{R}$. The integer exponents m and p are selected so that the integral does not diverge in the limit as $\mathcal{R} \to \infty$. The choice $m \geq 0$ and $p \geq 1$ is made. In order that the integral does not diverge in the limit as $\mathcal{R} \to \infty$ requires

$$m + 1 < 2p. \tag{20.106}$$

Evaluation of the contour integral yields

$$\int_{-\infty}^\infty \omega^m\{\varepsilon_0\varepsilon(\omega)^{-1} - 1\}^p\,\mathrm{d}\omega = 0. \tag{20.107}$$

The choice $m = 0,\ p = 1$ leads to Eq. (20.103), while the selection $m = 0,\ p = 2$ yields

$$\int_0^\infty \{\mathrm{Re}[\varepsilon_0\varepsilon(\omega)^{-1} - 1]\}^2\,\mathrm{d}\omega = \int_0^\infty \{\mathrm{Im}[\varepsilon_0\varepsilon(\omega)^{-1}]\}^2\,\mathrm{d}\omega. \tag{20.108}$$

This result also follows directly from the dispersion relations and the Parseval property of the Hilbert transform. The case $m = 1, p = 2$ yields the sum rule

$$\int_0^\infty \omega\,\mathrm{Im}[\varepsilon_0\varepsilon(\omega)^{-1}]\,\mathrm{Re}[\varepsilon_0\varepsilon(\omega)^{-1} - 1]\mathrm{d}\omega = 0, \tag{20.109}$$

which, in conjunction with Eq. (20.102), leads to the sum rule

$$\int_0^\infty \omega\,\mathrm{Im}[\varepsilon(\omega)^{-1}]\,\mathrm{Re}[\varepsilon(\omega)^{-1}]\mathrm{d}\omega = -\frac{\pi\omega_\mathrm{p}^2}{2\varepsilon_0^2}. \tag{20.110}$$

20.6 The permeability: dispersion relations

The permeability was introduced via Eq. (19.7). The magnetization **M** is defined by the relationship

$$\mathbf{M} = \chi_\mathrm{m}\mathbf{H}, \tag{20.111}$$

where χ_m denotes the magnetic susceptibility, which is related to the permeability as follows:

$$\mu = \mu_0(1 + \chi_m). \tag{20.112}$$

The magnetic susceptibility for a great many materials is very small, and a common approximation is to assume that the relative permeability $\mu\mu_0^{-1}$ (sometimes referred to as the relative magnetic permeability) is constant and assigned the value of unity. This approximation was employed in Chapter 19.

To obtain dispersion relations for the permeability and magnetic susceptibility, the arguments presented in Section 19.3 can be followed, with **D** and **E** replaced by **B** and **H**, respectively. The analog of Eq. (19.49) can be constructed, from which it can be deduced that the magnetic susceptibility is an analytical function in the upper half of the complex angular frequency plane. The appropriate crossing symmetry relationship is given by

$$\mu(-\omega_z^*) = \mu(\omega_z)^*, \tag{20.113}$$

from which it follows on writing

$$\mu(\omega) = \mu_r(\omega) + i\mu_i(\omega), \tag{20.114}$$

that

$$\mu_r(-\omega) = \mu_r(\omega) \tag{20.115}$$

and

$$\mu_i(-\omega) = -\mu_i(\omega). \tag{20.116}$$

On considering the integral

$$\oint_C \frac{\{\mu(\omega_z)\mu_0^{-1} - 1\}d\omega_z}{\omega - \omega_z},$$

where C denotes the contour shown in Figure 19.2, the Cauchy integral theorem yields, on taking the limits $\rho \to 0$ and $\mathcal{R} \to \infty$,

$$\frac{i}{\pi} P \int_{-\infty}^{\infty} \frac{\{\mu(\omega_r)\mu_0^{-1} - 1\}d\omega_r}{\omega - \omega_r} = \frac{\mu(\omega)}{\mu_0} - 1, \tag{20.117}$$

assuming for the moment that the integral does not diverge. Taking the real and imaginary parts and using the crossing symmetry relations for $\mu_r(\omega)$ and $\mu_i(\omega)$ leads to the Kramers–Kronig relations:

$$\mu_{\rm i}(\omega) = \frac{2\omega}{\pi} P \int_0^\infty \frac{[\mu_{\rm r}(\omega') - \mu_0]{\rm d}\omega'}{\omega^2 - \omega'^2} \tag{20.118}$$

and

$$\mu_{\rm r}(\omega) - \mu_0 = -\frac{2}{\pi} P \int_0^\infty \frac{\omega' \mu_{\rm i}(\omega'){\rm d}\omega'}{\omega^2 - \omega'^2}. \tag{20.119}$$

For the magnetic susceptibility, the Kramers–Kronig relations are often written as follows:

$$\chi''_{\rm m}(\omega) = \frac{1}{\pi} P \int_{-\infty}^\infty \frac{\{\chi'_{\rm m}(\omega') - \chi'_{\rm m}(\infty)\}{\rm d}\omega'}{\omega - \omega'} \tag{20.120}$$

and

$$\chi'_{\rm m}(\omega) - \chi'_{\rm m}(\infty) = -\frac{1}{\pi} P \int_{-\infty}^\infty \frac{\chi''_{\rm m}(\omega'){\rm d}\omega'}{\omega - \omega'}, \tag{20.121}$$

where $\chi_{\rm m}(\omega) = \chi'_{\rm m}(\omega) + {\rm i}\chi''_{\rm m}(\omega)$ has been employed to avoid excessive subscript labels, and the rather common practice of leaving the integration domain as $(-\infty, \infty)$ has been followed. The prime and double prime do not of course signify derivatives, but refer to the real and imaginary parts of $\chi_{\rm m}(\omega)$, respectively. To complete the derivation it is necessary to demonstrate that either $\mu(\omega_z)\mu_0^{-1} - 1$ or $\chi_{\rm m}(\omega_z)$ has the requisite asymptotic behavior as $\omega_z \to \infty$ in the upper half of the complex angular frequency plane.

The magnetization resulting from a system of spins is governed by the Bloch equations:

$$\frac{{\rm dM}_x}{{\rm d}t} = \omega_0 {\rm M}_y - \frac{{\rm M}_x}{T_2}, \tag{20.122}$$

$$\frac{{\rm dM}_y}{{\rm d}t} = -\omega_0 {\rm M}_x - \frac{{\rm M}_y}{T_2}, \tag{20.123}$$

and

$$\frac{{\rm dM}_z}{{\rm d}t} = -\frac{({\rm M}_z - {\rm M}_0)}{T_1}, \tag{20.124}$$

where ω_0 is the Larmor frequency, which is equal to $\gamma{\rm H}_0$, with γ the magnetogyric ratio, and ${\rm H}_0$ is a constant magnetic field. The parameters T_1 and T_2 are the spin–lattice relaxation time and the transverse relaxation time, respectively, and ${\rm M}_0$ is the steady magnetization resulting from the magnetic field ${\rm H}_0$ applied along the z-direction

($M_0 = \chi_{m_0} H_0$, where χ_{m_0} is the static magnetic susceptibility). The solution of these equations leads to the following results:

$$\chi'_m(\omega) = \frac{\chi_{m_0}\omega_0 T_2}{2} \frac{(\omega_0 - \omega)T_2}{1 + (\omega - \omega_0)^2 T_2^2} \tag{20.125}$$

and

$$\chi''_m(\omega) = \frac{\chi_{m_0}\omega_0 T_2}{2} \frac{1}{1 + (\omega - \omega_0)^2 T_2^2}. \tag{20.126}$$

The reader can find the details in many books on magnetic resonance. The chapter end-notes can be consulted for some appropriate references. If it is assumed that this model of the magnetization of the system holds at high frequency, then it follows that

$$\chi''_m(\omega) = O(\omega^{-2}), \quad \text{as } \omega \to \infty \tag{20.127}$$

and

$$\chi'_m(\omega) = O(\omega^{-1}), \quad \text{as } \omega \to \infty. \tag{20.128}$$

In this case, $\chi'_m(\infty)$ in Eqs. (20.120) and (20.121) would be zero.

20.7 The surface impedance: dispersion relations

The surface impedance function $Z_s(\omega)$ of a metal is commonly defined by the following relationship:

$$\mathbf{E}(\omega) = Z_s(\omega)\mathbf{H}(\omega) \times \mathbf{n}, \tag{20.129}$$

where $\mathbf{n}$ denotes a unit vector normal to the surface and directed into the metal. The direction of the plane electromagnetic wave incident on the metal is taken to be along $\mathbf{n}$; that is, the wave vector $\mathbf{k}$ is given by $\mathbf{k} = k\mathbf{n}$, where k is the magnitude of the vector $\mathbf{k}$. The fields $\mathbf{E}$ and $\mathbf{H}$ satisfy

$$\mathbf{E} \cdot \mathbf{n} = 0 \quad \text{and} \quad \mathbf{H} \cdot \mathbf{n} = 0. \tag{20.130}$$

These conditions define a transverse wave, where the fields $\mathbf{E}$ and $\mathbf{H}$ are perpendicular to the propagation direction. If $\mathbf{n}$ is set in the z-direction and the electric field is in the x-direction, then

$$Z_s(\omega) = \left.\frac{E_x(\omega)}{H_y(\omega)}\right|_0, \tag{20.131}$$

where the subscript zero indicates that the fields are taken at the metal surface. The surface impedance can be written as follows:

$$Z_s(\omega) = \sqrt{\left[\frac{\mu(\omega)}{\varepsilon(\omega)}\right]}. \tag{20.132}$$

To see how the boundary condition given in Eq. (20.129) arises and how the preceding formula for $Z_s(\omega)$ is obtained, start with Maxwell's equations in the absence of sources:

$$\nabla \cdot \mathbf{D} = 0, \tag{20.133}$$

$$\nabla \cdot \mathbf{B} = 0, \tag{20.134}$$

$$\nabla \times \mathbf{E} + \frac{\partial \mathbf{B}}{\partial t} = \mathbf{0}, \tag{20.135}$$

and

$$\nabla \times \mathbf{H} - \frac{\partial \mathbf{D}}{\partial t} = \mathbf{0}. \tag{20.136}$$

Employing the usual assumption of a harmonic time dependence of the form $e^{-i\omega t}$ allows the curl equations to be written in the following form:

$$\nabla \times \mathbf{E} - i\omega\mathbf{B} = \mathbf{0} \tag{20.137}$$

and

$$\nabla \times \mathbf{H} + i\omega\mathbf{D} = \mathbf{0}. \tag{20.138}$$

From the preceding results, the definitions $\mathbf{D} = \varepsilon(\omega)\mathbf{E}$ and $\mathbf{B} = \mu(\omega)\mathbf{H}$, and the vector identity

$$\nabla \times (\nabla \times \mathbf{a}) = \nabla(\nabla \cdot \mathbf{a}) - \nabla^2\mathbf{a}, \tag{20.139}$$

it follows that

$$(\nabla^2 + \mu(\omega)\varepsilon(\omega)\omega^2)\mathbf{E} = \mathbf{0} \tag{20.140}$$

and

$$(\nabla^2 + \mu(\omega)\varepsilon(\omega)\omega^2)\mathbf{B} = \mathbf{0}, \tag{20.141}$$

which are the Helmholtz wave equations. The fields take the following form:

$$\mathbf{E}(x,t) = \mathbf{E_0} e^{i(k\mathbf{n}\cdot\mathbf{x}-\omega t)} \tag{20.142}$$

and

$$\mathbf{B}(x,t) = \mathbf{B_0} e^{i(k\mathbf{n}\cdot\mathbf{x}-\omega t)}, \tag{20.143}$$

where $\mathbf{E_0}$ and $\mathbf{B_0}$ give the amplitude and directional information. Because the fields are real, the reader is reminded of the implied convention that the real part of the right-hand side of these equations is to be taken. From Eq. (20.141), it follows that

$$k = \surd[\mu(\omega)\varepsilon(\omega)]\,\omega. \tag{20.144}$$

From the definition of $\mathbf{n}$ in Eq. (20.129) it follows that

$$\frac{\mathbf{B} \times \mathbf{n}}{|\mathbf{B}|} = \frac{\mathbf{E}}{|\mathbf{E}|}, \tag{20.145}$$

and Eq. (20.137) yields

$$i\omega\mathbf{B} = \frac{|\mathbf{E}|}{|\mathbf{B}|}\nabla \times (\mathbf{B} \times \mathbf{n}), \tag{20.146}$$

which simplifies, on using the vector identity

$$\nabla \times (\mathbf{a} \times \mathbf{b}) = \mathbf{a}(\nabla \cdot \mathbf{b}) - \mathbf{b}(\nabla \cdot \mathbf{a}) + (\mathbf{b} \cdot \nabla)\mathbf{a} - (\mathbf{a} \cdot \nabla)\mathbf{b}, \tag{20.147}$$

to yield

$$\begin{aligned} i\omega\mathbf{B} &= \frac{|\mathbf{E}|}{|\mathbf{B}|}\{\mathbf{B}(\nabla \cdot \mathbf{n}) - \mathbf{n}(\nabla \cdot \mathbf{B}) + (\mathbf{n} \cdot \nabla)\mathbf{B} - (\mathbf{B} \cdot \nabla)\mathbf{n}\} \\ &= \frac{|\mathbf{E}|}{|\mathbf{B}|}(\mathbf{n} \cdot \nabla)\mathbf{B} \\ &= \frac{|\mathbf{E}|}{|\mathbf{B}|} ik\mathbf{B}. \end{aligned} \tag{20.148}$$

Employing Eq. (19.87) leads to

$$\frac{|\mathbf{E}|}{|\mathbf{B}|} = \frac{\omega}{k} = \frac{1}{\surd[\mu(\omega)\varepsilon(\omega)]}. \tag{20.149}$$

From Eq. (20.145),

$$\mathbf{E} = \frac{|\mathbf{E}|}{|\mathbf{B}|}\mathbf{B}\times\mathbf{n} = \frac{1}{\surd[\mu(\omega)\varepsilon(\omega)]}\mathbf{B}\times\mathbf{n} = \sqrt{\left[\frac{\mu(\omega)}{\varepsilon(\omega)}\right]}\mathbf{H}\times\mathbf{n}, \tag{20.150}$$

which is Eq. (20.129) with $Z_s(\omega)$ as given in Eq. (20.132).

The function $\varepsilon(\omega_z)$ is analytic in the upper half of the complex angular frequency plane and has no zeros there. This was discussed in Sections 19.3 and 19.6. The function $\mu(\omega_z)$ is also analytic in the upper half of the complex angular frequency plane, and this was discussed in the Section 20.6. Consider the integral

$$\oint_C \frac{\{Z_s(\omega_z) - Z_s(\infty)\}\mathrm{d}\omega_z}{\omega - \omega_z},$$

where C denotes the contour shown in Figure 19.2. The Cauchy integral theorem yields, on taking the limits $\rho \to 0$ and $\mathcal{R} \to \infty$,

$$\frac{\mathrm{i}}{\pi} P \int_{-\infty}^{\infty} \frac{\{Z_s(\omega') - Z_s(\infty)\}\mathrm{d}\omega'}{\omega - \omega'} = Z_s(\omega) - Z_s(\infty), \tag{20.151}$$

where for the moment the convergence of the integral is assumed. The surface impedance is sometimes written in terms of its real and imaginary parts as $Z_s(\omega) = R(\omega) + \mathrm{i}X(\omega)$, but this notation is avoided in the present chapter to minimize confusion with the reflectance introduced earlier in Section 20.2. Instead, the following notation is employed:

$$Z_s(\omega) = Z_s'(\omega) + \mathrm{i}Z_s''(\omega). \tag{20.152}$$

From the definition of $Z_s(\omega)$ and making use of Eqs. (20.113) and (19.53), it follows that

$$Z_s(-\omega) = \sqrt{\left[\frac{\mu(-\omega)}{\varepsilon(-\omega)}\right]} = \sqrt{\left[\frac{\mu(\omega)^*}{\varepsilon(\omega)^*}\right]}, \tag{20.153}$$

and hence

$$Z_s(-\omega) = Z_s(\omega)^*. \tag{20.154}$$

For the real and imaginary parts of $Z_s(\omega)$, the following crossing symmetry relations hold:

$$Z_s'(-\omega) = Z_s'(\omega) \tag{20.155}$$

and

$$Z_s''(-\omega) = -Z_s''(\omega). \tag{20.156}$$

Taking the real and imaginary parts of Eq. (20.151) and using the crossing symmetry relations just given leads to the following dispersion relations:

$$Z_s'(\omega) - Z_s'(\infty) = -\frac{2}{\pi} P \int_0^{\infty} \frac{\omega' Z_s''(\omega')\mathrm{d}\omega'}{\omega^2 - \omega'^2} \tag{20.157}$$

and

$$Z_s''(\omega) = \frac{2\omega}{\pi} P \int_0^\infty \frac{\{Z_s'(\omega') - Z_s'(\infty)\} d\omega'}{\omega^2 - \omega'^2}. \tag{20.158}$$

It remains to demonstrate that $Z_s(\omega)$ exhibits the requisite asymptotic behavior in order that the dispersion relations can be obtained. Employing Eqs. (19.60), (19.61), (20.112), (20.127), and (20.128) yields

$$Z_s(\omega) = \sqrt{\left(\frac{\mu_0}{\varepsilon_0}\right)} + \frac{a}{\omega} + \frac{b}{\omega^2} + O(\omega^{-3}), \quad \text{as } \omega \to \infty, \tag{20.159}$$

where a is a real constant and the constant b has $\operatorname{Im} b \neq 0$. The quantity $Z_s'(\infty)$ is given by

$$Z_s'(\infty) = Z_s(\infty) = c\mu_0 \approx 376.730\,313\,\Omega, \tag{20.160}$$

where $Z_s(\infty)$ denotes the impedance of free space, c is the speed of light in a vacuum, and the impedance is measured in units of ohms, denoted by Ω. From Eq. (20.159) and making use of the Phragmén–Lindelöf theorem, $Z_s(\omega_z) - Z_s(\infty)$ vanishes sufficiently rapidly as $\omega_z \to \infty$ to allow the dispersion relations to be obtained.

20.8 Anisotropic media

The discussion of optical properties carried out to this point has assumed that the media are isotropic. That is, the molecular species are assumed to be in the gas phase or solution, with appropriately rapid rotational motion, or that an isotropic solid is under consideration. In such cases the permittivity is characterized by a single parameter ε. When the medium is no longer isotropic, it is necessary to introduce the tensor character of the optical property. The case of the permittivity is considered in this section. The general relationship between the applied electric field and the resulting electric displacement can be written as follows:

$$\mathbf{D} = \boldsymbol{\varepsilon} \cdot \mathbf{E}, \tag{20.161}$$

which is the obvious extension of Eq. (19.4) to cover the cases that include anisotropic media. Tensors are given in bold type. For an isotropic medium, $\boldsymbol{\varepsilon}$ can be written as

$$\boldsymbol{\varepsilon} = \varepsilon \mathbf{I}, \tag{20.162}$$

where $\mathbf{I}$ denotes the unit second rank tensor whose elements are defined by

$$\delta_{ij} = 0, \quad i, j = x, y, z, \quad i \neq j, \tag{20.163}$$

and

$$\delta_{ij} = 1, \quad i, j = x, y, z, \quad i = j. \tag{20.164}$$

The individual components of the permittivity tensor are written as ε_{ij}. From Eq. (20.161) it follows that

$$\mathrm{D}_i = \varepsilon_{ij}\mathrm{E}_j, \tag{20.165}$$

where the indices i and j denote the coordinates x, y, or z. The standard Einstein convention in tensor analysis for repeated indices is employed, which implies a summation over the index j.

It is common to work in a coordinate system – called the principal axis system – where the permittivity tensor is diagonal. If the medium is a crystal with cubic symmetry, then $\varepsilon_{xx} = \varepsilon_{yy} = \varepsilon_{zz}$, and hence there is only one independent element necessary to characterize the permittivity tensor. The material is optically isotropic for linear optical properties. A crystal with hexagonal symmetry has $\varepsilon_{xx} = \varepsilon_{yy} \neq \varepsilon_{zz}$, and so two elements are required to characterize the permittivity tensor.

The dielectric tensor for a general medium is symmetric, that is

$$\varepsilon_{ij}(\omega) = \varepsilon_{ji}(\omega). \tag{20.166}$$

In the presence of an external magnetic field, this result is no longer true. The crossing symmetry relations for the elements of the dielectric tensor take the following form:

$$\varepsilon_{ij}(-\omega) = \varepsilon_{ij}(\omega)^*. \tag{20.167}$$

Proceeding from Eq. (20.165) in the same manner discussed in Section 19.3, the individual tensor elements $\varepsilon_{ij}(\omega)$ can be shown to be analytic functions in the upper half of the complex angular frequency plane. Consider the integral

$$\oint_C \frac{\{\varepsilon_{ij}(\omega_z) - \delta_{ij}\}\mathrm{d}\omega_z}{\omega - \omega_z},$$

where C denotes the contour shown in Figure 19.2. To avoid a proliferation of subscripts, $\varepsilon_{ij}(\omega)$ is written in terms of its real and imaginary parts as follows:

$$\varepsilon_{ij}(\omega) = \varepsilon'_{ij}(\omega) + \mathrm{i}\varepsilon''_{ij}(\omega). \tag{20.168}$$

Evaluating the contour integral and employing Eq. (20.167) leads to the following dispersion relations:

$$\varepsilon'_{ij}(\omega) - \varepsilon_0\delta_{ij} = -\frac{2}{\pi}P\int_0^\infty \frac{\omega'\varepsilon''_{ij}(\omega')\mathrm{d}\omega'}{\omega^2 - \omega'^2} \tag{20.169}$$

and

$$\varepsilon''_{ij}(\omega) = \frac{2\omega}{\pi} P \int_0^\infty \frac{\{\varepsilon'_{ij}(\omega') - \varepsilon_0 \delta_{ij}\} \mathrm{d}\omega'}{\omega^2 - \omega'^2}. \tag{20.170}$$

The Kronecker delta makes its appearance for the following reason. The elements of the dielectric tensor can be written as

$$\varepsilon_{ij}(\omega)\varepsilon_0^{-1} = \left\{1 - \frac{\omega_\mathrm{p}^2}{\omega^2}\right\} \delta_{ij} - \frac{\omega_\mathrm{p}^2 m}{\omega^2 \hbar} \chi_{ij}(\omega), \tag{20.171}$$

where $\chi_{ij}(\omega)$ denotes the ijth element of the current–current response tensor (Nozières and Pines, 1999, p. 255), and the reader is reminded that m is the electronic mass and $\hbar$ is Planck's constant divided by 2π. From this result and the functional form of $\chi_{ij}(\omega)$, it follows that the asymptotic behavior of the elements of the dielectric tensor is as follows:

$$\varepsilon_{ij}(\omega)\varepsilon_0^{-1} = \delta_{ij} - O(\omega^{-2})\delta_{ij}, \quad \text{as } \omega \to \infty. \tag{20.172}$$

It is therefore advantageous to start with $\varepsilon_{ij}(\omega_z) - \delta_{ij}$ to simplify the evaluation of the integral on the large semicircular section of the contour.

Sum rules can be derived directly from Eqs. (20.169) and (20.170). This is postponed to the following section, in which these two dispersion relations are generalized to incorporate the effects of spatial dispersion.

20.9 Spatial dispersion

The issue of including spatial dispersion in the optical constants is now investigated. The principal question of concern is how does the inclusion of spatial dispersion impact the derivation of the dispersion relations? In general, effects associated with spatial dispersion of the optical constants are less important than the normal dispersion associated with variations as a function of frequency.

The external sources $\rho(\mathbf{r}, t)$ and $\mathbf{J}(\mathbf{r}, t)$ that enter into Maxwell's equations are, respectively, the charge density and the current density. These functions are assumed to be sufficiently well behaved that a Fourier transform representation applies, that is

$$\mathbf{J}(\mathbf{r}, t) = \int_{-\infty}^{\infty} \mathbf{J}(\mathbf{k}, \omega) \mathrm{e}^{\mathrm{i}\mathbf{k}\cdot\mathbf{r} - \mathrm{i}\omega t} \, \mathrm{d}\mathbf{k} \, \mathrm{d}\omega, \tag{20.173}$$

and the inverse Fourier transform is given by

$$\mathbf{J}(\mathbf{k}, \omega) = \frac{1}{(2\pi)^4} \int_{-\infty}^{\infty} \mathbf{J}(\mathbf{r}, t) \mathrm{e}^{\mathrm{i}\omega t - \mathrm{i}\mathbf{k}\cdot\mathbf{r}} \, \mathrm{d}\mathbf{r} \, \mathrm{d}t. \tag{20.174}$$

The notation $\mathbf{dr}$ is shorthand for $dx\,dy\,dz$, and $\mathbf{dk}$ is shorthand for $dk_x\,dk_y\,dk_z$. The electric field $\mathbf{E}$ has the Fourier transform representation given by

$$\mathbf{E}(\mathbf{r},t) = \int_{-\infty}^{\infty} \mathbf{E}(\mathbf{k},\omega)\mathrm{e}^{\mathrm{i}\mathbf{k}\cdot\mathbf{r}-\mathrm{i}\omega t}\,\mathbf{dk}\,\mathrm{d}\omega \tag{20.175}$$

and

$$\mathbf{E}(\mathbf{k},\omega) = \frac{1}{(2\pi)^4}\int_{-\infty}^{\infty} \mathbf{E}(\mathbf{r},t)\mathrm{e}^{\mathrm{i}\omega t-\mathrm{i}\mathbf{k}\cdot\mathbf{r}}\,\mathbf{dr}\,\mathrm{d}t. \tag{20.176}$$

Similar relationships can be written for the quantities $\mathbf{D}$, $\mathbf{H}$, and $\mathbf{B}$.

The spatial-dependent electric displacement can be written as follows:

$$\mathbf{D}(\mathbf{k},\omega) = \boldsymbol{\varepsilon}(\mathbf{k},\omega)\cdot\mathbf{E}(\mathbf{k},\omega). \tag{20.177}$$

From Eq. (20.176), and noting that $\mathbf{E}(\mathbf{r},t)$ is real, it follows that

$$\mathbf{E}(-\mathbf{k},-\omega) = \mathbf{E}(\mathbf{k},\omega)^*, \tag{20.178}$$

and a similar result can be written for $\mathbf{D}(\mathbf{k},\omega)$. From Eq. (20.177) the crossing symmetry relation for $\boldsymbol{\varepsilon}(\mathbf{k},\omega)$ can be written as follows:

$$\boldsymbol{\varepsilon}(-\mathbf{k},-\omega) = \boldsymbol{\varepsilon}(\mathbf{k},\omega)^*. \tag{20.179}$$

The derivation of dispersion relations for a wave-vector- and frequency-dependent optical property is now examined. The initial focus is on the function $\boldsymbol{\varepsilon}(\mathbf{k},\omega)^{-1}$ rather than $\boldsymbol{\varepsilon}(\mathbf{k},\omega)$, for reasons that will become apparent shortly. The function $\boldsymbol{\varepsilon}(\mathbf{k},\omega)^{-1}$ can be expressed as follows (Nozières and Pines, 1999, p. 204):

$$\boldsymbol{\varepsilon}^{-1}(\mathbf{k},\omega)\varepsilon_0 = \mathsf{I} + \frac{e^2}{\varepsilon_0\hbar k^2}\boldsymbol{\chi}(\mathbf{k},\omega), \tag{20.180}$$

where $k = |\mathbf{k}|$, $-e$ is the electronic charge, and $\boldsymbol{\chi}(\boldsymbol{k},\omega)$ is called the density–density response function. This response measures the density fluctuations that arise when a perturbing density probe is introduced into an electronic system. The function $\boldsymbol{\chi}(\mathbf{k},\omega)$ can be expressed as follows:

$$\boldsymbol{\chi}(\mathbf{k},\omega) = \mathsf{I}\sum_n |\langle n|\rho_k^+|0\rangle|^2\left\{\frac{1}{\omega-\omega_{n0}+\mathrm{i}\eta} - \frac{1}{\omega+\omega_{n0}+\mathrm{i}\eta}\right\}. \tag{20.181}$$

In this formula, $|0\rangle$ and $|n\rangle$ denote the ground and excited eigenstates of the electronic system, respectively, ρ_k^+ represents a density fluctuation operator, and $\hbar\omega_{n0}$ is a transition energy between the states $|n\rangle$ and $|0\rangle$. The calculation leading to Eq. (20.181) is based on first-order time-dependent perturbation theory. The time-dependent perturbation is turned on very slowly – which is referred to as the adiabatic condition – and this is done by multiplying the perturbing interaction by a factor $\mathrm{e}^{\eta t}$, where η is a positive parameter. Readers with a background in time-dependent perturbation theory can consult Nozières and Pines (1999, Chap. 2) for further details on the derivation.

From the structure of $\boldsymbol{\chi}(\mathbf{k},\omega)$, it is clear that $\boldsymbol{\varepsilon}^{-1}(\mathbf{k},\omega_z)$ is an analytic function in the upper half of the complex angular frequency plane. This statement means that all the individual elements of the tensor $\boldsymbol{\varepsilon}^{-1}(\mathbf{k},\omega_z)$ are analytic in the upper half of the complex angular frequency plane. The poles of $\boldsymbol{\varepsilon}^{-1}(\mathbf{k},\omega_z)$ are located in the lower half of the complex angular frequency plane. Consider the integral

$$\oint_C \frac{\{\varepsilon_{ij}^{-1}(\mathbf{k},\omega_z)\varepsilon_0 - \delta_{ij}\}\mathrm{d}\omega_z}{\omega - \omega_z},$$

where C denotes the contour shown in Figure 19.2. The Cauchy integral theorem leads to

$$\int_{-\mathcal{R}}^{\omega-\rho} \frac{\{\varepsilon_{ij}^{-1}(\mathbf{k},\omega_\mathrm{r})\varepsilon_0 - \delta_{ij}\}\mathrm{d}\omega_\mathrm{r}}{\omega - \omega_\mathrm{r}} + \int_{\omega+\rho}^{\mathcal{R}} \frac{\{\varepsilon_{ij}^{-1}(\mathbf{k},\omega_\mathrm{r})\varepsilon_0 - \delta_{ij}\}\mathrm{d}\omega_\mathrm{r}}{\omega - \omega_\mathrm{r}}$$
$$+\,\mathrm{i}\int_0^{\pi} \{\varepsilon_{ij}^{-1}(\mathbf{k},\omega+\rho\mathrm{e}^{\mathrm{i}\phi})\varepsilon_0 - \delta_{ij}\}\mathrm{d}\phi + \mathrm{i}\mathcal{R}\int_0^{\pi} \frac{\{\varepsilon_{ij}^{-1}(\mathbf{k},\mathcal{R}\mathrm{e}^{\mathrm{i}\phi})\varepsilon_0 - \delta_{ij}\}\mathrm{e}^{\mathrm{i}\phi}\,\mathrm{d}\phi}{\omega - \mathcal{R}\mathrm{e}^{\mathrm{i}\phi}} = 0. \tag{20.182}$$

Taking the limits $\rho \to 0$ and $\mathcal{R} \to \infty$ in Eq. (20.182) leads to

$$P\int_{-\infty}^{\infty} \frac{\{\varepsilon_{ij}^{-1}(\mathbf{k},\omega_\mathrm{r})\varepsilon_0 - \delta_{ij}\}\mathrm{d}\omega_\mathrm{r}}{\omega - \omega_\mathrm{r}} + \mathrm{i}\pi\{\varepsilon_{ij}^{-1}(\mathbf{k},\omega)\varepsilon_0 - \delta_{ij}\} = 0. \tag{20.183}$$

To deal with the integral over the large semicircular section of the contour, the asymptotic behavior for large ω derived from Eq. (20.181) has been employed. The individual elements of the tensor $\boldsymbol{\varepsilon}(\mathbf{k},\omega)^{-1}$ are expressed in terms of the real and imaginary components in the following obvious notation:

$$\varepsilon_{ij}^{-1}(\mathbf{k},\omega) = \mathrm{Re}\,\varepsilon_{ij}^{-1}(\mathbf{k},\omega) + \mathrm{i}\,\mathrm{Im}\,\varepsilon_{ij}^{-1}(\mathbf{k},\omega), \tag{20.184}$$

and the elements of $\boldsymbol{\varepsilon}(\mathbf{k},\omega)$ are written as follows:

$$\varepsilon_{ij}(\mathbf{k},\omega) = \mathrm{Re}\,\varepsilon_{ij}(\mathbf{k},\omega) + \mathrm{i}\,\mathrm{Im}\,\varepsilon_{ij}(\mathbf{k},\omega). \tag{20.185}$$

On taking the real and imaginary parts of Eq. (20.183), the following dispersion relations are obtained:

$$\mathrm{Im}\,\varepsilon_{ij}^{-1}(\mathbf{k},\omega)\varepsilon_0 = \frac{1}{\pi}P\int_{-\infty}^{\infty} \frac{\{\mathrm{Re}\,\varepsilon_{ij}^{-1}(\mathbf{k},\omega')\varepsilon_0 - \delta_{ij}\}\mathrm{d}\omega'}{\omega - \omega'} \tag{20.186}$$

and

$$\operatorname{Re}\varepsilon_{ij}^{-1}(\mathbf{k},\omega)\varepsilon_0-\delta_{ij}=-\frac{\varepsilon_0}{\pi}P\int_{-\infty}^{\infty}\frac{\operatorname{Im}\varepsilon_{ij}^{-1}(\mathbf{k},\omega')\mathrm{d}\omega'}{\omega-\omega'}. \tag{20.187}$$

From Eq. (20.179) it follows that

$$\operatorname{Re}\varepsilon_{ij}(-\mathbf{k},-\omega)=\operatorname{Re}\varepsilon_{ij}(\mathbf{k},\omega) \tag{20.188}$$

and

$$\operatorname{Im}\varepsilon_{ij}(-\mathbf{k},-\omega)=-\operatorname{Im}\varepsilon_{ij}(\mathbf{k},\omega), \tag{20.189}$$

and hence the crossing symmetry relations for the inverse dielectric function can be written as follows:

$$\operatorname{Re}\varepsilon_{ij}^{-1}(\mathbf{k},-\omega)=\operatorname{Re}\varepsilon_{ij}^{-1}(-\mathbf{k},\omega) \tag{20.190}$$

and

$$\operatorname{Im}\varepsilon_{ij}^{-1}(\mathbf{k},-\omega)=-\operatorname{Im}\varepsilon_{ij}^{-1}(-\mathbf{k},\omega). \tag{20.191}$$

Using these relationships, Eqs. (20.186) and (20.187) can be written as follows:

$$\begin{aligned}\operatorname{Im}\varepsilon_{ij}^{-1}(\mathbf{k},\omega)\varepsilon_0=\frac{1}{\pi}\Bigg\{&P\int_0^{\infty}\frac{\{\operatorname{Re}\varepsilon_{ij}^{-1}(-\mathbf{k},\omega')\varepsilon_0-\delta_{ij}\}\mathrm{d}\omega'}{\omega+\omega'}\\&+P\int_0^{\infty}\frac{\{\operatorname{Re}\varepsilon_{ij}^{-1}(\mathbf{k},\omega')\varepsilon_0-\delta_{ij}\}\mathrm{d}\omega'}{\omega-\omega'}\Bigg\}\end{aligned} \tag{20.192}$$

and

$$\operatorname{Re}\varepsilon_{ij}^{-1}(\mathbf{k},\omega)\varepsilon_0-\delta_{ij}=\frac{\varepsilon_0}{\pi}\left\{P\int_0^{\infty}\frac{\operatorname{Im}\varepsilon_{ij}^{-1}(-\mathbf{k},\omega')\mathrm{d}\omega'}{\omega+\omega'}-P\int_0^{\infty}\frac{\operatorname{Im}\varepsilon_{ij}^{-1}(\mathbf{k},\omega')\mathrm{d}\omega'}{\omega-\omega'}\right\}. \tag{20.193}$$

Because of the reversal of sign for the wave vector in the crossing symmetry relations, the resulting dispersion relations do not appear in quite the same compact form as do the dispersion relations in the absence of spatial dispersion effects.

Consider the integral

$$\oint_C\{\varepsilon_{ij}^{-1}(\mathbf{k},\omega_z)\varepsilon_0-\delta_{ij}\}\mathrm{d}\omega_z,$$

where C denotes a semicircular contour in the upper half complex angular frequency plane of radius $\mathcal{R}$ and center the origin, and make use of Eq. (20.181); then it follows from the Cauchy integral theorem, in the limit $\mathcal{R} \to \infty$, that

$$\int_{-\infty}^{\infty} \{\varepsilon_{ij}^{-1}(\mathbf{k}, \omega)\varepsilon_0 - \delta_{ij}\}\mathrm{d}\omega = 0. \tag{20.194}$$

Taking the real and imaginary parts leads to the following sum rules:

$$\int_0^{\infty} \{\mathrm{Re}\, \varepsilon_{ij}^{-1}(\mathbf{k}, \omega) + \mathrm{Re}\, \varepsilon_{ij}^{-1}(-\mathbf{k}, \omega) - 2\varepsilon_0^{-1}\delta_{ij}\}\mathrm{d}\omega = 0 \tag{20.195}$$

and

$$\int_0^{\infty} \{\mathrm{Im}\, \varepsilon_{ij}^{-1}(\mathbf{k}, \omega) - \mathrm{Im}\, \varepsilon_{ij}^{-1}(-\mathbf{k}, \omega)\}\mathrm{d}\omega = 0. \tag{20.196}$$

These results can also be obtained directly from the dispersion relations in Eqs. (20.192) and (20.193) using the approach described in Section 19.9. From Eq. (20.181) it follows that

$$\boldsymbol{\chi}(\mathbf{k}, \omega) = \frac{2}{\omega^2}\mathsf{I}\sum_n \left\{ \frac{\omega_{n0}\left|\langle n|\rho_k^+|0\rangle\right|^2}{(1 - \omega_{n0}\omega^{-1} + \mathrm{i}\eta\omega^{-1})(1 + \omega_{n0}\omega^{-1} + \mathrm{i}\eta\omega^{-1})} \right\}, \tag{20.197}$$

and hence

$$\boldsymbol{\chi}(\mathbf{k}, \omega) = \frac{2}{\omega^2}\mathsf{I}\sum_n \omega_{n0}\left|\langle n|\rho_k^+|0\rangle\right|^2, \quad \text{as } \omega \to \infty. \tag{20.198}$$

The Thomas–Reiche–Kuhn sum rule takes the following form:

$$\frac{2m}{\hbar k^2}\sum_n \omega_{n0}\left|\langle n|\rho_k^+|0\rangle\right|^2 = \mathcal{N}, \tag{20.199}$$

where $\mathcal{N}$ is the number of particles in the system. Making use of this result and employing Eqs. (20.180) and (20.198) leads to

$$\boldsymbol{\varepsilon}^{-1}(\mathbf{k}, \omega)\varepsilon_0 = \left(1 + \frac{e^2\mathcal{N}}{\varepsilon_0 m\omega^2}\right)\mathsf{I}, \quad \text{as } \omega \to \infty; \tag{20.200}$$

that is,

$$\boldsymbol{\varepsilon}^{-1}(\mathbf{k}, \omega)\varepsilon_0 = \left(1 + \frac{\omega_\mathrm{p}^2}{\omega^2}\right)\mathsf{I}, \quad \text{as } \omega \to \infty. \tag{20.201}$$

From Eq. (20.193) it follows, in the limit $\omega \to \infty$, that

$$\frac{\omega_{\mathrm{p}}^2}{\omega^2}\delta_{ij} \approx \left\{ \frac{\varepsilon_0 \omega}{\pi} P \int_0^\infty \frac{\{\mathrm{Im}\, \varepsilon_{ij}^{-1}(-\mathbf{k}, \omega') - \mathrm{Im}\, \varepsilon_{ij}^{-1}(\mathbf{k}, \omega')\}\mathrm{d}\omega'}{\omega^2 - \omega'^2} \right.$$
$$\left. - \frac{\varepsilon_0}{\pi} P \int_0^\infty \frac{\omega'\{\mathrm{Im}\, \varepsilon_{ij}^{-1}(\mathbf{k}, \omega') + \mathrm{Im}\, \varepsilon_{ij}^{-1}(-\mathbf{k}, \omega')\}\mathrm{d}\omega'}{\omega^2 - \omega'^2} \right\}_{\omega \to \infty}. \tag{20.202}$$

Applying the approach of Section 19.9 leads to

$$\int_0^\infty \{\mathrm{Im}\, \varepsilon_{ij}^{-1}(\mathbf{k}, \omega) - \mathrm{Im}\, \varepsilon_{ij}^{-1}(-\mathbf{k}, \omega)\}\mathrm{d}\omega = 0 \tag{20.203}$$

and

$$\int_0^\infty \omega\{\mathrm{Im}\, \varepsilon_{ij}^{-1}(\mathbf{k}, \omega) + \mathrm{Im}\, \varepsilon_{ij}^{-1}(-\mathbf{k}, \omega)\}\mathrm{d}\omega = -\frac{\pi \omega_{\mathrm{p}}^2}{\varepsilon_0}\delta_{ij}. \tag{20.204}$$

The derivation of dispersion relations for $\boldsymbol{\varepsilon}(\mathbf{k}, \omega)$ is now considered. Here the situation is more complicated, as the reader will quickly appreciate. The function $\boldsymbol{\varepsilon}(\mathbf{k}, \omega)$ is analytic in the upper half of the complex angular frequency plane if $\boldsymbol{\varepsilon}^{-1}(\mathbf{k}, \omega)$ has no zeros in the same region. The initial focus here is centered on determining whether $\boldsymbol{\varepsilon}^{-1}(\mathbf{k}, \omega)$ has any zeros in the upper half complex angular frequency plane. The approach is based on a treatment in Nozières and Pines (1999, p. 206), which in turn is related to the ideas presented at the end of Section 19.10 (see Eq. (19.235)). To facilitate the discussion, the dynamic form factor (also called the dynamic structure factor) is introduced via the following definition:

$$S(\mathbf{k}, \omega) = \sum_n \left| \langle n | \, \rho_k^+ \, | 0 \rangle \right|^2 \delta(\omega - \omega_{n0}). \tag{20.205}$$

The quantity $S(\mathbf{k}, \omega)$ is real and positive. Making use of Eqs. (20.205) and (20.181), it follows that

$$\boldsymbol{\chi}(\mathbf{k}, \omega) = \mathrm{I} \int_0^\infty S(\mathbf{k}, \omega') \left\{ \frac{1}{\omega - \omega' + \mathrm{i}\eta} - \frac{1}{\omega + \omega' + \mathrm{i}\eta} \right\} \mathrm{d}\omega', \tag{20.206}$$

and hence

$$\boldsymbol{\varepsilon}^{-1}(\mathbf{k}, \omega)\varepsilon_0 = \mathrm{I} \left[1 + \frac{e^2}{\varepsilon_0 \hbar k^2} \int_0^\infty S(\mathbf{k}, \omega') \left\{ \frac{1}{\omega - \omega' + \mathrm{i}\eta} - \frac{1}{\omega + \omega' + \mathrm{i}\eta} \right\} \mathrm{d}\omega' \right]. \tag{20.207}$$

If the real angular frequency ω in this result is replaced by $\omega_r + i\omega_i$, the imaginary part is given by

$$\operatorname{Im} \boldsymbol{\varepsilon}^{-1}(\mathbf{k},\omega_z)\varepsilon_0 = -\mathsf{I}\frac{4e^2\omega_r(\omega_i+\eta)}{\varepsilon_0\hbar k^2} \times \int_0^\infty \frac{\omega' S(\mathbf{k},\omega')\mathrm{d}\omega'}{[(\omega_r-\omega')^2+(\omega_i+\eta)^2][(\omega_r+\omega')^2+(\omega_i+\eta)^2]}. \tag{20.208}$$

In Eq. (20.208) the integral is positive. Since $\omega_i > 0$ in the upper half complex angular frequency plane, the limit as $\eta \to 0+$ can be examined. The function $\operatorname{Im} \boldsymbol{\varepsilon}^{-1}(\mathbf{k},\omega_z)$ is zero if (i) $\omega_i = 0$ and $\omega_r = 0$, (ii) $\omega_i = 0$, and $\omega_r \neq 0$, or (iii) $\omega_i \neq 0$, and $\omega_r = 0$. The first two cases are excluded by the requirement that $\omega_i > 0$. Hence, the only possible zeros of $\boldsymbol{\varepsilon}^{-1}(\mathbf{k},\omega_z)$ in the upper half of the complex angular frequency plane must lie on the imaginary axis. On the imaginary axis it follows that

$$\boldsymbol{\varepsilon}^{-1}(\mathbf{k},i\omega_i)\varepsilon_0 = \mathsf{I}\left\{1 - \frac{2e^2}{\varepsilon_0\hbar k^2}\int_0^\infty \frac{\omega' S(\mathbf{k},\omega')\mathrm{d}\omega'}{\omega'^2+(\omega_i+\eta)^2}\right\}. \tag{20.209}$$

In Eq. (20.209) the integral is a monotonically decreasing function of ω_i, with the integral $\to 0$ as $\omega_i \to \infty$. Hence, there is a solution of

$$\boldsymbol{\varepsilon}^{-1}(\mathbf{k},\omega_z)\varepsilon_0 = 0 \tag{20.210}$$

if

$$\frac{2e^2}{\varepsilon_0\hbar k^2}\int_0^\infty \frac{S(\mathbf{k},\omega')}{\omega'}\mathrm{d}\omega' > 1. \tag{20.211}$$

Equivalently, if Eq. (20.211) holds, then from Eq. (20.207) it follows, in the limit $\eta \to 0$, that the static dielectric constant satisfies

$$\boldsymbol{\varepsilon}(\mathbf{k},0) < \mathbf{0}. \tag{20.212}$$

So, if $\boldsymbol{\varepsilon}(\mathbf{k},0) \geq \mathbf{0}$, then there are no zeros of $\boldsymbol{\varepsilon}^{-1}(\mathbf{k},\omega_z)$ in the upper half of the complex angular frequency plane. On the real axis the imaginary component of $\boldsymbol{\varepsilon}^{-1}(\mathbf{k},\omega)$ is non-zero for $|\omega| > 0$. Hence, it can be concluded that, for the majority of systems, $\boldsymbol{\varepsilon}(\mathbf{k},\omega_z)$ is analytic in the upper half complex angular frequency plane, with the possible exception of a pole at $\omega = 0$. If the latter situation occurs, then a subtraction from $\boldsymbol{\varepsilon}(\mathbf{k},\omega)$ is made to remove this singularity. From Eq. (20.201) it follows that

$$\boldsymbol{\varepsilon}(\mathbf{k},\omega)\varepsilon_0^{-1} = \left\{1 - \frac{\omega_p^2}{\omega^2}\right\}\mathsf{I}, \quad \text{as } \omega \to \infty. \tag{20.213}$$

Consider the integral

$$\oint_C \frac{\{\varepsilon_{ij}(\mathbf{k},\omega_z)\varepsilon_0^{-1} - \delta_{ij}\}\mathrm{d}\omega_z}{\omega - \omega_z},$$

where C denotes the contour shown in Figure 19.2. The following discussion is restricted to cases where there is no pole at $\omega = 0$. The Cauchy integral theorem leads, in the limits $\rho \to 0$ and $R \to \infty$, to

$$P\int_{-\infty}^{\infty} \frac{\{\varepsilon_{ij}(\mathbf{k},\omega_\mathrm{r}) - \varepsilon_0\delta_{ij}\}\mathrm{d}\omega_\mathrm{r}}{\omega - \omega_\mathrm{r}} + \mathrm{i}\pi\{\varepsilon_{ij}(\mathbf{k},\omega) - \varepsilon_0\delta_{ij}\} = 0. \tag{20.214}$$

Taking the real and imaginary parts gives the following dispersion relations:

$$\mathrm{Im}\,\varepsilon_{ij}(\mathbf{k},\omega) = \frac{1}{\pi}P\int_{-\infty}^{\infty} \frac{\{\mathrm{Re}\,\varepsilon_{ij}(\mathbf{k},\omega') - \varepsilon_0\delta_{ij}\}\mathrm{d}\omega'}{\omega - \omega'} \tag{20.215}$$

and

$$\mathrm{Re}\,\varepsilon_{ij}(\mathbf{k},\omega) - \varepsilon_0\delta_{ij} = -\frac{1}{\pi}P\int_{-\infty}^{\infty} \frac{\mathrm{Im}\,\varepsilon_{ij}(\mathbf{k},\omega')\mathrm{d}\omega'}{\omega - \omega'}. \tag{20.216}$$

Equations (20.215) and (20.216) apply for an insulator. The situation for conductors is treated later in Eqs. (20.225) and (20.226). Employing the symmetry relations given in Eqs. (20.188) and (20.189) leads to the following dispersion relations:

$$\begin{aligned}\mathrm{Im}\,\varepsilon_{ij}(\mathbf{k},\omega) = \frac{1}{\pi}\Bigg\{&\int_0^{\infty} \frac{\{\mathrm{Re}\,\varepsilon_{ij}(-\mathbf{k},\omega') - \varepsilon_0\delta_{ij}\}\mathrm{d}\omega'}{\omega + \omega'} \\ &+P\int_0^{\infty} \frac{\{\mathrm{Re}\,\varepsilon_{ij}(\mathbf{k},\omega') - \varepsilon_0\delta_{ij}\}\mathrm{d}\omega'}{\omega - \omega'}\Bigg\}\end{aligned} \tag{20.217}$$

and

$$\mathrm{Re}\,\varepsilon_{ij}(\mathbf{k},\omega) - \varepsilon_0\delta_{ij} = \frac{1}{\pi}\Bigg\{\int_0^{\infty} \frac{\mathrm{Im}\,\varepsilon_{ij}(-\mathbf{k},\omega')\mathrm{d}\omega'}{\omega + \omega'} - P\int_0^{\infty} \frac{\mathrm{Im}\,\varepsilon_{ij}(\mathbf{k},\omega')\mathrm{d}\omega'}{\omega - \omega'}\Bigg\}. \tag{20.218}$$

From the dispersion relations for $\boldsymbol{\varepsilon}(\mathbf{k},\omega)$, various sum rules can be derived using the same procedures previously indicated for $\boldsymbol{\varepsilon}^{-1}(\mathbf{k},\omega)$. Making use of Eq. (20.213), it follows from Eq. (20.218) that

$$\int_0^{\infty} \{\mathrm{Im}\,\varepsilon_{ij}(\mathbf{k},\omega) - \mathrm{Im}\,\varepsilon_{ij}(-\mathbf{k},\omega)\}\mathrm{d}\omega = 0 \tag{20.219}$$

and

$$\int_0^{\infty} \omega\{\mathrm{Im}\,\varepsilon_{ij}(\mathbf{k},\omega) + \mathrm{Im}\,\varepsilon_{ij}(-\mathbf{k},\omega)\}\mathrm{d}\omega = \varepsilon_0\pi\omega_\mathrm{p}^2\delta_{ij}. \tag{20.220}$$

From Eq. (20.217) the following sum rules are obtained:

$$\int_0^\infty \{\mathrm{Re}\, \varepsilon_{ij}(-\mathbf{k},\omega) + \mathrm{Re}\, \varepsilon_{ij}(\mathbf{k},\omega) - 2\varepsilon_0\delta_{ij}\}\mathrm{d}\omega = 0 \tag{20.221}$$

and

$$\int_0^\infty \omega\{\mathrm{Re}\, \varepsilon_{ij}(\mathbf{k},\omega) - \mathrm{Re}\, \varepsilon_{ij}(-\mathbf{k},\omega)\}\mathrm{d}\omega = 0. \tag{20.222}$$

For the case of a conducting material, the conductivity tensor $\boldsymbol{\sigma}(\mathbf{k},\omega)$ is introduced, which relates the applied electric field to the induced current density along the lines discussed at the start of Section 19.8, but with the modification indicated in Eq. (20.85), so that

$$\boldsymbol{\varepsilon}(\mathbf{k},\omega)\varepsilon_0^{-1} = \mathsf{I} + \mathrm{i}\omega^{-1}\varepsilon_0^{-1}\boldsymbol{\sigma}(\mathbf{k},\omega). \tag{20.223}$$

If $\boldsymbol{\sigma}(\mathbf{k},0) \neq 0$, then $\boldsymbol{\varepsilon}(\mathbf{k},\omega)$ has a pole at $\omega = 0$, which can be dealt with in the following way. Consider the integral

$$\oint_C \frac{\{\varepsilon_{ij}(\mathbf{k},\omega_z)\varepsilon_0^{-1} - \delta_{ij} - \mathrm{i}\omega_z^{-1}\varepsilon_0^{-1}\sigma_{ij}(\mathbf{k},0)\}\mathrm{d}\omega_z}{\omega - \omega_z},$$

where C is the same contour used for the non-conducting case. It follows that

$$P\int_{-\infty}^\infty \frac{\{\varepsilon_{ij}(\mathbf{k},\omega_\mathrm{r})/\varepsilon_0 - \delta_{ij} - \mathrm{i}(\sigma_{ij}(\mathbf{k},0)/\omega_\mathrm{r}\varepsilon_0)\}\mathrm{d}\omega_\mathrm{r}}{\omega - \omega_\mathrm{r}} + \mathrm{i}\pi\left\{\frac{\varepsilon_{ij}(\mathbf{k},\omega)}{\varepsilon_0} - \delta_{ij} - \mathrm{i}\frac{\sigma_{ij}(\mathbf{k},0)}{\omega\varepsilon_0}\right\} = 0. \tag{20.224}$$

Subtraction of the term $\mathrm{i}\omega^{-1}\sigma_{ij}(\mathbf{k},0)$ handles the pole at $\omega = 0$. The dc conductivity $\boldsymbol{\sigma}(\mathbf{k},0)$ is a real quantity. Taking the real and imaginary parts of Eq. (20.224) leads to

$$\mathrm{Im}\, \varepsilon_{ij}(\mathbf{k},\omega) = \omega^{-1}\sigma_{ij}(\mathbf{k},0) + \frac{1}{\pi}P\int_{-\infty}^\infty \frac{\mathrm{Re}\{\varepsilon_{ij}(\mathbf{k},\omega') - \varepsilon_0\delta_{ij}\}\mathrm{d}\omega'}{\omega - \omega'} \tag{20.225}$$

and

$$\mathrm{Re}\, \varepsilon_{ij}(\mathbf{k},\omega) - \varepsilon_0\delta_{ij} = -\frac{1}{\pi}P\int_{-\infty}^\infty \frac{\{\mathrm{Im}\, \varepsilon_{ij}(\mathbf{k},\omega') - \sigma_{ij}(\mathbf{k},0)/\omega'\}\mathrm{d}\omega'}{\omega - \omega'}. \tag{20.226}$$

Using the crossing symmetry relations yields the following dispersion relations:

$$\operatorname{Im}\varepsilon_{ij}(\mathbf{k},\omega)-\omega^{-1}\sigma_{ij}(\mathbf{k},0)=\frac{\omega}{\pi}P\int_0^\infty\frac{\{\operatorname{Re}\varepsilon_{ij}(\mathbf{k},\omega')+\operatorname{Re}\varepsilon_{ij}(-\mathbf{k},\omega')-2\varepsilon_0\delta_{ij}\}\mathrm{d}\omega'}{\omega^2-\omega'^2}$$
$$+\frac{1}{\pi}P\int_0^\infty\frac{\{\operatorname{Re}\varepsilon_{ij}(\mathbf{k},\omega')-\operatorname{Re}\varepsilon_{ij}(-\mathbf{k},\omega')\}\omega'\,\mathrm{d}\omega'}{\omega^2-\omega'^2} \tag{20.227}$$

and

$$\operatorname{Re}\varepsilon_{ij}(\mathbf{k},\omega)-\varepsilon_0\delta_{ij}=-\frac{\omega}{\pi}P\int_0^\infty\frac{\{\operatorname{Im}\varepsilon_{ij}(\mathbf{k},\omega')-\operatorname{Im}\varepsilon_{ij}(-\mathbf{k},\omega')\}\mathrm{d}\omega'}{\omega^2-\omega'^2}$$
$$-\frac{1}{\pi}P\int_0^\infty\frac{\left\{\begin{array}{c}\operatorname{Im}\varepsilon_{ij}(\mathbf{k},\omega')+\operatorname{Im}\varepsilon_{ij}(-\mathbf{k},\omega')\\-2\sigma_{ij}(\mathbf{k},0)/\omega'\end{array}\right\}\omega'\,\mathrm{d}\omega'}{\omega^2-\omega'^2}. \tag{20.228}$$

This pair of results represents the extension of Eqs. (20.217) and (20.218) to cover the conducting case. From Eqs. (20.227) and (20.228), the approach of Section 19.9, and the asymptotic condition given in Eq. (20.213), the following sum rules are obtained:

$$\int_0^\infty\{\operatorname{Re}\varepsilon_{ij}(\mathbf{k},\omega)-\operatorname{Re}\varepsilon_{ij}(-\mathbf{k},\omega)\}\omega\,\mathrm{d}\omega=0, \tag{20.229}$$

$$\int_0^\infty\{\operatorname{Re}\varepsilon_{ij}(\mathbf{k},\omega)+\operatorname{Re}\varepsilon_{ij}(-\mathbf{k},\omega)-2\varepsilon_0\delta_{ij}\}\mathrm{d}\omega=-\pi\sigma_{ij}(\mathbf{k},0), \tag{20.230}$$

$$\int_0^\infty\{\operatorname{Im}\varepsilon_{ij}(\mathbf{k},\omega)-\operatorname{Im}\varepsilon_{ij}(-\mathbf{k},\omega)\}\mathrm{d}\omega=0, \tag{20.231}$$

and

$$\int_0^\infty\{\operatorname{Im}\varepsilon_{ij}(\mathbf{k},\omega)+\operatorname{Im}\varepsilon_{ij}(-\mathbf{k},\omega)-2\sigma_{ij}(\mathbf{k},0)\omega^{-1}\}\omega\,\mathrm{d}\omega=\pi\varepsilon_0\omega_{\mathrm{p}}^2\delta_{ij}. \tag{20.232}$$

The preceding four results apply to both conductors and insulators.

A number of the dispersion relations and sum rules that have been given previously can be written in a modified form by employing the Onsager relationship,

$$\varepsilon_{ij}(\mathbf{k},\omega)=\varepsilon_{ji}(-\mathbf{k},\omega). \tag{20.233}$$

For example, from Eqs. (20.229)–(20.232) it follows that

$$\int_0^\infty \{\mathrm{Re}\,\varepsilon_{ij}(\mathbf{k},\omega) - \mathrm{Re}\,\varepsilon_{ji}(\mathbf{k},\omega)\}\omega\,\mathrm{d}\omega = 0, \tag{20.234}$$

$$\int_0^\infty \{\mathrm{Re}\,\varepsilon_{ij}(\mathbf{k},\omega) + \mathrm{Re}\,\varepsilon_{ji}(\mathbf{k},\omega) - 2\varepsilon_0\delta_{ij}\}\mathrm{d}\omega = -\pi\sigma_{ij}(\mathbf{k},0), \tag{20.235}$$

$$\int_0^\infty \{\mathrm{Im}\,\varepsilon_{ij}(\mathbf{k},\omega) - \mathrm{Im}\,\varepsilon_{ji}(\mathbf{k},\omega)\}\mathrm{d}\omega = 0, \tag{20.236}$$

and

$$\int_0^\infty \{\mathrm{Im}\,\varepsilon_{ij}(\mathbf{k},\omega) + \mathrm{Im}\,\varepsilon_{ji}(\mathbf{k},\omega) - 2\sigma_{ij}(\mathbf{k},0)\omega^{-1}\}\omega\,\mathrm{d}\omega = \pi\varepsilon_0\omega_{\mathrm{p}}^2\delta_{ij}. \tag{20.237}$$

This section is concluded with some remarks on the sign of the static dielectric constant and its bearing on the derivation of dispersion relations. Dispersion relations can still be derived for $\boldsymbol{\varepsilon}(\mathbf{k},\omega)$ for the case $\boldsymbol{\varepsilon}(\mathbf{k},0) < \mathbf{0}$, but the form is modified from that given previously. If $\boldsymbol{\varepsilon}(\mathbf{k},0) < \mathbf{0}$, then $\boldsymbol{\varepsilon}(\mathbf{k},\omega)$ has a pole at some complex frequency $\mathrm{i}\omega_0$ for ω_0 real and satisfying $\omega_0 > 0$. Making an appropriate subtraction of this pole behavior from $\boldsymbol{\varepsilon}(\mathbf{k},\omega)$ leads to a modified dispersion relation. The resulting formulas are less useful, since the frequency ω_0 must be located. The derivation of dispersion relations in this case is left as an exercise for the reader to pursue.

The situation $\boldsymbol{\varepsilon}(\mathbf{k},0) < \mathbf{0}$ contradicts no general principle of physics. When this case arises, it may in some problems be interpreted in terms of a specific system instability; however, this appears not to be a requirement. If the standard form of the Kramers–Kronig relations for the dielectric constant apply, then $\boldsymbol{\varepsilon}(\mathbf{k},0) > \mathbf{0}$; however, systems where $\boldsymbol{\varepsilon}(\mathbf{k},0) < \mathbf{0}$ applies lead to modified Kramers–Kronig relations. It is to be noted that a negative $\boldsymbol{\varepsilon}(\mathbf{k},0)$ does not lead to any contradiction of the causality principle.

20.10 Fourier series representation

The question of using a conjugate Fourier series approach to perform a Hilbert transform analysis on optical properties is now considered. The focus of this section is the dielectric constant, and the following section discusses a slightly more involved application to obtain the phase from the measured normal reflectance. Consider the conformal mapping depicted in Figure 20.3. This is accomplished by means of the following transformation:

$$w = \frac{\omega_z - \mathrm{i}}{\omega_z + \mathrm{i}}. \tag{20.238}$$

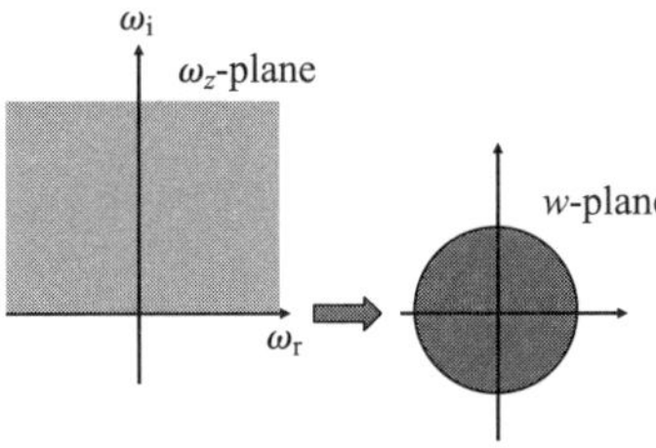

Figure 20.3. Conformal mapping from the upper half complex angular frequency plane to the interior of the unit disc in the w-plane.

The dielectric constant can then be expressed as follows:

$$\varepsilon(\omega_z)\varepsilon_0^{-1} - 1 \to \varepsilon(-\mathrm{i}(w+1)(w-1)^{-1})\varepsilon_0^{-1} - 1$$
$$= \sum_{m=0}^{\infty} c_m w^m, \quad \text{for } |w| \leq 1. \tag{20.239}$$

The preceding Laurent expansion follows from the fact that the dielectric constant is analytic within the interior of the unit disc in the complex w-plane. To handle the case of conductors, the factor $\varepsilon(\omega_z)\varepsilon_0^{-1} - 1 - \mathrm{i}\sigma(0)\omega_z^{-1}$ is used in place of $\varepsilon(\omega_z)\varepsilon_0^{-1} - 1$. The subtraction term $\mathrm{i}\sigma(0)\omega_z^{-1}$ handles the singularity as $\omega_z \to 0$.

Introducing the substitution

$$w = \mathrm{e}^{\mathrm{i}\Omega} \tag{20.240}$$

allows Eq. (20.239) to be written as follows:

$$\varepsilon(-\mathrm{i}(w+1)(w-1)^{-1})\varepsilon_0^{-1} - 1 = \sum_{m=0}^{\infty} c_m \cos m\Omega + \mathrm{i}\sum_{m=1}^{\infty} c_m \sin m\Omega. \tag{20.241}$$

The following definitions are introduced:

$$\varepsilon_1(\Omega) = \varepsilon_\mathrm{r}\left(\cot\frac{\Omega}{2}\right)\varepsilon_0^{-1} - 1 \tag{20.242}$$

and

$$\varepsilon_2(\Omega) = \varepsilon_\mathrm{i}\left(\cot\frac{\Omega}{2}\right)\varepsilon_0^{-1}. \tag{20.243}$$

On the real frequency axis, $\omega_z = \omega_\mathrm{r}$, and on setting $\omega_\mathrm{r} = \omega$ to simplify the notation, Eq. (20.241) becomes

$$\varepsilon(\omega)\varepsilon_0^{-1} - 1 \to \varepsilon\left(-\cot\frac{\Omega}{2}\right)\varepsilon_0^{-1} - 1 = \varepsilon_1(\Omega) - \mathrm{i}\varepsilon_2(\Omega). \tag{20.244}$$

Taking advantage of the crossing symmetry relations in Eqs. (19.54) and (19.55), the functions $\varepsilon_1(\Omega)$ and $\varepsilon_2(\Omega)$ can be expressed as Fourier series in the interval $(-\pi, \pi)$, so that

$$\varepsilon_1(\Omega) = \frac{1}{2}a_0 + \sum_{m=1}^{\infty} a_m \cos m\Omega \tag{20.245}$$

and

$$\varepsilon_2(\Omega) = \sum_{m=1}^{\infty} b_m \sin m\Omega, \tag{20.246}$$

where the a_m and b_m are expansion coefficients. Inserting the preceding pair of equations into Eq. (20.244) and comparing with Eq. (20.241) leads to

$$c_0 = \frac{1}{2}a_0, \quad c_m = a_m, \quad m \neq 0, \tag{20.247}$$

and

$$c_m = -b_m, \quad m \neq 0, \tag{20.248}$$

so that

$$b_m = -a_m, \quad m \neq 0. \tag{20.249}$$

The constant a_0 can be determined in the following manner. Employing the asymptotic condition given in Eq. (19.60) leads to

$$\lim_{\omega\to\infty}[\varepsilon_r(\omega)\varepsilon_0^{-1} - 1] = \lim_{\Omega\to 0}\left[\varepsilon_r\left(\cot\frac{\Omega}{2}\right)\varepsilon_0^{-1} - 1\right] = \frac{1}{2}a_0 + \sum_{m=1}^{\infty} a_m = 0, \tag{20.250}$$

and hence

$$a_0 = -2\sum_{m=1}^{\infty} a_m. \tag{20.251}$$

Equations (20.245) and (20.246) can be expressed as follows:

$$\varepsilon_1(\Omega) = \sum_{m=1}^{\infty} a_m\{\cos m\Omega - 1\} \tag{20.252}$$

and

$$\varepsilon_2(\Omega) = -\sum_{m=1}^{\infty} a_m \sin m\Omega, \tag{20.253}$$

which can be recognized as a conjugate pair of trigonometric series.

Equations (20.252) and (20.253) can be employed in the following manner. Suppose the dispersive mode of the complex dielectric constant has been determined experimentally. The measured data can be fitted to a Fourier cosine series for $\varepsilon_1(\Omega)$ using Eq. (20.252). In a practical calculation the series is truncated, so that $a_m = 0$ for m greater than some specified value. The dissipative mode is then directly determined from the conjugate sine series making use of Eq. (20.253). Since considerable attention has been devoted to the development of fast Fourier techniques, the numerical component of the calculations can be carried out using readily available algorithms.

As a very straightforward analytical example, suppose the dispersive component of the complex dielectric constant is given by

$$\varepsilon_r(\omega)\varepsilon_0^{-1} - 1 = \frac{1}{1+\omega^2}, \tag{20.254}$$

a choice made for reasons of simplicity so the preceding development can be illustrated in a straightforward manner. Angular frequencies have been divided by 1 Hz to render them dimensionless in this example. Equation (20.254) and the expression that will emerge for $\varepsilon_i(\omega)$ momentarily, are with the introduction of appropriate constants, the Debye equations used in the discussion of dielectrics. Note that Eq. (20.254) has obvious limitations; for example, the sum rule given in Eq. (19.225) is not satisfied. By inspection from Eq. (20.242), it follows that

$$\varepsilon_1(\Omega) = \sin^2\frac{\Omega}{2}, \tag{20.255}$$

and hence from Eq. (20.245) the a_m coefficients are determined as

$$a_0 = 1; \quad a_m = -\frac{1}{2}\delta_{m1}; \quad m \neq 0. \tag{20.256}$$

Equation (20.253) leads to

$$\varepsilon_2(\Omega) = \frac{1}{2}\sin\Omega = \frac{\cot\Omega/2}{1+\cot^2\Omega/2}, \tag{20.257}$$

and hence

$$\varepsilon_i(\omega)\varepsilon_0^{-1} = \frac{\omega}{1+\omega^2}. \tag{20.258}$$

The approach presented has the clear advantage that there is no longer the necessity to deal with a principal value integral evaluation. For band-limited data, which is a frequently occurring experimental circumstance, the same type of Fourier series representation can be carried out. In this case the conjugate series represents only an approximation to the Hilbert transform of the experimental data, unless the band-limited nature of the profile is exact, rather than just a limitation of the measurement process.

20.11 Fourier series approach to the reflectance

The ideas presented in Section 20.10 are now applied to treat reflectance data. From the definition in Eq. (20.1) it follows that

$$-\mathrm{i}\frac{\mathrm{d}\log\tilde{r}(\omega)}{\mathrm{d}\omega} = \frac{\mathrm{d}\theta(\omega)}{\mathrm{d}\omega} - \frac{\mathrm{i}}{2R(\omega)}\frac{\mathrm{d}R(\omega)}{\mathrm{d}\omega}. \tag{20.259}$$

In order to make a Fourier series expansion, the function must be bounded on the interval $[0,\infty)$. This requirement rules out working with $\log R(\omega)$ directly; however, the derivative function will serve the purpose. Recall from the developments of Section 20.2 that $\log\tilde{r}(\omega)$ is analytic in the upper half of the complex angular frequency plane, and hence $\mathrm{d}\log\tilde{r}(\omega)/\mathrm{d}\omega$ is analytic in the same region.

The conformal transformation

$$w = \frac{\omega_z - \mathrm{i}}{\omega_z + \mathrm{i}} \tag{20.260}$$

is employed again. From Eq. (20.259), the Laurent expansion for $-\mathrm{i}\,\mathrm{d}\log\tilde{r}(\omega_z)/\mathrm{d}\omega_z$ can be written as follows:

$$\begin{aligned} -\mathrm{i}\frac{\mathrm{d}\log\tilde{r}(\omega_z)}{\mathrm{d}\omega_z} &\to -\mathrm{i}\left[\frac{\mathrm{d}\log\tilde{r}(\omega_z)}{\mathrm{d}\omega_z}\right]_{\omega_z=-\mathrm{i}(w+1)(w-1)^{-1}} \\ &= \sum_{m=0}^{\infty} c_m w^m, \quad \text{for } |w| \le 1. \end{aligned} \tag{20.261}$$

The function

$$-\mathrm{i}\left[\frac{\mathrm{d}\log\tilde{r}(\omega_z)}{\mathrm{d}\omega_z}\right]_{\omega_z=-\mathrm{i}(w+1)(w-1)^{-1}}$$

is analytic in the unit disc. Introducing the substitution $w = \mathrm{e}^{\mathrm{i}\Omega}$ allows Eq. (20.261) to be written on the real frequency axis as follows:

$$-\mathrm{i}\left[\frac{\mathrm{d}\log\tilde{r}(\omega)}{\mathrm{d}\omega}\right]_{\omega=-\mathrm{i}(w+1)(w-1)^{-1}} = \sum_{m=0}^{\infty} c_m \cos m\Omega + \mathrm{i}\sum_{m=1}^{\infty} c_m \sin m\Omega. \tag{20.262}$$

Adopting the definitions

$$\overline{R}(\Omega) = \left[\frac{1}{2R(\omega)} \frac{\mathrm{d}R(\omega)}{\mathrm{d}\omega} \right]_{\omega=-\cot \Omega/2} \tag{20.263}$$

and

$$\overline{\theta}(\Omega) = \left[\frac{\mathrm{d}\theta(\omega)}{\mathrm{d}\omega} \right]_{\omega=-\cot \Omega/2} \tag{20.264}$$

leads to

$$-\mathrm{i} \left[\frac{\mathrm{d} \log \tilde{r}(\omega)}{\mathrm{d}\omega} \right]_{\omega=-\mathrm{i}(w+1)(w-1)^{-1}} = \overline{\theta}(\Omega) - \mathrm{i}\overline{R}(\Omega). \tag{20.265}$$

The Fourier series expansions of $\overline{R}(\Omega)$ and $\overline{\theta}(\Omega)$ in the interval $(-\pi, \pi)$ are given by

$$\overline{R}(\Omega) = \sum_{m=1}^{\infty} a_m \sin m\Omega \tag{20.266}$$

and

$$\overline{\theta}(\Omega) = \frac{1}{2} b_0 + \sum_{m=1}^{\infty} b_m \cos m\Omega. \tag{20.267}$$

The restriction to single trigonometric functions in the preceding two expansions follows from the crossing symmetry connections:

$$\overline{R}(-\Omega) = -\overline{R}(\Omega) \tag{20.268}$$

and

$$\overline{\theta}(-\Omega) = \overline{\theta}(\Omega). \tag{20.269}$$

These results can be deduced directly from Eqs. (20.32) and (20.33). On comparing Eqs. (20.262) and (20.265), and employing the series representations in Eqs. (20.266) and (20.267) yields

$$c_0 = \frac{1}{2} b_0, \quad c_m = b_m, \quad m \neq 0, \tag{20.270}$$

and

$$c_m = -a_m, \quad m \neq 0, \tag{20.271}$$

so that

$$b_m = -a_m, \quad m \neq 0. \tag{20.272}$$

The constant b_0 can be determined in the following manner. Employing the asymptotic condition

$$\lim_{\omega\to\infty} \frac{\mathrm{d}\theta(\omega)}{\mathrm{d}\omega} = 0, \tag{20.273}$$

where Eq. (20.54) has been utilized, leads to

$$\lim_{\omega\to\infty} \frac{\mathrm{d}\theta(\omega)}{\mathrm{d}\omega} = \overline{\theta}(0) = \frac{1}{2}b_0 + \sum_{m=1}^{\infty} b_m, \tag{20.274}$$

and hence

$$b_0 = -2\sum_{m=1}^{\infty} b_m. \tag{20.275}$$

This allows Eq. (20.267) to be rewritten as follows:

$$\overline{\theta}(\Omega) = \sum_{m=1}^{\infty} b_m\{\cos m\Omega - 1\}. \tag{20.276}$$

The results obtained are employed in the following fashion. Measurements of the reflectance allow the quantity $(1/2R(\omega))(\mathrm{d}R(\omega)/\mathrm{d}\omega)$ to be determined, and hence $\overline{R}(\Omega)$ can be calculated from Eq. (20.263). The Fourier coefficients a_m are determined in the standard way, and hence $\overline{\theta}(\Omega)$ is evaluated via Eq. (20.276). The derivative $\mathrm{d}\theta(\omega)/\mathrm{d}\omega$ is given by

$$\left[\frac{\mathrm{d}\theta(\omega)}{\mathrm{d}\omega}\right]_{\omega=-\cot\Omega/2} = \sum_{m=1}^{\infty} b_m \left\{2^{m-1}\cos^m\Omega - 1 + \sum_{k=1}^{[m/2]} \beta_{km}\cos^{m-2k}\Omega\right\}, \tag{20.277}$$

where

$$\beta_{km} = (-1)^k \frac{m}{k} 2^{m-2k-1} \binom{m-k-1}{k-1} \tag{20.278}$$

and $\binom{p}{q}$ denotes a binomial coefficient. Equation (20.277) can be rewritten as follows:

$$\frac{\mathrm{d}\theta(\omega')}{\mathrm{d}\omega'} = \sum_{m=1}^{\infty} b_m \left\{-1 + 2^{m-1}\left(\frac{\omega'^2-1}{\omega'^2+1}\right)^m + \sum_{k=1}^{[m/2]} \beta_{km}\left(\frac{\omega'^2-1}{\omega'^2+1}\right)^{m-2k}\right\}. \tag{20.279}$$

Integrating Eq. (20.279) with respect to ω' over the interval $(0, \omega)$, and recalling that $\theta(0) = 0$, leads to

$$\theta(\omega) = \sum_{m=1}^{\infty} b_m \left\{ -\omega + 2^{m-1} I_m(\omega) + \sum_{k=1}^{[m/2]} \beta_{km} I_{m-2k}(\omega) \right\}, \tag{20.280}$$

where

$$I_m(\omega) = \int_0^{\omega} \left(\frac{\omega'^2 - 1}{\omega'^2 + 1} \right)^m \mathrm{d}\omega'. \tag{20.281}$$

This latter integral can be expressed as follows:

$$I_m(\omega) = \omega - \alpha_m \tan^{-1} \omega - \omega \sum_{j=1}^{m-1} \frac{\alpha_{mj}}{(1+\omega^2)^j}, \quad m \geq 1, \tag{20.282}$$

and the coefficients α_m and α_{mj} are given by

$$\alpha_m = 2 \sum_{j=0}^{m-1} \binom{m}{j+1} \binom{2j}{j} \left(-\frac{1}{2} \right)^j = 2m\, {}_2F_1(1/2, 1-m; 2; 2), \quad m \geq 1, \tag{20.283}$$

and

$$\alpha_{mj} = \sum_{k=j}^{m-1} \binom{m}{k+1} (-1)^{k+j}\, 2^{2j-k} \sum_{n=j}^{k} \binom{2k}{k-n} \binom{n+j-1}{n-j} \frac{(-1)^n}{n}, \quad m \geq 1, \tag{20.284}$$

where ${}_2F_1(a, b; c; x)$ denotes a hypergeometric function.

This approach to obtaining the phase avoids the need to deal with a principal value integral evaluation. As a simple example, suppose a very simplistic model is adopted such that

$$\frac{1}{2R(\omega)} \frac{\mathrm{d}R(\omega)}{\mathrm{d}\omega} = -\frac{2\omega}{(1+\omega^2)^2}. \tag{20.285}$$

The fact that this model may not lead to the exact asymptotic behavior for $R(\omega)$ as $\omega \to \infty$, or perhaps not satisfy some of the required sum rules for the reflectance, is not a concern. This model is employed because it is sufficiently simple that it can be treated analytically. Equation (20.285) can be expressed as follows:

$$\left[\frac{1}{2R(\omega)} \frac{\mathrm{d}R(\omega)}{\mathrm{d}\omega} \right]_{\omega = -\cot \Omega/2} = \frac{1}{2} \sin \Omega - \frac{1}{4} \sin 2\Omega, \tag{20.286}$$

and hence from Eq. (20.266) the Fourier coefficients are determined by inspection to be

$$a_1 = \frac{1}{2}, \quad a_2 = -\frac{1}{4}, \quad a_m = 0, \quad m > 2. \tag{20.287}$$

Making use of Eqs. (20.272) and (20.280), with the aid of

$$I_0(\omega) = \omega, \tag{20.288}$$

$$I_1(\omega) = \omega - 2\tan^{-1}\omega, \tag{20.289}$$

and

$$I_2(\omega) = \omega - 2\tan^{-1}\omega + \frac{2\omega}{1+\omega^2}, \tag{20.290}$$

leads to the following expression for the phase:

$$\theta(\omega) = \frac{\omega}{1+\omega^2}. \tag{20.291}$$

The main computational component of this series approach is the determination of the expansion coefficients a_m in Eq. (20.266), and this can be accomplished by using standard algorithms; see, for example, Press *et al.* (1992).

20.12 Fourier and allied integral representation

In this section the allied integral representation is revisited. The real and imaginary parts of the permittivity for an isotropic medium can be written as follows:

$$\varepsilon_{\mathrm{i}}(\omega) = \frac{2}{\pi}\int_0^\infty \sin\omega t\,\mathrm{d}t \int_0^\infty \{\varepsilon_{\mathrm{r}}(\omega') - \varepsilon_0\}\cos\omega' t\,\mathrm{d}\omega' \tag{20.292}$$

and

$$\varepsilon_{\mathrm{r}}(\omega) - \varepsilon_0 = \frac{2}{\pi}\int_0^\infty \cos\omega t\,\mathrm{d}t \int_0^\infty \varepsilon_{\mathrm{i}}(\omega')\sin\omega' t\,\mathrm{d}\omega'. \tag{20.293}$$

These formulas are mathematically equivalent to the Kramers–Kronig relations given in Eqs. (19.34) and (19.35). The derivation follows directly the development given in Section 3.12. The crossing symmetry relations for the dielectric constant have been employed to restrict the integration interval to positive angular frequencies. One of the potential benefits of the allied integral form for the optical constants is the opportunity to take advantage of fast Fourier transform techniques to carry out the numerical evaluations. The discussion in this section will focus on obtaining sum rules from the allied integral equations.

Let $f(t) = \int_0^\infty \varepsilon_{\rm i}(\omega') \sin \omega' t \mathrm{d}\omega'$; then, from Eq. (20.293) it follows that

$$\begin{aligned}
\lim_{\lambda\to\infty} \int_0^\lambda \{\varepsilon_{\rm r}(\omega) - \varepsilon_0\}\mathrm{d}\omega &= \lim_{\lambda\to\infty} \frac{2}{\pi} \int_0^\lambda \mathrm{d}\omega \int_0^\infty \cos \omega t\, f(t)\mathrm{d}t \\
&= \lim_{\lambda\to\infty} \frac{2}{\pi} \int_0^\infty f(t)\mathrm{d}t \int_0^\lambda \cos \omega t \,\mathrm{d}\omega \\
&= \lim_{\lambda\to\infty} \frac{2}{\pi} \int_0^\infty \frac{\sin \lambda t\, f(t)}{t} \mathrm{d}t \\
&= \lim_{\lambda\to\infty} \frac{2}{\pi} \int_0^\infty \frac{\sin \lambda t \,\mathrm{d}t}{t} \int_0^\infty \varepsilon_{\rm i}(\omega') \sin \omega' t \,\mathrm{d}\omega' \\
&= \lim_{\lambda\to\infty} \frac{2}{\pi} \int_0^\infty \varepsilon_{\rm i}(\omega')\mathrm{d}\omega' \int_0^\infty \frac{\sin \lambda t \, \sin \omega' t \,\mathrm{d}t}{t} \\
&= 0. \qquad (20.294)
\end{aligned}$$

The final result follows from the Riemann–Lebesgue lemma, or, alternatively,

$$\begin{aligned}
\lim_{\lambda\to\infty} \int_0^\infty \frac{\sin \lambda t \, \sin \omega' t \,\mathrm{d}t}{t} &= \lim_{\lambda\to\infty} \frac{1}{2} \int_0^\infty \left\{ \frac{\cos(\lambda - \omega')t}{t} - \frac{\cos(\lambda + \omega')t}{t} \right\} \mathrm{d}t \\
&= \lim_{\lambda\to\infty} \frac{1}{2} \log \left(\frac{\lambda + \omega'}{\lambda - \omega'} \right) \\
&= 0. \qquad (20.295)
\end{aligned}$$

Hence, the following key sum rule is obtained:

$$\int_0^\infty \{\varepsilon_{\rm r}(\omega)\varepsilon_0^{-1} - 1\}\mathrm{d}\omega = 0. \qquad (20.296)$$

The reader is left to consider whether the f sum rule for $\varepsilon_{\rm i}(\omega)$ can be obtained in a similar fashion.

The corresponding allied integral formulas for the complex refractive index are given by

$$\kappa(\omega) = \frac{2}{\pi} \int_0^\infty \sin \omega t \,\mathrm{d}t \int_0^\infty \{n(\omega') - 1\} \cos \omega' t \,\mathrm{d}\omega' \qquad (20.297)$$

and

$$n(\omega) - 1 = \frac{2}{\pi} \int_0^\infty \cos \omega t \,\mathrm{d}t \int_0^\infty \kappa(\omega') \sin \omega' t \,\mathrm{d}\omega'. \qquad (20.298)$$

Using the same procedure described for the permittivity leads to the following result:

$$\int_0^\infty \{n(\omega) - 1\}\mathrm{d}\omega = 0. \qquad (20.299)$$

20.13 Integral inequalities

Integral inequalities for the optical constants have been less extensively employed for data analysis. In this section a few inequalities are considered that employ sum rules derived from previously developed dispersion relations. The principal idea is to find inequalities that might serve as suitable tests for experimentally determined data. The most desirable inequalities would be those that take the following form:

$$B(\omega_0, \omega_1, \omega_2) \leq \int_{\omega_1}^{\omega_2} w(\omega, \omega_0)\mathcal{O}(\omega)\mathrm{d}\omega \leq A(\omega_0, \omega_1, \omega_2), \tag{20.300}$$

where $\mathcal{O}(\omega)$ denotes a particular optical constant, $w(\omega, \omega_0)$ is a weight function depending on the angular frequency and perhaps some additional reference frequency ω_0, and $A(\omega_0, \omega_1, \omega_2)$ and $B(\omega_0, \omega_1, \omega_2)$ are bounds that can be readily determined. The spectral interval (ω_1, ω_2) is ideally not limited to the infinite interval. Additionally, it is desirable that any determined inequality of the form given in Eq. (20.300) is sharp. That is, the constant $A(\omega_0, \omega_1, \omega_2)$ gives, for a specified $\mathcal{O}(\omega), w(\omega, \omega_0)$, and interval (ω_1, ω_2), the lowest upper bound for the integral, and no other constant can be found for which a lower upper bound is possible. Similarly, the ideal constant $B(\omega_0, \omega_1, \omega_2)$ is the maximum lower bound for the stated $\mathcal{O}(\omega), w(\omega, \omega_0)$, and interval (ω_1, ω_2). Generally speaking, better bounds are more likely when the functions $w(\omega, \omega_0), \mathcal{O}(\omega)$, and the interval (ω_1, ω_2) are more restricted, rather than for cases where a very wide range of optical properties and weight functions is permitted. When $A(\omega_0, \omega_1, \omega_2)$ and $B(\omega_0, \omega_1, \omega_2)$ are very close to one another, then a useful inequality has been obtained. While knowing both $A(\omega_0, \omega_1, \omega_2)$ and $B(\omega_0, \omega_1, \omega_2)$ is the ideal situation, determination of integral inequalities where only tight upper bounds or tight lower bounds are known would serve as acceptable substitutes for Eq. (20.300).

The dispersion relations and most of the sum rules that have developed cover the infinite spectral range. A key question is whether or not it is possible to find inequalities based on weight functions that decay rapidly as $\omega \to \infty$, thereby minimizing errors associated with either missing data at high frequencies or errors associated with the extrapolation of data into this region. As an example, consider the following inequality:

$$\int_0^\infty \surd(\omega\kappa(\omega))\, \mathrm{e}^{-(\pi^3/128)(\omega^2/\omega_0^2)}\, \mathrm{d}\omega \leq \surd(\omega_0)\, \omega_\mathrm{p}. \tag{20.301}$$

In this inequality, ω_0 plays the role of an arbitrary angular frequency, and the Gaussian weight function damps out the contribution arising from $\surd(\omega\kappa(\omega))$ for large angular frequencies. A related inequality involving the refractive index can be written as follows:

$$\int_0^\infty \surd(\omega\kappa(\omega)n(\omega))\, \mathrm{e}^{-(\pi^3/128)(\omega^2/\omega_0^2)}\, \mathrm{d}\omega \leq \surd(\omega_0)\, \omega_\mathrm{p}. \tag{20.302}$$

The derivation of these two inequalities is straightforward. From the Cauchy–Schwarz–Buniakowski inequality,

$$\int_0^\infty \sqrt{(\omega\kappa(\omega))}\,\mathrm{e}^{-(\pi^3/128)(\omega^2/\omega_0^2)}\,\mathrm{d}\omega \le \left\{\int_0^\infty \omega\kappa(\omega)\,\mathrm{d}\omega \int_0^\infty \mathrm{e}^{-(\pi^3/64)(\omega^2/\omega_0^2)}\,\mathrm{d}\omega\right\}^{1/2}. \tag{20.303}$$

Employing the f sum rule, Eq. (19.242), and using

$$\int_0^\infty \mathrm{e}^{-a^2\omega^2}\,\mathrm{d}\omega = \frac{\sqrt{\pi}}{2a}, \quad \text{for } a > 0, \tag{20.304}$$

with $a = \pi^{3/2}(8\omega_0)^{-1}$, leads to Eq. (20.301). Making use of the sum rule given as entry (4) in Table 19.2, and following a similar derivation, leads to Eq. (20.302).

Suppose it is known that the optical constant of interest satisfies, on the angular frequency interval $[0, \infty)$, a bound of the following form:

$$0 \le \mathcal{O}(\omega) \le \mathcal{A}. \tag{20.305}$$

Let $w(\omega)$ denote a weight function which is decreasing on the same frequency interval; then it follows that

$$0 \le [\mathcal{A} - \mathcal{O}(\omega)][w(\omega) - w(\omega_0)], \quad \text{for } \omega \in [0, \omega_0], \tag{20.306}$$

and hence

$$\int_0^{\omega_0} [\mathcal{A} - \mathcal{O}(\omega)][w(\omega) - w(\omega_0)]\mathrm{d}\omega + \int_{\omega_0}^\infty \mathcal{O}(\omega)[w(\omega_0) - w(\omega)]\mathrm{d}\omega \ge 0. \tag{20.307}$$

Rearranging gives

$$F(\omega_0) \equiv \mathcal{A}\int_0^{\omega_0} w(\omega)\mathrm{d}\omega + w(\omega_0)\left[\int_0^\infty \mathcal{O}(\omega)\mathrm{d}\omega - \mathcal{A}\omega_0\right] \ge \int_0^\infty \mathcal{O}(\omega)w(\omega)\mathrm{d}\omega. \tag{20.308}$$

The optimal upper bound can be determined from the minimum of $F(\omega_0)$. So

$$\frac{\partial F(\omega_0)}{\partial\omega_0} = \frac{\partial w(\omega_0)}{\partial\omega_0}\left[\int_0^\infty \mathcal{O}(\omega)\mathrm{d}\omega - \mathcal{A}\omega_0\right] = 0, \tag{20.309}$$

and hence

$$\omega_0 = \frac{1}{\mathcal{A}}\int_0^\infty \mathcal{O}(\omega)\mathrm{d}\omega. \tag{20.310}$$

The second derivative satisfies $\partial^2 F(\omega_0)/\partial\omega_0^2 > 0$. Equation (20.308) can be written as follows:

$$\int_0^\infty \mathcal{O}(\omega)\, w(\omega)\mathrm{d}\omega \leq \mathcal{A} \int_0^{\omega_0} w(\omega)\mathrm{d}\omega, \tag{20.311}$$

with ω_0 given by Eq. (20.310). As an example, consider the case that $\mathcal{O}(\omega)$ is the reflectivity. In this case the upper bound of $R(\omega)$ is $\mathcal{A} = 1$. Suppose the following weight function is selected:

$$w(\omega) = \mathrm{e}^{-\omega/\omega_\mathrm{a}}, \tag{20.312}$$

where ω_a is an arbitrary angular frequency; then,

$$\omega_\mathrm{a}^{-1} \int_0^\infty R(\omega)\mathrm{e}^{-\omega/\omega_\mathrm{a}}\, \mathrm{d}\omega \leq 1 - \mathrm{e}^{-\omega_0/\omega_\mathrm{a}}, \tag{20.313}$$

with

$$\omega_0 = \int_0^\infty R(\omega)\mathrm{d}\omega. \tag{20.314}$$

Equation (20.313) is an improvement on the obvious inequality

$$\omega_\mathrm{a}^{-1} \int_0^\infty R(\omega)\mathrm{e}^{-\omega/\omega_\mathrm{a}}\, \mathrm{d}\omega \leq 1. \tag{20.315}$$

As a second example, suppose $\mathcal{O}(\omega) = \omega^{-1}\kappa(\omega)$, and denote the maximum of this function by $\mathcal{A}$. Let

$$w(\omega) = \mathrm{e}^{-\omega^2/\omega_\mathrm{a}^2}; \tag{20.316}$$

then,

$$\int_0^\infty \frac{\kappa(\omega)\mathrm{e}^{-\omega^2/\omega_\mathrm{a}^2}\, \mathrm{d}\omega}{\omega} \leq \omega_\mathrm{a}\mathcal{A} \int_0^{\omega_0/\omega_\mathrm{a}} \mathrm{e}^{-x^2}\, \mathrm{d}x, \tag{20.317}$$

and hence

$$\int_0^\infty \omega^{-1}\kappa(\omega)\mathrm{e}^{-\omega^2/\omega_\mathrm{a}^2}\, \mathrm{d}\omega \leq \frac{1}{2}\surd(\pi)\, \omega_\mathrm{a}\mathcal{A}\, \mathrm{erf}\left(\frac{[\pi(n(0)-1)]}{2\mathcal{A}\omega_\mathrm{a}}\right). \tag{20.318}$$

Note that this inequality is restricted to insulators, because the behavior $\kappa(\omega) \approx 1/\sqrt{\omega}$ for a conductor as $\omega \to 0$ would lead to a divergent integral in Eq. (20.318). This inequality would be harder to apply in practice because the maximum value $\mathcal{A}$ must be determined experimentally.

Notes

§20.2 For a discussion of the dispersion relations for the phase and reflectance the following references can be consulted: Toll (1956), Jahoda (1957), Gottlieb (1960), Velický (1961a), Stern (1963), Greenaway and Harbeke (1968), Ahrenkiel (1971), Jones and March (1973), Goedecke (1975a,b), Kröger (1975), Young (1976), Smith (1977), King (1979), Nash, Bell and Alexander (1995), Lee and Sindoni (1997), and Peiponen and Vartiainen (2006). Some of the historical papers on this topic are by Robinson (1952), Robinson and Price (1953), and Roessler (1965a,b, 1966). For some representative applications of the Kramers–Kronig relations to the determination of optical constants via reflectivity measurements, see the following: Rimmer and Dexter (1960), Velický (1961b), Philipp and Taft (1959, 1964), Bowlden and Wilmshurst (1963), Plaskett and Schatz (1963), Schatz, Maeda, and Kozima (1963), Philipp and Ehrenreich (1964), Andermann, Caron, and Dows (1965), Kozima *et al.* (1966), Andermann and Dows (1967), Berreman (1967), Andermann, Wu, and Duesler (1968), Verleur (1968), Wu and Andermann (1968), Scouler (1969), Neufeld and Andermann (1972), Klucker and Nielsen (1973), Veal and Paulikas (1974), Balzarotti *et al.* (1975), Chambers (1975), Leveque (1977), Rasigni and Rasigni (1977), Shiles *et al.* (1980), Bortz and French (1989), Grosse and Offermann (1991), Huang and Urban (1992), Tickanen, Tejedor–Tejedor, and Anderson (1992, 1997), Vartiainen, Peiponen, and Asakura (1993a), Bertie and Lan (1996), Peiponen, Vartiainen, and Asakura (1996, 1997b), Yamamoto and Ishida (1997), Palmer, Williams, and Budde (1998), and Räty, Peiponen, and Asakura (2004). For a cautionary note on some commercial software packages, see Lichvár, Liška, and Galusek (2002). For an application to thin surface films, see Plieth and Naegele (1975). For an application of subtractive Kramers–Kronig relations to deal with phase errors in terahertz reflection spectroscopy, see Lucarini *et al.* (2005), Peiponen *et al.* (2005), and Gornov *et al.* (2006). Blaschke factors are discussed further in Colwell (1985).

§20.3 Additional reading on the derivation of sum rules for the normal reflectance and related quantities can be found in Furuya, Villani, and Zimerman (1977), Inagaki *et al.* (1978), Inagaki, Kuwata, and Ueda (1979, 1980), King (1979), Smith and Manogue (1981), Kimel (1982b), and Smith (1985).

§20.4 For additional reading, see Tinkham (1956), Glover and Tinkham (1957), Tinkham and Ferrell (1959), Martin (1967), and Furuya and Zimerman (1976). For an application of the Kramers–Kronig relations to the determination of sum rules for the self-energy in alloys, see Scarfone and Chlipala (1975).

§20.5 Further discussion on the energy loss function can be found in Stern (1963), Nozières and Pines (1958, 1959), Kittel (1976), and Jackson (1999). The sum rule, Eq. (20.102), has been known for many years; see, for example, Fano (1956). For additional discussion on sum rules, see Stroud (1979).

§20.6 Gorter and Kronig (1936) appear to be the first authors to point out that the Kramers–Kronig relations also apply to the permeability. See also the work of Gorter (1938). For detailed derivations of the solutions of the Bloch equations, see Slichter (1963) and Carrington and McLachlan (1967), or Abragam (1961). An application of the Kramers–Kronig relations to the magnetic susceptibility of a colloidal suspension of magnetic particles can be found in Fannin, Molina, and Charles (1993).

§20.7 Additional reading on the surface impedance function can be found in Ginzburg (1956), Landau, Lifshitz, and Pitaevskiĭ (1984), and Jackson (1999).

§20.8 A concise discussion on the treatment of electromagnetic waves in anisotropic media can be found in Landau *et al.* (1984, chap. 11).

§20.9 A very useful starting source for further reading is Nozières and Pines (1999). Additional references that could be consulted are Leontovich (1961), Martin (1967), Melrose and Stoneham (1977), Dolgov, Kirzhnits, and Maksimov (1981), Mezincescu (1985), Musienko, Rudakov, and Solov'ev (1989), and, for a discussion on various sum rules including spatial dispersion, Altarelli *et al.* (1972). Mezincescu (1985) considers the derivation of dispersion relations for the situation where $\boldsymbol{\varepsilon}(\mathbf{k}, 0) < \mathbf{0}$, and Dolgov *et al.* (1981) review work on systems where this condition may apply.

§20.10 For further reading on the approach of this section, see Johnson (1975) and King (1978a) .

§20.11 The approach of this section can be found in King (1977).

§20.12 Early reference to this approach can be found in a brief abstract of Cole (1941), Gross (1941), and Kronig (1942). Additional discussion can be found in Peterson and Knight (1973) and King (1978a). For an application to thin-film spectroscopy and a comparison with the Kramers–Kronig approach, see Roth, Rao, and Dignam (1975).

§20.13 For some additional reading, see King (1981, 1982). Another approach to obtaining inequalities in a fairly straightforward fashion is given by Šachl (1963). Some techniques developed to derive inequalities for dispersion relations in elementary particle physics, for example Okubo (1974), have often not been fully exploited in optical data analysis.

Exercises

20.1 Let

$$f(x) = -\frac{x}{x^2+1} + \mathrm{i}\frac{x^2+2}{x^2+1}.$$

By considering the integral $\oint_C f(z)(\omega - z)^{-1}\,\mathrm{d}z$, where ω is real, and the contour is shown in Figure 19.2, what, if any, Hilbert transform relationships can be determined?

20.2 If the crossing symmetry relation $\tilde{r}(-\omega_z^*) = \tilde{r}(\omega_z)^*$ is to hold when a Blaschke factor is introduced, show how the individual terms of this factor must be constrained.

20.3 What (if any) sum rules can be derived by applying the Liu–Okubo technique to the function $\tilde{r}(\omega_z)$?

20.4 This question is a follow up to Exercise 19.7. How would the Kramers–Kronig relations for the reflectance be altered if a non-conducting material had a non-zero value for $\kappa(0)$?

20.5 Does examination of the limit $\omega \to \infty$ in Eq. (20.90) yield a sum rule? Explain your thinking.

20.6 Determine the sum rules that are obtained by examining $\oint_C \omega_z^m \{\sigma(\omega_z) + \mathrm{i}\omega_z\}^p \, \mathrm{d}\omega_z$, where C is a suitably chosen contour. Assume m and p are integer values and find the necessary constraints on these parameters.

20.7 Using the definition for the conductivity given in Eq. (20.85), derive sum rules for the moments of the conductivity, starting from the contour integral $\oint_C \omega_z^m \sigma(\omega_z)^p \, \mathrm{d}\omega_z$, where m and p are integer values. Determine the necessary constraints on these parameters.

20.8 Complete the details for the derivation of the sum rules for the energy loss function given in Eqs. (20.102) and (20.103).

20.9 How many independent elements are necessary to characterize the susceptibility tensor for crystals belonging to the following systems: (i) cubic, (ii) hexagonal, (iii) rhombohedral, (iv) tetragonal, (v) orthorhombic, (vi) monoclinic, and (vii) triclinic?

20.10 Suppose that $\boldsymbol{\varepsilon}(\mathbf{k}, \omega)$ has a pole at some complex frequency $\mathrm{i}\omega_0$ for ω_0 real and satisfying $\omega_0 > 0$. By making an appropriate subtraction of this pole behavior from $\boldsymbol{\varepsilon}(\mathbf{k}, \omega)$, derive the appropriate dispersion relations for this case.

20.11 Investigate the limits $\omega \to 0$ and $\omega \to \infty$ for $\theta(\omega)$ starting from Eq. (20.280).

20.12 Prove that Eq. (20.281) reduces to Eq. (20.282).

20.13 Determine a bound for the integral $\int_0^\infty (\kappa(\omega)/\sqrt{\omega}) \mathrm{d}\omega$. Does it apply to insulators or to both conductors and insulators?

20.14 Determine a bound for $\int_0^\infty R(\omega) \mathrm{e}^{-\omega^2/\omega_\mathrm{a}^2} \, \mathrm{d}\omega$, where ω_a is an arbitrary angular frequency.

20.15 Suppose the function f satisfies the following properties: (i) $f(\omega_z)$ is analytic in the upper half of the complex angular frequency plane, (ii) $f(\omega) = f(-\omega)^*$, (iii) $f(\omega_z) \to 0$ as $\omega_z \to \infty$. Determine $f(\mathrm{i}\omega)$, where ω is real. Hence, or otherwise, determine $\varepsilon_0 \varepsilon(\mathrm{i}\omega)^{-1} - 1$.

21

Dispersion relations for magneto-optical and natural optical activity

21.1 Introduction

The focus of this chapter is the derivation of dispersion relations arising in magneto-optical applications and for natural optical activity. A further topic considered is the development of sum rules for some of the quantities that are experimentally accessible. These include the optical rotatory dispersion (commonly abbreviated as ORD in the literature), circular dichroism (CD), the magnetic analogs, the Faraday dispersion, and magnetic circular dichroism (MCD). These terms will be defined shortly.

A number of the ideas that are employed are very similar to those developed in the preceding two chapters. There are, however, some major differences that arise for the cases of magneto-optical and natural optical activity. The dispersion relations for magneto-optical and natural optical activity have a different structure compared with what was found for the typical case of other optical properties discussed in Chapters 19 and 20. The difference arises primarily from the form of the crossing symmetry relations for the refractive indices and absorption coefficients for left and right circularly polarized light. The reason for the absence of simple dispersion relations for the individual modes corresponding to the refractive indices for left and right polarization, and the corresponding absorption coefficients for right and left circularly polarized light, is also discussed.

The Faraday effect describes the rotation of the plane of polarization of a beam of light as it traverses a medium under the influence of an imposed magnetic field in the direction of the beam. The rotation of the plane of polarization describes the situation of optical birefringence or double refraction, and in this situation it is referred to as circular birefringence. Linear birefringence can be induced by applying a static electric or magnetic field to a medium perpendicular to the direction of the electromagnetic beam. The former gives the *Kerr effect* and the latter is called the *Cotton–Mouton effect*.

The Faraday effect applies to all materials, irrespective of any particular symmetry the medium may have. Magnetic circular dichroism describes the difference in absorption of left and right circularly polarized beams as they traverse a medium acted upon by an external magnetic field parallel to the direction of propagation of the beam. Dispersion relations offer the potential to determine the magnetic circular dichroism from the Faraday dispersion, and conversely. In addition, the sum rules that are derived from the dispersion relations represent important constraints that the experimental data must satisfy for both the Faraday rotation and the magnetic circular dichroism.

For substances that have low intrinsic symmetry, the specific details of which are given later, the plane of polarization of the electromagnetic beam can be rotated as it propagates through the material. This occurs in the absence of any external applied magnetic field, and the phenomenon is termed optical rotatory dispersion. The frequency dependence of this rotation is linked by a dispersion relation to the corresponding difference of the absorption coefficients for left and right circularly polarized beams, and vice versa. This difference in absorption coefficients for left and right circularly polarized modes is called circular dichroism. A medium which exhibits optical rotatory dispersion or circular dichroism is called optically active. The term gyrotropy is used to denote the same phenomenon. The qualifier natural is often inserted, as in natural optical activity, to distinguish this from its magnetic analog. In applications, the experimental determination of one of these quantities allows the other one to be determined, provided the quantity measured is determined over a sufficiently wide spectral range. In some areas of modern chemistry, for example in natural product chemistry, organic species in solution can be characterized in part by their optical rotatory dispersion and circular dichroism curves. These techniques thus serve as "fingerprint" tools, which help in identifying or characterizing unknown molecules. Sum rules for natural optical activity provide useful checks on the quality of experimental data, when the measurements are performed over sufficiently wide spectral intervals.

21.2 Circular polarization

A short review on circular polarization modes for the electromagnetic field is discussed first. In the preceding two chapters, consideration has been restricted to linearly polarized light. This assumption is now dropped. Suppose $\boldsymbol{\epsilon}_1, \boldsymbol{\epsilon}_2$, and $\mathbf{n}$ are unit vectors along the axes of a right-handed axis system. For a wave traveling in the direction $\mathbf{n}$, the electric field vector $\mathbf{E}$ can be written in the following form:

$$\mathbf{E} = \boldsymbol{\epsilon}_1 E_1 + \boldsymbol{\epsilon}_2 E_2. \tag{21.1}$$

The quantities $\boldsymbol{\epsilon}_1$ and $\boldsymbol{\epsilon}_2$ are termed the polarization vectors, and E_1 and E_2 will in general be complex. Let

$$\tau = k\mathbf{n} \cdot \mathbf{r} - \omega t, \tag{21.2}$$

then E_1 and E_2 can be written as follows:

$$E_1 = a_1 \cos(\tau + \varphi_1), \quad a_1 > 0, \tag{21.3}$$

and

$$E_2 = a_2 \cos(\tau + \varphi_2), \quad a_2 > 0, \tag{21.4}$$

where φ_1 and φ_2 are phase factors. If the phase factors of E_1 and E_2 are the same, $\varphi_1 = \varphi_2 = \varphi$, then

$$\frac{E_1}{a_1} = \frac{E_2}{a_2}, \tag{21.5}$$

and hence

$$\mathbf{E} = \left(\boldsymbol{\epsilon}_1 + \frac{a_2}{a_1}\boldsymbol{\epsilon}_2\right) E_1. \tag{21.6}$$

This expression can be written as

$$\mathbf{E} = \boldsymbol{\epsilon} E_1, \tag{21.7}$$

where the wave is linearly polarized with polarization vector $\boldsymbol{\epsilon}$. If the phases φ_1 and φ_2 are different, then, from Eqs. (21.3) and (21.4), it follows that

$$\frac{E_1^2}{a_1^2} + \frac{E_2^2}{a_2^2} - 2\frac{E_1 E_2}{a_1 a_2}\cos\varphi = \sin^2\varphi, \tag{21.8}$$

where $\varphi = \varphi_2 - \varphi_1$. The case of linear polarization can be recovered by setting $\varphi = m\pi$, where $m \in \mathbb{Z}$. Introducing the simplifying assumptions that $a_1 = a_2 = a$ and $\varphi = \varphi_2 - \varphi_1 = (1/2)m\pi$, for $m = \pm 1, \pm 3, \pm 5 \ldots$, then Eq. (21.8) reduces to

$$E_1^2 + E_2^2 = a^2, \tag{21.9}$$

which corresponds to circular polarization. The polarization is called right-handed when $\sin\varphi > 0$, which means $\varphi = (1/2)\pi + 2m\pi$, with $m = 0, \pm 1, \pm 2, \ldots$ The polarization is termed left-handed when $\sin\varphi < 0$, and hence $\varphi = -(1/2)\pi + 2m\pi$, with $m = 0, \pm 1, \pm 2, \ldots$ Adopting a complex representation, and employing, for circular polarization, $a_1 = a_2$,

$$\frac{E_2}{E_1} = \frac{a_2 \mathrm{e}^{\mathrm{i}(\tau + \varphi_1 + \pi/2)}}{a_1 \mathrm{e}^{\mathrm{i}(\tau + \varphi_1)}} = \mathrm{i}, \tag{21.10}$$

and hence Eq. (21.1) can be written as follows:

$$\mathbf{E}_+ = (\boldsymbol{\epsilon}_1 + \mathrm{i}\boldsymbol{\epsilon}_2) E_1. \tag{21.11}$$

The other situation is given by

$$\frac{E_2}{E_1} = \frac{a_2 \mathrm{e}^{\mathrm{i}(\tau + \varphi_1 - \pi/2)}}{a_1 \mathrm{e}^{\mathrm{i}(\tau + \varphi_1)}} = -\mathrm{i}, \tag{21.12}$$

and so

$$\mathbf{E}_- = (\boldsymbol{\epsilon}_1 - \mathrm{i}\boldsymbol{\epsilon}_2)E_1. \tag{21.13}$$

It is common practice to introduce the orthogonal unit vector basis as follows:

$$\boldsymbol{\epsilon}_\pm = \frac{1}{\sqrt{2}}(\boldsymbol{\epsilon}_1 \pm \mathrm{i}\boldsymbol{\epsilon}_2); \tag{21.14}$$

the corresponding fields are thus given by

$$\mathbf{E}_\pm = \boldsymbol{\epsilon}_\pm E_1. \tag{21.15}$$

When the observer views the *oncoming* beam, the rotation of the electric field vector is *counter-clockwise* for $\boldsymbol{\epsilon}_+$, and this is called a left circularly polarized electric wave. For the case of $\boldsymbol{\epsilon}_-$, again observing the oncoming beam, the rotation of the electric vector is *clockwise*, and this is termed a right circularly polarized electric wave. Hence,

$$\mathbf{E}_\mathbf{L} = (\boldsymbol{\epsilon}_1 + \mathrm{i}\boldsymbol{\epsilon}_2)E_1 \tag{21.16}$$

and

$$\mathbf{E}_\mathbf{R} = (\boldsymbol{\epsilon}_1 - \mathrm{i}\boldsymbol{\epsilon}_2)E_1, \tag{21.17}$$

for the left and right circularly polarized electric fields. This is the convention commonly employed in optics (sometimes referred to as the optical convention). The reader needs to be alert that the opposite sign convention, that is associating $\mathbf{E}_\mathbf{R}$ with $\boldsymbol{\epsilon}_+$ and $\mathbf{E}_\mathbf{L}$ with $\boldsymbol{\epsilon}_-$, is also in use.

21.3 The complex refractive indices N_+ and N_-

It has been previously indicated in Section 20.8 that the elements of the dielectric tensor are symmetric, and in the principal axis system the tensor can be brought into diagonal form. In the presence of an external magnetic field, the dielectric tensor is no longer symmetric. Faraday made the first reported observation that the plane of polarization of a linearly polarized beam of electromagnetic radiation is rotated when passing through an optically inactive medium, in the presence of a magnetic field applied parallel to the direction of propagation of the beam. This observation is usually called the Faraday effect, but it is also referred to as magneto-optical rotation (or magnetic optical rotation, MOR.) When the plane of polarization is monitored as a function of the frequency of the electromagnetic radiation, the terminology employed is magnetic rotation spectra (MRS). The angle of rotation, ϕ, is proportional to the strength of the applied magnetic field H_z, which is applied in the z-direction, and to the thickness of the material, l, so that

$$\phi = VlH_z, \tag{21.18}$$

where V denotes the proportionality factor, which is called the Verdet constant. The angle of rotation can be directly related to the refractive indices of the two circularly polarized modes, which propagate through the material with different velocities. If the beam is also attenuated due to absorption processes as it transverses the medium, this is the phenomenon of magnetic circular dichroism.

A system having a C_n axis of symmetry is invariant under a rotation by 360°/n. For a system having at least a C_3 axis of symmetry, which means that the system is invariant under a rotation by 120°, then, on suppressing the angular frequency dependence, the dielectric tensor can be written in the following form:

$$\boldsymbol{\varepsilon} = \begin{pmatrix} \varepsilon_{xx} & \varepsilon_{xy} & 0 \\ -\varepsilon_{xy} & \varepsilon_{xx} & 0 \\ 0 & 0 & \varepsilon_{zz} \end{pmatrix}. \tag{21.19}$$

To see how this form for the dielectric tensor emerges from symmetry considerations alone, consider a rotation of the coordinate system by the angle θ about the z-axis. The matrix representation of this operation is given by

$$T = \begin{pmatrix} \cos\theta & \sin\theta & 0 \\ -\sin\theta & \cos\theta & 0 \\ 0 & 0 & 1 \end{pmatrix}. \tag{21.20}$$

The inverse transformation can be written as

$$T^{-1} = \begin{pmatrix} \cos\theta & -\sin\theta & 0 \\ \sin\theta & \cos\theta & 0 \\ 0 & 0 & 1 \end{pmatrix}. \tag{21.21}$$

If the dielectric tensor is invariant under a rotation by the angle θ, then

$$\boldsymbol{\varepsilon} = T\boldsymbol{\varepsilon}T^{-1}. \tag{21.22}$$

The following conditions on the elements of the dielectric tensor can be extracted from the preceding result:

$$(\varepsilon_{xx} - \varepsilon_{yy}) \sin^2\theta - (\varepsilon_{xy} + \varepsilon_{yx}) \sin\theta \cos\theta = 0, \tag{21.23}$$

$$(\varepsilon_{xx} - \varepsilon_{yy}) \sin\theta \cos\theta + (\varepsilon_{xy} + \varepsilon_{yx}) \sin^2\theta = 0, \tag{21.24}$$

$$\varepsilon_{xz}(\cos\theta - 1) + \varepsilon_{yz} \sin\theta = 0, \tag{21.25}$$

$$\varepsilon_{yz}(\cos\theta - 1) - \varepsilon_{xz} \sin\theta = 0, \tag{21.26}$$

$$\varepsilon_{zx}(\cos\theta - 1) + \varepsilon_{zy} \sin\theta = 0, \tag{21.27}$$

and

$$\varepsilon_{zy}(\cos\theta - 1) - \varepsilon_{zx} \sin\theta = 0. \tag{21.28}$$

These equations are satisfied in different cases either by a particular choice of the angle θ or by constraints on the elements of the dielectric tensor. For the case of a C_1 symmetry operation (system invariant under rotation about the z-axis by 2π), the preceding set of six equations is satisfied by inspection, without any constraints on the elements of the dielectric tensor. For a system with C_2 symmetry (invariant under rotation by π about the z-axis), then the dielectric tensor takes the following form:

$$\boldsymbol{\varepsilon} = \begin{pmatrix} \varepsilon_{xx} & \varepsilon_{xy} & 0 \\ \varepsilon_{yx} & \varepsilon_{yy} & 0 \\ 0 & 0 & \varepsilon_{zz} \end{pmatrix}. \tag{21.29}$$

For a system with C_3 symmetry about the z-axis, Eqs. (21.23)–(21.28) yield

$$\boldsymbol{\varepsilon} = \begin{pmatrix} \varepsilon_{xx} & \varepsilon_{xy} & 0 \\ -\varepsilon_{xy} & \varepsilon_{xx} & 0 \\ 0 & 0 & \varepsilon_{zz} \end{pmatrix}. \tag{21.30}$$

A system with C_n symmetry about the z-axis, that is a rotation by $2\pi/n$ leaves the system invariant, yields a dielectric tensor of the form given in Eq. (21.30) if $n \geq 3$. In what follows, the form indicated by Eq. (21.30) is selected to represent the dielectric tensor, since systems with high symmetry are quite prevalent.

On taking the curl of Eq. (20.135) and employing Eqs. (20.139) and (20.161), it follows that

$$\nabla^2\mathbf{E} - \frac{1}{c^2}\frac{\mu\boldsymbol{\varepsilon}\cdot}{\mu_0\varepsilon_0}\frac{\partial^2\mathbf{E}}{\partial t^2} = \mathbf{0}, \tag{21.31}$$

where it has been assumed that the components of the field, $\boldsymbol{\epsilon}_1$ and $\boldsymbol{\epsilon}_2$, are oriented along the x- and y-directions, and that the propagation direction is along the z-axis. The product of the tensor $\boldsymbol{\varepsilon}$ with the vector $\mathbf{E}$ is commonly written as $\boldsymbol{\varepsilon}\cdot\mathbf{E}$, which is sometimes abbreviated as $\boldsymbol{\varepsilon}\mathbf{E}$ when there is no risk of confusion as to what is implied.

The dot product $\mathbf{E} \cdot \mathbf{E}$ when it arises is not abbreviated as $\mathbf{EE}$, since the latter denotes a dyad. The material is assumed to be non-magnetic, so that $\mu\mu_0^{-1} = 1$. The electric fields are written as follows:

$$\mathbf{E}_+(\mathbf{r}, t) = E_0(\mathbf{i} + \mathrm{i}\,\mathbf{j})\mathrm{e}^{\mathrm{i}(\mathbf{k}_+ \cdot \mathbf{r} - \omega t)} \tag{21.32}$$

and

$$\mathbf{E}_-(\mathbf{r}, t) = E_0(\mathbf{i} - \mathrm{i}\,\mathbf{j})\mathrm{e}^{\mathrm{i}(\mathbf{k}_- \cdot \mathbf{r} - \omega t)}, \tag{21.33}$$

where $\mathbf{i}$ and $\mathbf{j}$ are unit vectors along the x- and y-directions, respectively.

Employing Eq. (21.30), the components of the electric displacement can be written in the following form:

$$D_x = \varepsilon_{xx}E_x + \varepsilon_{xy}E_y, \tag{21.34}$$

$$D_y = \varepsilon_{xx}E_y - \varepsilon_{xy}E_x, \tag{21.35}$$

and

$$D_z = \varepsilon_{zz}E_z. \tag{21.36}$$

From Eqs. (21.34) and (21.35) it follows that

$$D_x \pm \mathrm{i}D_y = (\varepsilon_{xx} \mp \mathrm{i}\varepsilon_{xy})(E_x \pm \mathrm{i}E_y), \tag{21.37}$$

which can be written in the following more compact form:

$$D_\pm = \varepsilon_\pm E_\pm. \tag{21.38}$$

From Eqs. (21.31), (21.32), and (21.33), it follows that

$$k_+^2\, E_0\mathrm{e}^{\mathrm{i}(\mathbf{k}_+ \cdot \mathbf{r} - \omega t)} = \frac{\omega^2}{c^2\varepsilon_0} E_0(\varepsilon_{xx} + \mathrm{i}\varepsilon_{xy})\mathrm{e}^{\mathrm{i}(\mathbf{k}_+ \cdot \mathbf{r} - \omega t)} \tag{21.39}$$

and

$$k_-^2\, E_0\mathrm{e}^{\mathrm{i}(\mathbf{k}_- \cdot \mathbf{r} - \omega t)} = \frac{\omega^2}{c^2\varepsilon_0} E_0(\varepsilon_{xx} - \mathrm{i}\varepsilon_{xy})\mathrm{e}^{\mathrm{i}(\mathbf{k}_- \cdot \mathbf{r} - \omega t)}. \tag{21.40}$$

These equations simplify to

$$k_+^2 = \frac{\omega^2}{c^2\varepsilon_0}(\varepsilon_{xx} + \mathrm{i}\varepsilon_{xy}) \tag{21.41}$$

and

$$k_-^2 = \frac{\omega^2}{c^2\varepsilon_0}(\varepsilon_{xx} - \mathrm{i}\varepsilon_{xy}). \tag{21.42}$$

For real ω, it follows from the preceding two results and Eq. (20.167) that

$$k_+^2(-\omega) = \{k_-^2(\omega)\}^* \tag{21.43}$$

and

$$k_-^2(-\omega) = \{k_+^2(\omega)\}^*. \tag{21.44}$$

The reader will notice immediately that these results are fundamentally different from the previous types of crossing symmetry relations encountered (see, for example, Eqs. (19.52) and (19.53)). The polarization modes switch over in these crossing symmetry relations. This has an important bearing on the form of the dispersion relations that can be derived. On taking the positive square root in Eqs. (21.43) and (21.44), it follows that

$$k_+(-\omega) = k_-(\omega)^* \tag{21.45}$$

and

$$k_-(-\omega) = k_+(\omega)^*. \tag{21.46}$$

Extending Eqs. (19.87) and (19.90), and assuming a non-magnetic medium, yields

$$N_\pm^2(\omega) = \frac{c^2 k_\pm^2(\omega)}{\omega^2} = \varepsilon_\pm \varepsilon_0^{-1} = (\varepsilon_{xx} \mp \mathrm{i}\varepsilon_{xy})\varepsilon_0^{-1}, \tag{21.47}$$

and then the following crossing symmetry relations are obtained:

$$N_+(-\omega) = N_-(\omega)^* \tag{21.48}$$

and

$$N_-(-\omega) = N_+(\omega)^*. \tag{21.49}$$

These relations will be employed to derive dispersion relations for the magneto-optical activity.

To derive dispersion relations, information is required about the asymptotic behavior of $N_+(\omega)$ and $N_-(\omega)$. The model developed in Section 19.2 (see Eq. (19.14)) can be modified to achieve this goal. The Lorentz force for a charge q in an electric field $\mathbf{E}$ and subjected to a magnetic induction $\mathbf{B}$ is given by

$$\mathbf{F} = q(\mathbf{E} + \mathbf{v} \times \mathbf{B}), \tag{21.50}$$

where $\mathbf{v}$ is the particle velocity. The equation of motion of an electron of charge $-e$, bound by a harmonic restoring force, subject to a damping force $-m\gamma\dot{\mathbf{r}}$, and acted upon by the preceding Lorentz force, is given by

$$m\ddot{\mathbf{r}} + m\gamma\dot{\mathbf{r}} + e\dot{\mathbf{r}} \times \mathbf{B} + m\omega_0^2\mathbf{r} = -e\mathbf{E}. \tag{21.51}$$

The following equations of motion can be written for the x- and y-components, assuming $\mathbf{B}$ is applied in the z-direction:

$$\ddot{x} + \gamma\dot{x} + \frac{eB_z}{m}\dot{y} + \omega_0^2 x = -\frac{e}{m}E_x \tag{21.52}$$

and

$$\ddot{y} + \gamma\dot{y} - \frac{eB_z}{m}\dot{x} + \omega_0^2 y = -\frac{e}{m}E_y. \tag{21.53}$$

Combining Eq. (21.52) with i times Eq. (21.53) leads to

$$\frac{\mathrm{d}^2(x+\mathrm{i}y)}{\mathrm{d}t^2} + \gamma\frac{\mathrm{d}(x+\mathrm{i}y)}{\mathrm{d}t} - \frac{\mathrm{i}eB_z}{m}\frac{\mathrm{d}(x+\mathrm{i}y)}{\mathrm{d}t} + \omega_0^2(x+\mathrm{i}y) = -\frac{e}{m}(E_x + \mathrm{i}E_y). \tag{21.54}$$

Assuming a harmonic time dependence for the fields of the form $\mathrm{e}^{-\mathrm{i}\omega t}$ leads to the following solution:

$$x + \mathrm{i}y = \frac{-e(E_x + \mathrm{i}E_y)}{m(\omega_0^2 - \omega^2 - \mathrm{i}\gamma\omega - \omega_\mathrm{c}\omega)}, \tag{21.55}$$

where the cyclotron frequency (in SI units) is introduced as

$$\omega_\mathrm{c} = \frac{eB_z}{m}. \tag{21.56}$$

It is common in some sources to see the cyclotron frequency defined for a charge q as $\omega_\mathrm{c} = qB_z/(mc)$. This latter definition reflects the use of different units. Introducing

$$F = \omega_0^2 - \omega^2 - \mathrm{i}\gamma\omega \tag{21.57}$$

and

$$\beta = \mathrm{i}\omega_\mathrm{c}\omega, \tag{21.58}$$

allows Eq. (21.55) to be written more compactly:

$$x + \mathrm{i}y = \frac{-e(E_x + \mathrm{i}E_y)}{m(F + \mathrm{i}\beta)}. \tag{21.59}$$

Combining Eq. (21.52) with −i times Eq. (21.53) leads, in a similar fashion, to

$$x - \mathrm{i}y = \frac{-e(E_x - \mathrm{i}E_y)}{m(F - \mathrm{i}\beta)}. \tag{21.60}$$

Let $x_\pm$ be defined by

$$x_\pm = x \pm \mathrm{i}y, \tag{21.61}$$

with analogous definitions for the electric fields $E_\pm$ and the electric polarizations $P_\pm$. Hence,

$$P_\pm = -\mathcal{N}ex_\pm = \frac{\mathcal{N}e^2 E_\pm}{m(F \pm \mathrm{i}\beta)} = \frac{\varepsilon_0 \omega_\mathrm{p}^2 E_\pm}{F \pm \mathrm{i}\beta}, \tag{21.62}$$

where $\mathcal{N}$ is the total number of electrons. From Eq. (21.62) it follows that

$$P_x = \frac{\varepsilon_0 \omega_\mathrm{p}^2}{2}\left\{\frac{E_+}{F + \mathrm{i}\beta} + \frac{E_-}{F - \mathrm{i}\beta}\right\}, \tag{21.63}$$

and

$$P_y = \frac{\varepsilon_0 \omega_\mathrm{p}^2}{2\mathrm{i}}\left\{\frac{E_+}{F + \mathrm{i}\beta} - \frac{E_-}{F - \mathrm{i}\beta}\right\}. \tag{21.64}$$

Making use of Eqs. (19.3) and (21.30) leads to

$$P_x = (\varepsilon_{xx} - \varepsilon_0)E_x + \varepsilon_{xy}E_y \tag{21.65}$$

and

$$P_y = -\varepsilon_{xy}E_x + (\varepsilon_{xx} - \varepsilon_0)E_y. \tag{21.66}$$

Comparing these two results with Eqs. (21.63) and (21.64) yields

$$(\varepsilon_{xx} - \varepsilon_0)E_x + \varepsilon_{xy}E_y = \frac{\varepsilon_0 \omega_\mathrm{p}^2}{2(F^2 + \beta^2)}\{E_+(F - \mathrm{i}\beta) + E_-(F + \mathrm{i}\beta)\} \tag{21.67}$$

and

$$-\varepsilon_{xy}E_x + (\varepsilon_{xx} - \varepsilon_0)E_y = \frac{\varepsilon_0 \omega_\mathrm{p}^2}{2\mathrm{i}(F^2 + \beta^2)}\{E_+(F - \mathrm{i}\beta) - E_-(F + \mathrm{i}\beta)\}. \tag{21.68}$$

Equating the coefficients of E_x and E_y for either of the preceding two equations leads to

$$\varepsilon_{xx}\varepsilon_0^{-1} = 1 + \frac{\omega_\mathrm{p}^2 F}{F^2 + \beta^2} \tag{21.69}$$

and

$$\varepsilon_{xy}\varepsilon_0^{-1} = \frac{\omega_{\rm p}^2\beta}{F^2+\beta^2}. \tag{21.70}$$

Using Eq. (21.47) leads to the following expressions:

$$N_+^2(\omega) - 1 = \frac{\omega_{\rm p}^2(F+\omega\omega_{\rm c})}{F^2+\beta^2} \tag{21.71}$$

and

$$N_-^2(\omega) - 1 = \frac{\omega_{\rm p}^2(F-\omega\omega_{\rm c})}{F^2+\beta^2}. \tag{21.72}$$

From Eqs. (21.69) and (21.70), the asymptotic behavior as $\omega \to \infty$ can be determined to be:

$$\varepsilon_{xx}\varepsilon_0^{-1} - 1 = -\frac{\omega_{\rm p}^2}{\omega^2}, \quad \text{as } \omega \to \infty, \tag{21.73}$$

and

$$\varepsilon_{xy}\varepsilon_0^{-1} = \frac{\omega_{\rm p}^2\beta}{\omega^4} = \frac{{\rm i}\omega_{\rm c}\omega_{\rm p}^2}{\omega^3}, \quad \text{as } \omega \to \infty. \tag{21.74}$$

Using Eqs. (21.71) and (21.72) leads to the following expressions:

$$N_\pm(\omega) - 1 = -\frac{\omega_{\rm p}^2}{2\omega^2}, \quad \text{as } \omega \to \infty, \tag{21.75}$$

and also

$$N_+(\omega) - N_-(\omega) = \frac{\omega_{\rm c}\omega_{\rm p}^2}{\omega^3}, \quad \text{as } \omega \to \infty. \tag{21.76}$$

21.4 Are there dispersion relations for the individual complex refractive indices N_+ and N_-?

The next objective is to address the question posed in the section title, and, during the course of the discussion, point to the essential features that distinguish the case of circular polarization modes from the case of linear polarization. If a strategy similar to the treatment of the complex refractive index for linear polarization is to be employed, then it is necessary to pin down the analytical properties of $N_+(\omega)$ and $N_-(\omega)$ in the upper half of the complex angular frequency plane.

For the moment assume that $N_+(\omega)$ is an analytic function in the upper half complex angular frequency plane. Consider the integral

$$\oint_C \frac{\{N_+(\omega_z) - 1\}\mathrm{d}\omega_z}{\omega - \omega_z},$$

where C denotes the contour shown in Figure 19.2. From the Cauchy integral theorem it follows that

$$P\int_{-\infty}^{\infty} \frac{\{N_+(\omega') - 1\}\mathrm{d}\omega'}{\omega - \omega'} + \mathrm{i}\pi\{N_+(\omega) - 1\} = 0. \tag{21.77}$$

In order to bring this result into the standard form, where the integration domain is over the positive spectral range, a difficulty is immediately encountered. The term $N_+(-\omega)$ does not transform into a function of $N_+(\omega)$, but rather is converted into the complex refractive index for the opposite helicity mode, $N_-(\omega)$. This is a direct outcome from the switching of modes that occurs in the crossing symmetry relations, Eqs. (21.48) and (21.49). So, clearly, the conventional approach to deriving dispersions relations for the optical constants fails for the case of $N_+(\omega)$. A similar situation applies for $N_-(\omega)$. In a sense, $N_+(\omega)$ is not a continuous function across the axis $\omega = 0$.

The problem point in the approach just indicated is the introduction of the crossing symmetry relations. The reader might wonder if it is possible to circumvent this difficulty by considering the integral

$$\oint_C \frac{\{N_+(\omega_z) - 1\}\mathrm{d}\omega_z}{\omega^2 - \omega_z^2},$$

where C denotes the contour shown in Figure 19.3. From the Cauchy integral theorem, and taking the limits $\rho \to 0$ and $R \to \infty$ yields

$$P\int_0^{\infty} \frac{\{N_+(\omega') - 1\}\mathrm{d}\omega'}{\omega^2 - \omega'^2} + \frac{\mathrm{i}\pi\{N_+(\omega) - 1\}}{2\omega} = \mathrm{i}\int_0^{\infty} \frac{\{N_+(\mathrm{i}\omega_\mathrm{i}) - 1\}\mathrm{d}\omega_\mathrm{i}}{\omega^2 + \omega_\mathrm{i}^2}. \tag{21.78}$$

The function $N_+(\omega)$ can be written as

$$N_+(\omega) = n_+(\omega) + \mathrm{i}\kappa_+(\omega), \tag{21.79}$$

where $n_+(\omega)$ and $\kappa_+(\omega)$ are the real and imaginary parts of $N_+(\omega)$, respectively; similarly, for $N_-(\omega)$,

$$N_-(\omega) = n_-(\omega) + \mathrm{i}\kappa_(\omega). \tag{21.80}$$

Taking the real and imaginary parts of Eq. (21.78) leads to

$$\kappa_+(\omega) = \frac{2\omega}{\pi}P\int_0^{\infty} \frac{\{n_+(\omega') - 1\}\mathrm{d}\omega'}{\omega^2 - \omega'^2} + \frac{2\omega}{\pi}\mathrm{Im}\int_0^{\infty} \frac{N_+(\mathrm{i}\omega_\mathrm{i})\mathrm{d}\omega_\mathrm{i}}{\omega^2 + \omega_\mathrm{i}^2} \tag{21.81}$$

and

$$n_+(\omega) - 1 = -\frac{2\omega}{\pi} P \int_0^\infty \frac{\kappa_+(\omega')\mathrm{d}\omega'}{\omega^2 - \omega'^2} + \frac{2\omega}{\pi} \int_0^\infty \frac{\{\mathrm{Re}\, N_+(\mathrm{i}\omega_\mathrm{i}) - 1\}\mathrm{d}\omega_\mathrm{i}}{\omega^2 + \omega_\mathrm{i}^2}. \quad (21.82)$$

The corresponding result for $N_-(\omega)$ follows by replacing $n_+(\omega)$ and $\kappa_+(\omega)$ by $n_-(\omega)$ and $\kappa_-(\omega)$, respectively.

To complete the derivation of Eqs. (21.81) and (21.82), it is necessary to establish that $N_+(\omega)$ is an analytic function in the first quadrant of the upper half of the complex angular frequency plane, and that $N_+(\omega_z) - 1$ has an appropriate asymptotic behavior for large ω_z. The latter issue can be dealt with by making use of the model developed in Section 21.3. From Eq. (21.73) it can be shown that the integral along the circular arc vanishes in the limit $R \to \infty$. The reader should check this assertion.

To deal with the issue of the analytic behavior of $N_+(\omega)$, it follows from Eq. (21.47) that

$$N_+(\omega) = \surd[(\varepsilon_{xx}(\omega) - \mathrm{i}\varepsilon_{xy}(\omega))\varepsilon_0^{-1}]. \quad (21.83)$$

The elements of the dielectric tensor are analytic in the upper half of the complex angular frequency plane and along the real axis for an insulator, and also for a conductor, provided the pole at $\omega = 0$ is excluded. This follows along the lines of the discussion presented in Section 19.3. Because of the square root relationship in Eq. (21.83), it is necessary to establish that there are no branch points for $N_+(\omega)$. This can be handled by establishing that $N_+^2(\omega)$ is not zero in the upper half of the complex frequency plane and on the real axis. If it is assumed that the system is in thermodynamic equilibrium, then it can be shown that the energy dissipation of a circularly polarized mode is proportional to $\omega\ \mathrm{Im}\, N_\pm^2(\omega)$. Hence, for $\omega > 0$, $\mathrm{Im}\, N_\pm^2(\omega) > 0$, which in turn implies that $N_\pm^2(\omega)$ has no zeros for $\omega > 0$ on the real axis. To complete the argument, it is possible to continue in a couple of different ways. By extending the results in Sections 19.3 and 19.10 (see in particular Eq. (19.48)), it can be shown that $\varepsilon_{ij}(\omega)$ has no zeros in the upper half of the complex angular frequency plane, and hence from Eq. (21.47) it follows that $N_\pm^2(\omega_z)$ does not have any zeros in the same region. Alternatively, consider the integral

$$\oint_C \frac{\{N_\pm^2(\omega_z) \pm N_\mp^2(\omega_z)\}\mathrm{d}\omega_z}{\omega^2 + \omega_z^2},$$

where C denotes the contour shown in Figure 19.5 and ω is real. The Cauchy residue theorem yields, on taking the limit $R \to \infty$,

$$\int_{-\infty}^\infty \frac{\{N_\pm^2(\omega_\mathrm{r}) \pm N_\mp^2(\omega_\mathrm{r})\}\mathrm{d}\omega_\mathrm{r}}{\omega^2 + \omega_\mathrm{r}^2} = \frac{\pi}{\omega}\{N_\pm^2(\mathrm{i}\omega) \pm N_\mp^2(\mathrm{i}\omega)\}. \quad (21.84)$$

Using the crossing symmetry relations for $N_\pm(\omega)$, expressions for $N_\pm^2(\mathrm{i}\omega)$ can be found in terms of integrals involving n_+, n_-, κ_+ and κ_-, which are in general non-zero. Hence there are no zeros for $N_\pm^2$ on the positive imaginary angular frequency axis. The argument can be generalized to all points in the upper half complex plane, with the result that $N_\pm^2(\omega_z)$ has no zeros in the upper half plane.

Equation (21.82) and its counterpart for $n_-(\omega) - 1$ are not particularly useful from a practical standpoint, since optical data are not measured at complex frequencies. These results could, however, be employed for checking theoretical models. So this approach fails to obtain the goal of connecting $n_+(\omega)$ with $\kappa_+(\omega)$ via a simple dispersion relation, one that is amenable to practical data analysis.

21.5 Magnetic optical activity: Faraday effect and magnetic circular dichroism

From the failure demonstrated in the preceding section, it should be apparent that it is necessary to take appropriate combinations of the functions $N_+(\omega)$ and $N_-(\omega)$. To derive dispersion relations, both the sum and difference of these two terms will work. The functions $N_+(\omega_z) \pm N_-(\omega_z)$ are both analytic in the upper half of the complex angular frequency plane. This follows directly from the discussion at the end of the preceding section. The asymptotic behavior of these two functions is required, and this can be deduced directly from Eqs. (21.75) and (21.76). Consider the integral

$$\oint_C \frac{\{N_+(\omega_z) + N_-(\omega_z) - 2\}\mathrm{d}\omega_z}{\omega - \omega_z},$$

where C denotes the contour shown in Figure 19.2. Application of the Cauchy integral theorem leads to

$$P\int_{-\infty}^{\infty} \frac{\{N_+(\omega') + N_-(\omega') - 2\}\mathrm{d}\omega'}{\omega - \omega'} = -\mathrm{i}\pi\{N_+(\omega) + N_-(\omega) - 2\}. \tag{21.85}$$

Employing the crossing symmetry connections in Eqs. (21.48) and (21.49) yields

$$N_+(\omega) + N_-(\omega) - 2 = \frac{\mathrm{i}}{\pi} P\int_0^{\infty} \left\{ \frac{\{N_+(\omega')^* + N_-(\omega')^* - 2\}}{\omega + \omega'} + \frac{\{N_+(\omega') + N_-(\omega') - 2\}}{\omega - \omega'} \right\} \mathrm{d}\omega', \tag{21.86}$$

which simplifies to

$$N_+(\omega) + N_-(\omega) - 2 = \frac{2\mathrm{i}}{\pi} P\int_0^{\infty} \left\{ \frac{\omega\{n_+(\omega') + n_-(\omega') - 2\}}{\omega^2 - \omega'^2} + \frac{\mathrm{i}\omega'\{\kappa_+(\omega') + \kappa_-(\omega')\}}{\omega^2 - \omega'^2} \right\} \mathrm{d}\omega'. \tag{21.87}$$

Taking the real and imaginary parts leads to the dispersion relations:

$$n_+(\omega) + n_-(\omega) - 2 = -\frac{2}{\pi} P \int_0^\infty \frac{\omega' \{\kappa_+(\omega') + \kappa_-(\omega')\} \mathrm{d}\omega'}{\omega^2 - \omega'^2} \tag{21.88}$$

and

$$\kappa_+(\omega) + \kappa_-(\omega) = \frac{2\omega}{\pi} P \int_0^\infty \frac{\{n_+(\omega') + n_-(\omega') - 2\} \mathrm{d}\omega'}{\omega^2 - \omega'^2}. \tag{21.89}$$

Consider the integral

$$\oint_C \frac{\{N_+(\omega_z) - N_-(\omega_z)\} \mathrm{d}\omega_z}{\omega - \omega_z},$$

where C denotes the contour shown in Figure 19.2; then, application of the Cauchy integral theorem and making use of the crossing symmetry relations, Eqs. (21.48) and (21.49), yields

$$N_+(\omega) - N_-(\omega) = \frac{\mathrm{i}}{\pi} P \int_0^\infty \left\{ \frac{\{N_-(\omega')^* - N_+(\omega')^*\}}{\omega + \omega'} + \frac{\{N_+(\omega') - N_-(\omega')\}}{\omega - \omega'} \right\} \mathrm{d}\omega', \tag{21.90}$$

which can be written as

$$N_+(\omega) - N_-(\omega) = \frac{2\mathrm{i}}{\pi} P \int_0^\infty \left\{ \frac{\omega' \{n_+(\omega') - n_-(\omega')\}}{\omega^2 - \omega'^2} + \frac{\mathrm{i}\omega \{\kappa_+(\omega') - \kappa_-(\omega')\}}{\omega^2 - \omega'^2} \right\} \mathrm{d}\omega'. \tag{21.91}$$

Taking the real and imaginary parts leads to the dispersion relations:

$$n_+(\omega) - n_-(\omega) = -\frac{2\omega}{\pi} P \int_0^\infty \frac{\{\kappa_+(\omega') - \kappa_-(\omega')\} \mathrm{d}\omega'}{\omega^2 - \omega'^2} \tag{21.92}$$

and

$$\kappa_+(\omega) - \kappa_-(\omega) = \frac{2}{\pi} P \int_0^\infty \frac{\omega' \{n_+(\omega') - n_-(\omega')\} \mathrm{d}\omega'}{\omega^2 - \omega'^2}. \tag{21.93}$$

The magneto-rotatory dispersion function, $\phi_\mathrm{F}(\omega)$, is introduced by the definition

$$\phi_\mathrm{F}(\omega) = \frac{\omega}{2c} \{n_+(\omega) - n_-(\omega)\}, \tag{21.94}$$

and the ellipticity function $\theta_\mathrm{F}(\omega)$ is defined as

$$\theta_\mathrm{F}(\omega) = \frac{\omega}{2c} \{\kappa_+(\omega) - \kappa_-(\omega)\}. \tag{21.95}$$

The most commonly employed symbols for the magneto-optical rotation and ellipticity are $\phi(\omega)$ and $\theta(\omega)$, respectively. These symbols are reserved for a description of natural optical activity, which is treated later in this chapter. A complex magneto-optical rotation function can be defined by

$$\Phi_{\mathrm{F}}(\omega) = \phi_{\mathrm{F}}(\omega) + \mathrm{i}\theta_{\mathrm{F}}(\omega). \tag{21.96}$$

Since

$$\Phi_{\mathrm{F}}(\omega) = \frac{\omega}{2c}\{N_+(\omega) - N_-(\omega)\}, \tag{21.97}$$

the complex optical rotation is analytic in the upper half complex angular frequency plane. The asymptotic behavior of $\Phi_{\mathrm{F}}(\omega)$ is given by

$$\Phi_{\mathrm{F}}(\omega) = \frac{\omega_{\mathrm{c}}\omega_{\mathrm{p}}^2}{2c\omega^2}, \quad \text{as } \omega \to \infty, \tag{21.98}$$

which follows directly from Eq. (21.76). The analytic behavior of $\Phi_{\mathrm{F}}(\omega)$ and knowledge of the asymptotic behavior allows dispersion relations for $\phi_{\mathrm{F}}(\omega)$ and $\theta_{\mathrm{F}}(\omega)$ to be readily determined. Alternatively, the definitions for $\phi_{\mathrm{F}}(\omega)$ and $\theta_{\mathrm{F}}(\omega)$ can be inserted into Eqs. (21.92) and (21.93), leading to

$$\phi_{\mathrm{F}}(\omega) = -\frac{2\omega^2}{\pi} P\int_0^\infty \frac{\theta_{\mathrm{F}}(\omega')\mathrm{d}\omega'}{\omega'(\omega^2 - \omega'^2)} \tag{21.99}$$

and

$$\theta_{\mathrm{F}}(\omega) = \frac{2\omega}{\pi} P\int_0^\infty \frac{\phi_{\mathrm{F}}(\omega')\mathrm{d}\omega'}{\omega^2 - \omega'^2}. \tag{21.100}$$

In order to derive sum rules for conducting media, it will be useful to give dispersion relations for $N_\pm(\omega)$ that avoid the pole behavior at $\omega = 0$. To handle this case, consider the integral

$$\oint_C \frac{\omega_z\{N_+(\omega_z) + N_-(\omega_z) - 2\}\mathrm{d}\omega_z}{\omega - \omega_z},$$

where C denotes the contour shown in Figure 19.2. From the Cauchy integral theorem it follows that

$$P\int_{-\infty}^\infty \frac{\omega'\{N_+(\omega') + N_-(\omega') - 2\}\mathrm{d}\omega'}{\omega - \omega'} = -\mathrm{i}\pi\omega\{N_+(\omega) + N_-(\omega) - 2\}. \tag{21.101}$$

The crossing symmetry relations allow this to be written as follows:

$$N_+(\omega) + N_-(\omega) - 2 = \frac{\mathrm{i}}{\pi\omega} P\int_0^\infty \omega' \left\{ \frac{\{N_+(\omega') + N_-(\omega') - 2\}}{\omega - \omega'} - \frac{\{N_+(\omega')^* + N_-(\omega')^* - 2\}}{\omega + \omega'} \right\} \mathrm{d}\omega', \tag{21.102}$$

which simplifies to

$$N_+(\omega) + N_-(\omega) - 2 = \frac{2\mathrm{i}}{\pi\omega} P \int_0^\infty \left\{ \frac{\omega'^2\{n_+(\omega') + n_-(\omega') - 2\}}{\omega^2 - \omega'^2} + \frac{\mathrm{i}\omega\omega'\{\kappa_+(\omega') + \kappa_-(\omega')\}}{\omega^2 - \omega'^2} \right\} \mathrm{d}\omega'. \tag{21.103}$$

Taking the real and imaginary parts leads to the dispersion relations:

$$n_+(\omega) + n_-(\omega) - 2 = -\frac{2}{\pi} P \int_0^\infty \frac{\omega'\{\kappa_+(\omega') + \kappa_-(\omega')\}\mathrm{d}\omega'}{\omega^2 - \omega'^2} \tag{21.104}$$

and

$$\kappa_+(\omega) + \kappa_-(\omega) = \frac{2}{\pi\omega} P \int_0^\infty \frac{\omega'^2\{n_+(\omega') + n_-(\omega') - 2\}\mathrm{d}\omega'}{\omega^2 - \omega'^2}. \tag{21.105}$$

Starting instead with the integral

$$\oint_C \frac{\omega_z\{N_+(\omega_z) - N_-(\omega_z)\}\mathrm{d}\omega_z}{\omega - \omega_z},$$

where C denotes the contour shown in Figure 19.2, yields

$$N_+(\omega) - N_-(\omega) = \frac{\mathrm{i}}{\pi\omega} P \int_0^\infty \omega' \left\{ \frac{\{N_+(\omega') - N_-(\omega')\}}{\omega - \omega'} - \frac{\{N_-(\omega')^* - N_+(\omega')^*\}}{\omega + \omega'} \right\} \mathrm{d}\omega', \tag{21.106}$$

which simplifies to

$$N_+(\omega) - N_-(\omega) = \frac{2\mathrm{i}}{\pi\omega} P \int_0^\infty \left\{ \frac{\omega\omega'\{n_+(\omega') - n_-(\omega')\}}{\omega^2 - \omega'^2} + \frac{\mathrm{i}\omega'^2\{\kappa_+(\omega') - \kappa_-(\omega')\}}{\omega^2 - \omega'^2} \right\} \mathrm{d}\omega'. \tag{21.107}$$

Taking the real and imaginary parts leads to the dispersion relations:

$$n_+(\omega) - n_-(\omega) = -\frac{2}{\pi\omega} P \int_0^\infty \frac{\omega'^2\{\kappa_+(\omega') - \kappa_-(\omega')\}\mathrm{d}\omega'}{\omega^2 - \omega'^2} \tag{21.108}$$

and

$$\kappa_+(\omega) - \kappa_-(\omega) = \frac{2}{\pi} P \int_0^\infty \frac{\omega'\{n_+(\omega') - n_-(\omega')\}\mathrm{d}\omega'}{\omega^2 - \omega'^2}. \tag{21.109}$$

In terms of $\phi_F(\omega)$ and $\theta_F(\omega)$, these preceding two dispersion relations are given by

$$\phi_F(\omega) = -\frac{2}{\pi} P \int_0^\infty \frac{\omega' \theta_F(\omega') d\omega'}{\omega^2 - \omega'^2} \tag{21.110}$$

and

$$\theta_F(\omega) = \frac{2\omega}{\pi} P \int_0^\infty \frac{\phi_F(\omega') d\omega'}{\omega^2 - \omega'^2}. \tag{21.111}$$

21.6 Sum rules for magneto-optical activity

A number of sum rules can be derived from the dispersion relations developed in the preceding section. The limit $\omega \to 0$ in Eqs. (21.88), (21.93), (21.105), and (21.108) leads to several sum rules. Making use of

$$n_+(0) = n_-(0) = n(0), \tag{21.112}$$

which applies for non-conducting media, then Eq. (21.88) yields

$$n(0) - 1 = \frac{1}{\pi} P \int_0^\infty \frac{\{\kappa_+(\omega) + \kappa_-(\omega)\} d\omega}{\omega}. \tag{21.113}$$

This result represents a generalization of Eq. (19.252). For non-conductors, κ_+ and κ_- satisfy

$$\kappa_+(0) = 0, \kappa_-(0) = 0, \tag{21.114}$$

so that Eq. (21.93) leads to the following sum rule:

$$\int_0^\infty \frac{\{n_+(\omega) - n_-(\omega)\} d\omega}{\omega} = 0. \tag{21.115}$$

Employing the definition for $\phi_F(\omega)$, Eq. (21.94), the preceding sum rule can be recast as follows:

$$\int_0^\infty \frac{\phi_F(\omega) d\omega}{\omega^2} = 0. \tag{21.116}$$

For a conductor or insulator, Eq. (21.105) in the limit $\omega \to 0$ leads to the sum rule

$$\int_0^\infty \left\{ \frac{n_+(\omega) + n_-(\omega)}{2} - 1 \right\} d\omega = 0, \tag{21.117}$$

which is a generalization of Eq. (19.247), and from Eq. (21.108) it follows that

$$\int_0^\infty \{\kappa_+(\omega) - \kappa_-(\omega)\} d\omega = 0. \tag{21.118}$$

Attention is now directed to a determination of the sum rules that can be obtained by examining the $\omega \to \infty$ behavior of the dispersion relations given previously. The key ideas discussed in Section 19.9 are employed to achieve this objective. The asymptotic behavior of $N_\pm(\omega)$ and $N_+(\omega) - N_-(\omega)$ given in Eqs. (21.75) and (21.76) is employed. From (21.73) it follows that

$$\kappa_\pm(\omega) = o(\omega^{-2}), \quad \text{as } \omega \to \infty, \tag{21.119}$$

and from Eq. (21.76)

$$\kappa_+(\omega) - \kappa_-(\omega) = o(\omega^{-3}), \quad \text{as } \omega \to \infty. \tag{21.120}$$

From the Lorentz model discussed at the end of Section 21.3, an expression for the precise fall-off of $\kappa_\pm(\omega)$ as $\omega \to \infty$ can be derived. In what follows it is sufficient to assume a conservative fall-off of $\kappa_\pm(\omega) = O(\omega^{-2-\delta})$ for $\delta > 0$, or the more conservative result $\kappa_\pm(\omega) = O(\omega^{-2}\log^{-\delta}\omega)$, with $\delta > 1$. In the limit $\omega \to \infty$, Eq. (21.104) yields

$$\int_0^\infty \omega\{\kappa_+(\omega) + \kappa_-(\omega)\}\mathrm{d}\omega = \frac{\pi\omega_\mathrm{p}^2}{2}, \tag{21.121}$$

which is a generalization of the Thomas–Reiche–Kuhn sum rule given in Eq. (19.242). Examining the limit $\omega \to \infty$ for Eq. (21.105) leads to Eq. (21.117), assuming explicitly that $\kappa_+(\omega) + \kappa_-(\omega) \neq O(\omega^{-3})$. The limit $\omega \to \infty$ for Eq. (21.92) yields Eq. (21.118) and the same limit for Eq. (21.93) yields Eq. (21.115), whereas Eq. (21.108) leads to the following sum rule:

$$\int_0^\infty \omega^2\{\kappa_+(\omega) - \kappa_-(\omega)\}\mathrm{d}\omega = -\frac{\pi\omega_\mathrm{c}\omega_\mathrm{p}^2}{2}, \tag{21.122}$$

or, in terms of $\theta_\mathrm{F}(\omega)$,

$$\int_0^\infty \omega\theta_\mathrm{F}(\omega)\mathrm{d}\omega = -\frac{\pi\omega_\mathrm{c}\omega_\mathrm{p}^2}{4c}. \tag{21.123}$$

A sum rule related to Eq. (21.116) can be obtained by applying the ideas of Section 19.9 to Eq. (21.100) and using Eq. (21.98), thus yielding the following result:

$$\int_0^\infty \phi_\mathrm{F}(\omega)\mathrm{d}\omega = 0. \tag{21.124}$$

Additional sum rules can be derived by considering some of the methods discussed in Chapter 19. These will be left for the reader to explore in the Exercises.

21.7 Magnetoreflectivity

The magnetoreflection can be defined via

$$\Delta R(\omega) = R_+(\omega) - R_-(\omega), \tag{21.125}$$

where $R_+(\omega)$ and $R_-(\omega)$ refer to the reflectivities for left and right circularly polarized light, respectively. There is an advantage to work instead with the quantity

$$\mathscr{R}(\omega) = \frac{\Delta R(\omega)}{R_a(\omega)} = \frac{R_+(\omega) - R_-(\omega)}{(1/2)\{R_+(\omega) + R_-(\omega)\}}, \tag{21.126}$$

which represents the magnetoreflectivity with a weighting factor of the averaged normal incident reflectivity for right and left circularly polarized modes.

Schnatterly (1969) proposed that the magnetoreflection could be written to first-order in the applied magnetic field for a conducting material as follows:

$$R_+(\omega) - R_-(\omega) = \omega_c \frac{\partial R(\omega)}{\partial \omega}, \tag{21.127}$$

where $R(\omega)$ is the intensity reflectivity defined in Eq. (20.2). Recall that the cyclotron frequency, ω_c, is given by

$$\omega_c = \frac{eB_0}{m}, \tag{21.128}$$

where m and e are the electronic mass and absolute value of the electronic charge, respectively, and B_0 denotes the static magnetic induction. Integration of Eq. (21.127) yields the sum rule

$$\int_0^\infty \{R_+(\omega) - R_-(\omega)\}\mathrm{d}\omega = -\omega_c, \tag{21.129}$$

where the result $R(0) = 1$ has been employed for conductors. A related result is the sum rule

$$\int_0^\infty \{\vartheta_+(\omega) - \vartheta_-(\omega)\}\mathrm{d}\omega = \pi\omega_c, \tag{21.130}$$

where $\vartheta_+(\omega)$ and $\vartheta_-(\omega)$ are the phases for left and right circularly polarized light, respectively. This result can be arrived at using the analogous result to Eq. (21.127):

$$\vartheta_+(\omega) - \vartheta_-(\omega) = \omega_c \frac{\partial \theta(\omega)}{\partial \omega}, \tag{21.131}$$

where $\theta(\omega)$ is the phase for the normal reflectivity. The results $\theta(0) = 0$ and $\theta(\infty) = \pi$ are also required (recall Eqs. (20.48) and 20.54)). Both the sum rules Eqs. (21.129) and (21.130) are approximate when arrived at from the corresponding expansions given in Eqs. (21.127) and (21.131). However, a rigorous derivation of Eq. (21.130) can be obtained from the appropriate dispersion relation. This approach will be shown momentarily.

The generalized reflectivity for circularly polarized modes is given by

$$\tilde{r}_\pm(\omega) = r_\pm(\omega)e^{i\vartheta_\pm(\omega)}. \tag{21.132}$$

The function $\log \tilde{r}_\pm(\omega)$ can be written in terms of its real and imaginary parts as follows:

$$\log \tilde{r}_\pm(\omega) = \log r_\pm(\omega) + i\vartheta_\pm(\omega). \tag{21.133}$$

The complex reflectivity for the different circular modes can be defined in terms of the corresponding complex refractive indices, so that

$$\tilde{r}_\pm(\omega) = \frac{N_\pm(\omega) - 1}{N_\pm(\omega) + 1}. \tag{21.134}$$

The function $\log\{\tilde{r}_+(\omega)/\tilde{r}_-(\omega)\}$ is analytic in the upper half complex angular frequency plane. This follows from the connection between $\tilde{r}_\pm(\omega)$ and $N_\pm(\omega)$, and the analytic properties of the latter functions. The absence of zeros of the permittivity in the upper half plane ensures an absence of branch points for $N_\pm(\omega)$. Consider the integral

$$\oint_C \frac{\log\{\tilde{r}_+(\omega_z)/\tilde{r}_-(\omega_z)\}d\omega_z}{\omega^2 - \omega_z^2},$$

where C denotes the contour shown in Figure 21.1. Evaluation of the integral in the limits $\rho \to 0, \delta \to 0$, and $\mathcal{R} \to \infty$ leads to

$$P\int_{-\infty}^{\infty} \frac{\log\{\tilde{r}_+(\omega')/\tilde{r}_-(\omega')\}}{\omega^2 - \omega'^2} d\omega' = -\frac{i\pi}{2\omega} \log\left\{\frac{\tilde{r}_+(\omega)\,\tilde{r}_-(-\omega)}{\tilde{r}_-(\omega)\,\tilde{r}_+(-\omega)}\right\}. \tag{21.135}$$

On making use of Eqs. (21.48) and (21.49), it follows that

$$\tilde{r}_+(-\omega) = \frac{N_+(-\omega) - 1}{N_+(-\omega) + 1} = \frac{N_-(\omega)^* - 1}{N_-(\omega)^* + 1} = \tilde{r}_-(\omega)^*, \tag{21.136}$$

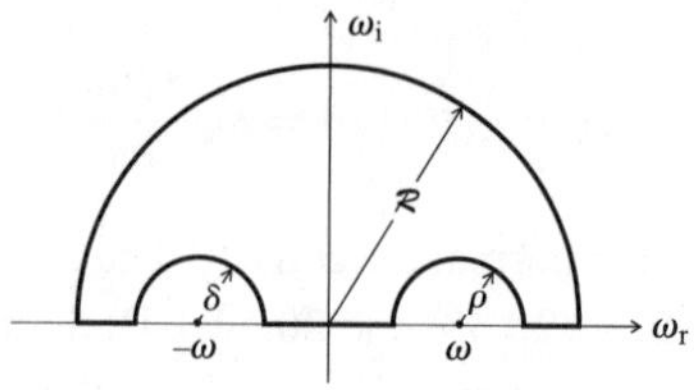

Figure 21.1. Contour for the determination of dispersion relations for the magnetoreflectivity.

and similarly for the other mode, so that

$$\tilde{r}_{\pm}(-\omega) = \tilde{r}_{\mp}(\omega)^*. \tag{21.137}$$

In the same manner as for the complex refractive indices for circularly polarized states, the crossing symmetry relations for the complex reflectivities for circularly polarized states lead to mode switching. From Eq. (21.137) it follows that

$$r_{\pm}(-\omega) = r_{\mp}(\omega) \tag{21.138}$$

and

$$\vartheta_{\pm}(-\omega) = -\vartheta_{\mp}(\omega). \tag{21.139}$$

Introducing the definition

$$\Delta\vartheta(\omega) = \vartheta_{+}(\omega) - \vartheta_{-}(\omega), \tag{21.140}$$

it follows that

$$\Delta\vartheta(-\omega) = \Delta\vartheta(\omega). \tag{21.141}$$

From Eq. (21.125) the magnetoreflectivity satisfies

$$\Delta R(-\omega) = -\Delta R(\omega), \tag{21.142}$$

and from Eq. (21.126) it follows that

$$\mathscr{R}(-\omega) = -\mathscr{R}(\omega). \tag{21.143}$$

Taking the imaginary part of Eq. (21.135) and using the crossing symmetry relations just given leads to the dispersion relation:

$$\log\left\{\frac{R_{+}(\omega)}{R_{-}(\omega)}\right\} = -\frac{4\omega}{\pi}P\int_0^\infty \frac{\Delta\vartheta(\omega')}{\omega^2 - \omega'^2}\,\mathrm{d}\omega'. \tag{21.144}$$

Because of the crossing symmetry relations, taking the real part of Eq. (21.135) does not lead to a dispersion relation.

Introducing the substitutions

$$R_{+}(\omega) = R_a(\omega) + \frac{1}{2}\Delta R(\omega) \tag{21.145}$$

and

$$R_{-}(\omega) = R_a(\omega) - \frac{1}{2}\Delta R(\omega), \tag{21.146}$$

allows Eq. (21.144) to be recast as follows:

$$\mathscr{R}(\omega)\left\{1+\frac{\mathscr{R}^2(\omega)}{12}+\frac{\mathscr{R}^4(\omega)}{80}+O(\mathscr{R}^6(\omega))\right\}=-\frac{4\omega}{\pi}P\int_0^\infty \frac{\Delta\vartheta(\omega')}{\omega^2-\omega'^2}\,\mathrm{d}\omega'. \tag{21.147}$$

For many situations of physical interest, $|\mathscr{R}(\omega)|$ is fairly small, of the order of 10^{-4} or smaller, and, to a very good approximation, the preceding result simplifies to

$$\mathscr{R}(\omega)=-\frac{4\omega}{\pi}P\int_0^\infty \frac{\Delta\vartheta(\omega')}{\omega^2-\omega'^2}\,\mathrm{d}\omega'. \tag{21.148}$$

The derivation leading to the dispersion relations for the reflectivity and phase given in Eqs. (20.43) and (20.44) can be adapted to the case of circularly polarized modes by considering the integral

$$\oint_C \{\log\tilde{r}_+(\omega_z)-\log\tilde{r}_-(\omega_z)\}\left(\frac{1}{\omega-\omega_z}-\frac{1}{\omega_0-\omega_z}\right)\mathrm{d}\omega_z,$$

where C denotes the contour shown in Figure 20.1. Application of the Cauchy integral theorem, and taking the limits $\delta\to 0$, $\rho\to 0$, and $\mathcal{R}\to\infty$, leads to

$$\mathrm{i}\pi\left\{\log\frac{\tilde{r}_+(\omega)}{\tilde{r}_+(\omega_0)}-\log\frac{\tilde{r}_-(\omega)}{\tilde{r}_-(\omega_0)}\right\}=-P\int_{-\infty}^{\infty}\{\log\tilde{r}_+(\omega')-\log\tilde{r}_-(\omega')\}\times\left(\frac{1}{\omega-\omega'}-\frac{1}{\omega_0-\omega'}\right)\mathrm{d}\omega'. \tag{21.149}$$

On taking the real and imaginary parts, it follows that

$$\Delta\vartheta(\omega)-\Delta\vartheta(\omega_0)=\frac{1}{\pi}P\int_{-\infty}^{\infty}\{\log r_+(\omega')-\log r_-(\omega')\}\left(\frac{1}{\omega-\omega'}-\frac{1}{\omega_0-\omega'}\right)\mathrm{d}\omega' \tag{21.150}$$

and

$$\log\frac{r_+(\omega)}{r_+(\omega_0)}-\log\frac{r_-(\omega)}{r_-(\omega_0)}=-\frac{1}{\pi}P\int_{-\infty}^{\infty}\Delta\vartheta(\omega')\left(\frac{1}{\omega-\omega'}-\frac{1}{\omega_0-\omega'}\right)\mathrm{d}\omega'. \tag{21.151}$$

Employing the crossing symmetry relation given in Eq. (21.141) yields

$$\log\frac{r_+(\omega)r_-(\omega_0)}{r_+(\omega_0)r_-(\omega)}=\frac{2}{\pi}P\int_0^\infty \Delta\vartheta(\omega')\left(\frac{\omega_0}{\omega_0^2-\omega'^2}-\frac{\omega}{\omega^2-\omega'^2}\right)\mathrm{d}\omega', \tag{21.152}$$

and from Eq. (21.150) it follows that

$$\Delta\vartheta(\omega) - \Delta\vartheta(\omega_0) = \frac{2}{\pi} P \int_0^\infty \omega' \log\left[\frac{r_+(\omega')}{r_-(\omega')}\right] \left(\frac{1}{\omega^2 - \omega'^2} - \frac{1}{\omega_0^2 - \omega'^2}\right) \mathrm{d}\omega'. \tag{21.153}$$

Starting with the simpler integral

$$\oint_C \frac{\omega_z\{\log\tilde{r}_+(\omega_z) - \log\tilde{r}_-(\omega_z)\}}{\omega^2 - \omega_z^2}\,\mathrm{d}\omega_z,$$

where C is the contour shown in Figure 21.1, yields the dispersion relation

$$\Delta\vartheta(\omega) = \frac{2}{\pi} P \int_0^\infty \frac{\omega' \log[r_+(\omega')/r_-(\omega')]\mathrm{d}\omega'}{\omega^2 - \omega'^2}. \tag{21.154}$$

For the integral in the preceding result to be convergent requires knowledge of the asymptotic behavior for $\log[r_+(\omega)/r_-(\omega)]$, and this will be addressed momentarily. Taking the limit $\omega \to 0$ in Eq. (21.154) and noting that $\Delta\vartheta(0) = 0$, yields the sum rule

$$\int_0^\infty \omega^{-1} \log\left[\frac{r_+(\omega)}{r_-(\omega)}\right] \mathrm{d}\omega = 0. \tag{21.155}$$

This can be recast into the approximate form

$$\int_0^\infty \omega^{-1} \mathscr{R}(\omega)\mathrm{d}\omega = 0. \tag{21.156}$$

A loose end from the derivation of Eq. (21.154) is the asymptotic behavior of $\log[r_+(\omega)/r_-(\omega)]$ as $\omega \to \infty$. This is addressed using Eqs. (21.75) and (21.76) and taking the real part of $\log[\tilde{r}_+(\omega)/\tilde{r}_-(\omega)]$, so that

$$\begin{aligned}
\lim_{\omega\to\infty} \log\left[\frac{r_+(\omega)}{r_-(\omega)}\right] &= \lim_{\omega\to\infty} \log\left\{\frac{(N_+(\omega) - 1)\,(N_-(\omega) + 1)}{(N_+(\omega) + 1)\,(N_-(\omega) - 1)}\right\} \\
&= \lim_{\omega\to\infty} \log\left\{\frac{(1 - \omega_\mathrm{p}^2/2\omega^2)^2 + (\omega_\mathrm{c}\omega_\mathrm{p}^2/\omega^3 - 1)}{(1 - \omega_\mathrm{p}^2/2\omega^2)^2 - (\omega_\mathrm{c}\omega_\mathrm{p}^2/\omega^3 - 1)}\right\} \\
&= \lim_{\omega\to\infty} \log\left\{\frac{1 - \omega_\mathrm{c}/\omega}{1 + \omega_\mathrm{c}/\omega}\right\} \\
&= -\frac{2\omega_\mathrm{c}}{\omega}, \quad \text{as } \omega \to \infty.
\end{aligned} \tag{21.157}$$

This section is concluded with the derivation of the sum rule given in Eq. (21.130). Using the crossing symmetry relations for $\tilde{r}_\pm(\omega)$ allows Eq. (21.135) to be written as follows:

$$P\int_0^\infty \frac{\log\{\tilde{r}_+(\omega')/\tilde{r}_-(\omega')\} - \log\{\tilde{r}_+(\omega')^*/\tilde{r}_-(\omega')^*\}}{\omega^2-\omega'^2}\,d\omega'$$

$$= -\frac{i\pi}{2\omega}\log\left\{\frac{\tilde{r}_+(\omega)}{\tilde{r}_-(\omega)}\frac{\tilde{r}_+^*(\omega)}{\tilde{r}_-^*(\omega)}\right\}. \tag{21.158}$$

This result simplifies to

$$\log\left\{\frac{r_+(\omega)}{r_-(\omega)}\right\} = -\frac{2\omega}{\pi}P\int_0^\infty \frac{\{\vartheta_+(\omega') - \vartheta_-(\omega')\}}{\omega^2-\omega'^2}\,d\omega'. \tag{21.159}$$

Taking the limit $\omega \to \infty$ in Eq. (21.159) and employing Eq. (21.157) yields Eq. (21.130).

21.8 Optical activity

Consider a medium that is non-magnetic, isotropic, and optically active, for example a dilute solution of an optically active species in a non-optically active solvent. In what follows, it is assumed for most formulas that the local field experienced by the individual molecules is the same as the external field. This assumption can be dropped by inclusion of the Lorentz approximation, and this issue is addressed later in Eq. (21.223). The term natural optical activity describes the situation where the plane of a linearly polarized beam of electromagnetic radiation is rotated as it passes through a solution of optically active molecules, or through a solid lacking certain symmetry elements. No external magnetic field is required. An optically active molecule cannot be superimposed on its mirror image. Such molecules lack planes of symmetry and a center of symmetry. In group-theoretic parlance, the molecules lack an S_n symmetry axis. The S_n symmetry element denotes a combined rotation–reflection operation, called an improper rotation axis, and corresponds to a rotation by $2\pi/n$ followed by reflection in a plane perpendicular to the rotation axis employed. For example, the methane molecule, CH_4, in which the hydrogen atoms are arranged in a tetrahedral configuration around the central carbon atom, has an S_4 axis as shown in Figure 21.2. Note that for the S_4 operation, the reflection plane is not a plane of symmetry for the molecule. An S_2 symmetry operation is identical to the inversion element (denoted by i) and an S_1 operation is equivalent to a plane of symmetry (denoted by the symbol σ).

The description of a dilute solution of optically active molecules in terms of electromagnetic fields is given by the following equations (Condon, 1937a):

$$\mathbf{D} = \varepsilon\mathbf{E} - g\frac{\partial \mathbf{H}}{\partial t} \tag{21.160}$$

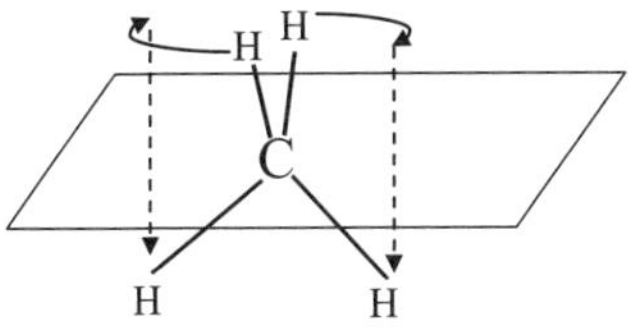

Figure 21.2. An improper S_4 symmetry operation denotes a rotation by $\pi/4$ followed by a reflection in a plane perpendicular to the rotation axis.

and

$$\mathbf{B} = \mu\mathbf{H} + g\frac{\partial \mathbf{E}}{\partial t}. \tag{21.161}$$

A quantum mechanical treatment leads to an additional term of the form $f\mathbf{H}$ for Eq. (21.160) and $f\mathbf{E}$ for Eq. (21.161). These extra terms do not contribute to the optical activity (Condon, 1937a), and are therefore discarded in the following discussion. The quantity g is termed the optical rotatory parameter, and it is a pseudoscalar. A pseudoscalar reverses sign under the inversion operation. A medium described by these two equations exhibits circular birefringence; that is, the plane of polarization for linearly polarized light is rotated as the beam traverses the medium. These equations are now solved to obtain an expression for the rotation of the plane of polarization as the beam propagates.

Equations (21.160) and (21.161) can be written as follows:

$$\mathbf{E} = \varepsilon^{-1}\mathbf{D} + \frac{g}{\mu\varepsilon}\frac{\partial \mathbf{B}}{\partial t} \tag{21.162}$$

and

$$\mathbf{H} = \mu^{-1}\mathbf{B} - \frac{g}{\mu\varepsilon}\frac{\partial \mathbf{D}}{\partial t}, \tag{21.163}$$

where terms of $O(g^2)$ are dropped since g is small. Let the unit vectors $\mathbf{i}$, $\mathbf{j}$, and $\mathbf{k}$ form a right-handed axis system, with $\mathbf{k}$ along the z-axis, and suppose the beam is propagating in the z-direction. The spatial and temporal dependence of the fields are of the form $e^{i\omega((N/c)\mathbf{k}\cdot\mathbf{r}-t)}$. Equations (21.162) and (21.163) can be rewritten as

$$\mathbf{E} = \varepsilon^{-1}\mathbf{D} - i\gamma\mathbf{B} \tag{21.164}$$

and

$$\mathbf{H} = \mu^{-1}\mathbf{B} + i\gamma\mathbf{D}, \tag{21.165}$$

where

$$\gamma = \frac{g\omega}{\mu\varepsilon}. \tag{21.166}$$

From the Maxwell equations,

$$\nabla \cdot \mathbf{D} = 0 \text{ and } \nabla \cdot \mathbf{B} = 0, \tag{21.167}$$

it follows that

$$\mathbf{k} \cdot \mathbf{B} = 0 \tag{21.168}$$

and

$$\mathbf{k} \cdot \mathbf{D} = 0. \tag{21.169}$$

The Maxwell curl equations,

$$\nabla \times \mathbf{E} + \frac{\partial \mathbf{B}}{\partial t} = 0 \text{ and } \nabla \times \mathbf{H} - \frac{\partial \mathbf{D}}{\partial t} = 0, \tag{21.170}$$

lead to

$$\frac{N}{c}\mathbf{k} \times \mathbf{H} = -\mathbf{D} \tag{21.171}$$

and

$$\frac{N}{c}\mathbf{k} \times \mathbf{E} = \mathbf{B}. \tag{21.172}$$

Applying the operation $\mathbf{k}\times$ to each of the preceding two equations, and making use of Eqs. (21.168) and (21.169), leads to the following expressions for **H** and **E**:

$$\frac{N}{c}\mathbf{H} = \mathbf{k} \times \mathbf{D} \tag{21.173}$$

and

$$\frac{N}{c}\mathbf{E} = -\mathbf{k} \times \mathbf{B}. \tag{21.174}$$

Employing Eqs. (21.164) and (21.165) yields

$$\frac{N}{c}\{\varepsilon^{-1}\mathbf{k} \times \mathbf{D} - \mathrm{i}\gamma\mathbf{k} \times \mathbf{B}\} = \mathbf{B} \tag{21.175}$$

and

$$\frac{N}{c}\{\mu^{-1}\mathbf{k} \times \mathbf{B} + \mathrm{i}\gamma\mathbf{k} \times \mathbf{D}\} = -\mathbf{D}. \tag{21.176}$$

Both **B** and **D** are transverse to the direction of propagation (recall Eqs. (21.168) and (21.169)), so the following results can be given:

$$\mathbf{B} = B_1\mathbf{i} + B_2\mathbf{j} \tag{21.177}$$

and

$$\mathbf{D} = D_1\mathbf{i} + D_2\mathbf{j}. \tag{21.178}$$

Inserting these expressions into Eqs. (21.175) and (21.176) leads to

$$\frac{N}{c}\{-\varepsilon^{-1}D_2 + \mathrm{i}\gamma B_2\} = B_1, \tag{21.179}$$

$$\frac{N}{c}\{\varepsilon^{-1}D_1 - \mathrm{i}\gamma B_1\} = B_2, \tag{21.180}$$

$$\frac{N}{c}\{\mu^{-1}B_2 + \mathrm{i}\gamma D_2\} = D_1, \tag{21.181}$$

and

$$\frac{N}{c}\{\mu^{-1}B_1 + \mathrm{i}\gamma D_1\} = -D_2. \tag{21.182}$$

Making use of a result from elementary linear algebra, the solution of this set of homogeneous simultaneous equations requires the determinant of the coefficients to vanish, that is

$$\begin{vmatrix} 1 & -a & 0 & b \\ -a & -1 & b & 0 \\ 0 & d & -1 & a \\ d & 0 & a & 1 \end{vmatrix} = 0, \tag{21.183}$$

where the following definitions have been employed:

$$a = \frac{\mathrm{i}N\gamma}{c}; \quad b = \frac{N}{c\varepsilon}; \quad d = \frac{N}{c\mu}. \tag{21.184}$$

Expanding the determinant yields

$$\frac{N^2}{c^2} = \varepsilon\mu\frac{(1 \pm \surd(\varepsilon\mu)\,\gamma)^2}{(1 - \varepsilon\mu\gamma^2)^2}, \tag{21.185}$$

which can be rewritten as follows:

$$N^2 = \frac{\varepsilon}{\varepsilon_0}\frac{\mu}{\mu_0}\frac{1}{(1 \mp \surd(\varepsilon\mu)\,\gamma)^2}. \tag{21.186}$$

Taking the positive root of this expression yields

$$N = \surd\left(\frac{\varepsilon}{\varepsilon_0}\frac{\mu}{\mu_0}\right)\frac{1}{(1 \mp \surd(\varepsilon\mu)\,\gamma)}. \tag{21.187}$$

For a non-magnetic material, $\mu\mu_0^{-1} \approx 1$ to a very good approximation, and since γ is a small correction, it follows, on introducing N_+ to denote the solution with a $+g$ component and N_- to designate the result with a $-g$ component, that

$$N_{\pm} = \surd\left(\frac{\varepsilon}{\varepsilon_0}\right)\left(1 \pm \frac{\omega g}{\surd(\varepsilon\mu)}\right). \tag{21.188}$$

From the preceding expression it follows, assuming a non-magnetic medium, that

$$N_+(\omega) - N_-(\omega) = 2\omega c g(\omega), \tag{21.189}$$

and an averaged complex refractive index can be written as

$$N_{\text{ave}}(\omega) = \frac{1}{2}\{N_+(\omega) + N_-(\omega)\} = \surd\left(\frac{\varepsilon}{\varepsilon_0}\right), \tag{21.190}$$

which is the expected result.

It is conventional to write the complex refractive indices for the medium as

$$N_{\pm}(\omega) = n_{\pm}(\omega) + \mathrm{i}\kappa_{\pm}(\omega), \tag{21.191}$$

which is similar to the case for magneto-optical activity. The optical rotatory dispersion, $\phi(\omega)$, is then defined by

$$\phi(\omega) = \frac{\omega}{2c}\{n_+(\omega) - n_-(\omega)\}. \tag{21.192}$$

The ellipticity function, $\Psi(\omega)$, is defined by the angle whose tangent is the ratio of the minor amplitude to the major amplitude of the emergent beam, and is given by

$$\tan\Psi(\omega) = \tanh\left[\frac{l\omega}{2c}\{\kappa_+(\omega) - \kappa_-(\omega)\}\right], \tag{21.193}$$

where l denotes the length of the medium traversed. Experimentally, the normal circumstance is that

$$\frac{l\omega}{2c}\{\kappa_+(\omega) - \kappa_-(\omega)\} \ll 1. \tag{21.194}$$

Introducing the ellipticity per unit length, defined as

$$\theta(\omega) = \Psi(\omega)/l, \tag{21.195}$$

allows Eq. (21.193) to be recast as

$$\theta(\omega) = \frac{\omega}{2c}\{\kappa_+(\omega) - \kappa_-(\omega)\}. \tag{21.196}$$

The circular dichroism of a system is determined by $\theta(\omega)$. The term *Cotton effect* is used to describe collectively circular dichroism bands and the associated optical rotatory dispersion occurring in the absorption regions.

A complex optical rotation function can be defined by

$$\Phi(\omega) = \phi(\omega) + \mathrm{i}\theta(\omega). \tag{21.197}$$

This is directly related to the function $g(\omega)$ by the result

$$\Phi(\omega) = \omega^2 g(\omega). \tag{21.198}$$

From Eqs. (21.189) and (21.198), it follows that

$$\Phi(\omega) = \frac{\omega}{2c}\{N_+(\omega) - N_-(\omega)\}. \tag{21.199}$$

The complex optical rotation function is analytic in the upper half of the complex angular frequency plane. This statement is supported using an argument presented by Thomaz and Nussenzveig (1982), which is very similar to the discussion presented in Section 19.3 for the permittivity. Suppose that the function $\omega g(\omega)$ is square integrable over the interval $(-\infty, \infty)$, that is

$$\int_{-\infty}^{\infty} |\omega g(\omega)|^2 < \infty. \tag{21.200}$$

The magnetization at some point $\mathbf{r}$ in the medium can be written, on making use of Eq. (21.161), assuming the medium to be non-magnetic and employing a harmonic time dependence for the fields of the form $\mathrm{e}^{-\mathrm{i}\omega t}$, as follows:

$$\begin{aligned}\mathbf{M}(\mathbf{r}, \omega) &= \frac{1}{\mu_0}\mathbf{B}(\mathbf{r}, \omega) - \mathbf{H}(\mathbf{r}, \omega) \\ &= -\mathrm{i}\omega\mu_0^{-1} g(\omega)\mathbf{E}(\mathbf{r}, \omega).\end{aligned} \tag{21.201}$$

The magnetization and electric field can be written as follows:

$$\mathbf{M}(\mathbf{r}, t) = \int_{-\infty}^{\infty} \mathbf{M}(\mathbf{r}, \omega)\mathrm{e}^{-\mathrm{i}\omega t}\, \mathrm{d}\omega \tag{21.202}$$

and

$$\mathbf{E}(\mathbf{r}, t) = \int_{-\infty}^{\infty} \mathbf{E}(\mathbf{r}, \omega)\mathrm{e}^{-\mathrm{i}\omega t}\, \mathrm{d}\omega, \tag{21.203}$$

and the inverse Fourier transform relationships are given by

$$\mathbf{M}(\mathbf{r}, \omega) = \frac{1}{2\pi}\int_{-\infty}^{\infty} \mathbf{M}(\mathbf{r}, t)\mathrm{e}^{\mathrm{i}\omega t}\, \mathrm{d}t \tag{21.204}$$

and

$$\mathbf{E}(\mathbf{r},\omega) = \frac{1}{2\pi}\int_{-\infty}^{\infty} \mathbf{E}(\mathbf{r},t)\mathrm{e}^{\mathrm{i}\omega t}\,\mathrm{d}t. \tag{21.205}$$

From Eq. (21.202) it follows that

$$\begin{aligned}\mathbf{M}(\mathbf{r},t) &= -\frac{\mathrm{i}\mu_0^{-1}}{2\pi}\int_{-\infty}^{\infty} \omega g(\omega)\mathrm{e}^{-\mathrm{i}\omega t}\,\mathrm{d}\omega \int_{-\infty}^{\infty} \mathbf{E}(\mathbf{r},t')\mathrm{e}^{\mathrm{i}\omega t'}\,\mathrm{d}t' \\ &= -\frac{\mathrm{i}\mu_0^{-1}}{2\pi}\int_{-\infty}^{\infty} \mathbf{E}(\mathbf{r},t')\mathrm{d}t' \int_{-\infty}^{\infty} \omega g(\omega)\mathrm{e}^{-\mathrm{i}(t-t')\omega}\,\mathrm{d}\omega \\ &= \int_{-\infty}^{\infty} \mathbf{E}(\mathbf{r},t')G(t-t')\mathrm{d}t',\end{aligned} \tag{21.206}$$

where

$$G(\tau) = \frac{1}{2\pi\mathrm{i}\mu_0}\int_{-\infty}^{\infty} \omega g(\omega)\mathrm{e}^{-\mathrm{i}\tau\omega}\,\mathrm{d}\omega. \tag{21.207}$$

The function $G(\tau)$ plays the role of a response function. It connects the incident electric field of the electromagnetic beam with the induced magnetization resulting from the optical activity of the medium. The causality principle can be expressed as

$$G(\tau) = 0, \quad \text{for } \tau < 0; \tag{21.208}$$

that is, from Eq. (21.206) there is no induced magnetization at the position **r** until the field **E** is applied. Application of Titchmarsh's theorem, and using Eqs. (21.200), (21.207), and (21.208), leads to the conclusion that $\omega g(\omega)$, and hence $\omega^{-1}\Phi(\omega)$, are analytic functions in the upper half of the complex angular frequency plane. Some refinements to this discussion will be presented momentarily.

Note from Eq. (21.189) that the difference $N_+(\omega) - N_-(\omega)$ is an analytic function, however, to establish that $N_+(\omega) + N_-(\omega)$ is an analytic function requires an examination of $\sqrt{(\varepsilon(\omega)\varepsilon_0^{-1})}$. The permittivity is an analytic function in the upper half of the complex angular frequency plane. It is also necessary to establish that there are no branch points for $N_+(\omega) + N_-(\omega)$ in this same region. This could be approached along the lines presented in Section 21.4 for magneto-optical rotation. The reader is invited to consider this as an exercise.

In order to derive dispersion relations for optical activity, information is required on the asymptotic behavior of $\Phi(\omega)$ as $\omega \to \infty$. From Eqs. (21.160), (21.161), and (21.170) it follows that

$$\mathbf{D} = \left(\varepsilon + \frac{g^2}{\mu}\frac{\partial^2}{\partial t^2}\right)\mathbf{E} + \frac{g}{\mu}\nabla \times \mathbf{E}, \tag{21.209}$$

and assuming the standard harmonic time dependence of the form $e^{-i\omega t}$ leads to

$$\mathbf{D} = \left(\varepsilon - \frac{\omega^2 g^2}{\mu}\right)\mathbf{E} + \frac{g}{\mu}\nabla \times \mathbf{E}. \tag{21.210}$$

The following shorthand notation for vector cross products is introduced:

$$(\mathbf{a} \times \mathbf{b})_k = \varepsilon_{ijk} a_i b_j, \tag{21.211}$$

where ε_{ijk} denotes the Levi-Civita pseudotensor, defined by

$$\varepsilon_{ijk} = \begin{cases} 1, & \text{if } ijk = 123, 231, 312 \\ -1, & \text{if } ijk = 321, 213, 132 \\ 0, & \text{otherwise,} \end{cases} \tag{21.212}$$

and 1, 2, and 3 denote the x-, y-, and z-components, respectively. From Eq. (21.210), and on comparing with $D_i = \varepsilon_{ij} E_j$, where ε_{ij} are the elements of the general permittivity tensor including the effects of optical activity, leads to

$$\varepsilon_{ij}(\mathbf{k}, \omega) = \left\{\varepsilon(0, \omega) - \frac{\omega^2 g^2}{\mu}\right\}\delta_{ij} + \frac{\mathrm{i}g}{\mu}\varepsilon_{ilj} k_l. \tag{21.213}$$

The contribution that is of the form $O(g^2)$ is usually very small and can be dropped. Linear response theory (Nozières and Pines, 1999, p. 255) leads to the following result:

$$\varepsilon_{ij}(\mathbf{k}, \omega)\varepsilon_0^{-1} = \left\{1 - \frac{\omega_\mathrm{p}^2}{\omega^2}\right\}\delta_{ij} - \frac{\omega_\mathrm{p}^2 m}{\omega^2 \hbar}\chi_{ij}(\mathbf{k}, \omega), \tag{21.214}$$

with the current–current response, $\chi_{ij}(\mathbf{k}, \omega)$, given by

$$\chi_{ij}(\mathbf{k}, \omega) = \sum_n \left\{\frac{\langle 0|J_{\mathbf{k}i}|n\rangle\langle n|J_{\mathbf{k}j}^{+}|0\rangle}{\omega - \omega_n + \mathrm{i}\delta} - \frac{\langle 0|J_{\mathbf{k}j}^{+}|n\rangle\langle n|J_{\mathbf{k}i}|0\rangle}{\omega + \omega_n + \mathrm{i}\delta}\right\}, \tag{21.215}$$

where δ plays the role of a positive infinitesimal, and the current operator is given by

$$\mathbf{J}_\mathbf{k} = \frac{1}{2m}\sum_j \{\mathbf{p}_j e^{-\mathrm{i}\mathbf{k}\cdot\mathbf{r}_j} + e^{-\mathrm{i}\mathbf{k}\cdot\mathbf{r}_j}\mathbf{p}_j\}, \tag{21.216}$$

and $\mathbf{p}_j$ denotes the momentum operator for electron j. Comparing Eqs. (21.213) and (21.214), and ignoring terms of order g^2, yields

$$\varepsilon(\omega)\varepsilon_0^{-1} = 1 - \frac{\omega_\mathrm{p}^2}{\omega^2}, \quad \text{as } \omega \to \infty, \tag{21.217}$$

and

$$\omega g(\omega) = \omega^{-1}\Phi(\omega) = o(\omega^{-2}), \quad \text{as } \omega \to \infty. \tag{21.218}$$

The preceding result leads to

$$\phi(\omega) = o(\omega^{-1}), \quad \text{as } \omega \to \infty, \tag{21.219}$$

and

$$\theta(\omega) = o(\omega^{-1}), \quad \text{as } \omega \to \infty. \tag{21.220}$$

An alternative approach to determine the asymptotic behavior of $\phi(\omega)$ and $\theta(\omega)$ will be presented shortly.

The connection between Eqs. (21.160) and (21.161) and the response of individual molecules to the electromagnetic field is made by means of the following results (Condon, 1937a):

$$\mathbf{p} = \alpha(\omega)\mathbf{E}' - \frac{\beta(\omega)\mu_0}{c}\frac{\partial \mathbf{H}}{\partial t} \tag{21.221}$$

and

$$\mathbf{m} = \frac{\beta(\omega)}{c}\frac{\partial \mathbf{E}'}{\partial t}, \tag{21.222}$$

where $\mathbf{p}$ and $\mathbf{m}$ denote, respectively, the induced electric dipole moment and the induced magnetic moment for an individual molecule. The quantity $\beta(\omega)$ is a measure of the natural optical activity of the medium and $\alpha(\omega)$ is the polarizability. The prime denotes the effective local field at the molecule, and this is often related to $\mathbf{E}$ by the Lorentz connection (in SI units):

$$\mathbf{E}' = \mathbf{E} + \frac{1}{3\varepsilon_0}\mathbf{P}. \tag{21.223}$$

The total electric and magnetic moments are given by

$$\mathbf{P} = \mathcal{N}\mathbf{p} \tag{21.224}$$

and

$$\mathbf{M} = \mathcal{N}\mathbf{m}, \tag{21.225}$$

where $\mathcal{N}$ is the average number of molecules per unit volume. Equations (21.221)–(21.225) lead directly to Eqs. (21.160) and (21.161). This allows a connection to be made between the molecular parameters α and β and the quantities ε and g. For the polarizability, it follows from Eqs. (19.1), (19.3), (19.5), (19.6), and (21.223), with $\mathbf{E}$

replaced by $\mathbf{E}'$ in Eq. (19.6), that

$$\alpha = \frac{3\varepsilon_0}{\mathcal{N}} \frac{\varepsilon\varepsilon_0^{-1} - 1}{\varepsilon\varepsilon_0^{-1} + 2}, \tag{21.226}$$

which is called the Clausius–Mossotti equation; when $\varepsilon\varepsilon_0^{-1}$ is replaced by n^2, it is referred to as the Lorentz–Lorenz equation. Starting from the expression

$$\mathbf{D} = \varepsilon_0 \mathbf{E} + \mathbf{P}, \tag{21.227}$$

substituting Eqs. (21.221) and (21.224), and comparing with Eq. (21.160), leads to

$$g(\omega) = \frac{\mathcal{N}\mu_0}{c}\beta(\omega). \tag{21.228}$$

Alternatively, starting with

$$\mathbf{B} = \mu_0 \mathbf{H} + \mu_0 \mathbf{M}, \tag{21.229}$$

and substituting Eqs. (21.225) and (21.222), ignoring local field effects, assuming the medium to be non-magnetic, and comparing with Eq. (21.161), also leads to Eq. (21.228). If the local field is not ignored in this sequence of steps, then Eq. (21.228) is replaced by

$$g(\omega) = \frac{\varepsilon\varepsilon_0^{-1} + 2}{3} \frac{\mathcal{N}\mu_0}{c}\beta(\omega). \tag{21.230}$$

The optical rotatory dispersion can be expressed directly in terms of β by the formula (in SI units)

$$\phi(\omega) = \frac{\varepsilon\varepsilon_0^{-1} + 2}{3} \frac{\omega^2 \mathcal{N}\mu_0}{c}\beta(\omega). \tag{21.231}$$

The next issue to address is how the asymptotic results for the complex optical rotatory dispersion given previously compare with the results obtained from a quantum mechanical treatment of optical activity. The connection with a quantum mechanical description of optical activity can be made via the Rosenfeld (1928) formula. In the Rosenfeld analysis, the parameter β is given by

$$\beta(\omega) = \frac{2c}{3\hbar} \sum_n \frac{R_{n0}}{\omega_{n0}^2 - \omega^2}, \tag{21.232}$$

where $\hbar\omega_{n0}$ denotes the transition energy from the ground state E_0 to the eigenstate E_n, and R_{n0}, called the rotational strength, is defined by

$$R_{n0} = \mathrm{Im}\{\langle 0|\mathbf{p}|n\rangle \cdot \langle n|\mathbf{m}|0\rangle\}. \tag{21.233}$$

The sum in Eq. (21.232) is over all the excited states. The rotational strengths satisfy the Kuhn sum rule, which takes the form

$$\sum_n R_{n0} = 0. \tag{21.234}$$

This result follows directly from the resolution of the identity (the closure theorem)

$$\sum_m |m\rangle\langle m| = 1, \tag{21.235}$$

where the summation is over all states. Hence,

$$\begin{aligned}\sum_n \operatorname{Im}\{\langle 0|\mathbf{p}|n\rangle\cdot\langle n|\mathbf{m}|0\rangle\} &= \operatorname{Im}\{\langle 0|\mathbf{p}\cdot\mathbf{m}|0\rangle - \langle 0|\mathbf{p}|0\rangle\cdot\langle 0|\mathbf{m}|0\rangle\}\\ &= 0. \end{aligned}\tag{21.236}$$

The last line of Eq. (21.236) makes use of the fact that the diagonal matrix element of a Hermitian operator is real. The Rosenfeld result applies only to dilute solutions or gases at low density. The independent particle model is assumed. This implies that the optical rotation from a sample of $\mathcal{N}$ molecules is $\mathcal{N}$ times the optical rotation resulting from one molecule. The average over Euler angles implicit in the formula implies that rotational motion must be fast, which implies that information on rotational states is lost. Of principal concern is the fact that the formula breaks down in the vicinity of an absorption band, that is when the resonance condition $\omega \approx \omega_{n0}$ applies. The Rosenfeld formula therefore provides no account of circular dichroism. It is common practice to estimate the asymptotic behavior for $\beta(\omega)$ in the limit $\omega \to \infty$ from the Rosenfeld equation in the following way:

$$\begin{aligned}\lim_{\omega\to\infty} \beta(\omega) &= \frac{2c}{3\hbar}\sum_n \frac{R_{n0}}{\omega_{n0}^2 - \omega^2}\\ &= -\frac{2c}{3\hbar\omega^2}\left\{\sum_n R_{n0} + \frac{1}{\omega^2}\sum_n \omega_{n0}^2 R_{n0} + O(\omega^{-4})\right\}\\ &= -\frac{2c}{3\hbar\omega^4}\sum_n \omega_{n0}^2 R_{n0} + O(\omega^{-6}), \end{aligned}\tag{21.237}$$

and hence

$$\beta(\omega) = O(\omega^{-4}), \quad \text{as } \omega \to \infty. \tag{21.238}$$

From this result and Eq. (21.228) it follows that

$$g(\omega) = O(\omega^{-4}), \quad \text{as } \omega \to \infty, \tag{21.239}$$

and, from Eq. (21.231),

$$\phi(\omega) = O(\omega^{-2}), \quad \text{as } \omega \to \infty. \tag{21.240}$$

While this estimate of the asymptotic behavior for the optical rotatory dispersion is consistent with Eq. (21.219), a word of caution is in order. One of the fundamental assumptions required in the derivation of the Rosenfeld formula, in the form given in Eq. (21.232), is that the wavelength of light must be much greater than the molecular dimensions of the optically active species. Taking the limit $\omega \to \infty$ in the Rosenfeld formula leads to an obvious contradiction with the previous assumption.

One of the difficulties with the Rosenfeld formula, namely the behavior near resonance, can be dealt with by employing the Condon (1937a) modification, given by

$$\beta(\omega) = \frac{2c}{3\hbar} \sum_n \frac{R_{n0}}{\omega_{n0}^2 - \omega^2 + \mathrm{i}\omega\gamma_n}, \tag{21.241}$$

where γ_n is a damping parameter associated with the line width of the transition from the ground state to the state n. Combining this result with Eqs. (21.198) and (21.231) yields (in SI units)

$$\Phi(\omega) = \frac{2\mathcal{N}\mu_0\omega^2}{3\hbar} \frac{\varepsilon(\omega)\varepsilon_0^{-1} + 2}{3} \sum_n \frac{R_{n0}}{\omega_{n0}^2 - \omega^2 + \mathrm{i}\omega\gamma_n}. \tag{21.242}$$

From this result it follows, on taking the real and imaginary parts, that

$$\phi(\omega) = \frac{2\mathcal{N}\mu_0\omega^2}{3\hbar} \frac{\varepsilon(\omega)\varepsilon_0^{-1} + 2}{3} \sum_n \frac{(\omega_{n0}^2 - \omega^2)R_{n0}}{(\omega_{n0}^2 - \omega^2)^2 + \omega^2\gamma_n^2} \tag{21.243}$$

and

$$\theta(\omega) = -\frac{2\mathcal{N}\mu_0\omega^3}{3\hbar} \frac{\varepsilon(\omega)\varepsilon_0^{-1} + 2}{3} \sum_n \frac{\gamma_n R_{n0}}{(\omega_{n0}^2 - \omega^2)^2 + \omega^2\gamma_n^2}. \tag{21.244}$$

The limit $\omega \to \infty$ in Eq. (21.243) yields

$$\phi(\omega) = O(\omega^{-2}), \quad \text{as } \omega \to \infty, \tag{21.245}$$

and from Eq. (21.244) the limit $\omega \to \infty$ leads to

$$\theta(\omega) = O(\omega^{-1}), \quad \text{as } \omega \to \infty. \tag{21.246}$$

This result is not in agreement with Eq. (21.220). If all the damping constants were the same, that is $\gamma_n = \Gamma$ for all n, then

$$\sum_n \gamma_n R_{n0} = \Gamma \sum_n R_{n0} = 0. \tag{21.247}$$

In this case, the limit $\omega \to \infty$ in Eq. (21.244) yields

$$\theta(\omega) = O(\omega^{-3}), \quad \text{as } \omega \to \infty. \tag{21.248}$$

However, the assumption that $\gamma_n = \Gamma$ for all n is unlikely to hold in general. Thomaz and Nussenzveig (1982) proposed an alternative form for $\Phi(\omega)$, with the local field effect ignored, which does not impact the basic order behavior for the asymptote of $\Phi(\omega)$:

$$\Phi(\omega) = \frac{2\mathcal{N}\mu_0}{3\hbar c} \sum_n \frac{(\omega + \mathrm{i}(\gamma_n/2))^2 R_{n0}}{\omega_{n0}^2 + (\gamma_n^2/4) - \omega^2 - \mathrm{i}\omega\gamma_n}, \tag{21.249}$$

which, on separating into real and imaginary parts and writing $\omega_{n0}'^2 = \omega_{n0}^2 + \gamma_n^2/4$, leads to

$$\phi(\omega) = \frac{2\mathcal{N}\mu_0}{3\hbar c} \sum_n \frac{[\omega^2(\omega_{n0}^2 - \omega^2 - \gamma_n^2/2) - \omega_{n0}'^2(\gamma_n^2/4)]R_{n0}}{(\omega_{n0}'^2 - \omega^2)^2 + \omega^2\gamma_n^2} \tag{21.250}$$

and

$$\theta(\omega) = \frac{2\mathcal{N}\mu_0\omega}{3\hbar c} \sum_n \frac{\omega_{n0}^2 \gamma_n R_{n0}}{(\omega_{n0}'^2 - \omega^2)^2 + \omega^2\gamma_n^2}. \tag{21.251}$$

The limit $\omega \to \infty$ in Eq. (21.250) yields Eq. (21.245), and the limit $\omega \to \infty$ in Eq. (21.251) yields (21.248), in contrast to the result obtained from Eq. (21.244), that is Eq. (21.246). A drawback of Eq. (21.250) is that the low-frequency limit does not lead to $\phi(0) = 0$.

From the preceding discussion it should be apparent that an entirely satisfactory quantum mechanical model has not been given, particularly in regards to obtaining a convincing explanation of the asymptotic behavior for the complex optical rotation as $\omega \to \infty$. Additionally, some of the simpler models, for example the Rosenfeld formula, do not lead to an appropriate description of the analytic behavior of $\Phi(\omega)$. In the following sections it is assumed that both Eqs. (21.245) and (21.248) apply. These assumptions lead to dispersion relations that appear to be consistent with experimental observations. A complete theoretical description of optical activity, one that treats both the molecule and the radiation field quantum mechanically, accounts for the analytic structure of the complex optical rotatory function, and provides a detailed description of the limits $\omega \to 0$ and particularly $\omega \to \infty$, still appears to be an unresolved problem.

The topic of the analytic behavior for the complex optical rotatory function is now reconsidered, and the previous discussion is refined using an argument due to Healy (1976). The induced electric dipole moment for an optically active molecule depends on both the electric field and the curl of the electric field. Ignoring local field effects, the induced dipole moment can be written as follows:

$$\mathbf{p}(\omega) = \alpha(\omega)\mathbf{E}(\omega) + \frac{\beta(\omega)\mu_0}{c\mu}\nabla \times \mathbf{E}(\omega). \tag{21.252}$$

This follows directly from Eqs. (21.221), (21.170), and (19.7). The important point of interest here is that, for the induced electric dipole moment to be zero for some time $t < \tau$, *both* $\mathbf{E}$ and $\nabla \times \mathbf{E}$ must be zero for $t < \tau$. In much the same manner as for the magnetization, in the time domain $\mathbf{p}(t)$ can be written as follows:

$$\mathbf{p}(t) = \int_{-\infty}^{\infty} \alpha(t-t')\mathbf{E}(t')\mathrm{d}t' + \frac{\mu_0}{c\mu}\int_{-\infty}^{\infty} \beta(t-t')\,\nabla \times \mathbf{E}(t')\mathrm{d}t'. \tag{21.253}$$

In order that $\mathbf{p}(t) = 0$ before the electric field and its curl are turned on requires

$$\alpha(\tau) = 0, \quad \text{for } \tau < 0, \tag{21.254}$$

and

$$\beta(\tau) = 0, \quad \text{for } \tau < 0. \tag{21.255}$$

That is,

$$E_k(t) \text{ and } \varepsilon_{ijk}\frac{\partial E_j}{\partial x_i} = 0, \quad \text{for } t < \tau, \Rightarrow p_k(t) = 0, \quad \text{for } t < \tau. \tag{21.256}$$

The important idea here is that the spatial variation of the field must be taken into account. An induced electric dipole can potentially occur at the molecular center when the field is applied anywhere in the molecule, independent of the appearance of a non-zero electric field at the center point. Note that an optically active molecule cannot be treated as being "point-like." Application of Eq. (21.256) and Titchmarsh's theorem leads to the analytic behavior of Φ in the upper half of the complex angular frequency plane. Since spatial effects are obviously important for discussing optical activity, the causality condition should be refined to deal with this phenomenon. For an electromagnetic beam moving in the z-direction, the relativistic version of the causality principle requires that the wave front cannot reach the point z, a positive position in the molecule measured from $z = 0$, before the time z/c, assuming that the beam passed the point $z = 0$ at time $t = 0$. This statement reflects the fact that the electric field does not instantaneously appear at all points across the molecule.

The final point considered in this section is the form of the crossing symmetry relations for the complex optical rotation. Two approaches are taken to deal with the complex optical rotatory function at negative angular frequencies. The first is simply to adopt, for an optically active medium, the following *definitions* for $\omega > 0$:

$$N_+(-\omega) = N_-(\omega)^* \tag{21.257}$$

and

$$N_-(-\omega) = N_+(\omega)^*, \tag{21.258}$$

from which it follows that

$$\Phi(-\omega) = \Phi(\omega)^*. \tag{21.259}$$

A direct consequence of this is that

$$\phi(-\omega) = \phi(\omega) \tag{21.260}$$

and

$$\theta(-\omega) = -\theta(\omega). \tag{21.261}$$

An alternative approach is to adopt the definition of the field of the electromagnetic beam as

$$E(z,t) = \int_{-\infty}^{\infty} \Big\{E_+(\omega)(\hat{\mathbf{x}} + \mathrm{i}\hat{\mathbf{y}})\mathrm{e}^{\mathrm{i}\omega N_+(\omega)z/c} + E_-(\omega)(\hat{\mathbf{x}} - \mathrm{i}\hat{\mathbf{y}})\mathrm{e}^{\mathrm{i}\omega N_-(\omega)z/c}\Big\}\,\mathrm{e}^{-\mathrm{i}\omega t}\,\mathrm{d}\omega, \tag{21.262}$$

where $\hat{\mathbf{x}}$ and $\hat{\mathbf{y}}$ are unit vectors in the x- and y-directions, respectively. For the field $E(z,t)$ to be real requires, for real ω, that

$$E_+(-\omega) = E_-(\omega)^*, \tag{21.263}$$

$$E_-(-\omega) = E_+(\omega)^*, \tag{21.264}$$

$$N_+(-\omega) = N_-(\omega)^*, \tag{21.265}$$

and

$$N_-(-\omega) = N_+(\omega)^*. \tag{21.266}$$

Equation (21.259) follows from the preceding two results and Eq. (21.199).

21.9 Dispersion relations for optical activity

Some of the dispersion relations that apply for optical activity will be derived in this section. Consider the integral

$$\oint_C \frac{\Phi(\omega_z)\mathrm{d}\omega_z}{\omega - \omega_z},$$

where C denotes the contour shown in Figure 19.2. The analytic behavior for $\Phi(\omega_z)$ in the upper half of the complex angular frequency plane and the asymptotic behavior were discussed in the previous section. Making use of this information and employing the Cauchy integral theorem leads to

$$P\int_{-\infty}^{\infty} \frac{\Phi(\omega')\mathrm{d}\omega'}{\omega - \omega'} = -\mathrm{i}\pi\,\Phi(\omega). \tag{21.267}$$

Taking the real and imaginary parts and using Eq. (21.197) yields the dispersion relations

$$\theta(\omega) = \frac{1}{\pi} P\int_{-\infty}^{\infty} \frac{\phi(\omega')\mathrm{d}\omega'}{\omega - \omega'} \tag{21.268}$$

and

$$\phi(\omega) = -\frac{1}{\pi} P\int_{-\infty}^{\infty} \frac{\theta(\omega')\mathrm{d}\omega'}{\omega - \omega'}. \tag{21.269}$$

Using the crossing symmetry relations for $\phi(\omega)$ and $\theta(\omega)$ given in Eqs. (21.260) and (21.261), the dispersion relations just given can be recast as

$$\theta(\omega) = \frac{2\omega}{\pi} P\int_{0}^{\infty} \frac{\phi(\omega')\mathrm{d}\omega'}{\omega^2 - \omega'^2} \tag{21.270}$$

and

$$\phi(\omega) = -\frac{2}{\pi} P\int_{0}^{\infty} \frac{\omega'\theta(\omega')\mathrm{d}\omega'}{\omega^2 - \omega'^2}. \tag{21.271}$$

The interpretation of these results should be apparent: the circular dichroism for the medium can be determined from the optical rotatory dispersion measured over a very wide spectral interval, and vice versa.

To obtain a different dispersion relation for the optical rotation, consider the integral

$$\oint_C \frac{\omega_z^{-1}\Phi(\omega_z)\mathrm{d}\omega_z}{\omega - \omega_z},$$

where C denotes the contour shown in Figure 19.2. Evaluation of the integral leads to

$$P\int_{-\infty}^{\infty}\frac{\Phi(\omega')\mathrm{d}\omega'}{\omega'(\omega-\omega')}=-\mathrm{i}\pi\omega^{-1}\Phi(\omega). \tag{21.272}$$

From Eq. (21.198) it is expected on physical grounds that $\omega g(\omega)\to 0$ as $\omega\to 0$. Alternatively, from Eq. (21.199) it is expected that $\{N_+(\omega)-N_-(\omega)\}\to 0$ as $\omega\to 0$. The latter condition is obvious for the case of insulators, where

$$n_+(0)=n(0),\quad n_-(0)=n(0), \tag{21.273}$$

and

$$\kappa_+(0)=0,\quad \kappa_-(0)=0. \tag{21.274}$$

For non-conductors there is no singularity for $\Phi(\omega)$ at $\omega=0$ in Eq. (21.272). For the case of a conductor, would the reader anticipate that the medium can or cannot differentiate between left and right circular polarization modes in the limit $\omega\to 0$? Is it expected that $\{N_+(\omega)-N_-(\omega)\}\to 0$ as $\omega\to 0$, keeping in mind, for a conductor, the presence of a $1/\sqrt{\omega}$ singularity for both $N_+(\omega)$ and $N_-(\omega)$? Taking the real and imaginary parts of Eq. (21.272) yields

$$\theta(\omega)=\frac{\omega}{\pi}P\int_{-\infty}^{\infty}\frac{\phi(\omega')\mathrm{d}\omega'}{\omega'(\omega-\omega')} \tag{21.275}$$

and

$$\phi(\omega)=-\frac{\omega}{\pi}P\int_{-\infty}^{\infty}\frac{\theta(\omega')\mathrm{d}\omega'}{\omega'(\omega-\omega')}, \tag{21.276}$$

which simplify, on application of the crossing symmetry relations, to yield

$$\phi(\omega)=-\frac{2\omega^2}{\pi}P\int_0^{\infty}\frac{\theta(\omega')\mathrm{d}\omega'}{\omega'(\omega^2-\omega'^2)}, \tag{21.277}$$

and Eq. (21.270) is also obtained. It is left as an exercise for the reader to decide what, if any, dispersion relations result if $\omega^{-1}\Phi(\omega)$ is replaced by $\omega^{-2}\Phi(\omega)$.

21.10 Sum rules for optical activity

Sum rules for the optical rotation and circular dichroism can be derived from the dispersion relations developed in the preceding section; they can be obtained by considering various contour integrals, or they may be obtained by direct application of some of the properties of the Hilbert transform. The approaches are similar to those first discussed in Sections 19.10 and 19.11.

The simplest sum rule to derive follows directly from Eq. (21.271), taking the limit $\omega \to 0$ and noting that $\phi(0) = 0$; hence,

$$\int_0^\infty \frac{\theta(\omega)\mathrm{d}\omega}{\omega} = 0. \tag{21.278}$$

If the approach of Section 19.9 is applied, and the asymptotic behavior given in Eqs. (21.240) and (21.220) employed, then from Eq. (21.270) it follows that

$$\int_0^\infty \phi(\omega)\mathrm{d}\omega = 0. \tag{21.279}$$

For an insulator it follows, on taking the limit $\omega \to 0$ in Eq. (21.270), that

$$\int_0^\infty \frac{\phi(\omega)\mathrm{d}\omega}{\omega^2} = 0. \tag{21.280}$$

For the preceding result to hold requires that

$$\lim_{\omega\to 0} \omega^{-1}\theta(\omega) = O(\omega^\alpha), \quad \text{for } \alpha > 0, \tag{21.281}$$

and this follows if the Condon extension of the Rosenfeld formula applies; see Eq. (21.244).

An alternative approach to obtain sum rules is to consider the evaluation of appropriately selected contour integrals with no singularities on the real axis. Some examples will illustrate this approach. Consider the integral

$$\oint_C \omega_z^m \Phi^p(\omega_z)\mathrm{d}\omega_z,$$

where C denotes a closed semicircular contour in the upper half of the complex angular frequency plane, center the origin and radius $\mathcal{R}$. The parameters m and p are taken to be integers, with $p \geq 1$, and m is allowed to be negative, selected so that the integrals are convergent. Because of the remarks made at the end of Section 21.8 concerning the asymptotic behavior of $\Phi(\omega)$ as $\omega \to \infty$, no attempt is made to pin down a precise condition relating m and p; however, a couple of simple examples are considered. The choice $m = 0$ and $p = 1$ leads to

$$\int_{-\mathcal{R}}^{\mathcal{R}} \Phi(\omega)\mathrm{d}\omega + \mathrm{i}\mathcal{R}\int_0^\pi \Phi(\mathcal{R}e^{\mathrm{i}\vartheta})e^{\mathrm{i}\vartheta}\mathrm{d}\vartheta = 0. \tag{21.282}$$

Using the result $\Phi(\omega) = o(\omega^{-1})$ as $\omega \to \infty$, and taking the limit $\mathcal{R} \to \infty$, leads to Eq. (21.279). Consider the case $m = -1$ and $p = 1$; then,

$$\int_{-\infty}^\infty \frac{\Phi(\omega)\mathrm{d}\omega}{\omega} = 0, \tag{21.283}$$

and hence Eq. (21.278) follows. The choice $m = 0$ and $p = 2$ leads to

$$\int_{-\infty}^{\infty} \Phi^2(\omega)\mathrm{d}\omega = 0, \tag{21.284}$$

from which it follows that

$$\int_0^{\infty} \phi^2(\omega)\mathrm{d}\omega = \int_0^{\infty} \theta^2(\omega)\mathrm{d}\omega, \tag{21.285}$$

which the reader will recognize as the Parseval-type identity applied to the Hilbert transform relations given in Eqs. (21.268) and (21.269).

The following can all be derived by appropriate choices of m and p:

$$\int_0^{\infty} \omega\phi(\omega)\theta(\omega)\mathrm{d}\omega = 0, \tag{21.286}$$

$$\int_0^{\infty} \frac{\phi(\omega)\theta(\omega)\mathrm{d}\omega}{\omega} = 0, \tag{21.287}$$

and

$$\int_0^{\infty} \frac{\phi^2(\omega)\mathrm{d}\omega}{\omega^2} = \int_0^{\infty} \frac{\theta^2(\omega)\mathrm{d}\omega}{\omega^2}. \tag{21.288}$$

The calculation of these sum rules is left as an exercise for the reader.

Notes

§21.2 Born and Wolf (1999) and Jackson (1999) are excellent sources for further reading on polarization modes.

§21.3 For additional reading, see Boswarva, Howard, and Lidiard (1962) and Smith (1980). Note that some sources work the classical model for a charge q, in place of the charge of the electron, $-e$. For an electron the cyclotron frequency is defined by $\omega_c = eB_z/m$ (the sense of motion is opposite for positively and negatively charged particles). However, some differences of sign will appear because of the negative charge. The reader also needs to be alert when making comparisons between different references as to which frequency dependence is employed for the field, $e^{i\omega t}$ or $e^{-i\omega t}$.

§21.4 Smith (1976a) addresses the key issues associated with deriving dispersion relations for circularly polarized modes, and this source should be consulted for further reading.

§21.5 Supplementary discussion on the topics of this section can be found in Henry, Schnatterly, and Slichter (1965), Schnatterly (1969), Smith (1976b, 1976c), and King (1978b). For some applications, see Boswarva *et al.* (1962) and Brown and Laramore (1967). For a chemistry perspective, see Buckingham and Stephens (1966).

§21.6 All of the sum rules in this section are discussed by Smith (1976c). For additional reading, see Furuya, Zimerman, and Villani (1976) and Mukhtarov (1979).

§21.7 Further discussion on magnetoreflectivity and dispersion relations can be found in Smith (1976b) and King (1978b). For a discussion of the rotation of the plane of polarization of reflected light from a metallic surface in the presence of a magnetic field, the polar reflection Faraday effect, see Stern, McGroddy, and Harte (1964).

§21.8 For additional background reading on optical activity, see Condon (1937a), Moscowitz (1962), Oosterhoff (1965), and Caldwell and Eyring (1971). Emeis, Oosterhoff, and De Vries (1967) (see also Healy (1974)) adopt Eqs. (21.257) and (21.258) as the starting definitions for the extension of the complex refractive indices for an optically active medium at negative frequencies, which then allows the determination of the domain of analyticity for the complex optical rotatory dispersion.

§21.9 For works discussing dispersion relations, see Moffitt and Moscowitz (1959), Moscowitz (1962, 1965), Emeis *et al.* (1967), Hutchinson (1968), Healy and Power (1974), Healy (1974, 1976), Smith (1976c), and Thomaz and Nussenzveig (1982). The latter two references address the issue of the validity of Eq. (21.272) for conductors. Parris and Van Der Walt (1975) discuss numerical aspects for the evaluation of the Kramers–Kronig relation for circular dichroism.

§21.10 Further reading on sum rules for natural optical activity can be found in Emeis *et al.* (1967), Smith (1976c), King (1980), Bokut, Penyaz, and Serdyukov (1981), Brockman and Moscowitz (1981), and Thomaz and Nussenzveig (1982).

Exercises

21.1 Prove that a system with C_n symmetry about the z-axis for $n \geq 3$ has a permittivity of the form

$$\boldsymbol{\varepsilon} = \begin{pmatrix} \varepsilon_{xx} & \varepsilon_{xy} & 0 \\ -\varepsilon_{xy} & \varepsilon_{xx} & 0 \\ 0 & 0 & \varepsilon_{zz} \end{pmatrix}.$$

21.2 By considering the integral

$$\oint_C \frac{\omega_z^m \Phi_F^p(\omega_z)\mathrm{d}\omega_z}{\omega - \omega_z},$$

where C is a suitably selected contour, determine the resulting dispersion relations assuming the parameters m and p are integers with $p \geq 1$.

21.3 Starting from the integral $\oint_C \omega_z^m \Phi_{\mathrm{F}}^p(\omega_z)\mathrm{d}\omega_z$, where C denotes a closed semicircular contour in the upper half of the complex angular frequency plane, center the origin and radius $\mathcal{R}$, and the parameters m and p are taken to be integers with $p \geq 1$, determine the resulting sum rules that can be obtained for the magneto-optical rotation and the magnetic circular dichroism.

21.4 What sum rules and integral inequalities can be derived by applying the common properties of the Hilbert transform to the pair Eqs. (21.99) and (21.100) and to Eqs. (21.110) and (21.111)?

21.5 What dispersion relations are obtained by considering the integral

$$\oint_C \frac{\omega_z^m \{r_+(\omega_z) - r_-(\omega_z)\}^p \, d\omega_z}{\omega - \omega_z},$$

where C is a suitably selected contour and the parameters m and p are integers with $p \geq 1$?

21.6 What sum rules can be derived from the dispersion relations obtained in Exercise 21.5?

21.7 Given that the optical activity parameter g is a pseudoscalar, explain what happens to the optical activity for a system composed of a dilute solution of molecules with inversion symmetry.

21.8 Determine the dispersion relations for the function $\Phi^2(\omega)$.

21.9 Derive the dispersion relations resulting from the choice $\omega^{-2}\Phi^2(\omega)$.

21.10 What sum rules can be derived from the dispersion relations obtained in Exercises 21.8 and 21.9?

21.11 Can any dispersion relations be derived starting with the function $\omega^{-2}\Phi(\omega)$?

21.12 If the answer to Exercise 21.11 is in the affirmative, what sum rules can be obtained?

21.13 Derive the sum rules given in Eqs. (21.286)–(21.288). Determine whether these sum rules apply for both conductors and insulators or only to insulators.

22

Dispersion relations for nonlinear optical properties

22.1 Introduction

The focus of this chapter is the derivation of dispersion relations for nonlinear optical properties. For reasons that will be soon apparent, this area has lagged significantly behind the developments for linear optical properties. The study of nonlinear optical properties essentially commences at the start of the 1960s, and the explosive growth in this area parallels the discovery, development, and exploitation of the laser. For linear optical properties, the polarization of the medium is given by

$$\mathbf{P} = \varepsilon_0 \boldsymbol{\chi}^{(1)} \cdot \mathbf{E}, \tag{22.1}$$

where $\boldsymbol{\chi}^{(1)}$ designates the linear electric susceptibility tensor. The $\boldsymbol{\chi}^{(1)}$ tensor is second-rank and it is often denoted simply by $\boldsymbol{\chi}$ when only linear phenomena are under discussion. To deal with nonlinear behavior, the standard approach is to expand the polarization in terms of powers of the applied electric field, so that

$$\mathbf{P} = \varepsilon_0 \{\boldsymbol{\chi}^{(1)} \cdot \mathbf{E} + \boldsymbol{\chi}^{(2)} : \mathbf{EE} + \boldsymbol{\chi}^{(3)} \vdots \mathbf{EEE} + \cdots\}, \tag{22.2}$$

where $\boldsymbol{\chi}^{(2)}$ and $\boldsymbol{\chi}^{(3)}$ are called the second-order and third-order nonlinear electric susceptibilities, respectively. The quantity $\boldsymbol{\chi}^{(2)}$ is a third-rank tensor and $\boldsymbol{\chi}^{(3)}$ is a fourth-rank tensor. The mode of writing the field terms as $: \mathbf{EE}$ and $\vdots \mathbf{EEE}$ is to alert the reader that the appropriate tensor product must be taken. The justification for the series expansion of the polarization is that the successive higher-order susceptibilities diminish in value fairly swiftly.

An alternative way to view nonlinear behavior is to define the polarization in terms of a general electric susceptibility via

$$\mathbf{P} = \varepsilon_0 \boldsymbol{\chi} \cdot \mathbf{E}, \tag{22.3}$$

where χ is now an electric-field-dependent tensor given by

$$\boldsymbol{\chi} = \boldsymbol{\chi}^{(1)} + \boldsymbol{\chi}^{(2)} \cdot \mathbf{E} + \boldsymbol{\chi}^{(3)} : \mathbf{EE} + \cdots. \tag{22.4}$$

In what follows it is more convenient to work with Eqs. (22.3) and (22.4), so that χ will denote the general nonlinear electric susceptibility. The term $\chi^{(n)}$ has units of $(\mathrm{m\,V^{-1}})^{n-1}$. There is a different convention employed where the factor ε_0 only precedes the linear susceptibility, in which case the units of $\chi^{(n)}$ become $\mathrm{m}^{n-2}\,\mathrm{V}^{-n}\,\mathrm{C}$ for $n \geq 2$. A detailed discussion of different unit systems employed in nonlinear optics can be found in Boyd (1992). The tensor component indices for the electric susceptibility are often dropped when they are not central to the topic under discussion. Some different conventions are used for displaying the frequency dependencies of $\chi^{(n)}$, and these are illustrated for the case of $\chi^{(2)}$. The general second-order electric susceptibility is often written as $\chi^{(2)}(\omega_1, \omega_2)$; however, it is also common to write it as the function $\chi^{(2)}(\omega_3, \omega_1, \omega_2)$, where the first given angular frequency is a combination of the other two frequencies. To signal that there are only two independent arguments, the notation $\chi^{(2)}(\omega_3; \omega_1, \omega_2)$ is commonly employed. Another convention is to write $\chi^{(2)}(\omega_3 = \omega_1 + \omega_2)$. In the discussion that follows, $\chi^{(2)}(\omega_1, \omega_2)$ is used unless the context makes $\chi^{(2)}(\omega_3; \omega_1, \omega_2)$ more informative. The former can be regarded as the general second-order susceptibility; the latter notation can be used to specify a particular type of second-order process, such as sum-frequency generation, difference-frequency generation, and so on. For harmonic generation, the notation will on occasion be simplified further, for example $\chi^{(2)}(2\omega)$ for second-harmonic generation. Some different nonlinear processes are briefly outlined in the following section. There are several conventions employed in the literature of nonlinear optics, and this sometimes makes comparisons between different formulas a more formidable exercise than it should be. The first thing to consider when making comparisons is the choice of units. Differences in unit systems often show up as the appearance of a 4π factor or the occurrence of the factor ε_0, the vacuum permittivity. Incident monochromatic fields are written commonly in the following forms:

$$\mathbf{E}(t) = \mathbf{E}_\omega \mathrm{e}^{-\mathrm{i}\omega t} + \mathbf{E}_\omega^* \mathrm{e}^{+\mathrm{i}\omega t} \tag{22.5}$$

or

$$\mathbf{E}(t) = \frac{1}{2}[\mathbf{E}_\omega \mathrm{e}^{-\mathrm{i}\omega t} + \mathbf{E}_\omega^* \mathrm{e}^{+\mathrm{i}\omega t}], \tag{22.6}$$

and $\mathbf{E}_\omega$ is sometimes written as $\mathbf{E}(\omega)$. The difference in these two choices will show up as various powers of 2 in a number of final formulas.

The total polarization can be expressed as follows:

$$\mathbf{P} = \mathbf{P}^{(1)} + \mathbf{P}^{(2)} + \mathbf{P}^{(3)} + \cdots, \tag{22.7}$$

where each term on the right-hand side has an obvious connection with the terms in the expansion in Eq. (22.4). The general time-dependent electric polarization can be expressed in the following form:

$$
\begin{aligned}
P_k(t) = \varepsilon_0 \Bigg\{ & \sum_l \int_0^\infty G_{kl}^{(1)}(t_1) E_l(t - t_1) \mathrm{d}t_1 \\
& + \sum_l \sum_m \int_0^\infty \mathrm{d}t_1 \int_0^\infty G_{klm}^{(2)}(t_1, t_2) E_l(t - t_1) E_m(t - t_2) \mathrm{d}t_2 + \cdots \\
& + \sum_{l_1} \cdots \sum_{l_n} \int_0^\infty \mathrm{d}t_1 \cdots \int_0^\infty G_{kl_1 \ldots l_n}^{(n)}(t_1, t_2, \ldots, t_n) \\
& \times E_{l_1}(t - t_1) \ldots E_{l_n}(t - t_n) \mathrm{d}t_n \Bigg\}.
\end{aligned} \tag{22.8}
$$

The subscripts designate the Cartesian coordinates x, y, and z. To simplify the notation, the summation convention is employed, which means sum over any repeated index. The function G plays the role of a response function. It is real and satisfies a causality condition in the following way:

$$
G_{kl_1 \cdots l_n}^{(n)}(t_1, t_2, \ldots, t_n) = 0, \quad \text{if } t_m < 0, \quad \text{for } m = 1, \ldots, n. \tag{22.9}
$$

For more complicated situations, where the function G has to be treated as a complex quantity, the approach outlined in Section 17.7.1 would be required. The first few terms in the expansion in Eq. (22.7) are now examined. Restricting to the linear response contribution, from Eq. (22.8) it follows that

$$
P_k^{(1)}(t) = \varepsilon_0 \int_0^\infty G_{kl}^{(1)}(t_1) E_l(t - t_1) \mathrm{d}t_1. \tag{22.10}
$$

Recall from Chapter 19 that the electric field can be written as

$$
E_l(t) = \int_{-\infty}^\infty E_l(\omega_1) \mathrm{e}^{-\mathrm{i}\omega_1 t} \, \mathrm{d}\omega_1. \tag{22.11}
$$

The reader is reminded about a common convention employed in the literature, where the time-dependent electric field and the frequency-dependent electric field have the same symbol; the variable dependence is meant to signify which quantity is under discussion. The same convention is also employed for the electric polarization. Equation (22.10) can be rewritten as follows:

$$
\begin{aligned}
P_k^{(1)}(t) &= \varepsilon_0 \int_0^\infty G_{kl}^{(1)}(t_1) \mathrm{d}t_1 \int_{-\infty}^\infty E_l(\omega_1) \mathrm{e}^{-\mathrm{i}\omega_1 (t - t_1)} \, \mathrm{d}\omega_1 \\
&= \varepsilon_0 \int_{-\infty}^\infty E_l(\omega_1) \mathrm{e}^{-\mathrm{i}\omega_1 t} \, \mathrm{d}\omega \int_0^\infty G_{kl}^{(1)}(t_1) \mathrm{e}^{\mathrm{i}\omega_1 t_1} \, \mathrm{d}t_1 \\
&= \varepsilon_0 \int_{-\infty}^\infty \chi_{kl}^{(1)}(\omega_1) E_l(\omega_1) \mathrm{e}^{-\mathrm{i}\omega_1 t} \, \mathrm{d}\omega_1,
\end{aligned} \tag{22.12}
$$

where the first-order electric susceptibility has been introduced as

$$\chi_{kl}^{(1)}(\omega) = \int_0^\infty G_{kl}^{(1)}(t)\mathrm{e}^{\mathrm{i}\omega t}\,\mathrm{d}t. \tag{22.13}$$

Taking the inverse Fourier transform of Eq. (22.12) yields

$$\frac{1}{2\pi}\int_{-\infty}^\infty P_k^{(1)}(t)\mathrm{e}^{\mathrm{i}\omega t}\,\mathrm{d}t = \varepsilon_0\frac{1}{2\pi}\int_{-\infty}^\infty \mathrm{e}^{\mathrm{i}\omega t}\,\mathrm{d}t\int_{-\infty}^\infty \chi_{kl}^{(1)}(\omega_1)E_l(\omega_1)\mathrm{e}^{-\mathrm{i}\omega_1 t}\,\mathrm{d}\omega_1; \tag{22.14}$$

that is,

$$P_k^{(1)}(\omega) = \varepsilon_0\frac{1}{2\pi}\int_{-\infty}^\infty \chi_{kl}^{(1)}(\omega_1)E_l(\omega_1)\mathrm{d}\omega_1\int_{-\infty}^\infty \mathrm{e}^{\mathrm{i}(\omega-\omega_1)t}\,\mathrm{d}t, \tag{22.15}$$

and hence

$$P_k^{(1)}(\omega) = \varepsilon_0\int_{-\infty}^\infty \chi_{kl}^{(1)}(\omega_1)E_l(\omega_1)\delta(\omega-\omega_1)\mathrm{d}\omega_1, \tag{22.16}$$

which is the required result. For the linear case, the integral can be evaluated to give

$$P_k^{(1)}(\omega) = \varepsilon_0\chi_{kl}^{(1)}(\omega)E_l(\omega). \tag{22.17}$$

From Eq. (22.8), an expression for $P_k^{(2)}(t)$ is given by

$$\begin{aligned}
P_k^{(2)}(t) &= \varepsilon_0\int_0^\infty \mathrm{d}t_1\int_0^\infty G_{klm}^{(2)}(t_1,t_2)E_l(t-t_1)E_m(t-t_2)\mathrm{d}t_2\\
&= \varepsilon_0\int_0^\infty \mathrm{d}t_1\int_0^\infty G_{klm}^{(2)}(t_1,t_2)\mathrm{d}t_2\int_{-\infty}^\infty E_l(\omega_1)\mathrm{e}^{-\mathrm{i}\omega_1(t-t_1)}\,\mathrm{d}\omega_1\\
&\quad\times\int_{-\infty}^\infty E_m(\omega_2)\mathrm{e}^{-\mathrm{i}\omega_2(t-t_2)}\,\mathrm{d}\omega_2\\
&= \varepsilon_0\int_{-\infty}^\infty E_l(\omega_1)\mathrm{e}^{-\mathrm{i}\omega_1 t}\,\mathrm{d}\omega_1\int_{-\infty}^\infty E_m(\omega_2)\mathrm{e}^{-\mathrm{i}\omega_2 t}\,\mathrm{d}\omega_2\\
&\quad\times\int_0^\infty \mathrm{d}t_1\int_0^\infty G_{klm}^{(2)}(t_1,t_2)\mathrm{e}^{\mathrm{i}\omega_1 t_1+\mathrm{i}\omega_2 t_2}\,\mathrm{d}t_2\\
&= \varepsilon_0\int_{-\infty}^\infty E_l(\omega_1)\mathrm{e}^{-\mathrm{i}\omega_1 t}\,\mathrm{d}\omega_1\int_{-\infty}^\infty E_m(\omega_2)\mathrm{e}^{-\mathrm{i}\omega_2 t}\chi_{klm}^{(2)}(\omega_1,\omega_2)\mathrm{d}\omega_2,
\end{aligned} \tag{22.18}$$

where the definition of the second-order electric susceptibility is introduced as follows:

$$\chi_{klm}^{(2)}(\omega_1,\omega_2) = \int_0^{\infty} \mathrm{d}t_1 \int_0^{\infty} G_{klm}^{(2)}(t_1,t_2)\mathrm{e}^{\mathrm{i}\omega_1 t_1+\mathrm{i}\omega_2 t_2}\,\mathrm{d}t_2. \tag{22.19}$$

Taking the inverse Fourier transform of Eq. (22.18) leads to

$$\frac{1}{2\pi}\int_{-\infty}^{\infty} P_k^{(2)}(t)\mathrm{e}^{\mathrm{i}\omega t}\,\mathrm{d}t = \varepsilon_0\frac{1}{2\pi}\int_{-\infty}^{\infty}\mathrm{e}^{\mathrm{i}\omega t}\,\mathrm{d}t\int_{-\infty}^{\infty} E_l(\omega_1)\mathrm{e}^{-\mathrm{i}\omega_1 t}\,\mathrm{d}\omega_1 \times \int_{-\infty}^{\infty} E_m(\omega_2)\mathrm{e}^{-\mathrm{i}\omega_2 t}\chi_{klm}^{(2)}(\omega_1,\omega_2)\mathrm{d}\omega_2; \tag{22.20}$$

that is,

$$P_k^{(2)}(\omega) = \varepsilon_0\frac{1}{2\pi}\int_{-\infty}^{\infty} E_l(\omega_1)\mathrm{d}\omega_1\int_{-\infty}^{\infty} E_m(\omega_2)\chi_{klm}^{(2)}(\omega_1,\omega_2)\mathrm{d}\omega_2 \times \int_{-\infty}^{\infty}\mathrm{e}^{\mathrm{i}(\omega-\omega_1-\omega_2)t}\,\mathrm{d}t, \tag{22.21}$$

and hence

$$P_k^{(2)}(\omega) = \varepsilon_0\int_{-\infty}^{\infty} E_l(\omega_1)\mathrm{d}\omega_1\int_{-\infty}^{\infty} E_m(\omega_2)\chi_{klm}^{(2)}(\omega_1,\omega_2)\delta(\omega-\omega_1-\omega_2)\mathrm{d}\omega_2, \tag{22.22}$$

which is the required result. The preceding arguments can be repeated to obtain the nth-order electric polarization $P_k^{(n)}(\omega)$ as follows:

$$P_k^{(n)}(\omega) = \varepsilon_0\int_{-\infty}^{\infty} E_{l_1}(\omega_1)\mathrm{d}\omega_1\int_{-\infty}^{\infty} E_{l_2}(\omega_2)\mathrm{d}\omega_2\cdots\int_{-\infty}^{\infty} E_{l_n}(\omega_n) \times \chi_{kl_1\cdots l_n}^{(n)}(\omega_1,\omega_2,\ldots,\omega_n)\delta(\omega-(\omega_1+\omega_2+\cdots+\omega_n))\mathrm{d}\omega_n, \tag{22.23}$$

where the nth-order electric susceptibility is defined by

$$\chi_{kl_1\ldots l_n}^{(n)}(\omega_1,\omega_2,\ldots,\omega_n)=\int_0^{\infty}\mathrm{d}t_1\ldots\int_0^{\infty} G_{kl_1\cdots l_2}^{(n)}(t_1,t_2,\ldots,t_n)\mathrm{e}^{\mathrm{i}(\omega_1 t_1+\omega_2 t_2+\cdots+\omega_n t_n)}\,\mathrm{d}t_n. \tag{22.24}$$

Different variations on the formula given in Eq. (22.23) can be found in the literature, some of which reflect the particular definition of the Fourier transform employed. These are recognized by the appearance of factors involving 2π.

Since the electric field and the polarization are real quantities, it follows from Eq. (22.8) that the various response functions are real. This in turn means that, from

Eq. (22.24), the susceptibilities satisfy

$$\chi^{(n)}_{kl_1\ldots l_n}(\omega_1,\omega_2,\ldots,\omega_n)^* = \chi^{(n)}_{kl_1\ldots l_n}(-\omega_1,-\omega_2,\ldots,-\omega_n). \tag{22.25}$$

This is the crossing symmetry relation for the generalized electric susceptibility. There are various permutation symmetries of the tensor indices, some of which depend on whether the medium is non-dissipative or dispersion free. For example, for a general medium,

$$\chi^{(2)}_{klm}(\omega_1,\omega_2) = \chi^{(2)}_{kml}(\omega_2,\omega_1). \tag{22.26}$$

This is referred to as an intrinsic permutation symmetry condition. Simultaneously interchanging $l \leftrightarrow m$ and $\omega_1 \leftrightarrow \omega_2$ in Eq. (22.22) leaves the electric polarization contribution $P^{(2)}_k(\omega)$ unchanged.

Spatial symmetry plays an important role in determining the number of independent tensor components that exist for a particular susceptibility tensor. For example, for an isotropic medium, $\boldsymbol{\chi}^{(1)}$ has only one unique tensor component, so $\boldsymbol{\chi}^{(1)}_{ij}(\omega) = \chi^{(1)}_{11}(\omega)\delta_{ij}$, and the third-order susceptibility can be expressed as follows:

$$\begin{aligned}\chi^{(3)}_{ijkl}(\omega_1,\omega_2,\omega_3) = {} & \chi^{(3)}_{1122}(\omega_1,\omega_2,\omega_3)\delta_{ij}\delta_{kl} + \chi^{(3)}_{1212}(\omega_1,\omega_2,\omega_3)\delta_{ik}\delta_{jl} \\ & + \chi^{(3)}_{1221}(\omega_1,\omega_2,\omega_3)\delta_{il}\delta_{jk},\end{aligned} \tag{22.27}$$

where the subscripts 1 and 2 denote the Cartesian components $\{x,y,z\}$. Results like this can be worked out by straightforward symmetry arguments. Equation (22.27) expresses the fact that of the eighty-one possible tensor components of $\boldsymbol{\chi}^{(3)}$, only three components are needed to specify the tensor for an isotropic system. The number of tensor components for each of the distinct crystal classes for the lower-order susceptibilities can be found in a number of sources, for example Butcher and Cotter (1990, pp. 298–302) or Boyd (1992, pp. 44–51).

The polarization terms in Eq. (22.7) are usually expressed in the following form:

$$\mathbf{P}^{(1)}(\omega) = \varepsilon_0\boldsymbol{\chi}^{(1)}(\omega)\cdot\mathbf{E}(\omega), \tag{22.28}$$

$$\mathbf{P}^{(2)}(\omega) = \varepsilon_0\boldsymbol{\chi}^{(2)}(\omega;\omega_1,\omega_2) \mathbin{\dot{:}} \mathbf{E}(\omega_1)\mathbf{E}(\omega_2), \tag{22.29}$$

with $\omega = \omega_1 + \omega_2$ in Eq. (22.29), and, in general,

$$\mathbf{P}^{(n)}(\omega) = \varepsilon_0\boldsymbol{\chi}^{(n)}(\omega;\omega_1,\omega_2,\ldots,\omega_n) \overset{(n)}{\vdots} \mathbf{E}(\omega_1)\mathbf{E}(\omega_2)\cdots\mathbf{E}(\omega_n), \tag{22.30}$$

where $\omega = \sum_{k=1}^{n}\omega_k$, and the appropriate tensor product is symbolized by $\overset{(n)}{\vdots}$.

22.2 Some types of nonlinear optical response

There are a number of optical processes that are classified as nonlinear. Some examples include second-harmonic generation (SHG), third-harmonic generation (THG), and sum-frequency and difference-frequency generation. A brief description of these four processes is now considered.

In second-harmonic generation, also referred to as frequency doubling, the incident light beam impinging on the material has an angular frequency of ω, and there is an emergent component with an angular frequency of 2ω. This is illustrated in the energy level diagram, Figure 22.1.

In the energy level diagram, the ground state is represented by a solid line, the dashed lines represent virtual levels of the system, and a wiggly line represents an incoming or outgoing photon of specified angular frequency. Virtual levels do not represent bound excited states of the system. The interaction of the light beam with the media is represented by a virtual process: a photon of angular frequency ω undergoes a virtual absorption; a virtual absorption of a second photon with an angular frequency of ω occurs, with the simultaneous release of a photon of angular frequency 2ω. The virtual levels represent a formal device for presenting energy changes on an energy level diagram. In this chapter it is assumed that the media under discussion are passive. The second-order nonlinear electric susceptibility $\boldsymbol{\chi}^{(2)}$ is the factor responsible for second-harmonic generation. SHG offers the possibility to switch the spectral region for an incident beam. For example, the visible 694.2 nm beam from a ruby laser can be converted by a quartz crystal to a beam in the ultraviolet at 347.1 nm. This particular case was the first observation of SHG (Franken *et al.*, 1961).

In third-harmonic generation the incident light beam impinging on the material has an angular frequency of ω, and there is an emergent component with an angular frequency of 3ω. This is indicated in the energy level diagram of Figure 22.2.

The third-order nonlinear electric susceptibility $\boldsymbol{\chi}^{(3)}$ is the factor giving rise to third-harmonic generation. THG was first observed in a study of crystalline calcite

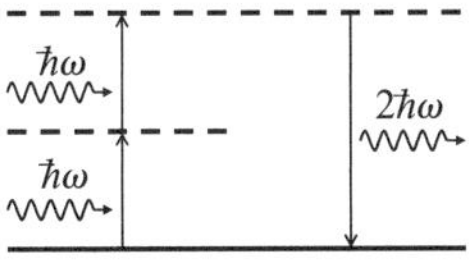

Figure 22.1. Energy level scheme for second-harmonic generation.

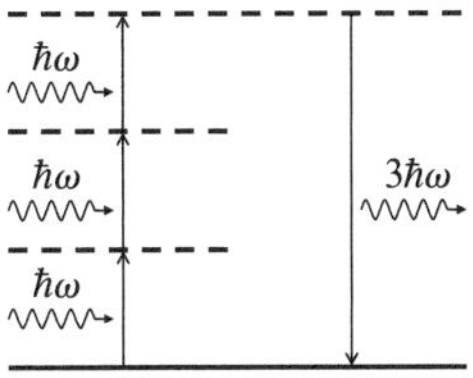

Figure 22.2. Energy level scheme for third-harmonic generation.

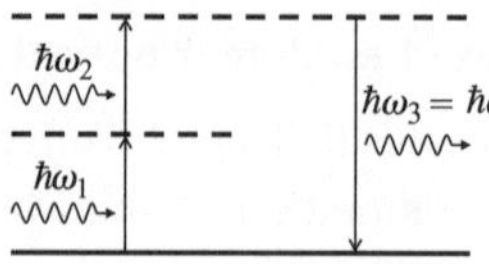

Figure 22.3. Energy level scheme for sum-frequency generation.

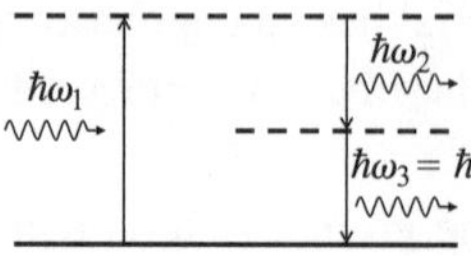

Figure 22.4. Energy level scheme for difference-frequency generation.

(Terhune, Maker, and Savage 1962). Typically the higher harmonics are harder to observe experimentally, though for the case of four of the rare gases, very high-order harmonic generation has been observed, and in particular, for neon, 135th-harmonic generation has been detected (L'Huillier and Balcou, 1993). For systems with an inversion center, that is the system is unchanged by the operation $\{x, y, z\} \rightarrow \{-x, -y, -z\}$ for the Cartesian coordinates, the second-order nonlinear electric susceptibility vanishes identically. Isotropic materials, including gases, normal liquids, and certain crystal classes do not exhibit second-order nonlinear effects. For these materials, the first nonlinear effects that can be observed experimentally depend on the third-order nonlinear electric susceptibility $\chi^{(3)}$. It is assumed that the electric susceptibilities are described in the electric dipole approximation; electric quadrupole and higher-order effects are ignored. At the interface between different media, each having an inversion center, or simply at the surface of a single medium, inversion symmetry is broken. In such cases, second-order nonlinear effects such as SHG are allowed.

Sum-frequency and difference-frequency generation are depicted in Figures 22.3 and 22.4, respectively. In difference-frequency generation, the incident light beams impinge on the media with angular frequencies ω_1 and ω_2. The two-photon emission observed has one photon at angular frequency ω_2, stimulated by the incident beam of angular frequency ω_2, and the second photon has the angular frequency difference $\omega_1 - \omega_2$. In sum-frequency generation, an exit photon has the angular frequency $\omega_1 + \omega_2$, the sum of the two incident angular frequencies ω_1 and ω_2. This was first detected by firing two ruby laser beams of different frequencies simultaneously at a crystal of triglycine sulfate (Bass *et al.*, 1962a). The different frequencies were obtained by running one laser at room temperature, the other at liquid nitrogen temperature (77 K), which leads to a wavelength difference of the two ruby laser beams of approximately 1.0 nm. Three closely spaced lines were detected in the ultraviolet region of the electromagnetic spectrum, corresponding to angular frequencies of $2\omega_1, 2\omega_2$, and $\omega_1 + \omega_2$.

22.3 Classical description: the anharmonic oscillator

The classical Lorentz model discussed in Section 19.2 can be extended to cover the case where the electron experiences, in addition to a harmonic restoring force and a damping term, an anharmonic perturbation of the form $m\alpha x^2$, so that

$$m\ddot{\mathbf{x}} + m\gamma\dot{\mathbf{x}} + m\omega_0^2\mathbf{x} + m\alpha\mathbf{x}^2 = -e\mathbf{E}. \tag{22.31}$$

The electric field is assumed to take the form

$$\mathbf{E} = \mathbf{E}_1\mathrm{e}^{-\mathrm{i}\omega_1 t} + \mathbf{E}_2\mathrm{e}^{-\mathrm{i}\omega_2 t} + c.c.\,, \tag{22.32}$$

where $c.c.$ denotes the complex conjugate of the terms that immediately precede this symbol. Vector notation is dropped at this point. To deal with the solution of Eq. (22.31), suppose that the anharmonic correction αx^2 is small, and develop the solution in the following form:

$$x = \sum_{j=1} \lambda^j x^{(j)}. \tag{22.33}$$

The parameter λ is employed to keep track of the order of the perturbation, and the field is written as λE. When the parameter λ takes the value zero, the electric field is switched off, and when $\lambda = 1$ the field is at full strength. Inserting Eq. (22.33) into Eq. (22.31), equating like powers of λ, and dropping vector notation, leads to the following equations:

$$m\ddot{x}^{(1)} + m\gamma\dot{x}^{(1)} + m\omega_0^2 x^{(1)} = -eE, \tag{22.34}$$

$$\ddot{x}^{(2)} + \gamma\dot{x}^{(2)} + \omega_0^2 x^{(2)} + \alpha\{x^{(1)}\}^2 = 0, \tag{22.35}$$

$$\ddot{x}^{(3)} + \gamma\dot{x}^{(3)} + \omega_0^2 x^{(3)} + 2\alpha x^{(1)}x^{(2)} = 0, \tag{22.36}$$

and so on. The solution of Eq. (22.34), which represents the linear problem, can be written as follows:

$$x^{(1)}(\omega_1,\omega_2,t) = x_1(\omega_1,t)E_1 + x_1(\omega_2,t)E_2 + c.c.\,, \tag{22.37}$$

leading to the familiar result

$$x_1(\omega,t) = -\frac{e}{mD(\omega)}\mathrm{e}^{-\mathrm{i}\omega t}, \tag{22.38}$$

where $D(\omega)$ is given by

$$D(\omega) = \omega_0^2 - \omega^2 - \mathrm{i}\gamma\omega. \tag{22.39}$$

To obtain the solution for $x^{(2)}$ it is necessary to insert Eq. (22.37) into Eq. (22.35). This approach continues, so that $x^{(3)}$ is determined after the solutions for $x^{(1)}$ and $x^{(2)}$ are inserted into Eq. (22.36), and so on. The second-order correction $x^{(2)}$ contains terms with an angular frequency dependence of $2\omega_1, 2\omega_2, \omega_1+\omega_2, \omega_1-\omega_2$, and a time-independent or dc component. To see why the preceding frequencies are expected, consider the square of the electric field, which takes the following form:

$$E^2 = E_1^2 \mathrm{e}^{-2\mathrm{i}\omega_1 t} + E_1^{*2} \mathrm{e}^{2\mathrm{i}\omega_1 t} + 2\,|E_1|^2 + E_2^2 \mathrm{e}^{-2\mathrm{i}\omega_2 t} + E_2^{*2} \mathrm{e}^{2\mathrm{i}\omega_2 t} + 2\,|E_2|^2$$
$$+ 2E_1E_2 \mathrm{e}^{-\mathrm{i}(\omega_1+\omega_2)t} + 2E_1^*E_2^* \mathrm{e}^{\mathrm{i}(\omega_1+\omega_2)t} + 2E_1E_2^* \mathrm{e}^{-\mathrm{i}(\omega_1-\omega_2)t}$$
$$+ 2E_1^*E_2 \mathrm{e}^{\mathrm{i}(\omega_1-\omega_2)t}. \tag{22.40}$$

In what follows, tensor notation is suppressed. The polarization according to Eqs. (22.3) and (22.4) is given by

$$P = \varepsilon_0 \Big\{ \chi^{(1)}\, [E_1 \mathrm{e}^{-\mathrm{i}\omega_1 t} + E_2 \mathrm{e}^{-\mathrm{i}\omega_2 t}] + \chi^{(2)} [E_1^2 \mathrm{e}^{-2\mathrm{i}\omega_1 t} + 2\{|E_1|^2 + |E_2|^2\}$$
$$+ E_2^2 \mathrm{e}^{-2\mathrm{i}\omega_2 t} + 2E_1E_2 \mathrm{e}^{-\mathrm{i}(\omega_1+\omega_2)t} + 2E_1E_2^* \mathrm{e}^{-\mathrm{i}(\omega_1-\omega_2)t}] + \cdots \Big\}. \tag{22.41}$$

The electric polarization to first-order, denoted by $P^{(1)}$, is given by

$$P^{(1)} = -\mathcal{N} e x^{(1)}(\omega_1, \omega_2, t)$$
$$= \frac{\mathcal{N} e^2}{m} \left\{ \frac{E_1 \mathrm{e}^{-\mathrm{i}\omega_1 t}}{D(\omega_1)} + \frac{E_2 \mathrm{e}^{-\mathrm{i}\omega_2 t}}{D(\omega_2)} \right\}. \tag{22.42}$$

Let $\omega_3 = \omega_1 + \omega_2, \omega_4 = \omega_1 - \omega_2$, and set

$$x_2(2\omega_1, t) = -\frac{\alpha e^2}{m^2 D(\omega_1)^2} \frac{\mathrm{e}^{-2\mathrm{i}\omega_1 t}}{D(2\omega_1)}, \tag{22.43}$$

$$x_{2\mathrm{s}}(\omega_1, \omega_2, \omega_3, t) = -\frac{2\alpha e^2}{m^2 D(\omega_1) D(\omega_2)} \frac{\mathrm{e}^{-\mathrm{i}\omega_3 t}}{D(\omega_3)}, \tag{22.44}$$

$$x_{2\mathrm{d}}(\omega_1, \omega_2, \omega_4, t) = -\frac{2\alpha e^2}{m^2 D(\omega_1) D(-\omega_2)} \frac{\mathrm{e}^{-\mathrm{i}\omega_4 t}}{D(\omega_4)}, \tag{22.45}$$

and

$$x_{2\mathrm{or}}(\omega_1) = -\frac{2\alpha e^2}{m^2 D(\omega_1) D(-\omega_1) D(0)}; \tag{22.46}$$

then, the solution of Eq. (22.35) is given by

$$x^{(2)}(\omega_1,\omega_2,t) = x_2(2\omega_1,t)E_1^2 + x_2(2\omega_2,t)E_2^2 + x_{2s}(\omega_1,\omega_2,\omega_3,t)E_1E_2 + x_{2d}(\omega_1,\omega_2,\omega_4,t)E_1E_2^* + x_{2or}(\omega_1)|E_1|^2 + x_{2or}(\omega_2)|E_2|^2 . \quad (22.47)$$

The subscripts s, d, and or refer to sum, difference, and optical rectification. Optical rectification is the generation of a dc field in the medium, the size of which is governed by $\chi^{(2)}(0;\omega,-\omega)$, assuming there is no inversion center present. This effect was first observed by Bass *et al.* (1962b) using a high-power ruby laser. The second-order polarization contribution, $P^{(2)}$, has contributions arising from the nonlinear susceptibility terms

$$P^{(2)} = \frac{\mathcal{N}\alpha e^3}{m^2}\left\{\frac{e^{-2i\omega_1 t}E_1^2}{D(\omega_1)^2D(2\omega_1)} + \frac{e^{-2i\omega_2 t}E_2^2}{D(\omega_2)^2D(2\omega_2)} + \frac{2e^{-i\omega_3 t}E_1E_2}{D(\omega_1)D(\omega_2)D(\omega_3)} + \frac{2e^{-i\omega_4 t}E_1E_2^*}{D(\omega_1)D(-\omega_2)D(\omega_4)} + \frac{2|E_1|^2}{D(\omega_1)D(-\omega_1)D(0)} + \frac{2|E_2|^2}{D(\omega_2)D(-\omega_2)D(0)}\right\}. \quad (22.48)$$

From this expression it follows that, for input angular frequencies of ω_1 and ω_2, there is a possibility for observing new frequency components at $2\omega_1, 2\omega_2, \omega_1 \pm \omega_2$, and 0. The electric susceptibility $\chi^{(2)}$ has components that depend on these different frequencies, and explicit expressions can be readily written down for each susceptibility term by reference to Eqs. (22.41) and (22.48). For example, for SHG (ignoring tensor components) $\chi^{(2)}(2\omega)$ is given by

$$\chi^{(2)}(2\omega) = \frac{\mathcal{N}\alpha e^3}{\varepsilon_0 m^2 D(\omega)^2 D(2\omega)}. \quad (22.49)$$

The value of $\chi^{(2)}(2\omega)$ depends directly on the size of the anharmonicity parameter α. For higher-order polarization contributions, the number of frequency combinations that can occur increases significantly, and the resulting expressions for $P^{(n)}$ are cumbersome.

Since a focus of this chapter is the derivation of dispersion relations for the nonlinear electric susceptibilities, the classical model expressions such as Eq. (22.49) are useful for determining the analytic structure when ω is allowed to become a complex variable. In addition, the asymptotic behavior as $\omega \to \infty$ can also be extracted in a rather straightforward manner from expressions like Eq. (22.49). Attention is now directed to a more rigorous description of the electric susceptibility via a quantum theoretic approach.

22.4 Density matrix treatment

As a prelude to a discussion of dispersion relations for the nonlinear susceptibility, a density matrix formulation of the electromagnetic interaction with the medium is examined. This allows the determination of the asymptotic behavior of the nonlinear susceptibility as $\omega \to \infty$, which is an essential element in the derivation of dispersion relations. The treatment is semiclassical, which means the electromagnetic field is not treated in a quantum field theoretic manner, but the medium is treated quantum mechanically. Only electronic contributions are considered.

In the following development, the wave function has a spatial and temporal dependence. The variables representing these quantities are suppressed to keep the notation compact. The wave function employed is normalized, so that

$$\langle \psi | \psi \rangle = 1, \tag{22.50}$$

and it is assumed that it describes a pure state. The density operator ρ, which is frequently symbolized by $\hat{\rho}$ to indicate its operator nature, is defined by

$$\rho = |\psi\rangle\langle\psi|, \tag{22.51}$$

and a one-line calculation shows that it is itempotent. In the following, the temporal dependence of ρ will frequently be suppressed to simplify the notation. When the system is not described by a single pure state ψ, ρ is given by the following expression:

$$\rho = \sum_n p_n |\psi_n\rangle\langle\psi_n|, \tag{22.52}$$

and p_n is a probability factor given by the Boltzmann distribution:

$$p_n = \frac{\mathrm{e}^{-(E_n/kT)}}{\sum_m \mathrm{e}^{-(E_m/kT)}}, \tag{22.53}$$

where k is the Boltzmann constant, T is the temperature on the Kelvin scale, and the E_n are energies corresponding to ψ_n.

The Hamiltonian for the system can be written as follows:

$$\mathscr{H} = \mathscr{H}_0 + \lambda \mathscr{H}_{\text{int}} + \mathscr{H}_{\text{random}}, \tag{22.54}$$

where $\mathscr{H}_0$ is the Hamiltonian in the absence of the electromagnetic field, $\mathscr{H}_{\text{int}}$ is the perturbation caused by the presence of the field, the term $\mathscr{H}_{\text{random}}$ treats the perturbation of the system by the thermal surroundings and accounts for the relaxation of excited state populations back to an equilibrium population, and λ is an order parameter. The lowest-order electric dipole approximation is adopted, so that

$$\mathscr{H}_{\text{int}} = -\boldsymbol{\mu} \cdot \mathbf{E}, \tag{22.55}$$

where $\boldsymbol{\mu}$ is the electric dipole operator, which takes the form

$$\boldsymbol{\mu} = \sum_{u=1}^{N} -e\mathbf{r}_u, \tag{22.56}$$

where N is the number of electrons, $\mathbf{r}_u$ is the position vector for electron u, and the reader is reminded that the charge of the electron is written as $-e$. The eigenfunctions of the unperturbed Hamiltonian $\mathscr{H}_0$ are taken to be the solutions of the Schrödinger equation,

$$\mathscr{H}_0\varphi_n = E_n\varphi_n, \tag{22.57}$$

where E_n is the energy corresponding to φ_n, and the φ_n are time-independent. The wave function ψ is expanded in the following manner:

$$\psi = \sum_{k=1}^{\infty} c_k(t)\varphi_k. \tag{22.58}$$

The functions φ_k are orthonormal and form a complete set, so that

$$\langle\varphi_k|\varphi_l\rangle = \delta_{kl} \tag{22.59}$$

and

$$\sum_{k=1} |\varphi_k\rangle\langle\varphi_k| = 1. \tag{22.60}$$

Matrix elements of the density operator are defined by

$$\rho_{mn} = \langle\varphi_m|\rho|\varphi_n\rangle, \tag{22.61}$$

which is frequently expressed in the more compact form

$$\rho_{mn} = \langle m|\rho|n\rangle. \tag{22.62}$$

The trace for a particular operator $\mathscr{O}$ is defined by

$$\mathrm{Tr}\,\mathscr{O} = \sum_{k=1} \langle\varphi_k|\mathscr{O}|\varphi_k\rangle. \tag{22.63}$$

The expectation value of the operator $\mathscr{O}$ is given by

$$\langle\mathscr{O}\rangle = \langle\psi|\mathscr{O}|\psi\rangle = \sum_{m=1}\sum_{n=1} c_m^* c_n\langle\varphi_m|\mathscr{O}|\varphi_n\rangle. \tag{22.64}$$

The expectation value of the operator $\mathscr{O}$ can be written in terms of $\mathrm{Tr}(\rho\mathscr{O})$ since

$$\begin{aligned}\mathrm{Tr}(\rho\mathscr{O}) &= \sum_{k=1} \langle\varphi_k|\rho\mathscr{O}|\varphi_k\rangle \\ &= \sum_{k=1} \langle\varphi_k|\psi\rangle\langle\psi|\mathscr{O}|\varphi_k\rangle \\ &= \sum_{k=1} c_k\langle\psi|\mathscr{O}|\varphi_k\rangle \\ &= \langle\psi|\mathscr{O}|\psi\rangle. \end{aligned} \tag{22.65}$$

From the time-dependent Schrödinger equation,

$$\mathrm{i}\hbar\frac{\partial|\psi\rangle}{\partial t} = \mathscr{H}|\psi\rangle, \tag{22.66}$$

the Hermitian conjugate is given by

$$-\mathrm{i}\hbar\frac{\partial\langle\psi|}{\partial t} = \langle\psi|\mathscr{H}. \tag{22.67}$$

Then it follows that

$$\begin{aligned}\mathrm{i}\hbar\frac{\partial\rho}{\partial t} &= \left(\mathrm{i}\hbar\frac{\partial|\psi\rangle}{\partial t}\right)\langle\psi| \quad + \quad |\psi\rangle\left(\mathrm{i}\hbar\frac{\partial\langle\psi|}{\partial t}\right) \\ &= \mathscr{H}|\psi\rangle\langle\psi| \quad - \quad |\psi\rangle\langle\psi|\mathscr{H}, \end{aligned} \tag{22.68}$$

and hence

$$\mathrm{i}\hbar\frac{\partial\rho}{\partial t} = [\mathscr{H}, \rho], \tag{22.69}$$

which is referred to as the quantum Liouville equation or the Liouville–von Neumann equation.

The solution of Eq. (22.69) is obtained in the following manner. The identification

$$\left(\frac{\partial\rho}{\partial t}\right)_{\text{relax}} = \frac{1}{\mathrm{i}\hbar}[\mathscr{H}_{\text{random}}, \rho] \tag{22.70}$$

is made, and a phenomenological damping term Γ_{nm} is introduced via

$$\left(\frac{\partial\rho_{nm}}{\partial t}\right)_{\text{relax}} = -\Gamma_{nm}\rho_{nm}, \tag{22.71}$$

with $\Gamma_{nm} = \Gamma_{mn}$. The term with a relax subscript denotes those relaxation processes, such as collisions in the gas phase, that allow the system to return to equilibrium.

The quantity $\rho(t)$ is expanded as the following series:

$$\rho(t) = \rho^{(0)}(t) + \lambda\rho^{(1)}(t) + \lambda^2\rho^{(2)}(t) + \cdots, \tag{22.72}$$

where λ is treated as an order parameter. Inserting this expression into Eq. (22.69) and evaluating the nmth matrix element leads to

$$\frac{\partial\rho_{nm}^{(0)}}{\partial t} = \frac{1}{\mathrm{i}\hbar}[\mathscr{H}_0, \rho^{(0)}]_{nm} - \Gamma_{nm}\rho_{nm}^{(0)}, \tag{22.73}$$

$$\frac{\partial\rho_{nm}^{(1)}}{\partial t} = \frac{1}{\mathrm{i}\hbar}[\mathscr{H}_0, \rho^{(1)}]_{nm} + \frac{1}{\mathrm{i}\hbar}[\mathscr{H}_{\mathrm{int}}, \rho^{(0)}]_{nm} - \Gamma_{nm}\rho_{nm}^{(1)}, \tag{22.74}$$

$$\frac{\partial\rho_{nm}^{(2)}}{\partial t} = \frac{1}{\mathrm{i}\hbar}[\mathscr{H}_0, \rho^{(2)}]_{nm} + \frac{1}{\mathrm{i}\hbar}[\mathscr{H}_{\mathrm{int}}, \rho^{(1)}]_{nm} - \Gamma_{nm}\rho_{nm}^{(2)}, \tag{22.75}$$

and so on for higher-order contributions. The solutions of these equations are now examined. From Eq. (22.73) it follows that

$$\frac{\partial\rho_{nm}^{(0)}}{\partial t} = \{-\mathrm{i}\omega_{nm} - \Gamma_{nm}\}\rho_{nm}^{(0)}, \tag{22.76}$$

and hence

$$\rho_{nm}^{(0)}(t) = \mathrm{e}^{-(\mathrm{i}\omega_{nm}+\Gamma_{nm})t}\rho_{nm}^{\mathrm{eq}}, \tag{22.77}$$

where ρ_{nm}^{eq} denotes the equilibrium value. The quantity ρ_{nm}^{eq} satisfies

$$\rho_{nm}^{\mathrm{eq}} = 0, \quad \text{for } n \neq m. \tag{22.78}$$

The off-diagonal elements are zero since there is an absence of coherence between different states of the ensemble.

For the solution of Eq. (22.74) it follows that

$$\frac{\partial\rho_{nm}^{(1)}}{\partial t} = (-\mathrm{i}\omega_{nm} - \Gamma_{nm})\rho_{nm}^{(1)} - \frac{\mathrm{i}}{\hbar}[\mathscr{H}_{\mathrm{int}}, \rho^{(0)}]_{nm}. \tag{22.79}$$

Let

$$\rho_{nm}^{(1)}(t) = g_{nm}(t)\mathrm{e}^{-(\mathrm{i}\omega_{nm}+\Gamma_{nm})t}, \tag{22.80}$$

then

$$\frac{\partial g_{nm}(t)}{\partial t} = -\frac{\mathrm{i}}{\hbar}[\mathscr{H}_{\mathrm{int}}, \rho^{(0)}]_{nm}\mathrm{e}^{(\mathrm{i}\omega_{nm}+\Gamma_{nm})t}, \tag{22.81}$$

and hence

$$g_{nm}(t) = -\frac{\mathrm{i}}{\hbar}\int_{-\infty}^{t}[\mathscr{H}_{\mathrm{int}}, \rho^{(0)}]_{nm}\mathrm{e}^{(\mathrm{i}\omega_{nm}+\Gamma_{nm})t'}\,\mathrm{d}t', \tag{22.82}$$

so that

$$\rho_{nm}^{(1)}(t) = -\frac{\mathrm{i}}{\hbar}\mathrm{e}^{-(\mathrm{i}\omega_{nm}+\Gamma_{nm})t}\int_{-\infty}^{t}[\mathcal{H}_{\text{int}}, \rho^{(0)}]_{nm}\mathrm{e}^{(\mathrm{i}\omega_{nm}+\Gamma_{nm})t'}\,\mathrm{d}t'. \tag{22.83}$$

Equation (22.75) can be written as follows:

$$\frac{\partial \rho_{nm}^{(2)}}{\partial t} = (-\mathrm{i}\omega_{nm} - \Gamma_{nm})\rho_{nm}^{(2)} - \frac{\mathrm{i}}{\hbar}[\mathcal{H}_{\text{int}}, \rho^{(1)}]_{nm}, \tag{22.84}$$

which is of the same form as Eq. (22.79), and can be solved in a similar fashion to yield

$$\rho_{nm}^{(2)}(t) = -\frac{\mathrm{i}}{\hbar}\mathrm{e}^{-(\mathrm{i}\omega_{nm}+\Gamma_{nm})t}\int_{-\infty}^{t}[\mathcal{H}_{\text{int}}, \rho^{(1)}]_{nm}\mathrm{e}^{(\mathrm{i}\omega_{nm}+\Gamma_{nm})t'}\,\mathrm{d}t'. \tag{22.85}$$

In fact, the kth-order solution is, by analogy, given by

$$\rho_{nm}^{(k)}(t) = -\frac{\mathrm{i}}{\hbar}\mathrm{e}^{-(\mathrm{i}\omega_{nm}+\Gamma_{nm})t}\int_{-\infty}^{t}[\mathcal{H}_{\text{int}}, \rho^{(k-1)}]_{nm}\mathrm{e}^{(\mathrm{i}\omega_{nm}+\Gamma_{nm})t'}\,\mathrm{d}t'. \tag{22.86}$$

These results are now employed to obtain the linear and second-order contributions to the electric susceptibility. To evaluate $[\mathcal{H}_{\text{int}}, \rho^{(0)}]_{nm}$, it follows from Eqs. (22.55) and (22.56), and (22.78) that

$$\begin{aligned}
[\mathcal{H}_{\text{int}}, \rho^{(0)}]_{nm} &= \left\langle n \middle| \mathcal{H}_{\text{int}}\rho^{(0)} - \rho^{(0)}\mathcal{H}_{\text{int}} \middle| m \right\rangle \\
&= e\sum_{u=1}^{N}\sum_{p=1}\left\{\langle n|\mathbf{r}_u|p\rangle\langle p|\rho^{(0)}|m\rangle - \langle n|\rho^{(0)}|p\rangle\langle p|\mathbf{r}_u|m\rangle\right\}\cdot\mathbf{E} \\
&= e\sum_{u=1}^{N}\sum_{p=1}\left\{\langle n|\mathbf{r}_u|p\rangle\rho_{pm}^{(0)} - \langle p|\mathbf{r}_u|m\rangle\rho_{np}^{(0)}\right\}\cdot\mathbf{E} \\
&= e\sum_{u=1}^{N}\left\{\langle n|\mathbf{r}_u|m\rangle\rho_{mm}^{(0)} - \langle n|\mathbf{r}_u|m\rangle\rho_{nn}^{(0)}\right\}\cdot\mathbf{E} \\
&= -\left\{\rho_{mm}^{(0)} - \rho_{nn}^{(0)}\right\}\boldsymbol{\mu}_{nm}\cdot\mathbf{E}.
\end{aligned} \tag{22.87}$$

The time-dependent electric field is assumed to have a harmonic time dependence of the following form:

$$\mathbf{E}(t) = \sum_{s}\mathbf{E}(\omega_s)\mathrm{e}^{-\mathrm{i}\omega_s t}, \tag{22.88}$$

and to simplify the following notation, Eq. (22.88) will be restricted to a single frequency term for the linear case, so that $\mathbf{E}(t) = \mathbf{E}(\omega)\mathrm{e}^{-\mathrm{i}\omega t}$. The reader should keep in mind the convention employed for the electric field (recall the comments following

Eq. (22.11)). It is necessary to determine $\rho_{nm}^{(1)}(t)$, which is required in the following development. Employing Eqs. (22.83) and (22.87) leads to

$$\begin{aligned}
\rho_{nm}^{(1)}(t) &= \frac{\mathrm{i}}{\hbar}\mathrm{e}^{-(\mathrm{i}\omega_{nm}+\Gamma_{nm})t}\{\rho_{mm}^{(0)} - \rho_{nn}^{(0)}\}\boldsymbol{\mu}_{nm} \cdot \int_{-\infty}^{t} \mathbf{E}(t')\mathrm{e}^{(\mathrm{i}\omega_{nm}+\Gamma_{nm})t'}\,\mathrm{d}t' \\
&= \frac{\mathrm{i}}{\hbar}\mathrm{e}^{-(\mathrm{i}\omega_{nm}+\Gamma_{nm})t}\{\rho_{mm}^{(0)} - \rho_{nn}^{(0)}\}\boldsymbol{\mu}_{nm} \cdot \mathbf{E}(\omega) \int_{-\infty}^{t} \mathrm{e}^{(\mathrm{i}\omega_{nm}-\mathrm{i}\omega+\Gamma_{nm})t'}\,\mathrm{d}t' \\
&= \frac{1}{\hbar}\{\rho_{mm}^{(0)} - \rho_{nn}^{(0)}\}\boldsymbol{\mu}_{nm} \cdot \mathbf{E}(\omega)\frac{\mathrm{e}^{-\mathrm{i}\omega t}}{\omega_{nm} - \omega - \mathrm{i}\Gamma_{nm}}.
\end{aligned} \tag{22.89}$$

The expectation value $\langle\boldsymbol{\mu}(t)\rangle$ is determined to first-order as follows:

$$\begin{aligned}
\langle\boldsymbol{\mu}(t)\rangle &= \mathrm{Tr}\left(\boldsymbol{\mu}(t)\rho^{(1)}\right) \\
&= \sum_m \left\langle m \left| \boldsymbol{\mu}(t)\rho^{(1)} \right| m \right\rangle \\
&= \sum_m \sum_n \langle m|\boldsymbol{\mu}(t)|n\rangle \left\langle n \left| \rho^{(1)} \right| m \right\rangle \\
&= \sum_m \sum_n \boldsymbol{\mu}_{mn}(t)\rho_{nm}^{(1)}.
\end{aligned} \tag{22.90}$$

It will be assumed that $\boldsymbol{\mu}(t)$ has a harmonic time dependence of the same form as the electric field, and to minimize the notation a restriction to a single frequency term is employed, so that

$$\langle\boldsymbol{\mu}(t)\rangle = \langle\boldsymbol{\mu}(\omega)\rangle\mathrm{e}^{-\mathrm{i}\omega t}. \tag{22.91}$$

Recall from Chapter 19 that the electric polarization of the medium is given by

$$\mathbf{P}(\omega) = \mathcal{N}\langle\boldsymbol{\mu}(\omega)\rangle, \tag{22.92}$$

where $\mathcal{N}$ is the number of molecules per unit volume. The polarization to first order in the field is expressed by

$$\mathbf{P}(\omega) = \varepsilon_0\boldsymbol{\chi}^{(1)} \cdot \mathbf{E}(\omega), \tag{22.93}$$

and the vector components of the polarization are given by

$$P_i(\omega) = \varepsilon_0 \sum_j \chi_{ij}^{(1)}(\omega)E_j(\omega). \tag{22.94}$$

Combining Eqs. (22.92), (22.91), (22.90), and (22.89), followed by comparison with Eq. (22.93), leads to the following expression for $\boldsymbol{\chi}^{(1)}$:

$$\boldsymbol{\chi}^{(1)} = \frac{\mathcal{N}}{\varepsilon_0 \hbar} \sum_m \sum_n \frac{\left\{\rho_{mm}^{(0)} - \rho_{nn}^{(0)}\right\} \boldsymbol{\mu}_{mn} \boldsymbol{\mu}_{nm}}{\omega_{nm} - \omega - \mathrm{i}\Gamma_{nm}}. \tag{22.95}$$

In this result $\boldsymbol{\mu}_{mn}\boldsymbol{\mu}_{nm}$ is a dyad. The tensor components of $\boldsymbol{\chi}^{(1)}$ are given by

$$\chi_{ij}^{(1)} = \frac{\mathcal{N}}{\varepsilon_0 \hbar} \sum_m \sum_n \frac{(\mu_i)_{mn}(\mu_j)_{nm} \left\{\rho_{mm}^{(0)} - \rho_{nn}^{(0)}\right\}}{\omega_{nm} - \omega - \mathrm{i}\Gamma_{nm}}, \tag{22.96}$$

which is more commonly written as follows:

$$\begin{aligned}
\chi_{ij}^{(1)} &= \frac{\mathcal{N}}{\varepsilon_0 \hbar} \sum_m \sum_n \left\{ \frac{(\mu_i)_{mn}(\mu_j)_{nm}\rho_{mm}^{(0)}}{\omega_{nm} - \omega - \mathrm{i}\Gamma_{nm}} - \frac{(\mu_i)_{mn}(\mu_j)_{nm}\rho_{nn}^{(0)}}{\omega_{nm} - \omega - \mathrm{i}\Gamma_{nm}} \right\} \\
&= \frac{\mathcal{N}}{\varepsilon_0 \hbar} \sum_m \sum_n \left\{ \frac{(\mu_i)_{mn}(\mu_j)_{nm}\rho_{mm}^{(0)}}{\omega_{nm} - \omega - \mathrm{i}\Gamma_{nm}} - \frac{(\mu_i)_{nm}(\mu_j)_{mn}\rho_{mm}^{(0)}}{\omega_{mn} - \omega - \mathrm{i}\Gamma_{mn}} \right\},
\end{aligned} \tag{22.97}$$

and hence

$$\chi_{ij}^{(1)} = \frac{\mathcal{N}}{\varepsilon_0 \hbar} \sum_m \sum_n \left\{ \frac{(\mu_i)_{nm}(\mu_j)_{mn}}{\omega_{nm} + \omega + \mathrm{i}\Gamma_{nm}} - \frac{(\mu_i)_{mn}(\mu_j)_{nm}}{\omega_{mn} + \omega + \mathrm{i}\Gamma_{nm}} \right\} \rho_{mm}^{(0)}. \tag{22.98}$$

To determine the second-order form of the electric susceptibility it is necessary to evaluate $[\mathscr{H}_{\mathrm{int}}, \rho^{(1)}]_{nm}$, which is determined as

$$\begin{aligned}
[\mathscr{H}_{\mathrm{int}}, \rho^{(1)}]_{nm} &= \left\langle n \middle| \mathscr{H}_{\mathrm{int}} \rho^{(1)} \middle| m \right\rangle - \left\langle n \middle| \rho^{(1)} \mathscr{H}_{\mathrm{int}} \middle| m \right\rangle \\
&= \sum_p \left\{ \langle n|\mathscr{H}_{\mathrm{int}}|p\rangle \left\langle p \middle| \rho^{(1)} \middle| m \right\rangle - \left\langle n \middle| \rho^{(1)} \middle| p \right\rangle \langle p|\mathscr{H}_{\mathrm{int}}|m\rangle \right\} \\
&= -\sum_p \left\{ \boldsymbol{\mu}_{np}\rho_{pm}^{(1)} - \boldsymbol{\mu}_{pm}\rho_{np}^{(1)} \right\} \cdot \mathbf{E}.
\end{aligned} \tag{22.99}$$

The p summation is over the complete set of states. To determine $\rho_{nm}^{(2)}(t)$, it follows from Eq. (22.85) that

$$\rho_{nm}^{(2)}(t) = \frac{\mathrm{i}}{\hbar} \mathrm{e}^{-(\mathrm{i}\omega_{nm} + \Gamma_{nm})t} \mathbf{E}(\omega_1) \cdot \sum_p \int_{-\infty}^{t} \{\boldsymbol{\mu}_{np}\rho_{pm}^{(1)} - \boldsymbol{\mu}_{pm}\rho_{np}^{(1)}\} \mathrm{e}^{(\mathrm{i}\omega_{nm} - \mathrm{i}\omega_1 + \Gamma_{nm})t'} \, \mathrm{d}t', \tag{22.100}$$

which simplifies, on using Eq. (22.89), to yield

$$\rho_{nm}^{(2)}(t) = \frac{\mathrm{i}}{\hbar^2}\mathrm{e}^{-(\mathrm{i}\omega_{nm}+\Gamma_{nm})t}\mathbf{E}(\omega_1)\cdot\sum_p \int_{-\infty}^{t} \mathrm{e}^{(\mathrm{i}\omega_{nm}-\mathrm{i}\omega_1+\Gamma_{nm})t'}\,\mathrm{d}t'$$
$$\times\left[\boldsymbol{\mu}_{np}\{\rho_{mm}^{(0)}-\rho_{pp}^{(0)}\}\boldsymbol{\mu}_{pm}\cdot\mathbf{E}(\omega_2)\frac{\mathrm{e}^{-\mathrm{i}\omega_2 t'}}{\omega_{pm}-\omega_2-\mathrm{i}\Gamma_{pm}}\right.$$
$$\left.-\boldsymbol{\mu}_{pm}\{\rho_{pp}^{(0)}-\rho_{nn}^{(0)}\}\boldsymbol{\mu}_{np}\cdot\mathbf{E}(\omega_2)\frac{\mathrm{e}^{-\mathrm{i}\omega_2 t'}}{\omega_{np}-\omega_2-\mathrm{i}\Gamma_{np}}\right]\mathrm{d}t'$$
$$=\frac{\mathrm{i}}{\hbar^2}\mathrm{e}^{-(\mathrm{i}\omega_{nm}+\Gamma_{nm})t}\mathbf{E}(\omega_1)\cdot$$
$$\times\sum_p\left[\frac{\boldsymbol{\mu}_{np}\{\rho_{mm}^{(0)}-\rho_{pp}^{(0)}\}\boldsymbol{\mu}_{pm}\cdot\mathbf{E}(\omega_2)}{\omega_{pm}-\omega_2-\mathrm{i}\Gamma_{pm}}\int_{-\infty}^{t}\mathrm{e}^{(\mathrm{i}\omega_{nm}-\mathrm{i}\omega_1-\mathrm{i}\omega_2+\Gamma_{nm})t'}\,\mathrm{d}t'\right.$$
$$\left.-\frac{\boldsymbol{\mu}_{pm}\{\rho_{pp}^{(0)}-\rho_{nn}^{(0)}\}\boldsymbol{\mu}_{np}\cdot\mathbf{E}(\omega_2)}{\omega_{np}-\omega_2-\mathrm{i}\Gamma_{np}}\int_{-\infty}^{t}\mathrm{e}^{(\mathrm{i}\omega_{nm}-\mathrm{i}\omega_1-\mathrm{i}\omega_2+\Gamma_{nm})t'}\,\mathrm{d}t'\right]$$
$$=\frac{\mathrm{e}^{-\mathrm{i}(\omega_1+\omega_2)t}}{\hbar^2(\omega_{nm}-\omega_1-\omega_2-\mathrm{i}\Gamma_{nm})}$$
$$\times\sum_p\left\{\frac{\boldsymbol{\mu}_{np}\cdot\mathbf{E}(\omega_1)\boldsymbol{\mu}_{pm}\cdot\mathbf{E}(\omega_2)(\rho_{mm}^{(0)}-\rho_{pp}^{(0)})}{\omega_{pm}-\omega_2-\mathrm{i}\Gamma_{pm}}\right.$$
$$\left.-\frac{\boldsymbol{\mu}_{pm}\cdot\mathbf{E}(\omega_1)\boldsymbol{\mu}_{np}\cdot\mathbf{E}(\omega_2)(\rho_{pp}^{(0)}-\rho_{nn}^{(0)})}{\omega_{np}-\omega_2-\mathrm{i}\Gamma_{np}}\right\}. \qquad (22.101)$$

The second-order contribution to $\langle\boldsymbol{\mu}(t)\rangle$ is given by

$$\langle\boldsymbol{\mu}(t)\rangle = \mathrm{Tr}(\boldsymbol{\mu}(t)\rho^{(2)})$$
$$=\sum_m\langle m|\boldsymbol{\mu}(t)\rho^{(2)}|m\rangle$$
$$=\sum_m\sum_n\boldsymbol{\mu}_{mn}(t)\rho_{nm}^{(2)}. \qquad (22.102)$$

The second-order susceptibility is given by

$$\boldsymbol{\chi}^{(2)}=\frac{\mathcal{N}}{\varepsilon_0\hbar^2}\sum_m\sum_n\frac{\boldsymbol{\mu}_{mn}}{(\omega_{nm}-\omega_1-\omega_2-\mathrm{i}\Gamma_{nm})}$$
$$\times\sum_{p=1}\left\{\frac{\boldsymbol{\mu}_{np}\boldsymbol{\mu}_{pm}(\rho_{mm}^{(0)}-\rho_{pp}^{(0)})}{\omega_{pm}-\omega_2-\mathrm{i}\Gamma_{pm}}-\frac{\boldsymbol{\mu}_{pm}\boldsymbol{\mu}_{np}(\rho_{pp}^{(0)}-\rho_{nn}^{(0)})}{\omega_{np}-\omega_2-\mathrm{i}\Gamma_{np}}\right\}. \qquad (22.103)$$

The tensor components of $\boldsymbol{\chi}^{(2)}$ can be written as follows:

$$\chi_{ijk}^{(2)} = \frac{\mathcal{N}}{\varepsilon_0 \hbar^2} \sum_m \sum_n \frac{(\mu_i)_{mn}}{(\omega_{nm} - \omega_1 - \omega_2 - \mathrm{i}\Gamma_{nm})}$$
$$\times \sum_p \left\{ \frac{(\mu_j)_{np}(\mu_k)_{pm}(\rho_{mm}^{(0)} - \rho_{pp}^{(0)})}{\omega_{pm} - \omega_2 - \mathrm{i}\Gamma_{pm}} - \frac{(\mu_j)_{pm}(\mu_k)_{np}(\rho_{pp}^{(0)} - \rho_{nn}^{(0)})}{\omega_{np} - \omega_2 - \mathrm{i}\Gamma_{np}} \right\}. \tag{22.104}$$

In order to ensure that the susceptibility is invariant under the appropriate permutation symmetry, one-half of the right-hand side of Eq. (22.104) is added to one-half of the same expression, but with the interchanges $\omega_2 \leftrightarrow \omega_1$ and $j \leftrightarrow k$ employed, which leads to

$$\chi_{ijk}^{(2)} = \frac{\mathcal{N}}{2\varepsilon_0 \hbar^2} \sum_m \sum_n \frac{(\mu_i)_{mn}}{(\omega_{nm} - \omega_1 - \omega_2 - \mathrm{i}\Gamma_{nm})}$$
$$\times \left[\sum_p \left\{ \frac{(\mu_j)_{np}(\mu_k)_{pm}(\rho_{mm}^{(0)} - \rho_{pp}^{(0)})}{\omega_{pm} - \omega_2 - \mathrm{i}\Gamma_{pm}} - \frac{(\mu_j)_{pm}(\mu_k)_{np}(\rho_{pp}^{(0)} - \rho_{nn}^{(0)})}{\omega_{np} - \omega_2 - \mathrm{i}\Gamma_{np}} \right\} \right.$$
$$\left. + \sum_p \left\{ \frac{(\mu_k)_{np}(\mu_j)_{pm}(\rho_{mm}^{(0)} - \rho_{pp}^{(0)})}{\omega_{pm} - \omega_1 - \mathrm{i}\Gamma_{pm}} - \frac{(\mu_k)_{pm}(\mu_j)_{np}(\rho_{pp}^{(0)} - \rho_{nn}^{(0)})}{\omega_{np} - \omega_1 - \mathrm{i}\Gamma_{np}} \right\} \right]. \tag{22.105}$$

Some authors do not include the factor of $1/2$, so the reader needs to be alert to this contribution when comparing formulas for $\chi_{ijk}^{(2)}$ from different sources. Equation (22.105) can be simplified in two stages. The first stage involves a summation variable interchange $n \leftrightarrow m$ in terms where $\rho_{nn}^{(0)}$ occurs, which leads to

$$\chi_{ijk}^{(2)} = \frac{\mathcal{N}}{2\varepsilon_0 \hbar^2}$$
$$\times \sum_m \sum_n \sum_p (\rho_{mm}^{(0)} - \rho_{pp}^{(0)}) \left[\frac{(\mu_i)_{mn}(\mu_j)_{np}(\mu_k)_{pm}}{(\omega_{nm} - \omega_1 - \omega_2 - \mathrm{i}\Gamma_{nm})(\omega_{pm} - \omega_2 - \mathrm{i}\Gamma_{pm})} \right.$$
$$\left. + \frac{(\mu_i)_{mn}(\mu_k)_{np}(\mu_j)_{pm}}{(\omega_{nm} - \omega_1 - \omega_2 - \mathrm{i}\Gamma_{nm})(\omega_{pm} - \omega_1 - \mathrm{i}\Gamma_{pm})} \right]$$
$$+ (\rho_{nn}^{(0)} - \rho_{pp}^{(0)}) \left[\frac{(\mu_i)_{mn}(\mu_j)_{pm}(\mu_k)_{np}}{(\omega_{nm} - \omega_1 - \omega_2 - \mathrm{i}\Gamma_{nm})(\omega_{np} - \omega_2 - \mathrm{i}\Gamma_{np})} \right.$$
$$\left. + \frac{(\mu_i)_{mn}(\mu_k)_{pm}(\mu_j)_{np}}{(\omega_{nm} - \omega_1 - \omega_2 - \mathrm{i}\Gamma_{nm})(\omega_{np} - \omega_1 - \mathrm{i}\Gamma_{np})} \right]$$

$$= \frac{\mathcal{N}}{2\varepsilon_0\hbar^2}\sum_m\sum_n\sum_p(\rho_{mm}^{(0)} - \rho_{pp}^{(0)})$$
$$\times\left[\frac{(\mu_i)_{mn}(\mu_j)_{np}(\mu_k)_{pm}}{(\omega_{nm} - \omega_1 - \omega_2 - \mathrm{i}\Gamma_{nm})(\omega_{pm} - \omega_2 - \mathrm{i}\Gamma_{pm})}\right.$$
$$+\frac{(\mu_i)_{mn}(\mu_k)_{np}(\mu_j)_{pm}}{(\omega_{nm} - \omega_1 - \omega_2 - \mathrm{i}\Gamma_{nm})(\omega_{pm} - \omega_1 - \mathrm{i}\Gamma_{pm})}$$
$$+\frac{(\mu_i)_{nm}(\mu_j)_{pn}(\mu_k)_{mp}}{(\omega_{mn} - \omega_1 - \omega_2 - \mathrm{i}\Gamma_{mn})(\omega_{mp} - \omega_2 - \mathrm{i}\Gamma_{mp})}$$
$$\left.+\frac{(\mu_i)_{nm}(\mu_k)_{pn}(\mu_j)_{mp}}{(\omega_{mn} - \omega_1 - \omega_2 - \mathrm{i}\Gamma_{mn})(\omega_{mp} - \omega_1 - \mathrm{i}\Gamma_{mp})}\right]. \tag{22.106}$$

The preceding expression can be reduced to a single factor $\rho_{pp}^{(0)}$ by making the interchange $m \leftrightarrow p$ for the terms involving $\rho_{mm}^{(0)}$ and the notation simplified by introducing $\omega_s = \omega_1 + \omega_2$, so that

$$\chi_{ijk}^{(2)} = \frac{\mathcal{N}}{2\varepsilon_0\hbar^2}\sum_m\sum_n\sum_p\rho_{pp}^{(0)}$$
$$\times\left[\frac{(\mu_i)_{pn}(\mu_j)_{nm}(\mu_k)_{mp}}{(\omega_{np} - \omega_s - \mathrm{i}\Gamma_{np})(\omega_{mp} - \omega_2 - \mathrm{i}\Gamma_{mp})} + \frac{(\mu_i)_{pn}(\mu_k)_{nm}(\mu_j)_{mp}}{(\omega_{np} - \omega_s - \mathrm{i}\Gamma_{np})(\omega_{mp} - \omega_1 - \mathrm{i}\Gamma_{mp})}\right.$$
$$+\frac{(\mu_i)_{np}(\mu_j)_{mn}(\mu_k)_{pm}}{(\omega_{pn} - \omega_s - \mathrm{i}\Gamma_{pn})(\omega_{pm} - \omega_2 - \mathrm{i}\Gamma_{pm})} + \frac{(\mu_i)_{np}(\mu_k)_{mn}(\mu_j)_{pm}}{(\omega_{pn} - \omega_s - \mathrm{i}\Gamma_{pn})(\omega_{pm} - \omega_1 - \mathrm{i}\Gamma_{pm})}$$
$$-\frac{(\mu_i)_{mn}(\mu_j)_{np}(\mu_k)_{pm}}{(\omega_{nm} - \omega_s - \mathrm{i}\Gamma_{nm})(\omega_{pm} - \omega_2 - \mathrm{i}\Gamma_{pm})} - \frac{(\mu_i)_{mn}(\mu_k)_{np}(\mu_j)_{pm}}{(\omega_{nm} - \omega_s - \mathrm{i}\Gamma_{nm})(\omega_{pm} - \omega_1 - \mathrm{i}\Gamma_{pm})}$$
$$\left.-\frac{(\mu_i)_{nm}(\mu_j)_{pn}(\mu_k)_{mp}}{(\omega_{mn} - \omega_s - \mathrm{i}\Gamma_{mn})(\omega_{mp} - \omega_2 - \mathrm{i}\Gamma_{mp})} - \frac{(\mu_i)_{nm}(\mu_k)_{pn}(\mu_j)_{mp}}{(\omega_{mn} - \omega_s - \mathrm{i}\Gamma_{mn})(\omega_{mp} - \omega_1 - \mathrm{i}\Gamma_{mp})}\right]. \tag{22.107}$$

By appropriate switches of the dummy variables m and n, Eq. (22.107) can be written with the matrix element indices aligned, so that

$$\chi_{ijk}^{(2)} = \frac{\mathcal{N}}{2\varepsilon_0\hbar^2}\sum_m\sum_n\sum_p\rho_{pp}^{(0)}$$
$$\times\left[\frac{(\mu_i)_{pn}(\mu_j)_{nm}(\mu_k)_{mp}}{(\omega_{np} - \omega_s - \mathrm{i}\Gamma_{np})(\omega_{mp} - \omega_2 - \mathrm{i}\Gamma_{mp})} + \frac{(\mu_i)_{pn}(\mu_k)_{nm}(\mu_j)_{mp}}{(\omega_{np} - \omega_s - \mathrm{i}\Gamma_{np})(\omega_{mp} - \omega_1 - \mathrm{i}\Gamma_{mp})}\right.$$
$$-\frac{(\mu_j)_{pn}(\mu_i)_{nm}(\mu_k)_{mp}}{(\omega_{mn} - \omega_s - \mathrm{i}\Gamma_{mn})(\omega_{pn} - \omega_1 - \mathrm{i}\Gamma_{pn})} - \frac{(\mu_j)_{pn}(\mu_i)_{nm}(\mu_k)_{mp}}{(\omega_{mn} - \omega_s - \mathrm{i}\Gamma_{mn})(\omega_{mp} - \omega_2 - \mathrm{i}\Gamma_{mp})}$$
$$+\frac{(\mu_j)_{pn}(\mu_k)_{nm}(\mu_i)_{mp}}{(\omega_{pm} - \omega_s - \mathrm{i}\Gamma_{pm})(\omega_{pn} - \omega_1 - \mathrm{i}\Gamma_{pn})} - \frac{(\mu_k)_{pn}(\mu_i)_{nm}(\mu_j)_{mp}}{(\omega_{mn} - \omega_s - \mathrm{i}\Gamma_{mn})(\omega_{pn} - \omega_2 - \mathrm{i}\Gamma_{pn})}$$
$$\left.-\frac{(\mu_k)_{pn}(\mu_i)_{nm}(\mu_j)_{mp}}{(\omega_{mn} - \omega_s - \mathrm{i}\Gamma_{mn})(\omega_{mp} - \omega_1 - \mathrm{i}\Gamma_{mp})} + \frac{(\mu_k)_{pn}(\mu_j)_{nm}(\mu_i)_{mp}}{(\omega_{pm} - \omega_s - \mathrm{i}\Gamma_{pm})(\omega_{pn} - \omega_2 - \mathrm{i}\Gamma_{pn})}\right]. \tag{22.108}$$

This is the quantum mechanical formula for the second-order contribution to the electric susceptibility. For those readers familiar with Feynman diagrams, one can appreciate that they are particularly useful as a book-keeping device for this type of calculation, and for the evaluation of the expressions for the more complex situations that arise for higher-order susceptibilities.

By treating ω_1 and ω_2 as complex variables, an examination of the denominator factors in Eq. (22.108) indicates that the poles lie in the lower half of the complex angular frequency planes, and therefore a quantum theoretic treatment confirms the analytic structure found from the anharmonic oscillator model. This formula can also be employed to ascertain the asymptotic behavior of $\chi_{ijk}^{(2)}$ as $\omega_1 \to \infty$ and $\omega_2 \to \infty$, a topic to be addressed in the following section.

22.5 Asymptotic behavior for the nonlinear susceptibility

Before proceeding to the derivation of the dispersion relations for the nonlinear susceptibilities, it is necessary to investigate the asymptotic behavior as $\omega \to \infty$. The linear case is considered first, to see the approach employed in its simplest setting.

From Eq. (22.98) it follows that

$$\begin{aligned}
\chi_{ij}^{(1)}(\omega) &= \frac{\mathcal{N}e^2}{\varepsilon_0 \hbar \omega^2} \sum_u \sum_v \rho_{uu}^{(0)} \left(1 + \frac{\omega_{vu}}{\omega} + \frac{\mathrm{i}\Gamma_{vu}}{\omega}\right)^{-1} \left(1 + \frac{\omega_{uv}}{\omega} + \frac{\mathrm{i}\Gamma_{uv}}{\omega}\right)^{-1} \\
&\quad \times [\omega\{\langle v|r_i|u\rangle\langle u|r_j|v\rangle - \langle u|r_i|v\rangle\langle v|r_j|u\rangle\} + \mathrm{i}\Gamma_{uv}\{\langle v|r_i|u\rangle\langle u|r_j|v\rangle \\
&\quad - \langle u|r_i|v\rangle\langle v|r_j|u\rangle\} + \omega_{uv}\{\langle v|r_i|u\rangle\langle u|r_j|v\rangle + \langle u|r_i|v\rangle\langle v|r_j|u\rangle\}] \\
&= \frac{\mathcal{N}e^2}{\varepsilon_0 \hbar \omega^2} \sum_u \sum_v \rho_{uu}^{(0)} [\omega\{\langle v|r_i|u\rangle\langle u|r_j|v\rangle - \langle u|r_i|v\rangle\langle v|r_j|u\rangle\} \\
&\quad + \mathrm{i}\Gamma_{uv}\{\langle v|r_i|u\rangle\langle u|r_j|v\rangle - \langle u|r_i|v\rangle\langle v|r_j|u\rangle\} \\
&\quad + \omega_{uv}\{\langle v|r_i|u\rangle\langle u|r_j|v\rangle + \langle u|r_i|v\rangle\langle v|r_j|u\rangle\} + O(\omega^{-1})].
\end{aligned} \tag{22.109}$$

The term $O(\omega^{-1})$ in the preceding result can be ignored. Making use of Eq. (22.60) yields

$$\begin{aligned}
&\sum_u \sum_v \rho_{uu}^{(0)} \{\langle v|r_i|u\rangle\langle u|r_j|v\rangle - \langle u|r_i|v\rangle\langle v|r_j|u\rangle\} \\
&\quad = \sum_u \sum_v \rho_{uu}^{(0)} \langle u|r_j|v\rangle\langle v|r_i|u\rangle - \sum_u \rho_{uu}^{(0)} \langle u|r_i r_j|u\rangle \\
&\quad = \sum_u \rho_{uu}^{(0)} \langle u|r_j r_i|u\rangle - \sum_u \rho_{uu}^{(0)} \langle u|r_i r_j|u\rangle \\
&\quad = 0.
\end{aligned} \tag{22.110}$$

The third term in Eq. (22.109) can be simplified by making use of the following result:

$$\langle v|[\mathscr{H}_0, r_i]|u\rangle = \langle v|\mathscr{H}_0 r_i - r_i \mathscr{H}_0|u\rangle = (\hbar\omega_v - \hbar\omega_u)\langle v|r_i|u\rangle; \tag{22.111}$$

employing

$$\langle v|[\mathscr{H}_0, r_i]|u\rangle = -\frac{\mathrm{i}\hbar}{m}\langle v|p_i|u\rangle \tag{22.112}$$

leads to

$$\omega_{vu}\langle v|r_i|u\rangle = -\frac{\mathrm{i}}{m}\langle v|p_i|u\rangle. \tag{22.113}$$

The third term in Eq. (22.109) can be rewritten in the following manner:

$$\begin{aligned}
&\sum_u \sum_v \rho_{uu}^{(0)}[-\omega_{vu}\{\langle v|r_i|u\rangle\langle u|r_j|v\rangle + \langle u|r_i|v\rangle\langle v|r_j|u\rangle\} \\
&\quad = -\sum_u \sum_v \rho_{uu}^{(0)}\{\omega_{vu}\langle v|r_i|u\rangle\langle u|r_j|v\rangle - \langle v|r_j|u\rangle\omega_{uv}\langle u|r_i|v\rangle\} \\
&\quad = \frac{\mathrm{i}}{m}\sum_u \sum_v \rho_{uu}^{(0)}\{\langle v|p_i|u\rangle\langle u|r_j|v\rangle - \langle v|r_j|u\rangle\langle u|p_i|v\rangle\} \\
&\quad = \frac{\mathrm{i}}{m}\sum_u \sum_v \rho_{uu}^{(0)}\{\langle u|r_j|v\rangle\langle v|p_i|u\rangle - \langle u|p_i|v\rangle\langle v|r_j|u\rangle\} \\
&\quad = \frac{\mathrm{i}}{m}\sum_u \rho_{uu}^{(0)}\{\langle u|r_j p_i|u\rangle - \langle u|p_i r_j|u\rangle\} \\
&\quad = \frac{\mathrm{i}}{m}\sum_u \rho_{uu}^{(0)}\{\langle u|r_j p_i|u\rangle - \langle u|(p_i r_j) + r_j p_i|u\rangle\} \\
&\quad = -\frac{\mathrm{i}}{m}\sum_u \rho_{uu}^{(0)}\langle u|(p_i r_j)|u\rangle \\
&\quad = -\frac{\hbar}{m}\delta_{ij}\sum_u \rho_{uu}^{(0)}\langle u|u\rangle.
\end{aligned} \tag{22.114}$$

Hence, from Eq. (22.109) it follows in the limit $\omega \to \infty$ that

$$\chi_{ij}^{(1)}(\omega) = -\frac{Ne^2}{\varepsilon_0 m}\frac{\delta_{ij}}{\omega^2}\sum_u \rho_{uu}^{(0)}\langle u|u\rangle. \tag{22.115}$$

There is an additional term $O(\omega^{-2})$ arising from the contribution involving Γ_{uv}. If all the Γ_{uv} factors are approximately constant, that is $\Gamma_{uv} \approx \Gamma$, then this additional $O(\omega^{-2})$ term can be deleted on using Eq. (22.110). At room temperature or below for a medium at thermal equilibrium, it is an excellent approximation for most systems to assume the excited electronic states are not significantly populated, which means

$$\rho_{uu}^{(0)} = \delta_{ug}, \tag{22.116}$$

where g is used to designate the ground electronic state. Hence Eq. (22.115) can be simplified as follows:

$$\chi_{ij}^{(1)}(\omega) = -\frac{Ne^2}{\varepsilon_0 m}\frac{\delta_{ij}}{\omega^2} + o(\omega^{-2}). \tag{22.117}$$

Inserting the definition of the plasma frequency, ω_{p}, yields

$$\chi_{ij}^{(1)}(\omega) = -\frac{\omega_{\mathrm{p}}^2\delta_{ij}}{\omega^2} + o(\omega^{-2}). \tag{22.118}$$

When spatial and relativistic effects are considered in first order, the asymptotic form of $\chi_{ij}^{(1)}(\mathbf{k},\omega)$ is given by

$$\chi_{ij}^{(1)}(\mathbf{k},\omega) = -\frac{\omega_{\mathrm{p}}^2\delta_{ij}}{\omega^2}\left(1 - \frac{2\langle T_0\rangle}{3mc^2}\right) + o(\omega^{-2}), \tag{22.119}$$

where $\langle T_0\rangle$ denotes the average ground state kinetic energy, m is the electronic mass, and c is the speed of light. For readers interested in pursuing the details leading to Eq. (22.119), see Scandolo *et al.* (2001).

The asymptotic behavior of $\chi_{ijk}^{(2)}(\omega;\omega_1,\omega_2)$ is now evaluated. Two cases are considered: the behavior of $\chi_{ijk}^{(2)}(\omega;\omega_1,\omega_2)$ as $\omega_1 \to \infty$ and the SHG case, where $\lim_{\omega\to\infty}\chi_{ijk}^{(2)}(2\omega;\omega,\omega)$. To simplify the writing, let

$$\mathsf{a} = (\mu_i)_{pn}(\mu_j)_{nm}(\mu_k)_{mp}, \tag{22.120}$$

$$\mathsf{b} = (\mu_i)_{pn}(\mu_k)_{nm}(\mu_j)_{mp}, \tag{22.121}$$

$$\mathsf{c} = (\mu_j)_{pn}(\mu_i)_{nm}(\mu_k)_{mp}, \tag{22.122}$$

$$\mathsf{d} = (\mu_j)_{pn}(\mu_k)_{nm}(\mu_i)_{mp}, \tag{22.123}$$

$$\mathsf{e} = (\mu_k)_{pn}(\mu_i)_{nm}(\mu_j)_{mp}, \tag{22.124}$$

and

$$\mathsf{f} = (\mu_k)_{pn}(\mu_j)_{nm}(\mu_i)_{mp}. \tag{22.125}$$

If the Γ_{uv} factors are ignored, then the expansion of Eq. (22.108) would be expected, on the basis of results from the anharmonic oscillator model, to take the following form:

$$\lim_{\omega_1\to\infty} \chi_{ijk}^{(2)}(\omega_s;\omega_1,\omega_2) = c_0 + \frac{c_1}{\omega_1} + \frac{c_2}{\omega_1^2} + \frac{c_3}{\omega_1^3} + O\left(\frac{1}{\omega_1^4}\right), \tag{22.126}$$

with $c_k = 0$, for $k = 0$, 1, 2, and 3. From Eq. (22.108) $c_0 = 0$, by inspection. The coefficient c_1 is given by

$$c_1 = S\left\{\frac{\mathsf{a}-\mathsf{c}}{(\omega_2-\omega_{mp})} + \frac{\mathsf{f}-\mathsf{e}}{(\omega_2+\omega_{np})}\right\}, \tag{22.127}$$

where S is a multiple summation operator represented by

$$S = \frac{N}{2\varepsilon_0\hbar^2}\sum_m\sum_n\sum_p \rho_{pp}^{(0)}. \tag{22.128}$$

If the sum over n is performed for the first term and over m for the second term in Eq. (22.127), then

$$\begin{aligned}
c_1 &= \frac{N}{2\varepsilon_0\hbar^2}\sum_p \rho_{pp}^{(0)}\left\{\sum_m\sum_n \frac{(\mu_i)_{pn}(\mu_j)_{nm}(\mu_k)_{mp} - (\mu_j)_{pn}(\mu_i)_{nm}(\mu_k)_{mp}}{(\omega_2-\omega_{mp})}\right. \\
&\quad \left. + \sum_m\sum_n \frac{(\mu_k)_{pn}(\mu_j)_{nm}(\mu_i)_{mp} - (\mu_k)_{pn}(\mu_i)_{nm}(\mu_j)_{mp}}{(\omega_2+\omega_{np})}\right\} \\
&= \frac{N}{2\varepsilon_0\hbar^2}\sum_p \rho_{pp}^{(0)}\left\{\sum_m \frac{(\mu_i\mu_j)_{pm}(\mu_k)_{mp} - (\mu_j\mu_i)_{pm}(\mu_k)_{mp}}{(\omega_2-\omega_{mp})}\right. \\
&\quad \left. + \sum_n \frac{(\mu_k)_{pn}(\mu_j\mu_i)_{np} - (\mu_k)_{pn}(\mu_i\mu_j)_{np}}{(\omega_2+\omega_{np})}\right\} \\
&= 0.
\end{aligned} \tag{22.129}$$

It is left as an exercise for the reader to examine c_k for $k = 2$, 3, and 4.

In a related manner, it is expected, also on the basis of the anharmonic oscillator model, that the asymptotic behavior for the SHG susceptibility would take the following form:

$$\lim_{\omega\to\infty} \chi_{ijk}^{(2)}(2\omega;\omega,\omega) = \sum_{k=0}^{5}\frac{c_k}{\omega^k} + O\left(\frac{1}{\omega^6}\right), \tag{22.130}$$

with $c_k = 0$ for $k = 0$ to 5. Equation (22.108) yields $c_0 = 0$ and $c_1 = 0$, by inspection. For the coefficient c_2 it follows that

$$c_2 = \frac{1}{2}S(\mathsf{a}+\mathsf{b}-2\mathsf{c}+\mathsf{d}-\mathsf{e}), \tag{22.131}$$

which simplifies, by carrying out the m and n summations, to yield

$$c_2 = 0. \tag{22.132}$$

Establishing that c_0, c_1, and c_2 vanish is sufficient to derive the standard form of the Kramers–Kronig relations for $\chi^{(2)}_{ijk}(2\omega; \omega, \omega)$, but is not sufficient to determine all the sum rules that must be satisfied by the real and imaginary parts of this optical property. The interested reader is left the task of examining c_k for $k = 3, 4, 5$, and 6.

The brute force technique just sketched clearly benefits from having access to one of the standard symbolic algebra software packages. Extending the brute force approach to deal with the general nth-order susceptibility has obvious limitations. Bassani and Scandolo (1991) gave a more direct approach. The basic idea is to represent the frequency-dependent optical constant as a Fourier transform, and then carry out an integration by parts, leading to an asymptotic series in the frequency variable of interest. This approach requires the evaluation of time derivatives of the response functions. This can be achieved by a density matrix formulation using an explicit formula for the response function. The final result is as follows:

$$\lim_{\omega_1 \to \infty} \chi^{(2)}(\omega_1, \omega_2) = \frac{\mathsf{C}}{\omega_1^4} + o(\omega_1^{-4}), \tag{22.133}$$

where C is determined as a somewhat complicated integral. This result is in agreement with the $O(\omega_1^{-4})$ behavior obtained from the simpler anharmonic oscillator model. The same authors (Scandolo and Bassani, 1995a) carried out an analysis for SHG and determined

$$\lim_{\omega \to \infty} \chi^{(2)}_{ijk}(\omega, \omega) = \frac{Ne^3}{8m^3} \left\langle \frac{\partial^3 V}{\partial x_i \partial x_j \partial x_k} \right\rangle_0 \frac{1}{\omega^6} + o(\omega^{-6}), \tag{22.134}$$

where V denotes the external potential, and the subscript zero signifies that the average is calculated for the ground state of the system. The asymptotic behavior $O(\omega^{-6})$ found for $\chi^{(2)}_{ijk}(\omega, \omega)$ coincides with that obtained from the anharmonic model. Rapapa and Scandolo (1996), using the same approach, determined the asymptotic behavior for third-harmonic generation as follows:

$$\lim_{\omega \to \infty} \chi^{(3)}_{ijkl}(\omega, \omega, \omega) = \frac{Ne^4}{27m^4} \left\langle \frac{\partial^4 V}{\partial x_i \partial x_j \partial x_k \partial x_l} \right\rangle_0 \frac{1}{\omega^8} + o(\omega^{-8}). \tag{22.135}$$

The analysis has been taken further by Bassani and Lucarini (2000), who found that

$$\lim_{\omega \to \infty} \chi^{(n)}_{ijkl\ldots}(n\omega; \omega, \omega, \ldots, \omega) = O\left(\frac{1}{\omega^{2n+2}}\right), \quad \text{for } n > 1. \tag{22.136}$$

Readers interested in the asymptotic behavior of the susceptibilities can pursue the intricacies of the derivations involved by consulting the references cited.

22.6 One-variable dispersion relations for the nonlinear susceptibility

This section considers the derivation of simple dispersion relations in one angular frequency, with all but one of the angular frequency variables held fixed. Tensor subscripts are omitted to simplify the notation. The nth-order susceptibility under discussion is $\chi^{(n)}(\omega_1, \omega_2, \ldots, \omega_k, \ldots, \omega_n)$, where it is assumed that all angular frequencies are distinct and positive. Attention is focused on the angular frequency ω_k, so that the function $\chi^{(n)}(\omega_1, \omega_2, \ldots, \omega_k, \ldots, \omega_n)$ is analytic in the upper half complex ω_k-plane, and has a suitable asymptotic behavior as $\omega_k \to \infty$. Consider the integral

$$\oint_C \frac{\chi^{(n)}(\omega_1, \omega_2, \ldots, \omega_z, \ldots, \omega_n)\mathrm{d}\omega_z}{\omega - \omega_z},$$

where C denotes the contour shown in Figure 19.2, and apply the Cauchy integral theorem. It follows, in the limits $R \to \infty$ and $\rho \to 0$, that

$$\chi^{(n)}(\omega_1, \omega_2, \ldots, \omega, \ldots, \omega_n) = -\frac{1}{\pi \mathrm{i}} P \int_{-\infty}^{\infty} \frac{\chi^{(n)}(\omega_1, \omega_2, \ldots, \omega_\mathrm{r}, \ldots, \omega_n)\mathrm{d}\omega_\mathrm{r}}{\omega - \omega_\mathrm{r}}. \tag{22.137}$$

If $\chi^{(n)}(\omega_1, \omega_2, \ldots, \omega, \ldots, \omega_n)$ is expressed in terms of its real and imaginary parts as follows:

$$\chi^{(n)}(\omega_1, \omega_2, \ldots, \omega, \ldots, \omega_n) = \mathrm{Re}\{\chi^{(n)}(\omega_1, \omega_2, \ldots, \omega, \ldots, \omega_n)\} + \mathrm{i}\,\mathrm{Im}\{\chi^{(n)}(\omega_1, \omega_2, \ldots, \omega, \ldots, \omega_n)\}, \tag{22.138}$$

then Eq. (22.137) can be written as

$$\mathrm{Im}\chi^{(n)}(\omega_1, \omega_2, \ldots, \omega, \ldots, \omega_n) = \frac{1}{\pi} P \int_{-\infty}^{\infty} \frac{\mathrm{Re}\,\chi^{(n)}(\omega_1, \omega_2, \ldots, \omega_\mathrm{r}, \ldots, \omega_n)\mathrm{d}\omega_\mathrm{r}}{\omega - \omega_\mathrm{r}} \tag{22.139}$$

and

$$\mathrm{Re}\chi^{(n)}(\omega_1, \omega_2, \ldots, \omega, \ldots, \omega_n) = -\frac{1}{\pi} P \int_{-\infty}^{\infty} \frac{\mathrm{Im}\,\chi^{(n)}(\omega_1, \omega_2, \ldots, \omega_\mathrm{r}, \ldots, \omega_n)\mathrm{d}\omega_\mathrm{r}}{\omega - \omega_\mathrm{r}}. \tag{22.140}$$

These relations correspond to single-variable versions of the conventional Kramers–Kronig relations applied to the nth-order susceptibility, with the previously stated conditions on the angular frequencies imposed. The preceding derivation rests on a similar set of assumptions to those outlined at the end of Section 19.2 for the derivation of the Kramers–Kronig relations for the linear susceptibility. In particular, the derivation is model-dependent, since the analytic structure of $\chi^{(n)}(\omega_1, \omega_2, \ldots, \omega_k, \ldots, \omega_n)$ and its asymptotic dependence as $\omega_k \to \infty$ are deduced directly from the anharmonic oscillator description of the physics involved.

A different approach is now explored for the development of the dispersion relations for the case of higher harmonic nonlinearities. Let p_k, for $k = 1, \ldots, n$, be the set of integers describing the harmonics which appear, and let the restriction that $p_k \geq 0$ for $k = 1, \ldots, n$ be imposed. For example, SHG has $n = 2$ and $p_1 = p_2 = 1$, THG has $n = 3$ and $p_1 = p_2 = p_3 = 1$. The nth-order susceptibility can be written as $\chi^{(n)}(p_1\omega, p_2\omega, \ldots, p_n\omega)$. Consider the integral

$$\oint_C \frac{\chi^{(n)}(p_1\omega_z, p_2\omega_z, \ldots, p_n\omega_z)\mathrm{d}\omega_z}{\omega - \omega_z},$$

where C denotes the contour shown in Figure 19.2, and apply the Cauchy integral theorem. It follows, after taking the limits $R \to \infty$ and $\rho \to 0$, that

$$\chi^{(n)}(p_1\omega, p_2\omega, \ldots, p_n\omega) = -\frac{1}{\pi \mathrm{i}} P \int_{-\infty}^{\infty} \frac{\chi^{(n)}(p_1\omega_\mathrm{r}, p_2\omega_\mathrm{r}, \ldots, p_n\omega_\mathrm{r})\mathrm{d}\omega_\mathrm{r}}{\omega - \omega_\mathrm{r}}. \quad (22.141)$$

The resulting dispersion relations then take the following form:

$$\mathrm{Im}\,\chi^{(n)}(p_1\omega, p_2\omega, \ldots, p_n\omega) = \frac{1}{\pi} P \int_{-\infty}^{\infty} \frac{\mathrm{Re}\,\chi^{(n)}(p_1\omega_\mathrm{r}, p_2\omega_\mathrm{r}, \ldots, p_n\omega_\mathrm{r})\mathrm{d}\omega_\mathrm{r}}{\omega - \omega_\mathrm{r}} \quad (22.142)$$

and

$$\mathrm{Re}\,\chi^{(n)}(p_1\omega, p_2\omega, \ldots, p_n\omega) = -\frac{1}{\pi} P \int_{-\infty}^{\infty} \frac{\mathrm{Im}\,\chi^{(n)}(p_1\omega_\mathrm{r}, p_2\omega_\mathrm{r}, \ldots, p_n\omega_\mathrm{r})\mathrm{d}\omega_\mathrm{r}}{\omega - \omega_\mathrm{r}}. \quad (22.143)$$

They can be put in a more useful form by employing the crossing symmetry relation Eq. (22.25), so that

$$\mathrm{Im}\,\chi^{(n)}(p_1\omega, p_2\omega, \ldots, p_n\omega) = \frac{2\omega}{\pi} P \int_{0}^{\infty} \frac{\mathrm{Re}\,\chi^{(n)}(p_1\omega_\mathrm{r}, p_2\omega_\mathrm{r}, \ldots, p_n\omega_\mathrm{r})\mathrm{d}\omega_\mathrm{r}}{\omega^2 - \omega_\mathrm{r}^2} \quad (22.144)$$

and

$$\mathrm{Re}\,\chi^{(n)}(p_1\omega, p_2\omega, \ldots, p_n\omega) = -\frac{2}{\pi} P \int_{0}^{\infty} \frac{\omega_\mathrm{r}\,\mathrm{Im}\,\chi^{(n)}(p_1\omega_\mathrm{r}, p_2\omega_\mathrm{r}, \ldots, p_n\omega_\mathrm{r})\mathrm{d}\omega_\mathrm{r}}{\omega^2 - \omega_\mathrm{r}^2}. \quad (22.145)$$

The preceding two results have the form of the conventional Kramers–Kronig relations. For SHG these equations can be expressed as

$$\mathrm{Im}\,\chi^{(2)}(2\omega; \omega, \omega) = \frac{2\omega}{\pi} P \int_{0}^{\infty} \frac{\mathrm{Re}\,\chi^{(2)}(2\omega_\mathrm{r}; \omega_\mathrm{r}, \omega_\mathrm{r})\mathrm{d}\omega_\mathrm{r}}{\omega^2 - \omega_\mathrm{r}^2} \quad (22.146)$$

and

$$\operatorname{Re}\chi^{(2)}(2\omega;\omega,\omega) = -\frac{2}{\pi}P\int_0^\infty \frac{\omega_r \operatorname{Im}\chi^{(2)}(2\omega_r;\omega_r,\omega_r)\mathrm{d}\omega_r}{\omega^2-\omega_r^2}. \qquad (22.147)$$

The previous discussion for harmonic generation dispersion relations can be generalized by considering the integral

$$\oint_C \frac{\omega_z^p\{\chi^{(n)}(p_1\omega_z,p_2\omega_z,\ldots,p_n\omega_z)\}^q\,\mathrm{d}\omega_z}{\omega-\omega_z},$$

where C denotes the contour shown in Figure 19.2, and p and q are integers subject to the constraints $q \geq 1$ and $0 \leq p < (2n+2)q$, with $n > 1$. This approach leads to the following result:

$$\begin{aligned}
&\omega^p\{\operatorname{Re}\chi^{(n)}(p_1\omega,p_2\omega,\ldots,p_n\omega)+\mathrm{i}\operatorname{Im}\chi^{(n)}(p_1\omega,p_2\omega,\ldots,p_n\omega)\}^q\\
&=\frac{\mathrm{i}}{\pi}P\int_0^\infty\left[\frac{\omega'^p\{\operatorname{Re}\chi^{(n)}(p_1\omega',p_2\omega',\ldots,p_n\omega')+\mathrm{i}\operatorname{Im}\chi^{(n)}(p_1\omega',p_2\omega',\ldots,p_n\omega')\}^q}{\omega-\omega'}\right.\\
&\left.+(-1)^p\frac{\omega'^p\{\operatorname{Re}\chi^{(n)}(p_1\omega',p_2\omega',\ldots,p_n\omega')-\mathrm{i}\operatorname{Im}\chi^{(n)}(p_1\omega',p_2\omega',\ldots,p_n\omega')\}^q}{\omega+\omega'}\right]\mathrm{d}\omega'.
\end{aligned} \qquad (22.148)$$

For $p = 0, q = 1$, and $n = 2$, Eq. (22.148) simplifies to Eqs. (22.146) and (22.147). A large number of the available sum rules for harmonic generation can be generated from Eq. (22.148).

The nonlinear susceptibility can also be expressed in the alternative following form:

$$\chi^{(n)} = \left|\chi^{(n)}\right| \mathrm{e}^{\mathrm{i}\phi}, \qquad (22.149)$$

where $\left|\chi^{(n)}\right|$ denotes the modulus and ϕ the phase shift, with both quantities depending on the appropriate frequencies. Since both of these are experimentally accessible, it is of interest to examine the dispersion relations that are obtained for the modulus and phase. The problem is very similar to that already covered in Section 20.2, and some of the details can be carried over directly to the present problem.

Let $\chi_u^{(n)}$ denote a dimensional factor having the same units as $\chi^{(n)}(\omega_1,\omega_2,\ldots,\omega_k,\ldots,\omega_n)$. In order to uncouple the modulus and phase, it is necessary to take the log of $\chi^{(n)}(\omega_1,\omega_2,\ldots,\omega_k,\ldots,\omega_n)$. Consider the integral

$$\oint_C \log\{\chi^{(n)}(\omega_1,\omega_2,\ldots,\omega_{z_k},\ldots,\omega_n)/\chi_u^{(n)}\}\left(\frac{1}{\omega-\omega_{z_k}}-\frac{1}{\omega_0-\omega_{z_k}}\right)\mathrm{d}\omega_{z_k},$$

where C designates the contour shown in Figure 20.1, and ω and ω_0 are two arbitrary frequencies. The factor $\chi_u^{(n)}$ keeps the argument of the logarithm dimensionless. To simplify the notation for the moment, all the frequency arguments except the one being integrated are suppressed, so that $\chi^{(n)}(\omega_1, \omega_2, \ldots, \omega_k, \ldots, \omega_n)$ is written as $\chi^{(n)}(\omega_k)$; the superscript (n) should remind the reader that it is the general nth-order nonlinear susceptibility under discussion. Employing the Cauchy integral theorem leads to

$$\begin{aligned}
&\int_{-\mathcal{R}}^{\omega-\delta} \log\{\chi^{(n)}(\omega_r)/\chi_u^{(n)}\} \left(\frac{1}{\omega-\omega_r} - \frac{1}{\omega_0-\omega_r}\right) d\omega_r \\
&+ \int_{\omega+\delta}^{\omega_0-\rho} \log\{\chi^{(n)}(\omega_r)/\chi_u^{(n)}\} \left(\frac{1}{\omega-\omega_r} - \frac{1}{\omega_0-\omega_r}\right) d\omega_r \\
&+ \int_{\omega_0+\rho}^{\mathcal{R}} \log\{\chi^{(n)}(\omega_r)/\chi_u^{(n)}\} \left(\frac{1}{\omega-\omega_r} - \frac{1}{\omega_0-\omega_r}\right) d\omega_r \\
&+ \int_0^{\pi} \log\{\chi^{(n)}(\omega+\delta e^{i\varphi})/\chi_u^{(n)}\} \left(\frac{1}{\delta e^{i\varphi}} + \frac{1}{\omega_0-\omega-\delta e^{i\varphi}}\right) i\delta e^{i\varphi}\, d\varphi \\
&- \int_0^{\pi} \log\{\chi^{(n)}(\omega_0+\rho e^{i\varphi})/\chi_u^{(n)}\} \left(\frac{1}{\omega-\omega_0-\rho e^{i\varphi}} + \frac{1}{\rho e^{i\varphi}}\right) i\rho e^{i\varphi}\, d\varphi \\
&+ \int_0^{\pi} \log\{\chi^{(n)}(\mathcal{R} e^{i\vartheta})/\chi_u^{(n)}\} \left(\frac{1}{\omega-\mathcal{R} e^{i\vartheta}} - \frac{1}{\omega_0-\mathcal{R} e^{i\vartheta}}\right) i\mathcal{R} e^{i\vartheta}\, d\vartheta = 0.
\end{aligned} \tag{22.150}$$

Taking the limits $\delta \to 0$ and $\rho \to 0$ in Eq. (22.150) leads to

$$\begin{aligned}
&P\int_{-\mathcal{R}}^{\mathcal{R}} \log\{\chi^{(n)}(\omega_r)/\chi_u^{(n)}\} \left(\frac{1}{\omega-\omega_r} - \frac{1}{\omega_0-\omega_r}\right) d\omega_r + i\pi \log\{\chi^{(n)}(\omega)/\chi^{(n)}(\omega_0)\} \\
&+ \int_0^{\pi} \log\{\chi^{(n)}(\mathcal{R} e^{i\vartheta})/\chi_u^{(n)}\} \left(\frac{1}{\omega-\mathcal{R} e^{i\vartheta}} - \frac{1}{\omega_0-\mathcal{R} e^{i\vartheta}}\right) i\mathcal{R} e^{i\vartheta}\, d\vartheta = 0.
\end{aligned} \tag{22.151}$$

Taking the limit $\mathcal{R} \to \infty$, the integral along the large semicircular arc vanishes, and hence

$$i\pi \log\{\chi^{(n)}(\omega)/\chi^{(n)}(\omega_0)\} = -P\int_{-\infty}^{\infty} \log\{\chi^{(n)}(\omega')/\chi_u^{(n)}\} \left(\frac{1}{\omega-\omega'} - \frac{1}{\omega_0-\omega'}\right) d\omega'. \tag{22.152}$$

From the real and imaginary parts of this result, it follows that

$$\begin{aligned}
&\phi(\omega_1, \omega_2, \ldots, \omega, \ldots, \omega_n) - \phi(\omega_1, \omega_2, \ldots, \omega_0, \ldots, \omega_n) \\
&= \frac{1}{\pi} P\int_{-\infty}^{\infty} \log\left|\chi^{(n)}(\omega_1, \omega_2, \ldots, \omega_k, \ldots, \omega_n)/\chi_u^{(n)}\right| \left(\frac{1}{\omega-\omega_k} - \frac{1}{\omega_0-\omega_k}\right) d\omega_k
\end{aligned} \tag{22.153}$$

and

$$\log\left|\frac{\chi^{(n)}(\omega_1,\omega_2,\ldots,\omega,\ldots,\omega_n)}{\chi^{(n)}(\omega_1,\omega_2,\ldots,\omega_0,\ldots,\omega_n)}\right| = -\frac{1}{\pi}P\int_{-\infty}^{\infty}\phi(\omega_1,\omega_2,\ldots,\omega_k,\ldots,\omega_n) \times\left(\frac{1}{\omega-\omega_k}-\frac{1}{\omega_0-\omega_k}\right)\mathrm{d}\omega_k. \quad (22.154)$$

Because of the multi-variable dependence, the crossing symmetry relations cannot be employed to convert the integration range to $[0,\infty)$.

The case of general harmonic generation is now considered. Equations (22.153) and (22.154) can be written as follows:

$$\phi(p_1\omega,p_2\omega,\ldots,p_n\omega)-\phi(p_1\omega_0,p_2\omega_0,\ldots,p_n\omega_0) = \frac{1}{\pi}P\int_{-\infty}^{\infty}\log\left|\chi^{(n)}(p_1\omega',p_2\omega',\ldots,p_n\omega')/\chi_u^{(n)}\right|\left(\frac{1}{\omega-\omega'}-\frac{1}{\omega_0-\omega'}\right)\mathrm{d}\omega' \quad (22.155)$$

and

$$\log\left|\frac{\chi^{(n)}(p_1\omega,p_2\omega,\ldots,p_n\omega)}{\chi^{(n)}(p_1\omega_0,p_2\omega_0,\ldots,p_n\omega_0)}\right| = -\frac{1}{\pi}P\int_{-\infty}^{\infty}\phi(p_1\omega',p_2\omega',\ldots,p_n\omega') \times\left(\frac{1}{\omega-\omega'}-\frac{1}{\omega_0-\omega'}\right)\mathrm{d}\omega'. \quad (22.156)$$

Employing Eq. (22.149) and the crossing relation Eq. (22.25) yields

$$\phi(-p_1\omega,-p_2\omega,\ldots,-p_n\omega) = -\phi(p_1\omega,p_2\omega,\ldots,p_n\omega) \quad (22.157)$$

and

$$\left|\chi^{(n)}(-p_1\omega,-p_2\omega,\ldots,-p_n\omega)\right| = \left|\chi^{(n)}(p_1\omega,p_2\omega,\ldots,p_n\omega)\right|, \quad (22.158)$$

so that Eqs. (22.155) and (22.156) simplify to

$$\phi(p_1\omega,p_2\omega,\ldots,p_n\omega)-\phi(p_1\omega_0,p_2\omega_0,\ldots,p_n\omega_0) = \frac{2}{\pi}P\int_{0}^{\infty}\log\left|\chi^{(n)}(p_1\omega',p_2\omega',\ldots,p_n\omega')/\chi_u^{(n)}\right|\left(\frac{\omega}{\omega^2-\omega'^2}-\frac{\omega_0}{\omega_0^2-\omega'^2}\right)\mathrm{d}\omega' \quad (22.159)$$

and

$$\log\left|\frac{\chi^{(n)}(p_1\omega,p_2\omega,\ldots,p_n\omega)}{\chi^{(n)}(p_1\omega_0,p_2\omega_0,\ldots,p_n\omega_0)}\right|$$

$$= \frac{2}{\pi}P\int_0^\infty \omega'\phi(p_1\omega',p_2\omega',\ldots,p_n\omega')\left(\frac{1}{\omega_0^2-\omega'^2}-\frac{1}{\omega^2-\omega'^2}\right)\mathrm{d}\omega'. \tag{22.160}$$

An alternative approach can be taken by considering the integral

$$\oint_C \frac{(\omega_0^2+\omega\omega_{z_k})\log\{\chi^{(n)}(\omega_{z_k})/\chi_u^{(n)}\}\mathrm{d}\omega_{z_k}}{(\omega_0^2+\omega_{z_k}^2)(\omega-\omega_{z_k})},$$

where C denotes the contour shown in Figure 19.2 and ω_0 designates an arbitrary frequency. The contribution $(\omega_0^2+\omega\omega_{z_k})(\omega_0^2+\omega_{z_k}^2)^{-1}$ serves as a weighting factor, inserted to improve the convergence behavior of the integrand as $\omega_z \to \infty$. There is a pole located at $\omega_z = \mathrm{i}\omega_0$ in the upper half of the complex frequency plane. The Cauchy residue theorem yields

$$P\int_{-\infty}^\infty \frac{(\omega_0^2+\omega\omega_\mathrm{r})\log\{\chi^{(n)}(\omega_\mathrm{r})/\chi_u^{(n)}\}\mathrm{d}\omega_\mathrm{r}}{(\omega_0^2+\omega_\mathrm{r}^2)(\omega-\omega_\mathrm{r})} + \mathrm{i}\pi\log\{\chi^{(n)}(\omega)/\chi_u^{(n)}\}$$

$$= \mathrm{i}\pi\log\{\chi^{(n)}(\mathrm{i}\omega_0)/\chi_u^{(n)}\}. \tag{22.161}$$

Taking the real and imaginary parts leads to

$$\phi(\omega_1,\omega_2,\ldots,\omega,\ldots,\omega_n)$$

$$= \frac{1}{\pi}P\int_{-\infty}^\infty \frac{(\omega_0^2+\omega\omega_\mathrm{r})\log\left|\chi^{(n)}(\omega_1,\omega_2,\ldots,\omega_\mathrm{r},\ldots,\omega_n)/\chi_u^{(n)}\right|\mathrm{d}\omega_\mathrm{r}}{(\omega_0^2+\omega_\mathrm{r}^2)(\omega-\omega_\mathrm{r})}$$

$$+ \mathrm{Im}[\log\{\chi^{(n)}(\omega_1,\omega_2,\ldots,\mathrm{i}\omega_0,\ldots,\omega_n)/\chi_u^{(n)}\}] \tag{22.162}$$

and

$$\log\left|\chi^{(n)}(\omega_1,\omega_2,\ldots,\omega,\ldots,\omega_n)/\chi_u^{(n)}\right|$$

$$= -\frac{1}{\pi}P\int_{-\infty}^\infty \frac{(\omega_0^2+\omega\omega_\mathrm{r})\phi(\omega_1,\omega_2,\ldots,\omega_\mathrm{r},\ldots,\omega_n)\mathrm{d}\omega_\mathrm{r}}{(\omega_0^2+\omega_\mathrm{r}^2)(\omega-\omega_\mathrm{r})}$$

$$+ \mathrm{Re}[\log\{\chi^{(n)}(\omega_1,\omega_2,\ldots,\mathrm{i}\omega_0,\ldots,\omega_n)/\chi_u^{(n)}\}]. \tag{22.163}$$

Restricting to the case of harmonic generation, then the two preceding dispersion relations can be written as follows:

$$\phi(p_1\omega, p_2\omega, \ldots, p_n\omega) = \frac{1}{\pi} P \int_{-\infty}^{\infty} \frac{(\omega_0^2 + \omega\omega') \log\left|\chi^{(n)}(p_1\omega', p_2\omega', \ldots, p_n\omega')/\chi_u^{(n)}\right| \mathrm{d}\omega'}{(\omega_0^2 + \omega'^2)(\omega - \omega')} \tag{22.164}$$

and

$$\log\left|\chi^{(n)}(p_1\omega, p_2\omega, \ldots, p_n\omega)/\chi_u^{(n)}\right| = -\frac{1}{\pi} P \int_{-\infty}^{\infty} \frac{(\omega_0^2 + \omega\omega')\phi(p_1\omega', p_2\omega', \ldots, p_n\omega')\mathrm{d}\omega'}{(\omega_0^2 + \omega'^2)(\omega - \omega')} + \log\{\chi^{(n)}(p_1\mathrm{i}\omega_0, p_2\mathrm{i}\omega_0, \ldots, p_n\mathrm{i}\omega_0)/\chi_u^{(n)}\}. \tag{22.165}$$

The crossing symmetry relations can be used to simplify Eqs. (22.164) and (22.165) to yield

$$\phi(p_1\omega, p_2\omega, \ldots, p_n\omega) = \frac{2\omega}{\pi} P \int_0^{\infty} \frac{\log\left|\chi^{(n)}(p_1\omega', p_2\omega', \ldots, p_n\omega')/\chi_u^{(n)}\right| \mathrm{d}\omega'}{(\omega^2 - \omega'^2)}, \tag{22.166}$$

and

$$\log\left|\chi^{(n)}(p_1\omega, p_2\omega, \ldots, p_n\omega)/\chi_u^{(n)}\right| = -\frac{2(\omega_0^2 + \omega^2)}{\pi} P \int_0^{\infty} \frac{\omega'\phi(p_1\omega', p_2\omega', \ldots, p_n\omega')\mathrm{d}\omega'}{(\omega_0^2 + \omega'^2)(\omega^2 - \omega'^2)} + \log\{\chi^{(n)}(p_1\mathrm{i}\omega_0, p_2\mathrm{i}\omega_0, \ldots, p_n\mathrm{i}\omega_0)/\chi_u^{(n)}\}. \tag{22.167}$$

The fact that the term $\chi^{(n)}(p_1\mathrm{i}\omega_0, p_2\mathrm{i}\omega_0, \ldots, p_n\mathrm{i}\omega_0)$ is real has been employed. To justify this statement, first note that the crossing symmetry relation given in Eq. (22.25) can be written in the following more general form:

$$\chi_{kl_1\ldots l_n}^{(n)}(\omega_{z1}, \omega_{z2}, \ldots, \omega_{zn})^* = \chi_{kl_1\ldots l_n}^{(n)}(-\omega_{z1}^*, -\omega_{z2}^*, \ldots, -\omega_{zn}^*). \tag{22.168}$$

This result follows directly from Eq. (22.24). On restricting to the case of harmonic generation, writing $\omega_z = \omega_\mathrm{r} + \mathrm{i}\omega_\mathrm{i}$, and setting the real part to zero, gives

$$\chi_{kl_1\ldots l_n}^{(n)}(\mathrm{i}\omega_\mathrm{i}, \mathrm{i}\omega_\mathrm{i}, \ldots, \mathrm{i}\omega_\mathrm{i})^* = \chi_{kl_1\ldots l_n}^{(n)}(\mathrm{i}\omega_\mathrm{i}, \mathrm{i}\omega_i, \ldots, \mathrm{i}\omega_\mathrm{i}), \tag{22.169}$$

from which it follows that the imaginary component of $\chi^{(n)}(p_1\mathrm{i}\omega_0, p_2\mathrm{i}\omega_0, \ldots, p_n\mathrm{i}\omega_0)$ is zero.

This section is concluded with a brief mention of subtracted dispersion relations for the nonlinear susceptibilities. From Eq. (22.144) it follows, on setting all $p_k = 1$, that

$$\operatorname{Im}\chi^{(n)}(\omega,\omega,\ldots,\omega) = \frac{2\omega}{\pi}P\int_0^\infty \left\{\frac{\operatorname{Re}\chi^{(n)}(\omega',\omega',\ldots,\omega')}{\omega^2-\omega'^2} - \frac{\operatorname{Re}\chi^{(n)}(\omega_0,\omega_0,\ldots,\omega_0)}{\omega_0^2-\omega'^2}\right\}\mathrm{d}\omega', \tag{22.170}$$

which can be rewritten in the following form:

$$\begin{aligned}\operatorname{Im}\chi^{(n)}(\omega,\omega,\ldots,\omega) = \frac{2\omega}{\pi}P\int_0^\infty &\{\omega'^2[\operatorname{Re}\chi^{(n)}(\omega_0,\omega_0,\ldots,\omega_0)\\ &- \operatorname{Re}\chi^{(n)}(\omega',\omega',\ldots,\omega')] + \omega_0^2\operatorname{Re}\chi^{(n)}(\omega',\omega',\ldots,\omega')\\ &- \omega^2\operatorname{Re}\chi^{(n)}(\omega_0,\omega_0,\ldots,\omega_0)\}\frac{\mathrm{d}\omega'}{(\omega^2-\omega'^2)(\omega_0^2-\omega'^2)}.\end{aligned} \tag{22.171}$$

The frequency ω_0 is sometimes referred to as an anchor point. For the case $\omega_0 = \omega$, Eq. (22.171) reduces to

$$\operatorname{Im}\chi^{(n)}(\omega,\omega,\ldots,\omega) = \frac{2\omega}{\pi}P\int_0^\infty \frac{[\operatorname{Re}\chi^{(n)}(\omega',\omega',\ldots,\omega') - \operatorname{Re}\chi^{(n)}(\omega,\omega,\ldots,\omega)]\mathrm{d}\omega'}{\omega^2-\omega'^2}. \tag{22.172}$$

Similarly, from Eq. (22.145) it follows that

$$\operatorname{Re}\chi^{(n)}(\omega,\omega,\ldots,\omega) = -\frac{2}{\pi}P\int_0^\infty \frac{[\omega'\operatorname{Im}\chi^{(n)}(\omega',\omega',\ldots,\omega') - \omega\operatorname{Im}\chi^{(n)}(\omega,\omega,\ldots,\omega)]\mathrm{d}\omega'}{\omega^2-\omega'^2}. \tag{22.173}$$

Multiple subtracted dispersion relations for the determination of both $\operatorname{Im}\chi^{(n)}(\omega, \omega,\ldots,\omega)$ and $\operatorname{Re}\chi^{(n)}(\omega,\omega,\ldots,\omega)$ can be derived. The form of the subtracted dispersion relations offer the potential for a more accurate numerical evaluation of the principal value integrals that arise.

22.7 Experimental verification of the dispersion relations for the nonlinear susceptibility

It is rather difficult to verify, even in an approximate manner, dispersion relations for the nonlinear susceptibility. The reader should keep in mind that nonlinear susceptibilities are only rendered accessible by laser sources, and these are typically monochromatic or tunable over rather narrow frequency intervals. The first effort attempting an experimental confirmation of the dispersion relations for the nonlinear susceptibility was carried out by Kishida *et al.* (1993), who focused their effort on verifying the dispersion relations for third-harmonic generation of the following form:

$$\phi(3\omega;\omega,\omega,\omega) = \frac{2\omega}{\pi} P \int_0^\infty \frac{\log\left|\chi^{(3)}(3\omega';\omega',\omega',\omega')/\chi_u^{(3)}\right| \mathrm{d}\omega'}{(\omega^2-\omega'^2)} \tag{22.174}$$

and

$$\log\left|\chi^{(3)}(3\omega;\omega,\omega,\omega)/\chi_u^{(3)}\right| = -\frac{2(\omega_0^2+\omega^2)}{\pi} P \int_0^\infty \frac{\omega'\phi(3\omega';\omega',\omega',\omega')\mathrm{d}\omega'}{(\omega_0^2+\omega'^2)(\omega^2-\omega'^2)} + \log\{\chi^{(3)}(3\mathrm{i}\omega_0;\mathrm{i}\omega_0,\mathrm{i}\omega_0,\mathrm{i}\omega_0)/\chi_u^{(3)}\}, \tag{22.175}$$

for samples of polymeric polydihexylsilane, and using $\omega_0 = 1$ Hz. Equations (22.174) and (22.175) are obtained as special cases of Eqs. (22.166) and (22.167) for $n = 3$, and all the p_k factors are set to unity. Kishida *et al.* (1993) measured both the modulus $|\chi^{(3)}|$ and the phase ϕ; however, the measurements were made over a relatively narrow spectral range of 0.75–1.90 eV. In the test of Eq. (22.175), the term $\log\{\chi^{(3)}(3\mathrm{i};\mathrm{i},\mathrm{i},\mathrm{i})/\chi_u^{(3)}\}$ was treated as an adjustable parameter. They found that the dispersion relation Eq. (22.174) using the experimental modulus $|\chi^{(3)}|$ gave a phase that was in satisfactory agreement with the measured phase. Using the experimental phase, the dispersion relation Eq. (22.175) gave a modulus that was also in satisfactory agreement with the experimentally determined modulus.

Kishida *et al.* also checked the dispersion relations,

$$\mathrm{Im}\,\chi^{(3)}(3\omega;\omega,\omega,\omega) = \frac{2\omega}{\pi} P \int_0^\infty \frac{\mathrm{Re}\,\chi^{(3)}(3\omega';\omega',\omega',\omega')\mathrm{d}\omega'}{\omega^2-\omega'^2} \tag{22.176}$$

and

$$\mathrm{Re}\,\chi^{(3)}(3\omega;\omega,\omega,\omega) = -\frac{2}{\pi} P \int_0^\infty \frac{\omega'\,\mathrm{Im}\,\chi^{(3)}(3\omega';\omega',\omega',\omega')\mathrm{d}\omega'}{\omega^2-\omega'^2}, \tag{22.177}$$

for polydihexylsilane. The values of $\mathrm{Im}\,\chi^{(3)}(3\omega;\omega,\omega,\omega)$ and $\mathrm{Re}\,\chi^{(3)}(3\omega;\omega,\omega,\omega)$ were obtained directly from the measured modulus and phase. The calculated $\mathrm{Im}\,\chi^{(3)}(3\omega;\omega,\omega,\omega)$ and $\mathrm{Re}\,\chi^{(3)}(3\omega;\omega,\omega,\omega)$ were found to be in reasonable agreement with the corresponding experimentally derived values.

Lucarini and Peiponen (2003) carried out an analysis of third-harmonic generation data for the polymers polysilane and polythiophene, for which experimental results are available over the intervals 0.4–2.5 eV and 0.5–2.0 eV, respectively. In place of examining the Kramers–Kronig relations for $\chi^{(3)}(3\omega;\omega,\omega,\omega)$ directly, they examined the more general dispersion relations based on the functions $\omega^{2\alpha}\chi^{(3)}(3\omega;\omega,\omega,\omega)$. Since the asymptotic behavior of $\chi^{(3)}(3\omega;\omega,\omega,\omega)$ as $\omega\to\infty$ is $O(\omega^{-8})$, then the integer parameter α is constrained to satisfy $0 \leq \alpha \leq 3$ to formulate Kramers–Kronig relations. A more generalized function, $\omega^{2\alpha}\{\chi^{(3)}(3\omega;\omega,\omega,\omega)\}^k$ for integer $k \geq 1$, was also studied. The authors found reasonable agreement between the calculated and experimental data: the low-energy part of the spectrum gave a poor comparison, but the high-energy correspondence was very good. The analysis of these authors did not employ various extrapolation models to extend the experimental data beyond the measured spectral interval.

A related study by Lucarini, Saarinen, and Peiponen (2003a, 2003b) considered a multiply subtractive form of the Kramers–Kronig relations applied to $\omega^{2\alpha}\chi^{(3)}(3\omega;\omega,\omega,\omega)$. A principal issue in this approach is the selection of anchor points, since experimental data for the real and imaginary components of a nonlinear susceptibility are not readily available. The improved convergence characteristics of the subtractive form of the Kramers–Kronig relations, due to the faster asymptotic fall-off of the integrands, gave improved results relative to the standard unsubtracted form of the Kramers–Kronig relations. Although very good data sets over wide spectral intervals for nonlinear susceptibilities are essentially unavailable, there is clear support for the validity of the Kramers–Kronig relations for harmonic generation susceptibilities based on the limited data sets that are accessible.

22.8 Dispersion relations in two variables

The case of dispersion relations for the electric susceptibility in two angular frequency variables is considered in this section. Tensor indices are ignored for notational simplicity, and attention is focused on the second-order electric susceptibility. Provided the signs of the frequency terms are positive, the electric susceptibility $\chi^{(2)}(\omega_1,\omega_2)$ is analytic in the upper half of the complex angular frequency plane for each variable separately. This follows directly from the analysis of the anharmonic model discussed in Section 22.3.

Applying the Cauchy integral theorem to the integral

$$\oint_{C_2} \frac{\chi^{(2)}(\omega_1,\omega_{z_2})\mathrm{d}\omega_{z_2}}{\omega_b-\omega_{z_2}},$$

where C_2 denotes the contour shown in Figure 19.2, which is located in the upper half of the complex plane of the variable ω_{z_2}, and taking the limits $R\to\infty$ and $\rho\to 0$, leads to

$$\chi^{(2)}(\omega_1,\omega_b) = -\frac{1}{\pi\mathrm{i}}P\int_{-\infty}^{\infty}\frac{\chi^{(2)}(\omega_1,\omega_\mathrm{r})\mathrm{d}\omega_\mathrm{r}}{\omega_b-\omega_\mathrm{r}}. \tag{22.178}$$

In a similar fashion, consider the contour integral

$$\oint_{C_1} \frac{\chi^{(2)}(\omega_{z_1},\omega_b)\mathrm{d}\omega_{z_1}}{\omega_a-\omega_{z_1}},$$

where C_1 denotes the contour shown in Figure 19.2, and the upper half complex plane now refers to the variable ω_{z_1}. It follows, on taking the limits $R \to \infty$ and $\rho \to 0$, that

$$\chi^{(2)}(\omega_a, \omega_b) = -\frac{1}{\pi \mathrm{i}} P \int_{-\infty}^{\infty} \frac{\chi^{(2)}(\omega_{\mathrm{r}}, \omega_b) \mathrm{d}\omega_{\mathrm{r}}}{\omega_a - \omega_{\mathrm{r}}}. \tag{22.179}$$

Substituting Eq. (22.178) into (22.179) yields

$$\chi^{(2)}(\omega_a, \omega_b) = -\frac{1}{\pi^2} P \int_{-\infty}^{\infty} \frac{\mathrm{d}\omega_{\mathrm{r}_1}}{\omega_a - \omega_{\mathrm{r}_1}} P \int_{-\infty}^{\infty} \frac{\chi^{(2)}(\omega_{\mathrm{r}_1}, \omega_{\mathrm{r}_2}) \mathrm{d}\omega_{\mathrm{r}_2}}{\omega_b - \omega_{\mathrm{r}_2}}. \tag{22.180}$$

Taking the real and imaginary parts of this result leads to the following:

$$\operatorname{Re} \chi^{(2)}(\omega_a, \omega_b) = -\frac{1}{\pi^2} P \int_{-\infty}^{\infty} \frac{\mathrm{d}\omega_{\mathrm{r}_1}}{\omega_a - \omega_{\mathrm{r}_1}} P \int_{-\infty}^{\infty} \frac{\operatorname{Re} \chi^{(2)}(\omega_{\mathrm{r}_1}, \omega_{\mathrm{r}_2}) \mathrm{d}\omega_{\mathrm{r}_2}}{\omega_b - \omega_{\mathrm{r}_2}}, \tag{22.181}$$

and

$$\operatorname{Im} \chi^{(2)}(\omega_a, \omega_b) = -\frac{1}{\pi^2} P \int_{-\infty}^{\infty} \frac{\mathrm{d}\omega_{\mathrm{r}_1}}{\omega_a - \omega_{\mathrm{r}_1}} P \int_{-\infty}^{\infty} \frac{\operatorname{Im} \chi^{(2)}(\omega_{\mathrm{r}_1}, \omega_{\mathrm{r}_2}) \mathrm{d}\omega_{\mathrm{r}_2}}{\omega_b - \omega_{\mathrm{r}_2}}. \tag{22.182}$$

The preceding two results are not useful for optical data analysis, because the same quantity appears on both the left- and right-hand sides of the equation. These results provide a constraint on any model employed to describe $\chi^{(2)}(\omega_1, \omega_2)$, and so serve as a necessary condition to be satisfied by any rigorous theoretical description of the second-order nonlinear susceptibility.

One of the standard approaches used to derive sum rules for a particular optical constant is to apply the superconvergence theorem (see Section 19.9) to the dispersion relations. A more fruitful approach is to work directly with contour integrals of functions such as $\chi^{(2)}(\omega_1, \omega_2)$, or extensions of this such as $\{\omega_1\omega_2\}^p\{\chi^{(2)}(\omega_1, \omega_2)\}^q$, for a suitable choice of the integer parameters p and q, and evaluate the integrals over the upper half of the complex ω_1- and ω_2-planes. The interested reader can explore the possibilities.

22.9 *n*-dimensional dispersion relations

The argument of the previous section can be extended in a straightforward fashion to yield

$$\chi^{(n)}(\omega_1, \omega_2, \ldots, \omega_n) = \frac{\mathrm{i}^n}{\pi^n} P \int_{-\infty}^{\infty} \cdots P \int_{-\infty}^{\infty} \frac{\chi^{(n)}(\omega_{\mathrm{r}_1}, \omega_{\mathrm{r}_2}, \ldots, \omega_{\mathrm{r}_n}) \mathrm{d}\omega_{\mathrm{r}_1} \cdots \mathrm{d}\omega_{\mathrm{r}_n}}{(\omega_1 - \omega_{\mathrm{r}_1}) \cdots (\omega_n - \omega_{\mathrm{r}_n})}. \tag{22.183}$$

For n even, both sides of the equation involve either $\mathrm{Re}\,\chi^{(n)}(\omega_1,\omega_2,\ldots,\omega_n)$ or $\mathrm{Im}\,\chi^{(n)}(\omega_1,\omega_2,\ldots,\omega_n)$, and the resulting dispersion relations provide necessary constraints that must be satisfied by any model for $\chi^{(n)}(\omega_1,\omega_2,\ldots,\omega_n)$. These results are not useful for optical data analysis. For n odd, it follows that

$$\mathrm{Re}\chi^{(n)}(\omega_1,\omega_2,\ldots,\omega_n)=\frac{\mathrm{i}^{n+1}}{\pi^n}P\int_{-\infty}^{\infty}\cdots P\int_{-\infty}^{\infty}\frac{\mathrm{Im}\,\chi^{(n)}(\omega_{\mathrm{r}_1},\omega_{\mathrm{r}_2},\ldots,\omega_{\mathrm{r}_n})\mathrm{d}\omega_{\mathrm{r}_1}\cdots\mathrm{d}\omega_{\mathrm{r}_n}}{(\omega_1-\omega_{\mathrm{r}_1})\cdots(\omega_n-\omega_{\mathrm{r}_n})} \tag{22.184}$$

and

$$\mathrm{Im}\chi^{(n)}(\omega_1,\omega_2,\ldots,\omega_n)=\frac{\mathrm{i}^{n-1}}{\pi^n}P\int_{-\infty}^{\infty}\cdots P\int_{-\infty}^{\infty}\frac{\mathrm{Re}\,\chi^{(n)}(\omega_{\mathrm{r}_1},\omega_{\mathrm{r}_2},\ldots,\omega_{\mathrm{r}_n})\mathrm{d}\omega_{\mathrm{r}_1}\cdots\mathrm{d}\omega_{\mathrm{r}_n}}{(\omega_1-\omega_{\mathrm{r}_1})\cdots(\omega_n-\omega_{\mathrm{r}_n})}. \tag{22.185}$$

The preceding two formulas have the potential to be more useful in optical data analysis, compared with the corresponding results for n even. More general dispersion relations can be derived by starting with the function

$$f(\omega_1,\omega_2,\ldots,\omega_n,\omega_{\alpha_1},\omega_{\alpha_2},\ldots,\omega_{\alpha_n},p,q)=\{\chi^{(n)}(\omega_{\mathrm{r}_1},\omega_{\mathrm{r}_2},\ldots,\omega_{\mathrm{r}_n})\}^q \times\prod_{k=1}^{n}\left\{\frac{\omega_k^p}{(\omega_{\alpha_k}-\omega_k)}\right\}. \tag{22.186}$$

Experimental data for susceptibilities with $n > 3$ are so limited that Eqs. (22.184) and (22.185) have not found any practical applications. Sum rules of an n-dimensional nature can be derived from the given dispersion relations, or derived starting from contour integrals using kernel functions of the form appearing in Eq. (22.186).

22.10 Situations where the dispersion relations do not hold

This section considers situations where the standard Kramers–Kronig relations cannot be derived for particular circumstances involving the nonlinear susceptibility. The case of difference-frequency generation, which is governed by the term $\chi^{(2)}(\omega_1-\omega_2;\omega_1,-\omega_2)$ in lowest order, is considered first. From the classical anharmonic model, Eq. (22.48), it follows that

$$\chi_{ijk}^{(2)}(\omega_1-\omega_2;\omega_1,-\omega_2)=\frac{P_i^{(2)}(\omega_1-\omega_2)}{\varepsilon_0E_j(\omega_1)E_k^*(\omega_2)}=\frac{2N\alpha e^3}{\varepsilon_0m^2D(\omega_1)D(-\omega_2)D(\omega_4)}, \tag{22.187}$$

where $P_i^{(2)}$ in this expression denotes the second-order polarization arising from difference-frequency generation. Recalling Eq. (22.39) and considering the complex ω_1-plane, it is clear from the denominator factors that $\chi_{ijk}^{(2)}(\omega_1-\omega_2;\omega_1,-\omega_2)$ has four simple poles, as depicted in Figure 22.5.

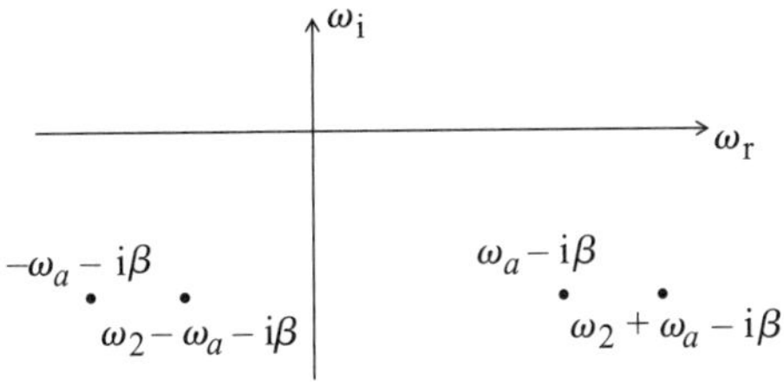

Figure 22.5. Pole structure for $\chi^{(2)}_{ijk}(\omega_1 - \omega_2; \omega_1, -\omega_2)$ in the ω_1-plane using $\beta = \gamma/2$ and $\omega_a = \surd(\omega_0^2 - \beta^2)$.

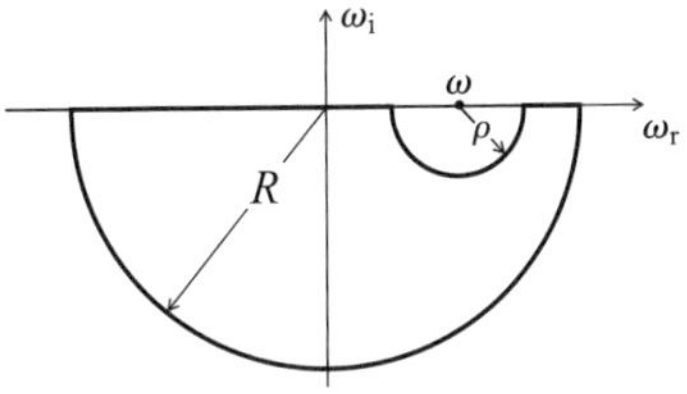

Figure 22.6. Contour for deriving a dispersion relation for $\chi^{(2)}(\omega_1 - \omega_2; \omega_1, -\omega_2)$ in the variable ω_2.

If the angular frequency ω_2 approaches zero, then pairs of these poles coalesce to form poles of order two. The important point is that all these poles are located in the lower half of the complex ω_1-plane. By considering the contour integral

$$\oint_{C_1} \frac{\chi^{(2)}(\omega_{z_1}, -\omega_2)\mathrm{d}\omega_{z_1}}{\omega_\alpha - \omega_{z_1}},$$

where C_1 denotes the contour shown in Figure 19.2; and applying the Cauchy integral theorem, leads to

$$\chi^{(2)}(\omega_\alpha - \omega_2; \omega_\alpha, -\omega_2) = -\frac{1}{\pi \mathrm{i}} P \int_{-\infty}^{\infty} \frac{\chi^{(2)}(\omega_r, -\omega_2)\mathrm{d}\omega_r}{\omega_\alpha - \omega_r}, \tag{22.188}$$

which corresponds to the dispersion relation for difference-frequency generation.

Consider the complex ω_2-plane. Here the situation is different: for this case the poles are located in the upper half of the complex ω_2-plane. A dispersion relation similar to Eq. (22.188) can be derived by considering the integral

$$\oint_{C_1} \frac{\chi^{(2)}(\omega_1, -\omega_{z_2})\mathrm{d}\omega_{z_2}}{\omega - \omega_{z_2}},$$

where the contour C_1 is shown in Figure 22.6. The preceding line of reasoning can be extended to $\chi^{(3)}_{ijkl}(\omega_1 + \omega_2 - \omega_3; \omega_1, \omega_2, -\omega_3)$, and so on.

Consideration is now turned to a more complicated situation, the third-order susceptibility $\chi^{(3)}_{ijkl}(\omega; \omega, \omega, -\omega)$, which corresponds to a particular case of $\chi^{(3)}_{ijkl}(\omega_1 +$

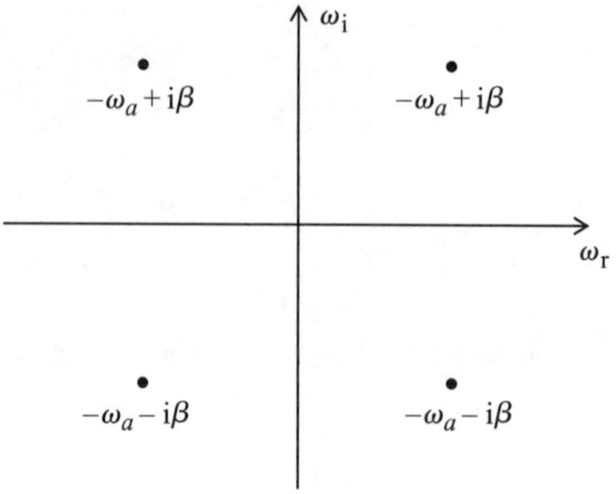

Figure 22.7. Pole structure for $\chi^{(3)}_{ijkl}(\omega;\omega,\omega,-\omega)$ in the ω_z-plane.

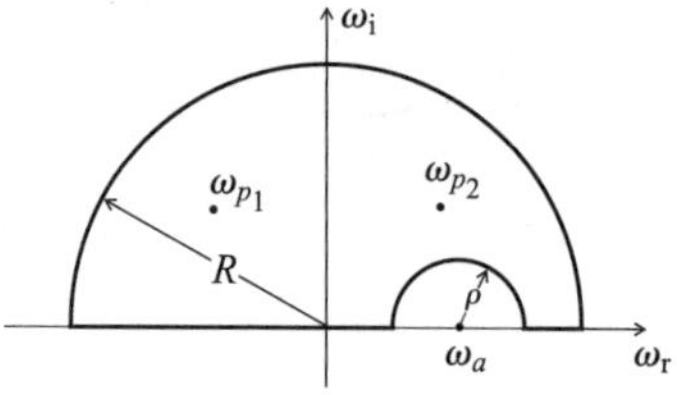

Figure 22.8. Contour for dealing with the kernel function $(\omega_\alpha-\omega)^{-1}\chi^{(3)}_{ijkl}(\omega;\omega,\omega,-\omega)$ in the ω_z-plane.

$\omega_2-\omega_3;\omega_1,\omega_2,-\omega_3)$. The poles of $\chi^{(3)}_{ijkl}(\omega;\omega,\omega,-\omega)$ are shown in Figure 22.7. In this case there are poles in both the upper and lower half complex angular frequency planes. There are no standard Kramers–Kronig relations for this particular type of nonlinear electric susceptibility function. The function $\chi^{(3)}_{ijkl}(\omega;\omega,\omega,-\omega)$ is not analytic in either the upper or lower half complex frequency planes. However, dispersion relations can be derived, but the pole structure in the upper half plane must be taken into consideration. Consider the integral

$$\oint_C \frac{\chi^{(3)}_{ijkl}(\omega_z;\omega_z,\omega_z,-\omega_z)\mathrm{d}\omega_z}{\omega_\alpha-\omega_z},$$

where C denotes the contour shown in Figure 22.8. Applying the Cauchy residue theorem leads to the following result involving the residues of $\chi^{(3)}_{ijkl}(\omega_z;\omega_z,\omega_z,-\omega_z)$:

$$\begin{aligned}P\int_{-\infty}^{\infty}\frac{\chi^{(3)}_{ijkl}(\omega;\omega,\omega,-\omega)\mathrm{d}\omega}{\omega_\alpha-\omega}&+\mathrm{i}\pi\chi^{(3)}_{ijkl}(\omega_\alpha;\omega_\alpha,\omega_\alpha,-\omega_\alpha)\\&=2\pi\mathrm{i}\sum_{\omega_p}\mathrm{Res}\{\chi^{(3)}_{ijkl}(\omega_z;\omega_z,\omega_z,-\omega_z)_{\omega_z=\omega_p}\},\end{aligned}\tag{22.189}$$

where the summation is over each of the poles in the upper half of the complex angular frequency plane. On letting $\omega_{p_1} = \mathrm{i}\beta + \surd(\omega_0^2 - \beta^2)$ and $\omega_{p_2} = \mathrm{i}\beta - \surd(\omega_0^2 - \beta^2)$, and making use of the crossing symmetry condition Eq. (22.25), Eq. (22.189) can be written in the following form:

$$\begin{aligned}
&\frac{2\omega_\alpha}{\pi} P \int_0^\infty \frac{\mathrm{Re}\,\chi_{ijkl}^{(3)}(\omega;\omega,\omega,-\omega)\mathrm{d}\omega}{\omega_\alpha^2 - \omega^2} + \frac{2\mathrm{i}}{\pi} P \int_0^\infty \frac{\omega\,\mathrm{Im}\,\chi_{ijkl}^{(3)}(\omega;\omega,\omega,-\omega)\mathrm{d}\omega}{\omega_\alpha^2 - \omega^2} \\
&\quad = -\mathrm{i}\,\mathrm{Re}\,\chi_{ijkl}^{(3)}(\omega_\alpha;\omega_\alpha,\omega_\alpha,-\omega_\alpha) + \mathrm{Im}\,\chi_{ijkl}^{(3)}(\omega_\alpha;\omega_\alpha,\omega_\alpha,-\omega_\alpha) \\
&\qquad + 2\mathrm{i}\,\mathrm{Re}[\mathrm{Res}\{\chi_{ijkl}^{(3)}(\omega_z;\omega_z,\omega_z,-\omega_z)_{\omega_z=\omega_{p_1}}\} \\
&\qquad + \mathrm{Res}\{\chi_{ijkl}^{(3)}(\omega_z;\omega_z,\omega_z,-\omega_z)_{\omega_z=\omega_{p_2}}\}] \\
&\qquad - 2\,\mathrm{Im}[\mathrm{Res}\{\chi_{ijkl}^{(3)}(\omega_z;\omega_z,\omega_z,-\omega_z)_{\omega_z=\omega_{p_1}}\} \\
&\qquad + \mathrm{Res}\{\chi_{ijkl}^{(3)}(\omega_z;\omega_z,\omega_z,-\omega_z)_{\omega_z=\omega_{p_2}}\}],
\end{aligned} \tag{22.190}$$

so that

$$\begin{aligned}
\mathrm{Im}\,\chi_{ijkl}^{(3)}(\omega_\alpha;\omega_\alpha,\omega_\alpha,-\omega_\alpha) = {}& \frac{2\omega_\alpha}{\pi} P \int_0^\infty \frac{\mathrm{Re}\,\chi_{ijkl}^{(3)}(\omega;\omega,\omega,-\omega)\mathrm{d}\omega}{\omega_\alpha^2 - \omega^2} \\
& + 2\,\mathrm{Im}[\mathrm{Res}\{\chi_{ijkl}^{(3)}(\omega_z;\omega_z,\omega_z,-\omega_z)_{\omega_z=\omega_{p_1}}\} \\
& + \mathrm{Res}\{\chi_{ijkl}^{(3)}(\omega_z;\omega_z,\omega_z,-\omega_z)_{\omega_z=\omega_{p_2}}\}]
\end{aligned} \tag{22.191}$$

and

$$\begin{aligned}
\mathrm{Re}\,\chi_{ijkl}^{(3)}(\omega_\alpha;\omega_\alpha,\omega_\alpha,-\omega_\alpha) = {}& -\frac{2}{\pi} P \int_0^\infty \frac{\omega\,\mathrm{Im}\,\chi_{ijkl}^{(3)}(\omega;\omega,\omega,-\omega)\mathrm{d}\omega}{\omega_\alpha^2 - \omega^2} \\
& + 2\,\mathrm{Re}[\mathrm{Res}\{\chi_{ijkl}^{(3)}(\omega_z;\omega_z,\omega_z,-\omega_z)_{\omega_z=\omega_{p_1}}\} \\
& + \mathrm{Res}\{\chi_{ijkl}^{(3)}(\omega_z;\omega_z,\omega_z,-\omega_z)_{\omega_z=\omega_{p_2}}\}].
\end{aligned} \tag{22.192}$$

Clearly, these expressions are more complicated than the standard form of the Kramers–Kronig relations. In terms of analyzing experimental optical data, these results would be of limited utility, since the pole structure has to be known precisely in order to pin down the values for the residue contributions.

An approximate model based on a two-level atom, which is sometimes employed to represent the nonlinear susceptibility, is given by Yariv (1989, p. 161) as follows:

$$\chi(\omega) = \frac{-\mu^2 \Delta N_0}{\varepsilon_0 \hbar} \frac{\omega - \omega_0 - \mathrm{i}\Gamma}{(\omega - \omega_0)^2 + \Gamma^2(1 + s^2)}, \tag{22.193}$$

where ΔN_0 is the population difference between the two levels at zero electric field, μ is the dipole transition moment for the resonance transition of angular frequency $\omega_0, \Gamma = T_2^{-1}$, where T_2 is a relaxation time, s denotes a saturation factor, and a sign change has been employed for the imaginary component of $\chi(\omega)$, consistent with the previous usage in this chapter. Saturation, as the term is used in spectroscopy, refers to the situation where the populations of the two states involved in a transition become equal. For the case of complete saturation, no net absorption of electromagnetic radiation occurs. For no saturation, $s = 0$, and it follows from Eq. (22.193) that $\chi(\omega)$ is an analytic function in the upper half of the complex angular frequency plane, since there is a single pole located in the lower half of the complex angular frequency plane at $\omega = \omega_0 - \mathrm{i}\Gamma$. In the presence of saturation, $s > 0$, there is a pole in both the upper and lower halves of the complex angular frequency plane. Clearly, in this case Kramers–Kronig relations in the standard form cannot be derived. Note also that in the case $s = 0$ the standard form of the Kramers–Kronig relations cannot be derived, since the $\chi(\omega)$ defined by the approximate model of Eq. (22.193) does not satisfy the crossing symmetry relations. However, a Hilbert transform connection can be written for $\chi(\omega)$. The Yariv model for the nonlinear susceptibity does not satisfy the standard dispersion relations on the frequency interval $[0, \infty)$.

22.11 Sum rules for the nonlinear susceptibilities

Consider the contour integral

$$\oint_C \omega_z^p \{\chi^{(n)}(\omega_1, \omega_2, \ldots, \omega_z, \ldots, \omega_n)\}^q \, \mathrm{d}\omega_z,$$

where C is the contour shown in Figure 22.9 and p and q are integers chosen so that $p \geq 0, q \geq 1$ and $1 + p \leq q$. This choice is based on the assumption that the asymptotic behavior of $\chi^{(n)}(\omega_1, \omega_2, \ldots, \omega_k, \ldots, \omega_n) = O(\omega_k^{-1-\delta})$, as $\omega_k \to \infty$ and $\delta > 0$. To simplify the notation slightly, the complex angular frequency ω_{z_k} and its real and imaginary components ω_{r_k} (which equals ω_k) and ω_{i_k}, are abbreviated by dropping the subscript k. All angular frequencies are assumed to be distinct and positive. The Cauchy integral theorem yields, in the limit $R \to \infty$,

$$\int_{-\infty}^{\infty} \omega_\mathrm{r}^p \{\chi^{(n)}(\omega_1, \omega_2, \ldots, \omega_\mathrm{r}, \ldots, \omega_n)\}^q \, \mathrm{d}\omega_\mathrm{r} = 0. \tag{22.194}$$

In terms of the real and imaginary parts,

$$\int_{-\infty}^{\infty} \omega_\mathrm{r}^p \{\mathrm{Re}\{\chi^{(n)}(\omega_1, \omega_2, \ldots, \omega_\mathrm{r}, \ldots, \omega_n)\} + \mathrm{i}\,\mathrm{Im}\{\chi^{(n)}(\omega_1, \omega_2, \ldots, \omega_\mathrm{r}, \ldots, \omega_n)\}^q \, \mathrm{d}\omega_\mathrm{r} = 0. \tag{22.195}$$

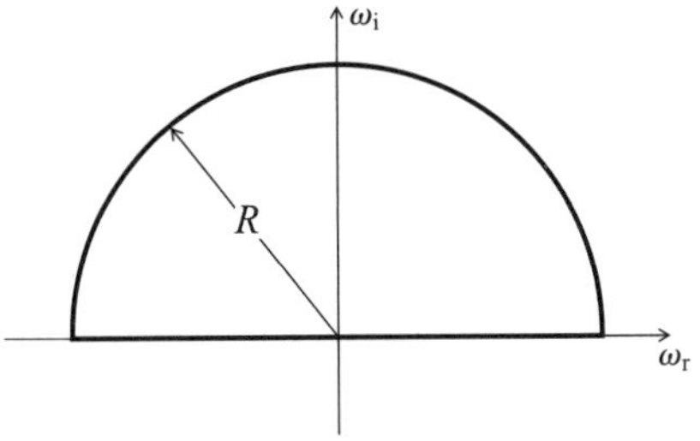

Figure 22.9. Contour for deriving sum rules for the function $\omega_z^p\{\chi^{(n)}(\omega_1,\omega_2,\ldots,\omega_z,\ldots,\omega_n)\}^q$.

Consider the integral

$$\oint_C \chi^{(n)}(p_1\omega_z, p_2\omega_z,\ldots,p_n\omega_z)\mathrm{d}\omega_z,$$

where C is the contour shown in Figure 22.9 and the p_k coefficients are defined in Section 22.6. It follows by the now familiar arguments that

$$\int_{-\infty}^{\infty} \chi^{(n)}(p_1\omega, p_2\omega,\ldots,p_n\omega)\mathrm{d}\omega = 0. \tag{22.196}$$

In this case, the integral over the interval $(-\infty,\infty)$ can be converted to an integral over the physically meaningful spectral range $[0,\infty)$ by taking advantage of the crossing symmetry relation for the nth-order nonlinear susceptibility (see Eq. (22.25)). Employing this result allows Eq. (22.196) to be written as the following sum rule:

$$\int_0^{\infty} \mathrm{Re}\{\chi^{(n)}(p_1\omega, p_2\omega,\ldots,p_n\omega)\}\mathrm{d}\omega = 0. \tag{22.197}$$

A special case of Eq. (22.197) occurs for second-harmonic generation, so that

$$\int_0^{\infty} \mathrm{Re}\{\chi^{(2)}(2\omega;\omega,\omega)\}\mathrm{d}\omega = 0. \tag{22.198}$$

With the tensor components inserted, the preceding two results take the following form:

$$\int_0^{\infty} \mathrm{Re}\{\chi^{(n)}_{kl_1\ldots l_n}(p_1\omega, p_2\omega,\ldots,p_n\omega)\}\mathrm{d}\omega = 0 \tag{22.199}$$

and

$$\int_0^{\infty} \mathrm{Re}\{\chi^{(2)}_{klm}(2\omega;\omega,\omega)\}\mathrm{d}\omega = 0. \tag{22.200}$$

Equation (22.197) can be generalized by considering the integral

$$\oint_C \omega_z^p \{\chi^{(n)}(p_1\omega_z, p_2\omega_z, \ldots, p_n\omega_z)\}^q \, \mathrm{d}\omega_z,$$

where C is the contour shown in Figure 22.9 and p and q are integers. To fix the values of p and q, recall that $\chi^{(n)}(p_1\omega, p_2\omega, \ldots, p_n\omega) = O(\omega^{-2n-2})$ as $\omega \to \infty$ for $n \geq 2$. This leads to the constraints

$$0 \leq p < (2n+2)q - 1, \tag{22.201}$$

for $q \geq 1$ and $n \geq 2$. Evaluating the contour integral leads to

$$\int_{-\infty}^{\infty} \omega^p [\chi^{(n)}(p_1\omega, p_2\omega, \ldots, p_n\omega)]^q \, \mathrm{d}\omega = 0, \tag{22.202}$$

which simplifies, on using the crossing symmetry constraint, to yield

$$\int_0^{\infty} \omega^p \{[\mathrm{Re}\,\chi^{(n)}(p_1\omega, p_2\omega, \ldots, p_n\omega) + \mathrm{i}\,\mathrm{Im}\,\chi^{(n)}(p_1\omega, p_2\omega, \ldots, p_n\omega)]^q + (-1)^p[\mathrm{Re}\,\chi^{(n)}(p_1\omega, p_2\omega, \ldots, p_n\omega) - \mathrm{i}\,\mathrm{Im}\,\chi^{(n)}(p_1\omega, p_2\omega, \ldots, p_n\omega)]^q\}\mathrm{d}\omega = 0. \tag{22.203}$$

For example, using $q = 1$ and considering SHG, it follows that

$$\int_0^{\infty} \omega^p(1 + (-1)^p)\,\mathrm{Re}\,\chi^{(2)}(2\omega; \omega, \omega)\mathrm{d}\omega = 0 \tag{22.204}$$

and

$$\int_0^{\infty} \omega^p(1 - (-1)^p)\,\mathrm{Im}\,\chi^{(2)}(2\omega; \omega, \omega)\mathrm{d}\omega = 0, \tag{22.205}$$

for $0 \leq p \leq 4$. The specific sum rules obtained are Eq. (22.198) and

$$\int_0^{\infty} \omega^2\,\mathrm{Re}\,\chi^{(2)}(2\omega; \omega, \omega)\mathrm{d}\omega = 0, \tag{22.206}$$

$$\int_0^{\infty} \omega^4\,\mathrm{Re}\,\chi^{(2)}(2\omega; \omega, \omega)\mathrm{d}\omega = 0, \tag{22.207}$$

$$\int_0^{\infty} \omega\,\mathrm{Im}\,\chi^{(2)}(2\omega; \omega, \omega)\mathrm{d}\omega = 0, \tag{22.208}$$

and

$$\int_0^{\infty} \omega^3\,\mathrm{Im}\,\chi^{(2)}(2\omega; \omega, \omega)\mathrm{d}\omega = 0. \tag{22.209}$$

A significant number of additional sum rules can be generated from Eq. (22.203) for larger powers of q and for the cases of higher-harmonic generation.

The strategy employed at the end of Section 19.13 can be applied directly to the nonlinear susceptibility for harmonic generation. For example, consider the integral

$$\oint_C \frac{\chi^{(n)}(\omega_z, \omega_z, \ldots, \omega_z)\mathrm{d}\omega_z}{\omega_z + \mathrm{i}\omega_0},$$

where C denotes the contour shown in Figure 19.7 and ω_0 is an arbitrary angular frequency satisfying $\omega_0 > 0$. Applying the Cauchy integral theorem and the crossing symmetry condition leads to

$$\int_0^\infty \frac{\begin{array}{l}\{\omega[\chi^{(n)}(\omega, \omega, \ldots, \omega) - \chi^{(n)}(\omega, \omega, \ldots, \omega)^*] - \mathrm{i}\omega_0[\chi^{(n)}(\omega, \omega, \ldots, \omega) \\ \quad +\chi^{(n)}(\omega, \omega, \ldots, \omega)^*]\}\mathrm{d}\omega\end{array}}{\omega^2 + \omega_0^2} = 0, \tag{22.210}$$

from which it follows that

$$\int_0^\infty \frac{\omega \operatorname{Im} \chi^{(n)}(\omega, \omega, \ldots, \omega)\mathrm{d}\omega}{\omega^2 + \omega_0^2} = \int_0^\infty \frac{\omega_0 \operatorname{Re} \chi^{(n)}(\omega, \omega, \ldots, \omega)\mathrm{d}\omega}{\omega^2 + \omega_0^2}. \tag{22.211}$$

This result applies for both the linear susceptibility and the nonlinear harmonic generation susceptibilities. Relations such as this can serve as necessary constraints on theoretical models. For further development of ideas along the lines of Eq. (22.211), including some generalizations, see Saarinen (2002).

22.12 Summary of sum rules for the nonlinear susceptibilities

This section provides a summary of the sum rules that have been derived for various forms of the nonlinear susceptibilities. Specific cases are given first in Table 22.1, and more general results applying to the nth-order susceptibility are given in the latter part of the table. The references given should be consulted for any restrictions that might apply. Tensor indices are not given to keep the notation as simple as possible.

22.13 The nonlinear refractive index and the nonlinear permittivity

A number of materials have a refractive index that exhibits a weak dependence on the intensity of the propagating electromagnetic pulse through the medium. For a general medium, the nonlinear refractive index $N^{\mathrm{NL}}(\omega)$, can be expressed as follows:

$$N^{\mathrm{NL}}(\omega) = N(\omega) + N_2(\omega)E^2(\omega), \tag{22.212}$$

where the tensor components are omitted to keep the notation as simple as possible. In this formula $N(\omega)$ is the normal complex refractive index and $N_2(\omega)$ is the

Table 22.1. *Summary of sum rules for the nonlinear susceptibilities*

Number	Sum rule	Reference
(1)	$\int_0^\infty \omega^{-1}\,\mathrm{Im}\,\chi^{(2)}(\omega,\omega)\mathrm{d}\omega = \chi^{(2)}(0,0)$	Scandolo and Bassani (1995a)
(2)	$\int_0^\infty \omega\,\mathrm{Im}\,\chi^{(2)}(\omega,\omega)\mathrm{d}\omega = 0$	Scandolo and Bassani (1995a)
(3)	$\int_0^\infty \mathrm{Re}\,\chi^{(2)}(\omega,\omega)\mathrm{d}\omega = 0$	Peiponen (1987c); Scandolo and Bassani (1995a)
(4)	$\int_0^\infty \omega^3\,\mathrm{Im}\,\chi^{(2)}(\omega,\omega)\mathrm{d}\omega = 0$	Scandolo and Bassani (1995a)
(5)	$\int_0^\infty \omega^2\,\mathrm{Re}\,\chi^{(2)}(\omega,\omega)\mathrm{d}\omega = 0$	Scandolo and Bassani (1995a)
(6)	$\int_0^\infty \omega^4\,\mathrm{Re}\,\chi^{(2)}(\omega,\omega)\mathrm{d}\omega = 0$	Scandolo and Bassani (1995a)
(7)	$\int_0^\infty \omega^5\,\mathrm{Im}\,\chi^{(2)}(\omega,\omega)\mathrm{d}\omega = -\frac{\pi N e^3}{16m^3}\left\langle \frac{\partial^3 V}{\partial x_i \partial x_j \partial x_k}\right\rangle_0$	Scandolo and Bassani (1995a)
(8)	$\int_{-\infty}^\infty \chi^{(3)}(\omega,\omega,-\omega)\mathrm{d}\omega = 0$	Peiponen, Vartiainen, and Asakura (1997a)
(9)	$\int_0^\infty \{\chi^{(2)}(\omega,\omega_1) + \chi^{(2)}(-\omega,\omega_1)\}\mathrm{d}\omega = 0$	Peiponen (1987c)
(10)	$\int_0^\infty \mathrm{Re}\,\chi^{(n)}(\omega,\omega,\ldots,\omega)\mathrm{d}\omega = 0$	Peiponen (1987c)

(11)	$\int_{-\infty}^{\infty} \omega_k^p \{\chi^{(n)}(\omega_1, \omega_2, \ldots, \omega_k, \ldots, \omega_n)\}^q \, \mathrm{d}\omega_k = 0,$ $p = 0, 1, \ldots, \quad q = 1, 2, \ldots, \quad 0 \le p \le q - 1$	Peiponen (1987a, c)
(12)	$\int_{-\infty}^{\infty} \omega_k^p \{\chi^{(n)}(\omega_1, \ldots, \omega_k, \ldots, -\omega_l, \ldots, \omega_n)\}^q \, \mathrm{d}\omega_k = 0,$ $p = 0, 1, \ldots, \quad q = 1, 2, \ldots, \quad 0 \le p \le q - 1$	Peiponen (1987c)
(13)	$\int_0^{\infty} \frac{\omega \operatorname{Im} \chi^{(3)}(3\omega; \omega, \omega, \omega) \mathrm{d}\omega}{\omega^2 + \omega_0^2} = \omega_0 \int_0^{\infty} \frac{\operatorname{Re} \chi^{(3)}(3\omega; \omega, \omega, \omega) \mathrm{d}\omega}{\omega^2 + \omega_0^2}$	Saarinen (2002)
(14)	$\omega_0 \int_0^{\infty} \frac{\omega^k \operatorname{Re}[\chi^{(n)}(n\omega; \omega, \ldots, \omega)]^l \, \mathrm{d}\omega}{\omega^2 + \omega_0^2} = \int_0^{\infty} \frac{\omega^{k+1} \operatorname{Im}[\chi^{(n)}(n\omega; \omega, \ldots, \omega)]^l \, \mathrm{d}\omega}{\omega^2 + \omega_0^2},$ $k = 0, 2, 4, \ldots, \quad l \in \mathbb{N}$	Saarinen (2002)
(15)	$\int_0^{\infty} \frac{\omega^{k+1} \operatorname{Re}[\chi^{(n)}(n\omega; \omega, \ldots, \omega)]^l \, \mathrm{d}\omega}{\omega^2 + \omega_0^2} = -\omega_0 \int_0^{\infty} \frac{\omega^{k+1} \operatorname{Im}[\chi^{(n)}(n\omega; \omega, \ldots, \omega)]^l \, \mathrm{d}\omega}{\omega^2 + \omega_0^2},$ $k = 1, 3, 5, \ldots, \quad l \in \mathbb{N}$	Saarinen (2002)
(16)	$\int_{-\infty}^{\infty} \omega_k^p \{\chi^{(n)}(\omega_1, \omega_2, \ldots, \omega_k, -\omega_k, \ldots, \omega_n)\}^q \, \mathrm{d}\omega_k = 0,$ $p = 0, 1, \ldots, \quad q = 1, 2, \ldots, \quad 0 \le p \le q - 1$	Peiponen (1987c)
(17)	$\int_{-\infty}^{\infty} \mathrm{d}\omega_1 \ldots \int_{-\infty}^{\infty} \{\omega_1 \ldots, \omega_n\}^p \{\chi^{(n)}(\omega_1, \ldots, \omega_n)\}^q \, \mathrm{d}\omega_n = 0,$ $p = 0, 1, \ldots, \quad q = 1, 2, \ldots, \quad 0 \le p \le q - 1$	Peiponen (1987b, 1988)

second-order refractive index. From Eq. (22.2) the electric polarization can be written, for a general medium, to the first nonlinear field term, again omitting tensor components, as follows:

$$P^{\mathrm{NL}}(\omega) = \varepsilon_0 \chi^{(1)}(\omega) E(\omega) + 3\varepsilon_0 \chi^{(3)}(\omega) E^3(\omega), \tag{22.213}$$

and it is assumed that a single laser frequency ω is employed. The nonlinear permittivity $\varepsilon^{\mathrm{NL}}(\omega)$ can be expressed in terms of the general susceptibility $\chi(\omega)$ by analogy to Eq. (19.5):

$$\varepsilon^{\mathrm{NL}}(\omega) = \varepsilon_0\{1 + \chi(\omega)\}. \tag{22.214}$$

When magnetic effects are ignored, it follows that

$$N^{\mathrm{NL}}(\omega)^2 = \frac{\varepsilon^{\mathrm{NL}}(\omega)}{\varepsilon_0} = 1 + \chi^{(1)}(\omega) + 3\chi^{(3)}(\omega) E^2(\omega), \tag{22.215}$$

so that the normal complex refractive index is related to the first-order electric susceptibility by

$$N^2(\omega) = 1 + \chi^{(1)}. \tag{22.216}$$

Ignoring higher-order field corrections, the contribution $N_2(\omega)$ can be written as follows:

$$N_2(\omega) = \frac{3\chi^{(3)}(\omega)}{2N(\omega)}. \tag{22.217}$$

The nonlinear refractive index $N^{\mathrm{NL}}(\omega)$ depends on the field strength, so, more formally, the appropriate notation is $N^{\mathrm{NL}}(\omega, E)$. The reader is expected to keep this field dependence in mind, even when it is not made explicit. In Section 19.6 it was established that, in the absence of spatial dispersion, the complex permittivity has no zeros in the upper half of the complex frequency plane. It follows that the same situation applies for the complex refractive index $N(\omega)$. The third-order electric susceptibility in Eq. (22.217) is actually $\chi^{(3)}(\omega; \omega, \omega, -\omega)$, and the discussion given in Section 22.10 established that $\chi^{(3)}(\omega; \omega, \omega, -\omega)$ is a meromorphic function, having poles in both the upper and lower halves of the complex frequency plane. Hence, $N_2(\omega)$ is a meromorphic function and $N^{\mathrm{NL}}(\omega)$ is also a meromorphic function with poles in both the upper and lower halves of the complex frequency plane. This means that the standard form of the Kramers–Kronig relations cannot be obtained for the functions $n^{\mathrm{NL}}(\omega)$ and $\kappa^{\mathrm{NL}}(\omega)$, the real and imaginary parts of $N^{\mathrm{NL}}(\omega)$. It is possible to obtain dispersion relations via the application of the Cauchy residue theorem, in a fashion similar to that indicated for the susceptibility in Section 22.10. However, the resulting formulas have limited practical value, since they depend on parameters governing the locations of the poles in the upper half of the complex frequency plane.

A second approach that can be employed involves two laser beams: one intense beam operating at the frequency ω, and the second probe beam at the frequency ω_1. The susceptibility factor that comes into play in this situation is $\chi^{(3)}(\omega_1;\omega_1,\omega,-\omega)$, so that

$$N^{\mathrm{NL}}(\omega_1)^2 = \frac{\varepsilon^{\mathrm{NL}}(\omega_1)}{\varepsilon_0} = 1 + \chi^{(1)}(\omega_1) + 6\chi^{(3)}(\omega_1;\omega_1,\omega,-\omega)E^2(\omega), \quad (22.218)$$

from which it follows, on ignoring higher-order field corrections, that

$$N_2(\omega_1) = \frac{3\chi^{(3)}(\omega_1;\omega_1,\omega,-\omega)}{N(\omega_1)}. \quad (22.219)$$

The quantity $N_2(\omega_1)$ is a fourth-rank tensor.

Consider the integral

$$\oint_C \frac{N_2(\omega_z)\mathrm{d}\omega_z}{\omega_\alpha - \omega_z},$$

where C denotes the contour shown in Figure 19.2. On applying the Cauchy integral theorem, it follows on taking the limits $R \to \infty$ and $\rho \to 0$ that

$$N_2(\omega_\alpha) = -\frac{1}{\pi \mathrm{i}} P \int_{-\infty}^{\infty} \frac{N_2(\omega)\mathrm{d}\omega}{\omega_\alpha - \omega}. \quad (22.220)$$

The asymptotic behavior employed is $N_2(\omega_1) = O(\omega_1^{-2})$ as $\omega \to \infty$. This allows the semicircular section of the contour integral to be evaluated. To proceed further the appropriate tensor subscripts are inserted:

$$N_{2_{ijkl}}(\omega_\alpha) = -\frac{1}{\pi \mathrm{i}} P \int_0^{\infty} \left\{ \frac{N_{2_{ijkl}}(\omega)}{\omega_\alpha - \omega} + \frac{N_{2_{ijkl}}(-\omega)}{\omega_\alpha + \omega} \right\} \mathrm{d}\omega. \quad (22.221)$$

Making use of the results

$$N_{ij}(-\omega_1) = N_{ij}(\omega_1)^* \quad (22.222)$$

and

$$\begin{aligned} \chi^{(3)}_{ijkl}(-\omega_1;-\omega_1,\omega,-\omega) &= \chi^{(3)}_{ijkl}(\omega_1;\omega_1,-\omega,\omega)^* \\ &= \chi^{(3)}_{ijlk}(\omega_1;\omega_1,\omega,-\omega)^*, \end{aligned} \quad (22.223)$$

where the preceding result employs the intrinsic permutation symmetry for $\chi^{(3)}_{ijkl}$, it follows that

$$N_{2_{ijkl}}(-\omega_1) = \frac{3\chi^{(3)}_{ijkl}(-\omega_1; -\omega_1, \omega, -\omega)}{N_{ij}(-\omega_1)}$$
$$= \frac{3\chi^{(3)}_{ijkl}(\omega_1; \omega_1, -\omega, \omega)^*}{N_{ij}(\omega_1)^*}$$
$$= \left[\frac{3\chi^{(3)}_{ijlk}(\omega_1; \omega_1, \omega, -\omega)}{N_{ij}(\omega_1)}\right]^*, \tag{22.224}$$

so that

$$N_{2_{ijkl}}(-\omega_1) = N_{2_{ijlk}}(\omega_1)^*. \tag{22.225}$$

Using this result, Eq. (22.221) can be expressed as follows:

$$N_{2_{ijkl}}(\omega_\alpha) = -\frac{1}{\pi \mathrm{i}} P \int_0^\infty \left\{ \frac{N_{2_{ijkl}}(\omega)}{\omega_\alpha - \omega} + \frac{N_{2_{ijlk}}(\omega)^*}{\omega_\alpha + \omega} \right\} \mathrm{d}\omega, \tag{22.226}$$

and on writing

$$N_{2_{ijkl}}(\omega) = n_{2_{ijkl}}(\omega) + \mathrm{i}\kappa_{2_{ijkl}}(\omega) \tag{22.227}$$

this yields

$$\kappa_{2_{ijkl}}(\omega_\alpha) = \frac{1}{\pi} P \int_0^\infty \left\{ \frac{n_{2_{ijkl}}(\omega)}{\omega_\alpha - \omega} + \frac{n_{2_{ijlk}}(\omega)}{\omega_\alpha + \omega} \right\} \mathrm{d}\omega, \tag{22.228}$$

and

$$n_{2_{ijkl}}(\omega_\alpha) = -\frac{1}{\pi} P \int_0^\infty \left\{ \frac{\kappa_{2_{ijkl}}(\omega)}{\omega_\alpha - \omega} - \frac{\kappa_{2_{ijlk}}(\omega)}{\omega_\alpha + \omega} \right\} \mathrm{d}\omega, \tag{22.229}$$

which represent the dispersion relations for the second-order correction to the nonlinear refractive index.

Dispersion relations can be given for $N^{\mathrm{NL}}(\omega_1)$ by ignoring higher-order susceptibility corrections beyond $\chi^{(3)}$. The argument of Section 19.6 can be repeated, to establish that $N^{\mathrm{NL}}(\omega_1) - 1$ is an analytic function in the upper half of the complex angular frequency plane. Following the same approach given previously for $N_{2_{ijkl}}(\omega)$ leads to

$$N^{\mathrm{NL}}_{ij}(\omega_\alpha) - \delta_{ij} = -\frac{1}{\pi \mathrm{i}} P \int_0^\infty \left\{ \frac{N^{\mathrm{NL}}_{ij}(\omega) - \delta_{ij}}{\omega_\alpha - \omega} + \frac{N^{\mathrm{NL}}_{ij}(-\omega) - \delta_{ij}}{\omega_\alpha + \omega} \right\} \mathrm{d}\omega. \tag{22.230}$$

If an isotopic medium is considered and attention focused on the diagonal tensor components $N_{ii}^{\rm NL}(\omega_\alpha)$, then it follows from Eq. (22.218) that

$$\begin{aligned} N_{ii}^{\rm NL}(-\omega_1)^2 &= 1 + \chi_{ii}^{(1)}(-\omega_1) + 6\chi_{iikl}^{(3)}(-\omega_1; -\omega_1, \omega, -\omega)E_k(\omega)E_l(-\omega) \\ &= 1 + \chi_{ii}^{(1)}(\omega_1)^* + 6\chi_{iikl}^{(3)}(\omega_1; \omega_1, -\omega, \omega)^* E_k(\omega)E_l(-\omega). \end{aligned} \quad (22.231)$$

Making use of the intrinsic permutation symmetry leads to

$$\chi_{iikl}^{(3)}(\omega_1; \omega_1, -\omega, \omega) = \chi_{iilk}^{(3)}(\omega_1; \omega_1, \omega, -\omega), \quad (22.232)$$

and employing Eq. (22.27) yields

$$\begin{aligned} \chi_{iikl}^{(3)}(\omega_1, \omega_2, \omega_3) = \chi_{1122}^{(3)}(\omega_1, \omega_2, \omega_3)\delta_{kl} + \{\chi_{1212}^{(3)}(\omega_1, \omega_2, \omega_3) \\ + \chi_{1221}^{(3)}(\omega_1, \omega_2, \omega_3)\}\delta_{ik}\delta_{il}. \end{aligned} \quad (22.233)$$

From Eqs. (22.232) and (22.233) it follows that

$$\chi_{iikl}^{(3)}(\omega_1; \omega_1, -\omega, \omega) = \chi_{iikl}^{(3)}(\omega_1; \omega_1, \omega, -\omega). \quad (22.234)$$

Hence, Eq. (22.231) can be written as follows:

$$\begin{aligned} N_{ii}^{\rm NL}(-\omega_1)^2 &= \{1 + \chi_{ii}^{(1)}(\omega_1) + 6\chi_{iikl}^{(3)}(\omega_1; \omega_1, \omega, -\omega)E_k(\omega)E_l(-\omega)\}^* \\ &= \{N_{ii}^{\rm NL}(\omega_1)^2\}^*. \end{aligned} \quad (22.235)$$

From the preceding equation it follows that the most plausible result is as follows:

$$N_{ii}^{\rm NL}(-\omega_1) = N_{ii}^{\rm NL}(\omega_1)^*, \quad (22.236)$$

which becomes consistent with the linear case when the $\chi^{(3)}$ contribution is omitted. Equation (22.236) can be obtained in a more rigorous fashion as follows. The function $N_{ii}^{\rm NL}(\omega)$ is analytic in the upper half of the complex angular frequency plane, and has a suitable asymptotic behavior there. Employing the previously derived results for the asymptotic behavior of $\chi_{ii}^{(1)}(\omega)$ and $\chi_{iikl}^{(3)}(\omega_1; \omega_1, \omega, -\omega)$ as $\omega \to \infty$, a Hilbert transform pair can be written for the real and imaginary parts of $N_{ii}^{\rm NL}(\omega)$. Titchmarsh's theorem can now be invoked, so that

$$N_{ii}^{\rm NL}(\omega) - 1 = \int_0^\infty R_{ii}(t)e^{i\omega t}\,dt \quad (22.237)$$

where $R_{ii}(t)$ can be viewed as a real response function. Equation (22.236) follows directly from Eq. (22.237).

If $N_{ii}^{\rm NL}(\omega)$ is written in terms of its real and imaginary parts as follows:

$$N_{ii}^{\rm NL}(\omega) = n_{ii}^{\rm NL}(\omega) + i\kappa_{ii}^{\rm NL}(\omega), \quad (22.238)$$

then Eq. (22.230) becomes

$$\kappa_{ii}^{\mathrm{NL}}(\omega_\alpha, E) = \frac{2\omega_\alpha}{\pi} P \int_0^\infty \frac{\{n_{ii}^{\mathrm{NL}}(\omega, E) - 1\}\mathrm{d}\omega}{\omega_\alpha^2 - \omega^2}, \tag{22.239}$$

and

$$n_{ii}^{\mathrm{NL}}(\omega_\alpha, E) - 1 = -\frac{2}{\pi} P \int_0^\infty \frac{\omega \kappa_{ii}^{\mathrm{NL}}(\omega, E)\mathrm{d}\omega}{\omega_\alpha^2 - \omega^2}, \tag{22.240}$$

and the field dependence has been made explicit in these final results. The preceding two equations have same structure as the Kramers–Kronig relations for the normal linear refractive index and the dissipative component of the complex refractive index for an isotropic medium.

A number of sum rules can be derived for $n_{ii}^{\mathrm{NL}}(\omega) - 1$ and $\kappa_{ii}^{\mathrm{NL}}(\omega)$. Dropping the tensor components, the following results can be derived:

$$\int_0^\infty \{n^{\mathrm{NL}}(\omega, E) - 1\}\mathrm{d}\omega = 0 \tag{22.241}$$

and

$$\int_0^\infty \omega \kappa^{\mathrm{NL}}(\omega, E)\mathrm{d}\omega = \frac{\pi}{4}\omega_{\mathrm{p}}^2. \tag{22.242}$$

The asymptotic behavior required to obtain these results comes directly from the asymptotic behavior of $\boldsymbol{\chi}^{(1)}$. Similar sum rules can be derived for the real and imaginary parts of the nonlinear permittivity, $\varepsilon_{\mathrm{r}}^{\mathrm{NL}}(\omega)$ and $\varepsilon_{\mathrm{i}}^{\mathrm{NL}}(\omega)$, respectively. For an isotropic medium, with the tensor components ignored, it follows that

$$\int_0^\infty [\varepsilon_{\mathrm{r}}^{\mathrm{NL}}(\omega, E) - \varepsilon_0]\mathrm{d}\omega = 0 \tag{22.243}$$

and

$$\int_0^\infty \omega \varepsilon_{\mathrm{i}}^{\mathrm{NL}}(\omega, E)\mathrm{d}\omega = \frac{\pi \varepsilon_0 \omega_{\mathrm{p}}^2}{2}. \tag{22.244}$$

Many additional sum rules can be derived based on the approaches discussed in Chapter 19.

Dispersion relations can be derived for the nonlinear reflectivity. This can be achieved by starting with the formula for the reflectivity in terms of either the nonlinear susceptibility or the nonlinear refractive index. To derive the dispersion relations it is necessary to deal with the meromorphic structure of the complex reflectivity in the upper half of the complex angular frequency plane. For readers interested in pursuing the details on this topic, consult Peiponen and Saarinen (2002).

Notes

§22.1 For general background on nonlinear optics, the texts by Bloembergen (1965), Shen (1984), Butcher and Cotter (1990), and Boyd (1992) are recommended. For concise accounts, see the chapters on this topic in Yariv (1989) and Brau (2004). Significant progress on the principal topics of this chapter has been made by the Finnish workers Peiponen, Vartiainen, and their colleagues and the Italians Bassani and Scandolo. Papers by these authors can be consulted for further general information.

§22.2 For detailed discussion on experimental issues, see Shen (1984) and Boyd (1992).

§22.3 Further discussion on the anharmonic oscillator model is given in the books cited in Section 22.1.

§22.4 The density matrix approach is discussed well in a number of sources. See, for example, Shen (1984), Butcher and Cotter (1990), and Boyd (1992); for a more advanced presentation, Mukamel (1995) is recommended. The seminal paper developing the ideas of this section is Kubo (1957).

§22.5 For a more detailed and general discussion of the asymptotic behavior of the susceptibilities, see Bassani and Scandolo (1991), Scandolo and Bassani (1995a), and Rapapa and Scandolo (1996). A discussion of the nth-order case is given by Bassani and Lucarini (2000).

§22.6 Early contributions to the study of dispersions relations for nonlinear systems are due to Kogan (1963), Price (1963), Caspers (1964), and Kawasaki (1971). Some later works on this topic are Ridener and Good (1974, 1975), Smet and van Groenendael (1979), Sen and Sen (1987), Peiponen (1987a, 1987b, 1987c), and the reviews by Peiponen *et al.* (1997a, 1997b, 1999). An alternative approach to the analysis of the nonlinear susceptibility involves the application of a maximum entropy model; see, for example, Vartiainen and Peiponen (1994) and Vartiainen *et al.* (1996).

§22.7 The first key effort to test dispersion relations for the nonlinear susceptibilities is due to Kishida *et al.* (1993). Recent work on this topic has been carried out by Lucarini and Peiponen (2003) and Lucarini *et al.* (2003b, 2005b).

§22.8 Double dispersion relations are discussed in Kawasaki (1971), Smet and Smet (1974), and Peiponen (1987b).

§22.9 For further reading on nth-order dispersion relations and the sum rules that can be derived from them, see Peiponen (1987b, 1988).

§22.10 For additional discussion, see Portis (1953), Troup (1971), Yariv (1989), Shore and Chan (1990), Hutchings *et al.* (1992), and Kircheva and Hadjichristov (1994). Extensive work on the meromorphic optical functions has been carried out by Peiponen and coworkers; see Peiponen, Vartiainen, and Asakura (1997c, 1998), Peiponen (2001), and Peiponen and Saarinen (2002).

§22.11 Further elaboration on the derivation of sum rules, testing, and the application of these results can be found in Peiponen (1987a, 1987b, 1987c), Bassani and Scandolo (1991, 1992), Peiponen, Vartiainen, and Asakura (1992), Scandolo and Bassani (1992a,b, 1995a,b), Vartiainen, Peiponen, and Asakura (1993b), Chernyak and Mukamel (1995), Rapapa and Scandolo (1996), Cataliotti *et al.* (1997), Bassani and Lucarini (1998, 1999), Scandolo *et al.* (2001), Saarinen (2002), Lucarini, *et al.* (2003c), Lucarini and Peiponen (2003), Peiponen, Saarinen, and Svirko (2004), and Peiponen *et al.* (2004a). For an application to the nonlinear Raman susceptibility, see Peiponen *et al.* (1990).

§22.12 In addition to the specific references cited in Table 22.1, Lucarini *et al.* (2005b) provide further extensions and elaborations on a number of the sum rules tabulated for the nonlinear susceptibility.

§22.13 Further discussion can be found in Boyd (1992). An investigation of sum rules for the nonlinear refractive index and dielectric constant can be found in Bassani and Scandolo (1991, 1992) and Scandolo and Bassani (1992b). For an application to semiconductors, where the quantity n_2 is determined, see Miller *et al.* (1981), and for an additional use see Kador (1995).

Exercises

22.1 By appealing to symmetry ideas, show that $\boldsymbol{\chi}^{(2)}$ vanishes for an isotopic medium.

22.2 Is the density operator for a non-pure state idempotent?

22.3 For readers with a knowledge of *Mathematica* (or any other symbolic algebra capable software), construct an automated code to evaluate the coefficients c_k in the asymptotic expansion of the susceptibility $\chi^{(2)}(\omega_1, \omega_2)$ as a sum of terms of the form $c_k\omega_1^{-k}$, for $k = 0, 1, 2, 3$, and 4, using the quantum formula Eq. (22.108).

22.4 Develop a *Mathematica* (or other software package) program to evaluate the coefficients c_k in the asymptotic expansion of the SHG susceptibility $\chi^{(2)}(\omega, \omega)$ as a series of terms of the form $c_k\omega^{-k}$, for $k = 0, 1, 2, 3, 4$, 5, and 6, using the quantum formula Eq. (22.108).

22.5 Can any dispersion relations be derived describing optical rectification?

22.6 Employing Eq. (22.203), derive the specific sum rules for THG for the cases $q = 1$ and 2, and for valid values of p.

22.7 What dispersion relations can be derived for the functions $n^{\mathrm{NL}}(\omega)$ and $\kappa^{\mathrm{NL}}(\omega)$, assuming the first nonlinear correction arises from $\chi^{(3)}(\omega; \omega, \omega, -\omega)$?

22.8 Can dispersion relations be derived for the susceptibilities connecting different phenomena, for example between SHG and optical rectification, or between SHG and sum-frequency or difference-frequency generation?

22.9 Explain the physical significance, if any, of the real and imaginary parts of $\boldsymbol{\chi}^{(n)}$ for $n > 1$.

22.10 What sum rules can be derived for the function

$$f(\omega) = \frac{\omega^p\{\chi^{(n)}(n\omega; \omega, \omega, \ldots, \omega)\}^q}{\omega_0^2 + \omega^2}? \tag{22.245}$$

Take ω_0 to be an arbitrary angular frequency. Assume p and q are integers and justify the range of these variables for the sum rules you derive.

22.11 Can a two-variable conjugate series approach be developed for $\chi^{(2)}(\omega_1, \omega_2)$ in a similar manner to the one-variable series technique applied to $\chi^{(1)}(\omega)$?

22.12 For the susceptibility $\chi^{(2)}(\omega_{z_1}, \omega_{z_2})$, can the product of the upper half complex ω_{z_1}- and ω_{z_2}-planes be mapped into the bidisc $\{(z_1, z_2) \in \mathbb{C}^2 : |z_1| < 1, |z_2| < 1\}$, and what relationships (if any) can be determined connecting the real and imaginary parts of the SHG susceptibility in the bidisc?

23

Some further applications of Hilbert transforms

23.1 Introduction

In this final chapter, some further applications of Hilbert transforms are considered. Several of these topics are of considerable significance and could easily be developed into rather lengthy chapters. The underlying theme of this chapter is that the applications all make use of the standard Hilbert transform, or one of its variants, as a substantial part of the technique or approach that is under discussion. Since problems from diverse areas are considered in this chapter, the reader is alerted to the fact that the same symbol may designate unrelated quantities in different sections.

23.2 Hilbert transform spectroscopy

The term *Hilbert transform spectroscopy* is used in the literature with two different meanings. Both techniques are discussed in this section. The first involves the use of the Josephson junction. Here the spectrum of the incident electromagnetic radiation on the junction is related to the Hilbert transform of the electrical response of the Josephson junction. The second development employs the conversion of a time domain signal into an absorption spectrum, using the Hilbert transform in a manner designed to try and improve spectral features, such as the identification of weak spectral lines and the minimization of spurious noise peaks.

23.2.1 The Josephson junction

The approach described in this section was proposed by Divin, Polyanskii, and Shul'man (1980), and called Hilbert transform spectroscopy by Divin *et al.* (1993). The terminology Hilbert spectroscopy has also been employed (Tarasov *et al.*, 1995). A Josephson junction consists of two superconducting layers separated by an insulator layer. In the dc Josephson effect, a current flows across the junction without the application of an external electric or magnetic field. The equation describing this behavior is given by

$$I = I_c \sin(\varphi_0), \tag{23.1}$$

where I is the superconducting current, φ_0 is a relative phase factor for the probability amplitudes of electron pairs on opposite sides of the junction, and the junction can pass a maximum zero-voltage current I_c, which is the critical current of the dc Josephson effect.

If a dc voltage is applied across the junction, then current oscillations in the radio-frequency region are set up across the junction, and this is termed the ac Josephson effect. The phase φ is now time-dependent and is given by

$$\frac{\mathrm{d}\varphi(t)}{\mathrm{d}t} = \frac{2eV}{\hbar}, \tag{23.2}$$

where V is the applied voltage, t is the time, and e is the absolute value of the electronic charge. A short calculation (see Kittel (1976)) shows that the superconducting current of Cooper pairs across the junction can be expressed as follows:

$$I(t) = I_c \sin\left(\varphi_0 + \frac{2eVt}{\hbar}\right). \tag{23.3}$$

The current has an angular frequency of oscillation given by

$$\omega = \frac{2eV}{\hbar}. \tag{23.4}$$

The frequency ν is related to the voltage by

$$\nu = 4.8355 \times 10^8 V (\mathrm{MHz\,V^{-1}}), \tag{23.5}$$

so that a 1 μV applied voltage produces a frequency of 483.55 MHz.

In the resistive-shunted junction model of a Josephson junction, the total current across the junction is given by the sum of the superconducting current and a normal resistive current. It is conventional to introduce relative dimensionless variables, defined by

$$\imath = I/I_c, \tag{23.6}$$

$$\mathrm{v} = V/V_c, \tag{23.7}$$

$$\varpi = \frac{\omega}{\omega_c}, \tag{23.8}$$

$$\tau = \omega_c t, \tag{23.9}$$

where

$$\omega_c = \frac{2eV_c}{\hbar} \tag{23.10}$$

and V_c is defined in terms of the resistance R by

$$V_c = RI_c. \tag{23.11}$$

In the resistive-shunted junction (RSJ) model the total current I is given by

$$I = \frac{V}{R} + I_c \sin\varphi(\tau). \tag{23.12}$$

Equations (23.2) and (23.12) can be expressed in terms of reduced variables as follows:

$$\frac{d\varphi(\tau)}{d\tau} = \mathrm{v}(\tau) \tag{23.13}$$

and

$$\frac{d\varphi(\tau)}{d\tau} + \sin\varphi(\tau) = \imath. \tag{23.14}$$

The solution of Eq. (23.14) for the case where the relative current $\imath$ is time-independent is given by

$$\varphi(\tau) = 2\tan^{-1}\left(\frac{1 + \surd(\imath^2 - 1)\tan\theta}{\imath}\right), \tag{23.15}$$

where

$$\theta = \frac{\surd(\imath^2 - 1)}{2}(\tau - c), \tag{23.16}$$

and c is an arbitrary constant. A short calculation shows that

$$\sin\varphi(\tau) = \frac{\imath\{1 + \cos 2\theta + \surd(\imath^2 - 1)\sin 2\theta\}}{\imath^2 + \cos 2\theta + \surd(\imath^2 - 1)\sin 2\theta} \tag{23.17}$$

and

$$\mathrm{v}(\tau) = \frac{d\varphi(\tau)}{d\tau} = \frac{\imath(\imath^2 - 1)}{\imath^2 + \cos 2\theta + \surd(\imath^2 - 1)\sin 2\theta}. \tag{23.18}$$

The time-averaged relative current, $\bar{\imath}$, is given in terms of the time-averaged relative voltage $\bar{\mathrm{v}}$ by

$$\bar{\imath} = \mathrm{sgn}\,\bar{\mathrm{v}}\,\surd\left(\bar{\mathrm{v}}^2 + 1\right). \tag{23.19}$$

Application of an additional external current $\tilde{\imath}(\varpi)$at the relative angular frequency ϖ to the junction leads to a change in the current–voltage behavior of the junction from $I_0(V)$ to $I(V)$. If the current response is denoted by $\Delta I(V) = I(V) - I_0(V)$ and reduced units adopted, replacing $\Delta I(V)$ by $\Delta\bar{\imath}$, then, for a given relative voltage $\bar{\mathrm{v}}$, it can be shown that

$$\Delta\bar{\imath}(\bar{\mathrm{v}},\varpi) = \frac{1}{4\bar{\imath}}\frac{\overline{\tilde{i}(\varpi)^2}}{\bar{\mathrm{v}}^2 - \varpi^2}. \tag{23.20}$$

The derivation of this result can be found in the work of Kanter and Vernon (1972). The preceding Kanter–Vernon formula (Eq. (23.20)), ignores the presence of any fluctuating current in the junction; when this is included in Eq. (23.14), the current response is found to be

$$\Delta\bar{\imath}(\bar{\mathrm{v}},\varpi) = \frac{\overline{\tilde{i}(\varpi)^2}}{8\bar{\imath}\bar{\mathrm{v}}}\left\{\frac{\bar{\mathrm{v}}+\varpi}{(\bar{\mathrm{v}}+\varpi)^2+\gamma^2} + \frac{\bar{\mathrm{v}}-\varpi}{(\bar{\mathrm{v}}-\varpi)^2+\gamma^2}\right\}, \tag{23.21}$$

where γ denotes a fluctuation induced line width of the Josephson oscillation. This result requires that $\bar{\mathrm{v}}$ and ϖ are much greater than γ. The frequency-integrated current response is given by

$$\begin{aligned}\Delta\bar{\imath}(\bar{\mathrm{v}}) &= \int_0^\infty \Delta\bar{\imath}(\bar{\mathrm{v}},\varpi)\mathrm{d}\varpi \\ &= \frac{1}{8\bar{\imath}\bar{\mathrm{v}}}\int_0^\infty \overline{\tilde{i}(\varpi)^2}\left\{\frac{\bar{\mathrm{v}}+\varpi}{(\bar{\mathrm{v}}+\varpi)^2+\gamma^2} + \frac{\bar{\mathrm{v}}-\varpi}{(\bar{\mathrm{v}}-\varpi)^2+\gamma^2}\right\}\mathrm{d}\varpi \\ &= \frac{1}{8\bar{\imath}\bar{\mathrm{v}}}\int_{-\infty}^\infty \overline{\tilde{i}(\varpi)^2}\frac{\bar{\mathrm{v}}-\varpi}{(\bar{\mathrm{v}}-\varpi)^2+\gamma^2}\,\mathrm{d}\varpi,\end{aligned} \tag{23.22}$$

which, in the limit that $\gamma \to 0$, leads to

$$\Delta\bar{\imath}(\bar{\mathrm{v}}) = \frac{1}{8\bar{\imath}\bar{\mathrm{v}}}P\int_{-\infty}^\infty \frac{\overline{\tilde{i}(\varpi)^2}\,\mathrm{d}\varpi}{\bar{\mathrm{v}}-\varpi}. \tag{23.23}$$

Introducing the definition

$$g(\bar{\mathrm{v}}) = 8\pi^{-1}\bar{\imath}\bar{\mathrm{v}}\Delta\bar{\imath}(\bar{\mathrm{v}}), \tag{23.24}$$

it follows that

$$g(\bar{\mathrm{v}}) = \frac{1}{\pi}P\int_{-\infty}^\infty \frac{\overline{\tilde{i}(\varpi)^2}\,\mathrm{d}\varpi}{\bar{\mathrm{v}}-\varpi} = H\left[\overline{\tilde{i}(\bar{\mathrm{v}})^2}\right], \tag{23.25}$$

and $g(\bar{\mathrm{v}})$ therefore represents the Hilbert transform of the time-averaged square of the external current. By analogy with the term interferogram used in Fourier spectroscopy, the designation *hilbertogram* has been employed to describe the function $g(\bar{\mathrm{v}})$ (Shul'man, Kosarev, and Tarasov 2003). The inversion property of the Hilbert transform can be employed to write the following:

$$\overline{\tilde{i}(\varpi)^2} = -\frac{1}{\pi}P\int_{-\infty}^\infty \frac{g(\bar{\mathrm{v}})\mathrm{d}\bar{\mathrm{v}}}{\varpi-\bar{\mathrm{v}}}. \tag{23.26}$$

If $S_i(\varpi)$ denotes the continuous spectrum associated with $\overline{\tilde{i}(\varpi)^2}$ and $S(\varpi)$ denotes the spectrum of electromagnetic radiation impinging on the Josephson junction, these two functions are connected by a transfer function $K(\varpi)$, so that

$$S_i(\varpi) = |K(\varpi)|^2 S(\varpi). \tag{23.27}$$

Hilbert transform spectroscopy has found application in the frequency range of approximately 100 GHz to the few terahertz region.

23.2.2 Absorption enhancement

Under the title Hartley/Hilbert transform spectroscopy, Williams and Marshall (1992) suggested a means by which enhanced absorption spectra might be obtained. The enhanced nature of the spectrum includes the possibility for slightly improved visual identification of weak peaks and the minimization of noise spikes that might be mischaracterized as weak absorption peaks. The basic idea is illustrated in the flow chart shown in Figure 23.1. The Hartley transform was employed by these authors to evaluate the required Hilbert transform. Other techniques such as the Fourier transform can be substituted for the Hartley transform, although the Hartley transform does have some features that recommend its use.

In the absence of noise, the scheme outlined in Figure 23.1 would lead to the same peak heights for absorption bands as would be found from a direct analysis of the time-domain data. Since the time-domain data are discrete, a discrete cosine Fourier transform of N data points yields N frequency-domain points, of which $N/2$ are unique – the second half of the frequency-domain data is a repeat of the first half. A similar situation holds for the sine Fourier transform. It is possible to work with just the cosine Fourier transform using $2N$ time-domain data points, of which N have been obtained by appropriate zero filling. When this is done, both the normal absorption spectral profile and the so-called enhanced profile should be the same. In the absence of zero-filling, the enhanced absorption spectrum shows improvements over the normal absorption spectrum, when spectral noise is taken into consideration.

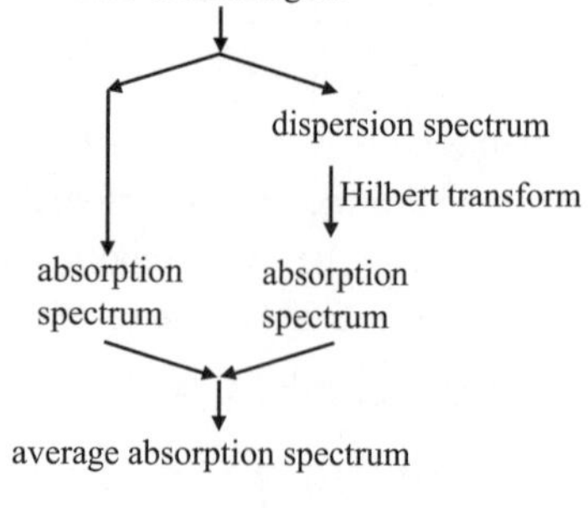

Figure 23.1. Scheme proposed to obtain an enhanced absorption spectrum.

Padding the original N-point time-domain signal with N zeros, leads, after Fourier transformation and application of the scheme given in Figure 23.1, to a spectrum that is not enhanced relative to the normal absorption spectrum. The difference between the two approaches amounts to one of computational convenience. Since the FFT approach can be executed very quickly, even on very large data sets, probably little is gained by combing the FFT with a Hilbert transform evaluation.

23.3 The phase retrieval problem

It is common in many experiments involving the interaction of electromagnetic radiation with matter to measure intensity data. Phase information for the electromagnetic wave is lost in this measurment process. It is often impractical to determine directly phase information experimentally. The phase problem, or synonymously the phase retrieval problem, deals with the determination of phase information from measurements of the modulus. Intensity measurements are proportional to the square of the modulus. Some different approaches that have been applied to deal with this rather venerable problem are examined in this section. A key idea that forms the basis of various procedures is to set up the dispersion relation connection between the phase and the modulus.

Let the intensity of some physical measurement be represented as follows:

$$I \propto |F(\omega)|^2 \tag{23.28}$$

and

$$F(\omega) = |f(\omega)|\, e^{i\phi(\omega)}, \tag{23.29}$$

where $\phi(\omega)$ denotes the phase and $|f(\omega)|$ is the modulus. It is assumed that the function F is an analytic function in the upper half of the complex angular frequency plane. In order to set up a dispersion relation connection for the phase, it is necessary to separate $\phi(\omega)$ and $f(\omega)$. This is accomplished by working with the function $\log F(\omega)$, so that

$$\log F(\omega) = \log|f(\omega)| + i\phi(\omega). \tag{23.30}$$

The material presented for the reflectivity in Section 20.2 can be applied directly, leading to the following dispersion relation:

$$\phi(\omega) = \frac{1}{\pi} P \int_{-\infty}^{\infty} \frac{\log|f(\omega')|\,d\omega'}{\omega - \omega'}. \tag{23.31}$$

It is assumed that $f(\omega) = O(\omega^{-\lambda})$ with $\lambda > 1$, as $\omega \to \infty$. To examine the convergence of the preceding integral in the asymptotic region, select an angular frequency $\omega_0 >> \omega$, and then consider the sum of the two integrals

$$\int_{-T}^{-\omega_0} \frac{\log|\omega'|\mathrm{d}\omega'}{\omega - \omega'} + \int_{\omega_0}^{T} \frac{\log|\omega'|\mathrm{d}\omega'}{\omega - \omega'}.$$

This pair of integrals gives a convergent result in the limit $T \to \infty$. The two important factors that are required for arriving at this result are: (i) that $|f(\omega)|$ has no zeros in the upper half of the complex angular frequency plane, and (ii) the absence of Blaschke factors. For the latter issue, if $B(\omega)$ is a Blaschke factor (recall Eq. (20.19)) then

$$|f(\omega)B(\omega)|^2 = |f(\omega)|^2. \tag{23.32}$$

The presence of zeros for $|f(\omega)|$ leads to logarithmic divergences for $\log|f(\omega)|$.

One approach (Burge *et al.*, 1974) that has been employed to avoid the problem of zeros for the function $f(\omega)$ is to write

$$g(\omega) = f(\omega) + C, \tag{23.33}$$

where C is a complex constant selected so that the function $|g|$ has no zeros in the upper half complex angular frequency plane. Let the modulus and phase of g be denoted by $|g(\omega)|$ and $\varphi(\omega)$, respectively, so that

$$g(\omega) = |g(\omega)|\, \mathrm{e}^{\mathrm{i}\varphi(\omega)}, \tag{23.34}$$

with

$$\tan\varphi(\omega) = \frac{g_{\mathrm{i}}(\omega)}{g_{\mathrm{r}}(\omega)}, \tag{23.35}$$

where $g_{\mathrm{r}}(\omega)$ and $g_{\mathrm{i}}(\omega)$ are the real and imaginary parts of $g(\omega)$, respectively.

If it is assumed that $f(\omega_z)$ satisfies $f(\omega_z) = O(\omega_z^{-1-\delta})$ for $\delta > 0$, as $\omega_z \to \infty$, then the following result can be derived:

$$f(\omega) = -\frac{1}{\pi\mathrm{i}} P \int_{-\infty}^{\infty} \frac{f(\omega')\mathrm{d}\omega'}{\omega - \omega'}. \tag{23.36}$$

The resulting dispersion relations then take the following form:

$$f_{\mathrm{i}}(\omega) = \frac{1}{\pi} P \int_{-\infty}^{\infty} \frac{f_{\mathrm{r}}(\omega')\mathrm{d}\omega'}{\omega - \omega'} \tag{23.37}$$

and

$$f_{\mathrm{r}}(\omega) = -\frac{1}{\pi} P \int_{-\infty}^{\infty} \frac{f_{\mathrm{i}}(\omega')\mathrm{d}\omega'}{\omega - \omega'}, \tag{23.38}$$

which can be modified by the addition of a constant C, with real and imaginary parts C_r and C_i, respectively, so that

$$f_\mathrm{i}(\omega) + C_\mathrm{i} = \frac{1}{\pi} P \int_{-\infty}^{\infty} \frac{\{f_\mathrm{r}(\omega') + C_\mathrm{r}\}\mathrm{d}\omega'}{\omega - \omega'} + C_\mathrm{i} \tag{23.39}$$

and

$$f_\mathrm{r}(\omega) + C_\mathrm{r} = -\frac{1}{\pi} P \int_{-\infty}^{\infty} \frac{\{f_\mathrm{i}(\omega') + C_\mathrm{i}\}\mathrm{d}\omega'}{\omega - \omega'} + C_\mathrm{r}. \tag{23.40}$$

These equations can be rewritten as follows:

$$g_\mathrm{i}(\omega) = \frac{1}{\pi} P \int_{-\infty}^{\infty} \frac{g_\mathrm{r}(\omega')\mathrm{d}\omega'}{\omega - \omega'} + C_\mathrm{i} \tag{23.41}$$

and

$$g_\mathrm{r}(\omega) = -\frac{1}{\pi} P \int_{-\infty}^{\infty} \frac{g_\mathrm{i}(\omega')\mathrm{d}\omega'}{\omega - \omega'} + C_\mathrm{r}. \tag{23.42}$$

From Eqs. (23.35) and (23.42), it follows that

$$\frac{g_\mathrm{i}(\omega)}{\tan\varphi(\omega)} = -\frac{1}{\pi} P \int_{-\infty}^{\infty} \frac{g_\mathrm{i}(\omega')\mathrm{d}\omega'}{\omega - \omega'} + C_\mathrm{r}. \tag{23.43}$$

Solving this integral equation offers the possibility for obtaining the phase. The problem can be approached in the following way making use of the Sokhotsky–Plemelj formulas (Peřina, 1985, p. 48). If the function $\psi(\omega_z)$ is introduced by the following definition (for $\omega_\mathrm{i} \neq 0$):

$$\psi(\omega_z) = \frac{1}{2\pi} \int_{-\infty}^{\infty} \frac{g_\mathrm{i}(\omega')\mathrm{d}\omega'}{\omega' - \omega_z} + \frac{1}{2} C_\mathrm{r}, \tag{23.44}$$

then examining the limit $\omega_\mathrm{i} \to 0$ and using Eqs. (10.7) and (10.8) yields

$$\psi^+(\omega) = \frac{1}{2\pi} P \int_{-\infty}^{\infty} \frac{g_\mathrm{i}(\omega')\mathrm{d}\omega'}{\omega' - \omega} + \frac{1}{2}\mathrm{i} g_\mathrm{i}(\omega) + \frac{1}{2} C_\mathrm{r} \tag{23.45}$$

and

$$\psi^-(\omega) = \frac{1}{2\pi} P \int_{-\infty}^{\infty} \frac{g_\mathrm{i}(\omega')\mathrm{d}\omega'}{\omega' - \omega} - \frac{1}{2}\mathrm{i} g_\mathrm{i}(\omega) + \frac{1}{2} C_\mathrm{r}, \tag{23.46}$$

where the limits are taken as $\omega_\mathrm{i} \to 0+$ for ψ^+ and $\omega_\mathrm{i} \to 0-$ for ψ^-. Employing Eq. (23.42) yields

$$\psi^+(\omega) = \frac{1}{2} g(\omega) \tag{23.47}$$

and

$$\psi^-(\omega) = \frac{1}{2}g^*(\omega), \tag{23.48}$$

from which it follows that

$$\psi^+(\omega) + \psi^-(\omega) = g_r(\omega) \tag{23.49}$$

and

$$\psi^+(\omega) - \psi^-(\omega) = \mathrm{i}g_i(\omega). \tag{23.50}$$

Equation (23.34) leads to

$$\log\left(\frac{g^2(\omega)}{|g(\omega)|^2}\right) = \log\left(\frac{|g(\omega)|^2\,\mathrm{e}^{2\mathrm{i}\varphi(\omega)}}{|g(\omega)|^2}\right) = 2\mathrm{i}\varphi(\omega) \tag{23.51}$$

and

$$\frac{\psi^+(\omega)}{\psi^-(\omega)} = \frac{g(\omega)}{g(\omega)^*} = \frac{g^2(\omega)}{|g^2(\omega)|} = \mathrm{e}^{2\mathrm{i}\varphi(\omega)}, \tag{23.52}$$

so that

$$\log\psi^+(\omega) - \log\psi^-(\omega) = 2\mathrm{i}\varphi(\omega). \tag{23.53}$$

The solution of Eq. (23.53) is given by

$$\log\psi(\omega_z) = \frac{1}{2\pi\mathrm{i}}\int_{-\infty}^{\infty}\frac{2\mathrm{i}\varphi(\omega')\mathrm{d}\omega'}{\omega' - \omega_z} + \log c, \tag{23.54}$$

where c is a constant, and hence

$$\psi(\omega_z) = c\exp\left\{\frac{1}{\pi}\int_{-\infty}^{\infty}\frac{\varphi(\omega')\mathrm{d}\omega'}{\omega' - \omega_z}\right\}. \tag{23.55}$$

From the preceding result it follows, on using Eqs. (10.7) and (10.8), that

$$\psi^+(\omega) = c\exp\left\{\frac{1}{\pi}P\int_{-\infty}^{\infty}\frac{\varphi(\omega')\mathrm{d}\omega'}{\omega' - \omega} + \mathrm{i}\varphi(\omega)\right\}, \tag{23.56}$$

and

$$\psi^-(\omega) = c\exp\left\{\frac{1}{\pi}P\int_{-\infty}^{\infty}\frac{\varphi(\omega')\mathrm{d}\omega'}{\omega' - \omega} - \mathrm{i}\varphi(\omega)\right\}. \tag{23.57}$$

Equations (23.49) and (23.50) yield

$$g_i(\omega) = 2c \sin \varphi(\omega) \exp\left\{\frac{1}{\pi} P \int_{-\infty}^{\infty} \frac{\varphi(\omega')\mathrm{d}\omega'}{\omega' - \omega}\right\}, \tag{23.58}$$

$$g_r(\omega) = 2c \cos \varphi(\omega) \exp\left\{\frac{1}{\pi} P \int_{-\infty}^{\infty} \frac{\varphi(\omega')\mathrm{d}\omega'}{\omega' - \omega}\right\}, \tag{23.59}$$

and hence

$$|g(\omega)| = \surd\left(g_r^2(\omega) + g_i^2(\omega)\right) = 2c \exp\left\{\frac{1}{\pi} P \int_{-\infty}^{\infty} \frac{\varphi(\omega')\mathrm{d}\omega'}{\omega' - \omega}\right\}, \tag{23.60}$$

so that

$$\log\left|\frac{g(\omega)}{2c}\right| = \frac{1}{\pi} P \int_{-\infty}^{\infty} \frac{\varphi(\omega')\mathrm{d}\omega'}{\omega' - \omega}. \tag{23.61}$$

Take the Hilbert transform of both sides of the preceding result, assuming both integrals converge. Recall that if $\varphi \in L^p(\mathbb{R})$ for $1 < p < \infty$, then $H\varphi \in L^p(\mathbb{R})$, and from Eq. (23.61) this is not expected to be the case for functions likely to arise in practical applications. However, functions still exist where taking the Hilbert transform leads to convergent results. So if the integrals exist, it follows that

$$\frac{1}{\pi} P \int_{-\infty}^{\infty} \frac{\log|g(\omega)/2c|\mathrm{d}\omega}{\omega'' - \omega} = \frac{1}{\pi} P \int_{-\infty}^{\infty} \frac{\mathrm{d}\omega}{\omega'' - \omega} \frac{1}{\pi} P \int_{-\infty}^{\infty} \frac{\varphi(\omega')\mathrm{d}\omega'}{\omega' - \omega}, \tag{23.62}$$

and hence

$$\varphi(\omega) = \frac{1}{\pi} P \int_{-\infty}^{\infty} \frac{\log|g(\omega')|\mathrm{d}\omega'}{\omega - \omega'}. \tag{23.63}$$

Recalling the iteration property for the Hilbert transform, the simplification of the right-hand side of Eq. (23.62) is straightforward for functions $\varphi \in L^p(\mathbb{R})$ for $1 < p < \infty$, but can also be carried out for particular choices of other functions. From Eqs. (23.33) and (23.34) it follows that

$$f_r(\omega) + C_r = |g(\omega)| \cos \varphi(\omega) \tag{23.64}$$

and

$$f_i(\omega) + C_i = |g(\omega)| \sin \varphi(\omega), \tag{23.65}$$

so that

$$\tan \phi = \frac{f_i}{f_r} = \frac{|g(\omega)| \sin \varphi(\omega) - C_i}{|g(\omega)| \cos \varphi(\omega) - C_r}. \tag{23.66}$$

The preceding result can be simplified by noting that a zero of $|f(\omega_z)|$ must have both the real and imaginary components equal to zero, so that it is sufficient to add a real constant C to f and let $C_i = 0$ in Eq. (23.66). The computational approach to be employed is as follows. Fix the constant C, for example taking $C > \max|f|$, and evaluate $\varphi(\omega)$ from Eq. (23.63). It is assumed that $|f|$ is available over a sufficiently wide spectral range for this to be done accurately. Then determine ϕ from Eq. (23.66). In situations where the Hilbert transform of Eq. (23.61) cannot be taken, an alternative avenue is to employ a subtraction for the function f, based at some selected frequency. Results in this direction were indicated in Section 20.2.

Let $f(\omega) = |f(\omega)|\, e^{i\phi(\omega)}$ and consider the case where $|f(0)|$ is non-zero. A modified approach can be employed to obtain dispersion relations connecting $|f(\omega)|$ with $\phi(\omega)$. Consider the integral

$$\oint_C \frac{\log f(\omega_z) d\omega_z}{\omega_z(\omega_z - \omega)},$$

where the contour C is taken to be the familiar semicircular contour in the upper half of the complex angular frequency plane with suitable indentations of the contour into the upper half plane at the angular frequencies $\omega_z = 0$ and $\omega_z = \omega$. The extra factor of ω_z^{-1} improves the convergence. The resulting dispersion relations are readily found to be as follows:

$$\varphi(\omega) = \frac{\omega}{\pi} P \int_{-\infty}^{\infty} \frac{\log |f(\omega')| d\omega'}{\omega'(\omega - \omega')} + \varphi(0) \tag{23.67}$$

and

$$\log|f(\omega)| = -\frac{\omega}{\pi} P \int_{-\infty}^{\infty} \frac{\varphi(\omega') d\omega'}{\omega'(\omega - \omega')} + \log|f(0)|. \tag{23.68}$$

The derivation explicitly requires the absence of zeros for the function $f(\omega_z)$ in the upper half of the complex angular frequency plane.

Suppose the function $f(\omega_z)$ has N zeros, located at the positions $\omega_k, k = 1\ldots, N$, in the upper half of the complex angular frequency plane, and their locations are known to be at the positions $\omega_k, k = 1, \ldots, N$. To keep the discussion simple, it is assumed that all the zeros are of order one. The function $f(\omega_z)$ can be modified by introducing the new function $g(\omega_z)$ defined in the following way:

$$g(\omega_z) = f(\omega_z) \prod_{k=1}^{N} \frac{\omega_z - \omega_k^*}{\omega_z - \omega_k}, \tag{23.69}$$

where the second function is a Blaschke factor. On the real frequency axis it follows that

$$|g(\omega)|^2 = |f(\omega)|^2. \tag{23.70}$$

Hoenders (1975) considered the class of functions which are band-limited to the finite interval $[a, b]$, with the Fourier transform of f given by

$$F(\omega) = \int_a^b f(t)\mathrm{e}^{-\mathrm{i}\omega t}\,\mathrm{d}t. \tag{23.71}$$

In the preceding formula let ω be the complex variable ω_z. A function $F(\omega_z)$ defined by Eq. (23.71) is an entire function of exponential type. By taking advantage of this characterization, a product-type expansion for the entire function can be written (see Levin (1964), chap. 1). Under relatively mild assumptions, Hoenders (1975) derived dispersion relations starting with a contour integral of the function $(\omega_z - \omega)^{-1}\omega_z \mathrm{e}^{\mathrm{i}\omega_z} g(\omega_z)$, where $g(\omega_z)$, defined in Eq. (23.69), is assumed to satisfy a condition of the form given in Eq. (23.71). The choice of contour is a semicircle, with a suitable semicircular indentation at $\omega_z = \omega$ and at any zeros that may happen to be located on the real frequency axis. The reader is left to explore the outcome. The major issue to resolve is the actual location of the zeros. The end-notes provide some sources for further reading.

23.4 X-ray crystallography

The scattering of a wave by a point-sized scattering center, located at a point P with position vector $\mathbf{r}$ relative to the origin, is given by a scattering factor $F_{\mathbf{h}}$, which can be expressed as follows:

$$F_{\mathbf{h}} = f\mathrm{e}^{2\pi \mathrm{i}\mathbf{r}\cdot\mathbf{h}}, \tag{23.72}$$

where f is the amplitude factor of the scattered wave and $\mathbf{h}$ is a wave vector representing the difference between the wave vectors for the scattered and incident radiation. The preceding result can be generalized to include N scattering centers, so that

$$F_{\mathbf{h}} = \sum_{k=1}^{N} f_k \mathrm{e}^{2\pi \mathrm{i}\mathbf{r}_k\cdot\mathbf{h}}. \tag{23.73}$$

The notation F_{hkl} in place of $F_{\mathbf{h}}$ is common. The indices h, k, and l, called the Miller indices, are used to characterize the various lattice planes or faces in a crystal. They are usually written in the form (*hkl*), and to indicate lattice points the notation *hkl* is used. A transition from the set of discrete scattering centers to the continuous situation can be made by replacing the sum in Eq. (23.73) by an integral. Let $\rho(\mathbf{r})$ denote the electron density function at the point $\mathbf{r}$, then the structure factor $F_{\mathbf{h}}$ is given by the Fourier transform of $\rho(\mathbf{r})$ as follows:

$$F_{\mathbf{h}} = \int_{-\infty}^{\infty} \rho(\mathbf{r})\mathrm{e}^{2\pi \mathrm{i}\mathbf{r}\cdot\mathbf{h}}\,\mathrm{d}V, \tag{23.74}$$

where the integral is over a volume in real space and $\mathbf{h}$ denotes a coordinate in the reciprocal space (also called the Fourier space by some authors). The inverse Fourier transform of Eq. (23.74) leads to the following expression for the electronic density:

$$\rho(\mathbf{r}) = \int_{-\infty}^{\infty} F_{\mathbf{h}} e^{-2\pi i \mathbf{r}\cdot\mathbf{h}} \, dV_{\mathbf{h}}, \tag{23.75}$$

and the integration is over a the volume in reciprocal space. In X-ray crystallography the principal item of interest is the location of the scattering centers, that is, the positions of the nuclei in the structure under investigation. The electron density is also a key item, and this can be extracted from an X-ray scattering experiment. Knowledge of the electronic density distribution provides access to information on chemical bonding within the crystal structure. The discrete analog of Eq. (23.75) is given by

$$\rho(\mathbf{r}) = \frac{1}{V} \sum_h \sum_k \sum_l F_{hkl} e^{-2\pi i(hx+ky+lz)}, \tag{23.76}$$

where V is the volume of the unit cell of the crystal.

The quantity that is experimentally accessible is the intensity I_{hkl}, which satisfies

$$I_{hkl} \propto |F_{hkl}|^2 . \tag{23.77}$$

The structure factor can be written in the form

$$F_{hkl} = |F_{hkl}| \, e^{i\phi_{hkl}}, \tag{23.78}$$

where ϕ_{hkl} is the phase. In order to carry out the Fourier synthesis of the electronic density using Eq. (23.76), it is necessary to determine the phase factors ϕ_{hkl}, keeping in mind that the intensity measurement only furnishes the modulus of the structure factor. This is called the *phase problem* in crystallography.

There is a long history of work devoted to the resolution of the phase problem in crystallography, and some clever schemes have been developed that are successful to varying degrees. A couple of approaches related to the Hilbert transform, which for the most part have been a more recent advance, are considered in this section.

Ramachandran (1969) developed an approach based on the discrete Hilbert transform. To simplify, a one-dimensional version of the theory is considered. The structure factor is written in the following form:

$$F_h = |F_h| \, e^{i\phi_h} = A(x) + iB(x). \tag{23.79}$$

Assume for the one-dimensional case that $\rho(x)$ is periodic and

$$\rho(x) = \sum_{h=-\infty}^{\infty} F_h e^{-2\pi i hx}, \tag{23.80}$$

with

$$F_h = \int_0^1 \rho(x) \mathrm{e}^{2\pi \mathrm{i} hx}\, \mathrm{d}x. \tag{23.81}$$

Treating the structure factor as a function of the continuous variable h, and differentiating Eq. (23.81) with respect to this variable, leads to a pair of formulas that are related to discrete Hilbert transforms. One result, expressing A in terms of B and its derivative, and the other giving B in terms of A and its derivative. There are two difficulties. The first is that the derivatives of the structure factors are not readily accessible. The second issue concerns the fact that the structure factor F_h is not a continuous function; it vanishes everywhere except at the lattice points. For this reason the standard form of the Hilbert transform relations connecting the real and imaginary parts of F_h do not apply. Also, this raises issues associated with taking the derivatives of the structure factors. This approach has not found any practical application.

A second scheme is now considered. Suppose the structure factor, considered as a function of the continuous variable s in reciprocal space, is analytic in the upper half of the complex s plane. Further assume that $F(s) = O(s^{-\lambda})$ with $\lambda > 1$, as $s \to \infty$. In this case it follows that

$$F(s) = -\frac{1}{\pi \mathrm{i}} P \int_{-\infty}^{\infty} \frac{F(s')\mathrm{d}s'}{s - s'}. \tag{23.82}$$

The structure factor may be written, for the one-dimensional case, in the following form:

$$F(s/a) = \int_0^a \rho(x) \mathrm{e}^{2\pi \mathrm{i} sx/a}\, \mathrm{d}x, \tag{23.83}$$

where the electronic density is assumed to satisfy

$$\rho(x) = 0, \quad x < 0,\ x > a. \tag{23.84}$$

Since $F(s/a)$ has an inverse Fourier transform with compact support in the interval $[0, a]$, the Whittaker–Shannon–Kotel'nikov sampling theorem (recall Section 7.4) can be applied to write

$$F(s/a) = \sum_{k=-\infty}^{\infty} F(k/2a)\, \mathrm{sinc}\,(2s - k), \quad \text{for } s \in \mathbb{R}, \tag{23.85}$$

and recall that the sinc function is defined by

$$\mathrm{sinc}\, x = \frac{\sin \pi x}{\pi x}. \tag{23.86}$$

Inserting Eq. (23.85) into Eq. (23.82) and recalling that the Hilbert transform of the sinc function is $H[\operatorname{sinc} x] = (1 - \cos \pi x)/(\pi x)$, leads to the following:

$$F(s/a) = \mathrm{i} \sum_{k=-\infty}^{\infty} F(k/2a) \left\{ \frac{1 - \cos[\pi(2s - k)]}{\pi(2s - k)} \right\}. \tag{23.87}$$

Combining this result with Eq. (23.85) yields

$$\sum_{k=-\infty}^{\infty} F(k/2a) \left\{ \operatorname{sinc}\{(2s - k)\} - \mathrm{i} \left\{ \frac{1 - \cos[\pi(2s - k)]}{\pi(2s - k)} \right\} \right\} = 0. \tag{23.88}$$

Writing the structure factor in terms of its real and imaginary parts as

$$F(s) = A(s) + \mathrm{i}B(s), \tag{23.89}$$

and setting $s = h/2$ allows expressions for A and B to be obtained from Eq. (23.87) as follows:

$$A\left(\frac{h}{2a}\right) = -\frac{1}{\pi} \sum_{k=-\infty}^{\infty}{}' B\left(\frac{k}{2a}\right) \frac{[1 - (-1)^{h-k}]}{h - k} \tag{23.90}$$

and

$$B\left(\frac{h}{2a}\right) = \frac{1}{\pi} \sum_{k=-\infty}^{\infty}{}' A\left(\frac{k}{2a}\right) \frac{[1 - (-1)^{h-k}]}{h - k}, \tag{23.91}$$

where the prime on the summation is used to signify the omission of the term in the sum with $k = h$. Equations (23.90) and (23.91) represent the discrete Hilbert transforms for the real and imaginary components of the structure factor. The preceding development is due to Mishnev (1993), who also considered the extension to the three-dimensional case, which follows in a rather straightforward manner. The three-dimensional analog of Eq. (23.82) can be written as follows:

$$F(s_1, s_2, s_3) = -\frac{\mathrm{i}}{\pi^3} P \int_{-\infty}^{\infty} P \int_{-\infty}^{\infty} P \int_{-\infty}^{\infty} \frac{F(s_1', s_2', s_3')\, \mathrm{d}s_1', \mathrm{d}s_2', \mathrm{d}s_3'}{(s_1 - s_1')(s_2 - s_2')(s_3 - s_3')}. \tag{23.92}$$

The three-dimensional analog of the Whittaker–Shannon–Kotel'nikov sampling theorem is given by

$$F\left(\frac{s_1}{a_1}, \frac{s_2}{a_2}, \frac{s_3}{a_3}\right) = \sum_{h=-\infty}^{\infty} \sum_{k=-\infty}^{\infty} \sum_{l=-\infty}^{\infty} F\left(\frac{h}{2a_1}, \frac{k}{2a_2}, \frac{l}{2a_3}\right) \times \operatorname{sinc}\{(2s_1 - h)\} \operatorname{sinc}\{(2s_2 - k)\} \operatorname{sinc}\{(2s_3 - l)\}. \tag{23.93}$$

To simplify the formulas, the restriction $a_1 = a_2 = a_3 = 1$ is applied. Substituting this result into Eq. (23.92) yields

$$F(s_1, s_2, s_3) = -\mathrm{i} \sum_{h=-\infty}^{\infty} \sum_{k=-\infty}^{\infty} \sum_{l=-\infty}^{\infty} F\left(\frac{h}{2}, \frac{k}{2}, \frac{l}{2}\right) \left\{ \frac{1 - \cos[\pi(2s_1 - h)]}{\pi(2s_1 - h)} \right\} \times \left\{ \frac{1 - \cos[\pi(2s_2 - k)]}{\pi(2s_2 - k)} \right\} \left\{ \frac{1 - \cos[\pi(2s_3 - l)]}{\pi(2s_3 - l)} \right\}. \tag{23.94}$$

Introducing the change of variables $s_1 = h'/2, s_2 = k'/2$, and $s_3 = l'/2$ yields

$$F\left(\frac{h'}{2}, \frac{k'}{2}, \frac{l'}{2}\right) = -\frac{\mathrm{i}}{\pi^3} \sum_{h=-\infty}^{\infty}{}' \sum_{k=-\infty}^{\infty}{}' \sum_{l=-\infty}^{\infty}{}' F\left(\frac{h}{2}, \frac{k}{2}, \frac{l}{2}\right) \times \frac{\{1 - (-1)^{h'-h}\}\{1 - (-1)^{k'-k}\}\{1 - (-1)^{l'-l}\}}{(h' - h)(k' - k)(l' - l)}. \tag{23.95}$$

Writing $F = A + \mathrm{i}B$ leads to the following expressions for A and B:

$$A\left(\frac{h'}{2}, \frac{k'}{2}, \frac{l'}{2}\right) = \frac{1}{\pi^3} \sum_{h=-\infty}^{\infty}{}' \sum_{k=-\infty}^{\infty}{}' \sum_{l=-\infty}^{\infty}{}' B\left(\frac{h}{2}, \frac{k}{2}, \frac{l}{2}\right) \times \frac{\{1 - (-1)^{h'-h}\}\{1 - (-1)^{k'-k}\}\{1 - (-1)^{l'-l}\}}{(h' - h)(k' - k)(l' - l)} \tag{23.96}$$

and

$$B\left(\frac{h'}{2}, \frac{k'}{2}, \frac{l'}{2}\right) = -\frac{1}{\pi^3} \sum_{h=-\infty}^{\infty}{}' \sum_{k=-\infty}^{\infty}{}' \sum_{l=-\infty}^{\infty}{}' A\left(\frac{h}{2}, \frac{k}{2}, \frac{l}{2}\right) \times \frac{\{1 - (-1)^{h'-h}\}\{1 - (-1)^{k'-k}\}\{1 - (-1)^{l'-l}\}}{(h' - h)(k' - k)(l' - l)}. \tag{23.97}$$

Zanotti *et al.* (1996) discuss one implementation procedure, and their approach is as follows. Suppose that the reflections are divided into two groups: those with known phases, call this set Γ_{n}, and those with unknown phases, denote this set by Γ_{u}. The first of these two groups might be determined by *direct methods*, which involve various probabilistic procedures, inequality techniques for the structure factors, and others (see Stout and Jensen (1989), chap. 11). The computational steps employed to implement the approach are as follows. (i) Using Eqs. (23.96) and (23.97) and the data from Γ_{n}, evaluate the A and B terms for the set Γ_{u} to some level of accuracy. A key ingredient of this calculation is knowledge of $F(0,0,0)$. (ii) From all the available A and B terms, recalculate the A and B terms to an improved accuracy. (iii) Renormalize the determined A and B values using the known values of the structure factors $|F|^2$, noting that $|F|^2 = A^2 + B^2$. A recycling procedure through steps (ii) and (iii) can be employed. Once the A and B values have been determined, the phases can be determined. Some examples of the approach can be found in Zanotti *et al.* (1996).

An alternative strategy that takes advantage of the discrete Hilbert transform relations can be applied when the phases for all the reflections are known, but with varying degrees of uncertainty. Let m denote a *figure of merit* for a reflection, and let this quantity be assigned in the following manner. For a reflection with a known phase that is expected to be correct, assign for that reflection $m = 1$, and for a reflection for which the phase is expected to be totally incorrect, use $m = 0$. Now modify the right-hand side of Eqs. (23.96) and (23.97) by making the following replacements:

$$B\left(\frac{h}{2}, \frac{k}{2}, \frac{l}{2}\right) \to mB\left(\frac{h}{2}, \frac{k}{2}, \frac{l}{2}\right) \text{ and } A\left(\frac{h}{2}, \frac{k}{2}, \frac{l}{2}\right) \to mA\left(\frac{h}{2}, \frac{k}{2}, \frac{l}{2}\right),$$

respectively. This amounts to weighting more heavily the more accurately determined A and B terms. A cycling procedure can be carried out, where only the phases with low m values are modified. In each step of the cycling, modifications of the assigned figures of merit can be made. The calculation is terminated when some preset level of convergence has been achieved.

23.5 Electron–atom scattering

Dispersion relations are concisely discussed in this section for two types of scattering processes: potential scattering and electron–hydrogen atom scattering. Both problems have been of interest for a considerable period of time. The chapter end-notes provide some references to the historical development of the subject, as well as the more recent progress. The focus of this section is electron–hydrogen atom scattering, and in particular, the analytic structure of the scattering amplitude in the complex energy plane. Some introductory background from potential scattering is reviewed first.

23.5.1 Potential scattering

Scattering experiments have a very long history in the physical sciences. Scattering a particle off a target provides an opportunity to investigate the forces in play between the target and scattering particle, the internal structure of the target, information on decay times and hence stability, and other factors. The common variables that are measured are energy, momentum transfer, various angular distributions and the associated intensities, together with different polarization factors in a number of experiments.

Consider the collision of two spinless particles of masses m_1 and m_2, with the reduced mass for the system given by $m = m_1m_2/(m_1+m_2)$. If the interaction potential V depends on the difference in coordinates of the two particles, the Schrödinger equation for the system is separable into two parts in the center-of-mass coordinate system: one of the resulting equations describes the translation of the system; the other describes the internal dynamics of the system. It is the latter dynamics that is of interest. It is assumed that the interaction potential has an asymptotic fall-off at large distances that is much faster than r^{-1}, where r is the distance between the center of mass of the particle system and the origin of the potential. The Schrödinger equation in the center-of-mass coordinate system takes the following form:

$$-\frac{\hbar^2}{2m}\nabla^2\psi + V(\mathbf{r})\psi = E\psi, \tag{23.98}$$

where ∇ is the del operator and E is the energy. At very large separation distances, where the potential can be approximately neglected, the energy is given by

$$E = \frac{p^2}{2m} = \frac{\hbar^2 k^2}{2m}, \tag{23.99}$$

where the momentum $\mathbf{p} = \hbar\mathbf{k}$. In a system of units where $\hbar$ is set to the value unity (that is, atomic units, a.u.), $\mathbf{k}$ denotes the momentum vector of the particle. In the SI unit system, multiply $\mathbf{k}$ by $\hbar$ to get the momentum. Loosely speaking, $\mathbf{k}$ is often referred to as the momentum, even when atomic units are not in use. In the asymptotic region the solution of the Schrödinger equation behaves as follows:

$$\psi^+(\mathbf{r}, \mathbf{k}) \to \mathrm{e}^{ikz} + f(\theta, \phi)\frac{\mathrm{e}^{ikr}}{r}, \quad \text{as } r \to \infty. \tag{23.100}$$

The interpretation is that, at large distances from the scattering center, the first part of the solution is a plane wave; more generally, it comprises a superposition of plane waves traveling in the z-direction with momentum $\hbar k$. The second part of the solution is an outgoing wave with the angular amplitude $f(\theta, \phi)$, which corresponds to the scattered beam. The quantity f is referred to as the scattering amplitude. A key quantity of interest in any scattering experiment is the differential cross-section, which is given by

$$\sigma(\theta, \phi) = |f(\theta, \phi)|^2. \tag{23.101}$$

It is conventional to simplify the notation by employing

$$U(\mathbf{r}) = \frac{2m}{\hbar^2}V(\mathbf{r}), \tag{23.102}$$

so that Eq. (23.98) can be written as follows:

$$(\nabla^2 + k^2)\psi = U(\mathbf{r})\psi, \tag{23.103}$$

and the factor $U(\mathbf{r})\psi$ can be regarded as an inhomogeneity term. The corresponding homogeneous problem is

$$(\nabla^2 + k^2)\varphi(\mathbf{r}) = 0, \tag{23.104}$$

and if the direction of the outgoing particle is fixed as the z-direction, then the solution of Eq. (23.104) can be written as follows:

$$\varphi(\mathbf{r}) = e^{i\mathbf{k}\cdot\mathbf{r}} = e^{ikz}. \tag{23.105}$$

The solution of Eq. (23.103) is given by

$$\psi(\mathbf{r}) = \varphi(\mathbf{r}) - \int G(\mathbf{r}-\mathbf{r}')U(\mathbf{r}')\psi(\mathbf{r}')\mathrm{d}\mathbf{r}', \tag{23.106}$$

where the integration is over three-dimensional space and the Green function G satisfies

$$(\nabla^2 + k^2)G(\mathbf{r}) = -\delta(\mathbf{r}). \tag{23.107}$$

For a free particle a standard calculation (see, for example, Schiff (1955), pp. 163–164) leads to the result that

$$G(\mathbf{r}) = \frac{1}{4\pi r}e^{ikr}. \tag{23.108}$$

Insertion of Eq. (23.108) into Eq. (23.106) and examination of the asymptotic region $r \to \infty$, leads to the result given in Eq. (23.100), with

$$f(\theta,\phi) = -\frac{1}{4\pi}\int U(\mathbf{r}')e^{-i\mathbf{k}_{\mathrm{d}}\cdot\mathbf{r}'}\psi^+(\mathbf{r}',\mathbf{k})\mathrm{d}\mathbf{r}', \tag{23.109}$$

where $\mathbf{k}_{\mathrm{d}}$ is a vector whose direction indicates the trajectory towards the detector.

The integral in Eq. (23.109) can be simplified by developing $\psi^+(\mathbf{r}',\mathbf{k})$ in a series expansion. As a zeroth approximation, let $\psi^+(\mathbf{r}',\mathbf{k})$ be represented by just the first term, Eq. (23.106), so that

$$\psi^+_{(0)}(\mathbf{r}',\mathbf{k}) = \varphi(\mathbf{r}) = e^{ikz}. \tag{23.110}$$

Inserting this result in to Eq. (23.109) yields the following approximation for the scattering amplitude:

$$f(\theta,\phi) = -\frac{1}{4\pi}\int U(\mathbf{r}')e^{-i\mathbf{k}_{\mathrm{d}}\cdot\mathbf{r}'+ikz'}\,\mathrm{d}\mathbf{r}'. \tag{23.111}$$

This is referred to as the first Born approximation for the scattering amplitude. The momentum transfer vector $\mathbf{K}$ is introduced by the definition

$$\mathbf{K} = \mathbf{k}_0 - \mathbf{k}_{\mathrm{d}}, \tag{23.112}$$

where $\mathbf{k}_0$ is the momentum vector of the incident particle. Then Eq. (23.111) can be written as follows:

$$f(\theta,\phi) = -\frac{1}{4\pi}\int U(\mathbf{r}')\mathrm{e}^{\mathrm{i}\mathbf{K}\cdot\mathbf{r}'}\,\mathrm{d}\mathbf{r}', \tag{23.113}$$

which can be simplified, when U is a radial function, to yield

$$f(\theta,\phi) = -\frac{1}{K}\int_0^\infty r\sin KrU(r)\mathrm{d}r. \tag{23.114}$$

In this case, the scattering amplitude is observed to be independent of ϕ. It is common to write the scattering amplitude as a function of the magnitude of the momentum $k = |\mathbf{k}_\mathrm{d}|$ and of the momentum transfer vector $K = |\mathbf{K}|$. In the literature the symbol Δ is often used in place of K, which can cause some confusion since ∇^2 is often written using this symbol, and the opposite sign convention is also employed for the definition of $\mathbf{K}$.

23.5.2 Dispersion relations for potential scattering

The focus of attention in this subsection is elastic scattering in the forward direction. The scattering amplitude can be written as

$$f(k,\theta) = -\frac{1}{4\pi}\int U(\mathbf{r})\mathrm{d}\mathbf{r} - \frac{1}{4\pi}\int\int U(\mathbf{r})G(\mathbf{r},\mathbf{r}',k)U(\mathbf{r}')\mathrm{e}^{\mathrm{i}\mathbf{k}\cdot(\mathbf{r}'-\mathbf{r})}\,\mathrm{d}\mathbf{r}\,\mathrm{d}\mathbf{r}', \tag{23.115}$$

where the Green function $G(\mathbf{r},\mathbf{r}',k)$ is given by (Roman, 1965, p. 211)

$$G(\mathbf{r},\mathbf{r}',E) = \sum_{j=1}^{N}\frac{\varphi_j(\mathbf{r})\varphi_j^*(\mathbf{r}')}{E-E_j} + \int\frac{\psi_{k'}(\mathbf{r})\psi_{k'}^*(\mathbf{r}')}{E-E'}\,\mathrm{d}\mathbf{k}'. \tag{23.116}$$

In Eq. (23.116), $\varphi_j(\mathbf{r})$ designates an eigenfunction for a bound state of the system, and the corresponding energy is denoted by E_j. The integral term corresponds to the continuum contribution, and it is useful to keep in mind the connection between k and E given in Eq. (23.99). The first term on the right-hand side of Eq. (23.115) corresponds to the first Born approximation to the scattering amplitude in the forward direction. It is implicitly assumed that potentials are restricted to the class for which the integral $\int U(\mathbf{r})\mathrm{d}\mathbf{r}$ does not diverge. The scattering amplitude in the forward direction can be written as a function of energy in the following form:

$$f(E,0) = f_\mathrm{B}(E,0) - \frac{1}{4\pi}\int\int U(\mathbf{r})G(\mathbf{r},\mathbf{r}',k)U(\mathbf{r}')\mathrm{e}^{\mathrm{i}\mathbf{k}\cdot(\mathbf{r}'-\mathbf{r})}\mathrm{d}\mathbf{r}\,\mathrm{d}\mathbf{r}', \tag{23.117}$$

where the first Born approximation to the scattering amplitude, $f_\mathrm{B}(E,0)$, is identified with $-(1/4\pi)\int U(\mathbf{r})\mathrm{d}\mathbf{r}$.

In the complex E-plane, the elastic scattering amplitude in the forward direction has a sequence of first-order poles on the negative real energy axis at the locations E_j, and it is assumed there are N such poles. These correspond to the bound states of the system. There is a branch cut in the complex plane, which is taken along the positive real energy axis. This corresponds to the positive energy continuum states. It is assumed that the scattering amplitude has the following asymptotic behavior in the complex energy plane:

$$f(E,0) - f_{\mathrm{B}}(E,0) = O(E^{-1-\delta}), \quad \text{for } E \to \infty, \tag{23.118}$$

where $\delta > 0$. Consider the integral

$$\oint_C \frac{\{f(E',0) - f_{\mathrm{B}}(E',0)\}\mathrm{d}E'}{E' - E},$$

where C denotes the contour shown in Figure 23.2; then it follows that

$$\begin{aligned}\oint_C \frac{\{f(E',0) - f_{\mathrm{B}}(E',0)\}\mathrm{d}E'}{E' - E} &= \int_{\Gamma_{AB}} \{\}\mathrm{d}E' + \int_{\Gamma} \{\}\mathrm{d}E' \\ &\quad + \int_{\Gamma_{CD}} \{\}\mathrm{d}E' + \int_{\Gamma_{DA}} \{\}\mathrm{d}E' \\ &= 2\pi\mathrm{i}\{f(E,0) - f_{\mathrm{B}}(E,0)\} \\ &\quad + 2\pi\mathrm{i}\sum_{j=1}^{N} \mathrm{Res}\left[\frac{\{f(E',0) - f_{\mathrm{B}}(E',0)\}}{E' - E}\right]_{E'=E_j}.\end{aligned} \tag{23.119}$$

Employing the asymptotic condition given in Eq. (23.118), and writing $E' = R\mathrm{e}^{\mathrm{i}\theta}$, leads to

$$\begin{aligned}\lim_{R\to\infty} \int_{\Gamma} \frac{\{f(E',0) - f_{\mathrm{B}}(E',0)\}\mathrm{d}E'}{E' - E} &= \lim_{R\to\infty} \int_{\varepsilon}^{2\pi-\varepsilon} \frac{\{f(R\mathrm{e}^{\mathrm{i}\theta},0) - f_{\mathrm{B}}(R\mathrm{e}^{\mathrm{i}\theta},0)\}\mathrm{i}R\mathrm{e}^{\mathrm{i}\theta}\,\mathrm{d}\theta}{R\mathrm{e}^{\mathrm{i}\theta} - E} \\ &= 0,\end{aligned} \tag{23.120}$$

with $\varepsilon > 0$. For the integral around the origin, it follows, on using the substitution $E' = \rho\mathrm{e}^{\mathrm{i}\theta}$, that

$$\begin{aligned}\lim_{\rho\to 0} \int_{\Gamma_{DA}} \frac{\{f(E',0) - f_{\mathrm{B}}(E',0)\}\mathrm{d}E'}{E' - E} &= \lim_{\rho\to 0} \int_{2\pi-\varepsilon}^{\varepsilon} \frac{\{f(\rho\mathrm{e}^{\mathrm{i}\theta},0) - f_{\mathrm{B}}(\rho\mathrm{e}^{\mathrm{i}\theta},0)\}\mathrm{i}\rho\mathrm{e}^{\mathrm{i}\theta}\,\mathrm{d}\theta}{\rho\mathrm{e}^{\mathrm{i}\theta} - E} \\ &= 0.\end{aligned} \tag{23.121}$$

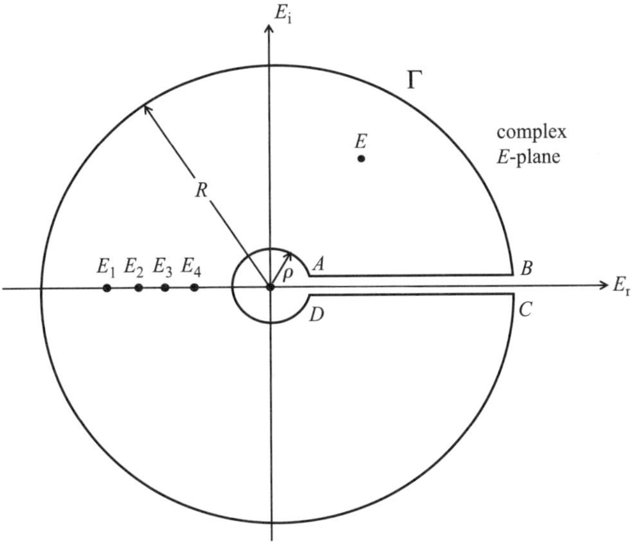

Figure 23.2. Contour for the evaluation of dispersion relations for the forward elastic scattering amplitude.

The remaining two integrals lead to

$$\int_{\Gamma_{AB}} \{\} \mathrm{d}E' + \int_{\Gamma_{CD}} \{\} \mathrm{d}E' = \int_{\rho}^{R} \frac{\{f(E_x + \mathrm{i}\varepsilon, 0) - f_{\mathrm{B}}(E_x + \mathrm{i}\varepsilon, 0)\} \mathrm{d}E_x}{E_x + \mathrm{i}\varepsilon - E} - \int_{\rho}^{R} \frac{\{f(E_x - \mathrm{i}\varepsilon, 0) - f_{\mathrm{B}}(E_x - \mathrm{i}\varepsilon, 0)\} \mathrm{d}E_x}{E_x - \mathrm{i}\varepsilon - E}, \quad (23.122)$$

where $E' = E_x + \mathrm{i}\varepsilon$ has been used above the cut and $E' = E_x - \mathrm{i}\varepsilon$ has been used below the cut. Taking the limit $R \to \infty$ and letting $\rho \to 0$ leads to

$$\lim_{\substack{R\to\infty \\ \rho\to 0}} \left[\int_{\Gamma_{AB}} \{\} \mathrm{d}E' + \int_{\Gamma_{CD}} \{\} \mathrm{d}E' \right] = \lim_{\varepsilon\to 0} \int_0^{\infty} (E_x - E)^{-1} [\{f(E_x + \mathrm{i}\varepsilon, 0) - f_{\mathrm{B}}(E_x + \mathrm{i}\varepsilon, 0)\} - \{f(E_x - \mathrm{i}\varepsilon, 0) - f_{\mathrm{B}}(E_x - \mathrm{i}\varepsilon, 0)\}] \mathrm{d}E_x. \quad (23.123)$$

The preceding integral can be simplified by using

$$\lim_{\varepsilon\to 0} \{f(E_x + \mathrm{i}\varepsilon, 0) - f_{\mathrm{B}}(E_x + \mathrm{i}\varepsilon, 0)\} - \{f(E_x - \mathrm{i}\varepsilon, 0) - f_{\mathrm{B}}(E_x - \mathrm{i}\varepsilon, 0)\} = 2\mathrm{i}\,\mathrm{Im}\{f(E_x, 0) - f_{\mathrm{B}}(E_x, 0)\} = 2\mathrm{i}\,\mathrm{Im}\, f(E_x, 0). \quad (23.124)$$

This last line is obtained using the fact that the potential is taken to be real. From Eq. (23.119) it follows that

$$f(E,0) = f_{\mathrm{B}}(E,0) + \frac{1}{\pi}\int_0^\infty \frac{\operatorname{Im} f(E',0)\mathrm{d}E'}{E'-E} - \sum_{j=1}^{N} \operatorname{Res}\left[\frac{\{f(E',0) - f_{\mathrm{B}}(E',0)\}}{E'-E}\right]_{E'=E_j}. \tag{23.125}$$

If the real part of each of the residues is denoted by $R(E_j)$, and the point E is moved onto the real axis, so that $\operatorname{Im} E = 0$, then the real part of the preceding result can be written as follows:

$$\operatorname{Re} f(E,0) = f_{\mathrm{B}}(E,0) + \frac{1}{\pi} P \int_0^\infty \frac{\operatorname{Im} f(E',0)\mathrm{d}E'}{E'-E} - \sum_{j=1}^{N} R(E_j), \tag{23.126}$$

which represents the dispersion relation for the elastic forward scattering amplitude.

The generalization to the case of non-forward scattering for fixed θ can be accomplished in a straightforward manner. A much more difficult problem is the derivation of the two-dimensional dispersion relations in both the variables k and Δ. The references cited in the end-notes direct the reader to the appropriate sources for both of these problems.

23.5.3 Dispersion relations for electron–hydrogen atom scattering

The dispersion relation for electron scattering from a hydrogen atom in the ground state is examined in this subsection. Attention is focused on elastic scattering in the forward direction. The discussion of the preceding subsection can be carried over to the present problem with some key changes. The most important modification takes into account that the system has two electrons, which requires the proper symmetry properties of the wave function to be employed. If all the spin-dependent terms in the Hamiltonian are ignored, then the total wave function can be expressed as a product of a spatial-dependent wave function and a spin-dependent wave function. The latter does not appear explicitly in the calculations, although it is implicitly assumed that it is antisymmetric under interchange of the two electron spins. The spatial part of the wave function is symmetric with respect to interchange of the spatial coordinates of the two electrons. This situation applies for singlet states.

The atomic two-electron Hamiltonian is non-separable, and consequently no exact solutions are known for this problem, although highly accurate numerical calculations are available for many properties of these atomic species. The simplest approach is to describe the system with one electron occupying the ground state of the hydrogen atom, for which an exact solution of the Schrödinger equation is available, and represent the other electron by an appropriate plane wave eigenfunction. Antisymmetrization of the product of these two eigenfunctions leads to the so-called exchange term, where the spatial coordinates $\mathbf{r}_1$ and $\mathbf{r}_2$ are interchanged. Gerjuoy and Krall (1960) gave the dispersion relation for the forward elastic scattering amplitude for electron–hydrogen atom scattering by analogy with the result for potential scattering, as follows:

$$\mathrm{Re}\{f(E,0) \pm g(E,0)\} = f_{\mathrm{B}}(E,0) \pm g_{\mathrm{B}}(E,0) + \frac{1}{\pi} P\int_0^{\infty} \frac{\mathrm{Im}\{f(E',0) \pm g(E',0)\}\mathrm{d}E'}{E'-E} - \sum_{j=1}^{N_{\mathrm{s}}} R(E_j) - \sum_{j=1}^{N_{\mathrm{t}}} R(E_j), \tag{23.127}$$

where $f(E,0)$ denotes the direct contribution to the scattering amplitude and $g(E,0)$ is the exchange contribution. The sums are taken over the bound state singlets and bound state triplets, respectively, of the H^- ion. The $+$ combination of amplitudes corresponds to the singlet situation and the $-$ combination denotes the triplet state. The residues are determined from the simple poles occurring at the bound states. The connection with experimental results can be made by use of the following result:

$$\mathrm{Im}\{f(E,0) \pm g(E,0)\} = \frac{k}{4\pi}\sigma_{\pm}(E), \tag{23.128}$$

where $\sigma_+(E)$ denotes the total cross-section, integrated over all angles, for scattering in a singlet combination with an electron in the ground state of the hydrogen atom, and $\sigma_-(E)$ is the corresponding cross-section for scattering an electron forming a triplet combination. Most negative atomic ions have a small number of bound states. The H^- ion has one bound state of ^1S symmetry with an energy of $-0.527\,751\,016\,5\ldots$ a.u. and a bound state of ^3P symmetry with an energy of $-0.125\,350\ldots$ a.u. The dispersion relation given in Eq. (23.127) has also been applied to deal with target atoms having more than one electron.

Tests of Eq. (23.127) revealed inconsistencies with available experimental data. Interestingly, this dispersion relation gave fairly satisfactory results for positron–atom scattering. In positron–atom scattering there is no exchange term. This pointed the way to resolving the possible problems with the Gerjuoy–Krall formulation of the dispersion relation. The analytic structure assumed by Gerjuoy and Krall for the exchange term was over-simplified. To give some idea of what is involved, a part of the exchange term is examined in detail to see what type of singularities might arise.

The first Born term that contributes to the exchange amplitude and depends on the electron–electron potential, is given by

$$g_{12}^{\mathrm{Born}}(\mathbf{k},\mathbf{k}') = -\frac{m_{\mathrm{e}}}{2\pi\hbar^2}\int\!\!\int \chi(\mathbf{k}',\mathbf{r}_2)^*\varphi(\mathbf{r}_1)^*\frac{e^2}{r_{12}}\chi(\mathbf{k},\mathbf{r}_1)\varphi(\mathbf{r}_2)\,\mathbf{dr}_1\,\mathbf{dr}_2, \tag{23.129}$$

where r_{12} is the separation distance between the incoming electron and the electron of the hydrogen atom, m_{e} is the mass of the electron, $\hbar$ is Planck's constant divided by 2π,

$$\chi(\mathbf{k},\mathbf{r}) = \mathrm{e}^{\mathrm{i}\mathbf{k}\cdot\mathbf{r}}, \tag{23.130}$$

and $\varphi(\mathbf{r})$ is the ground-state eigenfunction of the hydrogen atom, given by

$$\varphi(\mathbf{r}) = \frac{1}{\sqrt{(\pi a_0^3)}} \mathrm{e}^{-r/a_0}, \tag{23.131}$$

where a_0 is the Bohr radius. The case of forward elastic scattering is now considered, and hence $\mathbf{k}' = \mathbf{k}$. The exchange term $g_{12}^{\mathrm{Born}}(\mathbf{k}, \mathbf{k})$ has first-, second-, and third-order pole terms. To see this, Eq. (23.129) is evaluated in the following manner, taking advantage of Fourier transforms. The integral can be calculated in other ways. Let

$$f(\mathbf{r}_1) = \mathrm{e}^{-r_1/a_0}, \tag{23.132}$$

$$h(\mathbf{r}_{12}) = \frac{\mathrm{e}^{\mathrm{i}\mathbf{k}\cdot\mathbf{r}_{12}}}{r_{12}}, \tag{23.133}$$

and denote the Fourier transforms of $f(\mathbf{r}_1)$ and $h(\mathbf{r}_{12})$ by $F(\mathbf{k}_1)$ and $H(\mathbf{k}_3)$, respectively. The Fourier transform $F(\mathbf{k}_1)$ is worked out in spherical polar coordinates using the change of variables $r = a_0 s$ and $\kappa_1 = a_0 k_1$, and to simplify the calculation the polar axis has been placed along the vector $\mathbf{k}_1$, so that

$$\begin{aligned} F(\mathbf{k}_1) &= \int f(\mathbf{r})\mathrm{e}^{-\mathrm{i}\mathbf{k}_1\cdot\mathbf{r}}\,\mathrm{d}\mathbf{r} \\ &= \int_0^\infty r^2 \mathrm{e}^{-r/a_0}\,\mathrm{d}r \int_0^\pi \sin\theta\,\mathrm{e}^{-\mathrm{i}k_1 r\cos\theta}\,\mathrm{d}\theta \int_0^{2\pi} \mathrm{d}\phi \\ &= 2\pi a_0^3 \int_0^\infty s^2 \mathrm{e}^{-s}\,\mathrm{d}s \int_{-1}^1 \mathrm{e}^{-\mathrm{i}\kappa_1 s t}\,\mathrm{d}t \\ &= \frac{4\pi a_0^3}{\kappa_1}\int_0^\infty s\sin(\kappa_1 s)\mathrm{e}^{-s}\,\mathrm{d}s \\ &= \frac{8\pi a_0^3}{(1+\kappa_1^2)^2}. \end{aligned} \tag{23.134}$$

The Fourier transform of $h(\mathbf{r}_{12})$ does not exist in the conventional sense. This issue can be dealt with by modifying Eq. (23.133) using the replacement $\mathrm{i}\mathbf{k}\cdot\mathbf{r}_{12} \to -\alpha\,|\mathbf{r}_{12}| + \mathrm{i}\mathbf{k}\cdot\mathbf{r}_{12}$, for $\alpha > 0$, and taking the limit $\alpha \to 0+$ after evaluating the Fourier transform, but before performing the final integration. Spherical polar coordinates are employed with the polar axis placed along the vector $\mathbf{k} - \mathbf{k}_3$, and the substitutions $r = r_{12}$ and $\kappa_3 = |\mathbf{k} - \mathbf{k}_3|$ are used, so that

$$\begin{aligned} H(\mathbf{k}_3) &= \int h(\mathbf{r}_{12})\mathrm{e}^{-\mathrm{i}\mathbf{k}_3\cdot\mathbf{r}_{12}}\,\mathrm{d}\mathbf{r}_{12} \\ &= \int \frac{\mathrm{e}^{\mathrm{i}\mathbf{k}\cdot\mathbf{r}_{12} - \mathrm{i}\mathbf{k}_3\cdot\mathbf{r}_{12} - \alpha r_{12}}\,\mathrm{d}\mathbf{r}_{12}}{r_{12}} \\ &= 2\pi \int_0^\infty r\mathrm{e}^{-\alpha r}\,\mathrm{d}r \int_0^\pi \sin\theta\,\mathrm{e}^{-\mathrm{i}\kappa_3 r\cos\theta}\,\mathrm{d}\theta \end{aligned}$$

$$= 2\pi \int_0^\infty r\mathrm{e}^{-\alpha r}\,\mathrm{d}r \int_{-1}^{1} \mathrm{e}^{-\mathrm{i}\kappa_3 r t}\,\mathrm{d}t$$

$$= \frac{4\pi}{\kappa_3}\int_0^\infty \sin(\kappa_3 r)\mathrm{e}^{-\alpha r}\,\mathrm{d}r$$

$$= \frac{4\pi}{\kappa_3^2 + \alpha^2}. \tag{23.135}$$

An alternative approach to evaluate $H(\mathbf{k}_3)$ which avoids the insertion of the convergence factor, is to take note of the distributional nature of the integral involved, so that

$$\int_0^\infty \sin\kappa r\,\mathrm{d}r = \frac{1}{2\mathrm{i}}\left\{\int_0^\infty \mathrm{e}^{\mathrm{i}\kappa r}\,\mathrm{d}r - \int_0^\infty \mathrm{e}^{-\mathrm{i}\kappa r}\,\mathrm{d}r\right\}$$

$$= \frac{1}{2\mathrm{i}}\{2\pi\delta^+(\kappa) - 2\pi\delta^-(\kappa)\}$$

$$= \frac{\pi}{\mathrm{i}}\left\{-\frac{1}{2\pi\mathrm{i}}\lim_{\varepsilon\to 0+}\frac{1}{\kappa+\mathrm{i}\varepsilon} - \frac{1}{2\pi\mathrm{i}}\lim_{\varepsilon\to 0+}\frac{1}{\kappa-\mathrm{i}\varepsilon}\right\}$$

$$= \frac{1}{2}\lim_{\varepsilon\to 0+}\left\{\frac{1}{\kappa+\mathrm{i}\varepsilon} + \frac{1}{\kappa-\mathrm{i}\varepsilon}\right\}$$

$$= \frac{\kappa}{\kappa^2 + 0+}, \tag{23.136}$$

where the Heisenberg delta functions given in Eqs. (10. 9) and (10. 10) have been employed. From Eq. (23.129) it follows, on inserting the Fourier transforms and using the substitutions $k_3 = xk$ and $\lambda = a_0 k$, that

$$-(\pi a_0^3)\left(\frac{2\pi\hbar^2}{m_\mathrm{e}e^2}\right) g_{12}^{\mathrm{Born}}(\mathbf{k},\mathbf{k}) = \iint f(\mathbf{r}_1) f(\mathbf{r}_2) h(\mathbf{r}_{12})\mathrm{d}\mathbf{r}_1\,\mathrm{d}\mathbf{r}_2$$

$$= \frac{1}{(2\pi)^9}\iiint F(\mathbf{k}_1)\mathrm{e}^{\mathrm{i}\mathbf{k}_1\cdot\mathbf{r}_1}\,\mathrm{d}\mathbf{k}_1 \int F(\mathbf{k}_2)\mathrm{e}^{\mathrm{i}\mathbf{k}_2\cdot\mathbf{r}_2}\,\mathrm{d}\mathbf{k}_2$$

$$\times \int H(\mathbf{k}_3)\mathrm{e}^{\mathrm{i}\mathbf{k}_3\cdot\mathbf{r}_{12}}\,\mathrm{d}\mathbf{k}_3\,\mathrm{d}\mathbf{r}_1\,\mathrm{d}\mathbf{r}_2$$

$$= \frac{1}{(2\pi)^9}\iiint F(\mathbf{k}_1)F(\mathbf{k}_2)H(\mathbf{k}_3)\mathrm{d}\mathbf{k}_1\,\mathrm{d}\mathbf{k}_2\,\mathrm{d}\mathbf{k}_3$$

$$\times \int \mathrm{e}^{\mathrm{i}\mathbf{k}_1\cdot\mathbf{r}_1}\mathrm{e}^{\mathrm{i}\mathbf{k}_2\cdot\mathbf{r}_2}\mathrm{e}^{\mathrm{i}\mathbf{k}_3\cdot\mathbf{r}_{12}}\,\mathrm{d}\mathbf{r}_1\,\mathrm{d}\mathbf{r}_2$$

$$= \frac{1}{(2\pi)^3}\iiint F(\mathbf{k}_1)F(\mathbf{k}_2)H(\mathbf{k}_3)\delta(\mathbf{k}_1+\mathbf{k}_3)$$

$$\times\,\delta(\mathbf{k}_2-\mathbf{k}_3)\mathrm{d}\mathbf{k}_1\,\mathrm{d}\mathbf{k}_2\,\mathrm{d}\mathbf{k}_3$$

$$= \frac{1}{(2\pi)^3}\int F(-\mathbf{k}_3)F(\mathbf{k}_3)H(\mathbf{k}_3)\mathrm{d}\mathbf{k}_3$$

$$
\begin{aligned}
&= \frac{256\pi^3 a_0^6}{(2\pi)^3} \int \frac{1}{(1+a_0^2k_3^2)^4}\frac{1}{\kappa_3^2}\, \mathrm{d}\mathbf{k}_3 \\
&= 32a_0^6 \int_0^\infty \frac{k_3^2\, \mathrm{d}k_3}{(1+a_0^2k_3^2)^4} \\
&\quad \times \int_0^\pi \frac{\sin\theta_3}{(k_3^2+k^2-2\mathbf{k}_3\cdot\mathbf{k})}\, \mathrm{d}\theta_3 \int_0^{2\pi} \mathrm{d}\phi_3 \\
&= 64\pi a_0^6 \int_0^\infty \frac{k_3^2\, \mathrm{d}k_3}{(1+a_0^2k_3^2)^4} \int_{-1}^1 \frac{\mathrm{d}t}{k_3^2+k^2-2kk_3t} \\
&= -\frac{32\pi a_0^6}{k} \int_0^\infty \frac{k_3}{(1+a_0^2k_3^2)^4} \log\left(\frac{k-k_3}{k+k_3}\right)^2 \mathrm{d}k_3 \\
&= -32\pi k a_0^6 \int_0^\infty \frac{x}{(1+\lambda^2x^2)^4} \log\left(\frac{1-x}{1+x}\right)^2 \mathrm{d}x \\
&= -32\pi k a_0^6 \left\{ -\frac{\pi(15+10\lambda^2+3\lambda^4)}{24\lambda(1+\lambda^2)^3} \right\} \\
&= \frac{4\pi^2 a_0^5(15+10\lambda^2+3\lambda^4)}{3(1+\lambda^2)^3}.
\end{aligned} \tag{23.137}
$$

Making use of $a_0 = \hbar^2/(m_e e^2)$, where e is the absolute value of the electronic charge, it follows that

$$
\begin{aligned}
g_{12}^{\mathrm{Born}}(\mathbf{k},\mathbf{k}) &= -\frac{2a_0(15+10\lambda^2+3\lambda^4)}{3(1+\lambda^2)^3} \\
&= -\frac{a_0}{3}\left\{ \frac{6}{1+\lambda^2} + \frac{8}{(1+\lambda^2)^2} + \frac{16}{(1+\lambda^2)^3} \right\},
\end{aligned} \tag{23.138}
$$

which is the desired result. The Born approximation to the exchange contribution arising from the electron–electron potential is found to have first-, second-, and third-order poles at $k^2 = -a_0^{-2}$. In atomic units, which are popular in atomic physics, $\hbar = 1$, $m_e = 1$, and $e = 1$. Consequently, $a_0 = 1$, and hence the poles occur at $k^2 = -1$, or at an energy of $E = -0.5$ a.u.

The pole structure of $g_{12}^{\mathrm{Born}}(\mathbf{k},\mathbf{k})$ is not cancelled by higher-order contributions in the Born expansion, nor is it cancelled by considering the other term in the potential. For these reasons the original derivation of the dispersion relation by Gerjuoy and krall is known to be incorrect. Gerjuoy and Krall assumed a particular form for the singular behavior of the term $g(\mathbf{k},\mathbf{k}) - g^{\mathrm{Born}}(\mathbf{k},\mathbf{k})$ which does not match what has just been derived for $g_{12}^{\mathrm{Born}}(\mathbf{k},\mathbf{k})$. Determination of the singular structure of the exchange term has proved to be a rather intricate problem, and the last word has not been said on this topic. The references in the end-notes will give the reader a glimpse of the evolution of the efforts on this important problem.

23.6 Magnetic resonance applications

In classical nuclear magnetic resonance spectroscopy (NMR), a radiofrequency source is used to induce a nuclear spin moment reorientation for certain nuclei in the presence of a magnetic field. The traditional practice was to plot the resulting absorption versus frequency. Since nuclei with non-zero nuclear spin are sensitive to the local magnetic environment, NMR has become the most important instrumental technique in chemistry. A limitation of the classical experiment was the problem of saturation. To observe weak peaks it is desirable to work with higher incident powers; however, this can readily lead to saturation, where the populations of molecules having the lower- and higher-energy nuclear spin moment orientations are equal. When this condition arises, the absorption signal is lost. In some situations, the dispersion mode saturates less quickly in comparison to the absorption. In such cases, the dispersion mode can be utilized; taking the Hilbert transform produces the traditional absorption mode format.

The magnetic susceptibility χ_m appropriate to a discussion of NMR, when conditions are far from saturation, is given by $\chi_m = \chi'_m + i\chi''_m$, where the real component χ'_m and the imaginary part χ''_m take the following forms:

$$\chi'_m(\omega) = \frac{1}{2}\omega_0 \chi_{m_0} \frac{(\omega - \omega_0)T_2^2}{1 + (\omega - \omega_0)^2 T_2^2} \tag{23.139}$$

and

$$\chi''_m(\omega) = \frac{1}{2}\omega_0 \chi_{m_0} \frac{T_2}{1 + (\omega - \omega_0)^2 T_2^2}, \tag{23.140}$$

where ω_0 is the resonant angular frequency, χ_{m_0} is the static magnetic susceptibility, and T_2 is the spin–spin (also termed the transverse) relaxation time. The magnetic susceptibility components satisfy the following dispersion relations:

$$\chi'_m(\omega) - \chi'_m(\infty) = -\frac{1}{\pi} P \int_{-\infty}^{\infty} \frac{\chi''_m(\omega')\mathrm{d}\omega'}{\omega - \omega'} \tag{23.141}$$

and

$$\chi''_m(\omega) = \frac{1}{\pi} P \int_{-\infty}^{\infty} \frac{\{\chi'_m(\omega') - \chi'_m(\infty)\}\mathrm{d}\omega'}{\omega - \omega'}. \tag{23.142}$$

In the presence of saturation, the susceptibility components take the following forms (Carrington and McLachlan, 1967, p. 181):

$$\chi'_m(\omega) = \frac{1}{2}\omega_0 \chi_{m_0} \frac{(\omega - \omega_0)T_2^2}{1 + (\omega - \omega_0)^2 T_2^2 + \gamma^2 H_1^2 T_1 T_2} \tag{23.143}$$

and

$$\chi''_{\rm m}(\omega) = \frac{1}{2}\omega_0 \chi_{\rm m_0} \frac{T_2}{1 + (\omega - \omega_0)^2 T_2^2 + \gamma^2 H_1^2 T_1 T_2}, \tag{23.144}$$

where H_1 is the applied alternating magnetic field, T_1 is the spin–lattice (also termed the longitudinal) relaxation time, and γ is the nuclear gyromagnetic ratio. Modified Hilbert transform relations can be derived connecting the components $\chi'_{\rm m}$ and $\chi''_{\rm m}$ in the presence of saturation.

A second application of the Hilbert transform in NMR arises in making phase corrections. This is done by constructing a suitable combination of the original signal with the Hilbert transform of the signal. The discrete Hilbert transform is required since the experimental data set is discrete. Different methods exist to fix the phase angle θ. One approach involves calculating the phase via

$$\theta = \tan^{-1}\left(\frac{I_{\rm H}}{I_0}\right), \tag{23.145}$$

where I_0 is the integrated intensity of the original data and $I_{\rm H}$ is the integrated intensity determined from the Hilbert-transformed signal. The interested reader can pursue the details in Ernst (1969).

In the modern approach to NMR spectrometry, a relatively intense radiofrequency pulse is employed to populate the excited energy levels. The free induction decay (FID) observed as the excited energy levels depopulate is Fourier-transformed to produce the standard NMR absorption profile. Techniques where the Fourier transform is employed to convert from the time domain to the frequency domain are collectively referred to as Fourier spectroscopy, or Fourier transform spectroscopy. The distinguishing feature for the different techniques is the spectral range that is covered. In the standard Fourier transform NMR (FTNMR) technique, a periodic short duration pulse is applied with a time delay between each pulse. In general, the Fourier transform approach does not yield dispersive and absorptive components equal to the low-power spectra obtained via the classical continuous wave approach, although in practice it is convenient to interpret the absorptive mode from the FT technique as the normal absorption spectrum.

Two issues that arise in the FTNMR technique are as follows. The first is whether the nuclear spin system can be approximated as a linear system in the presence of a strong periodic radiofrequency pulse sequence. If the magnetization of the system is examined after a single pulse has been applied, and there is no radiofrequency field, the equations of motion can be treated as arising from a linear system. The initial state of the spin system is not linear during the application of the pulse. The pulse affects the initial state of the prepared system. For multi-pulse experiments, nonlinear effects must be taken into consideration.

The second issue concerns causality. Since the applied pulse is periodic, the standard application of the causality statement that the impulse response vanishes for negative times needs to be modified. The real and imaginary components of the Fourier transform of the free induction decay curve are independent. Bartholdi and Ernst (1973) proposed a procedure to force the Fourier transform to have its real and imaginary components connected by a discrete Hilbert transform, and to ensure a causality condition. This is achieved by modifying the FID curve from each pulse by adding at the end of the decay, and just prior to the initiation of the next pulse, an equal time segment of data having zero amplitude. In effect, the system is forced into a true equilibrium state; that is, the decay of the system from the first pulse is complete before the second pulse strikes the system. These authors provide a discrete Hilbert transform analysis of the response in this experiment. The interested reader can pursue the details in the work of Bartholdi and Ernst.

This section is concluded with a very brief reference to two-dimensional NMR. Several approaches are available to obtain two-dimensional NMR; a detailed discussion can be found in Ernst, Bodenhausen, and Wokaun (1987). In the time domain the signal depends on two time variables. After a two-dimensional Fourier transform, the resulting signal $S(\omega_1, \omega_2)$ depends on two frequency variables. A standard Hilbert transform connection can be established between the real and imaginary components of $S(\omega_1, \omega_2)$ in the separate variables ω_1 and ω_2. In general, it may not be possible to resolve $S(\omega_1, \omega_2)$ into pure dispersive and absorptive mode spectra. A number of approaches have been suggested for extracting pure two-dimensional absorption peaks. For further reading on this topic, see Ernst *et al.* (1987).

23.7 DISPA analysis

A technique that has been employed for the analysis of spectral information is considered in this section. The acronym of the section heading stands for a plot of dispersion versus absorption for any of a number of different spectroscopic methods. The idea was initially developed by Cole and Cole (1941, 1942). These authors considered plots where the dispersive and absorptive components of the dielectric constant were plotted versus each other.

Let $A(\omega)$ and $D(\omega)$ denote the absorption and dispersion curves, respectively. If the absorption profile is a Lorentzian curve, recall Eq. (14.107), then

$$A(x_\omega) = \frac{1}{1 + x_\omega^2}, \tag{23.146}$$

with

$$x_\omega = \frac{\omega - \omega_0}{a}, \tag{23.147}$$

where the angular frequency ω_0 is the center of the Lorentzian and $2a$ is the width of the curve at half height of the curve. The absorption curve has been normalized to give

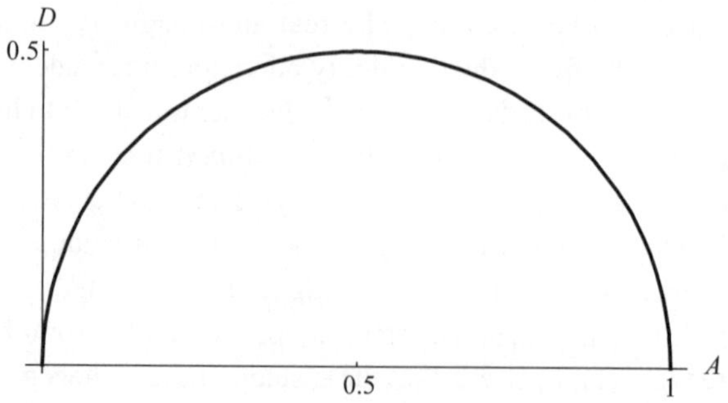

Figure 23.3. DISPA plot of dispersion versus absorption for a Lorentzian profile.

unit height at the peak maximum. The corresponding dispersion curve is given by

$$D(x_\omega) = H[A(x_\omega)] = \frac{x_\omega}{1 + x_\omega^2}. \tag{23.148}$$

A simple observation is that

$$\left\{A(x_\omega) - \frac{1}{2}\right\}^2 + D^2(x_\omega) = \frac{1}{2^2}, \tag{23.149}$$

and so a plot of $D(x_\omega)$ versus $A(x_\omega)$ gives a circle centered at (1/2, 0) with a radius of 1/2. Since the plot is symmetric about the x_ω-axis, it is sufficient to plot the positive semicircular section. The situation is illustrated in Figure 23.3, which is called a DISPA plot by some authors; other authors refer to it as a Cole–Cole plot.

The DISPA plot presentation of data has several uses. The simplest and most obvious application is to use the DISPA plot to detect departures from a Lorentzian profile. For example, if the time-domain data from an NMR experiment lead to a single absorption line, but the DISPA plot shows distortion from the expected semicircular behavior, then it can be concluded that the line profile is not a simple Lorentzian, provided certain other factors can also be excluded.

Complications in line-shape analysis can arise in several ways. The simplest two situations that can occur are the following. Two lines with different line widths fortuitously coincide, with the outcome that the resulting profile corresponds to an inhomogeneously broadened Lorentzian profile. A second possibility is that two Lorentzian absorption curves that have slightly offset maxima overlap, to produce a profile that corresponds to an inhomogeneously broadened Lorentzian line shape. A DISPA presentation of the spectral data for each of these cases offers the possibility to detect departures from what would be expected from a single Lorentzian profile.

Another application of the DISPA plot technique arises in NMR. A chemical system undergoing a rearrangement to a modified species affects the line shape of the absorption profile. The DISPA approach can discern departures from pure Lorentzian behavior, and can therefore help detect chemical exchange processes.

A similar type of DISPA analysis can be carried out when the absorption profiles are expected to be Gaussian in nature. A Gaussian profile takes the following form:

$$A(\omega) = \frac{1}{\sqrt{(2\pi)}\,\sigma} e^{-(\omega-\omega_0)^2/(2\sigma^2)}, \tag{23.150}$$

where the angular frequency ω_0 corresponds to the center of the absorption line and the width at half height is directly proportional to the parameter σ. This form is normalized to unit area when integrated over the interval $(-\infty, \infty)$. See the Exercises at the end of this chapter for some problems investigating the DISPA outcome when the profile is a Gaussian.

23.8 Electrical circuit analysis

Consider the simple circuit composed of a resistance R, a capacitor C, and an inductor L, connected in series to a voltage source $\mathcal{E}(t)$, which has the Fourier transform representation given by

$$\mathcal{E}(t) = \int_{-\infty}^{\infty} E(\omega) e^{-i\omega t}\, d\omega. \tag{23.151}$$

Employing the opposite sign choice in the exponent of the Fourier transform, that is using $e^{i\omega t}$ in place of $e^{-i\omega t}$ leads to some sign switches in the following formulas, and these alternative expressions are often seen in the literature. The circuit is shown in Figure 23.4.

The current in the circuit is denoted by $\mathcal{I}(t)$, and this has a Fourier transform representation given by

$$\mathcal{I}(t) = \int_{-\infty}^{\infty} I(\omega) e^{-i\omega t}\, d\omega. \tag{23.152}$$

Standard circuit analysis leads to

$$\mathcal{E}(t) = R\mathcal{I}(t) + L\frac{d\mathcal{I}(t)}{dt} + \frac{q}{C}, \tag{23.153}$$

where q is the charge on the capacitor, and the current is given by dq/dt. Taking the derivative of the preceding result leads to

$$\frac{d\mathcal{E}(t)}{dt} = L\frac{d^2\mathcal{I}(t)}{dt^2} + R\frac{d\mathcal{I}(t)}{dt} + \frac{1}{C}\mathcal{I}(t). \tag{23.154}$$

Inserting Eqs. (23.151) and (23.152) into Eq. (23.154), and taking the inverse Fourier transform, leads to

$$E(\omega) = \left\{R - i\omega L + \frac{i}{\omega C}\right\} I(\omega). \tag{23.155}$$

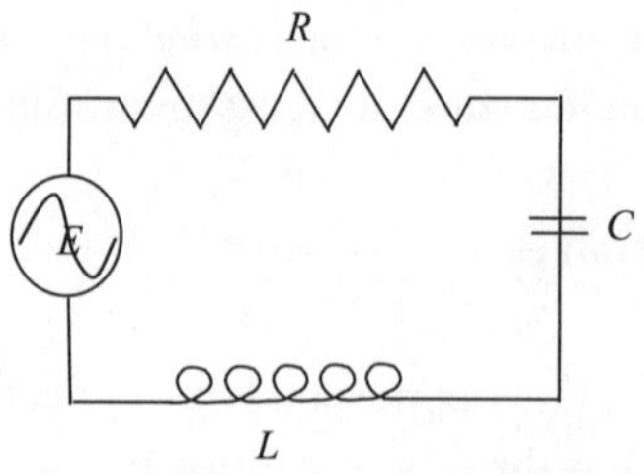

Figure 23.4. Circuit composed of a resistor, inductor, and capacitor in series with a voltage source.

The impedance is introduced by the following equation:

$$Z(\omega) = R + \mathrm{i}X(\omega), \tag{23.156}$$

where $X(\omega)$ is the reactance, which, for the circuit under consideration, is given by

$$X(\omega) = -\omega L + \frac{1}{\omega C}. \tag{23.157}$$

A more general expression for $Z(\omega)$ is given later (see Eq. (23.174)). Equation (23.155) becomes

$$E(\omega) = Z(\omega)I(\omega), \tag{23.158}$$

which can be rewritten as

$$I(\omega) = Y(\omega)E(\omega), \tag{23.159}$$

where the admittance function $Y(\omega)$ is given by

$$Y(\omega) = \frac{1}{R - \mathrm{i}\omega L + \mathrm{i}/\omega C}. \tag{23.160}$$

Sometimes the term *immittance* is used to describe collectively either the impedance or the admittance when discussing relations that are satisfied by both functions. For $4L/C \geq R^2$, the admittance function has singularities in the lower half of the complex angular frequency plane, located at

$$\omega = \pm\frac{1}{2L}\surd\left(\frac{4L}{C} - R^2\right) - \mathrm{i}\frac{R}{2L}; \tag{23.161}$$

for $4L/C < R^2$ the poles of $Y(\omega)$ are located in the lower half of the complex plane at

$$\omega = -\mathrm{i}\frac{R}{2L}\left[1 \pm \surd\left(1 - \frac{4L}{CR^2}\right)\right]. \tag{23.162}$$

The admittance function is analytic in the upper half of the complex angular frequency plane. If the opposite sign convention had been employed in Eq. (23.151), the poles are located in the upper half of the complex angular frequency plane, and the admittance function is analytic in the lower half of the complex angular frequency plane. Consider the integral

$$\oint_C \frac{Y(\omega_z)\mathrm{d}\omega_z}{\omega - \omega_z},$$

where C denotes the contour shown in Figure 19.2, then

$$Y(\omega) = -\frac{1}{\mathrm{i}\pi} P \int_{-\infty}^{\infty} \frac{Y(\omega')\mathrm{d}\omega'}{\omega - \omega'}, \tag{23.163}$$

which can be rewritten in terms of the real and imaginary parts of the admittance, $Y_\mathrm{r}(\omega)$ and $Y_\mathrm{i}(\omega)$ respectively, as follows:

$$Y_\mathrm{r}(\omega) = -\frac{1}{\pi} P \int_{-\infty}^{\infty} \frac{Y_\mathrm{i}(\omega')\mathrm{d}\omega'}{\omega - \omega'} \tag{23.164}$$

and

$$Y_\mathrm{i}(\omega) = \frac{1}{\pi} P \int_{-\infty}^{\infty} \frac{Y_\mathrm{r}(\omega')\mathrm{d}\omega'}{\omega - \omega'}. \tag{23.165}$$

From Eq. (23.160) it follows that

$$Y(-\omega) = Y(\omega)^*; \tag{23.166}$$

therefore

$$Y_\mathrm{r}(-\omega) = Y_\mathrm{r}(\omega) \tag{23.167}$$

and

$$Y_\mathrm{i}(-\omega) = -Y_\mathrm{i}(\omega). \tag{23.168}$$

Using these results, Eqs. (23.164) and (23.165) can be written as follows:

$$Y_\mathrm{r}(\omega) = -\frac{2}{\pi} P \int_{0}^{\infty} \frac{\omega' Y_\mathrm{i}(\omega')\mathrm{d}\omega'}{\omega^2 - \omega'^2} \tag{23.169}$$

and

$$Y_\mathrm{i}(\omega) = \frac{2\omega}{\pi} P \int_{0}^{\infty} \frac{Y_\mathrm{r}(\omega')\mathrm{d}\omega'}{\omega^2 - \omega'^2}. \tag{23.170}$$

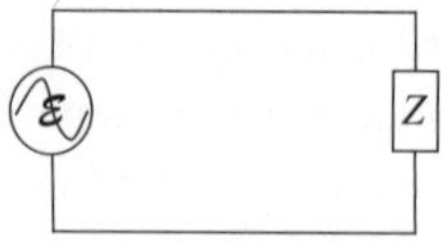

Figure 23.5. Replacement of the components L, C, and R by an impedance Z.

These are the standard dispersion relations for the admittance of the circuit under consideration. The circuit admittance is sometimes expressed as follows:

$$Y(\omega) = G(\omega) + \mathrm{i}B(\omega), \tag{23.171}$$

where $G(\omega)$ is the circuit conductance and $B(\omega)$ is the circuit susceptance. The dispersion relations for the circuit in question can be therefore written as follows:

$$G(\omega) = -\frac{2}{\pi} P \int_0^\infty \frac{\omega' B(\omega')\mathrm{d}\omega'}{\omega^2 - \omega'^2} \tag{23.172}$$

and

$$B(\omega) = \frac{2\omega}{\pi} P \int_0^\infty \frac{G(\omega')\mathrm{d}\omega'}{\omega^2 - \omega'^2}. \tag{23.173}$$

When the circuit conductance does not vanish as $\omega \to \infty$, the preceding dispersion relations are modified by replacing $G(\omega)$ by $G(\omega) - G(\infty)$.

The circuit indicated in Figure 23.4 can be replaced by the setup shown in Figure 23.5. Consider a generalization of the first circuit, for which the impedance of the circuit is given by

$$Z(\omega) = \mathrm{R}(\omega) + \mathrm{i}X(\omega), \tag{23.174}$$

where $\mathrm{R}(\omega)$ and $X(\omega)$ are used to denote, respectively, the resistance and reactance of the *circuit*. Note that, in general, the resistance of a circuit involves components other than resistor elements. The reader needs to be alert to the notation distinguishing the resistance of the circuit and a resistor resistance. Two conditions are imposed: (1) the zeros of $Z(\omega)$ for the circuit lie in the lower half of the complex angular frequency plane, and (2) the asymptotic behavior of $Y(\omega)$ as $|\omega| \to \infty$ is of the form $Y(\omega) = O(\omega^{-1-\delta})$ for $\delta \geq 0$. With these conditions in place, Eqs. (23.169) and (23.170) can be derived in the manner previously outlined. Not all circuits lead to an admittance function that satisfies the dispersion relations given in Eqs. (23.164) and (23.165). If the circuit impedance has a zero on the real frequency axis, then the standard dispersion relations for the admittance do not apply, but in this case it is possible to construct modified dispersion relations by the subtraction of a suitable non-zero contribution to the admittance, evaluated at the angular frequency for which the zero occurs. In the event that the circuit admittance does not belong to $L^2(\mathbb{R})$

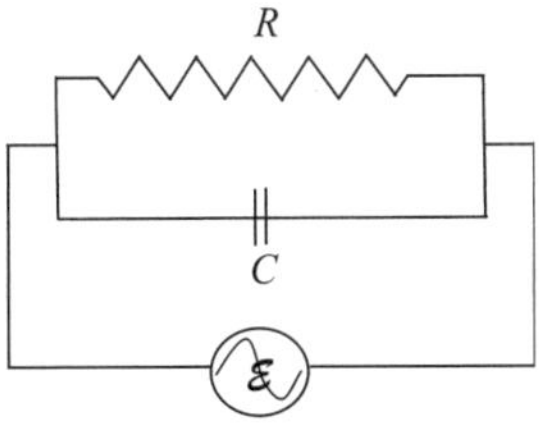

Figure 23.6. A sample RC circuit for which dispersion relations can be derived for the impedance.

because of an unsuitable asymptotic behavior as $|\omega| \to \infty$, it is usually possible to replace the standard dispersion relations by subtracted dispersion relations. That is, suppose the asymptotic behavior of the impedance as $|\omega| \to \infty$ is given by $Y(\omega) = Y_\infty + O(\omega^{-1-\delta})$, where Y_∞ is a constant, equal to the value $Y(\infty)$. In this case, dispersion relations can be derived by considering the function $Y(\omega) - Y_\infty$, in much the same way as the dielectric constant and refractive index were treated in Chapter 19.

In a number of situations it is possible to determine dispersion relations directly for the circuit impedance. Consider, for example, the circuit indicated in Figure 23.6. The circuit impedance satisfies

$$\frac{1}{Z(\omega)} = \frac{1}{R} + \mathrm{i}\omega C, \tag{23.175}$$

so that

$$Z(\omega) = \frac{R}{1 + \mathrm{i}\omega RC}. \tag{23.176}$$

The impedance for this circuit is not an analytic function in the upper half of the complex angular frequency plane: there is a simple pole at $\omega = \mathrm{i}(RC)^{-1}$. Employing the definition of Eq. (23.174), dispersion relations for the resistance and reactance of the circuit can be written as follows:

$$\mathrm{R}(\omega) = \frac{1}{\pi} P \int_{-\infty}^{\infty} \frac{X(\omega')\mathrm{d}\omega'}{\omega - \omega'} \tag{23.177}$$

and

$$X(\omega) = -\frac{1}{\pi} P \int_{-\infty}^{\infty} \frac{\mathrm{R}(\omega')\mathrm{d}\omega'}{\omega - \omega'}. \tag{23.178}$$

From Eq. (23.176) it follows that

$$Z(-\omega) = Z(\omega)^*, \tag{23.179}$$

so that

$$\mathrm{R}(-\omega) = \mathrm{R}(\omega) \tag{23.180}$$

and

$$X(-\omega) = -X(\omega). \tag{23.181}$$

Using these results, Eqs. (23.177) and (23.178) can be written as follows:

$$\mathrm{R}(\omega) = \frac{2}{\pi} P \int_0^\infty \frac{\omega' X(\omega')\mathrm{d}\omega'}{\omega^2 - \omega'^2} \tag{23.182}$$

and

$$X(\omega) = -\frac{2\omega}{\pi} P \int_0^\infty \frac{\mathrm{R}(\omega')\mathrm{d}\omega'}{\omega^2 - \omega'^2}. \tag{23.183}$$

Consider a general circuit with an impedance having no singularities in the lower half of the complex angular frequency plane and satisfying the crossing symmetry relation of Eq. (23.179). If the impedance has a suitable behavior as $|\omega| \to \infty$, allowing for the possibility that the resistance of the circuit behaves at large angular frequencies like a constant, $\mathrm{R}(\infty)$, which may be non-zero, the corresponding dispersion relations take the following form:

$$\mathrm{R}(\omega) - \mathrm{R}(\infty) = \frac{2}{\pi} P \int_0^\infty \frac{\omega' X(\omega')\mathrm{d}\omega'}{\omega^2 - \omega'^2} \tag{23.184}$$

and

$$X(\omega) = -\frac{2\omega}{\pi} P \int_0^\infty \frac{\{\mathrm{R}(\omega') - \mathrm{R}(\infty)\}\mathrm{d}\omega'}{\omega^2 - \omega'^2}. \tag{23.185}$$

The following sum rule is obtained from the latter result:

$$\int_0^\infty \{\mathrm{R}(\omega) - \mathrm{R}(\infty)\}\mathrm{d}\omega = -\frac{\pi}{2} \lim_{\omega\to\infty} \omega X(\omega). \tag{23.186}$$

Equation (23.184) yields the following sum rule:

$$\int_0^\infty \frac{X(\omega)\mathrm{d}\omega}{\omega} = \frac{\pi}{2}\{\mathrm{R}(\infty) - \mathrm{R}(0)\}. \tag{23.187}$$

Equation (23.186) is referred to as the resistance integral theorem, and Eq. (23.185) is the reactance integral theorem.

In practical applications, the connection between the magnitude and phase of the admittance function is of interest. In this case the admittance is written in the following form:

$$Y(\omega) = M_Y(\omega)\mathrm{e}^{\mathrm{i}\theta_Y(\omega)}. \tag{23.188}$$

The problem of determining the phase in terms of the modulus $M_Y(\omega)$ corresponds to the subject treated in Section 23.3 (see also Section 20.2). If the dispersion relations for the admittance, Eqs. (23.164) and (23.165), are satisfied for a linear passive device, then it follows automatically that the circuit impedance has no zeros in the upper half of the complex angular frequency plane. If the impedance had zeros in this region of the complex angular frequency plane, the admittance would have poles in the upper half plane. Hence, the dispersion relations would not be obtained in the standard form presented. For example, for the series *LCR* circuit considered previously, the system impedance can be expressed as follows:

$$Z(\omega) = \frac{L}{\mathrm{i}\omega}(\omega - \omega_1)(\omega - \omega_2), \tag{23.189}$$

where ω_1 and ω_2 denote the two roots given in Eq. (23.161), and they correspond to the zeros of the impedance. They are both located in the lower half of the complex angular frequency plane.

Relying on the previous developments (recall Section 20.2), the phase and modulus of an electric circuit satisfy the following dispersion relations:

$$\theta_Y(\omega) = \frac{2\omega}{\pi} P \int_0^\infty \frac{\log M_Y(\omega')\mathrm{d}\omega'}{\omega^2 - \omega'^2} \tag{23.190}$$

and

$$\log M_Y(\omega) - \log M_Y(\omega_0) = \frac{2}{\pi} P \int_0^\infty \omega' \theta_Y(\omega') \left(\frac{1}{\omega_0^2 - \omega'^2} - \frac{1}{\omega^2 - \omega'^2} \right) \mathrm{d}\omega'. \tag{23.191}$$

In a similar fashion to Eq. (23.188), the impedance can be expressed as follows:

$$Z(\omega) = M_Z(\omega) \mathrm{e}^{\mathrm{i}\theta_Z(\omega)}. \tag{23.192}$$

The resulting dispersion relations are given by

$$\theta_Z(\omega) = \frac{2\omega}{\pi} P \int_0^\infty \frac{\log M_Z(\omega')\mathrm{d}\omega'}{\omega^2 - \omega'^2} \tag{23.193}$$

and

$$\log M_Z(\omega) - \log M_Z(\omega_0) = \frac{2}{\pi} P \int_0^\infty \omega' \theta_Z(\omega') \left(\frac{1}{\omega_0^2 - \omega'^2} - \frac{1}{\omega^2 - \omega'^2} \right) \mathrm{d}\omega', \tag{23.194}$$

which apply for circuits where the impedance has no zeros in the upper half of the complex angular frequency plane.

23.9 Applications in acoustics

In analogy with the propagation of electromagnetic waves in a medium, there is a related development for the derivation of dispersion relations of the Kramers–Kronig type for the transmission of sound waves. The applications and general progress on this topic have been relatively limited in comparison with the growth of developments in the electromagnetic situation.

The derivation of the dispersion relations for the acoustic case is now examined. The medium is assumed to be linear and the causality principle satisfied. The key variable that is of interest is the complex wave number, $K(\omega)$, which takes the following form:

$$K(\omega) = \frac{\omega}{c(\omega)} + i\alpha(\omega). \tag{23.195}$$

In Eq. (23.195) $c(\omega)$ is the phase velocity in the medium and $\alpha(\omega)$, which is the conventional symbol employed in the literature, is the attenuation coefficient. The reader should not get this latter notation confused with the polarizability, which shares the same symbol. In order to discuss acoustic waves, some definitions are needed. Attention is focused on the propagation of acoustic waves in a fluid. The condensation s is defined by

$$s = \frac{\rho - \rho_0}{\rho_0}, \tag{23.196}$$

where ρ is the instantaneous density and ρ_0 is the equilibrium density of the fluid. The condensation is a measure of the density change relative to the equilibrium density as an acoustic wave propagates in a medium. The time development of the condensation can be related to the pressure p via

$$s(t) = \int_{-\infty}^{\infty} \kappa_s(t - t')p(t')\mathrm{d}t', \tag{23.197}$$

where $\kappa_s(t)$ is the adiabatic compressibility. With the imposition of causality, $s(t)$ depends only on the past history, and not on the future, so that

$$\kappa_s(t - t') = 0, \quad \text{if } t - t' < 0. \tag{23.198}$$

If the Fourier transforms of $s(t)$, $\kappa_s(t)$, and $p(t)$ exist and are denoted by $S(\omega)$, $\mathrm{K}_s(\omega)$, and $P(\omega)$, respectively, then it follows that

$$S(\omega) = \mathrm{K}_s(\omega)P(\omega). \tag{23.199}$$

From the relationship

$$\mathrm{K}_s(\omega) = \int_{-\infty}^{\infty} \kappa_s(t)\mathrm{e}^{-i\omega t}\,\mathrm{d}t, \tag{23.200}$$

and the fact that the adiabatic compressibility is a real quantity, leads to

$$\mathrm{K}_s(-\omega) = \mathrm{K}_s(\omega)^*, \tag{23.201}$$

which represents the crossing symmetry relation for the frequency-domain compressibility. The quantity $\mathrm{K}_s(\omega)$ can be analytically continued into the upper half of the complex angular frequency plane. By considering the integral

$$\oint_C \frac{\{\mathrm{K}_s(\omega_z) - \mathrm{K}_s(\infty)\}\mathrm{d}\omega_z}{\omega - \omega_z},$$

where C represents the contour shown in Figure 19.2 and $\mathrm{K}_s(\infty)$ denotes the compressibility at infinite frequency, it follows that

$$\frac{\mathrm{i}}{\pi} P \int_{-\infty}^{\infty} \frac{\{\mathrm{K}_s(\omega_\mathrm{r}) - \mathrm{K}_s(\infty)\}\mathrm{d}\omega_\mathrm{r}}{\omega - \omega_\mathrm{r}} = \mathrm{K}_s(\omega) - \mathrm{K}_s(\infty). \tag{23.202}$$

Writing $\mathrm{K}_s(\omega)$ in terms of its real and imaginary parts as

$$\mathrm{K}_s(\omega) = \mathrm{K}_s'(\omega) + \mathrm{i}\mathrm{K}_s''(\omega), \tag{23.203}$$

and employing Eq. (23.201), yields

$$\mathrm{K}_s''(\omega) - \mathrm{K}_s''(\infty) = \frac{2\omega}{\pi} P \int_0^{\infty} \frac{\{\mathrm{K}_s'(\omega_\mathrm{r}) - \mathrm{K}_s'(\infty)\}\mathrm{d}\omega_\mathrm{r}}{\omega^2 - \omega_\mathrm{r}^2} \tag{23.204}$$

and

$$\mathrm{K}_s'(\omega) - \mathrm{K}_s'(\infty) = -\frac{2}{\pi} P \int_0^{\infty} \frac{\{\omega_\mathrm{r}\mathrm{K}_s''(\omega_\mathrm{r}) - \omega\mathrm{K}_s''(\infty)\}\mathrm{d}\omega_\mathrm{r}}{\omega^2 - \omega_\mathrm{r}^2}. \tag{23.205}$$

The link between the attenuation coefficient and phase velocity can be made via the dispersion formula for acoustic propagation:

$$K^2(\omega) = \omega^2 \rho_0 \mathrm{K}_s(\omega), \tag{23.206}$$

where $K(\omega)$ is the acoustic wave number, and the reader is reminded that the term dispersion relation is used to denote an integral connection between the real and imaginary parts of a particular property, and that a dispersion formula is used to give the frequency behavior of a quantity. From Eq. (23.206) it follows that

$$K(\omega) = \omega\sqrt{(\rho_0 \mathrm{K}_s(\omega))}, \tag{23.207}$$

and, using an argument closely related to that presented in Section 19.6 (Eqs. (19.104)–(19.106)) for the refractive index, it follows that K is an analytic function in the upper half of the complex angular frequency plane. Recall from Section 19.6 that the refractive index function $N(\omega) - 1$ is not the Fourier transform of a time-dependent system response function that is physically realizable. In a similar manner, the function $K(\omega)$ is not the Fourier transform of a physically realizable time-dependent system response function. Let $f = \omega^{-1}K(\omega)$, where f denotes the reciprocal of the complex velocity of sound and satisfies a crossing symmetry relationship of the form $f(-\omega) = f(\omega)^*$. Consider the integral

$$\oint_C \frac{\{f(\omega_z) - f(\infty)\}\mathrm{d}\omega_z}{\omega - \omega_z},$$

with Figure 19.2 defining the contour C; then

$$\frac{\alpha(\omega)}{\omega} - \left(\frac{\alpha(\omega)}{\omega}\right)_{\omega=\infty} = \frac{2\omega}{\pi} P \int_0^\infty \frac{\{c(\omega_\mathrm{r})^{-1} - c(\infty)^{-1}\}\mathrm{d}\omega_\mathrm{r}}{\omega^2 - \omega_\mathrm{r}^2} \tag{23.208}$$

and

$$\frac{1}{c(\omega)} - \frac{1}{c(\infty)} = -\frac{2}{\pi} P \int_0^\infty \frac{\{\alpha(\omega_\mathrm{r}) - \omega(\alpha(\omega)/\omega)_{\omega=\infty}\}\mathrm{d}\omega_\mathrm{r}}{\omega^2 - \omega_\mathrm{r}^2}. \tag{23.209}$$

These results represent the dispersion relations connecting the phase velocity and the attenuation coefficient for an acoustic wave.

An alternative approach to these dispersion relations (Angel and Achenbach, 1991) starts with the following set of assumptions: $\alpha(\omega)$ and its first derivative are assumed to vanish at $\omega = 0$, and $c(\omega)$ is non-zero at the same point. Both $\alpha(\omega)$ and $c(\omega)$ are bounded as $\omega \to \infty$, with $c(\infty)$ being non-zero. The function $\omega^{-1}K(\omega)$ satisfies $\omega^{-1}K(\omega) = O(\omega^{-1})$ as $\omega \to \infty$, and this function is differentiable at $\omega = 0$. The dispersion relations for $\alpha(\omega)$ and $c(\omega)$ can be obtained from these statements. Different dispersion relations can be obtained by modifying some of these assumptions. For example, the attenuation of acoustic waves propagating in a number of materials are modeled with an empirical frequency power law of the following form (Szabo, 1994; He, 1998; Waters *et al.*, 2000a):

$$\alpha(\omega) = \alpha_0 |\omega|^y, \tag{23.210}$$

where α_0 and y are non-negative constants, and the spectral interval is assumed to have some finite bandwidth. Typically, the exponent satisfies $0 < y \leq 2$, but for many materials $1 \leq y \leq 2$. For the situations where Eq. (23.210) applies, then $\omega^{-1}K(\omega) = O(\omega^{y-1})$. In the case $y < 1$, Eqs. (23.208) and (23.209) apply, but for larger values of y it is necessary to insert some convergence factor into the choice of function before performing the contour integration, thereby leading to modified dispersion relations involving subtractions. The number of subtractions to be employed can be connected back to the value of y (see the discussion in Waters *et al.* (2000b)).

The results stated in Eqs. (23.208) and (23.209) were first given by Ginzberg (1955), who noted that the upper limit of the integration is actually an approximation. An underlying assumption of the derivation is that sound is propagating in a continuous medium. In the case of gases, this requires that the wavelength λ of the sound wave has to exceed the mean free path of the molecules. For liquids and solids the mean free path is replaced by the interatomic spacing. Clearly, letting $\omega \to \infty$ leads to a fundamental conceptual problem. Ginzberg deals with this issue by supposing the upper limit of infinity is replaced by some large but finite angular frequency. He suggests upper bounds for the angular frequency of the order of $10^{13} - 10^{14}$ Hz.

A different approach can be taken by writing

$$\mathrm{H}(\omega) = \mathrm{e}^{\mathrm{i}K(\omega)d} = A(\omega)\mathrm{e}^{\mathrm{i}\omega d/c(\omega)}, \tag{23.211}$$

where d denotes a propagation distance and $A(\omega) = \mathrm{e}^{-\alpha(\omega)d}$. By considering the function $\log\{\mathrm{H}(\omega)/d\}$, with d appropriately scaled to bring it into dimensionless form by dividing by the units of distance, and assuming the absence of zeros of $\mathrm{H}(\omega)$ in the upper half of the complex angular frequency plane, then dispersion relations can be constructed connecting the real and imaginary parts of $\log\{\mathrm{H}(\omega)/d\}$, using the approaches discussed in Sections 20.2 and 23.3. The real and imaginary parts of this functions are directly related to $\alpha(\omega)$ and $c(\omega)$, respectively.

The surface acoustic impedance $Z_{\mathrm{s}}(\omega)$ is defined by

$$Z_{\mathrm{s}}(\omega) = \frac{P(\omega)}{V_{\mathrm{N}}(\omega)}, \tag{23.212}$$

where $P(\omega)$ is the pressure at the surface and $V_{\mathrm{N}}(\omega)$ is the normal velocity component into the surface. The velocity can be written in terms of the surface admittance function $Y_{\mathrm{s}}(\omega)$ via

$$V_{\mathrm{N}}(\omega) = Y_{\mathrm{s}}(\omega)P(\omega). \tag{23.213}$$

Depending on the asymptotic behavior of the surface admittance function or surface acoustic impedance, dispersion relations can be obtained in much the same manner as was done in Section 23.8. This is left as an exercise for the reader to consider.

23.10 Viscoelastic behavior

The general connection between the dynamic stress σ_{ij} and the strain ε_{kl} components of the stress and strain tensors are given by

$$\sigma_{ij}(t) = \int_{-\infty}^{t} c_{ijkl}(t-\tau)\varepsilon_{kl}(\tau)\mathrm{d}\tau, \tag{23.214}$$

where c_{ijkl} are elements of the stiffness tensor. The function $c_{ijkl}(t)$ is also referred to as a memory function. In the following development, the tensor indices are dropped to simplify the notation. The Fourier transforms of $\sigma(t), c(t)$, and $\varepsilon(t)$ are denoted by $\bar{\sigma}(\omega), M(\omega)$, and $\bar{\varepsilon}(\omega)$, respectively, and, assuming that these quantities exist, it follows that

$$\bar{\sigma}(\omega) = M(\omega)\bar{\varepsilon}(\omega), \tag{23.215}$$

where $M(\omega)$ is a modulus of elasticity. The quantity $M(\omega)$ can be written as follows:

$$M(\omega) = M_{\rm d}(\omega) + {\rm i}M_{\rm l}(\omega), \tag{23.216}$$

where $M_{\rm d}(\omega)$ denotes the dynamic modulus of elasticity and $M_{\rm l}(\omega)$ designates the loss modulus. The modulus at zero frequency is denoted by M_0, and this is a real quantity. By considering the integral

$$\oint_C \frac{\omega_z^{-1}\{M(\omega_z) - M_0\}{\rm d}\omega_z}{\omega - \omega_z},$$

where C represents the contour shown in Figure 19.2, the following dispersion relations are obtained:

$$M_{\rm d}(\omega) = M_0 - \frac{2\omega^2}{\pi} P \int_0^\infty \frac{M_{\rm l}(\omega'){\rm d}\omega'}{\omega'(\omega^2 - \omega'^2)} \tag{23.217}$$

and

$$M_{\rm l}(\omega) = \frac{2\omega}{\pi} P \int_0^\infty \frac{\{M_{\rm d}(\omega') - M_0\}{\rm d}\omega'}{\omega^2 - \omega'^2}. \tag{23.218}$$

The derivation of these results has made use of the crossing symmetry condition:

$$M(-\omega) = M(\omega)^*, \tag{23.219}$$

which follows directly from the Fourier transform connection between $M(\omega)$ and $c(t)$, and on noting that $c(t)$ is a real quantity. It has also been assumed that $\omega_z^{-1}\{M(\omega_z) - M_0\}$ vanishes at least like $O(\omega_z^{-1})$ as $\omega_z \to \infty$ in the upper half of the complex angular frequency plane. This assumption is actually problematic for the following reason. At very high frequencies there is an issue similar to the one indicated in the preceding section, namely the difficulty associated with attaching a meaning to the strain in this spectral region. The way to deal with this matter is to assume some upper cutoff point for integrals over frequency.

23.11 Epilog

Other applications beyond those covered in this chapter and elsewhere in the book can be found in the literature. Primary examples include problems dealing with more advanced scattering situations, particularly those involving analytic function theory of more than one variable.

Sufficient theoretical developments involving the Hilbert transform have been covered in the early chapters for the reader to attack many problems likely to arise in practical settings. Good luck!

Notes

§23.2.1 Hilbert transform spectroscopy based on the Josephson junction is discussed in a number of papers by Divin and coworkers; see Divin *et al.* (1980, 1983, 1993, 1995, 1996, 1997a, 1997b, 1999, 2001) and Volkov *et al.* (1999). For further reading, see Larkin, Anischenko, and Khabayer (1994), Tarasov *et al.* (1995), Larkin *et al.* (1997), Ludwig *et al.* (2001), and Shul'man *et al.* (2003). Discussion of the Josephson effect can be found in a number of sources; see, for example, Barone and Paternò (1982) and Likharev (1986).

§23.2.2 For some related reading, see Verdun, Giancaspro, and Marshall (1988) and Liang and Marshall (1990). Some useful background can be found in Bartholdi and Ernst (1973).

§23.3 The literature on the phase retrieval problem is extensive. Some selective further reading can be found in Walther (1962), Roman and Marathay (1963), Saxton (1974), Burge *et al.* (1974, 1976), Misell and Greenaway (1974a, 1974b), Misell, Burge, and Greenaway (1974), Hoenders (1975), De Heer *et al.* (1976), Tip (1977), Ross *et al.* (1978), Ross, Fiddy, and Moezzi (1980), Nieto-Vesperinas (1980), Andersson, Johansson, and Eklund (1981), Taylor (1981), Peřina (1985), and Nakajima (1988). For suitable sources for further exploration on the issue of the location of zeros, see the work by Ross *et al.* (1980), and for some mathematical background consult Levin (1964). The problem of phase determination in Raman spectroscopy has been an active research area; see, for example, Cable and Albrecht (1986), Joo and Albrecht (1993), Lee (1995), and Lee, Feng, and Yeo (1997). For applications of the Hilbert transform to the study of phase synchronization, see Rosenblum, Pikovsky, and Kurths (1996), and in neuroscience see Le Van Quyen *et al.* (2001).

§23.4 Further reading on applications in crystallography can be found in Ramachandran (1969), Kaufmann (1985), Tang and Chang (1990), Mishnev (1993, 1996), Nikulin, Zaumseil, and Petrashen (1996), Nikulin (1997), and Giacovazzo, Siliqi, and Fernández–Castaño (1999). Mishnev's (1993) work examined in this section is related to an investigation of Kramer (1973). Some additional comments on the sampling theorem can be found in Kohlenberg (1953). For discussion on the importance of the electronic density, its connection with chemical bonding, and its determination from X-ray diffraction studies, see Coppens (1997).

§23.5.1 For further reading on potential scattering, see Schiff (1955), Goldberger and Watson (1964), Landau and Lifshitz (1965), and Roman (1965).

§23.5.2 For additional reading on dispersion relations for potential scattering, see Khuri (1957), Klein and Zemach (1959), Goldberger (1960), Landau and Lifshitz (1965), and Roman (1965). For more advanced discussions on dispersion relations, see Goldberger and Watson (1964), Barton (1965), Nussenzveig (1972), and

Nishijima (1974). An alternative approach that has been developed involves examining derivative analyticity relations. Some references to work on the differential form of the dispersion relations are Bronzan, Kane, and Sukhatme (1974), Sukhatme *et al.* (1975), Kolář and Fischer (1984), Fischer and Kolář (1987), Martini *et al.* (1999), and Menon, Motter, and Pimentel (1999).

§23.5.3 Reflecting the importance of the topic of this subsection, there has been considerable work published on dispersion relations for electron–atom scattering, and in particular for the specific case of electron–hydrogen atom scattering. A selection of papers for further reading is: Gerjuoy and Krall (1960, 1962), Krall and Gerjuoy (1960), Byron, De Heer, and Joachain (1975), Hutt *et al.* (1976), Blum and Burke (1977), Byron and Joachain (1977, 1978), McDowell and Farmer (1977), Gerjuoy and Lee (1978), Kuchiev and Amusia (1978), Dumbrajs and Martinis (1981), Amusia and Kuchiev (1982), Kuchiev (1985), Temkin, Bhatia, and Kim (1986), Bessis, Haffad, and Msezane (1994), Bessis and Temkin (2000), Temkin and Drachman (2000), Vrinceanu *et al.* (2001), and Sucher (2002).

§23.6 A good source for a classical discussion of NMR is Abragam (1961), and FTNMR is described in many books; see, for example, Drago (1992) and, for a more mathematical account, Ernst *et al.* (1987). An early application of the Kramers–Kronig relations in magnetic resonance is given in Pake and Purcell (1948). Further discussion of applications in magnetic resonance can be found in the works by Bolton, Troup, and Wilson (1964), Bolton (1969a, 1969b), Ernst (1969), and Bartholdi and Ernst (1973). The latter paper in particular amplifies on the discussion of this section.

§23.7 The idea for the topic of this section can be found in Cole and Cole (1941, 1942). Some additional discussion can be found in Fang (1961, 1965) and Bolton (1969b). Marshall and coworkers have discussed the DISPA approach in several papers; see Marshall and Roe (1978), Roe, Marshall, and Smallcombe (1978), Herring *et al.* (1980), Marshall (1982), and Wang and Marshall (1983). The first of the aforementioned Marshall papers gives a summary of the impact of various line broadening mechanisms and the resulting appearance of the DISPA curve.

§23.8 For further reading, see Carson (1926), Lee (1932), Gross (1943), Bode (1945), Murakami and Corrington (1948), Guillemin (1949), Brachman (1955), Page (1955), Tuttle (1958), and Hamilton (1960). For a discussion of polarization resistance, see Mansfeld and Kendig (1999). The singular structure of various circuits is considered in Gross and Braga (1961). In a number of sources the analytic structure of the impedance and admittance is discussed in the complex angular p-plane, which is connected to the complex angular frequency plane by the change of variable $p = \mathrm{i}\omega$. For an application of the multiply subtractive form of the Kramers–Kronig relations for the impedance function of concrete, see Peiponen (2005).

§23.9 For further reading on a number of the topics in this section, see Mangulis (1964), Horton (1974), O'Donnell, Jaynes, and Miller (1978, 1981), Weaver and Pao (1981), Brauner and Beltzer (1985), Weaver (1986), Lee, Lahham, and Martin (1990), Angel and Achenbach (1991), Szabo (1994, 1995), Audoin and Roux (1996), He (1998), Waters *et al.* (2000a, 2000b, 2003), Mobley, Waters, and Miller (2005), and Waters, Mobley, and Miller (2005). Berthelot (2001) can be consulted for the case of the surface acoustic impedance function. Waters *et al.* (2003) discuss an approach based on differential forms of the Kramers–Kronig relations.

§23.10 Additional discussion can be found in Gross (1948, 1968), Schwarzl and Struik (1967–1968), Ferry (1970), Booij and Thoone (1982), Tschoegl (1989), and Pritz (1999).

Exercises

23.1 Evaluate the exchange contribution

$$g_1^{\mathrm{Born}}(\mathbf{k},\mathbf{k}') = -\frac{m_e}{2\pi\hbar^2}\iint \chi(\mathbf{k}',\mathbf{r}_2)^*\varphi(\mathbf{r}_1)^* V_1(\mathbf{r}_1)\chi(\mathbf{k},\mathbf{r}_1)\varphi(\mathbf{r}_2)\mathrm{d}\mathbf{r}_1\,\mathrm{d}\mathbf{r}_2$$

for $\mathbf{k} = \mathrm{k}'$, where the one-electron potential is $V_1(\mathbf{r}_1) = -e^2/r_1$. Determine the pole structure that arises, and indicate if there is any cancellation with the pole structure of $g_{12}^{\mathrm{Born}}(\mathbf{k},\mathbf{k})$.

23.2 Can a logarithmic Hilbert transform approach (in the absence of zeros in the appropriate regions of the complex angular frequency planes) lead to the determination of the phase for the general two-dimensional phase retrieval problem? State any assumptions that are necessary to arrive at your result.

23.3 Suppose Exercise 23.2 is reconsidered, but the assumption is made that the two-dimensional function $F(x_1,x_2)$ of interest has the form $F(x_1,x_2) = F_1(x_1)F_2(x_2)$. What, if any, simplification arises? Can the phase be determined from knowledge of $|F_1(x_1)|$ and $|F_2(x_2)|$?

23.4 Consider the two functions

$$f_1(t) = \mathrm{sinc}(t+1) + \mathrm{sinc}\, t + \mathrm{sinc}(t-1)$$

and

$$f_2(t) = -\mathrm{i}\,\mathrm{sinc}(t+1) + \mathrm{sinc}\, t + \mathrm{i}\,\mathrm{sinc}(t-1).$$

How are the functions related? How do the moduli of the functions compare at the points $t = -1$, 0, and 1? How do the moduli of the functions compare for general t? What is the support of the Fourier transform of each function? What is the relationship between the zeros of $f_1(t)$ and the zeros of $f_2(t)$?

23.5 Construct a DISPA plot for a single absorption peak assuming a Gaussian line profile.

23.6 Construct the DISPA plot obtained using a Gaussian absorption profile with the line width factor $\sigma = 1$. Compare this with an experimental absorption curve which is obtained from two Gaussians centered at the same angular frequency, but having line width factors of 1.05σ and 1.1σ.

23.7 Repeat Exercise 23.6 with the experimental curve replaced by a sum of two Gaussian functions with the same line width factor σ, but centered at angular frequencies $0.95\omega_0$ and $1.05\omega_0$.

23.8 Consider a circuit consisting of a resistor R in parallel with a capacitor C, which is then linked in series with an inductor L and a voltage source $\mathcal{E}(t)$. Discuss the analytic behavior as a function of complex frequencies of the impedance and admittance functions for the circuit.

23.9 If the circuit of Figure 23.5 is modified so that a capacitor with capacitance C is placed in parallel with the impedance (call it Z_0), do the dispersion relations for the admittance given in Eqs. (23.169) and (23.170) hold?

23.10 If the capacitor of the previous question is replaced by an inductor of inductance L, do the dispersion relations Eqs. (23.169) and (23.170) hold? If not, can they be altered in any suitable manner to give a pair of modified dispersion relations for the admittance?

23.11 If the circuit of Figure 23.5 is modified by adding a series combination of a capacitor and an inductor in parallel with the impedance Z_0, what dispersion relations (if any) can be obtained for the admittance function?

Appendix 1

Tables of selected Hilbert transforms

The tables are laid out in the following order.

(1) General properties.
(2) Powers and algebraic functions.
(3) Exponential functions.
(4) Hyperbolic functions.
(5) Trigonometric functions.
(6) Logarithmic functions.
(7) Inverse trigonometric and hyperbolic functions.
(8) Special functions:

- (A) Legendre polynomials;
- (B) Hermite polynomials;
- (C) Laguerre polynomials;
- (D) Bessel functions of the first kind of integer order;
- (E) Bessel functions of the first kind of fractional order;
- (F) Bessel functions of the second kind of fractional order;
- (G) product of Bessel functions of the first kind of fractional order;
- (H) modified Bessel functions of the first kind;
- (I) modified Bessel functions of the second kind;
- (J) spherical Bessel functions of the first kind;
- (K) spherical Bessel functions of the second kind;
- (L) cosine integral function;
- (M) sine integral function;
- (N) Struve functions;
- (O) Anger functions;
- (P) miscellaneous special functions.

(9) Pulse and wave forms.
(10) Distributions.
(11) Multiple Hilbert transforms.

(12) Finite Hilbert transforms:

  (A) the interval $[-1, 1]$;
  (B) the interval $[0, 1]$;
  (C) the interval $[a, b]$.

(13) Miscellaneous cases:

  (A) the cosine form;
  (B) the one-sided Hilbert transform;
  (C) the cotangent form;
  (D) the Hilbert transforms H_e and H_o.

To find an entry quickly, search under the part of the functional form mentioned furthest down the above list. For example, to find $(Hf)(x)$ with $f(x) = \sin ax\, J_n(ax)$, search under "Special functions", and not under "Trigonometric functions". A number of special cases of more general formulas have been included to make the table quick and easy to employ. All the entries have been checked numerically to verify their accuracy. When a generic function f is employed to indicate a particular property, it is assumed the appropriate Hilbert transform exists, that is $f \in L^p(a, b)$ for $p \geq 1$, (or for some properties $p > 1$), where (a, b) denotes the range of the Hilbert transform or one of its variants. Other sources giving tables of Hilbert transforms are Erdélyi *et al.* (1954, Vol. II, p. 239), MacDonald and Brachman (1956), Alavi-Sereshki and Prabhakar (1972), Hahn (p. 397; 1996a, 1996b), and Poularikas (1999). The opposite sign convention to that employed in the present table is used in Erdélyi *et al.* (1954). Individual Hilbert transforms can be found scattered across different sections of Bierens de Haan (1867) and Gradshteyn and Ryzhik (1965).

Special symbols

The following special functions and symbols are employed in the tables.

Pochhammer symbol: $(a)_k = a(a+1)(a+2)\cdots(a+k-1) = \Gamma(a+k)/\Gamma(a)$;

beta function (Euler's integral of the first kind): $B(a,b) = \int_0^1 t^{a-1}(1-t)^{b-1}\,dt$, $\operatorname{Re} a > 0, \operatorname{Re} b > 0$;

Fresnel cosine integral: $C(z) = \int_0^z \cos\left(\pi t^2/2\right)dt$;

Hartley cas function: $\operatorname{cas} x = \sin x + \cos x$;

cosine integral: $\operatorname{Ci}(x) = \gamma + \log x + \int_0^x \frac{\cos y - 1}{y}\,dy = -\int_x^\infty \frac{\cos y\,dy}{y} \equiv -\operatorname{ci}(x)$;

cosine–exponential integral: $\operatorname{cie}(\alpha, \beta) = \int_\alpha^\infty \frac{\cos y\, e^{-\beta y}\,dy}{y}, \quad \alpha > 0, \ \beta > 0$;

cosine–exponential integral: $\operatorname{Cie}(\alpha, \beta) = P\int_{-\alpha}^\infty \frac{\cos y\, e^{-\beta y}dy}{y}, \quad \alpha > 0, \ \beta > 0$;

Clausen function: $\mathrm{Cl}_2(x) = \sum_{n=1}^{\infty} \frac{\sin nx}{n^2}, \quad \mathrm{Cl}_2(x) = -\int_0^x \log\{2\sin(t/2)\}\mathrm{d}t$;
Gegenbauer polynomial (ultraspherical polynomial):

$$C_n^{\lambda}(x) = \frac{1}{\Gamma(\lambda)} \sum_{m=0}^{[n/2]} \frac{(-1)^m \Gamma(\lambda + n - m)(2x)^{n-2m}}{m!(n-2m)!}, \quad \lambda > -1/2, \quad \lambda \neq 0;$$

ultraspherical function of the second kind: $D_n^{(\lambda)}(x)$, see Eq. (11.299);
exponential integral: $E_n(z) = \int_1^{\infty} \frac{\mathrm{e}^{-zy}\,\mathrm{d}y}{y^n}, \quad n = 0, 1, 2, \ldots$, for $\mathrm{Re}\, z > 0$;
Weber function: $\mathbf{E}_{\upsilon}(z) = \frac{1}{\pi} \int_0^{\pi} \sin(\upsilon\theta - z\sin\theta)\mathrm{d}\theta$;
exponential integral: $\mathrm{Ei}(x) = -P \int_{-x}^{\infty} \frac{\mathrm{e}^{-y}\,\mathrm{d}y}{y}$, for $x > 0$;
error function: $\mathrm{erf}(z) = \frac{2}{\sqrt{\pi}} \int_0^z \mathrm{e}^{-s^2}\,\mathrm{d}s$;
Kummer's confluent hypergeometric function: ${}_1F_1(\alpha; \beta; z) = \sum_{k=0}^{\infty} \frac{(\alpha)_k z^k}{(\beta)_k k!}$;
Gauss' hypergeometric function: ${}_2F_1(a, b; c; z) = \sum_{k=0}^{\infty} \frac{(a)_k (b)_k z^k}{(c)_k k!}$;
Heaviside distribution: $\mathrm{H}(x) = \begin{cases} 0, & x < 0 \\ 1, & x > 0; \end{cases}$
Hermite polynomial: $H_n(x) = n! \sum_{m=0}^{[n/2]} \frac{(-1)^m (2x)^{n-2m}}{(n-2m)!m!}$;
Struve function: $\mathbf{H}_{\nu}(z) = \frac{2(z/2)^{\nu}}{\sqrt{(\pi)}\,\Gamma(\nu + 1/2)} \int_0^{\pi/2} \sin(z\sin\theta)\sin^{2\nu}\theta\,\mathrm{d}\theta$;

$$\mathbf{H}_{\nu}(z) = (z/2)^{\nu+1} \sum_{m=0}^{\infty} \frac{(-1)^m (z/2)^{2m}}{\Gamma(m + 3/2)\Gamma(m + \nu + 3/2)};$$

modified Bessel function of the first kind: $I_{\nu}(z) = (z/2)^{\nu} \sum_{m=0}^{\infty} \frac{(z/2)^{2m}}{m!\Gamma(\nu + m + 1)}$;
Bessel function of the first kind: $J_{\nu}(z) = (z/2)^{\nu} \sum_{m=0}^{\infty} \frac{(-1)^m (z/2)^{2m}}{m!\Gamma(\nu + m + 1)}$;
spherical Bessel function of the first kind: $j_n(x) = (-x)^n \left(\frac{1}{x}\frac{\mathrm{d}}{\mathrm{d}x}\right)^n \left(\frac{\sin x}{x}\right)$, $n \in \mathbb{Z}^+$;
Anger function: $\mathbf{J}_{\upsilon}(z) = \frac{1}{\pi} \int_0^{\pi} \cos(\upsilon\theta - z\sin\theta)\mathrm{d}\theta$;

modified Bessel function of the second kind (Basset's function), also called the modified Bessel function of the third kind:

$$K_n(x) = -(-1)^n \log(z/2) I_n(z) + \frac{1}{2}(z/2)^{-n} \sum_{m=0}^{n-1} \frac{(-1)^m (n-m-1)!(z/2)^{2m}}{m!}$$
$$+ \frac{(-z)^n}{2^{n+1}} \sum_{m=0}^{\infty} \frac{\{\psi(n+m+1)+\psi(m+1)\}(z/2)^{2m}}{m!(n+m)!};$$

Laguerre polynomial: $L_n(x) = \sum_{m=0}^{n} (-1)^m \binom{n}{n-m} \frac{x^m}{m!}$;

modified Struve function: $\mathbf{L}_v(z) = \frac{2(z/2)^v}{\sqrt{(\pi)}\,\Gamma(v+1/2)} \int_0^{\pi/2} \sinh(z\cos\theta)\sin^{2v}\theta\, \mathrm{d}\theta$, $\operatorname{Re} v > -1/2$;

polylogarithm function: $\mathrm{Li}_n(z) = \sum_{k=1}^{\infty} k^{-n} z^k$, $|z| \le 1$,

$$\mathrm{Li}_n(z) = \int_0^z \frac{\mathrm{Li}_{n-1}(z)}{z}\, \mathrm{d}z, \quad n \ge 3;$$

dilogarithm function: $\mathrm{Li}_2(z) = \sum_{k=1}^{\infty} \frac{z^k}{k^2}$, $|z| \le 1$, $\mathrm{Li}_2(z) = -\int_0^z \frac{\log(1-z)}{z}\, \mathrm{d}z$;

Legendre polynomial: $P_n(x) = \frac{1}{2^n} \sum_{m=0}^{[n/2]} (-1)^m \binom{n}{m} \binom{2n-2m}{n} x^{n-2m}$;

associated Legendre function of the first kind: $P_v^m(x) = (-1)^m (1-x^2)^{m/2} \frac{\mathrm{d}^m P_v(x)}{\mathrm{d}x^m}$;

Jacobi polynomial: $P_n^{(\alpha,\beta)}(x) = 2^{-n} \sum_{m=0}^{n} \binom{n+\alpha}{m} \binom{n+\beta}{n-m} (x-1)^{n-m}(x+1)^m$;

Legendre function of the second kind:

$$Q_n(x) = \frac{1}{2} P_n(x) \log\left(\frac{1+x}{1-x}\right) - \sum_{j=1}^{n} \frac{P_{j-1}(x) P_{n-j}(x)}{j};$$

associated Legendre function of the second kind:

$$Q_v^m(x) = (-1)^m (1-x^2)^{m/2} (\mathrm{d}^m Q_v(x)/\mathrm{d}x^m);$$

Jacobi function of the second kind:

$$Q_n^{(\alpha,\beta)}(x) = \frac{1}{2}(x-1)^{-\alpha}(x+1)^{-\beta} \int_{-1}^{1} \frac{(1-t)^\alpha (1+t)^\beta P_n^{(\alpha,\beta)}(t)\mathrm{d}t}{x-t};$$

Fresnel sine integral: $S(z) = \int_0^z \sin\left(\pi t^2/2\right)\mathrm{d}t$ (other conventions are also commonly employed in the literature for the Fresnel integrals);

signum function ("sign" function): $\operatorname{sgn} x = \begin{cases} 1, & \text{for } x > 0 \\ 0, & \text{for } x = 0 \\ -1, & \text{for } x < 0; \end{cases}$

hyperbolic sine integral function: $\mathrm{Shi}(z) = \int_0^z \frac{\sinh t \, \mathrm{d}t}{t}$;

sine integral: $\mathrm{si}(x) = -\int_x^\infty \frac{\sin y \, \mathrm{d}y}{y}$;

sine integral: $\mathrm{Si}(x) = \int_0^x \frac{\sin y \, \mathrm{d}y}{y} = \frac{\pi}{2} + \mathrm{si}(x)$;

sine–exponential integral: $\mathrm{sie}(\alpha, \beta) = -\int_\alpha^\infty \frac{\sin y \, \mathrm{e}^{-\beta y} \, \mathrm{d}y}{y}, \quad \beta > 0$;

sinc function: $\mathrm{sinc}\, x = \frac{\sin \pi x}{\pi x}$;

Chebyshev polynomial of the first kind: $T_n(x) = \frac{n}{2} \sum_{m=0}^{[n/2]} \frac{(-1)^m (n-m-1)!(2x)^{n-2m}}{(n-2m)!m!}$;

Chebyshev polynomial of the second kind: $U_n(x) = \sum_{m=0}^{[n/2]} \frac{(-1)^m (n-m)!(2x)^{n-2m}}{(n-2m)!m!}$;

Bessel function of the second kind (Neumann's function):

$$Y_n(x) = -n!(z/2)^{-n}\pi^{-1} \sum_{m=0}^{\infty} \frac{(z/2)^m J_m(z)}{(n-m)m!} - 2\pi^{-1} \sum_{m=0}^{\infty} \frac{(-1)^m (n+2m) J_{n+2m}(z)}{(n+m)m}$$
$$+ 2\pi^{-1}\{\log(z/2) - \psi(n+1)\} J_n(z)$$

(the symbol $N_n(x)$ is also employed in place of $Y_n(x)$);

spherical Bessel function of the second kind: $y_n(x) = x^n(-1)^{n+1}\left(\frac{1}{x}\frac{\mathrm{d}}{\mathrm{d}x}\right)^n \left(\frac{\cos x}{x}\right)$, $n \in \mathbb{Z}^+$.

Greek symbols

Catalan's constant: $\beta(2) = \sum_{k=0}^{\infty} \frac{(-1)^k}{(2k+1)^2} \approx 0.915\,965\,594\,177\,219\,015\,1\ldots$;

Euler's constant: $\gamma \approx 0.577\,215\,664\,901\,532\,860\,6\ldots$

gamma function: $\Gamma(z) = \int_0^\infty t^{z-1} \mathrm{e}^{-t} \mathrm{d}t$, for $\mathrm{Re}\, z > 0$;

incomplete gamma function: $\Gamma(a, x) = \int_x^\infty t^{a-1} \mathrm{e}^{-t} \, \mathrm{d}t$;

Dirac delta "function": $\delta(x)$;

Lerch function: $\Phi(z, s, v) = \sum_{n=0}^{\infty} (n+v)^{-s} z^n, \ v \neq 0, -1, -2, \ldots$;

digamma function: $\psi(z) = \frac{\Gamma'(z)}{\Gamma(z)}, \psi(z+1) = -\gamma + \sum_{n=1}^{\infty} \frac{z}{(z+n)n}, z \neq -1, -2, \ldots$;

polygamma function: $\psi^{(n)}(z) = \frac{\mathrm{d}^n \psi(z)}{\mathrm{d}z^n} = \frac{\mathrm{d}^{n+1} \ln \psi(z)}{\mathrm{d}z^{n+1}}$.

Miscellaneous

Summation convention: $\sum_{j=n}^{m} a_j = 0$, for $m < n$;

floor symbol: $\lfloor x \rfloor$, greatest integer less than or equal to x.

Table 1.1. *General properties*

Number	Function $f(x)$	H[function] $g(x) = \frac{1}{\pi} P \int_{-\infty}^{\infty} \frac{f(s)\mathrm{d}s}{x-s}$
(1.1)	$f(x)$	$g(x) = (Hf)(x)$
(1.2)	$g(x)$	$f(x) = -(Hg)(x)$
(1.3)	$f_1(x) + f_2(x)$	$H\{f_1(x) + f_2(x)\} = (Hf_1)(x) + (Hf_2)(x)$
(1.4)	$(Hf)(x)$	$\{H(Hf)\}(x) = (H^2 f)(x) = -f(x)$
(1.5)	$(H^n f)(x)$, integer $n \geq 0$	$(H^{n+1}f)(x) = \begin{cases} (-1)^{(n+1)/2} f(x), & \text{for } n \text{ odd} \\ (-1)^{n/2} g(x), & \text{for } n \text{ even} \end{cases}$
(1.6)	$f(x+a)$	$g(x+a)$
(1.7)	$f(ax),\ a > 0$	$g(ax)$
(1.8)	$f(-ax),\ a > 0$	$-g(-ax)$
(1.9)	$f(ax+b),\ b \in \mathbb{R}$	$\operatorname{sgn} a\, g(ax+b)$
(1.10)	$f(ax^{-1}),\ a > 0$	$g(0) - g(ax^{-1})$
(1.11)	$f(a/(x+b)),\ a > 0$	$g(0) - g(a/(x+b))$
(1.12)	$f\left(a + \frac{b}{x+c}\right),\ b > 0$	$g(a) - g\left(a + \frac{b}{x+c}\right)$
(1.13)	$f(ax - bx^{-1})$, $a > 0, b > 0$	$g(ax - bx^{-1})$
(1.14)	$xf(x)$	$xg(x) - \frac{1}{\pi}\int_{-\infty}^{\infty} f(t)\mathrm{d}t$
(1.15)	$(x+a)f(x)$	$(x+a)g(x) - \frac{1}{\pi}\int_{-\infty}^{\infty} f(t)\mathrm{d}t$
(1.16)	$x^2 f(x)$	$x^2 g(x) - \frac{1}{\pi}\int_{-\infty}^{\infty} (t+x)f(t)\mathrm{d}t$
(1.17)	$x^n f(x)$, integer $n \geq 0$	$H\{x^n f(x)\} = x^n g(x) - \frac{1}{\pi}\sum_{k=0}^{n-1} x^k \int_{-\infty}^{\infty} t^{n-1-k} f(t)\mathrm{d}t$
(1.18)	$\frac{f(x)}{x},\ x^{-1}f(x) \in L^1_{\text{loc}}(\mathbb{R})$	$\frac{g(x) - g(0)}{x}$
(1.19)	$\frac{f(x)}{x^2},\ x^{-2}f(x) \in L^1_{\text{loc}}(\mathbb{R})$	$\frac{g(x) - g(0) - xH\{t^{-1}f(t)\}(0)}{x^2}$
(1.20)	$f'(x)$	$g'(x)$
(1.21)	$\frac{\mathrm{d}^n f(x)}{\mathrm{d}x^n}$	$\frac{\mathrm{d}^n g(x)}{\mathrm{d}x^n}$
(1.22)	$f(x)$ even function	$\frac{2x}{\pi} P \int_0^{\infty} \frac{f(s)\mathrm{d}s}{x^2 - s^2}$
(1.23)	$f(x)$ odd function	$\frac{2}{\pi} P \int_0^{\infty} \frac{sf(s)\mathrm{d}s}{x^2 - s^2}$
(1.24)	$f(x) * h(x)$	$g(x) * h(x)$
(1.25)	$h(x) * f(x)$	$h(x) * g(x)$

$*$ denotes convolution.

Table 1.2. *Powers and algebraic functions*

Number	$f(x)$	$\frac{1}{\pi} P \int_{-\infty}^{\infty} \frac{f(s)\mathrm{d}s}{x-s}$
(2.1)	a (a constant)	0
(2.2)	$(x+a)^{-1}, \quad \mathrm{Im}\, a > 0$	$-\mathrm{i}(x+a)^{-1}$
(2.3)	$(x+a)^{-1}, \quad \mathrm{Im}\, a < 0$	$\mathrm{i}(x+a)^{-1}$
(2.4)	$a(x^2+a^2)^{-1}, \quad \mathrm{Re}\, a > 0$	$x(x^2+a^2)^{-1}$
(2.5)	$x(x^2+a^2)^{-1}, \quad \mathrm{Re}\, a > 0$	$-a(x^2+a^2)^{-1}$
(2.6)	$(bx+c)(x^2+a^2)^{-1}, \quad \mathrm{Re}\, a > 0$	$\frac{(cx-ba^2)}{a(x^2+a^2)}$
(2.7)	$\frac{b}{(x+a)^2+b^2}, \quad b > 0$	$\frac{x+a}{(x+a)^2+b^2}$
(2.8)	$\frac{x+a}{(x+a)^2+b^2}, \quad b > 0$	$-\frac{b}{(x+a)^2+b^2}$
(2.9)	$\begin{cases} 0, & x < 0 \\ \frac{1}{x^3+1}, & x > 0 \end{cases}$	$\frac{1}{9\pi(1+x^3)}\{9\log\lvert x\rvert + 2\surd(3)\,\pi x(x+1)\}$
(2.10)	$\begin{cases} 0, & x < 0 \\ \frac{x}{x^3+1}, & x > 0 \end{cases}$	$\frac{1}{9\pi(1+x^3)}\{9x\log\lvert x\rvert + 2\surd(3)\,\pi(x^2-1)\}$
(2.11)	$\begin{cases} 0, & x < 0 \\ \frac{x^2}{x^3+1}, & x > 0 \end{cases}$	$\frac{1}{9\pi(1+x^3)}\{9x^2\log\lvert x\rvert - 2\surd(3)\,\pi(x+1)\}$
(2.12)	$\frac{1}{(x^4+a^4)}, \quad a > 0$	$\frac{x(x^2+a^2)}{(x^4+a^4)a^3\surd 2}$

Table 1.2. (*Cont.*)

Number	$f(x)$	$\dfrac{1}{\pi} P \displaystyle\int_{-\infty}^{\infty} \frac{f(s)\mathrm{d}s}{x-s}$
(2.13)	$\dfrac{1}{(x^2+a^2)^2}, \quad a>0$	$\dfrac{x(x^2+3a^2)}{2a^3(x^2+a^2)^2}$
(2.14)	$\dfrac{1}{(x^2+a^2)(x^2+b^2)}, \quad a>0, b>0$	$\dfrac{x(x^2+a^2+b^2+ab)}{ab(a+b)(x^2+a^2)(x^2+b^2)}$
(2.15)	$\dfrac{x}{(x^2+a^2)^2}, \quad a>0$	$\dfrac{x^2-a^2}{2a(x^2+a^2)^2}$
(2.16)	$\dfrac{x}{(x^2+a^2)(x^2+b^2)}, \quad a>0, b>0$	$\dfrac{x^2-ab}{(a+b)(x^2+a^2)(x^2+b^2)}$
(2.17)	$\dfrac{x}{(x^4+a^4)}, \quad a>0$	$\dfrac{x^2-a^2}{(x^4+a^4)a\sqrt{2}}$
(2.18)	$\dfrac{x^2}{(x^4+a^4)}, \quad a>0$	$\dfrac{x(x^2-a^2)}{(x^4+a^4)a\sqrt{2}}$
(2.19)	$\dfrac{x^2}{(x^2+a^2)^2}, \quad a>0$	$\dfrac{x(x^2-a^2)}{2a(x^2+a^2)^2}$
(2.20)	$\dfrac{x^2}{(x^2+a^2)(x^2+b^2)}, \quad a>0, b>0$	$\dfrac{x(x^2-ab)}{(a+b)(x^2+a^2)(x^2+b^2)}$
(2.21)	$\dfrac{x^3}{(x^4+a^4)}, \quad a>0$	$-\dfrac{a(x^2+a^2)}{\sqrt{(2)}\,(x^4+a^4)}$
(2.22)	$\dfrac{x^3}{(x^2+a^2)^2}, \quad a>0$	$-\dfrac{a(3x^2+a^2)}{2(x^2+a^2)^2}$
(2.23)	$\dfrac{x^3}{(x^2+a^2)(x^2+b^2)}, \quad a>0, b>0$	$\dfrac{-x^2\{a^2+b^2+ab\}-a^2b^2}{(a+b)(x^2+a^2)(x^2+b^2)}$

Number	$f(x)$	$\frac{1}{\pi}P\int_{-\infty}^{\infty}\frac{f(s)\mathrm{d}s}{x-s}$
(2.24)	$\frac{ax^3+bx^2+cx+d}{(1+x^4)}$	$[(x^4+1)\sqrt{2}]^{-1}\{x^3(b+d)-x^2(a-c)-x(b-d)-a-c\}$
(2.25)	$\frac{ax^3+bx^2+cx+d}{(1+x^2)^2}$	$[2(1+x^2)^2]^{-1}\{x^3(b+d)+x^2(c-3a)+x(3d-b)-a-c\}$
(2.26)	$\frac{x(1-x^2)}{x^4+(a^2-2)x^2+1}, \quad a>0$	$\frac{ax^2}{x^4+(a^2-2)x^2+1}$
(2.27)	$\frac{1}{(x^6+a^6)}, \quad a>0$	$\frac{x(2x^4+a^2x^2+2a^4)}{3a^5(x^6+a^6)}$
(2.28)	$\frac{1}{(x^2+a^2)^3}, \quad a>0$	$\frac{x(3x^4+10a^2x^2+15a^4)}{8a^5(x^2+a^2)^3}$
(2.29)	$\frac{1}{(x^4+a^4)(x^2+b^2)}, \quad a>0, \quad b>0$	$[2a^3b(a^4+b^4)(x^4+a^4)(x^2+b^2)]^{-1}\{a^2(2a^5+\sqrt{(2)}\,a^2b^3+\sqrt{(2)}\,b^5)x$ $+b\sqrt{(2)}\,(a^4+b^4)x^3+(2a^3-\sqrt{(2)}\,a^2b+\sqrt{(2)}\,b^3)x^5\}$
(2.30)	$\frac{x}{(x^4+1)(x^2+1)},$	$\frac{\sqrt{2}\,x^2(x^2+1)-(x^4+1)}{2(x^2+1)(x^4+1)}$
(2.31)	$[(x^2+a^2)(x^2+b^2)(x^2+c^2)]^{-1},$ $a>0, \quad b>0, \quad c>0$	$[abc(a+b)(b+c)(c+a)(x^2+a^2)(x^2+b^2)(x^2+c^2)]^{-1}x[(a+b+c)x^4$ $+(a^3+b^3+c^3+a^2b+a^2c+b^2a+b^2c+c^2a+c^2b+abc)x^2+a^3b^2$ $+a^3c^2+b^3a^2+b^3c^2+c^3a^2+c^3b^2+a^3bc+b^3ac+c^3ab+2(a^2b^2c$ $+a^2c^2b+b^2c^2a)]$
(2.32)	$\frac{x}{(x^6+a^6)}, \quad a>0$	$\frac{x^4+2a^2x^2-2a^4}{3a^3(x^6+a^6)}$
(2.33)	$\frac{x}{(x^2+a^2)^3}, \quad a>0$	$\frac{x^4+6a^2x^2-3a^4}{8a^3(x^2+a^2)^3}$

Table 1.2. (*Cont.*)

Number	$f(x)$	$\frac{1}{\pi} P \int_{-\infty}^{\infty} \frac{f(s)\mathrm{d}s}{x-s}$
(2.34)	$\frac{x^2}{(x^6+a^6)}, \quad a>0$	$\frac{x(x^4+2a^2x^2-2a^4)}{3a^3(x^6+a^6)}$
(2.35)	$\frac{x^2}{(x^2+a^2)^3}, \quad a>0$	$\frac{x(x^4+6a^2x^2-3a^4)}{8a^3(x^2+a^2)^3}$
(2.36)	$\frac{a(3x^2-a^2)}{(a^2+x^2)^3}, \quad \mathrm{Re}\, a>0$	$\frac{(x^3-3a^2x)}{(a^2+x^2)^3}$
(2.37)	$\frac{x^3}{(x^6+a^6)}, \quad a>0$	$\frac{2x^4-2a^2x^2-a^4}{3a(x^6+a^6)}$
(2.38)	$\frac{x^3}{(x^2+a^2)^3}, \quad a>0$	$\frac{3x^4-6a^2x^2-a^4}{8a(x^2+a^2)^3}$
(2.39)	$\frac{x^4}{(x^6+a^6)}, \quad a>0$	$\frac{x(2x^4-2a^2x^2-a^4)}{3a(x^6+a^6)}$
(2.40)	$\frac{x^4}{(x^2+a^2)^3}, \quad a>0$	$\frac{x(3x^4-6a^2x^2-a^4)}{8a(x^2+a^2)^3}$
(2.41)	$\frac{x^5}{(x^6+a^6)}, \quad a>0$	$-\frac{a(2x^4+a^2x^2+2a^4)}{3(x^6+a^6)}$
(2.42)	$\frac{x^5}{(x^2+a^2)^3}, \quad a>0$	$-\frac{a(15x^4+10a^2x^2+3a^4)}{8(x^2+a^2)^3}$
(2.43)	$(1+x^6)^{-1}\{ax^5+bx^4 + cx^3+dx^2+ex+f\}$	$[3(1+x^6)]^{-1}\{(2b+d+2f)x^5+(2c-2a+e)x^4+(2d-2b+f)x^3 +(2e-a-2c)x^2+(2f-b-2d)x-2a-c-2e\}$
(2.44)	$[(1+x^2)^{-3}]\{ax^5+bx^4 + cx^3+dx^2+ex+f\}$	$[8(1+x^2)]^{-3}\{(3b+d+3f)x^5-(15a-3c-e)x^4-(6b-6d-10f)x^3 -(10a+6c-6e)x^2-(b+3d-15f)x-(3a+c+3e)\}$

Number	$f(x)$	$\frac{1}{\pi}P\int_{-\infty}^{\infty}\frac{f(s)\mathrm{d}s}{x-s}$
(2.45)	$[(1+x^4)(1+x^2)]^{-1}\{ax^5+bx^4 + cx^3+dx^2+ex+f\}$	$[2(1+x^4)(1+x^2)]^{-1}\{(b+(\surd(2)-1)d+f)x^5+(c-(1+\surd(2))a + (\surd(2)-1)e)x^4+\surd(2)(d+f-b)x^3+\surd(2)(e-a-c)x^2 + ((1-\surd(2))b-d+(1+\surd2)f)x-a-(\surd(2)-1)c-e\}$
(2.46)	$\frac{1}{x^8+a^8},\quad a>0$	$\frac{x(2+\surd 2)\surd[2-\surd 2]}{4a(x^8+a^8)}\left\{1+(\surd(2)-1)\left(\frac{x^2}{a^2}+\frac{x^4}{a^4}\right)+\frac{x^6}{a^6}\right\}$
(2.47)	$\frac{x}{x^8+a^8},\quad a>0$	$\frac{a(2+\surd 2)\surd[2-\surd 2]}{4(x^8+a^8)}\left\{-1+\frac{x^2}{a^2}+(\surd(2)-1)\left(\frac{x^4}{a^4}+\frac{x^6}{a^6}\right)\right\}$
(2.48)	$\frac{x^2}{x^8+a^8},\quad a>0$	$\frac{ax(2+\surd 2)\surd[2-\surd 2]}{4(x^8+a^8)}\left\{-1+\frac{x^2}{a^2}+(\surd(2)-1)\left(\frac{x^4}{a^4}+\frac{x^6}{a^6}\right)\right\}$
(2.49)	$\frac{x^3}{x^8+a^8},\quad a>0$	$\frac{a^3(2+\surd 2)\surd[2-\surd 2]}{4(x^8+a^8)}\left\{1-\surd(2)-\frac{x^2}{a^2}+\frac{x^4}{a^4}+(\surd(2)-1)\frac{x^6}{a^6}\right\}$
(2.50)	$\frac{4ax(a^2-x^2)}{(a^2+x^2)^4},\quad \mathrm{Re}\, a>0$	$\frac{-(x^4-6a^2x^2+a^4)}{(a^2+x^2)^4}$
(2.51)	$\frac{x^4}{x^8+a^8},\quad a>0$	$\frac{a^3x(2+\surd 2)\surd[2-\surd 2]}{4(x^8+a^8)}\left\{1-\surd(2)-\frac{x^2}{a^2}+\frac{x^4}{a^4}+(\surd(2)-1)\frac{x^6}{a^6}\right\}$
(2.52)	$\frac{x^5}{x^8+a^8},\quad a>0$	$\frac{a^5(2+\surd 2)\surd[2-\surd 2]}{4(x^8+a^8)}\left\{(1-\surd 2)\left(1+\frac{x^2}{a^2}\right)-\frac{x^4}{a^4}+\frac{x^6}{a^6}\right\}$
(2.53)	$\frac{a(a^4-10a^2x^2+5x^4)}{(a^2+x^2)^5},\quad \mathrm{Re}\, a>0$	$\frac{(x^5-10a^2x^3+5a^4x)}{(a^2+x^2)^5}$
(2.54)	$\frac{1}{\surd(x^2+a^2)},\quad a>0$	$\frac{2\sinh^{-1}(x/a)}{\pi\surd(x^2+a^2)}$
(2.55)	$\frac{1}{\surd(\lvert x\rvert+a)},\quad a>0$	$\frac{2\,\mathrm{sgn}\, x}{\pi}\left[\frac{\sec^{-1}(\surd(\lvert x\rvert/a))}{\surd(\lvert x\rvert-a)}+\frac{\log(\surd\lvert x\rvert/[\surd(\lvert x\rvert+a)+\surd a])}{\surd(\lvert x\rvert+a)}\right]$

Table 1.2. (*Cont.*)

Number	$f(x)$	$\frac{1}{\pi} P \int_{-\infty}^{\infty} \frac{f(s)\mathrm{d}s}{x-s}$
(2.56)	$(x^2+a^2)^{-v}, \ a>0, v>0$	$\frac{2x\Gamma(v+1/2)}{\sqrt{(\pi)}\,\Gamma(v)a^{1+2v}}\, {}_2F_1\left(v+1/2, 1; 3/2; -x^2/a^2\right)$
(2.57)	$\frac{\sqrt{\lvert x\rvert}}{a^2+x^2}, \ a>0$	$\frac{\sqrt{(2a^{-1})}\,x-\sqrt{\lvert x\rvert}}{(a^2+x^2)}$
(2.58)	$\sqrt{(\lvert a-x\rvert)}-\sqrt{(\lvert b-x\rvert)}, \ a>0, \ b>0$	$-\sqrt{(b-x)}+\sqrt{(a-x)}, \quad -\infty<x<a$ $-\sqrt{(b-x)}-\sqrt{(x-a)}, \quad a<x<b$ $\sqrt{(x-b)}-\sqrt{(x-a)}, \quad b<x<\infty$
(2.59)	$\lvert x\rvert^{\mu}, -1<\mathrm{Re}\,\mu<0$	$-\tan(\mu\pi/2)\,\mathrm{sgn}\,x\,\lvert x\rvert^{\mu}$
(2.60)	$\mathrm{sgn}\,x\,\lvert x\rvert^{\mu}, \ -1<\mathrm{Re}\,\mu<0$	$\cot(\mu\pi/2)\ \lvert x\rvert^{\mu}$
(2.61)	x^{-1} (interpreted in the distributional sense, that is $P(x^{-1})$ or $p.v.(x^{-1})$; a similar comment applies to the following four entries)	$-\pi\,\delta(x)$
(2.62)	x^{-2}	$\pi\,\delta'(x)$
(2.63)	x^{-3}	$-\frac{\pi\,\delta''(x)}{2}$
(2.64)	x^{-4}	$\frac{\pi\,\delta'''(x)}{6}$
(2.65)	$x^{-n-1}, \ n=0,1,2,\ldots$	$\frac{(-1)^{n+1}\pi\,\delta^{n'}(x)}{n!}$ (n' signifies the nth derivative)

Table 1.3. *Exponential functions*

Number	$f(x)$	$\frac{1}{\pi}P\int_{-\infty}^{\infty}\frac{f(s)\mathrm{d}s}{x-s}$
(3.1)	e^{iax}	$-\mathrm{i}\,\mathrm{sgn}\,a\,\mathrm{e}^{iax}$
(3.2)	$\frac{\mathrm{e}^{iax}}{x}$	$-\pi\delta(x)+\frac{\mathrm{i}\,\mathrm{sgn}\,a}{x}(1-\mathrm{e}^{iax})$
(3.3)	$\mathrm{e}^{-a\|x\|},\ a>0$	$\frac{\mathrm{sgn}\,x}{\pi}[\mathrm{e}^{a\|x\|}E_1(a\,\|x\|)+\mathrm{e}^{-a\|x\|}\mathrm{Ei}(a\|x\|)]$
(3.4)	$\mathrm{sgn}\,x\,\mathrm{e}^{-a\|x\|},\ a>0$	$\frac{1}{\pi}[\mathrm{e}^{-a\|x\|}\mathrm{Ei}(a\,\|x\|)-\mathrm{e}^{a\|x\|}E_1(a\|x\|)]$
(3.5)	$\mathrm{e}^{-ax^2},\ a\geq 0$	$-\mathrm{i}\mathrm{e}^{-ax^2}\mathrm{erf}\,(\mathrm{i}\sqrt{(a)}\,x)$ $=2\sqrt{\left(\frac{a}{\pi}\right)}x\mathrm{e}^{-ax^2}{}_1F_1(1/2;3/2;ax^2)$ $=2\sqrt{\left(\frac{a}{\pi}\right)}x{}_1F_1(1;3/2;-ax^2)$
(3.6)	$f(x)=\mathrm{e}^{-ax^{2n}}$, $n=1,2,\ldots$, and $a>0$,	$\frac{2a}{\pi}\mathrm{e}^{-ax^{2n}}\sum_{k=0}^{n-1}M_{nk}(a)E_{n(n-k-1)}(a,x)$, $M_{nk}(a)=\frac{\Gamma\,((2k+1)/2n)}{a^{(2k+1)/2n}}$, $E_{nk}(a,x)=\int_0^x t^{2k}\mathrm{e}^{at^{2n}}\,\mathrm{d}t$
(3.7)	$x^{2n}\mathrm{e}^{-ax^2},\ a>0$	$x^{2n}G(a,x)-\frac{1}{\pi\sqrt{a}}\sum_{k=0}^{n-1}\frac{x^{2n-2k-1}\Gamma(k+1/2)}{a^k}$, where $G(a,x)=H[\mathrm{e}^{-ax^2}]$
(3.8)	$x^{2n+1}\mathrm{e}^{-ax^2},\ a>0$	$x^{2n+1}G(a,x)-\frac{1}{\pi\sqrt{a}}\sum_{k=0}^{n}\frac{x^{2n-2k}\Gamma(k+1/2)}{a^k}$, where $G(a,x)=H[\mathrm{e}^{-ax^2}]$
(3.9)	$\mathrm{e}^{-ax^2-bx},\ a>0$	$-\,\mathrm{i}\mathrm{e}^{b^2/4a}\mathrm{e}^{-ax^2}\,\mathrm{erf}[\mathrm{i}\sqrt{(a)}\,(x+b/2a)]$
(3.10)	$\frac{\mathrm{e}^{-ax^2}}{b^2+x^2},\ a\geq 0,\ b>0$	$(x^2+b^2)^{-1}[x\mathrm{e}^{ab^2}b^{-1}\{1-\mathrm{erf}\,(b\sqrt{a})\}$ $-\,\mathrm{i}\mathrm{e}^{-ax^2}\,\mathrm{erf}\,(\mathrm{i}x\sqrt{a})]$
(3.11)	$\mathrm{e}^{iax^{-1}}$	$\mathrm{i}\,\mathrm{sgn}\,a\{\mathrm{e}^{iax^{-1}}-1\}$
(3.12)	$\mathrm{e}^{\mathrm{i}(ax-bx^{-1})},\ a>0,\ b>0$	$-\mathrm{i}\mathrm{e}^{\mathrm{i}(ax-bx^{-1})}$

Table 1.4. *Hyperbolic functions*

In the following order: coth, csch, (coth, csch), sech, tanh, (sech, tanh)

Number	$f(x)$	$\frac{1}{\pi}P\int_{-\infty}^{\infty}\frac{f(s)\mathrm{d}s}{x-s}$
(4.1)	$3x^{-3}+x^{-1}-3x^{-2}\coth x$	$-\frac{6}{\pi}\sum_{k=1}^{\infty}\frac{1}{k(\pi^2k^2+x^2)}$ $=-\frac{3}{\pi x^2}\left\{2\gamma+\psi\left(1+\frac{\mathrm{i}x}{\pi}\right)+\psi\left(1-\frac{\mathrm{i}x}{\pi}\right)\right\}$
(4.2)	$3x^{-4}+x^{-2}-3x^{-3}\coth x$	$\frac{6x}{\pi^3}\sum_{k=1}^{\infty}\frac{1}{k^3(\pi^2k^2+x^2)}$
(4.3)	$\operatorname{csch} x-x^{-1}$	$-2\pi\sum_{k=1}^{\infty}\frac{(-1)^k k}{\pi^2k^2+x^2}$ $=-\mathrm{i}\{\operatorname{csch} x+x^{-1}\}$ $+\pi^{-1}\left\{\psi\left(\frac{\mathrm{i}x}{2\pi}\right)-\psi\left(\frac{\mathrm{i}x}{2\pi}+\frac{1}{2}\right)\right\}$
(4.4)	$\frac{x\operatorname{csch} x-1}{x^2}$	$\frac{2x}{\pi}\sum_{k=1}^{\infty}\frac{(-1)^k}{k(\pi^2k^2+x^2)}$ $=\frac{1}{\pi x}\left[\psi\left(\frac{\mathrm{i}x}{2\pi}\right)-\psi\left(\frac{\mathrm{i}x}{2\pi}+\frac{1}{2}\right)-\log 4\right.$ $\left.-\pi\mathrm{i}\{x^{-1}+\operatorname{csch} x\}\right]$
(4.5)	$6x^{-3}-x^{-1}-6x^{-2}\operatorname{csch} x$	$-\frac{12}{\pi}\sum_{k=1}^{\infty}\frac{(-1)^k}{k(\pi^2k^2+x^2)}$ $=\frac{6}{\pi x^2}\left[\psi\left(\frac{\mathrm{i}x}{2\pi}+\frac{1}{2}\right)-\psi\left(\frac{\mathrm{i}x}{2\pi}\right)\right.$ $\left.+\log 4+\pi\mathrm{i}\{x^{-1}+\operatorname{csch} x\}\right]$
(4.6)	$6x^{-4}-x^{-2}-6x^{-3}\operatorname{csch} x$	$\frac{12x}{\pi^3}\sum_{k=1}^{\infty}\frac{(-1)^k}{k^3(\pi^2k^2+x^2)}$ $=\frac{6\log 4}{\pi x^3}-\frac{9\varsigma(3)}{\pi^3 x}+\frac{6}{\pi x^3}\left\{\psi\left(\frac{1}{2}+\frac{\mathrm{i}x}{2\pi}\right)\right.$ $\left.-\psi\left(\frac{\mathrm{i}x}{2\pi}\right)+\pi\mathrm{i}\left[x^{-1}+\operatorname{csch} x\right]\right\}$
(4.7)	$x^{-1}\coth x-\operatorname{csch}^2 x$	$\frac{2x}{\pi}\sum_{k=1}^{\infty}\frac{x^2+3\pi^2k^2}{k(\pi^2k^2+x^2)^2}$
(4.8)	$x^{-2}-\operatorname{csch} x\coth x$	$4\pi x\sum_{k=1}^{\infty}\frac{(-1)^k k}{(\pi^2k^2+x^2)^2}$ $=\mathrm{i}\{x^{-2}+\operatorname{csch} x\coth x\}$ $+\frac{\mathrm{i}}{2\pi^2}\left\{\psi^{(1)}\left(\frac{\mathrm{i}x}{2\pi}\right)-\psi^{(1)}\left(\frac{1}{2}+\frac{\mathrm{i}x}{2\pi}\right)\right\}$

Table 1.4. (*Cont.*)

Number	$f(x)$	$\frac{1}{\pi}P\int_{-\infty}^{\infty}\frac{f(s)\mathrm{d}s}{x-s}$
(4.9)	$\operatorname{csch} x\{1 - x\coth x\}$	$2\pi\sum_{k=1}^{\infty}\frac{(-1)^k k(x^2-\pi^2k^2)}{(\pi^2k^2+x^2)^2}$
(4.10)	$x^{-1}\operatorname{csch} x\{1 - x\coth x\}$	$\frac{2x}{\pi}\sum_{k=1}^{\infty}\frac{(-1)^k(x^2+3\pi^2k^2)}{k(\pi^2k^2+x^2)^2}$
(4.11)	$x^{-1}\coth x + \operatorname{csch}^2 x$ $-2x\operatorname{csch}^2 x\coth x$	$\frac{2x}{\pi}\sum_{k=1}^{\infty}\frac{x^4+6\pi^2k^2x^2-3\pi^4k^4}{k(\pi^2k^2+x^2)^3}$
(4.12)	$x^{-2}\coth x + x^{-1}\operatorname{csch}^2 x$ $-2\operatorname{csch}^2 x\coth x$	$\frac{2}{\pi}\sum_{k=1}^{\infty}\frac{x^4+6\pi^2k^2x^2-3\pi^4k^4}{k(\pi^2k^2+x^2)^3}$
(4.13)	$x^{-3}\coth x + x^{-2}\operatorname{csch}^2 x$ $-2x^{-1}\operatorname{csch}^2 x\coth x$	$\frac{2x}{\pi^3}\sum_{k=1}^{\infty}\frac{3x^4+10\pi^2k^2x^2+15\pi^4k^4}{k^3(\pi^2k^2+x^2)^3}$
(4.14)	$\operatorname{csch} x\{x + x^2\coth x$ $-x^3 - 2x^3\operatorname{csch}^2 x\}$	$2\pi x\sum_{k=1}^{\infty}(-1)^k k(\pi^2k^2+x^2)^{-3}$ $\times(3x^4 - 6\pi^2k^2x^2 - \pi^4k^4)$
(4.15)	$\operatorname{csch} x\{1 + x\coth x$ $-x^2 - 2x^2\operatorname{csch}^2 x\}$	$2\pi\sum_{k=1}^{\infty}(-1)^k k(\pi^2k^2+x^2)^{-3}$ $\times(3x^4-6\pi^2k^2x^2-\pi^4k^4)$
(4.16)	$\operatorname{csch} x\{x^{-1} + \coth x$ $-x - 2x\operatorname{csch}^2 x\}$	$\frac{2x}{\pi}\sum_{k=1}^{\infty}\frac{(-1)^k(x^4+6\pi^2k^2x^2-3\pi^4k^4)}{k(\pi^2k^2+x^2)^3}$
(4.17)	$\operatorname{csch} x\{x^{-2} + x^{-1}\coth x$ $-1 - 2\operatorname{csch}^2 x\}$	$\frac{2}{\pi}\sum_{k=1}^{\infty}\frac{(-1)^k(x^4+6\pi^2k^2x^2-3\pi^4k^4)}{k(\pi^2k^2+x^2)^3}$
(4.18)	$\operatorname{csch} x\{x^{-3} + x^{-2}\coth x$ $-x^{-1} - 2x^{-1}\operatorname{csch}^2 x\}$	$\frac{2x}{\pi^3}\sum_{k=1}^{\infty}(-1)^k k^{-3}(\pi^2k^2+x^2)^{-3}$ $\times(3x^4+10\pi^2k^2x^2+15\pi^4k^4)$
(4.19)	$2x^{-1} + 12x^{-3}$ $-9x^{-2}\coth x$ $-3x^{-1}\operatorname{csch}^2 x$	$-\frac{6}{\pi}\sum_{k=1}^{\infty}\frac{3x^2+\pi^2k^2}{k(\pi^2k^2+x^2)^2}$
(4.20)	$2x^{-2} + 12x^{-4}$ $-9x^{-3}\coth x$ $-3x^{-2}\operatorname{csch}^2 x$	$\frac{6x}{\pi^3}\sum_{k=1}^{\infty}\frac{x^2-\pi^2k^2}{k^3(\pi^2k^2+x^2)^2}$
(4.21)	$2x^{-3} + 12x^{-5}$ $-9x^{-4}\coth x$ $-3x^{-3}\operatorname{csch}^2 x$	$\frac{6}{\pi^3}\sum_{k=1}^{\infty}\frac{x^2-\pi^2k^2}{k^3(\pi^2k^2+x^2)^2}$

Table 1.4. (*Cont.*)

Number	$f(x)$	$\frac{1}{\pi}P\int_{-\infty}^{\infty}\frac{f(s)\mathrm{d}s}{x-s}$
(4.22)	$2x^{-4}+12x^{-6}$ $-9x^{-5}\coth x$ $-3x^{-4}\operatorname{csch}^2 x$	$\frac{6x}{\pi^5}\sum_{k=1}^{\infty}\frac{3\pi^2k^2+x^2}{k^5(\pi^2k^2+x^2)^2}$
(4.23)	$12x^{-3}-x^{-1}$ $-9x^{-2}\operatorname{csch} x$ $-3x^{-1}\operatorname{csch} x\coth x$	$-\frac{6}{\pi}\sum_{k=1}^{\infty}\frac{(-1)^k(\pi^2k^2+3x^2)}{k(\pi^2k^2+x^2)^2}$
(4.24)	$12x^{-4}-x^{-2}$ $-9x^{-3}\operatorname{csch} x$ $-3x^{-2}\operatorname{csch} x\coth x$	$\frac{6x}{\pi^3}\sum_{k=1}^{\infty}\frac{(-1)^k(x^2-\pi^2k^2)}{k^3(\pi^2k^2+x^2)^2}$
(4.25)	$12x^{-5}-x^{-3}$ $-9x^{-4}\operatorname{csch} x$ $-3x^{-3}\operatorname{csch} x\coth x$	$\frac{6}{\pi^3}\sum_{k=1}^{\infty}\frac{(-1)^k(x^2-\pi^2k^2)}{k^3(\pi^2k^2+x^2)^2}$
(4.26)	$12x^{-6}-x^{-4}$ $-9x^{-5}\operatorname{csch} x$ $-3x^{-4}\operatorname{csch} x\coth x$	$\frac{6x}{\pi^5}\sum_{k=1}^{\infty}\frac{(-1)^k(3\pi^2k^2+x^2)}{k^5(\pi^2k^2+x^2)^2}$
(4.27)	$\operatorname{sech} x$	$8x\sum_{k=0}^{\infty}\frac{(-1)^k}{(2k+1)^2\pi^2+4x^2}$ $=\frac{\mathrm{i}}{\pi}\left\{\psi\left(\frac{1}{4}-\frac{\mathrm{i}x}{2\pi}\right)-\psi\left(\frac{1}{4}+\frac{\mathrm{i}x}{2\pi}\right)\right\}-\tanh x$
(4.28)	$\operatorname{sech}^2 x$	$32\pi x\sum_{k=0}^{\infty}\frac{2k+1}{[(2k+1)^2\pi^2+4x^2]^2}$
(4.29)	$\frac{1-\operatorname{sech} x}{x}$	$-8\sum_{k=0}^{\infty}\frac{(-1)^k}{(2k+1)^2\pi^2+4x^2}$ $=\frac{\mathrm{i}}{\pi x}\left\{\psi\left(\frac{1}{4}+\frac{\mathrm{i}x}{2\pi}\right)-\psi\left(\frac{1}{4}-\frac{\mathrm{i}x}{2\pi}\right)\right\}$ $+x^{-1}\tanh x$
(4.30)	$\frac{1-\operatorname{sech} x}{x^2}$	$\frac{32x}{\pi^2}\sum_{k=0}^{\infty}\frac{(-1)^k}{(2k+1)^2[(2k+1)^2\pi^2+4x^2]}$ $=\frac{\mathrm{i}}{\pi x^2}\left\{\pi\operatorname{sech} x+\psi\left(\frac{1}{4}+\frac{\mathrm{i}x}{2\pi}\right)\right.$ $\left.-\psi\left(\frac{3}{4}+\frac{\mathrm{i}x}{2\pi}\right)\right\}$ $+\frac{8\beta(2)}{\pi^2 x}$

Table 1.4. (*Cont.*)

Number	$f(x)$	$\frac{1}{\pi}P\int_{-\infty}^{\infty}\frac{f(s)\mathrm{d}s}{x-s}$
(4.31)	$\frac{\tanh x}{x}$	$\frac{16x}{\pi}\sum_{k=0}^{\infty}\frac{1}{(2k+1)\{(2k+1)^2\pi^2+4x^2\}} = \frac{2}{\pi x}\left\{\gamma+\log 4+\psi\left(\frac{1}{2}+\frac{\mathrm{i}x}{\pi}\right) - \frac{\mathrm{i}\pi}{2}\tanh x\right\}$
(4.32)	$\frac{x-\tanh x}{x^2}$	$-\frac{16}{\pi}\sum_{k=0}^{\infty}\frac{1}{(2k+1)\{(2k+1)^2\pi^2+4x^2\}} = -\frac{2}{\pi x^2}\left\{\gamma+\log 4+\psi\left(\frac{1}{2}+\frac{\mathrm{i}x}{\pi}\right) - \frac{\mathrm{i}\pi}{2}\tanh x\right\}$
(4.33)	$\frac{x-\tanh x}{x^3}$	$\frac{64x}{\pi^3}\sum_{k=0}^{\infty}\frac{1}{(2k+1)^3\{(2k+1)^2\pi^2+4x^2\}}$
(4.34)	$x\operatorname{sech} x\tanh x$	$8x\sum_{k=0}^{\infty}\frac{(-1)^k(4x^2-(2k+1)^2\pi^2)}{\{(2k+1)^2\pi^2+4x^2\}^2}$
(4.35)	$\operatorname{sech} x\tanh x$	$-8\sum_{k=0}^{\infty}\frac{(-1)^k\{(2k+1)^2\pi^2-4x^2\}}{\{(2k+1)^2\pi^2+4x^2\}^2}$
(4.36)	$\frac{\operatorname{sech} x\tanh x}{x}$	$\frac{32x}{\pi^2}\sum_{k=0}^{\infty}\frac{(-1)^k\{3(2k+1)^2\pi^2+4x^2\}}{(2k+1)^2\{(2k+1)^2\pi^2+4x^2\}^2}$
(4.37)	$x^{-1}\tanh x+\operatorname{sech}^2 x$	$\frac{16x}{\pi}\sum_{k=0}^{\infty}\frac{3(2k+1)^2\pi^2+4x^2}{(2k+1)\{(2k+1)^2\pi^2+4x^2\}^2}$
(4.38)	$\frac{\tanh x-x\operatorname{sech}^2 x}{x}$	$\frac{16x}{\pi}\sum_{k=0}^{\infty}\frac{4x^2-(2k+1)^2\pi^2}{(2k+1)\{(2k+1)^2\pi^2+4x^2\}^2}$
(4.39)	$\frac{\tanh x-x\operatorname{sech}^2 x}{x^2}$	$\frac{16}{\pi}\sum_{k=0}^{\infty}\frac{4x^2-(2k+1)^2\pi^2}{(2k+1)\{(2k+1)^2\pi^2+4x^2\}^2}$
(4.40)	$\frac{\tanh x-x\operatorname{sech}^2 x}{x^3}$	$\frac{64x}{\pi^3}\sum_{k=0}^{\infty}\frac{3(2k+1)^2\pi^2+4x^2}{(2k+1)^3\{(2k+1)^2\pi^2+4x^2\}^2}$

Table 1.4. (*Cont.*)

Number	$f(x)$	$\frac{1}{\pi}P\int_{-\infty}^{\infty}\frac{f(s)\mathrm{d}s}{x-s}$
(4.41)	$x^{-3}\{\tanh x - x\,\mathrm{sech}^2 x + 2x^2\,\mathrm{sech}^2 x \tanh x\}$	$\frac{64x}{\pi^3}\sum_{k=0}^{\infty}[(2k+1)\{\pi^2(2k+1)^2+4x^2\}]^{-3} \times\{48x^4+40\pi^2(2k+1)^2x^2 + 15\pi^4(2k+1)^4\}$
(4.42)	$2x^{-1}(1-\mathrm{sech}\,x) - \mathrm{sech}\,x\tanh x$	$-8\sum_{k=0}^{\infty}\frac{(-1)^k\{12x^2+\pi^2(2k+1)^2\}}{\{(2k+1)^2\pi^2+4x^2\}^2}$
(4.43)	$2x^{-2}(1-\mathrm{sech}\,x) - x^{-1}\,\mathrm{sech}\,x\tanh x$	$\frac{32x}{\pi^2}\sum_{k=0}^{\infty}[(2k+1)\{(2k+1)^2\pi^2 + 4x^2\}]^{-2}(-1)^k\{4x^2-\pi^2(2k+1)^2\}$
(4.44)	$2x^{-3}(1-\mathrm{sech}\,x) - x^{-2}\,\mathrm{sech}\,x\tanh x$	$\frac{32}{\pi^2}\sum_{k=0}^{\infty}[(2k+1)\{(2k+1)^2\pi^2 + 4x^2\}]^{-2}(-1)^k\{4x^2-\pi^2(2k+1)^2\}$
(4.45)	$2x^{-4}(1-\mathrm{sech}\,x) - x^{-3}\,\mathrm{sech}\,x\tanh x$	$\frac{128x}{\pi^4}\sum_{k=0}^{\infty}[(2k+1)^2\{(2k+1)^2\pi^2 + 4x^2\}]^{-2}(-1)^k\{4x^2+3\pi^2(2k+1)^2\}$
(4.46)	$2x^{-1} - 3x^{-2}\tanh x + x^{-1}\,\mathrm{sech}^2 x$	$-\frac{16}{\pi}\sum_{k=0}^{\infty}\frac{12x^2+\pi^2(2k+1)^2}{(2k+1)\{(2k+1)^2\pi^2+4x^2\}^2}$
(4.47)	$2x^{-2} - 3x^{-3}\tanh x + x^{-2}\,\mathrm{sech}^2 x$	$\frac{64x}{\pi^3}\sum_{k=0}^{\infty}\frac{4x^2-\pi^2(2k+1)^2}{(2k+1)^3\{(2k+1)^2\pi^2+4x^2\}^2}$
(4.48)	$2x^{-3} - 3x^{-4}\tanh x + x^{-3}\,\mathrm{sech}^2 x$	$\frac{64}{\pi^3}\sum_{k=0}^{\infty}\frac{4x^2-\pi^2(2k+1)^2}{(2k+1)^3\{(2k+1)^2\pi^2+4x^2\}^2}$
(4.49)	$2x^{-4} - 3x^{-5}\tanh x + x^{-4}\,\mathrm{sech}^2 x$	$\frac{256x}{\pi^5}\sum_{k=0}^{\infty}(2k+1)^{-5}\{(2k+1)^2\pi^2 + 4x^2\}^{-2}\{4x^2+3\pi^2(2k+1)^2\}$

Table 1.5. *Trigonometric functions*

Number	$f(x)$	$\frac{1}{\pi}P\int_{-\infty}^{\infty}\frac{f(s)\mathrm{d}s}{x-s}$
(5.1)	$\sin ax$	$-\operatorname{sgn} a\cos ax$
(5.2)	$\cos ax$	$\operatorname{sgn} a\sin ax = \sin\lvert a\rvert x$
(5.3)	$\operatorname{cas} ax$	$\operatorname{sgn} a\,(\sin ax - \cos ax) = \surd(2)\operatorname{sgn} a\sin(ax - \pi/4)$
(5.4)	$\sin(ax+b)$	$-\operatorname{sgn} a\cos(ax+b)$
(5.5)	$\cos(ax+b)$	$\operatorname{sgn} a\sin(ax+b)$
(5.6)	$\cos ax\cos bx$	$\frac{1}{2}\{\operatorname{sgn}(a+b)\sin[(a+b)x] + \operatorname{sgn}(a-b)\sin[(a-b)x]\}$ $= \begin{cases}\sin ax\cos bx, & a>b>0\\ \dfrac{\operatorname{sgn} a\sin 2ax}{2}, & a=b\\ \cos ax\sin bx, & b>a>0\end{cases}$
(5.7)	$\sin ax\sin bx$	$\frac{1}{2}\{\operatorname{sgn}(a-b)\sin[(a-b)x] - \operatorname{sgn}(a+b)\sin[(a+b)x]\}$ $= \begin{cases}-\cos ax\sin bx, & a>b>0\\ -\dfrac{\operatorname{sgn} a\sin 2ax}{2}, & a=b\\ -\sin ax\cos bx, & b>a>0\end{cases}$

Table 1.5. (*Cont.*)

Number	$f(x)$	$\frac{1}{\pi} P \int_{-\infty}^{\infty} \frac{f(s)\mathrm{d}s}{x-s}$
(5.8)	$\sin ax \cos bx$	$-\frac{1}{2}\{\operatorname{sgn}(a-b)\cos[(a-b)x] + \operatorname{sgn}(a+b)\cos[(a+b)x]\}$ $= \begin{cases} -\cos ax \cos bx, & a > b > 0 \\ -\frac{\operatorname{sgn} a \cos 2ax}{2}, & a = b \\ \sin ax \sin bx, & b > a > 0 \end{cases}$
(5.9)	$\cos(ax+c)\cos(bx+d)$	$\frac{1}{2}\{\operatorname{sgn}(a+b)\sin[(a+b)x+c+d] + \operatorname{sgn}(a-b)\sin[(a-b)x+c-d]\}$
(5.10)	$\sin(ax+c)\sin(bx+d)$	$\frac{1}{2}\{\operatorname{sgn}(a-b)\sin[(a-b)x+c-d] - \operatorname{sgn}(a+b)\sin[(a+b)x+c+d]\}$
(5.11)	$\sin(ax+c)\cos(bx+d)$	$-\frac{1}{2}\{\operatorname{sgn}(a+b)\cos[(a+b)x+c+d] + \operatorname{sgn}(a-b)\cos[(a-b)x+c-d]\}$
(5.12)	$\sin^2 ax$	$-\frac{1}{2}\operatorname{sgn} a \sin 2ax$
(5.13)	$\cos^2 ax$	$\frac{1}{2}\operatorname{sgn} a \sin 2ax$
(5.14)	$\sin^3 ax$	$\frac{\operatorname{sgn} a}{4}[\cos 3ax - 3\cos ax]$
(5.15)	$\cos^3 ax$	$\frac{\operatorname{sgn} a}{4}[\sin 3ax + 3\sin ax]$

Number	$f(x)$	$\frac{1}{\pi} P \int_{-\infty}^{\infty} \frac{f(s)\mathrm{d}s}{x-s}$
(5.16)	$\sin^4 ax$	$\frac{\operatorname{sgn} a}{8}[\sin 4ax - 4\sin 2ax]$
(5.17)	$\cos^4 ax$	$\frac{\operatorname{sgn} a}{8}[\sin 4ax + 4\sin 2ax]$
(5.18)	$\sin^5 ax$	$-\frac{\operatorname{sgn} a}{16}[\cos 5ax - 5\cos 3ax + 10\cos ax]$
(5.19)	$\cos^5 ax$	$\frac{\operatorname{sgn} a}{16}[\sin 5ax + 5\sin 3ax + 10\sin ax]$
(5.20)	$\sin^6 ax$	$-\frac{\operatorname{sgn} a}{32}[\sin 6ax - 6\sin 4ax + 15\sin 2ax]$
(5.21)	$\cos^6 ax$	$\frac{\operatorname{sgn} a}{32}[\sin 6ax + 6\sin 4ax + 15\sin 2ax]$
(5.22)	$\sin^{2n} ax, \quad n = 1, 2, \ldots$	$\frac{\operatorname{sgn} a}{2^{2n-1}} \sum_{k=0}^{n-1} (-1)^{n-k} \binom{2n}{k} \sin[2(n-k)ax]$
(5.23)	$\sin^{2n+1} ax, \quad n = 0, 1, 2, \ldots$	$\frac{\operatorname{sgn} a}{2^{2n}} \sum_{k=0}^{n} (-1)^{n+k+1} \binom{2n+1}{k} \cos[(2n-2k+1)ax]$
(5.24)	$\sin^{2n} ax \cos ax, \quad n = 0, 1, 2, \ldots$	$\frac{\operatorname{sgn} a}{4^n(2n+1)} \sum_{k=0}^{n} (-1)^k \binom{2n+1}{n-k} (2k+1)\sin[(2k+1)ax]$
(5.25)	$\sin^{2n} ax \cos bx, \quad n = 0, 1, 2, \ldots$	$4^{-n}\binom{2n}{n} \operatorname{sgn} b \sin bx + 4^{-n} \sum_{k=0}^{n-1} (-1)^{n+k} \binom{2n}{k} \{\operatorname{sgn}[b-2(n-k)a]$ $\times \sin[bx - 2(n-k)ax] + \operatorname{sgn}[b+2(n-k)a]\sin[bx+2(n-k)ax]\}$
(5.26)	$\sin^{2n-1} ax \cos ax, \quad n = 1, 2, \ldots$	$\frac{\operatorname{sgn} a}{2^{2n-1} n} \sum_{k=1}^{n} (-1)^k k \binom{2n}{n-k} \cos(2kax)$

Table 1.5. (*Cont.*)

Number	$f(x)$	$\frac{1}{\pi} P \int_{-\infty}^{\infty} \frac{f(s)\,ds}{x-s}$
(5.27)	$\sin^{2n-1} ax \cos bx, \quad n = 1, 2, \ldots$	$\frac{1}{2^{2n-1}} \sum_{k=0}^{n-1} (-1)^{n+k} \binom{2n-1}{k} \{\operatorname{sgn}[(2n-2k-1)a - b]$ $\times \cos[(2n-2k-1)ax - bx] + \operatorname{sgn}[(2n-2k-1)a+b] \cos[(2n-2k-1)ax+bx]\}$
(5.28)	$\cos^{2n} ax, \quad n = 1, 2, \ldots$	$\frac{\operatorname{sgn} a}{2^{2n-1}} \sum_{k=0}^{n-1} \binom{2n}{k} \sin[2(n-k)ax]$
(5.29)	$\cos^{2n-1} ax, \quad n = 1, 2, \ldots$	$\frac{\operatorname{sgn} a}{2^{2n-2}} \sum_{k=0}^{n-1} \binom{2n-1}{k} \sin[(2n-2k-1)ax]$
(5.30)	$\cos^{2n} ax \sin ax, \quad n = 0, 1, 2, \ldots$	$-\frac{\operatorname{sgn} a}{(2n+1)4^n} \sum_{k=0}^{n} (2k+1) \binom{2n+1}{n-k} \cos[(2k+1)ax]$
(5.31)	$\cos^{2n} ax \sin bx, \quad n = 0, 1, 2, \ldots$	$-\frac{1}{4^n} \Big\{ \binom{2n}{n} \operatorname{sgn} b \cos bx + \sum_{k=0}^{n-1} \{\operatorname{sgn}[b - 2(n-k)a] \cos[bx - 2(n-k)ax]$ $+ \operatorname{sgn}[b + 2(n-k)a] \cos[bx + 2(n-k)ax]\} \binom{2n}{k} \Big\}$
(5.32)	$\cos^{2n-1} ax \sin ax, \quad n = 1, 2, \ldots$	$-\frac{\operatorname{sgn} a}{2^{2n-1} n} \sum_{k=1}^{n} k \binom{2n}{n-k} \cos(2kax)$

Number	$f(x)$	$\frac{1}{\pi}P\int_{-\infty}^{\infty}\frac{f(s)\mathrm{d}s}{x-s}$
(5.33)	$\cos^{2n-1} ax \sin bx, \quad n=1,2,\dots$	$-\frac{1}{2^{2n-1}}\sum_{k=0}^{n-1}\binom{2n-1}{k}\{\operatorname{sgn}[b+a-2a(n-k)]\cos[\{b+a-2a(n-k)\}x]$ $+\operatorname{sgn}[b-a+2a(n-k)]\cos[\{b-a+2a(n-k)\}x]\}$
(5.34)	$\cos^{2n} ax \sin 2nax, \quad n=1,2,\dots$	$-\frac{\operatorname{sgn} a}{4^n}\left\{\binom{2n}{n}\cos 2nax+\cos 4nax+\sum_{k=1}^{n-1}\{\cos 2kax+\cos[(2n-k)2ax]\}\binom{2n}{k}\right\}$
(5.35)	$\cos^{2n-1} ax \sin(2n-1)ax,$ $n=1,2,\dots$	$-\frac{\operatorname{sgn} a}{2^{2n-1}}\left\{\cos(2n-1)2ax+\sum_{k=1}^{n-1}\binom{2n-1}{k}\{\cos 2kax+\cos[(2n-1-k)2ax]\}\right\}$
(5.36)	$\cos^{2n} ax \cos 2nax, \quad n=1,2,\dots$	$\frac{\operatorname{sgn} a}{4^n}\left\{\binom{2n}{n}\sin 2nax+\sin 4nax+\sum_{k=1}^{n-1}\{\sin 2kax+\sin[(2n-k)2ax]\}\binom{2n}{k}\right\}$
(5.37)	$\cos^{2n-1} ax \times \cos(2n-1)ax,$ $n=1,2,\dots$	$\frac{\operatorname{sgn} a}{2^{2n-1}}\left\{\sin(2n-1)2ax+\sum_{k=1}^{n-1}\binom{2n-1}{k}\{\sin 2kax+\sin[(2n-1-k)2ax]\}\right\}$
(5.38)	$\sin(a\|x\|)$	$\frac{2\operatorname{sgn}(ax)}{\pi}\{\sin(\|ax\|)\operatorname{Ci}(\|ax\|)-\cos(ax)\operatorname{Si}(\|ax\|)\}$
(5.39)	$\operatorname{sgn} x \cos ax$	$\frac{2}{\pi}\{\cos(ax)\operatorname{Ci}(\|ax\|)+\sin(\|ax\|)\operatorname{Si}(\|ax\|)\}$
(5.40)	$\operatorname{sgn} x \sin ax$	$\frac{2\operatorname{sgn}(ax)}{\pi}\{\sin(\|ax\|)\operatorname{Ci}(\|ax\|)-\cos(ax)\operatorname{Si}(\|ax\|)\}$
(5.41)	$\operatorname{sgn} x \sin(a\sqrt{\|x\|}), \quad a>0$	$-\cos(a\sqrt{\|x\|})-e^{-a\sqrt{\|x\|}}$

Table 1.5. (*Cont.*)

Number	$f(x)$	$\frac{1}{\pi} P \int_{-\infty}^{\infty} \frac{f(s)\mathrm{d}s}{x-s}$
(5.42)	$\operatorname{sinc} x$	$\frac{1-\cos \pi x}{\pi x}$
(5.43)	$\sin ax \operatorname{sinc} bx, \quad 0 < b\pi < a$	$-\cos ax \operatorname{sinc} bx$
(5.44)	$\cos ax \operatorname{sinc} bx, \quad 0 < b\pi < a$	$\sin ax \operatorname{sinc} bx$
(5.45)	$\sin ax \operatorname{sinc} bx$	$-\frac{1}{2}[\operatorname{sgn}(a+\pi b) + \operatorname{sgn}(a-\pi b)] \cos ax \operatorname{sinc} bx$ $- \frac{1}{2\pi bx}[\operatorname{sgn}(a+\pi b) - \operatorname{sgn}(a-\pi b)] \sin ax \cos b\pi x$
(5.46)	$\cos ax \operatorname{sinc} bx$	$\frac{1}{2\pi bx}[\operatorname{sgn}(\pi b + a)\{1-\cos[(\pi b + a)x]\}$ $+ \operatorname{sgn}(\pi b - a)\{1 - \cos[(\pi b - a)x]\}]$
(5.47)	$\frac{\sin ax}{x}$	$\frac{\operatorname{sgn} a}{x}(1-\cos ax)$
(5.48)	$\frac{\sin ax \sin bx}{x}, \quad 0 < b < a$	$-\frac{\cos ax \sin bx}{x}$
(5.49)	$\frac{\cos ax \sin bx}{x}, \quad 0 < b < a$	$\frac{\sin ax \sin bx}{x}$
(5.50)	$\frac{\sin^2 ax}{x}$	$-\frac{\operatorname{sgn} a \sin 2ax}{2x}$

Number	$f(x)$	$\frac{1}{\pi}P\int_{-\infty}^{\infty}\frac{f(s)\mathrm{d}s}{x-s}$
(5.51)	$\frac{\sin^2 ax}{x^2}$	$\frac{\operatorname{sgn} a}{2x^2}(2ax - \sin 2ax)$
(5.52)	$\frac{\sin ax \sin^2 bx}{x^2},\quad 0 < 2b < a$	$-\frac{\cos ax \sin^2 bx}{x^2}$
(5.53)	$\frac{\cos ax \sin^2 bx}{x^2},\quad 0 < 2b < a$	$\frac{\sin ax \sin^2 bx}{x^2}$
(5.54)	$\frac{\sin ax \sin \pi x}{1-x^2},\quad \pi < a$	$-\frac{\cos ax \sin \pi x}{1-x^2}$
(5.55)	$\frac{\cos ax \sin \pi x}{1-x^2},\quad \pi < a$	$\frac{\sin ax \sin \pi x}{1-x^2}$
(5.56)	$\frac{\sin^3 ax}{x}$	$\frac{\operatorname{sgn} a}{4x}(\cos 3ax - 3\cos ax + 2)$
(5.57)	$\frac{\sin ax \sin^2 bx}{x},\quad 0 < 2b < a$	$-\frac{\cos ax \sin^2 bx}{x}$
(5.58)	$\frac{\cos ax \sin^2 bx}{x},\quad 0 < 2b < a$	$\frac{\sin ax \sin^2 bx}{x}$
(5.59)	$\frac{\sin ax \sin bx \sin cx}{x},\quad 0 < c \le b,\ b + c < a$	$-\frac{\cos ax \sin bx \sin cx}{x}$
(5.60)	$\frac{\cos ax \sin bx \sin cx}{x},\quad 0 < c \le b,\ b + c < a$	$\frac{\sin ax \sin bx \sin cx}{x}$

Table 1.5. (*Cont.*)

Number	$f(x)$	$\frac{1}{\pi} P \int_{-\infty}^{\infty} \frac{f(s)\mathrm{d}s}{x-s}$
(5.61)	$\frac{\sin^3 ax}{x^2}$	$\frac{\operatorname{sgn} a}{4x^2}(\cos 3ax - 3\cos ax + 2)$
(5.62)	$\frac{\sin^3 ax}{x^3}$	$\frac{\operatorname{sgn} a}{4x^3}(\cos 3ax - 3\cos ax + 2 + 3a^2x^2)$
(5.63)	$\frac{\sin ax \sin^3 bx}{x^3}, \quad 0 < 3b < a$	$-\frac{\cos ax \sin^3 bx}{x^3}$
(5.64)	$\frac{\cos ax \sin^3 bx}{x^3}, \quad 0 < 3b < a$	$\frac{\sin ax \sin^3 bx}{x^3}$
(5.65)	$\frac{\sin^4 ax}{x}$	$\frac{\operatorname{sgn} a}{8x}(\sin 4ax - 4\sin 2ax)$
(5.66)	$\frac{\sin^4 ax}{x^2}$	$\frac{\operatorname{sgn} a}{8x^2}(\sin 4ax - 4\sin 2ax + 4ax)$
(5.67)	$\frac{\sin^4 ax}{x^3}$	$\frac{\operatorname{sgn} a}{8x^3}(\sin 4ax - 4\sin 2ax + 4ax)$
(5.68)	$\frac{\sin^4 ax}{x^4}$	$\frac{\operatorname{sgn} a}{8x^4}\left(\sin 4ax - 4\sin 2ax + 4ax + \frac{16}{3}a^3x^3\right)$
(5.69)	$\frac{\sin^5 ax}{x}$	$\frac{\operatorname{sgn} a}{16x}(6 - 10\cos ax + 5\cos 3ax - \cos 5ax)$

Number	$f(x)$	$\frac{1}{\pi} P \int_{-\infty}^{\infty} \frac{f(s)\mathrm{d}s}{x-s}$
(5.70)	$\frac{\sin^5 ax}{x^2}$	$\frac{\operatorname{sgn} a}{16x^2}(6 - 10\cos ax + 5\cos 3ax - \cos 5ax)$
(5.71)	$\frac{\sin^5 ax}{x^3}$	$\frac{\operatorname{sgn} a}{16x^3}(6 + 5a^2x^2 - 10\cos ax + 5\cos 3ax - \cos 5ax)$
(5.72)	$\frac{\sin^5 ax}{x^4}$	$\frac{\operatorname{sgn} a}{16x^4}(6 + 5a^2x^2 - 10\cos ax + 5\cos 3ax - \cos 5ax)$
(5.73)	$\frac{\sin^5 ax}{x^5}$	$\frac{\operatorname{sgn} a}{16x^5}\left(6 + 5a^2x^2 + \frac{115}{12}a^4x^4 - 10\cos ax + 5\cos 3ax - \cos 5ax\right)$
(5.74)	$\frac{\sin^6 ax}{x}$	$-\frac{\operatorname{sgn} a}{32x}(\sin 6ax - 6\sin 4ax + 15\sin 2ax)$
(5.75)	$\frac{\sin^6 ax}{x^2}$	$-\frac{\operatorname{sgn} a}{32x^2}(\sin 6ax - 6\sin 4ax + 15\sin 2ax - 12ax)$
(5.76)	$\frac{\sin^6 ax}{x^3}$	$-\frac{\operatorname{sgn} a}{32x^3}(\sin 6ax - 6\sin 4ax + 15\sin 2ax - 12ax)$
(5.77)	$\frac{\sin^{2n} ax}{x^p}$, integer n, p, $2n \geq p \geq 1$	$\frac{\operatorname{sgn} a}{2^{2n-1}x^p}\sum_{k=0}^{n-1}(-1)^{n-k}\binom{2n}{k}\sin[2(n-k)ax]$ $+ \frac{\operatorname{sgn} a}{x^p}\sum_{k=1}^{[\frac{p}{2}]}\frac{(-4)^{k-n}(ax)^{2k-1}}{(2k-1)!}\sum_{j=0}^{n-1}(-1)^j\binom{2n}{j}(n-j)^{2k-1}$, with $\left[\frac{p}{2}\right] = p/2$ for p even and $(p-1)/2$ for p odd

Table 1.5. (*Cont.*)

Number	$f(x)$	$\frac{1}{\pi} P \int_{-\infty}^{\infty} \frac{f(s)\mathrm{d}s}{x-s}$
(5.78)	$\frac{\sin^{2n+1} ax}{x^p}$, integer n, p, $2n+1 \geq p$ $p \geq 1$	$\frac{-(-1)^n \operatorname{sgn} a}{4^n x^p} \sum_{k=0}^{n} (-1)^k \binom{2n+1}{k} \cos[(2n-2k+1)ax]$ $+ \frac{\operatorname{sgn} a}{4^n x^p} \sum_{k=0}^{\left[\frac{p-1}{2}\right]} \frac{(-1)^{k+n}(ax)^{2k}}{(2k)!} \sum_{j=0}^{n} (-1)^j \binom{2n+1}{j} (2n-2j+1)^{2k}$
(5.79)	$\frac{\sin ax \sin^n x}{x^n}$, $n = 0, 1, 2, \ldots,\ n < a$	$-\frac{\cos ax \sin^n x}{x^n}$
(5.80)	$\frac{\cos ax \sin^n x}{x^n}$, $n = 0, 1, 2, \ldots,\ n < a$	$\frac{\sin ax \sin^n x}{x^n}$
(5.81)	$\frac{\sin a\sqrt{\lvert x\rvert}}{\sqrt{\lvert x\rvert}}$, $a > 0$	$\frac{\operatorname{sgn} x}{\pi\sqrt{\lvert x\rvert}} \{\mathrm{e}^{-a\sqrt{\lvert x\rvert}} \operatorname{Ei}(a\sqrt{\lvert x\rvert}) - \mathrm{e}^{a\sqrt{\lvert x\rvert}} \operatorname{Ei}(-a\sqrt{\lvert x\rvert})$ $+ 2\sin(a\sqrt{\lvert x\rvert}) \operatorname{Ci}(a\sqrt{\lvert x\rvert}) - 2\cos(a\sqrt{\lvert x\rvert}) \operatorname{Si}(a\sqrt{\lvert x\rvert})\}$
(5.82)	$\frac{\sin ax}{\sqrt{(x^2+c^2)}} \sin b\sqrt{(x^2+c^2)}$, $0 < b < a$	$-\frac{\cos ax \sin b\sqrt{(x^2+c^2)}}{\sqrt{(x^2+c^2)}}$
(5.83)	$\frac{\cos ax}{\sqrt{(x^2+c^2)}} \sin b\sqrt{(x^2+c^2)}$, $0 < b < a$	$\frac{\sin ax \sin b\sqrt{(x^2+c^2)}}{\sqrt{(x^2+c^2)}}$

Number	$f(x)$	$\frac{1}{\pi} P \int_{-\infty}^{\infty} \frac{f(s)\mathrm{d}s}{x-s}$
(5.84)	$\frac{1}{1 - b\cos ax}, \quad \lvert b \rvert < 1$	$\frac{b \operatorname{sgn} a \sin ax}{(1 - b\cos ax)\sqrt{(1-b^2)}}$
(5.85)	$(5 - 4\cos x)^{-1}$	$\frac{4 \sin x}{3(5 - 4\cos x)}$
(5.86)	$(5 - 4\cos x)^{-1} \sin x$	$\frac{1 - 2\cos x}{2(5 - 4\cos x)}$
(5.87)	$(5 - 4\cos x)^{-1} \cos x$	$\frac{5 \sin x}{3(5 - 4\cos x)}$
(5.88)	$\frac{1}{1 - 2a\cos bx + a^2}, \ a^2 < 1, b > 0$	$\frac{2a \sin bx}{(1-a^2)(1 - 2a\cos bx + a^2)}$
(5.89)	$\frac{\sin bx}{1 - 2a\cos bx + a^2}, \ a^2 < 1, b > 0$	$\frac{a - \cos bx}{1 - 2a\cos bx + a^2}$
(5.90)	$\frac{\sin bx}{1 - 2a\cos 2bx + a^2}, \ a^2 < 1, b > 0$	$\frac{(a-1)\cos bx}{(a+1)(1 - 2a\cos 2bx + a^2)}$
(5.91)	$\frac{\sin bcx}{1 - 2a\cos bx + a^2}, \ a^2 < 1, b > 0, 0 < c$	$\frac{2a \sin bcx \sin bx}{(1-a^2)(1 - 2a\cos bx + a^2)} - \frac{\cos bcx}{(1-a^2)}$ $- \frac{2}{(1-a^2)} \sum_{k=1}^{\lfloor c \rfloor} a^k \cos(c-k)bx$
(5.92)	$\frac{\sin bx}{x(1 - 2a\cos bx + a^2)}, \ a^2 < 1, b > 0$	$\frac{(1+a)(1 - \cos bx)}{x(1-a)(1 - 2a\cos 2bx + a^2)}$

Table 1.5. (*Cont.*)

Number	$f(x)$	$\frac{1}{\pi} P \int_{-\infty}^{\infty} \frac{f(s)\mathrm{d}s}{x-s}$
(5.93)	$\frac{\sin bcx}{x(1-2a\cos bx+a^2)},\ a^2<1,\ b>0,\ 0<c$	$\frac{1-\cos bcx}{x(1-a^2)} + \frac{2a\sin bcx \sin bx}{x(1-a^2)(1-2a\cos bx+a^2)} + \frac{2}{x(1-a^2)} \times \sum_{k=1}^{\lfloor c\rfloor} a^k\{1-\cos(c-k)bx\}$
(5.94)	$\frac{\cos bx}{1-2a\cos bx+a^2},\ a^2<1,\ b>0$	$\frac{(1+a^2)\sin bx}{(1-a^2)(1-2a\cos bx+a^2)}$
(5.95)	$\frac{\cos bx}{1-2a\cos 2bx+a^2},\ a^2<1,\ b>0$	$\frac{(1+a)\sin bx}{(1-a)(1-2a\cos 2bx+a^2)}$
(5.96)	$\frac{1-a\cos bx}{1-2a\cos 2bx+a^2},\ a^2<1,\ b>0$	$\frac{a\sin bx}{1-2a\cos 2bx+a^2}$
(5.97)	$\frac{\cos bcx}{1-2a\cos bx+a^2},\ a^2<1,\ b>0,\ 0<c$	$\frac{2a\cos bcx\sin bx}{(1-a^2)(1-2a\cos bx+a^2)} + \frac{\sin bcx}{(1-a^2)} + \frac{2}{(1-a^2)} \times \sum_{k=1}^{\lfloor c\rfloor} a^k \sin(c-k)bx$
(5.98)	$\frac{\sin bcx \sin bdx}{1-2a\cos bx+a^2},\ a^2<1,\ b>0,\ c>d>0$	$\frac{2a\sin bx\sin bcx\sin bdx}{(1-a^2)(1-2a\cos bx+a^2)} - \frac{\cos bcx\sin bdx}{(1-a^2)} + \frac{1}{(1-a^2)}\left\{\sum_{k=1}^{\lfloor c-d\rfloor} a^k\sin(c-d-k)bx - \sum_{k=1}^{\lfloor c+d\rfloor} a^k \sin(c+d-k)bx\right\}$

Number	$f(x)$	$\frac{1}{\pi} P \int_{-\infty}^{\infty} \frac{f(s)\mathrm{d}s}{x-s}$
(5.99)	$\frac{\sin bcx \cos bdx}{1 - 2a\cos bx + a^2}, \ a^2 < 1, \ b > 0, \ c > 0, \ d > 0$	$\frac{2a \sin bx \sin bcx \cos bdx}{(1-a^2)(1-2a\cos bx + a^2)} - \frac{\cos bcx \cos bdx}{(1-a^2)} - \frac{1}{(1-a^2)} \left\{ \sum_{k=1}^{\lfloor c+d \rfloor} a^k \cos(c+d-k)bx + \sum_{k=1}^{\lfloor c-d \rfloor} a^k \cos(c-d-k)bx \right\}$, for $c > d$ $\frac{2a \sin bx \sin bcx \cos bdx}{(1-a^2)(1-2a\cos bx + a^2)} + \frac{\sin bcx \sin bdx}{(1-a^2)} - \frac{1}{(1-a^2)} \times \left\{ \sum_{k=1}^{\lfloor c+d \rfloor} a^k \cos(c+d-k)bx - \sum_{k=1}^{\lfloor d-c \rfloor} a^k \cos(d-c-k)bx \right\}$, for $d > c$
(5.100)	$\frac{\cos bcx \cos bdx}{1 - 2a\cos bx + a^2}, \ a^2 < 1, \ b > 0, \ c > d > 0$	$\frac{2a \sin bx \cos bcx \cos bdx}{(1-a^2)(1-2a\cos bx + a^2)} + \frac{\sin bcx \cos bdx}{(1-a^2)} + \frac{1}{(1-a^2)} \left\{ \sum_{k=1}^{\lfloor c+d \rfloor} a^k \sin(c+d-k)bx + \sum_{k=1}^{\lfloor c-d \rfloor} a^k \sin(c-d-k)bx \right\}$
(5.101)	$\frac{\sin(2abx)}{\sin(bx)}, \quad b > 0$	$\frac{2\sin^2(abx)}{\sin(bx)}$, for $a \in \mathbb{Z}^+$ $2\sum_{k=1}^{n} \sin(2kbx)$, for $a = n + 1/2, \ n \in \mathbb{Z}^+$

Table 1.5. (*Cont.*)

Number	$f(x)$	$\frac{1}{\pi}P\int_{-\infty}^{\infty}\frac{f(s)\mathrm{d}s}{x-s}$
(5.102)	$\frac{\sin^2(nbx)}{\sin(bx)}, \quad b>0,\ n\in\mathbb{Z}^+$	$-\frac{\sin(2nbx)}{2\sin(bx)}$
(5.103)	$\frac{\sin^2(nbx)}{x\sin(bx)}, \quad b>0,\ n\in\mathbb{Z}^+$	$\frac{1}{2x}\left[2n-\frac{\sin(2nbx)}{\sin(bx)}\right]$
(5.104)	$\frac{\cos ax}{x}$	$-\pi\delta(x)+\operatorname{sgn} a\frac{\sin ax}{x}$
(5.105)	$\frac{\sin ax}{x^2}$	$-a\pi\delta(x)+\operatorname{sgn} a\left(\frac{1-\cos ax}{x^2}\right)$
(5.106)	$\frac{\sin ax - ax}{x^2}$	$\operatorname{sgn} a\left(\frac{1-\cos ax}{x^2}\right)$
(5.107)	$\frac{\cos ax}{x^2}$	$\pi\delta'(x)-\operatorname{sgn} a\left(\frac{ax-\sin ax}{x^2}\right)$
(5.108)	$\frac{\sin ax}{x^3}$	$a\pi\delta'(x)-\frac{a^2\operatorname{sgn} a}{2x}+\operatorname{sgn} a\left(\frac{1-\cos ax}{x^3}\right)$
(5.109)	$\frac{\cos ax}{x^3}$	$\frac{\pi}{2}[a^2\delta(x)-\delta''(x)]-\operatorname{sgn} a\left(\frac{ax-\sin ax}{x^3}\right)$
(5.110)	$\frac{ax\cos ax-\sin ax}{x}$	$\frac{\operatorname{sgn} a}{x}(ax\sin ax+\cos ax-1)$

Number	$f(x)$	$\frac{1}{\pi} P \int_{-\infty}^{\infty} \frac{f(s)\mathrm{d}s}{x-s}$
(5.111)	$\frac{ax\cos ax - \sin ax}{x^2}$	$\frac{\operatorname{sgn} a}{x^2}(ax\sin ax + \cos ax - 1)$
(5.112)	$\frac{ax\cos ax - \sin ax}{x^3}$	$\frac{\operatorname{sgn} a}{2x^3}(2ax\sin ax + 2\cos ax - 2 - a^2x^2)$
(5.113)	$\frac{1-\cos ax}{x}$	$-\frac{\operatorname{sgn} a \sin ax}{x} = -x^{-1}\sin(\lvert a\rvert x)$
(5.114)	$\frac{1-\cos ax}{x^2}$	$\frac{\operatorname{sgn} a}{x^2}(ax - \sin ax)$
(5.115)	$\frac{\sin ax(1-\cos bx)}{x^2}, \quad 0 < b < a$	$-\frac{\cos ax(1-\cos bx)}{x^2}$
(5.116)	$\frac{\cos ax(1-\cos bx)}{x^2}, \quad 0 < b < a$	$\frac{\sin ax(1-\cos bx)}{x^2}$
(5.117)	$\frac{(1-\cos ax)^2}{x}$	$\frac{\operatorname{sgn} a}{2x}(\sin 2ax - 4\sin ax)$
(5.118)	$\left(\frac{1-\cos ax}{x}\right)^2$	$\frac{\operatorname{sgn} a}{2x^2}(\sin 2ax - 4\sin ax + 2ax)$
(5.119)	$\frac{(1-\cos ax)^2}{x^3}$	$\frac{\operatorname{sgn} a}{2x^3}(\sin 2ax - 4\sin ax + 2ax)$
(5.120)	$\frac{(1-\cos ax)^2}{x^4}$	$\frac{\operatorname{sgn} a}{2x^4}\left(\sin 2ax - 4\sin ax + 2ax + \frac{2}{3}a^3x^3\right)$

Table 1.5. (*Cont.*)

Number	$f(x)$	$\frac{1}{\pi} P \int_{-\infty}^{\infty} \frac{f(s)\mathrm{d}s}{x-s}$
(5.121)	$\frac{(1-\cos ax)^3}{x}$	$-\frac{\operatorname{sgn} a}{4x}(\sin 3ax - 6\sin 2ax + 15\sin ax)$
(5.122)	$\frac{(1-\cos ax)^3}{x^2}$	$-\frac{\operatorname{sgn} a}{4x^2}(\sin 3ax - 6\sin 2ax + 15\sin ax - 6ax)$
(5.123)	$\left(\frac{1-\cos ax}{x}\right)^3$	$-\frac{\operatorname{sgn} a}{4x^3}(\sin 3ax - 6\sin 2ax + 15\sin ax - 6ax)$
(5.124)	$\frac{(1-\cos ax)^n}{x^p}$, integer $n, p, 2n \geq p \geq 1$	$\frac{\operatorname{sgn} a}{2^{n-1}x^p} \sum_{k=0}^{n-1} (-1)^{n-k} \binom{2n}{k} \sin[(n-k)ax]$ $+ \frac{\operatorname{sgn} a}{2^{n-1}x^p} \sum_{k=1}^{\left[\frac{p}{2}\right]} \frac{(-1)^{k-n}(ax)^{2k-1}}{(2k-1)!} \sum_{j=0}^{n-1} (-1)^j \binom{2n}{j} (n-j)^{2k-1}$, with $\left[\frac{p}{2}\right] = p/2$ for p even and $(p-1)/2$ for p odd
(5.125)	$\operatorname{sgn}[\sin ax]$, $a > 0$	$\frac{2}{\pi} \log \lvert\tan(ax/2)\rvert$
(5.126)	$\operatorname{sgn}[\cos ax]$, $a > 0$	$\frac{2}{\pi} \log \left\lvert \tan\left(\frac{ax}{2} + \frac{\pi}{4}\right)\right\rvert$
(5.127)	$\sin(ax^2)$, $a > 0$	$-\operatorname{sgn} x\{S(\surd(2a/\pi)\,\lvert x\rvert)[\cos ax^2 + \sin ax^2] + C(\surd(2a/\pi)\,\lvert x\rvert) \times [\cos ax^2 - \sin ax^2]\}$

Number	$f(x)$	$\frac{1}{\pi}P\int_{-\infty}^{\infty}\frac{f(s)\mathrm{d}s}{x-s}$
(5.128)	$\cos(ax^2),\ a>0$	$\operatorname{sgn} x\{S(\sqrt{(2a/\pi)}\,\lvert x\rvert)[\sin ax^2-\cos ax^2]+C(\sqrt{(2a/\pi)}\,\lvert x\rvert)$ $\times[\sin ax^2+\cos ax^2]\}$
(5.129)	$\sin(ax^{-1})$	$\operatorname{sgn} a[\cos(ax^{-1})-1]$
(5.130)	$\cos(ax^{-1})$	$-\operatorname{sgn} a\sin(ax^{-1})$
(5.131)	$\sin\left(ax-\frac{a}{x}\right),\ a>0$	$-\cos\left(ax-\frac{a}{x}\right)$
(5.132)	$\cos\left(ax-\frac{a}{x}\right),\ a>0$	$\sin\left(ax-\frac{a}{x}\right)$
(5.133)	$\sin(ax^{-2}),\ a>0$	$\operatorname{sgn}(x^{-1})\{S(\sqrt{(2a/\pi)}\,\lvert x^{-1}\rvert)[\cos ax^{-2}+\sin ax^{-2}]$ $+C(\sqrt{(2a/\pi)}\,\lvert x^{-1}\rvert)[\cos ax^{-2}-\sin ax^{-2}]\}$
(5.134)	$\cos(ax^{-2}),\ a>0$	$-\operatorname{sgn}(x^{-1})\{S(\sqrt{(2a/\pi)}\,\lvert x^{-1}\rvert)[\sin ax^{-2}-\cos ax^{-2}]$ $+C(\sqrt{(2a/\pi)}\,\lvert x^{-1}\rvert)[\sin ax^{-2}+\cos ax^{-2}]\}$
(5.135)	$\frac{1}{x+a-\cot(x^{-1})},\ \operatorname{Im} a>0$	$\frac{-\mathrm{i}[x-\cot(x^{-1})]}{a[x+a-\cot(x^{-1})]}$
(5.136)	$\cos(a\cot x),\ a>0$	$-\sin(a\cot x)$
(5.137)	$\mathrm{e}^{-ax^2}\cos bx,\ a\geq 0,\ b\geq 0$	$-\frac{\mathrm{i}}{2}\mathrm{e}^{-ax^2}\left\{\mathrm{e}^{\mathrm{i}bx}\operatorname{erf}\left[\sqrt{(a)}\left(\mathrm{i}x+\frac{b}{2a}\right)\right]+\mathrm{e}^{-\mathrm{i}bx}\operatorname{erf}\left[\sqrt{(a)}\left(\mathrm{i}x-\frac{b}{2a}\right)\right]\right\}$ $=\mathrm{e}^{-ax^2}\operatorname{Im}\left\{\mathrm{e}^{\mathrm{i}bx}\operatorname{erf}\left[\sqrt{(a)}\left(\frac{b}{2a}+\mathrm{i}x\right)\right]\right\}$

Table 1.5. (*Cont.*)

Number	$f(x)$	$\frac{1}{\pi} P \int_{-\infty}^{\infty} \frac{f(s)\mathrm{d}s}{x-s}$
(5.138)	$\mathrm{e}^{-ax^2} \sin bx,\ a \geq 0,\ b \geq 0$	$-\frac{1}{2}\mathrm{e}^{-ax^2}\left\{\mathrm{e}^{\mathrm{i}bx}\,\mathrm{erf}\left[\sqrt{(a)}\left(\frac{b}{2a}+\mathrm{i}x\right)\right]+\mathrm{e}^{-\mathrm{i}bx}\,\mathrm{erf}\left[\sqrt{(a)}\left(\frac{b}{2a}-ix\right)\right]\right\}$ $= -\mathrm{e}^{-ax^2}\mathrm{Re}\left\{\mathrm{e}^{\mathrm{i}bx}\,\mathrm{erf}\left[\sqrt{(a)}\left(\frac{b}{2a}+\mathrm{i}x\right)\right]\right\}$
(5.139)	$\cos ax\, \mathrm{e}^{-b\|x\|},\ a > 0,\ b \geq 0$	$\frac{\sin ax}{\pi}\left\{\mathrm{sie}\left(ax, \frac{b}{a}\right)\mathrm{e}^{bx} + \mathrm{sie}\left(-ax, \frac{b}{a}\right)\mathrm{e}^{-bx}\right\}$ $+ \frac{\mathrm{sgn}\ x \cos ax}{\pi}\left\{\mathrm{Cie}\left(a\|x\|, \frac{b}{a}\right)\mathrm{e}^{b\|x\|} - \mathrm{Cie}\left(a\|x\|, \frac{b}{a}\right)\mathrm{e}^{-b\|x\|}\right\}$
(5.140)	$\sin ax\, \mathrm{e}^{-b\|x\|},\ a > 0,\ b \geq 0$	$\frac{\cos ax}{\pi}\left\{\mathrm{sie}\left(ax, \frac{b}{a}\right)\mathrm{e}^{bx} + \mathrm{sie}\left(-ax, \frac{b}{a}\right)\mathrm{e}^{-bx}\right\}$ $+ \frac{\mathrm{sgn}\ x \sin ax}{\pi}\left\{\mathrm{Cie}\left(a\|x\|, \frac{b}{a}\right)\mathrm{e}^{b\|x\|} - \mathrm{Cie}\left(a\|x\|, \frac{b}{a}\right)\mathrm{e}^{-b\|x\|}\right\}$
(5.141)	$\mathrm{e}^{\mathrm{i}ax} \sin bx$	$\frac{1}{2}\left[\mathrm{sgn}(a-b)\mathrm{e}^{\mathrm{i}(a-b)x} - \mathrm{sgn}(a+b)\mathrm{e}^{\mathrm{i}(a+b)x}\right]$
(5.142)	$\mathrm{e}^{\mathrm{i}ax} \cos bx$	$-\frac{\mathrm{i}}{2}\left[\mathrm{sgn}(a+b)\mathrm{e}^{\mathrm{i}(a+b)x} + \mathrm{sgn}(a-b)\mathrm{e}^{\mathrm{i}(a-b)x}\right]$
(5.143)	$\mathrm{e}^{a\cos bx} \sin(a \sin bx)$	$\mathrm{sgn}\, b\left[1 - \mathrm{e}^{a\cos bx}\cos(a\sin bx)\right]$
(5.144)	$\mathrm{e}^{a\cos bx} \cos(a \sin bx)$	$\mathrm{sgn}\, b\, \mathrm{e}^{a\cos bx}\sin(a\sin bx)$

Number	$f(x)$	$\frac{1}{\pi}P\int_{-\infty}^{\infty}\frac{f(s)\mathrm{d}s}{x-s}$
(5.145)	$\mathrm{e}^{a\cos bx}\cos(cx+a\sin bx)$, $b>0, c\geq 0$	$\mathrm{e}^{a\cos bx}\sin(cx+a\sin bx)$
(5.146)	$\frac{\cos ax}{x^2+b^2}$, $a>0, b>0$	$\frac{x\mathrm{e}^{-ab}+b\sin ax}{b(x^2+b^2)}$
(5.147)	$\frac{\sin ax}{x^2+b^2}$, $a>0, b>0$	$\frac{\mathrm{e}^{-ab}-\cos ax}{(x^2+b^2)}$
(5.148)	$\frac{x\cos ax}{x^2+b^2}$	$\frac{x \operatorname{sgn} a \sin ax - \lvert b\rvert \mathrm{e}^{-\lvert ab\rvert}}{b^2+x^2}$
(5.149)	$\frac{x\sin ax}{x^2+b^2}$, $a>0, b>0$	$\frac{x(\mathrm{e}^{-ab}-\cos ax)}{x^2+b^2}$
(5.150)	$\frac{(\alpha x^2+\beta x+\gamma)\cos ax}{x^2+b^2}$, $a>0, b>0$	$\frac{\mathrm{e}^{-ab}[x(\gamma-\alpha b^2)-\beta b^2]+b(\alpha x^2+\beta x+\gamma)\sin ax}{b(x^2+b^2)}$
(5.151)	$\frac{(\alpha x^2+\beta x+\gamma)\sin ax}{x^2+b^2}$, $a>0, b>0$	$\frac{\mathrm{e}^{-ab}(x\beta-\alpha b^2+\gamma)-(\alpha x^2+\beta x+\gamma)\cos ax}{x^2+b^2}$
(5.152)	$\frac{\sin ax}{x^4+4b^4}$, $a>0, b>0$	$\frac{1}{x^4+4b^4}\left[\frac{\mathrm{e}^{-ab}}{2b^2}(2b^2\cos ab+x^2\sin ab)-\cos ax\right]$
(5.153)	$\frac{\cos ax}{x^4+4b^4}$, $a>0, b>0$	$\frac{1}{x^4+4b^4}\left\{\sin ax+\frac{x\mathrm{e}^{-ab}}{4b^3}[(x^2+2b^2)\cos ab+(x^2-2b^2)\sin ab]\right\}$

Table 1.5. (*Cont.*)

Number	$f(x)$	$\frac{1}{\pi} P \int_{-\infty}^{\infty} \frac{f(s)\mathrm{d}s}{x-s}$
(5.154)	$\frac{\sinh x + \sin x}{x(\cosh x - \cos x)} - 2x^{-2}$	$\frac{2x}{\pi} \sum_{k=1}^{\infty} \frac{(x^2 - 2\pi^2 k^2)}{k(4\pi^4 k^4 + x^4)}$
(5.155)	$\frac{\sinh x + \sin x}{x^2(\cosh x - \cos x)} - 2x^{-3}$	$\frac{2}{\pi} \sum_{k=1}^{\infty} \frac{(x^2 - 2\pi^2 k^2)}{k(4\pi^4 k^4 + x^4)}$
(5.156)	$\frac{\sinh x + \sin x}{x^3(\cosh x - \cos x)} - 2x^{-4}$	$\frac{x}{\pi^3} \sum_{k=1}^{\infty} \frac{(2\pi^2 k^2 + x^2)}{k^3(4\pi^4 k^4 + x^4)}$
(5.157)	$\frac{\sinh x + \sin x}{\cosh x - \cos x} - \frac{\sinh 2x + \sin 2x}{\cosh 2x - \cos 2x} - x^{-1}$	$2\pi \sum_{k=1}^{\infty} \frac{(-1)^{k+1} k(\pi^2 k^2 + 2x^2)}{(\pi^4 k^4 + 4x^4)}$
(5.158)	$\frac{\sinh x + \sin x}{x(\cosh x - \cos x)} - \frac{\sinh 2x + \sin 2x}{x(\cosh 2x - \cos 2x)} - x^{-2}$	$\frac{4x}{\pi} \sum_{k=1}^{\infty} \frac{(-1)^k (2x^2 - \pi^2 k^2)}{k(\pi^4 k^4 + 4x^4)}$
(5.159)	$\frac{\sinh x + \sin x}{x^2(\cosh x - \cos x)} - \frac{\sinh 2x + \sin 2x}{x^2(\cosh 2x - \cos 2x)} - x^{-3}$	$\frac{4}{\pi} \sum_{k=1}^{\infty} \frac{(-1)^k (2x^2 - \pi^2 k^2)}{k(\pi^4 k^4 + 4x^4)}$

Number	$f(x)$	$\frac{1}{\pi}P\int_{-\infty}^{\infty}\frac{f(s)\mathrm{d}s}{x-s}$
(5.160)	$\frac{\sinh x+\sin x}{x^3(\cosh x-\cos x)} - \frac{\sinh 2x+\sin 2x}{x^3(\cosh 2x-\cos 2x)} - x^{-4}$	$\frac{8x}{\pi^3}\sum_{k=1}^{\infty}\frac{(-1)^k(2x^2+\pi^2k^2)}{k^3(\pi^4k^4+4x^4)}$
(5.161)	$\frac{\mathrm{e}^a-\cos x}{\cosh a-\cos x},\ a>0$	$\frac{\sin x}{\cosh a-\cos x}$
(5.162)	$\frac{\sin x}{\cosh a-\cos x},\ a>0$	$\frac{\mathrm{e}^{-a}-\cos x}{\cosh a-\cos x}$
(5.163)	$\cosh(a\sin bx)\sin(a\cos bx),\ b>0$	$\sinh(a\sin bx)\cos(a\cos bx)$
(5.164)	$\sinh(a\sin bx)\sin(a\cos bx),\ b>0$	$\cosh(a\sin bx)\cos(a\cos bx)-1$
(5.165)	$\cosh(a\sin bx)\cos(a\cos bx),\ b>0$	$-\sinh(a\sin bx)\sin(a\cos bx)$
(5.166)	$\sinh(a\sin bx)\cos(a\cos bx),\ b>0$	$-\cosh(a\sin bx)\sin(a\cos bx)$
(5.167)	$\sin ax\, f(x)$, with $\mathcal{F}[f(x)]=0$, for $\|x\|>b$ and $0<b<a$	$-\cos ax\, f(x)$
(5.168)	$\cos ax\, f(x)$, with $\mathcal{F}[f(x)]=0$, for $\|x\|>b$ and $0<b<a$	$\sin ax\, f(x)$

Table 1.6. *Logarithmic functions*

Note that log stands for $\log_e$ in this table

Number	$f(x)$	$\frac{1}{\pi} P \int_{-\infty}^{\infty} \frac{f(s)\mathrm{d}s}{x-s}$
(6.1)	$\log\left\lvert\frac{b-x}{a-x}\right\rvert,\quad b > a$	$0,\quad -\infty < x < a$ $\pi,\quad a < x < b$ $0,\quad b < x < \infty$
(6.2)	$x^{-1}\log\left\lvert\frac{a+x}{b-x}\right\rvert,\quad a > 0,\ b > 0$	$\pi x^{-1},\quad -\infty < x < -a$ $0,\quad -a < x < b$ $\pi x^{-1},\quad b < x < \infty$
(6.3)	$\log\left\lvert\frac{x^2-a^2}{x^2-b^2}\right\rvert,\quad 0 < a < b$	$\pi,\quad -b < x < -a$ $-\pi,\quad a < x < b$ $0,\quad \lvert x\rvert < a$ or $b < \lvert x\rvert$
(6.4)	$\log(a^2 + b^2x^2),\quad a > 0,\ b > 0$	$-2\tan^{-1}(bx/a)$
(6.5)	$\frac{\log(1+a^2x^2)}{x},\quad a > 0$	$-2x^{-1}\tan^{-1}(ax)$
(6.6)	$\log\left(\frac{a^2+b^2x^2}{x^2}\right),\quad a > 0,\ b > 0$	$2\cot^{-1}(bx/a)$
(6.7)	$\log(1-a^2x^{-2})^2\log\lvert 1-b^2x^{-2}\rvert,$ $a > 0,\ b > 0$	$2\pi\{\log\lvert b^2x^{-2}-1\rvert + 2\log\lvert 1+ax^{-1}\rvert\},\quad x < a < b$ $2\pi\{\log\lvert a^2x^{-2}-1\rvert + 2\log\lvert 1+bx^{-1}\rvert\},\quad x < b < a$ $4\pi\log\lvert 1+ax^{-1}\rvert,\quad a < x < b$ $2\pi\log\left\lvert\frac{x+a}{x-a}\right\rvert,\quad a < b < x$ $4\pi\log\lvert 1+bx^{-1}\rvert,\quad b < x < a$ $2\pi\log\left\lvert\frac{x+b}{x-b}\right\rvert,\quad b < a < x$

Table 1.7. *Inverse trigonometric and inverse hyperbolic functions*

Number	$f(x)$	$\frac{1}{\pi} P \int_{-\infty}^{\infty} \frac{f(s)\mathrm{d}s}{x-s}$
(7.1)	$\frac{\tan^{-1}(x)}{x}$	$\frac{\log(1+x^2)}{2x}$
(7.2)	$\cot^{-1}(x),\ x > 0$	$\log\left(\frac{x}{\sqrt{(1+x^2)}}\right)$
(7.3)	$x\cot^{-1}(ax),\ a > 0$	$\frac{x}{2}\log\left(\frac{a^2x^2}{1+a^2x^2}\right)$
(7.4)	$\frac{\cos[v\tan^{-1}(x/a)]}{\sqrt{[(x^2+a^2)^v]}},\ v > 0,\ a > 0$	$\frac{\sin[v\tan^{-1}(x/a)]}{\sqrt{[(x^2+a^2)^v]}}$
(7.5)	$\frac{\sin[v\tan^{-1}(x/a)]}{\sqrt{[(x^2+a^2)^v]}},\ v > 0,\ a > 0$	$-\frac{\cos[v\tan^{-1}(x/a)]}{\sqrt{[(x^2+a^2)^v]}}$
(7.6)	$\frac{\sinh^{-1}(x/a)}{\sqrt{(x^2+a^2)}},\ a > 0$	$-\frac{\pi}{2\sqrt{(x^2+a^2)}}$
(7.7)	$\tan^{-1}\left(\frac{a\sin bx}{1+a\cos bx}\right),\ a^2 < 1,\ b > 0$	$-(1/2)\log(1+2a\cos bx+a^2)$
(7.8)	$\tan^{-1}\left(\frac{2a\cos x}{1-a^2}\right),\ a^2 < 1$	$(1/2)\log\left(\frac{1+2a\sin x+a^2}{1-2a\sin x+a^2}\right)$

Table 1.8A. *Special functions: Legendre polynomials*

Number	$f(x)$	$\frac{1}{\pi} P \int_{-\infty}^{\infty} \frac{f(s)ds}{x-s}$
(8A.1)	$P_0(\cos x)$	0
(8A.2)	$P_1(\cos x)$	$\sin x$
(8A.3)	$P_2(\cos x)$	$\frac{3}{4}\sin 2x$
(8A.4)	$P_3(\cos x)$	$\frac{1}{8}(3\sin x + 5\sin 3x)$
(8A.5)	$P_4(\cos x)$	$\frac{5}{64}(4\sin 2x + 7\sin 4x)$
(8A.6)	$P_5(\cos x)$	$\frac{1}{128}(30\sin x + 35\sin 3x + 63\sin 5x)$
(8A.7)	$P_6(\cos x)$	$\frac{21}{512}(5\sin 2x + 6\sin 4x + 11\sin 6x)$
(8A.8)	$P_7(\cos x)$	$\frac{1}{1024}(175\sin x + 189\sin 3x + 231\sin 5x + 429\sin 7x)$
(8A.9)	$P_n(\cos x)$	$\frac{1}{2^{2n-1}}\sum_{m=1}^{n/2}\binom{n+2m}{\frac{n}{2}+m}\binom{n-2m}{\frac{n}{2}-m}\sin 2mx,$ for n even $\frac{1}{2^{2n-1}}\sum_{m=0}^{(n-1)/2}\binom{n+2m+1}{\frac{n+1}{2}+m}\binom{n-2m-1}{\frac{n-1}{2}-m}\sin[(2m+1)x],$ for n odd
(8A.10)	$P_0(\sin x)$	0
(8A.11)	$P_1(\sin x)$	$-\cos x$
(8A.12)	$P_2(\sin x)$	$-\frac{3}{4}\sin 2x$
(8A.13)	$P_3(\sin x)$	$\frac{1}{8}(5\cos 3x - 3\cos x)$
(8A.14)	$P_4(\sin x)$	$\frac{5}{64}(7\sin 4x - 4\sin 2x)$
(8A.15)	$P_5(\sin x)$	$-\frac{1}{128}(63\cos 5x - 35\cos 3x + 30\cos x)$

Table 1.8A. (*Cont.*)

Number	$f(x)$	$\frac{1}{\pi}P\int_{-\infty}^{\infty}\frac{f(s)\mathrm{d}s}{x-s}$
(8A.16)	$P_6(\sin x)$	$-\frac{1}{512}(231\sin 6x - 126\sin 4x + 105\sin 2x)$
(8A.17)	$P_7(\sin x)$	$\frac{1}{1024}(429\cos 7x - 231\cos 5x + 189\cos 3x - 175\cos x)$
(8A.18)	$P_{2n}(\sin x),\ n\in\mathbb{Z}^+$	$\frac{1}{2^{4n-1}}\sum_{m=1}^{n}(-1)^m\binom{2n+2m}{n+m}\binom{2n-2m}{n-m}\sin 2mx$
(8A.19)	$P_{2n+1}(\sin x),\ n\in\mathbb{Z}^+$	$\frac{1}{2^{4n+1}}\sum_{m=0}^{n}(-1)^{m+1}\binom{2n+2m+2}{n+1+m}\binom{2n-2m}{n-m} \times \cos[(2m+1)x]$
(8A.20)	$P_0(x)\Pi_2(x),\ \lvert x\rvert\neq 1$	$\pi^{-1}\log\left\lvert\frac{1+x}{1-x}\right\rvert$
(8A.21)	$P_1(x)\Pi_2(x),\ \lvert x\rvert\neq 1$	$\pi^{-1}\left(x\log\left\lvert\frac{1+x}{1-x}\right\rvert - 2\right)$
(8A.22)	$P_2(x)\Pi_2(x),\ \lvert x\rvert\neq 1$	$\pi^{-1}\left(\left[\frac{3}{2}x^2 - 1/2\right]\log\left\lvert\frac{1+x}{1-x}\right\rvert - 3x\right)$
(8A.23)	$P_3(x)\Pi_2(x),\ \lvert x\rvert\neq 1$	$\pi^{-1}\left(\left[\frac{5}{2}x^3 - \frac{3}{2}x\right]\log\left\lvert\frac{1+x}{1-x}\right\rvert - 5x^2 + \frac{4}{3}\right)$
(8A.24)	$P_4(x)\Pi_2(x),\ \lvert x\rvert\neq 1$	$\pi^{-1}\left(\left[\frac{35}{8}x^4 - \frac{15}{4}x^2 + \frac{3}{8}\right]\log\left\lvert\frac{1+x}{1-x}\right\rvert - \frac{35}{4}x^3 + \frac{55}{12}x\right)$
(8A.25)	$P_5(x)\Pi_2(x),\ \lvert x\rvert\neq 1$	$\pi^{-1}\left(\left[\frac{63}{8}x^5 - \frac{35}{4}x^3 + \frac{15}{8}x\right]\log\left\lvert\frac{1+x}{1-x}\right\rvert - \frac{63}{4}x^4 + \frac{49}{4}x^2 - \frac{16}{15}\right)$
(8A.26)	$P_6(x)\Pi_2(x),\ \lvert x\rvert\neq 1$	$\pi^{-1}\left(\left[\frac{231}{16}x^6 - \frac{315}{16}x^4 + \frac{105}{16}x^2 - \frac{5}{16}\right] \times\log\left\lvert\frac{1+x}{1-x}\right\rvert - \frac{231}{8}x^5 + \frac{119}{4}x^3 - \frac{231}{40}x\right)$
(8A.27)	$P_7(x)\Pi_2(x),\ \lvert x\rvert\neq 1$	$\pi^{-1}\left(\left[\frac{429}{16}x^7 - \frac{693}{16}x^5 + \frac{315}{16}x^3 - \frac{35}{16}x\right]\log\left\lvert\frac{1+x}{1-x}\right\rvert - \frac{429}{8}x^6 + \frac{275}{4}x^4 - \frac{849}{40}x^2 + \frac{32}{35}\right)$

Table 1.8A. (*Cont.*)

Number	$f(x)$	$\frac{1}{\pi}P\int_{-\infty}^{\infty}\frac{f(s)\mathrm{d}s}{x-s}$
(8A.28)	$P_8(x)\Pi_2(x)$, $\lvert x\rvert\neq 1$	$\pi^{-1}\Big(\Big[\frac{6435}{128}x^8-\frac{3003}{32}x^6+\frac{3465}{64}x^4-\frac{315}{32}x^2+\frac{35}{128}\Big]\log\left\lvert\frac{1+x}{1-x}\right\rvert-\frac{6435}{64}x^7+\frac{9867}{64}x^5-\frac{4213}{64}x^3+\frac{15\,159}{2240}x\Big)$
(8A.29)	$P_9(x)\Pi_2(x)$, $\lvert x\rvert\neq 1$	$\pi^{-1}\Big(\Big[\frac{12\,155}{128}x^9-\frac{6435}{32}x^7+\frac{9009}{64}x^5-\frac{1155}{32}x^3+\frac{315}{128}x\Big]\log\left\lvert\frac{1+x}{1-x}\right\rvert-\frac{12155}{64}x^8+\frac{65\,065}{192}x^6-\frac{11\,869}{64}x^4+\frac{14\,179}{448}x^2-\frac{256}{315}\Big)$
(8A.30)	$P_{10}(x)\Pi_2(x)$, $\lvert x\rvert\neq 1$	$\pi^{-1}\Big(\Big[\frac{46\,189}{256}x^{10}-\frac{109\,395}{256}x^8+\frac{45\,045}{128}x^6-\frac{15\,015}{128}x^4+\frac{3465}{256}x^2-\frac{63}{256}\Big]\log\left\lvert\frac{1+x}{1-x}\right\rvert-\frac{46\,189}{128}x^9+\frac{70\,499}{96}x^7-\frac{157\,157}{320}x^5+\frac{26\,741}{224}x^3-\frac{61\,567}{8064}x\Big)$
(8A.31)	$P_{2j}(x)\Pi_2(x)$, $\lvert x\rvert\neq 1$	$\frac{1}{\pi}\log\left\lvert\frac{1+x}{1-x}\right\rvert\sum_{k=0}^{j}a_{jk}x^{2k}-\frac{2}{\pi}\sum_{k=1}^{j}x^{2k-1}\sum_{m=k}^{j}\frac{a_{jm}}{(2m-2k+1)}$, $a_{jk}=4^{-j}(-1)^{k+j}\binom{2j+2k}{2j}\binom{2j}{j-k}$
(8A.32)	$P_{2j+1}(x)\Pi_2(x)$, $\lvert x\rvert\neq 1$	$\frac{1}{\pi}\log\left\lvert\frac{1+x}{1-x}\right\rvert\sum_{k=0}^{j}b_{jk}x^{2k+1}-\frac{2}{\pi}\sum_{k=0}^{j}x^{2k}\sum_{m=k}^{j}\frac{b_{jm}}{(2m-2k+1)}$, $b_{jk}=2^{-2j-1}(-1)^{k+j}\binom{2j+2k+2}{2j+1}\binom{2j+1}{j-k}$

Table 1.8B. *Special functions: Hermite polynomials*

Number	$f(x)$	$\frac{1}{\pi}P\int_{-\infty}^{\infty}\frac{f(s)\mathrm{d}s}{x-s}$
(8B.1)	$H_0(x)\mathrm{e}^{-x^2}$	$G(x) \equiv -\mathrm{i}\mathrm{e}^{-x^2}\,\mathrm{erf}\,(\mathrm{i}x)$
(8B.2)	$H_1(x)\mathrm{e}^{-x^2}$	$2xG(x) - \frac{2}{\sqrt{\pi}}$
(8B.3)	$H_2(x)\mathrm{e}^{-x^2}$	$(4x^2-2)G(x) - \frac{4x}{\sqrt{\pi}}$
(8B.4)	$H_3(x)\mathrm{e}^{-x^2}$	$(8x^3-12x)G(x) + \frac{8(1-x^2)}{\sqrt{\pi}}$
(8B.5)	$H_4(x)\mathrm{e}^{-x^2}$	$(16x^4-48x^2+12)G(x) + \frac{8(5x-2x^3)}{\sqrt{\pi}}$
(8B.6)	$H_5(x)\mathrm{e}^{-x^2}$	$(32x^5-160x^3+120x)G(x) - \frac{16(4-9x^2+2x^4)}{\sqrt{\pi}}$
(8B.7)	$H_6(x)\mathrm{e}^{-x^2}$	$(64x^6-480x^4+720x^2-120)G(x) - \frac{(528x-448x^3+64x^5)}{\sqrt{\pi}}$
(8B.8)	$H_7(x)\mathrm{e}^{-x^2}$	$(128x^7-1344x^5+3360x^3-1680x)G(x) + \frac{(768-2784x^2+1280x^4-128x^6)}{\sqrt{\pi}}$
(8B.9)	$H_8(x)\mathrm{e}^{-x^2}$	$(256x^8-3584x^6+13\,440x^4-13\,440x^2+1680)G(x) + \frac{(8928x-11\,840x^3+3456x^5-256x^7)}{\sqrt{\pi}}$
(8B.10)	$H_9(x)\mathrm{e}^{-x^2}$	$(512x^9-9216x^7+48\,384x^5-80\,640x^3 + 30\,240x)G(x) - \frac{1}{\sqrt{\pi}}(12\,288-62\,400x^2 + 44\,160x^4-8960x^6+512x^8)$
(8B.11)	$H_{10}(x)\mathrm{e}^{-x^2}$	$(1024x^{10}-23\,040x^8+161\,280x^6-403\,200x^4 + 302\,400x^2-30\,240)G(x) - \frac{1}{\sqrt{\pi}}(185\,280x - 337\,920x^3+150\,528x^5-22\,528x^7+1024x^9)$
(8B.12)	$H_{2j}(x)\mathrm{e}^{-x^2}$, integer $j \geq 0$	$G(x)\sum_{m=0}^{j} a_{jm}x^{2m} - \sum_{m=1}^{j} b_{jm}x^{2m-1}$, $a_{jm} = \frac{4^m(-1)^{j+m}(2j)!}{(2m)!(j-m)!}$, $b_{jm} = 4^j\pi^{-1}(2j)!\sum_{k=0}^{j-m}\frac{\Gamma(j-m-k+1/2)}{(-4)^k\,k!(2j-2k)!}$

Table 1.8B. (*Cont.*)

Number	$f(x)$	$\frac{1}{\pi}P\int_{-\infty}^{\infty}\frac{f(s)\mathrm{d}s}{x-s}$
(8B.13)	$H_{2j+1}(x)\mathrm{e}^{-x^2}$, integer $j \geq 0$	$\sum_{m=0}^{j}\left\{\frac{(4j+2)a_{jm}G(x)}{(2m+1)}x^{2m+1} - c_{jm}x^{2m}\right\}$, $c_{jm} = 2^{2j+1}\pi^{-1}(2j+1)!\sum_{k=0}^{j-m}\frac{\Gamma(j-m-k+1/2)}{(-4)^k\, k!(2j-2k+1)!}$, a_{jm} as in (8B.12)

Table 1.8C. *Special functions: Laguerre polynomials*

Number	$f(x)$	$\frac{1}{\pi}P\int_{-\infty}^{\infty}\frac{f(s)\mathrm{d}s}{x-s}$
(8C.1)	$L_0(x)\mathrm{H}(x)\mathrm{e}^{-x}$	$E(x) = \pi^{-1}\mathrm{e}^{-x}\mathrm{Ei}(x)$
(8C.2)	$L_1(x)\mathrm{H}(x)\mathrm{e}^{-x}$	$(1-x)E(x) + \pi^{-1}$
(8C.3)	$L_2(x)\mathrm{H}(x)\mathrm{e}^{-x}$	$[(2-4x+x^2)E(x) + \pi^{-1}(3-x)]/2$
(8C.4)	$L_3(x)\mathrm{H}(x)\mathrm{e}^{-x}$	$[(6-18x+9x^2-x^3)E(x)+\pi^{-1}(11-8x+x^2)]/6$
(8C.5)	$L_4(x)\mathrm{H}(x)\mathrm{e}^{-x}$	$[(24-96x+72x^2-16x^3+x^4)E(x) + \pi^{-1}(50-58x+15x^2-x^3)]/24$
(8C.6)	$L_5(x)\mathrm{H}(x)\mathrm{e}^{-x}$	$[(120-600x+600x^2-200x^3+25x^4-x^5)E(x) + \pi^{-1}(274-444x+177x^2-24x^3+x^4)]/120$
(8C.7)	$L_6(x)\mathrm{H}(x)\mathrm{e}^{-x}$	$[(720-4320x+5400x^2-2400x^3+450x^4 - 36x^5+x^6)E(x)+\pi^{-1}(1764-3708x+2016x^2 - 416x^3+35x^4-x^5)]/720$
(8C.8)	$L_7(x)\mathrm{H}(x)\mathrm{e}^{-x}$	$[(5040-35\,280x+52\,920x^2-29\,400x^3 + 7350x^4-882x^5+49x^6-x^7)E(x) + \pi^{-1}(13\,068-33\,984x+23\,544x^2-6560x^3 + 835x^4-48x^5+x^6)]/5040$
(8C.9)	$L_8(x)\mathrm{H}(x)\mathrm{e}^{-x}$	$[(40\,320-322\,560x+564\,480x^2-376\,320x^3 + 117\,600x^4-18\,816x^5+1568x^6-64x^7 + x^8)E(x)+\pi^{-1}(109\,584-341\,136x + 288\,360x^2-101\,560x^3+17\,370x^4-1506x^5 + 63x^6-x^7)]/40\,320$

Table 1.8C. (*Cont.*)

Number	$f(x)$	$\frac{1}{\pi}P\int_{-\infty}^{\infty}\frac{f(s)\mathrm{d}s}{x-s}$
(8C.10)	$L_9(x)\mathrm{H}(x)\mathrm{e}^{-x}$	$[(362\,880 - 3\,265\,920x + 6\,531\,840x^2 - 5\,080\,320x^3 + 1\,905\,120x^4 - 381\,024x^5 + 42\,336x^6 - 2592x^7 + 81x^8 - x^9)E(x) + \pi^{-1}(1\,026\,576 - 3\,733\,920x + 3\,736\,440x^2 - 1\,595\,040x^3 + 343\,410x^4 - 39\,900x^5 + 2513x^6 - 80x^7 + x^8)]/362\,880$

Table 1.8D. *Special functions: Bessel functions of the first kind of integer order*

Number	$f(x)$	$\frac{1}{\pi}P\int_{-\infty}^{\infty}\frac{f(s)\mathrm{d}s}{x-s}$
(8D.1)	$\sin ax\, J_0[b\surd(x^2+c^2)],\ 0<b<a,\ c\geq 0$	$-\cos ax\, J_0[b\surd(x^2+c^2)]$
(8D.2)	$\cos ax\, J_0[b\surd(x^2+c^2)],\ 0<b<a,\ c\geq 0$	$\sin ax\, J_0[b\surd(x^2+c^2)]$
(8D.3)	$\sin ax\, J_0[b\surd(x^2-c^2)],\ 0<b<a,\ c\geq 0$	$-\cos ax\, J_0[b\surd(x^2-c^2)]$
(8D.4)	$\cos ax\, J_0[b\surd(x^2-c^2)],\ 0<b<a,\ c\geq 0$	$\sin ax\, J_0[b\surd(x^2-c^2)]$
(8D.5)	$\cos ax\, J_1 ax,\ a>0$	$\sin ax\, J_1(ax)$
(8D.6)	$\sin ax\, J_n ax, a>0,\ n=0,1,2,\ldots$	$-\cos ax\, J_n ax$
(8D.7)	$\sin ax\, J_n bx, 0<b<a,\ n=0,1,2,\ldots$	$-\cos ax\, J_n bx$
(8D.8)	$\cos ax\, J_n bx, 0<b<a,\ n=0,1,2,\ldots$	$\sin ax\, J_n(bx)$

Table 1.8E. *Special functions: Bessel functions of the first kind of fractional order*

Number	$f(x)$	$\frac{1}{\pi} P \int_{-\infty}^{\infty} \frac{f(s)\mathrm{d}s}{x-s}$
(8E.1)	$\lvert x\rvert^{v} J_v(a\lvert x\rvert),\ a>0,$ $-1/2<\operatorname{Re} v<3/2$	$\operatorname{sgn} x\,\lvert x\rvert^{v}\,[\sec(v\pi)\mathbf{H}_{-v}(a\lvert x\rvert) - \tan(v\pi)J_v(a\lvert x\rvert)]$
(8E.2)	$\operatorname{sgn} x\,\lvert x\rvert^{v} J_v(a\lvert x\rvert),\ a>0,$ $-1/2<\operatorname{Re} v<3/2$	$\lvert x\rvert^{v}\,Y_v(a\,\lvert x\rvert)$
(8E.3)	$\lvert x\rvert^{-v} J_v(a\,\lvert x\rvert),\ a>0,$ $-1/2<\operatorname{Re} v$	$\operatorname{sgn} x\,\lvert x\rvert^{-v}\,\mathbf{H}_v(a\,\lvert x\rvert)$
(8E.4)	$\frac{\sin ax}{\sqrt{\lvert x\rvert}} J_{n+1/2}(\lvert x\rvert), 1<a,$ $n=0,1,2,\ldots$	$-\frac{\cos ax\,J_{n+1/2}(\lvert x\rvert)}{\sqrt{\lvert x\rvert}}$
(8E.5)	$\frac{\cos ax}{\sqrt{\lvert x\rvert}} J_{n+1/2}(\lvert x\rvert), 1<a,$ $n=0,1,2,\ldots$	$\frac{\sin ax\,J_{n+1/2}(\lvert x\rvert)}{\sqrt{\lvert x\rvert}}$
(8E.6)	$\frac{\sin ax}{\sqrt{\lvert x\rvert}} J_{2n+1/2}(b\lvert x\rvert), 0<b<a,$ $n=0,1,2,\ldots$	$-\frac{\cos ax\,J_{2n+1/2}(b\lvert x\rvert)}{\sqrt{\lvert x\rvert}}$
(8E.7)	$\frac{\cos ax}{\sqrt{\lvert x\rvert}} J_{2n+1/2}(b\lvert x\rvert), 0<b<a,$ $n=0,1,2,\ldots$	$\frac{\sin ax\,J_{2n+1/2}(b\,\lvert x\rvert)}{\sqrt{\lvert x\rvert}}$
(8E.8)	$\frac{\sin ax}{\lvert x\rvert^{v}} J_v(bx), 0<b<a,$ $\operatorname{Re} v>-1/2$	$-\frac{\cos ax}{\lvert x\rvert^{v}} J_v(bx)$
(8E.9)	$\frac{\cos ax}{\lvert x\rvert^{v}} J_v(bx), 0<b<a,$ $\operatorname{Re} v>-1/2$	$\frac{\sin ax}{\lvert x\rvert^{v}} J_v(bx)$
(8E.10)	$\frac{\sin ax}{\lvert x\rvert^{v}} J_{v+1}(bx), 0<b<a,$ $\operatorname{Re} v>-1/2$	$-\frac{\cos ax}{\lvert x\rvert^{v}} J_{v+1}(bx)$
(8E.11)	$\frac{\cos ax}{\lvert x\rvert^{v}} J_{v+1}(bx), 0<b<a,$ $\operatorname{Re} v>-1/2$	$\frac{\sin ax}{\lvert x\rvert^{v}} J_{v+1}(bx)$
(8E.12)	$\frac{\sin ax}{\lvert x\rvert^{v}} J_{2n+v+1}(bx), 0<b<a,$ $\operatorname{Re} v>-1/2,\quad n=0,1,2,\ldots$	$-\frac{\cos ax}{\lvert x\rvert^{v}} J_{2n+v+1}(bx)$
(8E.13)	$\frac{\cos ax}{\lvert x\rvert^{v}} J_{2n+v+1}(bx), 0<b<a,$ $\operatorname{Re} v>-1/2,\quad n=0,1,2,\ldots$	$\frac{\sin ax}{\lvert x\rvert^{v}} J_{2n+v+1}(bx)$

Table 1.8E. (*Cont.*)

Number	$f(x)$	$\frac{1}{\pi}P\int_{-\infty}^{\infty}\frac{f(s)\mathrm{d}s}{x-s}$
(8E.14)	$\frac{\sin ax}{\lvert x\rvert^{v+1/2}}J_{n+v+1}(x),\ 1<a,$ $\operatorname{Re}v>-1,\quad n=0,1,2,\ldots$	$-\frac{\cos ax}{\lvert x\rvert^{v+1/2}}J_{n+v+1}(x)$
(8E.15)	$\frac{\cos ax}{\lvert x\rvert^{v+1/2}}J_{n+v+1}(x),\ 1<a,$ $\operatorname{Re}v>-1,\quad n=0,1,2,\ldots$	$\frac{\sin ax}{\lvert x\rvert^{v+1/2}}J_{n+v+1}(x)$
(8E.16)	$\frac{\sin ax\ \cos x}{\lvert x\rvert^{v}}J_{v}(x),\ 2<a,$ $\operatorname{Re}v>-1/2$	$-\frac{\cos ax\ \cos x}{\lvert x\rvert^{v}}J_{v}(x)$
(8E.17)	$\frac{\cos ax\ \cos x}{\lvert x\rvert^{v}}J_{v}(x),\ 2<a,$ $\operatorname{Re}v>-1/2$	$\frac{\sin ax\ \cos x}{\lvert x\rvert^{v}}J_{v}(x)$
(8E.18)	$\frac{\sin ax\ \sin x}{\lvert x\rvert^{v}}J_{v}(x),\ 2<a,$ $\operatorname{Re}v>-1/2$	$-\frac{\cos ax\ \sin x}{\lvert x\rvert^{v}}J_{v}(x)$
(8E.19)	$\frac{\cos ax\ \sin x}{\lvert x\rvert^{v}}J_{v}(x),\ 2<a,$ $\operatorname{Re}v>-1/2$	$\frac{\sin ax\ \sin x}{\lvert x\rvert^{v}}J_{v}(x)$
(8E.20)	$\frac{\sin ax\ \sin x}{\lvert x\rvert^{v}}J_{v+1}(x),\ 2<a,$ $\operatorname{Re}v>-1/2$	$-\frac{\cos ax\ \sin x}{\lvert x\rvert^{v}}J_{v+1}(x)$
(8E.21)	$\frac{\cos ax\ \sin x}{\lvert x\rvert^{v}}J_{v+1}(x),\ 2<a,$ $\operatorname{Re}v>-1/2$	$\frac{\sin ax\ \sin x}{\lvert x\rvert^{v}}J_{v+1}(x)$
(8E.22)	$\frac{\sin ax\ \cos x}{\lvert x\rvert^{v}}J_{v+1}(x),\ 2<a,$ $\operatorname{Re}v>-1/2$	$-\frac{\cos ax\ \cos x}{\lvert x\rvert^{v}}J_{v+1}(x)$
(8E.23)	$\frac{\cos ax\ \cos x}{\lvert x\rvert^{v}}J_{v+1}(x),\ 2<a,$ $\operatorname{Re}v>-1/2$	$\frac{\sin ax\ \cos x}{\lvert x\rvert^{v}}J_{v+1}(x)$
(8E.24)	$\lvert x\rvert^{-v}\sin(ax)J_{v}(a\lvert x\rvert),\ a>0,$ $\operatorname{Re}v>-1/2$	$-\lvert x\rvert^{-v}\cos(ax)J_{v}(a\lvert x\rvert)$
(8E.25)	$\lvert x\rvert^{-v}\cos(ax)J_{v}(a\lvert x\rvert),\ a>0,$ $\operatorname{Re}v>-1/2$	$\lvert x\rvert^{-v}\sin(ax)J_{v}(a\lvert x\rvert)$

Table 1.8E. (*Cont.*)

Number	$f(x)$	$\frac{1}{\pi}P\int_{-\infty}^{\infty}\frac{f(s)\mathrm{d}s}{x-s}$
(8E.26)	$\frac{\sin ax}{(x^2+c^2)^{v/2}}J_v[b\surd(x^2+c^2)]$, $0<b<a,\ c>0,\ \operatorname{Re} v>-1/2$	$-\frac{\cos ax}{(x^2+c^2)^{v/2}}J_v[b\surd(x^2+c^2)]$
(8E.27)	$\frac{\cos ax}{(x^2+c^2)^{v/2}}J_v[b\surd(x^2+c^2)]$, $0<b<a,\ c>0,\ \operatorname{Re} v>-1/2$	$\frac{\sin ax}{(x^2+c^2)^{v/2}}J_v[b\surd(x^2+c^2)]$

Table 1.8F. *Special functions: Bessel functions of the second kind of fractional order*

Number	$f(x)$	$\frac{1}{\pi}P\int_{-\infty}^{\infty}\frac{f(s)\mathrm{d}s}{x-s}$
(8F.1)	$\lvert x\rvert^v Y_v(a\lvert x\rvert),\ a>0,$ $-1/2<\operatorname{Re} v<3/2$	$-\operatorname{sgn} x\,\lvert x\rvert^v J_v(a\lvert x\rvert)$
(8F.2)	$\lvert x\rvert^v Y_v(a\lvert x\rvert)\cos bx,\ 0<b<a,$ $-1/2<\operatorname{Re} v<1/2$	$-\operatorname{sgn} x\,\lvert x\rvert^v Y_v(a\lvert x\rvert)\cos bx$
(8F.3)	$\lvert x\rvert^v Y_v(a\lvert x\rvert)\sin bx,\ 0<b<a,$ $-1/2<\operatorname{Re} v<1/2$	$-\operatorname{sgn} x\,\lvert x\rvert^v Y_v(a\lvert x\rvert)\sin bx$

Table 1.8G. *Special functions: product of Bessel functions of the first kind of fractional order*

Number	$f(x)$	$\frac{1}{\pi}P\int_{-\infty}^{\infty}\frac{f(s)ds}{x-s}$
(8G.1)	$\sin ax\, J_{n+1/2}(bx) \times J_{n+1/2}(cx)$, $n = 0, 1, 2, \ldots$, $b > 0, c > 0$, $a > b + c$	$-\cos ax\, J_{n+1/2}(bx)J_{n+1/2}(cx)$
(8G.2)	$\cos ax\, J_{n+1/2}(bx) \times J_{n+1/2}(cx)$, $n = 0, 1, 2, \ldots$, $b > 0, c > 0$, $a > b + c$	$\sin ax\, J_{n+1/2}(bx)J_{n+1/2}(cx)$
(8G.3)	$\sin ax\, J_\nu(x)J_{-\nu}(x)$, $a > 2$	$-\cos ax\, J_\nu(x)J_{-\nu}(x)$
(8G.4)	$\cos ax\, J_\nu(x)J_{-\nu}(x)$, $a > 2$	$\sin ax\, J_\nu(x)J_{-\nu}(x)$
(8G.5)	$\sin ax\, J_{\mu+x}(b)J_{\nu-x}(b)$, $\mathrm{Re}(\mu + \nu) > 1$, $a > \pi$	$-\cos ax\, J_{\mu+x}(b)J_{\nu-x}(b)$
(8G.6)	$\cos ax\, J_{\mu+x}(b)J_{\nu-x}(b)$, $\mathrm{Re}(\mu + \nu) > 1$, $a > \pi$	$\sin ax\, J_{\mu+x}(b)J_{\nu-x}(b)$
(8G.7)	$\sqrt{(x)}\sin ax\, \{J_{-1/4}(bx)\}^2$, $0 < 2b < a$	$-\sqrt{(x)}\cos ax\, \{J_{-1/4}(bx)\}^2$
(8G.8)	$\sqrt{(x)}\cos ax\, \{J_{-1/4}(bx)\}^2$, $0 < 2b < a$	$\sqrt{(x)}\sin ax\, \{J_{-1/4}(bx)\}^2$
(8G.9)	$\sqrt{(x)}\sin ax\, \{J_{1/4}(bx)\}^2$, $0 < 2b < a$	$-\sqrt{(x)}\cos ax\, \{J_{1/4}(bx)\}^2$
(8G.10)	$\sqrt{(x)}\cos ax\, \{J_{1/4}(bx)\}^2$, $0 < 2b < a$	$\sqrt{(x)}\sin ax\, \{J_{1/4}(bx)\}^2$
(8G.11)	$\sqrt{(x)}\sin ax\, J_{\nu-1/4}(bx) \times J_{-\nu-1/4}(bx)$, $0 < 2b < a$	$-\sqrt{(x)}\cos ax\, J_{\nu-1/4}(bx)\, J_{-\nu-1/4}(bx)$
(8G.12)	$\sqrt{(x)}\cos ax\, J_{\nu-1/4}(bx) \times J_{-\nu-1/4}(bx)$, $0 < 2b < a$	$\sqrt{(x)}\sin ax\, J_{\nu-1/4}(bx)\, J_{-\nu-1/4}(bx)$
(8G.13)	$\sqrt{(x)}\sin ax\, J_{\nu+1/4}(bx) \times J_{-\nu+1/4}(bx)$, $0 < 2b < a$	$-\sqrt{(x)}\cos ax\, J_{\nu+1/4}(bx)\, J_{-\nu+1/4}(bx)$
(8G.14)	$\sqrt{(x)}\cos ax\, J_{\nu+1/4}(bx) \times J_{-\nu+1/4}(bx)$, $0 < 2b < a$	$\sqrt{(x)}\sin ax\, J_{\nu+1/4}(bx)J_{-\nu+1/4}(bx)$
(8G.15)	$\sin ax\, J_{\nu-x}(b)J_{\nu+x}(b)$, $\pi < a$, $\nu > 1/2$	$-\cos ax\, J_{\nu-x}(b)J_{\nu+x}(b)$
(8G.16)	$\cos ax\, J_{\nu-x}(b)\, J_{\nu+x}(b)$, $\pi < a$, $\nu > 1/2$	$\sin ax\, J_{\nu-x}(b)\, J_{\nu+x}(b)$

Table 1.8H. *Special functions: modified Bessel functions of the first kind*

Number	$f(x)$	$\frac{1}{\pi}P\int_{-\infty}^{\infty}\frac{f(s)\mathrm{d}s}{x-s}$
(8H.1)	$\mathrm{e}^{-a\lvert x\rvert}I_0(ax), \quad a>0$	$2\pi^{-1}\sinh(ax)K_0(a\lvert x\rvert)$
(8H.2)	$\operatorname{sgn} x\, \mathrm{e}^{-a\lvert x\rvert}I_0(ax), a>0$	$-2\pi^{-1}\cosh(ax)K_0(a\lvert x\rvert)$
(8H.3)	$\begin{cases} 0, & x<0 \\ \mathrm{e}^{-ax}I_0(ax), & x>0, \quad a>0 \end{cases}$	$-\frac{1}{\pi}\mathrm{e}^{-ax}K_0(a\lvert x\rvert)$
(8H.4)	$\begin{cases} \mathrm{e}^{ax}I_0(ax), & x<0 \\ 0, & x>0, \quad a>0 \end{cases}$	$\frac{1}{\pi}\mathrm{e}^{ax}K_0(a\lvert x\rvert)$
(8H.5)	$\sin ax\, I_{\nu-x}(b)I_{\nu+x}(b), \ \pi<a$	$-\cos ax\, I_{\nu-x}(b)I_{\nu+x}(b)$
(8H.6)	$\cos ax\, I_{\nu-x}(b)I_{\nu+x}(b), \ \pi<a$	$\sin ax\, I_{\nu-x}(b)I_{\nu+x}(b)$

Table 1.8I. *Special functions: modified Bessel functions of the second kind*

Number	$f(x)$	$\frac{1}{\pi}P\int_{-\infty}^{\infty}\frac{f(s)\mathrm{d}s}{x-s}$
(8I.1)	$K_0(a\lvert x\rvert), \ a>0$	$\frac{\pi}{2}\{\operatorname{sgn} x\, I_0(ax)-\mathbf{L}_0(ax)\}$
(8I.2)	$\lvert x\rvert^{\nu}K_\nu(a\lvert x\rvert), \ a>0,$ $-1/2<\operatorname{Re}\nu<1/2$	$\frac{\pi\sec(\pi\nu)}{2}\operatorname{sgn} x\,\lvert x\rvert^{\nu}\{I_\nu(a\lvert x\rvert)-\mathbf{L}_{-\nu}(a\lvert x\rvert)\}$
(8I.3)	$\mathrm{e}^{ax}K_0(a\lvert x\rvert), \ a>0$	$-\pi\mathrm{e}^{ax}I_0(ax), \ -\infty<x<0,$ $0, \ 0<x<\infty$
(8I.4)	$\mathrm{e}^{-ax}K_0(a\lvert x\rvert),$ $a>0$	$0, \ -\infty<x<0,$ $\pi\mathrm{e}^{-ax}I_0(ax), \ 0<x<\infty$
(8I.5)	$\sinh(ax)K_0(a\lvert x\rvert),$	$-\frac{\pi}{2}\mathrm{e}^{-a\lvert x\rvert}I_0(ax)$
(8I.6)	$\cosh(ax)K_0(a\lvert x\rvert),$ $a>0$	$\frac{\pi}{2}\operatorname{sgn} x\,\mathrm{e}^{-a\lvert x\rvert}I_0(ax)$
(8I.7)	$\mathrm{e}^{-a\lvert x\rvert}\sinh(ax)$ $\times I_0(ax)K_0(a\lvert x\rvert),$ $a>0$	$\frac{1}{\pi}\sinh^2(ax)\{K_0(a\lvert x\rvert)\}^2-\frac{\pi}{4}\mathrm{e}^{-2a\lvert x\rvert}\{I_0(ax)\}^2$

Table 1.8J. *Special functions: spherical Bessel functions of the first kind*

Number	$f(x)$	$\frac{1}{\pi}P\int_{-\infty}^{\infty}\frac{f(s)\mathrm{d}s}{x-s}$
(8J.1)	$j_0(x)$	$\frac{1-\cos x}{x}$
(8J.2)	$j_1(x)$	$-\frac{\sin x}{x}+\frac{1-\cos x}{x^2}$
(8J.3)	$j_2(x)$	$\frac{1}{2x}+\frac{3}{x^3}+\left(\frac{1}{x}-\frac{3}{x^3}\right)\cos x-\frac{3\sin x}{x^2}$
(8J.4)	$j_{n+1}(x)$	$H[j_{n+1}(x)]=\frac{(2n+1)}{x}H[j_n(x)]-H[j_{n-1}(x)] +\frac{1}{\pi x}\int_{-\infty}^{\infty}[j_{n+1}(t)+j_{n-1}(t)]\mathrm{d}t$

Table 1.8K. *Special functions: spherical Bessel functions of the second kind*

Number	$f(x)$	$\frac{1}{\pi}P\int_{-\infty}^{\infty}\frac{f(s)\mathrm{d}s}{x-s}$
(8K.1)	$y_0(x)$	$\pi\delta(x)-\frac{\sin x}{x}$
(8K.2)	$y_1(x)$	$-\pi\delta'(x)+\frac{\cos x}{x}-\frac{\sin x}{x^2}$
(8K.3)	$y_2(x)$	$\frac{\pi}{2}(\delta(x)+3\delta''(x))+\frac{3\cos x}{x^2}+\left(\frac{1}{x}-\frac{3}{x^3}\right)\sin x$

Table 1.8L. *Special functions: cosine integral function*

Number	$f(x)$	$\frac{1}{\pi}P\int_{-\infty}^{\infty}\frac{f(s)\mathrm{d}s}{x-s}$
(8L.1)	$\mathrm{Ci}(a\lvert x\rvert),\quad a>0$	$\operatorname{sgn} x\,\mathrm{si}(a\lvert x\rvert)$
(8L.2)	$\mathrm{Ci}(a\lvert x\rvert)\sin bx, a>0$	$\operatorname{sgn} x\ \sin bx[\mathrm{Si}(a\lvert x\rvert)-\pi/2],$ $a\ge b>0$ $\operatorname{sgn} x\sin bx[\mathrm{Si}(b\lvert x\rvert)-\pi/2]$ $+\cos bx[\mathrm{Ci}(b\lvert x\rvert)-\mathrm{Ci}(a\lvert x\rvert)],$ $b\ge a$ $\operatorname{sgn} x\sin bx[\mathrm{Si}(\lvert bx\rvert)-\pi/2]$ $+\cos bx[\mathrm{Ci}(a\lvert x\rvert)-\mathrm{Ci}(\lvert bx\rvert)],$ $b<0$ and $\lvert b\rvert\ge a,$ $\operatorname{sgn} x\sin bx[\mathrm{Si}(a\lvert x\rvert)-\pi/2],$ $b<0$ and $\lvert b\rvert\le a$
(8L.3)	$\mathrm{Ci}(a\lvert x\rvert)\cos bx,\quad a>0$	$\cos bx\ \operatorname{sgn} x[\mathrm{Si}(a\lvert x\rvert)-\pi/2],$ $a\ge b>0$ $\cos bx\ \operatorname{sgn} x[\mathrm{Si}(b\lvert x\rvert)-\pi/2]$ $+\sin bx[\mathrm{Ci}(a\lvert x\rvert)-\mathrm{Ci}(b\lvert x\rvert)],$ $b\ge a$ $\cos bx\ \operatorname{sgn} x[\mathrm{Si}(\lvert bx\rvert)-\pi/2]$ $+\sin bx[\mathrm{Ci}(\lvert bx\rvert)-\mathrm{Ci}(a\lvert x\rvert)],$ $b<0$ and $\lvert b\rvert\ge a$ $\cos bx\ \operatorname{sgn} x[\mathrm{Si}(a\lvert x\rvert)-\pi/2],$ $b<0$ and $\lvert b\rvert\le a$

Table 1.8M. *Special functions: sine integral function*

Number	$f(x)$	$\frac{1}{\pi}P\int_{-\infty}^{\infty}\frac{f(s)\mathrm{d}s}{x-s}$
(8M.1)	$\operatorname{sgn} x\ \mathrm{si}(a\lvert x\rvert),\ a>0$	$-\mathrm{Ci}(a\lvert x\rvert)$
(8M.2)	$\mathrm{Si}(ax)\sin bx$	$-\cos bx\,\mathrm{Si}(ax),\ b\ge a>0$ $-\cos bx\,\mathrm{Si}(bx)+\sin bx[\mathrm{Ci}(b\lvert x\rvert)$ $-\mathrm{Ci}(a\lvert x\rvert)],\quad a\ge b>0$
(8M.3)	$\mathrm{Si}(ax)\cos bx$	$\sin bx\,\mathrm{Si}(ax),\ b\ge a>0$ $\sin bx\,\mathrm{Si}(bx)-\cos bx[\mathrm{Ci}(a\lvert x\rvert)$ $-\mathrm{Ci}(b\lvert x\rvert)],\quad a\ge b>0$
(8M.4)	$\sin a\lvert x\rvert\,\mathrm{si}(a\lvert x\rvert),\ a>0$	$-\sin(ax)\mathrm{Ci}(a\lvert x\rvert)$
(8M.5)	$\sin(ax)\,\mathrm{Ci}(a\lvert x\rvert)$ $-\operatorname{sgn} x\cos\ ax\,\mathrm{si}(a\lvert x\rvert),\ a>0$	$\sin(a\lvert x\rvert)\mathrm{si}(a\lvert x\rvert)+\cos(ax)\mathrm{Ci}(a\lvert x\rvert)$

Table 1.8N. *Special functions: Struve functions*

Number	$f(x)$	$\frac{1}{\pi}P\int_{-\infty}^{\infty}\frac{f(s)\mathrm{d}s}{x-s}$
(8N.1)	$\mathbf{H}_0(ax),\ 0<a$	$-J_0(ax)$
(8N.2)	$\frac{\sin ax}{x^\nu}\mathbf{H}_\nu(bx),\ 0<b<a,\quad \mathrm{Re}\,\nu>-1/2$	$-\frac{\cos ax}{x^\nu}\mathbf{H}_\nu(bx)$
(8N.3)	$\frac{\cos ax}{x^\nu}\mathbf{H}_\nu(bx),\ 0<b<a,\quad \mathrm{Re}\,\nu>-1/2$	$\frac{\sin ax}{x^\nu}\mathbf{H}_\nu(bx)$

Table 1.8O. *Special functions: Anger functions*

Number	$f(x)$	$\frac{1}{\pi}P\int_{-\infty}^{\infty}\frac{f(s)\mathrm{d}s}{x-s}$
(8O.1)	$\mathbf{J}_\nu(ax),\quad 0<a$	$-\mathbf{E}_\nu(ax)$
(8O.2)	$\sin ax[\mathbf{J}_\nu(bx)-\mathbf{J}_{-\nu}(bx)],\ 0<b<a$	$-\cos ax[\mathbf{J}_\nu(bx)-\mathbf{J}_{-\nu}(bx)]$
(8O.3)	$\cos ax[\mathbf{J}_\nu(bx)-\mathbf{J}_{-\nu}(bx)],\ 0<b<a$	$\sin ax[\mathbf{J}_\nu(bx)-\mathbf{J}_{-\nu}(bx)]$

Table 1.8P. *Special functions: miscellaneous special functions*

Number	$f(x)$	$\frac{1}{\pi}P\int_{-\infty}^{\infty}\frac{f(s)\mathrm{d}s}{x-s}$
(8P.1)	$x\,{}_2F_1\left(v+1/2,1;3/2;-\frac{x^2}{a^2}\right)$ $v>0,\ a>0,$	$-\frac{\sqrt{(\pi)}\,\Gamma(v)a^{1+2v}}{2\Gamma(v+1/2)(x^2+a^2)^v}$
(8P.2)	$\mathbf{E}_\nu(ax),\quad 0<a$	$\mathbf{J}_\nu(ax)$
(8P.3)	$\mathbf{L}_0(\mathrm{i}a\,x),\quad 0<a$	$-\mathrm{i}\,J_0(ax)$
(8P.4)	$\mathrm{Li}_2(\mathrm{e}^{\mathrm{i}x})$	$-\mathrm{i}\,\mathrm{Li}_2(\mathrm{e}^{\mathrm{i}x})$
(8P.5)	$\mathrm{Cl}_2(x)$	$-\sum_{n=1}^{\infty}\frac{\cos nx}{n^2}=-\frac{\pi^2}{6}+\frac{\pi x}{2}-\frac{x^2}{4},$ $0\le x<2\pi$

Table 1.9. *Pulse and wave forms*

Number	$f(x)$	$\frac{1}{\pi}P\int_{-\infty}^{\infty}\frac{f(s)\mathrm{d}s}{x-s}$
(9.1)	$0, -\infty < x < a$, $1, a < x < b$, $0, b < x < \infty$, rectangular pulse "boxcar pulse"	$\frac{1}{\pi}\log\left\lvert\frac{a-x}{b-x}\right\rvert$
(9.2)	$\Pi_{2a}(x)$ unit rectangular step function	$\frac{1}{\pi}\log\left\lvert\frac{a+x}{a-x}\right\rvert$
(9.3)	$\Pi_{2a}(x-a)$ one-sided rectangular pulse	$\frac{1}{\pi}\log\left\lvert\frac{x}{x-2a}\right\rvert$
(9.4)	$0, -\infty < x < 0$ $\frac{hx}{a}, 0 \le x < a$ $h, a \le x \le b$ $0, b < x < \infty$ ramp(x)	$-\frac{h}{\pi}\left\{1+\frac{x}{a}\log\left\lvert\frac{a-x}{x}\right\rvert+\log\left\lvert\frac{b-x}{x-a}\right\rvert\right\}$
(9.5)	$0, -\infty < x < a$ $h\frac{(x-a)}{(b-a)}, a < x < b$ $h\frac{(c-x)}{(c-b)}, b < x < c$ $0, c < x < \infty$ triangular pulse	$-\frac{h}{\pi}\left\{\frac{x-a}{b-a}\log\left\lvert\frac{x-b}{x-a}\right\rvert+\frac{c-x}{c-b}\log\left\lvert\frac{x-c}{x-b}\right\rvert\right\}$

Number	$f(x)$	$\frac{1}{\pi}P\int_{-\infty}^{\infty}\frac{f(s)\mathrm{d}s}{x-s}$
(9.6)	$1-(\lvert x\rvert/a), \lvert x\rvert \le a$ $0, \lvert x\rvert > a$ $a>0$ triangular pulse	$-\frac{1}{\pi}\left\{\log\left\lvert\frac{x-a}{x+a}\right\rvert+\frac{x}{a}\log\left\lvert\frac{x^2}{x^2-a^2}\right\rvert\right\}$
(9.7)	$0, \ -\infty<x<0$ $a^{-1}b(a-x), \ 0\le x\le a$ $0, \ a\le x<\infty$ one-sided triangular pulse	$\frac{b}{a\pi}\left\{a-(a-x)\log\left\lvert\frac{x-a}{x}\right\rvert\right\}$
(9.8)	$\operatorname{sgn} x\,\Pi_{2a}(x)$	$-\frac{1}{\pi}\log\left\lvert\frac{x^2-a^2}{x^2}\right\rvert$
(9.9)	$0, -\infty<x\le -b$ $\frac{b+x}{b-a}, -b<x<-a$ $1, \ -a<x<a$ $\frac{b-x}{b-a}, \ a<x<b$ $0, \ b\le x<\infty$ for $0<a<b$ trapezoid pulse	$-\frac{1}{\pi}\left\{\frac{b}{b-a}\log\left\lvert\frac{(a+x)(b-x)}{(a-x)(b+x)}\right\rvert+\frac{x}{b-a}\log\left\lvert\frac{a^2-x^2}{b^2-x^2}\right\rvert+\log\left\lvert\frac{a-x}{a+x}\right\rvert\right\}$
(9.10)	$1-(x/a)^2, \quad \lvert x\rvert\le a$ $0, \quad \lvert x\rvert>a$ parabolic pulse	$\frac{1}{\pi}\left\{\frac{2x}{a}-[1-(x/a)^2]\log\left\lvert\frac{x-a}{x+a}\right\rvert\right\}$
(9.11)	$\cos^2(\pi x/2a), \quad \lvert x\rvert\le a$ $0, \quad \lvert x\rvert>a$ $a>0$ video pulse	$\frac{1}{2\pi}\log\left\lvert\frac{x+a}{x-a}\right\rvert+\frac{\cos(\pi x/a)}{2\pi}\left\{\mathrm{Ci}\left[\frac{\pi}{a}\lvert x+a\rvert\right]-\mathrm{Ci}\left[\frac{\pi}{a}\lvert a-x\rvert\right]\right\}$ $+\frac{\sin(\pi x/a)}{2\pi}\left\{\mathrm{Si}\left[\frac{\pi}{a}(x+a)\right]-\mathrm{Si}\left[\frac{\pi}{a}(x-a)\right]\right\}$

Table 1.9. (*Cont.*)

Number	$f(x)$	$\frac{1}{\pi} P \int_{-\infty}^{\infty} \frac{f(s)\mathrm{d}s}{x-s}$
(9.12)	$0, \quad -\infty < x < a$ $x^{-1}, \quad a < x < \infty$ $a > 0$	$\frac{1}{\pi x} \log\left\|\frac{a-x}{a}\right\|, \quad x \neq 0, a$
(9.13)	$x^{-1}, \quad -\infty < x < a$ $0, \quad a < x < b$ $x^{-1}, \quad b < x < \infty$ $a < 0, \; b > 0$	$\frac{1}{\pi x} \log\left\|\frac{a(b-x)}{b(x-a)}\right\|, \quad x \neq 0, a, b$
(9.14)	$0, \quad -\infty < x < a$ $x^{-2}, a < x < \infty$ $a > 0$	$\frac{1}{\pi x^2}\left[\log\left\|\frac{a-x}{a}\right\| + a^{-1}x\right], \quad x \neq 0, a$
(9.15)	$0, -\infty < x < 0$ $x^{\mu}, \quad 0 < x < \infty$ $-1 < \mathrm{Re}\,\mu < 0$	$(-x)^{\mu}\csc(\mu\pi), \quad -\infty < x < 0$ $x^{\mu}\cot(\mu\pi), \quad 0 < x < \infty$
(9.16)	$0, \quad -\infty < x < 0$ $1/(ax+b), \quad 0 < x < \infty$ $a > 0, \; b > 0$	$\frac{1}{\pi(ax+b)} \log\|ax/b\|, \quad x \neq 0, -b/a$
(9.17)	$0, \quad -\infty < x < 0$ $1/(ax+b)^2, \quad 0 < x < \infty$ $a > 0, \; b > 0$	$\frac{1}{\pi(ax+b)^2} \log\left\|\frac{ax}{b}\right\| + \frac{1}{\pi b(ax+b)}, \quad x \neq 0, -b/a$

Number	$f(x)$	$\frac{1}{\pi} P \int_{-\infty}^{\infty} \frac{f(s)\mathrm{d}s}{x-s}$
(9.18)	$0, \quad -\infty < x < 0$ $1/(x^2+a^2), \quad 0 < x < \infty$ $a > 0$	$\frac{\{x + 2a\pi^{-1} \log(\lvert x\rvert / a)\}}{2a(x^2+a^2)}$
(9.19)	$0, \quad -\infty < x < 0$ $x/x^2+a^2, \quad 0 < x < \infty$ $a > 0$	$\frac{\left\{2x\pi^{-1} \log\left(\frac{\lvert x\rvert}{a}\right) - a\right\}}{2(x^2+a^2)}$
(9.20)	$0, \quad -\infty < x < 0$ $(cx+d)/(ax+b)^2, \quad 0 < x < \infty$ $a > 0, b > 0, \quad x \neq 0, -b/a$	$\frac{cx+d}{\pi(ax+b)^2} \log\left\lvert\frac{ax}{b}\right\rvert + \frac{ad-bc}{\pi ab(ax+b)}$
(9.21)	$0, \quad -\infty < x < -a$ $\sqrt{(a^2-x^2)}, \quad -a < x < a$ $0, \quad a < x < \infty$	$x + \sqrt{(x^2-a^2)}, \quad -\infty < x < -a$ $x, \quad -a < x < a$ $x - \sqrt{(x^2-a^2)}, \quad a < x < \infty$
(9.22)	$0, \quad -\infty < x < -a$ $x\sqrt{(a^2-x^2)}, \quad -a < x < a$ $0, \quad a < x < \infty$	$x^2 + x\sqrt{(x^2-a^2)} - a^2/2, \quad -\infty < x < -a$ $x^2 - a^2/2, \quad -a < x < a$ $x^2 - x\sqrt{(x^2-a^2)} - a^2/2, \quad a < x < \infty$

Table 1.9. (*Cont.*)

Number	$f(x)$	$\frac{1}{\pi}P\int_{-\infty}^{\infty}\frac{f(s)\mathrm{d}s}{x-s}$
(9.23)	$0,\quad -\infty < x < 0$	$\frac{a}{\pi}+\frac{x}{2}+\sqrt{(x^2-a^2)}\left\{\frac{1}{2}+\frac{1}{\pi}\sin^{-1}(a/x)\right\}, -\infty < x < -a$
	$\sqrt{(a^2-x^2)},\quad 0 < x < a$	$\frac{a}{\pi}+\frac{x}{2}-\frac{\sqrt{(a^2-x^2)}}{\pi}\log\left\lvert\frac{a+\sqrt{(a^2-x^2)}}{x}\right\rvert, -a < x < a$
	$0,\quad a < x < \infty$	$\frac{a}{\pi}+\frac{x}{2}-\sqrt{(x^2-a^2)}\left\{\frac{1}{2}+\frac{1}{\pi}\sin^{-1}(a/x)\right\}, a < x < \infty$
		principal value of arcsin is taken (between $-\pi/2$ and $\pi/2$)
(9.24)	$0,\quad -\infty < x < -a$	$-\frac{1}{\sqrt{(x^2-a^2)}},\quad -\infty < x < -a$
	$\frac{1}{\sqrt{(a^2-x^2)}},\quad -a < x < a$	$0, -a < x < a$
	$0,\quad a < x < \infty$	$\frac{1}{\sqrt{(x^2-a^2)}},\quad a < x < \infty$
(9.25)	$-\frac{1}{\sqrt{(x^2-a^2)}},\quad -\infty < x < -a$	$0, -\infty < x < -a$
	$0,\quad -a < x < a$	$-\frac{1}{\sqrt{(a^2-x^2)}},\quad -a < x < a$
	$\frac{1}{\sqrt{(x^2-a^2)}},\quad a < x < \infty$	$0, a < x < \infty$
(9.26)	$0,\quad -\infty < x < -a$	$1-\sqrt{\left(\frac{a-x}{\lvert x+a\rvert}\right)},\quad -\infty < x < -a$
	$\sqrt{\left(\frac{a-x}{a+x}\right)},\quad -a < x < a$	$1, -a < x < a$
	$0,\quad a < x < \infty$	$1-\sqrt{\left(\frac{x-a}{x+a}\right)},\quad a < x < \infty$

Number	$f(x)$	$\frac{1}{\pi} P \int_{-\infty}^{\infty} \frac{f(s)\mathrm{d}s}{x-s}$
(9.27)	$0, \quad -\infty < x < 0$ $\sqrt{\left(\frac{a-x}{a+x}\right)}, \quad 0 < x < a$ $0, \quad a < x < \infty$	$1/2 - \frac{1}{\pi}\sqrt{\left(\left\|\frac{a-x}{a+x}\right\|\right)} \cos^{-1}(-a/x), \quad a < \|x\| < \infty$ $1/2 - \frac{1}{\pi}\sqrt{\left(\frac{a-x}{a+x}\right)} \log\left\|\frac{a+\sqrt{(a^2-x^2)}}{x}\right\|, \quad \|x\| < a,$ principal value of arccos is taken (between 0 and π)
(9.28)	$0, \quad -\infty < x < a$ $\frac{(x-a)^\mu}{(b-x)^\mu}, \quad a < x < b$ $0, \quad b < x < \infty$ $-1 < \operatorname{Re}\mu < 1$	$\csc(\mu\pi)\left\{(a-x)^\mu(b-x)^{-\mu} - 1\right\}, \quad -\infty < x < a$ $\csc(\mu\pi)\{(x-a)^\mu(b-x)^{-\mu}\cos(\mu\pi) - 1\}, a < x < b$ $\csc(\mu\pi)\left\{(x-a)^\mu(x-b)^{-\mu} - 1\right\}, \quad b < x < \infty$
(9.29)	$0, \quad -\infty < x < a$ $\frac{(x-a)^{\mu-1}}{(b-x)^\mu}, \quad a < x < b$ $0, \quad b < x < \infty$ $0 < \operatorname{Re}\mu < 1$	$\csc(\mu\pi)(x-b)^{-1}\left\|(a-x)(b-x)^{-1}\right\|^{\mu-1}, -\infty < x < a$ or $b < x < \infty$ $(x-a)^{\mu-1}(b-x)^{-\mu}\cot(\mu\pi), \quad a < x < b$
(9.30)	$0, \quad -\infty < x < a$ $(x-a)^\mu(b-x)^\mu, \quad a < x < b$ $0, \quad b < x < \infty$ $-1 < \operatorname{Re}\mu$	$\frac{(b-a)^{1+2\mu}\Gamma(\mu+1)_2F_1\left(1,\ 1+\mu;\ 2+2\mu;\ \frac{b-a}{x-a}\right)}{2^{1+2\mu}\sqrt{(\pi)}\,(x-a)\Gamma(\mu+3/2)},$ $-\infty < x < a, \quad b < x < \infty$ $(x-a)^\mu(b-x)^\mu\cot\pi\mu$ $-\frac{(b-a)^{2\mu}\Gamma(\mu)_2F_1\left(1, -2\mu; 1-\mu; \frac{x-a}{b-a}\right)}{4^\mu\sqrt{\pi}\,\Gamma(\mu+1/2)},$ $a < x < b$
(9.31)	$0, \quad -\infty < x < -a$ $x\sqrt{\left(\frac{a-x}{a+x}\right)}, \quad -a < x < a$ $0, \quad a < x < \infty$	$x - a - x\sqrt{\left(\left\|\frac{a-x}{a+x}\right\|\right)}, \quad -\infty < x < -a$ $x - a, \quad -a < x < a$ $x - a - x\sqrt{\left(\frac{x-a}{x+a}\right)}, \quad a < x < \infty$

Table 1.9. (*Cont.*)

Number	$f(x)$	$\frac{1}{\pi} P \int_{-\infty}^{\infty} \frac{f(s)\mathrm{d}s}{x-s}$
(9.32)	$0, \quad -\infty < x < a$ $\sqrt{(x-a)}, \quad a < x < b$ $\sqrt{(x-a)} - \sqrt{(x-b)}, \quad b < x < \infty$	$\sqrt{(a-x)} - \sqrt{(b-x)}, \quad -\infty < x < a$ $-\sqrt{(b-x)}, \quad a < x < b$ $0, \quad b < x < \infty$
(9.33)	$\sqrt{(a-x)} - \sqrt{(b-x)}, -\infty < x < a$ $-\sqrt{(b-x)}, \quad a < x < b$ $0, \quad b < x < \infty$	$0, -\infty < x < a$ $-\sqrt{(x-a)}, \quad a < x < b$ $\sqrt{(x-b)} - \sqrt{(x-a)}, \quad b < x < \infty$
(9.34)	$0, \quad -\infty < x < 0$ $\frac{1}{\sqrt{(ax+b)}}, \quad 0 < x < \infty$ $a, b > 0$	$-\frac{2\tan^{-1}[\sqrt{(\lvert ax+b\rvert/b)}]}{\pi\sqrt{(\lvert ax+b\rvert)}}, \quad -\infty < x < -b/a$ $\frac{1}{\pi\sqrt{(ax+b)}} \log\left\lvert\frac{\sqrt{b}-\sqrt{(ax+b)}}{\sqrt{b}+\sqrt{(ax+b)}}\right\rvert, \quad -b/a < x < \infty$
(9.35)	$0, \quad -\infty < x < 0$ $\frac{1}{\sqrt{(a^2-x^2)}}, \quad 0 < x < a$ $0, \quad a < x < \infty$	$\frac{-\cos^{-1}(-a/x)}{\pi\sqrt{(x^2-a^2)}}, \quad -\infty < x < -a$ $\frac{\log\left\lvert\frac{x}{\sqrt{(a^2-x^2)}+a}\right\rvert}{\pi\sqrt{(a^2-x^2)}}, \quad -a < x < a$ $\frac{\cos^{-1}(-a/x)}{\pi\sqrt{(x^2-a^2)}}, \quad a < x < \infty$ principal value of arccos is taken (between 0 and π)

Table 1.9. (*Cont.*)

Number	$f(x)$	$\frac{1}{\pi}P\int_{-\infty}^{\infty}\frac{f(s)\mathrm{d}s}{x-s}$
(9.36)	$0, \quad -\infty < x < a$ $\frac{1}{\sqrt{(x^2-a^2)}}, \quad a < x < \infty$ $a > 0$	$\frac{\log\left\lvert\frac{a}{\sqrt{(x^2-a^2)}-x}\right\rvert}{\pi\sqrt{(x^2-a^2)}}, \quad -\infty < x < -a, \quad a < x < \infty$ $\frac{-\cos^{-1}(-x/a)}{\pi\sqrt{(a^2-x^2)}}, \quad -a < x < a$ principal value of arccos is taken (between 0 and π)
(9.37)	$\frac{x}{\sqrt{(a^2-x^2)}}, \quad \lvert x\rvert < a$ $0, \quad \lvert x\rvert > a$	$-1, \lvert x\rvert < a$ $-1+\frac{\lvert x\rvert}{\sqrt{(x^2-a^2)}}, \quad \lvert x\rvert > a$
(9.38)	$0, -\infty < x < a$ $\mathrm{e}^{-bx}, \quad a < x < \infty$ $b > 0$	$-\pi^{-1}\mathrm{e}^{-bx}E_1(ab-bx), \quad -\infty < x < a$ $\pi^{-1}\mathrm{e}^{-bx}\mathrm{Ei}(bx-ab), \quad a < x < \infty$
(9.39)	$0, \quad -\infty < x < 0$ $\mathrm{e}^{-ax}, \quad 0 < x < \infty$ $a > 0$	$-\pi^{-1}\mathrm{e}^{-ax}E_1(-ax), \quad -\infty < x < 0$ $\pi^{-1}\mathrm{e}^{-ax}\mathrm{Ei}(ax), \quad 0 < x < \infty$
(9.40)	$0, \quad -\infty < x < 0$ $xe^{-ax}, \quad 0 < x < \infty$ $a > 0$	$-\pi^{-1}\{x\mathrm{e}^{-ax}E_1(-ax)+a^{-1}\}, \quad -\infty < x < 0$ $\pi^{-1}\{x\mathrm{e}^{-ax}\mathrm{Ei}(ax)-a^{-1}\}, \quad 0 < x < \infty$
(9.41)	$0, \quad -\infty < x < 0$ $\sin ax, \quad 0 < x < \infty,$ $a > 0$	$\frac{1}{\pi}\{\sin ax\,\mathrm{Ci}(a\lvert x\rvert)-\cos ax[\mathrm{Si}(ax)+\pi/2]\}$
(9.42)	$0, -\infty < x < 0$ $\cos ax, \quad 0 < x < \infty,$ $a > 0$	$\frac{1}{\pi}\{\cos ax\,\mathrm{Ci}(a\lvert x\rvert)+\sin ax[\pi/2+\mathrm{Si}(ax)]\}$

Table 1.9. (*Cont.*)

Number	$f(x)$	$\frac{1}{\pi} P \int_{-\infty}^{\infty} \frac{f(s)\mathrm{d}s}{x-s}$
(9.43)	$0, \quad -\infty < x < 0$ $\sin(a\sqrt{x}), \quad 0 < x < \infty$	$-\mathrm{sgn}\, a\, \mathrm{e}^{-\lvert a\rvert\sqrt{\lvert x\rvert}}, \quad -\infty < x < 0$ $-\mathrm{sgn}\, a\, \cos(a\sqrt{x}), \quad 0 < x < \infty$
(9.44)	$\sin \pi x\, \mathrm{e}^{-\pi\lvert x\rvert}\mathrm{H}(x)$	$\frac{\mathrm{e}^{-\pi x}}{\pi}[\cos \pi x\, \mathrm{sie}(-\pi x, 1) - \sin \pi x\, \mathrm{Cie}(\pi x, 1)], \ x > 0$ $\frac{\mathrm{e}^{-\pi x}}{\pi}[\cos \pi x\, \mathrm{sie}(-\pi x, 1) - \sin \pi x\, \mathrm{cie}(-\pi x, 1)], \ x < 0$
(9.45)	$\cos \pi x\, \mathrm{e}^{-\pi\lvert x\rvert}\mathrm{H}(x)$	$-\frac{\mathrm{e}^{-\pi x}}{\pi}[\sin \pi x\, \mathrm{sie}(-\pi x, 1) + \cos \pi x\, \mathrm{Cie}(\pi\, x, 1)], \ x > 0$ $-\frac{\mathrm{e}^{-\pi x}}{\pi}[\sin \pi x\, \mathrm{sie}(-\pi x, 1) + \cos \pi x\, \mathrm{cie}(-\pi x, 1)], \ x < 0$

Table 1.10. *Distributions*

Some cases for which the result for *Hf* includes a delta function can also be found in Tables 1.2 and 1.5.

Number	$f(x)$	$\frac{1}{\pi}P\int_{-\infty}^{\infty}\frac{f(s)ds}{x-s}$
(10.1)	$\delta(x)$	$p.v.\frac{1}{\pi x}$
(10.2)	$\delta(x+a)$	$p.v.\frac{1}{\pi(x+a)}$
(10.3)	$\delta(ax+b)$	$\frac{\operatorname{sgn} a}{\pi}p.v.\frac{1}{(ax+b)}$
(10.4)	$\delta(x^2-a^2), \quad a \neq 0$	$\frac{x}{\pi\|a\|}p.v.\frac{1}{(x^2-a^2)}$
(10.5)	$\delta'(x)$	$-p.v.\frac{1}{\pi x^2}$
(10.6)	$\delta^{n'}(x), n = 0, 1, \ldots,$ n' signifies the nth derivative	$\frac{(-1)^n n!}{\pi}p.v.\frac{1}{x^{n+1}}$
(10.7)	$p.v.(1/x)$	$-\pi\delta(x)$

Table 1.11. *Multiple Hilbert transforms*

Notation: $\mathbf{a}\cdot\mathbf{x} = \{a_1x_1 + a_2x_2 + \cdots + a_nx_n\}$; $\mathbf{s} \equiv \{s_1, s_2, s_3, \ldots, s_n\}$.

Number	$f(\mathbf{x})$	$\frac{1}{\pi^n} P \int_{-\infty}^{\infty} f(\mathbf{s}) d\mathbf{s} \prod_{k=1}^{n} \frac{1}{(x_k - s_k)}$
(11.1)	$\sin(\mathbf{a}\cdot\mathbf{x})$	$(-1)^{n/2}\sin(\mathbf{a}\cdot\mathbf{x})\prod_{k=1}^{n}\operatorname{sgn} a_k$, for n even
		$(-1)^{(n+1)/2}\cos(\mathbf{a}\cdot\mathbf{x})\prod_{k=1}^{n}\operatorname{sgn} a_k$, for n odd
(11.2)	$\cos(\mathbf{a}\cdot\mathbf{x})$	$(-1)^{n/2}\cos(\mathbf{a}\cdot\mathbf{x})\prod_{k=1}^{n}\operatorname{sgn} a_k$, for n even
		$(-1)^{(n-1)/2}\sin(\mathbf{a}\cdot\mathbf{x})\prod_{k=1}^{n}\operatorname{sgn} a_k$, for n odd
(11.3)	$\sin(\mathbf{a}\cdot\mathbf{x})\sin(\mathbf{b}\cdot\mathbf{x})$	$(1/2)(-1)^{n/2}\{\cos\{(\mathbf{a}-\mathbf{b})\cdot\mathbf{x}\}\prod_{k=1}^{n}\operatorname{sgn}(a_k - b_k)$
		$-\cos\{(\mathbf{a}+\mathbf{b})\cdot\mathbf{x}\}\prod_{k=1}^{n}\operatorname{sgn}(a_k + b_k)\}$, for n even
		$(1/2)(-1)^{(n-1)/2}\{\sin\{(\mathbf{a}-\mathbf{b})\cdot\mathbf{x}\}\prod_{k=1}^{n}\operatorname{sgn}(a_k - b_k)$
		$-\sin\{(\mathbf{a}+\mathbf{b})\cdot\mathbf{x}\}\prod_{k=1}^{n}\operatorname{sgn}(a_k + b_k)\}$, for n odd
(11.4)	$\cos(\mathbf{a}\cdot\mathbf{x})\cos(\mathbf{b}\cdot\mathbf{x})$	$(1/2)(-1)^{n/2}\{\cos\{(\mathbf{a}-\mathbf{b})\cdot\mathbf{x}\}\prod_{k=1}^{n}\operatorname{sgn}(a_k - b_k)$
		$+\cos\{(\mathbf{a}+\mathbf{b})\cdot\mathbf{x}\}\prod_{k=1}^{n}\operatorname{sgn}(a_k + b_k)\}$, for n even
		$(1/2)(-1)^{(n-1)/2}\{\sin\{(\mathbf{a}-\mathbf{b})\cdot\mathbf{x}\}\prod_{k=1}^{n}\operatorname{sgn}(a_k - b_k)$
		$+\sin\{(\mathbf{a}+\mathbf{b})\cdot\mathbf{x}\}\prod_{k=1}^{n}\operatorname{sgn}(a_k + b_k)\}$, for n odd
(11.5)	$\sin(\mathbf{a}\cdot\mathbf{x})\cos(\mathbf{b}\cdot\mathbf{x})$	$(1/2)(-1)^{n/2}\{\sin\{(\mathbf{a}-\mathbf{b})\cdot\mathbf{x}\}\prod_{k=1}^{n}\operatorname{sgn}(a_k - b_k)$
		$+\sin\{(\mathbf{a}+\mathbf{b})\cdot\mathbf{x}\}\prod_{k=1}^{n}\operatorname{sgn}(a_k + b_k)\}$, for n even
		$(1/2)(-1)^{(n+1)/2}\{\cos\{(\mathbf{a}-\mathbf{b})\cdot\mathbf{x}\}\prod_{k=1}^{n}\operatorname{sgn}(a_k - b_k)$
		$+\cos\{(\mathbf{a}+\mathbf{b})\cdot\mathbf{x}\}\prod_{k=1}^{n}\operatorname{sgn}(a_k + b_k)\}$, for n odd

Table 1.11. (*Cont.*)

Number	$f(\mathbf{x})$	$\frac{1}{\pi^n} P \int_{-\infty}^{\infty} f(\mathbf{s}) \mathrm{d}\mathbf{s} \prod_{k=1}^{n} \frac{1}{(x_k - s_k)}$
(11.6)	$\mathrm{e}^{\mathbf{ia}\cdot\mathbf{x}}$	$(-\mathrm{i})^n \mathrm{e}^{\mathbf{ia}\cdot\mathbf{x}} \prod_{k=1}^{n} \operatorname{sgn} a_k$
(11.7)	$\delta(x_1, x_2, \ldots, x_n)$	$\frac{1}{\pi^n} p.v. \frac{1}{x_1 x_2 \cdots x_n}$
(11.8)	$p.v. \frac{1}{x_1 x_2 \cdots x_n}$	$(-\pi)^n \delta(x_1, x_2, \ldots, x_n)$

Table 1.12A. *Finite Hilbert transforms: the interval* $[-1, 1]$

Number	$f(x)$	$g(x) = \frac{1}{\pi} P \int_{-1}^{1} \frac{f(s)\mathrm{d}s}{x - s}, \ -1 < x < 1$
(12A.1)	c (constant)	$c\pi^{-1} \log \left\| \frac{1+x}{1-x} \right\|$
(12A.2)	x	$\frac{x}{\pi} \log \left\| \frac{1+x}{1-x} \right\| - \frac{2}{\pi}$
(12A.3)	x^n, integer $n \geq 0$	$\frac{x^n}{\pi} \log \left\| \frac{1+x}{1-x} \right\| - \frac{2}{\pi} \sum_{j=0}^{[(n-1)/2]} \frac{x^{n-1-2j}}{2j+1}$
(12A.4)	$(x+a)^{-1}, \quad \|a\| > 1$	$\frac{1}{\pi(x+a)} \log \left(\frac{(a+1)(1+x)}{(a-1)(1-x)} \right)$
(12A.5)	$1/\sqrt{(1-x^2)}$	0
(12A.6)	$x^n/\sqrt{(1-x^2)}$, integer $n > 0$	$-x^{n-1} - \sum_{k=1}^{N} \frac{(2k-1)!!\, x^{n-1-2k}}{(2k)!!}$ $N = (n-2)/2, \quad n$ even, $N = (n-1)/2, \quad n$ odd
(12A.7)	$\frac{1}{(x^2+a^2)\sqrt{(1-x^2)}}$, $a > 0$	$\frac{x}{a\sqrt{(1+a^2)}\,(x^2+a^2)}$
(12A.8)	$\sqrt{(1-x^2)}$	x
(12A.9)	$x^n\sqrt{(1-x^2)}$, integer $n \geq 1$	$x^{n+1} - (1/2)x^{n-1} - \sum_{j=1}^{[(n-1)/2]} \frac{(2j-1)!!\, x^{n-1-2j}}{2^{j+1}(j+1)!}$

Table 1.12A. (*Cont.*)

Number	$f(x)$	$g(x)=\dfrac{1}{\pi}P\displaystyle\int_{-1}^{1}\frac{f(s)ds}{x-s},\ -1<x<1$
(12A.10)	$\sqrt{(1-x^2)}\log(1+x)$	$1-x\log 2-\sqrt{(1-x^2)}\,(\pi/2-\sin^{-1}x)$
(12A.11)	$\dfrac{\log(1-x)}{\sqrt{(1-x^2)}}$	$\dfrac{\pi-\cos^{-1}x}{\sqrt{(1-x^2)}}$
(12A.12)	$\dfrac{\log(1+x)}{\sqrt{(1-x^2)}}$	$-\dfrac{\cos^{-1}x}{\sqrt{(1-x^2)}}$
(12A.13)	$\dfrac{x\log(1-x)}{\sqrt{(1-x^2)}}$	$\log 2+\dfrac{x(\pi-\cos^{-1}x)}{\sqrt{(1-x^2)}}$
(12A.14)	$\dfrac{x\log(1+x)}{\sqrt{(1-x^2)}}$	$\log 2-\dfrac{x\cos^{-1}x}{\sqrt{(1-x^2)}}$
(12A.15)	$\dfrac{x^2\log(1-x)}{\sqrt{(1-x^2)}}$	$1+x\log 2+\dfrac{x^2(\pi-\cos^{-1}x)}{\sqrt{(1-x^2)}}$
(12A.16)	$\dfrac{x^2\log(1+x)}{\sqrt{(1-x^2)}}$	$-1+x\log 2-\dfrac{x^2\cos^{-1}x}{\sqrt{(1-x^2)}}$
(12A.17)	$\sqrt{(1-x^2)}\log(1-x)$	$-1-x\log 2+\sqrt{(1-x^2)}\,(\pi-\cos^{-1}x)$
(12A.18)	$\sqrt{(1-x^2)}\log(1+x)$	$1-x\log 2-\sqrt{(1-x^2)}\cos^{-1}x$
(12A.19)	$\left(\dfrac{1-x}{1+x}\right)^a,\ 0<\lvert a\rvert<1$	$-\left(\dfrac{1-x}{1+x}\right)^a\cot a\pi+\csc a\pi$
(12A.20)	$x^n\sqrt{\left(\dfrac{1+x}{1-x}\right)}$, integer $n>0$	$-\dfrac{(n-1)!!}{n!!}$ $-(x+1)x^{n-1}\left\{1+\displaystyle\sum_{k=1}^{[(n-2)/2]}\frac{(2k-1)!!}{(2k)!!\,x^{2k}}\right\}$, n even; $-(x+1)x^{n-1}\left\{1+\displaystyle\sum_{k=1}^{[(n-1)/2]}\frac{(2k-1)!!}{(2k)!!\,x^{2k}}\right\}$, n odd
(12A.21)	$x^n\sqrt{[(1-x)/(1+x)]}$, integer $n>0$	$\dfrac{(n-1)!!}{n!!}+(x-1)x^{n-1}$ $\times\left\{1+\displaystyle\sum_{k=1}^{[(n-2)/2]}\frac{(2k-1)!!}{(2k)!!\,x^{2k}}\right\}$, n even; $(x-1)x^{n-1}\left\{1+\displaystyle\sum_{k=1}^{[(n-1)/2]}\frac{(2k-1)!!}{(2k)!!\,x^{2k}}\right\}$, n odd

Table 1.12A. (*Cont.*)

Number	$f(x)$	$g(x) = \frac{1}{\pi} P \int_{-1}^{1} \frac{f(s)\mathrm{d}s}{x-s}, \ -1 < x < 1$
(12A.22)	$\frac{(1-x)^{a-1}}{(1+x)^a}$, $0 < a < 1$	$-\frac{(1-x)^{a-1}}{(1+x)^a} \cot a\pi$
(12A.23)	$\frac{(1+x)^{a-1}}{(1-x)^a}$, $0 < a < 1$	$\frac{(1+x)^{a-1}}{(1-x)^a} \cot a\pi$
(12A.24)	$(1-x)^a(1+x)^b$, $a > -1, b > -1$	$\frac{2^{a+b}\Gamma(a)\Gamma(b+1)}{\pi\Gamma(a+b+1)} {}_2F_1(1, \ -a-b;$ $1-a; \ (1-x)/2) - (1-x)^a(1+x)^b \cot a\pi$
(12A.25)	$\frac{\surd(1-x^2)}{(1-ax)}$, $-1 < a < 0$	$\frac{\surd(1-a^2)}{a(1-ax)} - a^{-1}$
(12A.26)	$\frac{\pi}{2} P_n(x)$	$Q_n(x)$
(12A.27)	$\frac{\pi}{2} P_n(x)p(x)$, where p is a polynomial of order $\leq n$	$Q_n(x)p(x)$
(12A.28)	$\frac{\pi}{2}(1-x^2)^{m/2} P_n^m(x)$	$(1-x^2)^{m/2} Q_n^m(x)$
(12A.29)	$(1-x)^\alpha(1+x)^\beta \times P_n^{(\alpha,\beta)}(x)$, $\alpha > -1, \beta > -1$, $\alpha \neq 0, 1, 2, \ldots$	$-\cot \alpha\pi(1-x)^\alpha(1+x)^\beta P_n^{(\alpha,\beta)}(x) + \frac{2^{\alpha+\beta}\Gamma(\alpha)\Gamma(n+1+\beta)}{\pi\Gamma(n+\alpha+\beta+1)} \times {}_2F_1(n+1, -n-\alpha-\beta; 1-\alpha; (1-x)/2)$
(12A.30)	$\frac{1}{\surd(1-2rx+r^2)}$, $-1 < r < 1$	$\frac{2}{\pi} \sum_{n=0}^{\infty} r^n Q_n(x)$
(12A.31)	$T_n(x), n = 0, 1, \ldots$	$\frac{1}{\pi} T_n(x) \log\left(\frac{1+x}{1-x}\right) - \frac{2^n n}{\pi} \sum_{j=0}^{[(n-1)/2]} a_{nj} x^{n-2j-1}$, with $a_{nj} = \sum_{v=0}^{j} \frac{(-4)^{-v}(n-v-1)!}{v!(n-2v)!(2j+1-2v)}$
(12A.32)	$\frac{T_0(x)}{\surd(1-x^2)}$	0

Table 1.12A. (*Cont.*)

Number	$f(x)$	$g(x) = \frac{1}{\pi} P \int_{-1}^{1} \frac{f(s)\mathrm{d}s}{x-s}, \ -1 < x < 1$
(12A.33)	$\frac{T_1(x)}{\sqrt{(1-x^2)}}$	-1
(12A.34)	$\frac{T_n(x)}{\sqrt{(1-x^2)}}$, $n = 1, 2, \ldots$	$-U_{n-1}(x)$
(12A.35)	$U_n(x)$, $n = 0, 1, 2, \ldots$	$\frac{1}{\pi} U_n(x) \log\left(\frac{1+x}{1-x}\right) - \frac{2^{n+1}}{\pi} \sum_{j=0}^{[(n-1)/2]} a_{nj} x^{n-2j-1}$, with $a_{nj} = \sum_{v=0}^{j} \frac{(-4)^{-v}(n-v)!}{v!(n-2v)!(2j+1-2v)}$
(12A.36)	$\sqrt{(1-x^2)}\, U_{n-1}(x)$, $n = 1, 2, \ldots$	$T_n(x)$
(12A.37)	$f_n(x) = -\frac{T_n(x)}{\sqrt{(1-x^2)}}$, $n = 0, 1, 2, \ldots$	$g_n(x) = \frac{1}{\pi} P \int_{-1}^{1} \frac{f_n(s)\mathrm{d}s}{x-s}$, $g_0(x) = 0; g_1(x) = 1;$ $g_n(x) = 2x g_{n-1}(x) - g_{n-2}(x), n \geq 2$
(12A.38)	$\sin ax, \quad a > 0$	$\frac{\sin ax}{\pi}\{\mathrm{Ci}[a(1+x)] - \mathrm{Ci}[a(1-x)]\} - \frac{\cos ax}{\pi}\{\mathrm{Si}[a(1+x)] + \mathrm{Si}[a(1-x)]\}$
(12A.39)	$\cos ax, \quad a > 0$	$\frac{\sin ax}{\pi}\{\mathrm{Si}[a(1+x)] + \mathrm{Si}[a(1-x)]\} + \frac{\cos ax}{\pi}\{\mathrm{Ci}[a(1+x)] - \mathrm{Ci}[a(1-x)]\}$
(12A.40)	e^{ax}	$\frac{\mathrm{e}^{ax}}{\pi}\{E_1(a(1-x)) - E_1(a(1+x)) - 2\mathrm{Shi}(a(1-x))\}, \quad a > 0,$ $\frac{\mathrm{e}^{ax}}{\pi}\{E_1(-a(1-x)) - E_1(-a(1+x)) + 2\mathrm{Shi}(-a(1+x))\}, \quad a < 0$
(12A.41)	$x\mathrm{e}^{ax}$	$\frac{x\mathrm{e}^{ax}}{\pi}\{E_1(a(1-x)) - E_1(a(1+x)) - 2\mathrm{Shi}(a(1-x))\} + \frac{1}{a\pi}(\mathrm{e}^{-a} - \mathrm{e}^{a}), \quad a > 0,$ $\frac{x\mathrm{e}^{ax}}{\pi}\{E_1(-a(1-x)) - E_1(-a(1+x)) + 2\mathrm{Shi}(-a(1+x))\} + \frac{1}{a\pi}(\mathrm{e}^{-a} - \mathrm{e}^{a}), \quad a < 0$
(12A.42)	$(1-x^2)^{\lambda-1/2} C_n^{(\lambda)}(x)$	$(1-x^2)^{\lambda-1/2} D_n^{(\lambda)}(x)$

Table 1.12B. *Finite Hilbert transforms: the interval* [0, 1]

Number	$f(x)$	$\frac{1}{\pi} P \int_0^1 \frac{f(s)}{x-s} \, \mathrm{d}s, \quad 0 < x < 1$
(12B.1)	$\frac{1}{\surd(x(1-x))}$	0
(12B.2)	$\surd\left(\frac{x}{1-x}\right)$	-1
(12B.3)	$\surd\left(\frac{1-x}{x}\right)$	1
(12B.4)	$\surd(x(1-x))$	$x - 1/2$
(12B.5)	$\frac{x^2}{\surd(x(1-x))}$	$-x - 1/2$
(12B.6)	$\surd(1-x)$	$\frac{1}{\pi}\left[2 + \surd(1-x)\log\left(\frac{1-\surd(1-x)}{1+\surd(1-x)}\right)\right]$
(12B.7)	$\surd x$	$\frac{1}{\pi}\left[\surd(x)\log\left(\frac{1+\surd x}{1-\surd x}\right) - 2\right]$

Table 1.12B. (*Cont.*)

Number	$f(x)$	$\frac{1}{\pi} P \int_0^1 \frac{f(s)}{x-s}\,\mathrm{d}s,\ 0 < x < 1$
(12B.8)	$\surd(1+x)$	$\frac{\surd(1+x)}{\pi} \log \left\lvert \frac{(\surd(1+x)-1)(\surd(1+x)+\surd 2)}{(\surd(1+x)+1)(\surd(1+x)-\surd 2)} \right\rvert - \frac{2(\surd(2)-1)}{\pi}$
(12B.9)	$\surd(1-x^2)$	$\frac{\surd(1-x^2)}{\pi} \log \left\lvert \frac{(\surd(1+x)+\surd(1-x))(1-\surd(1-x^2))}{(\surd(1+x)-\surd(1-x))(1+\surd(1-x^2))} \right\rvert + \frac{x}{2} + \frac{1}{\pi}$
(12B.10)	$\surd\left(\frac{1+x}{1-x}\right)$	$\frac{1}{\pi} \surd\left(\frac{1+x}{1-x}\right) \log \left\lvert \frac{(\surd(1+x)+\surd(1-x))(1-\surd(1-x^2))}{(\surd(1+x)-\surd(1-x))(1+\surd(1-x^2))} \right\rvert - \frac{1}{2}$
(12B.11)	$\surd\left(\frac{1-x}{1+x}\right)$	$\frac{1}{\pi} \surd\left(\frac{1-x}{1+x}\right) \log \left\lvert \frac{(\surd(1+x)+\surd(1-x))(1-\surd(1-x^2))}{(\surd(1+x)-\surd(1-x))(1+\surd(1-x^2))} \right\rvert + \frac{1}{2}$
(12B.12)	$\frac{x^n}{\surd(x(1-x))},\quad n \in \mathbb{N}$	$-x^{n-1} - \sum_{k=1}^{n-1} \frac{(2k-1)!!\, x^{n-k-1}}{(2k)!!}$

Table 1.12C. *Finite Hilbert transforms: the interval* $[a,\ b]$

Number	$f(x)$	$\dfrac{1}{\pi}P\displaystyle\int_a^b \frac{f(s)}{x-s}\,\mathrm{d}s, \quad a<x<b$
(12C.1)	$\dfrac{1}{\sqrt{[(b-x)(x-a)]}}$	0
(12C.2)	$\dfrac{x}{\sqrt{[(b-x)(x-a)]}}$	-1
(12C.3)	$\dfrac{x^2}{\sqrt{[(b-x)(x-a)])}}$	$-x-\dfrac{b+a}{2}$
(12C.4)	$\sqrt{[(b-x)(x-a)]}$	$x-\dfrac{b+a}{2}$
(12C.5)	$\dfrac{1}{x\sqrt{[(b-x)(x-a)]}}$, $0<a$	$\dfrac{1}{x\sqrt{(ab)}}$
(12C.6)	$\dfrac{\sqrt{[(b-x)(x-a)]}}{x}$, $0<a$	$1-\dfrac{\sqrt{(ab)}}{x}$
(12C.7)	$(b-x)^{-\mu}(x-a)^{\mu-1}$, $0<\operatorname{Re}\mu<1$	$(b-x)^{-\mu}(x-a)^{\mu-1}\cot(\pi\mu)$
(12C.8)	$(b-x)^{-\mu}(x-a)^{\mu}$, $\lvert\operatorname{Re}\mu\rvert<1$	$\csc(\pi\mu)\left\{\cos(\pi\mu)\dfrac{(x-a)^{\mu}}{(b-x)^{\mu}}-1\right\}$
(12C.9)	$\sin cx, \quad c>0$	$\dfrac{\sin cx}{\pi}\{\mathrm{Ci}[c(x-a)]-\mathrm{Ci}[c(b-x)]\}$ $-\dfrac{\cos cx}{\pi}\{\mathrm{Si}[c(x-a)]+\mathrm{Si}[a(b-x)]\}$
(12C.10)	$\cos cx, \quad c>0$	$\dfrac{\sin cx}{\pi}\{\mathrm{Si}[c(x-a)]+\mathrm{Si}[c(b-x)]\}$ $+\dfrac{\cos cx}{\pi}\{\mathrm{Ci}[c(x-a)]-\mathrm{Ci}[c(b-x)]\}$

Table 1.13A. *Miscellaneous cases: the cosine form*

Number	$f(x)$	$\dfrac{1}{\pi}P\displaystyle\int_0^{\pi}\frac{f(y)\mathrm{d}y}{\cos x-\cos y}$
(13A.1)	c (constant)	0
(13A.2)	x	$\dfrac{4\csc x}{\pi}\displaystyle\sum_{k=0}^{\infty}\frac{\sin(2k+1)x}{(2k+1)^2}$ $=\dfrac{\mathrm{i}\csc x}{2\pi}\{\mathrm{e}^{-\mathrm{i}x}\Phi(\mathrm{e}^{-2\mathrm{i}x},2,1/2)$ $-\mathrm{e}^{\mathrm{i}x}\Phi(\mathrm{e}^{2\mathrm{i}x},2,\ 1/2)\}$
(13A.3)	$\cos nx$, integer $n\geq 0$	$-\dfrac{\sin nx}{\sin x}$

Table 1.13A. (*Cont.*)

Number	$f(x)$	$\frac{1}{\pi}P\int_0^{\pi}\frac{f(y)\mathrm{d}y}{\cos x-\cos y}$
(13A.4)	$\sin x$	$\frac{2\log\{\cot(x/2)\}}{\pi}$
(13A.5)	$\sin x\sin nx$, integer $n\ge 1$	$\cos nx$
(13A.6)	$\cot x\sin nx$, integer $n\ge 0$	$\csc^2 x\{\cos nx\cos x-\cos^2(x/2)$ $+(-1)^n\sin^2(x/2)\}$
(13A.7)	$\cos x\cos nx$, integer $n\ge 1$	$\cos nx-\csc x\sin(n+1)x$
(13A.8)	$\sin x\cos x$	$\frac{2}{\pi}\{\cos x\log[\cot(x/2)]-1\}$
(13A.9)	$\sin^2(x/2)$	$1/2$
(13A.10)	$\cos^2(x/2)$	$-1/2$
(13A.11)	$\frac{\sin nx}{\sin x}$, integer $n\ge 0$	$\csc^2 x\{\cos nx-\cos^2(x/2)$ $-(-1)^n\sin^2(x/2)\}$
(13A.12)	$\sin\left[\left(\frac{1-a}{2}\right)x\right]\sin\left(\frac{x}{2}\right)$ $\times\sec^a\left(\frac{x}{2}\right)$, $a<1$	$\frac{\cos[(1-a)(x/2)]}{2\cos^{1+a}(x/2)}$
(13A.13)	$\cos\left[\frac{(1-a)x}{2}\right]\cos^{1-a}\left(\frac{x}{2}\right)$, $a<1$	$-\frac{\sin[(1-a)(x/2)]}{2\cos^a(x/2)\sin(x/2)}$
(13A.14)	$\cos(ax/2)\sec^a(x/2)$, $a<1$	$\frac{1}{2}\csc(x/2)\sin(ax/2)(\sec(x/2))^{1+a}$
(13A.15)	$\log\lvert\cos x-\cos a\rvert$	$(\pi-a)\csc x,\quad 0<a<x<\pi$ $-x\csc x,\quad 0<x<a<\pi$

Table 1.13B. *Miscellaneous cases: the one-sided Hilbert transform*

Number	$f(x)$	$\frac{1}{\pi}P\int_0^{\infty}\frac{f(s)\mathrm{d}s}{x-s},\quad x>0$
(13B.1)	$(x+a)^{-1},\quad a>0$	$\frac{\log(x/a)}{\pi(x+a)}$
(13B.2)	$(x^2+a^2)^{-1},\quad a>0$	$\frac{\{x+2a\pi^{-1}\log(x/a)\}}{2a(x^2+a^2)}$

Table 1.13B. (*Cont.*)

Number	$f(x)$	$\frac{1}{\pi}P\int_0^\infty \frac{f(s)\mathrm{d}s}{x-s},\ x>0$
(13B.3)	$x(x^2+a^2)^{-1}, \quad a>0$	$\frac{\{2x\pi^{-1}\log(x/a)-a\}}{2(x^2+a^2)}$
(13B.4)	$\frac{cx+d}{(ax+b)^2}, \ a>0, \quad b>0$	$\frac{(cx+d)\log(ax/b)}{\pi(ax+b)^2}+\frac{ad-bc}{\pi ab(ax+b)}$
(13B.5)	$\frac{1}{\sqrt{(ax+b)}}, \ a>0, \quad b>0$	$\frac{2\log\{\sqrt{(ax)}/(\sqrt{(b)}+\sqrt{(ax+b)})\}}{\pi\sqrt{(ax+b)}}$
(13B.6)	$x^\mu, -1<\mathrm{Re}\,\mu<0$	$x^\mu\cot(\pi\mu)$
(13B.7)	$\frac{x^\mu}{x+a}, \ -1<\mathrm{Re}\,\mu<1,\ a>0$	$\frac{1}{x+a}\{x^\mu\cot(\pi\mu)-a^\mu\csc(\pi\mu)\}$
(13B.8)	$x^{b-1}(x+a)^{1-c}, \ a>0, \ 0<\mathrm{Re}\,b<\mathrm{Re}\,c$	$\frac{a^{b+1-c}}{(x+a)\pi}B(c-b-1,b) \times {}_2F_1\left(2-c,\ 1;\ 2-c+b;\ \frac{a}{x+a}\right) - x^{b-1}(x+a)^{1-c}\cot[(c-b)\pi]$
(13B.9)	$x^\mu\log x, -1<\mathrm{Re}\,\mu<0$	$x^\mu\{\cot(\pi\mu)\log x-\pi\csc^2(\pi\mu)\}$
(13B.10)	$\frac{\log x}{x+a}, \quad a>0$	$\frac{\{\log^2 x-\log^2 a-\pi^2\}}{2\pi(x+a)}$
(13B.11)	$\frac{\log(x/a)}{(x+a)\sqrt{x}}, \quad a>0$	$-\frac{\pi}{(x+a)\sqrt{x}}$
(13B.12)	$\sin ax$	$\frac{\mathrm{sgn}\,a}{\pi}\{\sin\lvert a\rvert x\,\mathrm{Ci}(\lvert a\rvert x) - \cos ax[\mathrm{Si}(\lvert a\rvert x)+\pi/2]\}$
(13B.13)	$\cos ax$	$\pi^{-1}\cos ax\,\mathrm{Ci}(\lvert a\rvert x) + (2\pi)^{-1}[\pi+2\mathrm{Si}(\lvert a\rvert x)]\,\mathrm{sgn}\,a\sin ax$
(13B.14)	$\mathrm{e}^{-ax}, \quad a>0$	$\pi^{-1}\mathrm{e}^{-ax}\mathrm{Ei}(ax)$
(13B.15)	$x^n\mathrm{e}^{-ax}$, integer $n\ge 0, \quad a>0$	$\frac{x^n}{\pi}\left\{\mathrm{e}^{-ax}\mathrm{Ei}(ax)-\sum_{k=1}^{n}\frac{(k-1)!}{(ax)^k}\right\}$
(13B.16)	$\mathrm{e}^{-a\sqrt{x}}, \quad a>0$	$\pi^{-1}\{\mathrm{e}^{-a\sqrt{x}}\mathrm{Ei}(a\sqrt{x})-\mathrm{e}^{a\sqrt{x}}E_1(a\sqrt{x})\}$
(13B.17)	$x^{-b}\mathrm{e}^{-ax}, \ 0<b<1, \ 0\le a$	$-\frac{\mathrm{e}^{-ax}}{\pi x^b}\{\mathrm{i}\pi+\mathrm{e}^{-\mathrm{i}b\pi}\Gamma(1-b)\Gamma(b,-ax)\}$
(13B.18)	$\sin(a\sqrt{x}), \quad 0<a$	$-\cos(a\sqrt{x})$
(13B.19)	$\mathrm{e}^{-ax}I_0(ax), \quad 0<a$	$-\pi^{-1}\mathrm{e}^{-ax}K_0(ax)$

Table 1.13C. *Miscellaneous cases: the cotangent form*

Number	$f(x)$	$g(x) = \frac{1}{2\pi} P \int_{-\pi}^{\pi} f(s) \cot\left(\frac{x-s}{2}\right) ds$
(13C.1)	c (constant)	0
(13C.2)	$g(x)$	$-f(x) + \frac{1}{2\pi} \int_{-\pi}^{\pi} f(\theta) d\theta$
(13C.3)	$f(x+a)$	$g(x+a)$
(13C.4)	x	$-2\log\{2\cos(x/2)\}$
(13C.5)	$\|x\|$	$-\frac{4}{\pi} \sum_{n=0}^{\infty} \frac{\sin(2n+1)x}{(2n+1)^2}$
(13C.6)	x^2	$4\sum_{n=1}^{\infty} \frac{(-1)^n \sin nx}{n^2}$ $= 2i\{\mathrm{Li}_2(-e^{-ix}) - \mathrm{Li}_2(-e^{ix})\}$
(13C.7)	$x(\pi^2 - x^2)$	$12\sum_{n=1}^{\infty} \frac{(-1)^n \cos nx}{n^3}$ $= 6\mathrm{Li}_3(-e^{-ix}) + 6\mathrm{Li}_3(-e^{ix})$
(13C.8)	$\sin nx$, integer $n \geq 1$	$-\cos nx$
(13C.9)	$\sin ax, a \neq$ integer	$\frac{2\sin a\pi}{\pi} \sum_{n=1}^{\infty} \frac{(-1)^n n \cos nx}{n^2 - a^2}$, for $\|a\| < 1$, $= -\frac{\sin a\pi\, e^{-ix}}{2\pi} \{\Phi(-e^{-ix}, 1, 1-a)$ $+ \Phi(-e^{-ix}, 1, 1+a)$ $+ e^{2ix}\Phi(-e^{ix}, 1, 1-a)$ $+ e^{2ix}\Phi(-e^{ix}, 1, 1+a)\}$
(13C.10)	$\|\sin x\|$	$-\frac{4}{\pi} \sum_{n=1}^{\infty} \frac{\sin 2nx}{4n^2 - 1}$ $= -\frac{2}{\pi} \sin x \{\tanh^{-1}(e^{ix}) + \tanh^{-1}(e^{-ix})\}$
(13C.11)	$x \sin x$	$-2\sin x \log\{2\cos(x/2)\}$
(13C.12)	$\sin^2 nx$, integer $n \geq 0$	$-(1/2)\sin 2nx$
(13C.13)	$\cos nx$, integer $n \geq 0$	$\sin nx$
(13C.14)	$\cos ax, a \neq$ integer	$-\frac{2a\sin a\pi}{\pi} \sum_{n=1}^{\infty} \frac{(-1)^n \sin nx}{n^2 - a^2}$

Table 1.13C. (*Cont.*)

Number	$f(x)$	$g(x) = \frac{1}{2\pi} P \int_{-\pi}^{\pi} f(s) \cot\left(\frac{x-s}{2}\right) ds$
(13C.15)	$\cos^2 nx$, integer $n \geq 0$	$(1/2) \sin 2nx$
(13C.16)	$x \cos x$	$1 - 2\cos x \log\{2\cos(x/2)\}$
(13C.17)	e^{inx}, $n \in \mathbb{Z}$	$-i \operatorname{sgn} n \, e^{inx}$
(13C.18)	e^{ax}	$\frac{2\sinh a\pi}{\pi} \sum_{n=1}^{\infty} \frac{(-1)^n\{a \sin nx + n \cos nx\}}{n^2 + a^2}$
(13C.19)	$\sinh ax$	$\frac{2\sinh a\pi}{\pi} \sum_{k=1}^{\infty} \frac{(-1)^k k \cos kx}{k^2 + a^2}$
(13C.20)	$\cosh ax$	$\frac{2a\sinh a\pi}{\pi} \sum_{k=1}^{\infty} \frac{(-1)^k \sin kx}{k^2 + a^2}$
(13C.21)	$\sin 2nx \cot x$, integer $n \geq 0$	$(1 - \cos 2nx) \cot x$
(13C.22)	$\sin 2nx \csc x$, integer $n \geq 0$	$2\sin^2 nx \csc x$
(13C.23)	$\sin^2 nx \csc x$, integer $n \geq 0$	$-(1/2) \sin 2nx \csc x$
(13C.24)	$\sin 2nx \csc^2 x - 2n \cos(2n-1)x \csc x$, integer $n \geq 0$	$(1 - \cos 2nx) \csc^2 x - 2n \sin(2n-1)x \csc x$
(13C.25)	$\sin(n+1)x \sin nx \csc x$, integer $n \geq 0$	$-\cos(n+1)x \sin nx \csc x$
(13C.26)	$\cos(n+1)x \sin nx \csc x$, integer $n \geq 0$	$\sin(n+1)x \sin nx \csc x$
(13C.27)	$\cos^n x \cos(n+1)x$, integer $n \geq 0$	$\cos^n x \sin(n+1)x$
(13C.28)	$\cos^n x \sin(n+1)x$, integer $n \geq 0$	$-\cos^n x \cos(n+1)x$
(13C.29)	$\cos^n x \cos nx$, integer $n \geq 0$	$\cos^n x \sin nx$
(13C.30)	$\cos^n x \sin nx$, integer $n \geq 0$	$2^{-n} - \cos^n x \cos nx$
(13C.31)	$\sin^n x \sin nx$, integer $n \geq 0$	$\begin{cases} -\sin^n x \cos nx, & n \text{ odd} \\ 2^{-n}(-1)^{n/2} - \sin^n x \cos nx, & n \text{ even} \end{cases}$

Table 1.13C. (*Cont.*)

Number	$f(x)$	$g(x) = \frac{1}{2\pi} P \int_{-\pi}^{\pi} f(s) \cot\left(\frac{x-s}{2}\right) ds$
(13C.32)	$\sin^n x \cos nx$, integer $n \geq 0$	$\begin{cases} \sin^n x \sin nx, & n \text{ even} \\ 2^{-n}(-1)^{(n+1)/2} + \sin^n x \sin nx, & n \text{ odd} \end{cases}$
(13C.33)	$\cot\left(\frac{x-a}{2}\right), \quad x \neq a$	1
(13C.34)	$e^{a \cos x} \cos(a \sin x)$	$e^{a \cos x} \sin(a \sin x)$
(13C.35)	$e^{a \cos x} \sin(a \sin x)$	$1 - e^{a \cos x} \cos(a \sin x)$
(13C.36)	$\sinh(a \cos x) \sin(a \sin x)$	$1 - \cosh(a \cos x) \cos(a \sin x)$
(13C.37)	$\cosh(a \cos x) \cos(a \sin x)$	$\sinh(a \cos x) \sin(a \sin x)$
(13C.38)	$\sinh(a \cos x) \cos(a \sin x)$	$\cosh(a \cos x) \sin(a \sin x)$
(13C.39)	$\cosh(a \cos x) \sin(a \sin x)$	$-\sinh(a \cos x) \cos(a \sin x)$
(13C.40)	$(5 - 4\cos x)^{-1}$	$\frac{4 \sin x}{3(5 - 4\cos x)}$
(13C.41)	$(5 - 4\cos x)^{-1} \sin x$	$\frac{1 - 2\cos x}{2(5 - 4\cos x)}$
(13C.42)	$(5 - 4\cos x)^{-1} \cos x$	$\frac{5 \sin x}{3(5 - 4\cos x)}$
(13C.43)	$\frac{1 - r^2}{1 - 2r\cos x + r^2}, \quad \lvert r \rvert < 1$	$\frac{2r \sin x}{1 - 2r\cos x + r^2}$
(13C.44)	$\frac{\sin x}{1 - 2a\cos x + a^2}, \quad \lvert a \rvert < 1$	$\frac{a - \cos x}{1 - 2a\cos x + a^2}$
(13C.45)	$\frac{(1 - a^2)\sin x}{(1 - 2a\cos x + a^2)^2}$, $\lvert a \rvert < 1, \quad x \neq n\pi$, for integer n	$\frac{2a - (1 + a^2)\cos x}{(1 - 2a\cos x + a^2)^2}$
(13C.46)	$\frac{a\cos x - 1}{1 - 2a\cos x + a^2}, \quad \lvert a \rvert > 1$	$\frac{a \sin x}{1 - 2a\cos x + a^2}$
(13C.47)	$\frac{1 - a\cos x}{1 - 2a\cos x + a^2}, \quad \lvert a \rvert < 1$	$\frac{a \sin x}{1 - 2a\cos x + a^2}$
(13C.48)	$\frac{\cos x}{1 - 2a\cos 2x + a^2}, \quad \lvert a \rvert < 1$	$\frac{(1 - a^2)^{-1}(1 + a^2)\sin x}{1 - 2a\cos 2x + a^2}$
(13C.49)	$\frac{\sin x}{(1 - 2a\cos x + a^2)^3}, \quad \lvert a \rvert < 1$	$-\frac{1}{2a(1 - a^2)^3} \sum_{k=1}^{\infty} k a^k \cos kx \times \{k(1 - a^2) + a^2 + 1\}$

Table 1.13C. (*Cont.*)

Number $f(x)$	$g(x) = \frac{1}{2\pi} P \int_{-\pi}^{\pi} f(s) \cot\left(\frac{x-s}{2}\right) ds$
	$= -\frac{(1+a^2)\{(1+a)^2 \cos x - 2a\}}{2(1-a^2)^3(1-2a\cos x + a^2)^2} - \frac{\{(1+a)^2 \cos x + 2a(\cos^2 x - 2)\}}{2(1-a^2)(1-2a\cos x + a^2)^3}$
(13C.50) $\frac{\sin x}{(1-2a\cos x + a^2)^3}$, $\lvert a\rvert > 1$	$-\frac{1}{2a(a^2-1)^3} \sum_{k=1}^{\infty} ka^{-k} \cos kx \times \{k(a^2-1) + 1 + a^2\}$ $= -\frac{(1+a^2)\{(1+a)^2 \cos x - 2a\}}{2(a^2-1)^3(1-2a\cos x + a^2)^2} - \frac{\{(1+a)^2 \cos x + 2a(\cos^2 x - 2)\}}{2(a^2-1)(1-2a\cos x + a^2)^3}$
(13C.51) $\frac{\sin\{[(2n+1)x]/2\}}{\sin(x/2)}$, integer $n \geq 0$	$\cot(x/2) - \frac{\cos\{[(2n+1)x]/2\}}{\sin(x/2)}$ $= \sin nx + (1 - \cos nx)\cot(x/2)$
(13C.52) $\frac{\left[\frac{\sin\{(n+1)x/2\}}{\sin(x/2)}\right]^2}{(n+1)}$, integer $n \geq 0$	$\cot(x/2) - \frac{\sin\{(n+1)x\}}{2(n+1)\sin^2(x/2)}$
(13C.53) $\log\{1 - 2a\cos x + a^2\}$, $\lvert a\rvert < 1$	$-2\tan^{-1}\left[\frac{a \sin x}{1 - a\cos x}\right]$
(13C.54) $\log\{1 - 2a\cos x + a^2\}$, $\lvert a\rvert > 1$	$-2\tan^{-1}\left[\frac{\sin x}{a - \cos x}\right]$
(13C.55) $\log\left\{\frac{1 + 2a\sin x + a^2}{1 - 2a\sin x + a^2}\right\}$, $\lvert a\rvert < 1$	$-2\tan^{-1}\left[\frac{2a\cos x}{1 - a^2}\right]$
(13C.56) $\log\left\{\frac{1 + 2a\cos x + a^2}{1 - 2a\cos x + a^2}\right\}$, $\lvert a\rvert < 1, x \neq n\pi$, for integer n	$2\tan^{-1}\left[\frac{2a\sin x}{1 - a^2}\right]$
(13C.57) $\log\lvert\tan(x/2)\rvert$	$-\frac{\pi}{2}\operatorname{sgn} x$
(13C.58) $\log\{2\cos(x/2)\}$	$x/2$
(13C.59) $\log\lvert\sin(x/2)\rvert$, $x \neq 0$	$\frac{1}{2}(x - \pi \operatorname{sgn} x)$

Table 1.13C. (*Cont.*)

Number	$f(x)$	$g(x) = \frac{1}{2\pi} P \int_{-\pi}^{\pi} f(s) \cot\left(\frac{x-s}{2}\right) ds$
(13C.60)	$\int_0^x \log\|2\cos(t/2)\| dt$	$x^2/4 - \pi^2/12$
(13C.61)	$\operatorname{sgn}(\sin x)$	$\frac{2}{\pi} \log\|\tan(x/2)\|$
(13C.62)	$\operatorname{sgn}(\cos x)$	$\frac{2}{\pi} \log\|\tan(x/2 + \pi/4)\|$
(13C.63)	$\operatorname{sgn} x, \quad x \neq 0$	$\frac{2}{\pi} \log\|\tan(x/2)\|$
(13C.64)	$x(\pi - x \operatorname{sgn} x)$	$-\frac{8}{\pi} \sum_{n=1}^{\infty} \frac{\cos(2n-1)x}{(2n-1)^3}$ $= -\frac{1}{2\pi} \left\{ e^{ix} \Phi\left(e^{2ix}, 3, \frac{1}{2}\right) + e^{-ix} \Phi\left(e^{-2ix}, 3, \frac{1}{2}\right) \right\}$
(13C.65)	$\tan^{-1}\left[\frac{2a\sin x}{1-a^2}\right], \quad \|a\| < 1,$ $x \neq n\pi$, for integer n	$-2 \sum_{n=1}^{\infty} \frac{a^{2n-1}\cos(2n-1)x}{(2n-1)}$ $= -\tanh^{-1}(ae^{ix}) - \tanh^{-1}(ae^{-ix})$
(13C.66)	$\tan^{-1}\left[\frac{2a\cos x}{1-a^2}\right], \quad \|a\| < 1$	$-2 \sum_{n=1}^{\infty} \frac{(-1)^n a^{2n-1} \sin(2n-1)x}{(2n-1)}$ $= i\{\tan^{-1}(ae^{-ix}) - \tan^{-1}(ae^{ix})\}$
(13C.67)	$\tan^{-1}\left[\frac{a\sin x}{1 - a\cos x}\right],$ $\|a\| < 1, \quad x \neq n\pi,$ for integer n	$\frac{1}{2} \log\{1 - 2a\cos x + a^2\}$
(13C.68)	$\tan^{-1}(\operatorname{csch} a \cos x), \quad 0 < a$	$2 \sum_{k=1}^{\infty} \frac{e^{-a(2k-1}(-1)^{k+1}\sin(2k-1)x}{2k-1}$ $= i\{\tan^{-1}(e^{-a-ix}) - \tan^{-1}(e^{-a+ix})\}$

Table 1.13D. *Miscellaneous cases: the Hilbert transforms H_e and H_o*

Additional entries for H_e and H_o can be worked out directly from the other sections of this table by noting the even or odd character of the integrand.

Number	$f(x)$	$\frac{2}{\pi}P\int_0^\infty \frac{f(s)}{x^2-s^2}\,ds$
(13D.1)	$f(x), \quad f$ even	$x^{-1}Hf(x)$
(13D.2)	$xf(x), \quad f$ odd	$(Hf)(x)$
(13D.3)	c (constant), $\quad x \neq 0$	0
(13D.4)	$\frac{1}{x^2+a^2}, \quad a > 0$	$\frac{1}{a(x^2+a^2)}$
(13D.5)	$\frac{x}{x^2+a^2}, \quad a > 0$	$\frac{2\log(\lvert x\rvert/a)}{\pi(x^2+a^2)}$
(13D.6)	$\sin ax, \quad a > 0$	$\frac{2}{\pi x}\{\sin ax\,\mathrm{Ci}(a\lvert x\rvert) - \cos ax\,\mathrm{Si}(ax)\}$
(13D.7)	$x\sin ax$	$-\operatorname{sgn} a \cos ax$
(13D.8)	$x^{-1}\sin ax$	$x^{-2}\operatorname{sgn} a(1-\cos ax)$
(13D.9)	$\sin(ax+b), \quad a > 0$	$\frac{2\cos b}{\pi x}\{\sin ax\,\mathrm{Ci}(a\lvert x\rvert) - \cos ax\,\mathrm{Si}(ax)\} + \frac{\sin b \sin ax}{x}$
(13D.10)	$\cos ax$	$x^{-1}\operatorname{sgn} a \sin ax$
(13D.11)	$x\cos ax, \quad a > 0$	$\frac{2}{\pi}\{\cos ax\,\mathrm{Ci}(a\lvert x\rvert) + \sin ax\,\mathrm{Si}(ax)\}$
(13D.12)	$\cos(ax+b), \quad a > 0$	$\frac{\cos b \sin ax}{x} - \frac{2\sin b}{\pi x}\{\sin ax\,\mathrm{Ci}(a\lvert x\rvert) - \cos ax\,\mathrm{Si}(ax)\}$
(13D.13)	$e^{-ax}, \quad a > 0$	$(\pi\lvert x\rvert)^{-1}\{e^{a\lvert x\rvert}E_1(a\lvert x\rvert) + e^{-a\lvert x\rvert}\mathrm{Ei}(a\lvert x\rvert)\}$
(13D.14)	$e^{iax}, \quad a > 0$	$\frac{\sin ax}{x}\left\{1 + \frac{2i\,\mathrm{Ci}(a\lvert x\rvert)}{\pi}\right\} - \frac{2i\cos ax\,\mathrm{Si}(ax)}{\pi x}$
(13D.15)	$e^{-ax^2}, \quad a > 0$	$-ix^{-1}e^{-ax^2}\operatorname{erf}(i\sqrt{(a)}\,x) = 2\sqrt{\left(\frac{a}{\pi}\right)}e^{-ax^2}\,{}_1F_1\left(1/2;\ 3/2;\ ax^2\right)$

Appendix 2

Atlas of Hilbert transform pairs

1 $f = \Pi(x)$ $\qquad$ $Hf = \dfrac{1}{\pi}\log\left|\dfrac{2x+1}{2x-1}\right|$

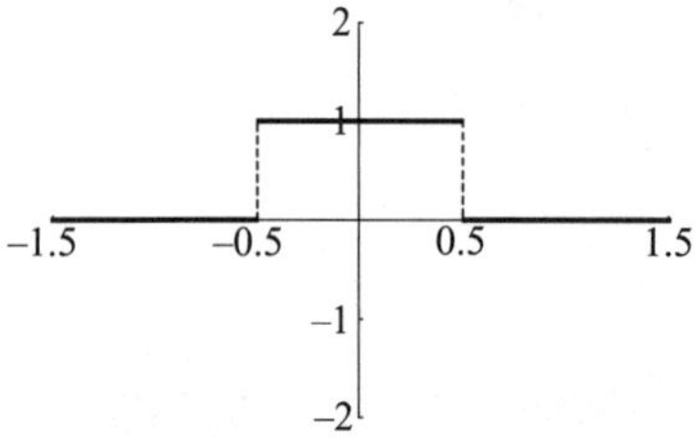

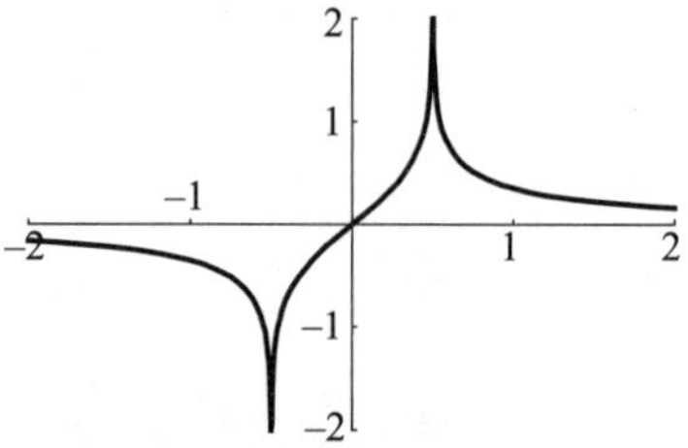

2 $f = \Pi(x - 1/2)$ $\qquad$ $Hf = \dfrac{1}{\pi}\log\left|\dfrac{x}{x-1}\right|$

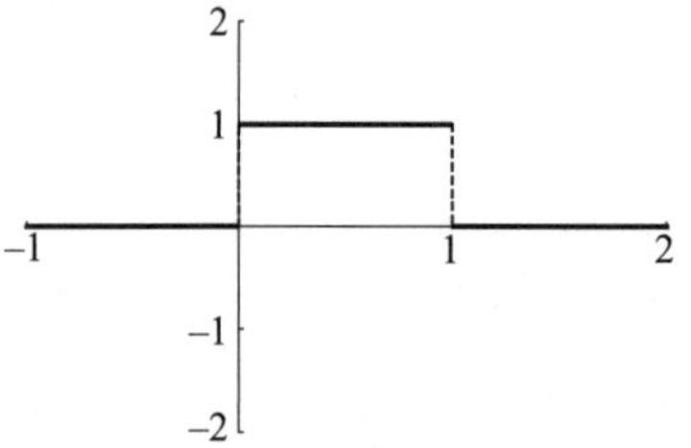

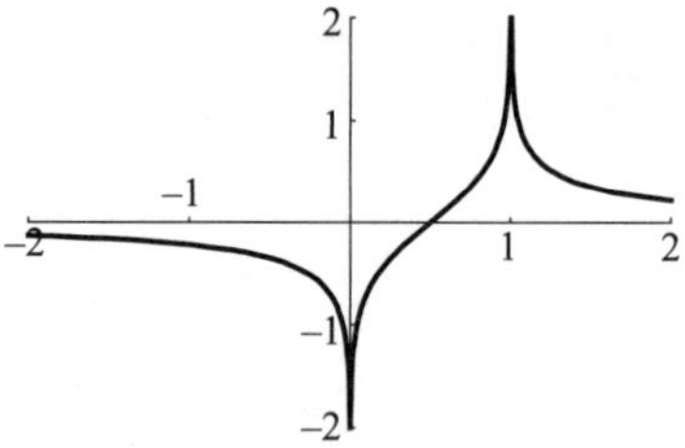

3 $f = \operatorname{sgn} x\,\Pi_2(x)$ $\qquad$ $Hf = \dfrac{1}{\pi}\log\left|\dfrac{x^2}{x^2-1}\right|$

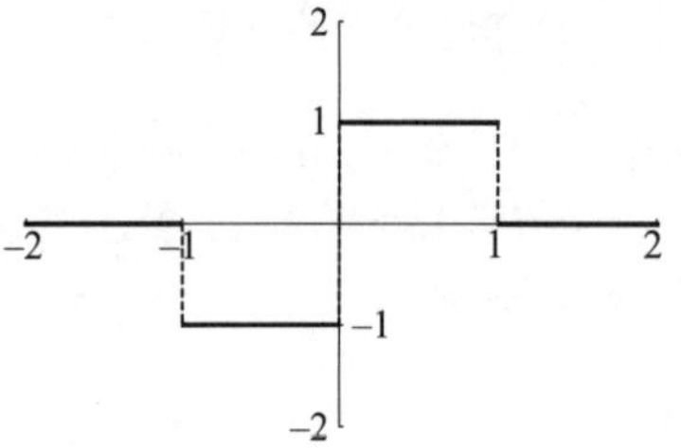

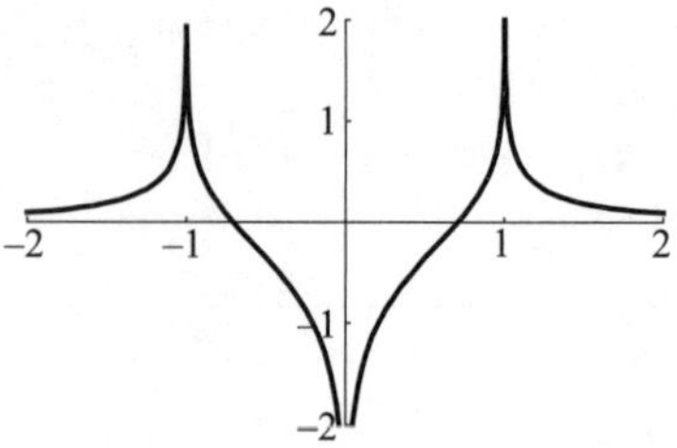

4 $f = \Pi(x-1) - \Pi(x+1)$ $\qquad$ $Hf = \dfrac{1}{\pi}\log\left|\dfrac{4x^2-1}{4x^2-9}\right|$

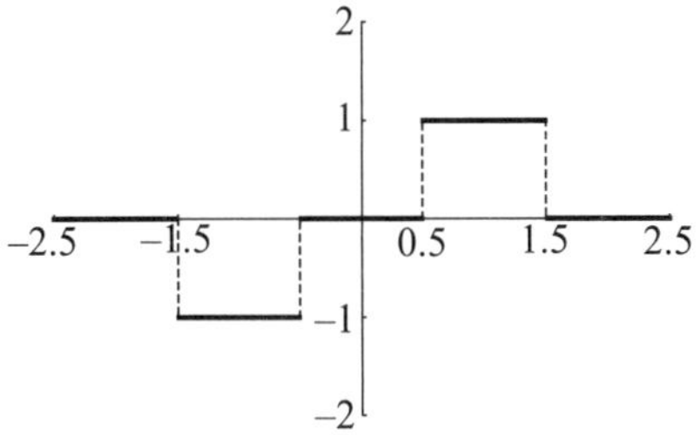

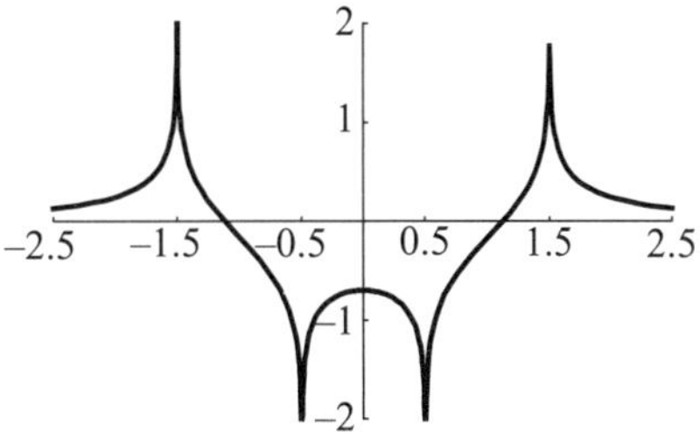

5 $f = \dfrac{1}{2}\{1 + \operatorname{sgn}(\cos \pi x)\}$ $\qquad$ $Hf = \dfrac{1}{\pi}\log\left|\tan\left\{\dfrac{\pi}{4}(2x+1)\right\}\right|$

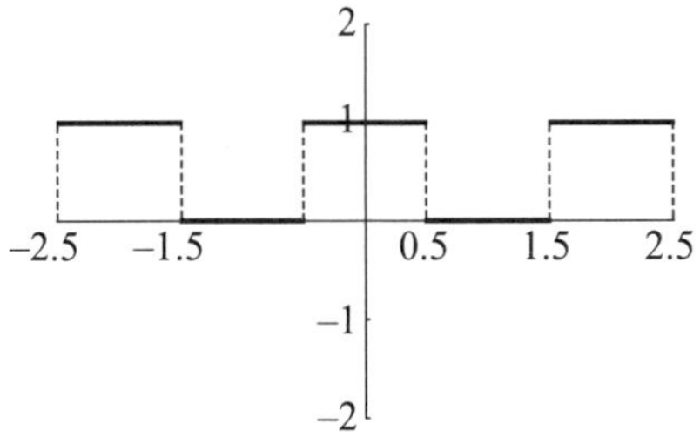

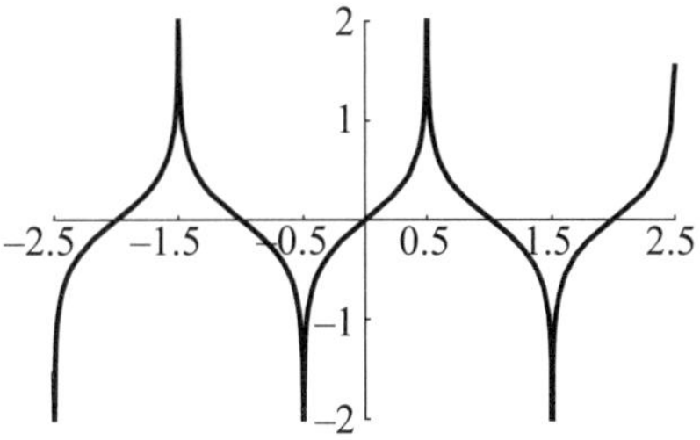

6 $f = \dfrac{1}{2}\{1 + \operatorname{sgn}(\sin \pi x)\}$ $\qquad$ $Hf = \dfrac{1}{\pi}\log\left|\tan\left\{\dfrac{\pi x}{4}\right\}\right|$

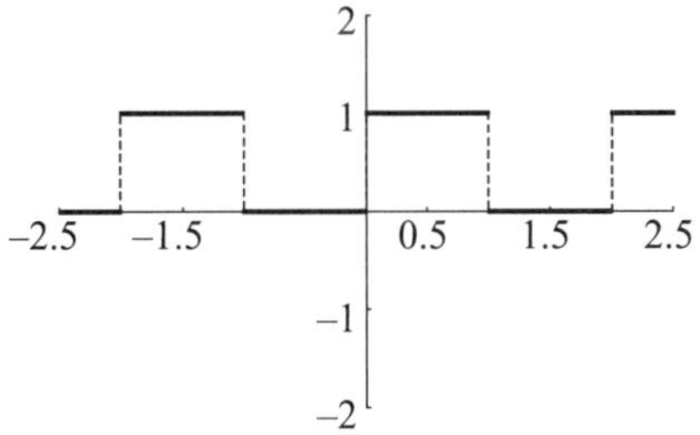

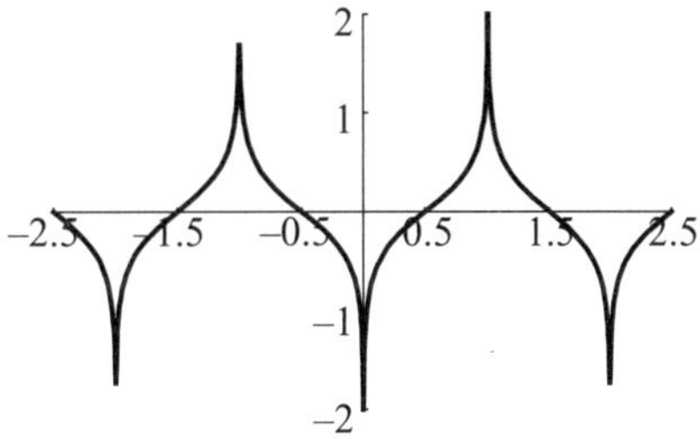

7 $f = \operatorname{sgn}(\cos \pi x)$ $\qquad$ $Hf = \dfrac{2}{\pi}\log\left|\tan\left\{\dfrac{\pi}{4}(2x+1)\right\}\right|$

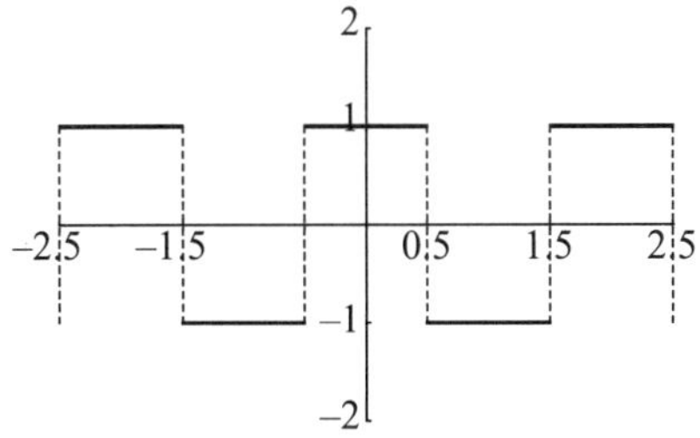

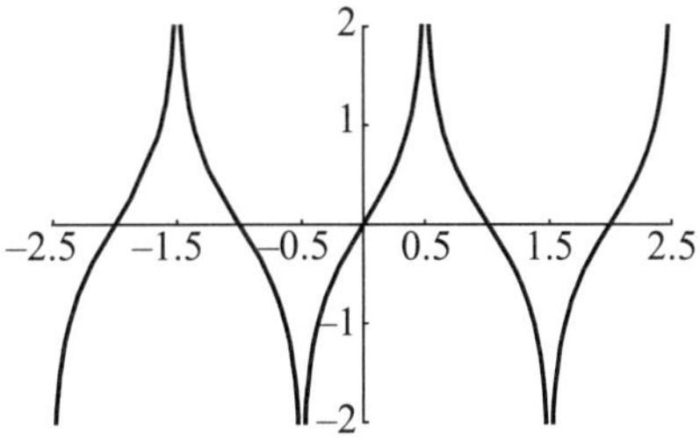

8 $f = \operatorname{sgn}(\sin \pi x)$ $\qquad$ $Hf = \dfrac{2}{\pi} \log \left| \tan \left\{ \dfrac{\pi x}{2} \right\} \right|$

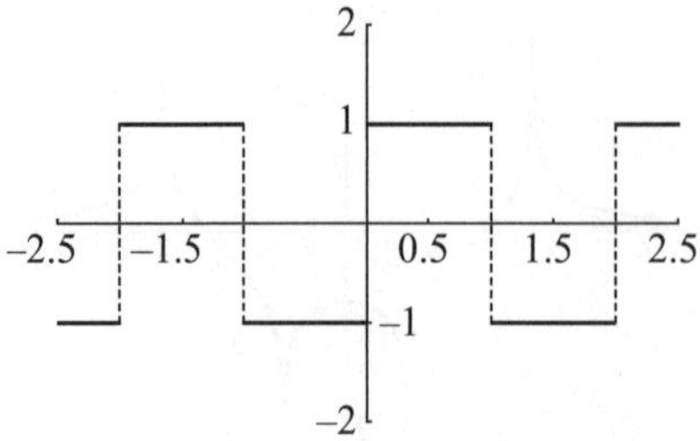

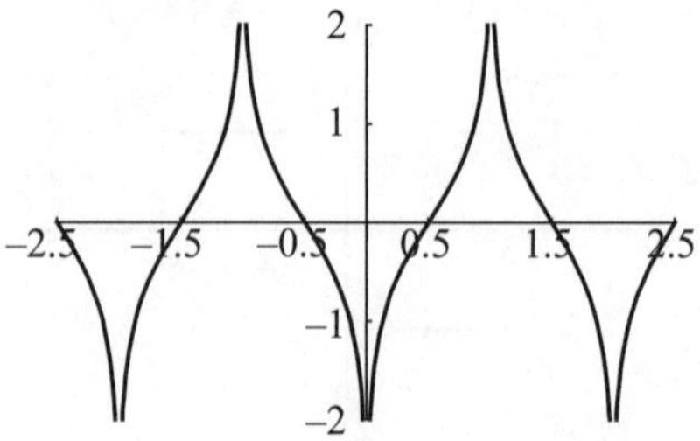

9 $f = \sin \pi x$ $\qquad$ $Hf = -\cos \pi x$

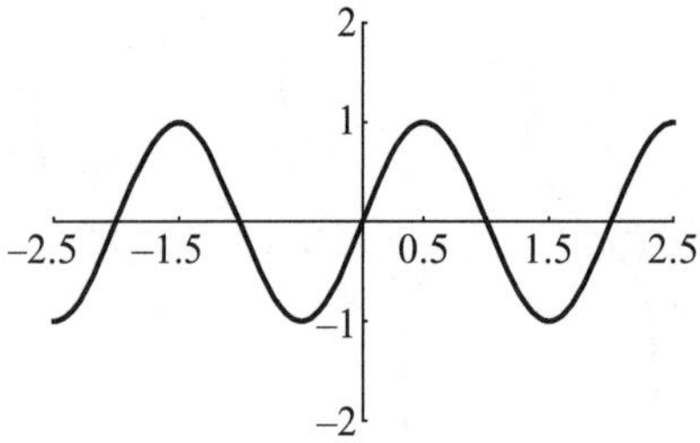

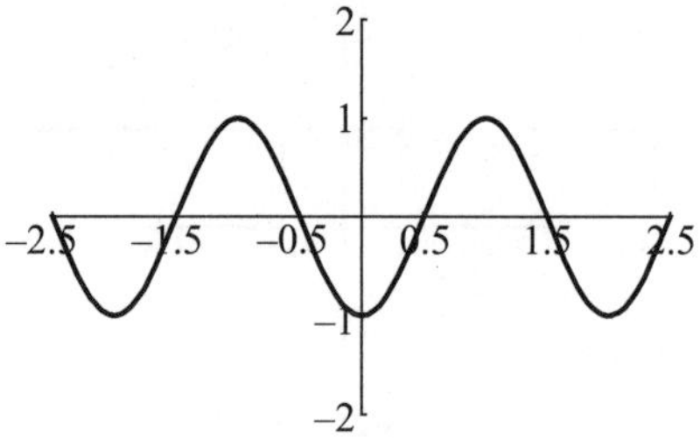

10 $f = \sin^2 \pi x$ $\qquad$ $Hf = -\dfrac{1}{2} \sin 2\pi x$

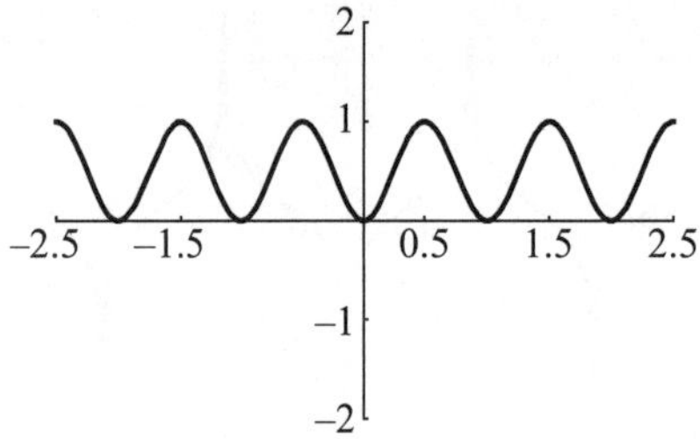

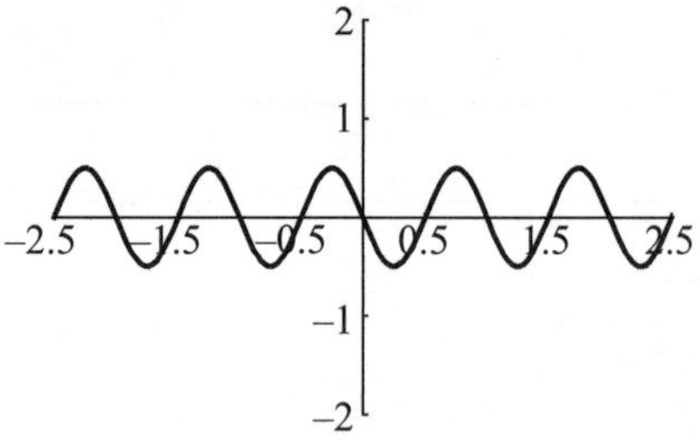

11 $f = \cos \pi x$ $\qquad$ $Hf = \sin \pi x$

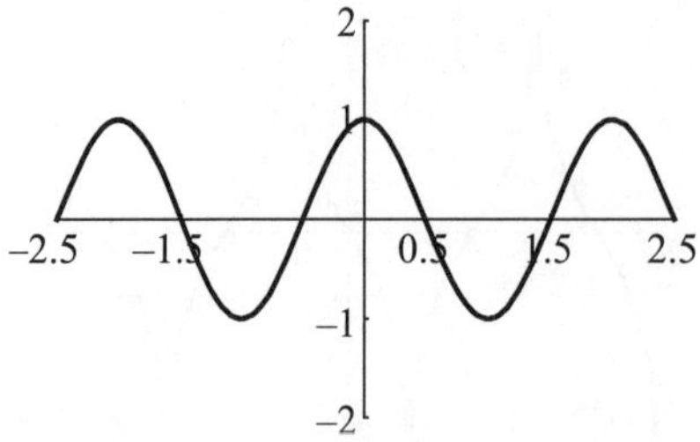

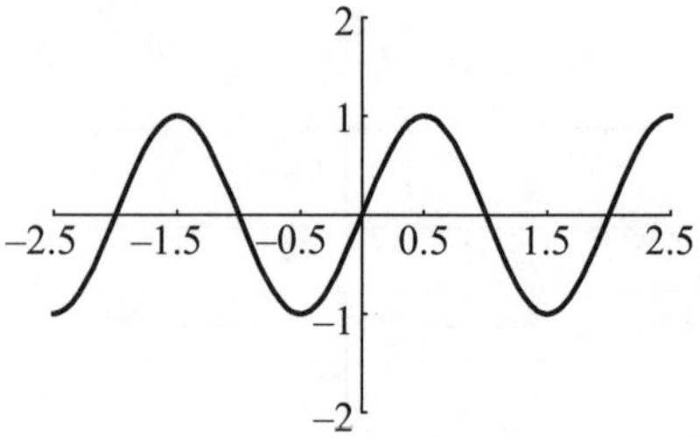

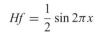

12 $f = \cos^2 \pi x$ $\qquad$ $Hf = \frac{1}{2} \sin 2\pi x$

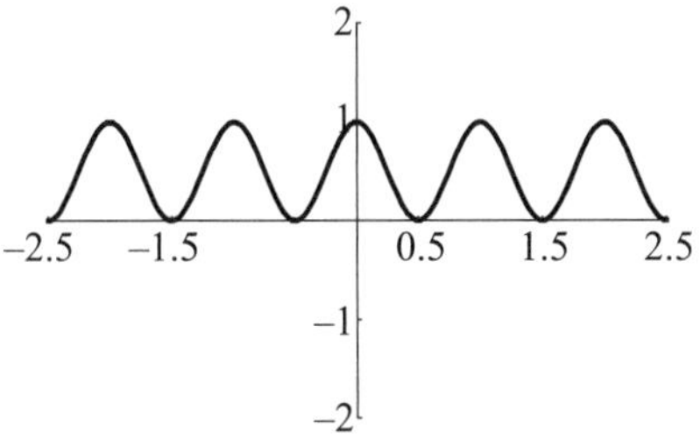

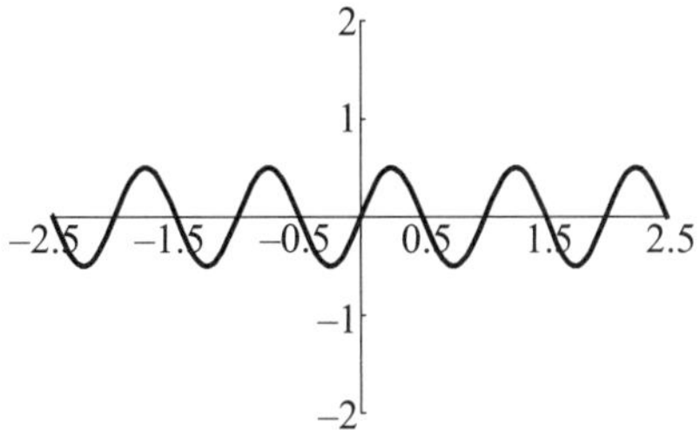

13 $f = \operatorname{cas} \pi x$ $\qquad$ $Hf = \surd(2) \sin \frac{\pi}{4}(4x - 1)$

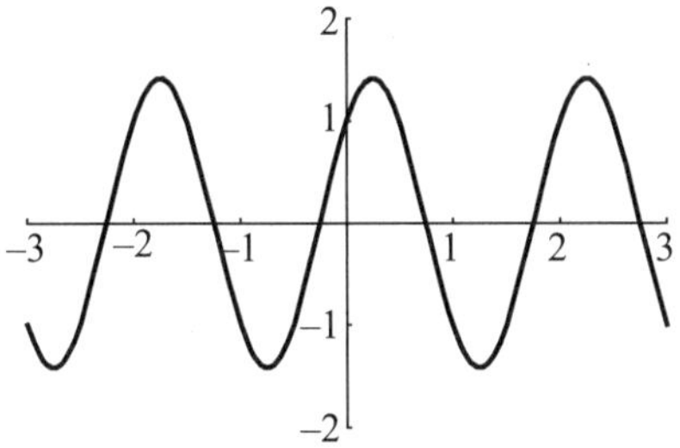

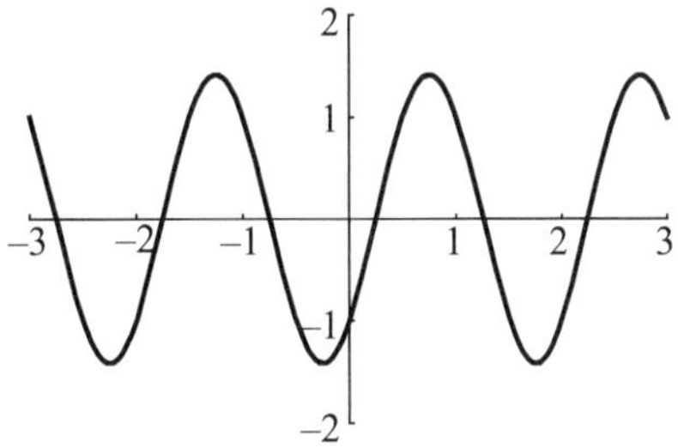

14 $f = \operatorname{sinc} x$ $\qquad$ $Hf = \frac{1 - \cos \pi x}{\pi x}$

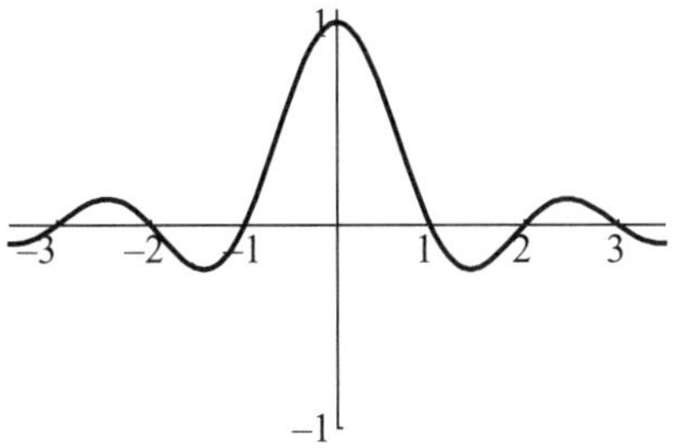

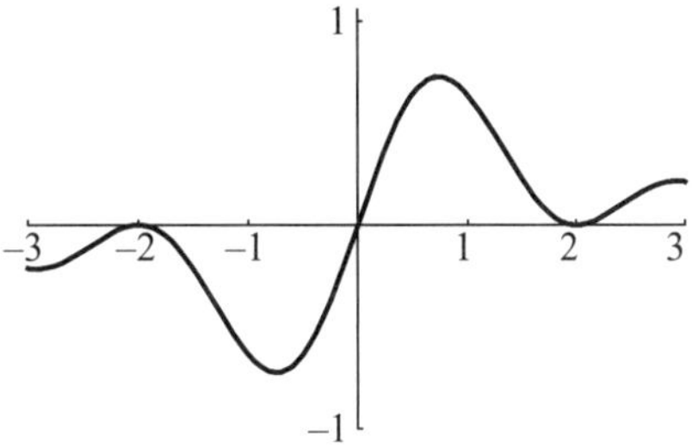

15 $f = \operatorname{sinc}^2 x$ $\qquad$ $Hf = \frac{2\pi x - \sin 2\pi x}{2\pi^2 x^2}$

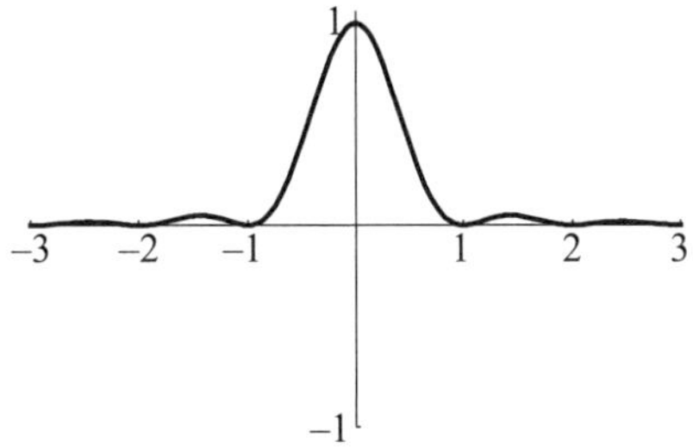

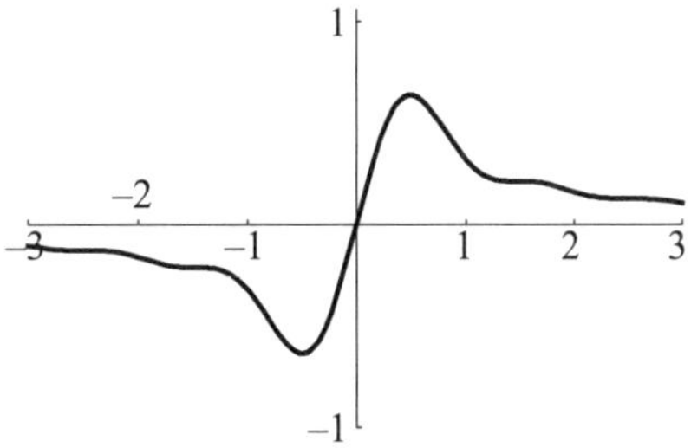

16 $f = \dfrac{1 - \cos \pi x}{x}$ $\qquad$ $Hf = -\pi \operatorname{sinc} x$

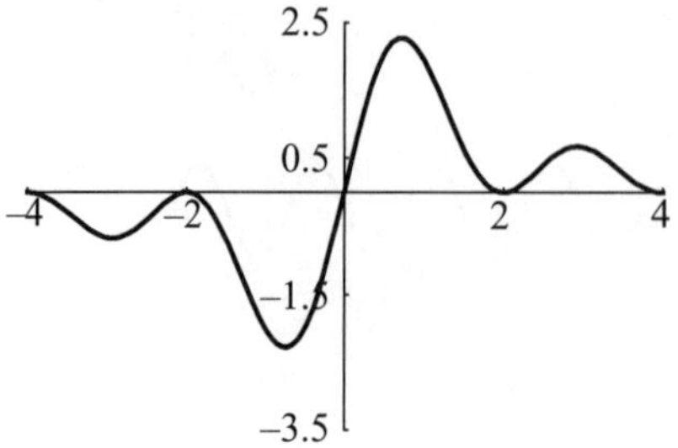

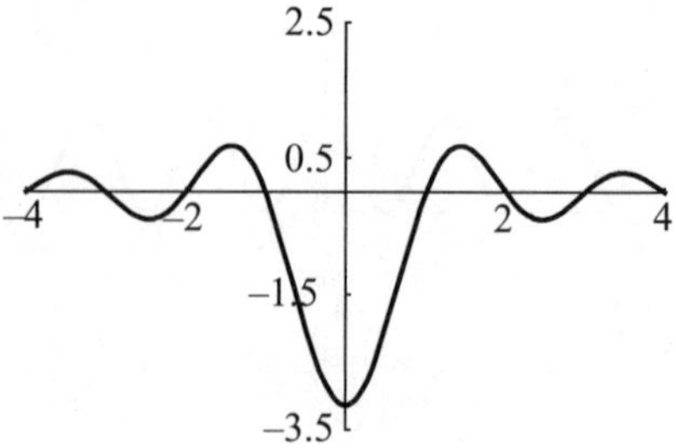

17 $f = \cos \dfrac{\pi x}{2} \Pi_2(x)$

$$Hf = \frac{1}{\pi}\left\{\cos\frac{\pi x}{2}\left[\operatorname{Ci}\left(\frac{\pi}{2}|x+1|\right) - \operatorname{Ci}\left(\frac{\pi}{2}|x-1|\right)\right] + \sin\frac{\pi x}{2}\left[\operatorname{Si}\left(\frac{\pi}{2}(x+1)\right) - \operatorname{Si}\left(\frac{\pi}{2}(x-1)\right)\right]\right\}$$

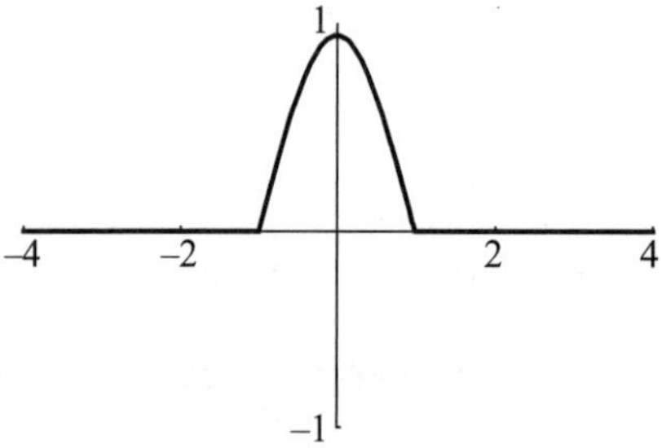

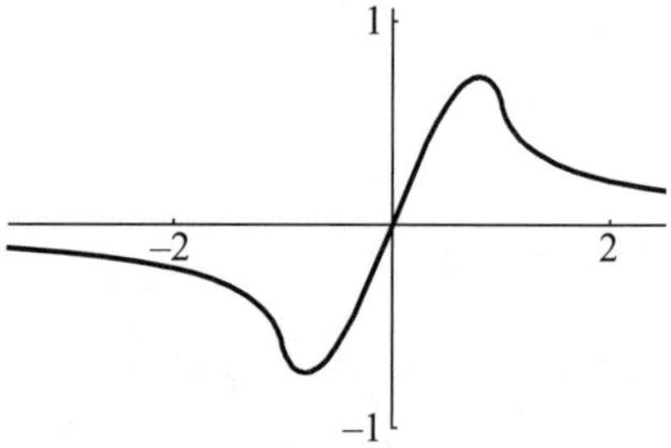

18 $f = \cos^2 \dfrac{\pi x}{2} \Pi_2(x)$

$$Hf = \frac{1}{2\pi}\log\left|\frac{x+1}{x-1}\right| + \frac{\cos \pi x}{2\pi} \times [\operatorname{Ci}(\pi|x+1|) - \operatorname{Ci}(\pi|x-1|)] + \frac{\sin \pi x}{2\pi}[\operatorname{Si}(\pi(x+1)) - \operatorname{Si}(\pi(x-1))]$$

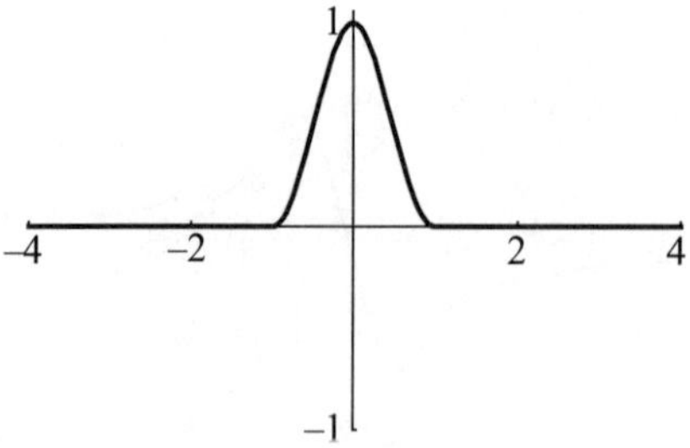

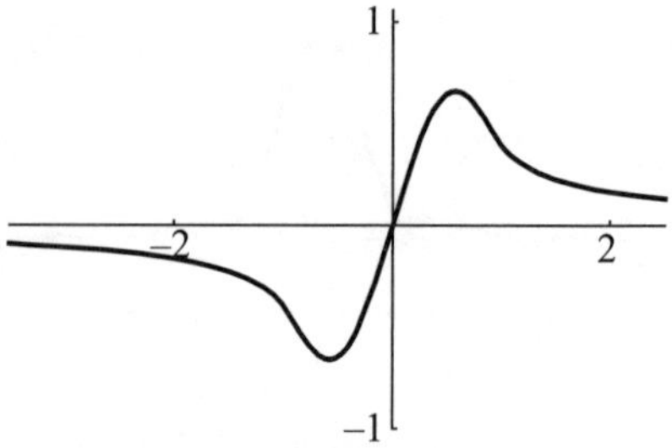

19 $f = \cos \frac{\pi x}{2} \mathrm{H}(x)\Pi_2(x)$

$$Hf = \frac{1}{\pi}\left\{\cos\frac{\pi x}{2}\left[\mathrm{Ci}\left(\frac{\pi|x|}{2}\right) - \mathrm{Ci}\left(\frac{\pi|x-1|}{2}\right)\right] + \sin\frac{\pi x}{2}\left[\mathrm{Si}\left(\frac{\pi x}{2}\right) - \mathrm{Si}\left(\frac{\pi(x-1)}{2}\right)\right]\right\}$$

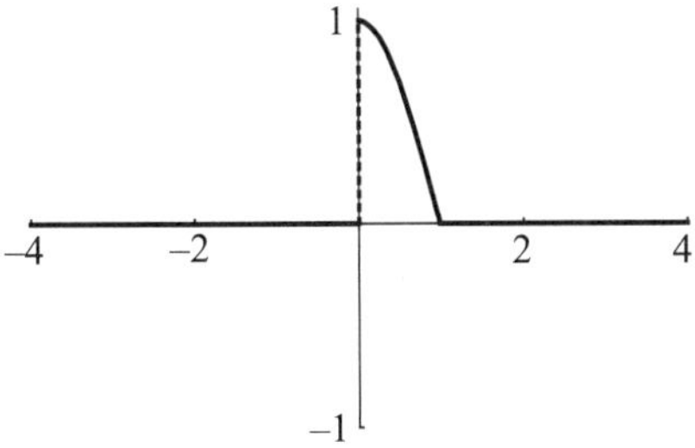

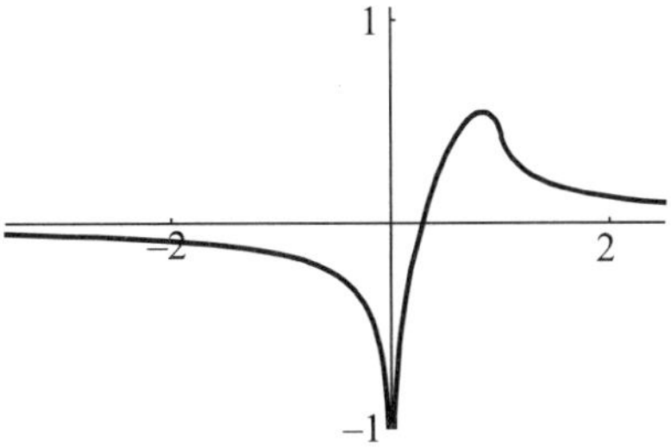

20 $f = \sin \pi x\, \mathrm{e}^{-\pi|x|}$

$$Hf = \pi^{-1}\cos\pi x[\mathrm{sie}(\pi x, 1)\mathrm{e}^{\pi x} + \mathrm{sie}(-\pi x, 1)\mathrm{e}^{-\pi x}] + \pi^{-1}\mathrm{sgn}\, x \sin\pi x[\mathrm{cie}(\pi|x|, 1)\mathrm{e}^{\pi|x|} - \mathrm{Cie}(\pi|x|, 1)\mathrm{e}^{-\pi|x|}]$$

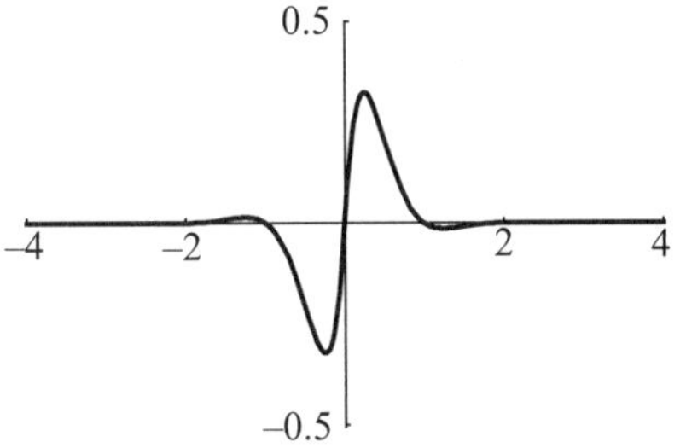

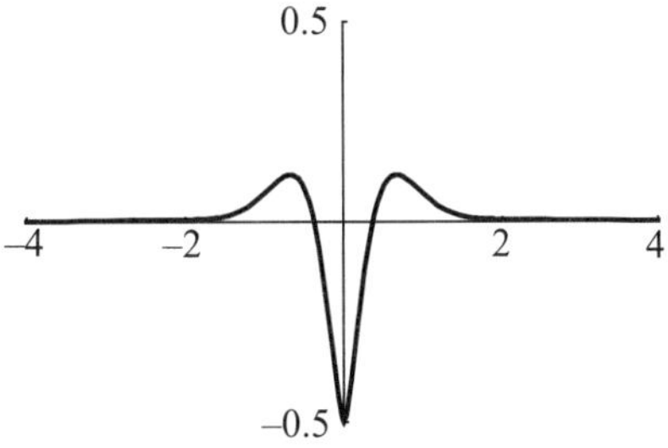

21 $f = \sin \pi x\, \mathrm{e}^{-\pi|x|}\mathrm{H}(x)$

$$Hf = \begin{cases} \frac{\mathrm{e}^{-\pi x}}{\pi}[\cos\pi x\, \mathrm{sie}(-\pi x, 1) - \sin\pi x\, \mathrm{Cie}(\pi x, 1)], & x > 0 \\ \frac{\mathrm{e}^{-\pi x}}{\pi}[\cos\pi x\, \mathrm{sie}(-\pi x, 1) - \sin\pi x\, \mathrm{cie}(-\pi x, 1)], & x < 0 \end{cases}$$

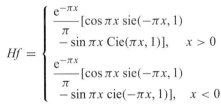

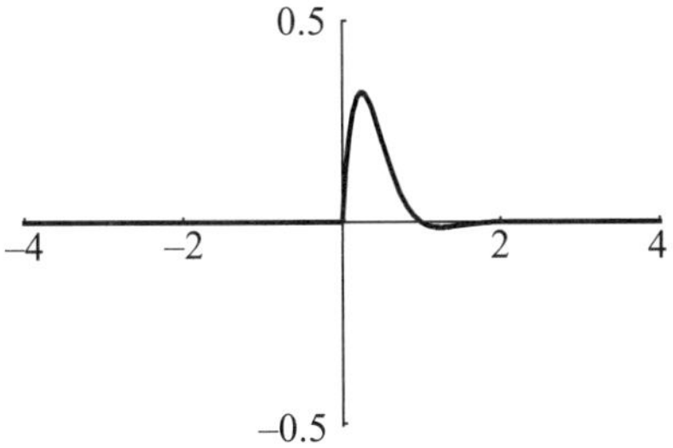

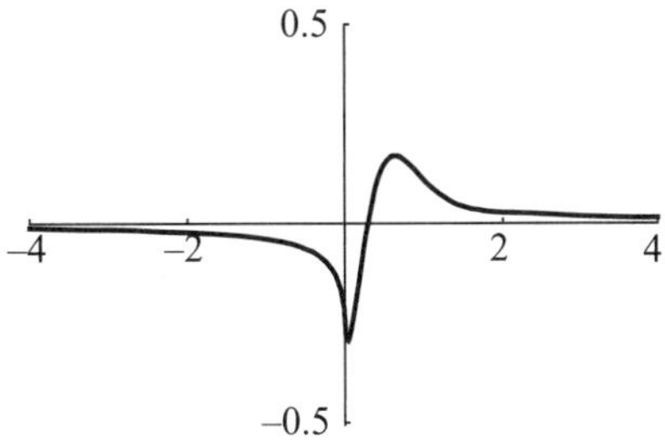

22 $f = \cos \pi x \, e^{-\pi|x|}$

$$Hf = -\pi^{-1} \sin \pi x [\mathrm{sie}(\pi x, 1) e^{\pi x} + \mathrm{sie}(-\pi x, 1) e^{-\pi x}] + \pi^{-1} \operatorname{sgn} x \cos \pi x [\mathrm{cie}(\pi |x|, 1) e^{\pi |x|} - \mathrm{Cie}(\pi |x|, 1) e^{-\pi |x|}]$$

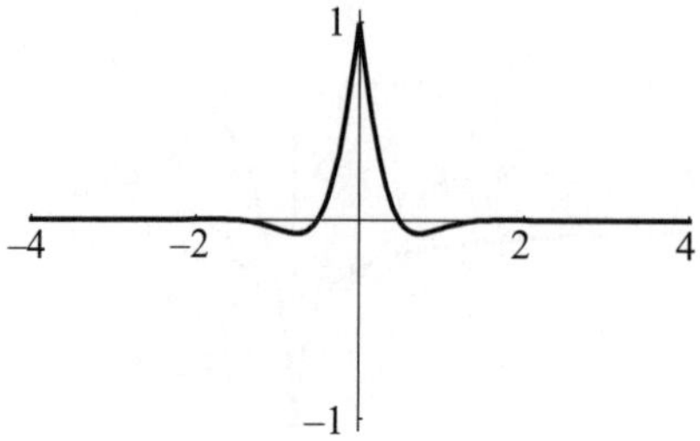

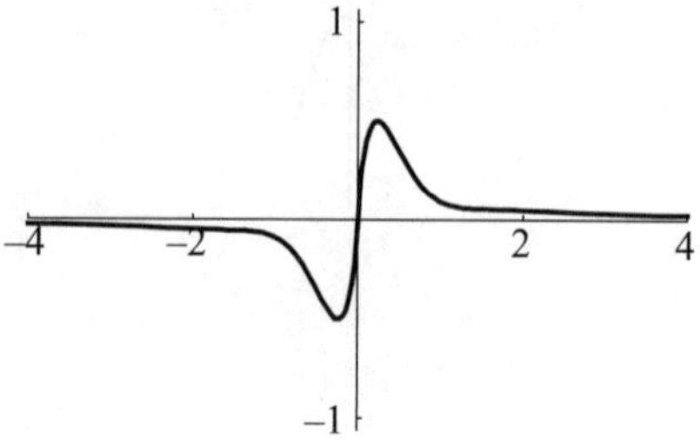

23 $f = \cos \pi x \, e^{-\pi|x|} \mathrm{H}(x)$

$$Hf = \begin{cases} -\dfrac{e^{-\pi x}}{\pi} [\sin \pi x \, \mathrm{sie}(-\pi x, 1) + \cos \pi x \, \mathrm{Cie}(\pi x, 1)], & x > 0 \\ -\dfrac{e^{-\pi x}}{\pi} [\sin \pi x \, \mathrm{sie}(-\pi x, 1) + \cos \pi x \, \mathrm{cie}(-\pi x, 1)], & x < 0 \end{cases}$$

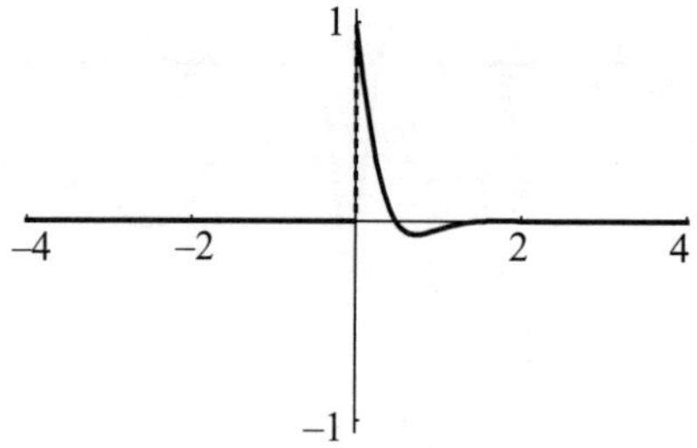

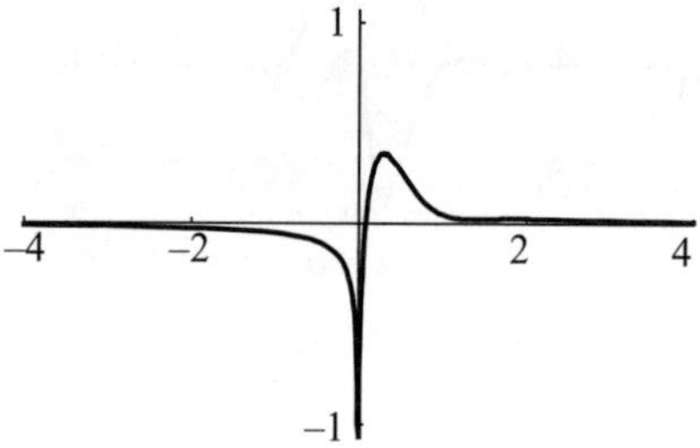

24 $f = e^{-|x|}$

$$Hf = \frac{\operatorname{sgn} x}{\pi} \left[e^{|x|} E_1(|x|) + e^{-|x|} \mathrm{Ei}(|x|) \right]$$

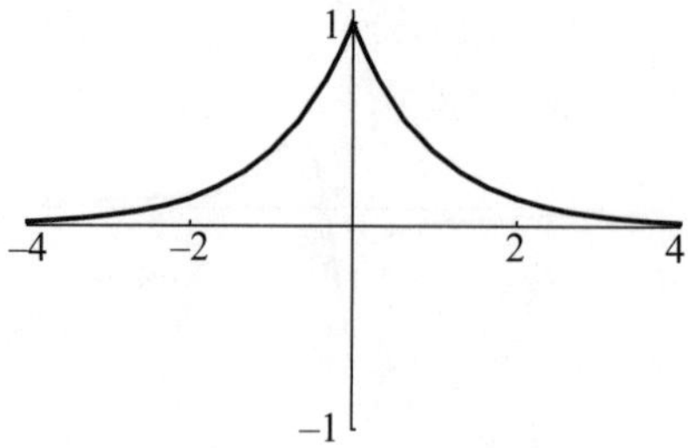

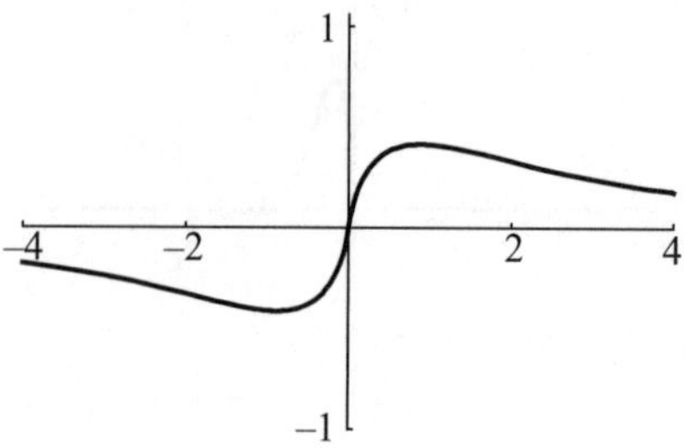

25 $f = \mathrm{e}^{-|x|}\mathrm{H}(x)$

$$Hf = \begin{cases} \pi^{-1}\mathrm{e}^{-x}\mathrm{Ei}(x), & x > 0 \\ -\pi^{-1}\mathrm{e}^{-x}E_1(-x), & x < 0 \end{cases}$$

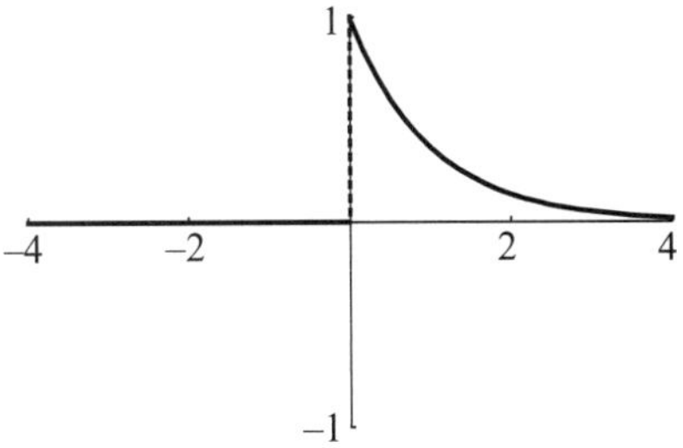

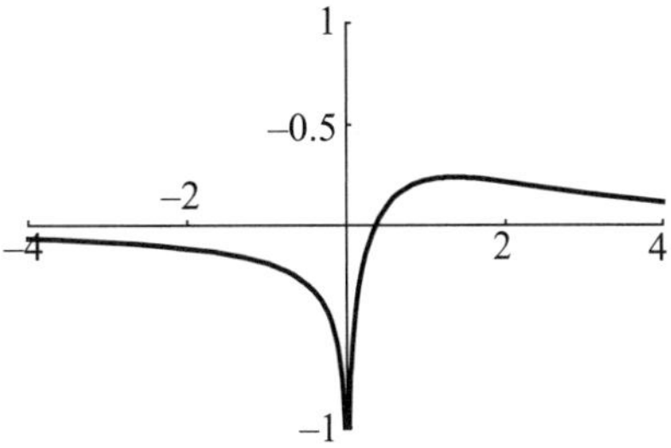

26 $f = \operatorname{sgn} x\, \mathrm{e}^{-|x|}$

$$Hf = \frac{1}{\pi}\left[\mathrm{e}^{-|x|}\mathrm{Ei}(|x|) - \mathrm{e}^{|x|}E_1(|x|)\right]$$

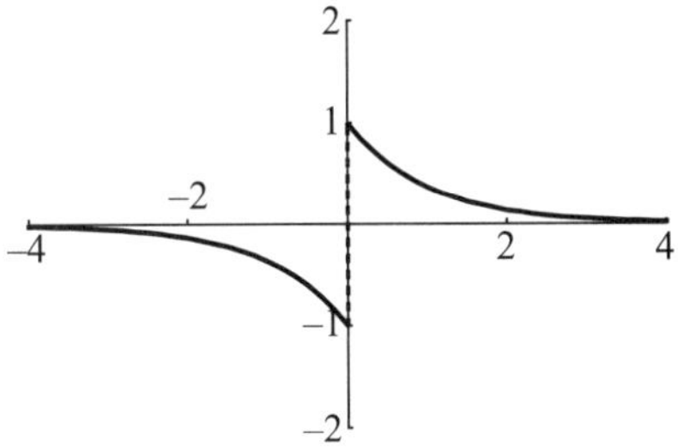

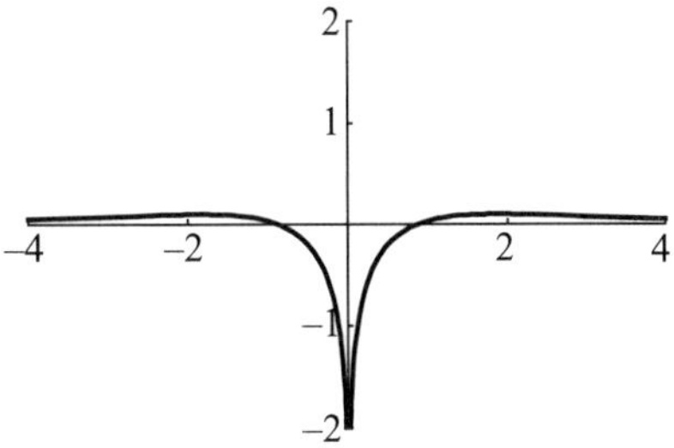

27 $f = x\mathrm{e}^{-|x|}\mathrm{H}(x)$

$$Hf = \begin{cases} \pi^{-1}[x\mathrm{e}^{-x}\mathrm{Ei}(x) - 1], & x > 0 \\ -\pi^{-1}[x\mathrm{e}^{-x}E_1(-x) + 1], & x < 0 \end{cases}$$

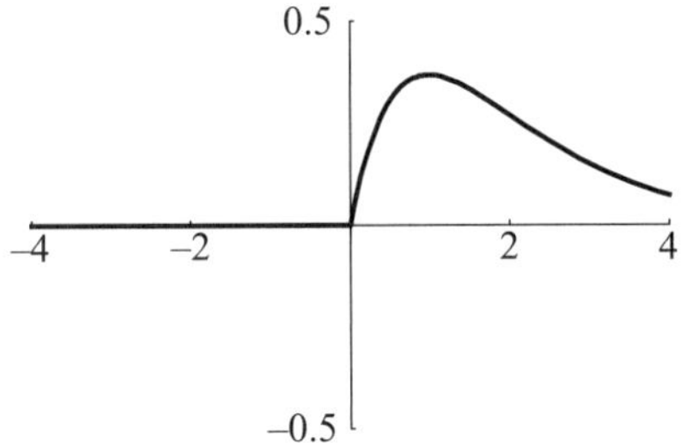

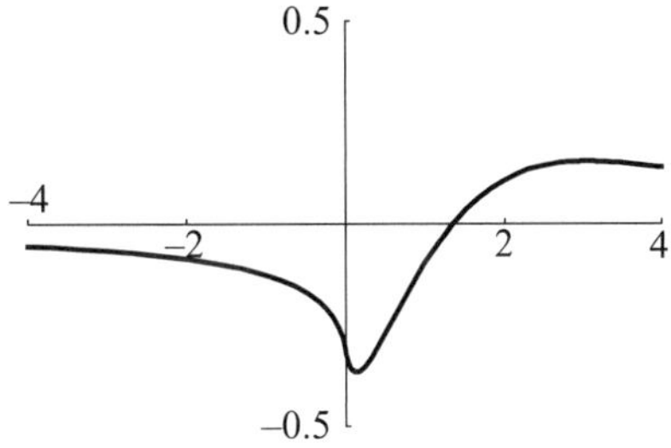

28 $f = \mathrm{e}^{-x^2}$

$$Hf = -\mathrm{i}\mathrm{e}^{-x^2}\mathrm{erf}(\mathrm{i}x)$$

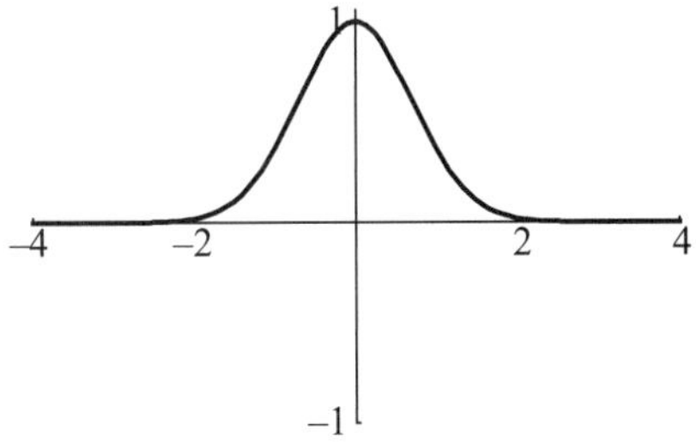

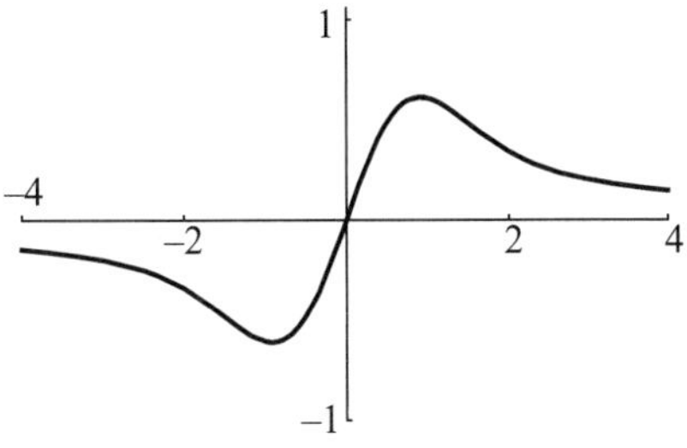

29 $f = \cos \pi x \, e^{-x^2}$ $\qquad Hf = e^{-x^2} \operatorname{Im}\left\{e^{i\pi x} \operatorname{erf}\left(\frac{\pi}{2} + ix\right)\right\}$

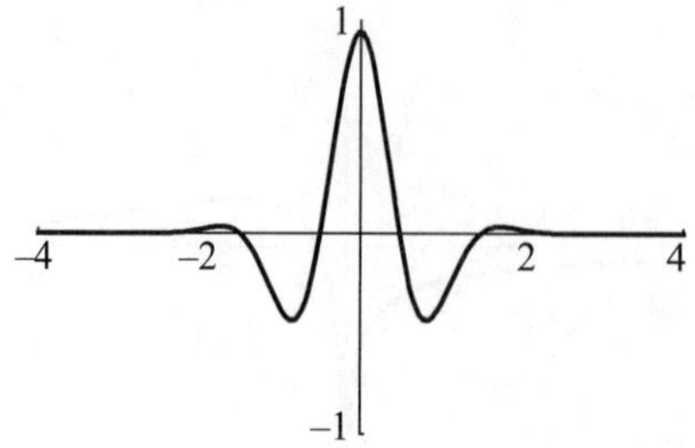

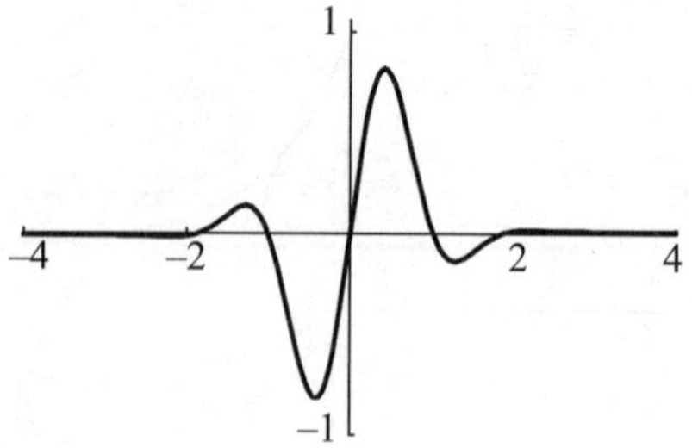

30 $f = \operatorname{sech} 2\pi x$ $\qquad Hf = \frac{i}{\pi}\left[\psi\left(\frac{1}{4} - ix\right) - \psi\left(\frac{1}{4} + ix\right)\right] - \tanh 2\pi x$

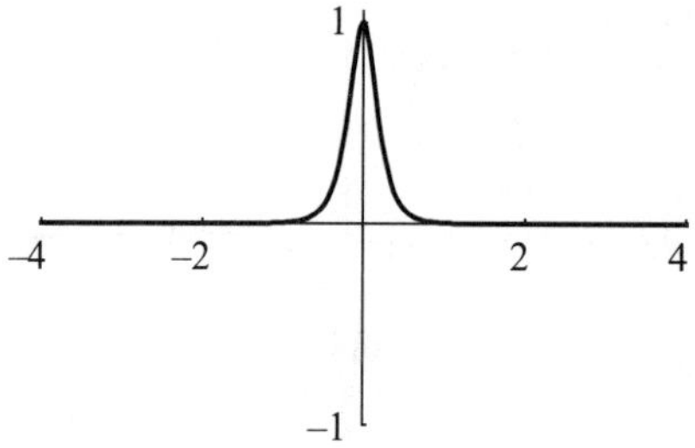

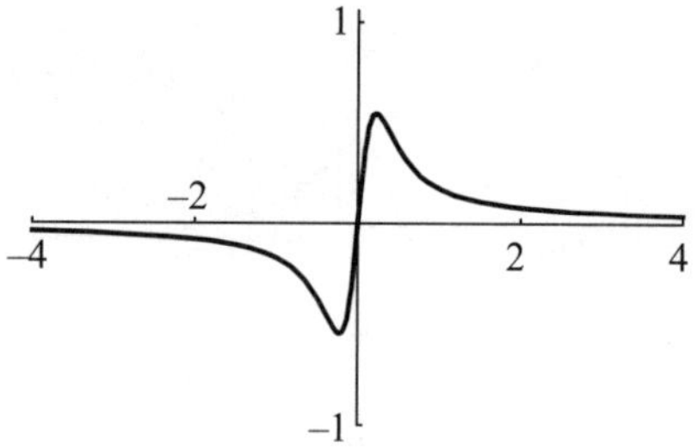

31 $f = \operatorname{sech}^2 x$ $\qquad Hf = 32\pi x \sum_{k=0}^{\infty} \frac{2k+1}{[(2k+1)^2\pi^2 + 4x^2]^2}$

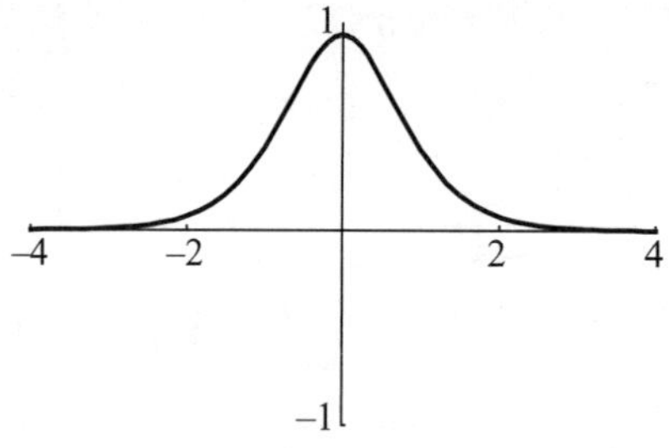

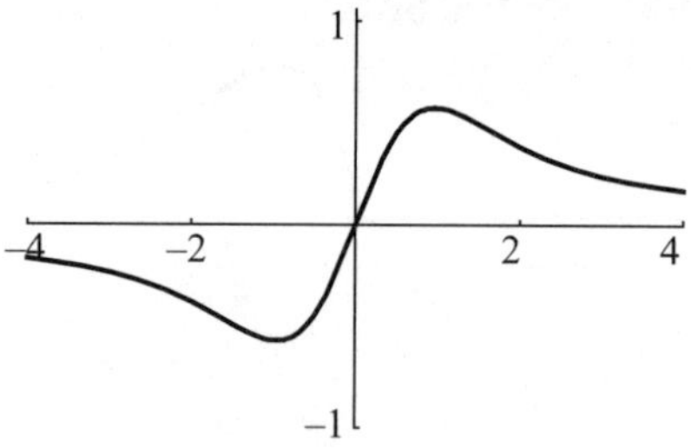

32 $f = x\Pi_2(x-1)$ $\qquad Hf = \frac{1}{\pi}\left\{\log\left|\frac{x}{x-2}\right| - 2\right\}$

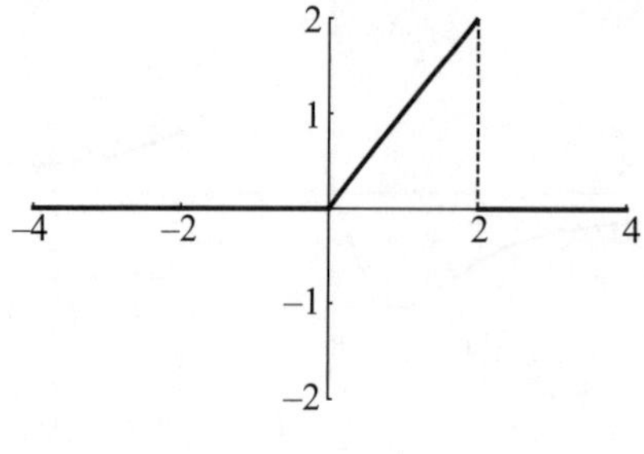

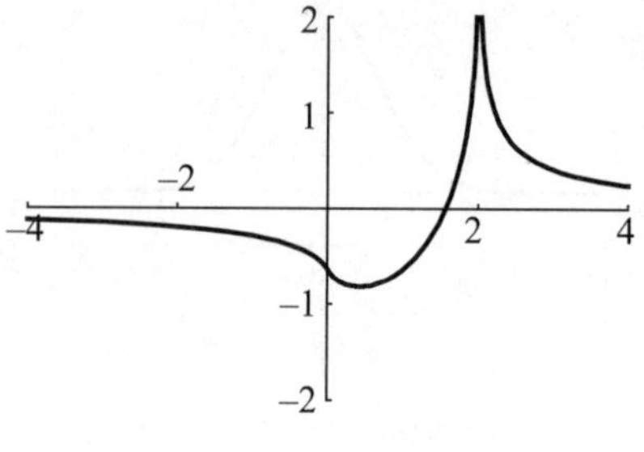

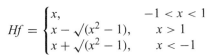

33 $f = \sqrt{(1 - x^2)}\,\Pi_2(x)$ $\qquad$ $Hf = \begin{cases} x, & -1 < x < 1 \\ x - \sqrt{(x^2 - 1)}, & x > 1 \\ x + \sqrt{(x^2 - 1)}, & x < -1 \end{cases}$

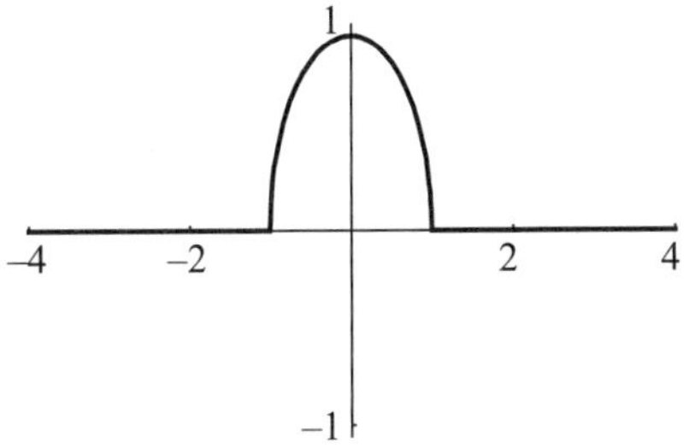

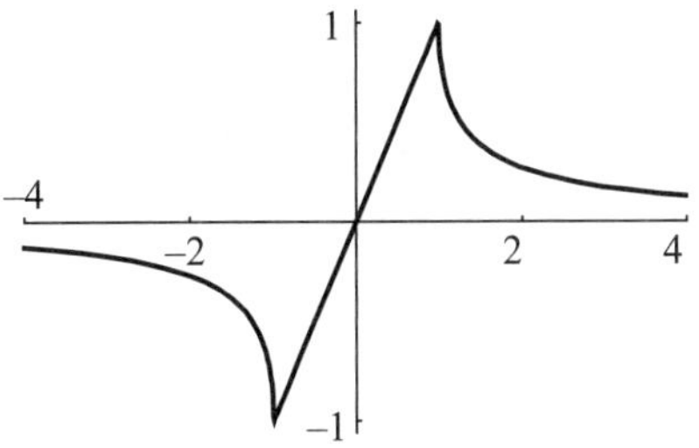

34 $f = (1 - x^2)\Pi_2(x)$ $\qquad$ $Hf = \dfrac{1}{\pi}\left\{2x + (1 - x^2)\log\left|\dfrac{1 + x}{1 - x}\right|\right\}$

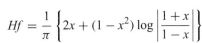

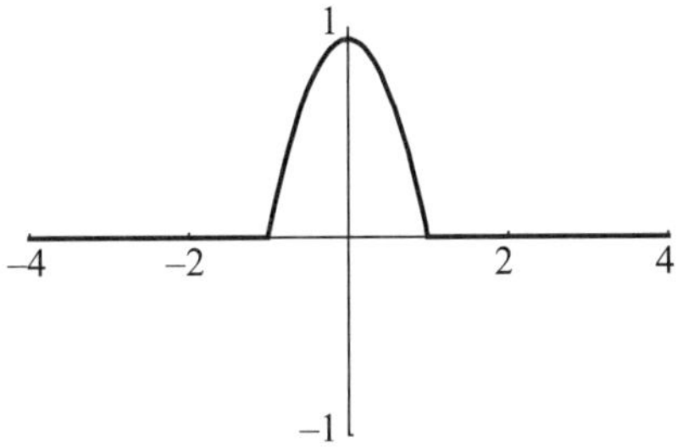

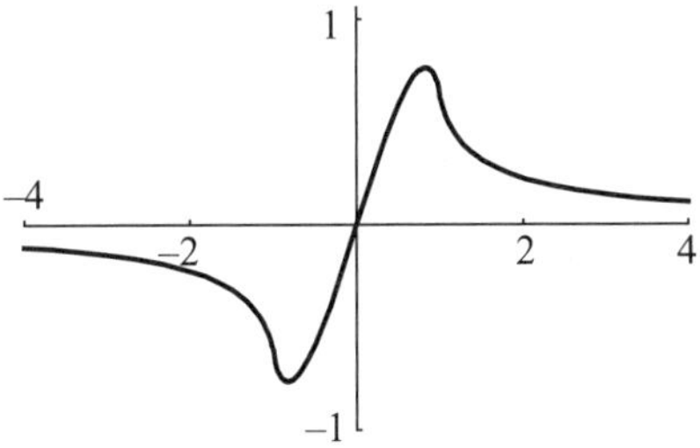

35 $f = \Lambda(x)$ $\qquad$ $Hf = \dfrac{1}{\pi}\left\{\log\left|\dfrac{1 + x}{1 - x}\right| + x\log\left|\dfrac{x^2 - 1}{x^2}\right|\right\}$

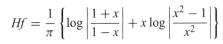

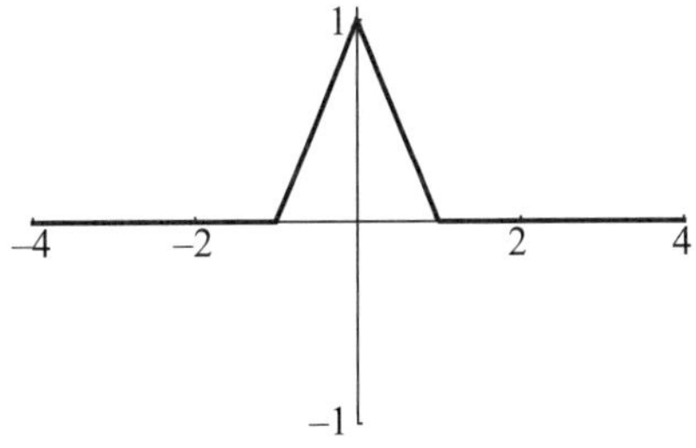

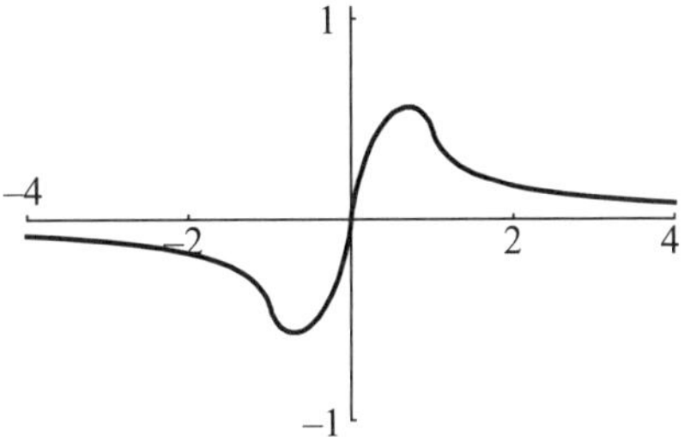

36 $f = \operatorname{sgn} x\,\Lambda(x)$ $\qquad$ $Hf = \dfrac{1}{\pi}\left\{2 + \log\left|\dfrac{x^2}{x^2 - 1}\right| + x\log\left|\dfrac{x - 1}{x + 1}\right|\right\}$

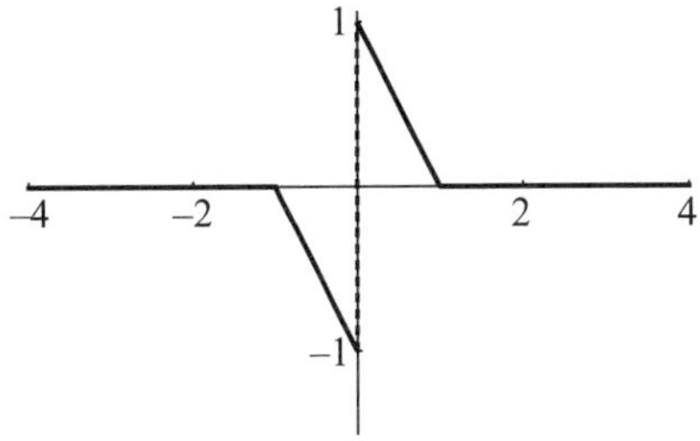

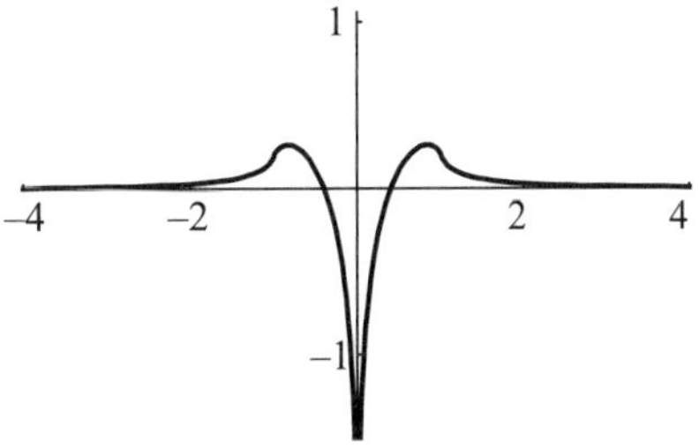

37 $f = \sum_{k=-\infty}^{\infty} \Lambda(x - 2k)$ $\qquad Hf = \frac{4}{\pi^2} \sum_{k=0}^{\infty} \frac{\sin[(2k+1)\pi x]}{(2k+1)^2}$

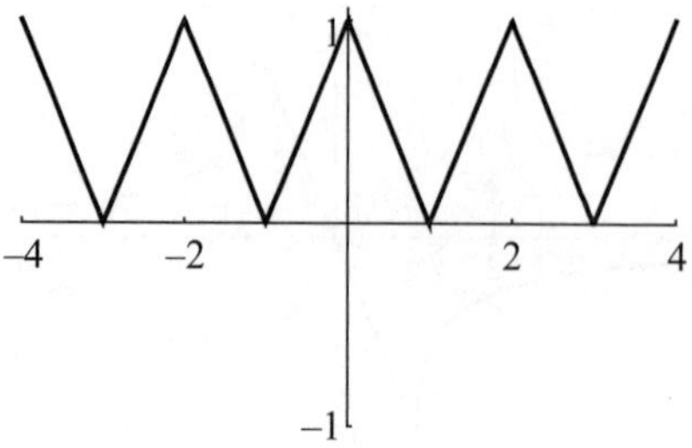

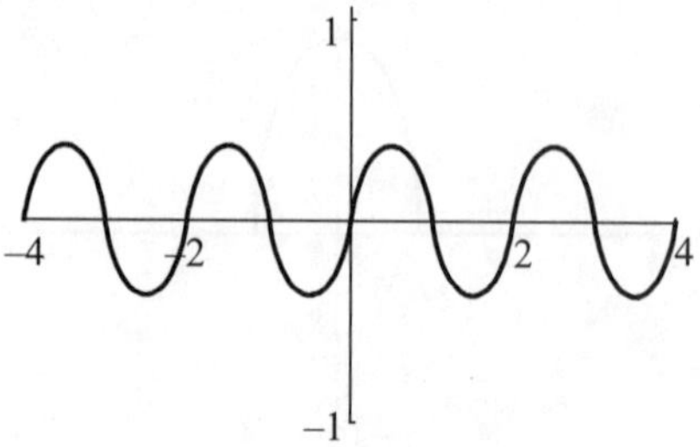

38 $f = \sum_{k=-\infty}^{\infty} \Lambda(x - 2k - 1)$ $\qquad Hf = -\frac{4}{\pi^2} \sum_{k=0}^{\infty} \frac{\sin[(2k+1)\pi x]}{(2k+1)^2}$

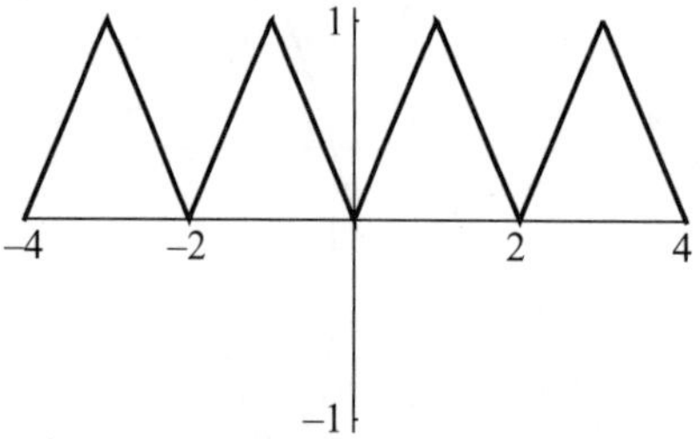

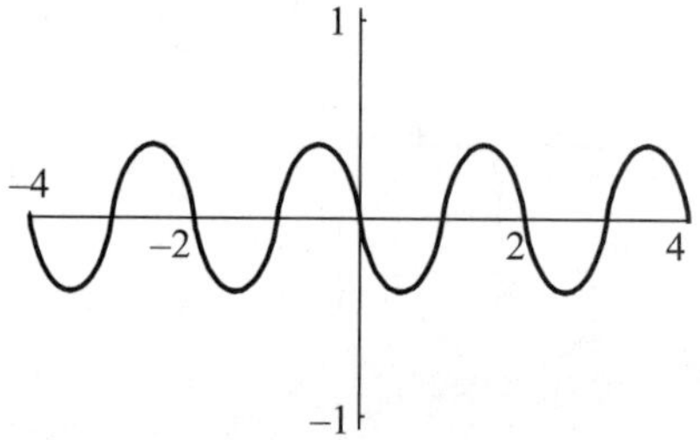

39 $f = \frac{1}{2} + \frac{4}{\pi^2} \sum_{k=0}^{\infty} \frac{\cos[(2k+1)\pi x] + 2\cos\left[\frac{(2k+1)\pi x}{2}\right]}{(2k+1)^2}$

$$Hf = \frac{4}{\pi^2} \sum_{k=0}^{\infty} \frac{\sin[(2k+1)\pi x] + 2\sin[\frac{(2k+1)\pi x}{2}]}{(2k+1)^2}$$

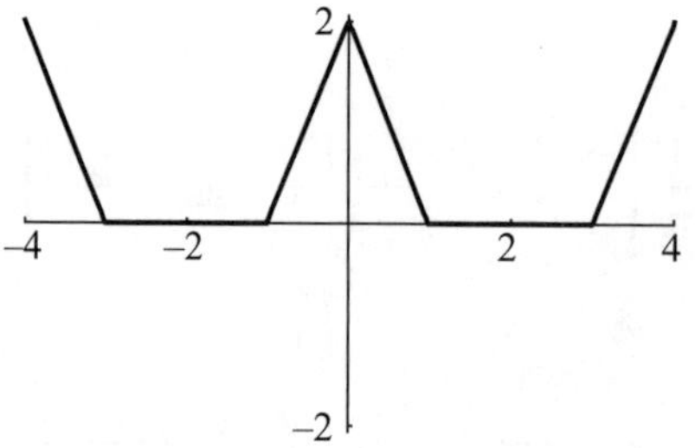

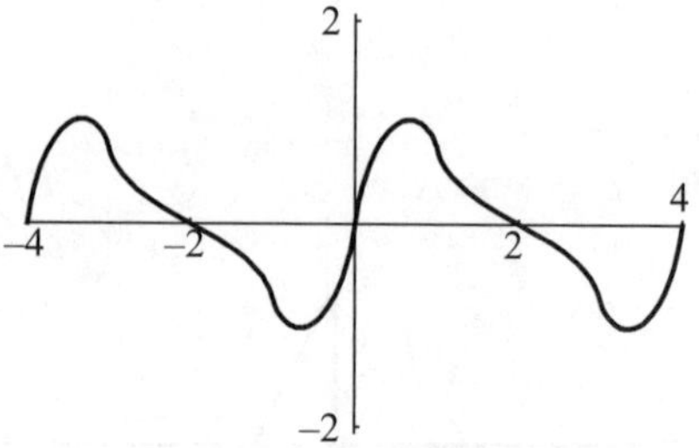

40

$$f = \frac{1}{2} - \frac{4}{\pi^2} \sum_{k=0}^{\infty} \frac{\cos[(2k+1)\pi x] + 2(-1)^k \sin\left[\frac{(2k+1)\pi x}{2}\right]}{(2k+1)^2}$$

$$Hf = -\frac{4}{\pi^2} \sum_{k=0}^{\infty} \frac{\sin[(2k+1)\pi x] - 2(-1)^k \cos\left[\frac{(2k+1)\pi x}{2}\right]}{(2k+1)^2}$$

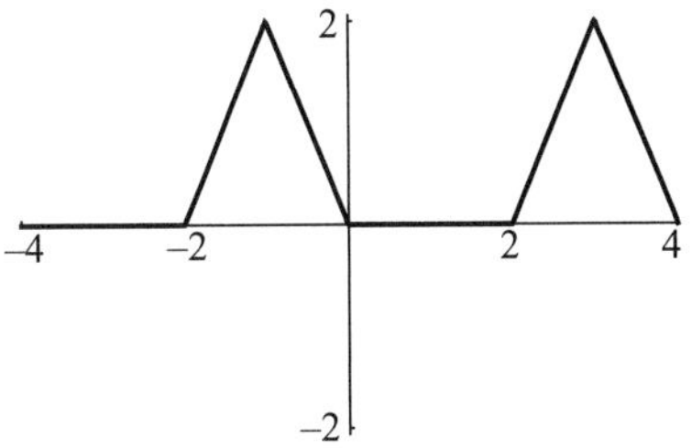

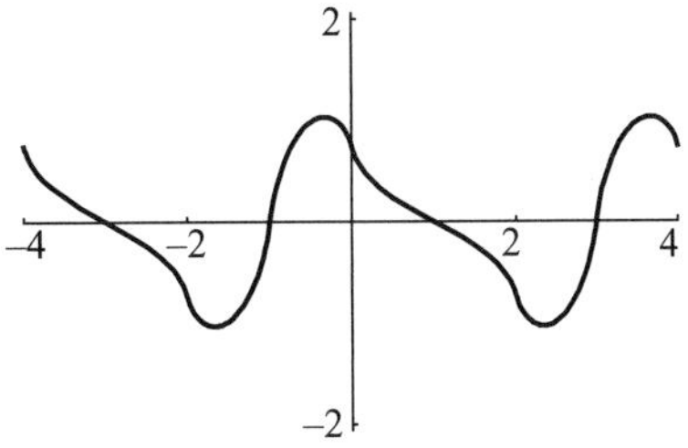

41 $f = \frac{1}{x^2+1}$ $Hf = \frac{x}{x^2+1}$

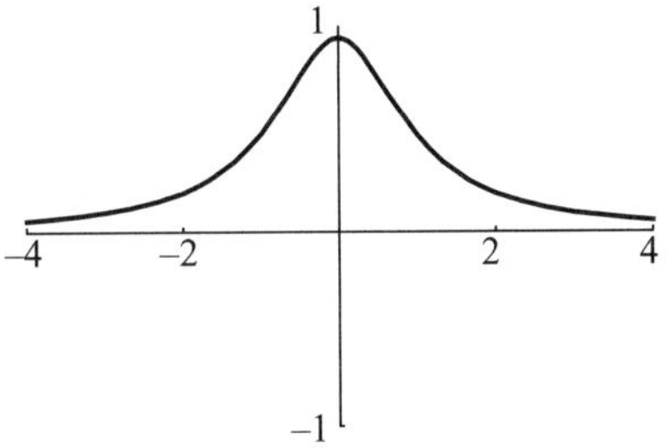

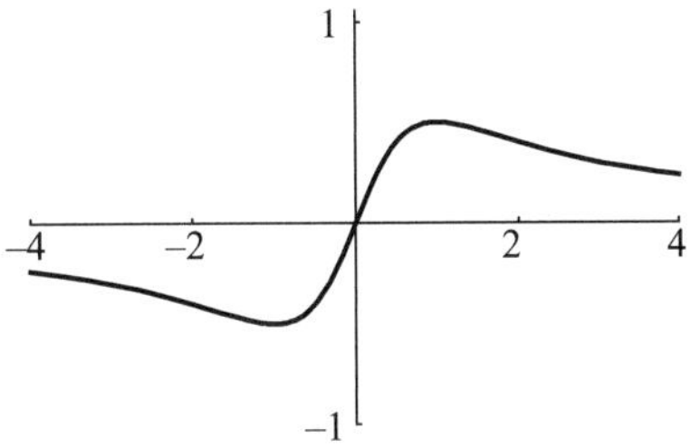

42 $f = \frac{1}{\pi} \log\left|\frac{x-1}{x+1}\right|$ $Hf = \Pi_2(x)$

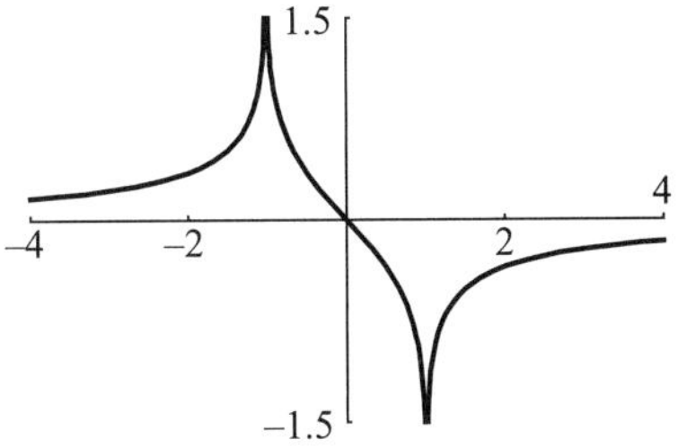

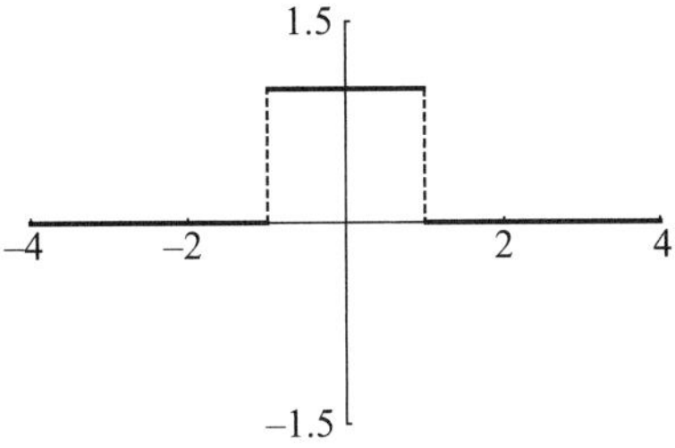

43 $f = \operatorname{sgn} x \operatorname{si}(|x|)$ $\qquad$ $Hf = -\operatorname{Ci}(|x)|$

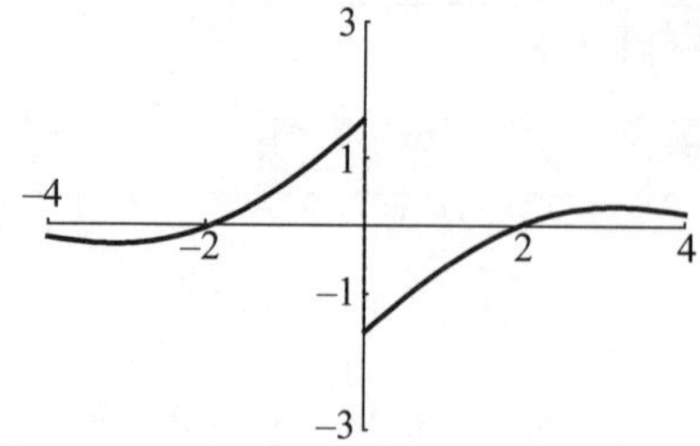

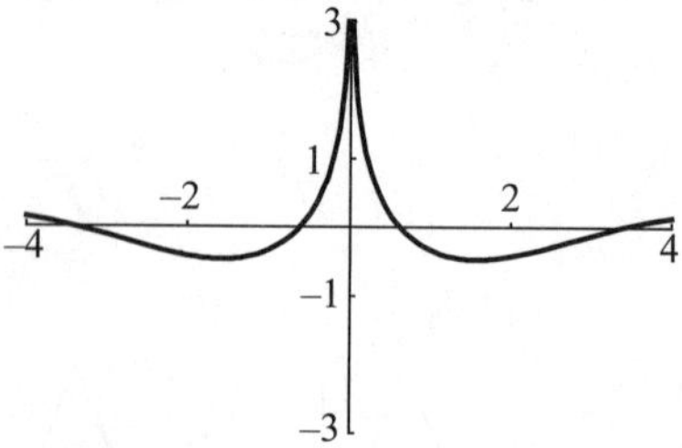

44 $f = e^{-|x-1|}\,H(x-1)$ $\qquad$ $Hf = \begin{cases} -\dfrac{e^{1-x}}{\pi} E_1(1-x), & x < 1 \\ \dfrac{e^{1-x}}{\pi} \operatorname{Ei}(x-1), & x > 1 \end{cases}$

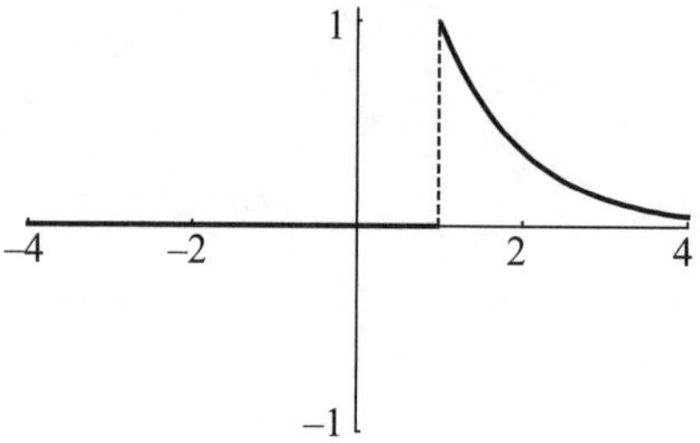

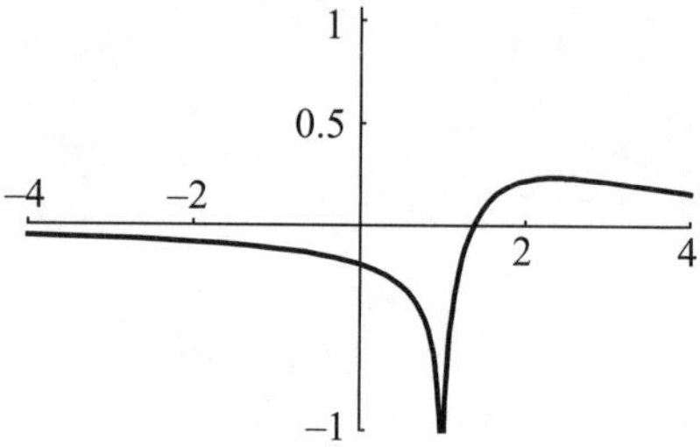

45 $f = \begin{cases} 0, & |x| > 2 \\ 1, & |x| < 1 \\ 2+x, & -2 < x < -1 \\ 2-x, & 1 < x < 2 \end{cases}$ $\qquad$ $Hf = \dfrac{1}{\pi}\left\{\log\left|\dfrac{(x+2)^2(x-1)}{(x-2)^2(x+1)}\right| + x\log\left|\dfrac{x^2-4}{x^2-1}\right|\right\}$

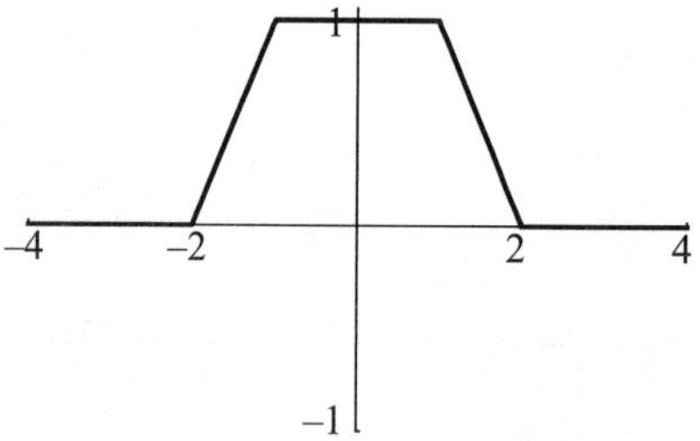

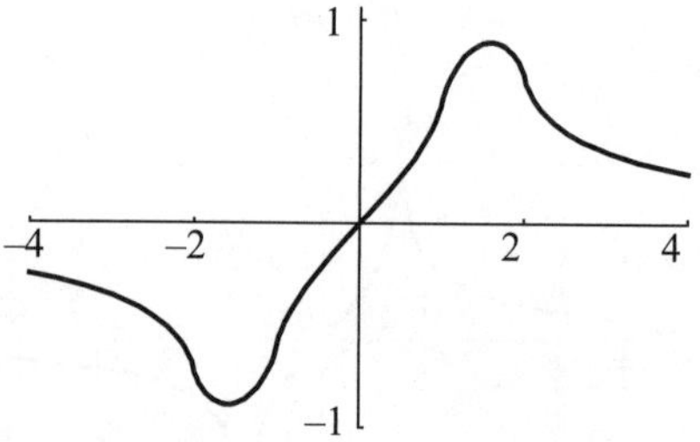

References

Abdelouhab, L., Bona, J. L., Felland, M., and Saut, J.-C. (1989). Nonlocal models for nonlinear, dispersive waves, *Physica D* **40**, 360–392.

Ablowitz, M. J. and Clarkson P. A. (1991). *Solitons, Nonlinear Evolution Equations and Inverse Scattering* (Cambridge: Cambridge University Press).

Abragam, A. (1961). *The Principles of Nuclear Magnetism* (London: Oxford University Press).

Abramowitz, M. and Stegun, I. A. (1965). *Handbook of Mathematical Functions* (New York: Dover).

Achieser, N. I. (1956). *Theory of Approximation* (New York: Frederick Ungar Publ. Co.).

Adams, E. (1982). On weighted norm inequalities for the Hilbert transform of functions with moments zero, *Trans. Am. Math. Soc.* **272**, 487–500.

Afshar, R., Mueller, F. M., and Shaffer, J. C. (1973). Hilbert transformation of densities of states using Hermite functions, *J. Comput. Phys.* **11**, 190–209.

Agneni, A. (1990). Hilbert and Hartley transforms in the time domain system identification, *Aero. Missili E Spazio* **69**, 26–32.

Aheizer, N. I. and Krein, M. (1962). *Some Questions in the Theory of Moments* (Providence, RI: American Mathematical Society).

Ahrenkiel, R. K. (1971). Modified Kramers-Kronig analysis of optical spectra, *J. Opt. Soc. Am.* **61**, 1651–1655. Erratum: *J. Opt. Soc. Am.* **62**, 1009 (1972).

Akhiezer, N. I. (1965). *The Classical Moment Problem* (New York: Hafner Publishing Company).

Aki, K. and Richards, P. G. (1980). *Quantitative Seismology Theory and Methods*, Vols. I and II (San Francisco: W. H. Freeman).

Alavi-Sereshki, M. M. and Prabhakar, J. C. (1972). A tabulation of Hilbert transforms for electrical engineers, *IEEE Trans. Commun.* **20**, 1194–1198.

Alfimov, G. L., Usero, D., and Vázquez, L. (2000). On complex singularities of solutions of the equation $Hu_x - u + u^p = 0$, *J. Phys. A: Math. Gen.* **33**, 6707–6720.

Alieva, T., Bastiaans, M. J., and Calvo M. L. (2005). Fractional transforms in optical information processing, *EURASIP J. Appl. Signal Process.* **2005**(10), 1498–1519.

Almeida, L. B. (1994). The fractional Fourier transform and time-frequency representations, *IEEE Trans. Signal Process.* **42**, 3084–3091.

Altarelli, M. and Smith, D. Y. (1974). Superconvergence and sum rules for the optical constants: Physical meaning, comparison with experiment, and generalization, *Phys. Rev. B* **9**, 1290–1298. Erratum: *Phys. Rev. B* **12**, 3511 (1975).

Altarelli, M., Dexter, D. L. Nussenzveig, H. M., and Smith D. Y. (1972). Superconvergence and sum rules for the optical constants, *Phys. Rev. B* **6**, 4502–4509.

Alzer, H. (1997). A new refinement of the arithmetic mean-geometric mean inequality, *Rocky Mountain J. Math.* **27**, 663–667.

Amari, S. (1994). Evaluation of Cauchy principal value integrals using modified Simpson rules, *Appl. Math. Lett.* **7**, 19–23.

Amusia, M. Ya. and Kuchiev, M. Yu. (1982). Multiple pole in the electron-hydrogen atom scattering amplitude, *Phys. Rev. Lett.* **48**, 1726–1729.

Andermann, G. and Dows, D. D. (1967). Infrared reflectance spectrum and optical constants of sodium chlorate, *J. Phys. Chem. Solids* **28**, 1307–1315.

Andermann, G., Caron, A., and Dows, D. A. (1965). Kramers-Kronig dispersion analysis of infrared reflectance bands, *J. Opt. Soc. Am.* **55**, 1210–1216.

Andermann, G., Wu, C. K., and Duesler, E. (1968). Kramers-Kronig phase-angle partitioning method for disclosing systematic errors in infrared reflectance data, *J. Opt. Soc. Am.* **58**, 1663–1664.

Andersen, K. F. (1976a). Weighted norm inequalities for Hilbert transforms and conjugate functions of even and odd functions, *Proc. Am. Math. Soc.* **56**, 99–107.

Andersen, K. F. (1976b). Discrete Hilbert transforms and rearrangement invariant sequence spaces, *Appl. Analysis* **5**, 193–200.

Andersen, K. F. (1977a). On an inequality for the Hilbert transform, *J. London Math. Soc.* (2) **16**, 290–296.

Andersen, K. F. (1977b). Inequalities with weights for discrete Hilbert transforms, *Can. Math. Bull.* **20**, 9–16.

Andersen, K. F. (1995). Weighted inequalities for convolutions, *Proc. Am. Math. Soc.* **123**, 1129–1136.

Andersen, K. F. and Muckenhoupt, B. (1982). Weighted weak type Hardy inequalities with applications to Hilbert transforms and maximal functions, *Studia Math.* **72**, 9–26.

Anderson, B. D. O. and Green, M. (1988). Hilbert transform and gain/phase error bounds for rational functions, *IEEE Trans. Circuits Syst.* **35**, 528–535.

Anderson, D. G. (1965). Gaussian quadrature formula for $\int_0^1 -\ln(x) f(x) \mathrm{d}x$, *Math. Comp.* **19**, 477–481.

Andersson, T., Johansson, J., and Eklund, H. (1981). Numerical solution of the Hilbert transform for phase calculation from an amplitude spectrum, *Math. Computers Simul.* **23**, 262–266.

Andrews, G. E., Askey, R., and Roy, R. (1999). *Special Functions* (Cambridge: Cambridge University Press).

Angel, Y. C. and Achenbach, J. D. (1991). Attenuation and speed of antiplane waves in a cracked solid using the Kramers-Kronig relations, *J. Acoust. Soc. Am.* **90**, 2757–2762.

Ansari, R. (1987). IIR discrete-time Hilbert transformers, *IEEE Trans Acoustics, Speech, Signal Process.* **35**, 1116–1119.

Arcozzi, N. and Fontana, L. (1998). A characterization of the Hilbert transform, *Proc. Am. Math. Soc.* **126**, 1747–1749.

Arnison, M. R., Cogswell, C. J., Smith, N. I., Fekete, P. W., and Larkin, K. G. (2000). Using the Hilbert transform for 3D visualization interference contrast microscopic images, *J. Microscopy* **199**, 79–84.

Arocena, R., Cotlar, M., and Sadosky, C. (1981). Weighted inequalities in L^2 and lifting properties. In *Mathematical Analysis and Applications, Part A*, L. Nachbin, ed. (New York: Academic Press), pp. 95–128.

Artiaga, L. (1964). Generalized Hilbert kernels, *Duke Math. J.* **31**, 471–478.

Asmar, N. and Hewitt, E. (1988). Marcel Riesz's theorem on conjugate Fourier series and its descendants. In *Proceedings of the Analysis Conference, Singapore 1986*, S. T. L. Choy, J. P. Jesudason, and P. Y. Lee, eds. (Amsterdam: North-Holland).

Aspnes, D. E. (1985). The accurate determination of optical properties by ellipsometry. In *Handbook of Optical Constants of Solids*, E. D. Palik, ed. (Orlando, FL: Academic Press), pp. 89–112.

Assani, I. and Petersen, K. (1992). The helical transform as a connection between ergodic theory and harmonic analysis, *Trans. Am. Math. Soc.* **331**, 131–142.

Astala, K., Päivärinta, L., and Saksman, E. (1996). The finite Hilbert transform in weighted spaces, *Proc. Roy. Soc. Edinburgh* **126A**, 1157–1167.

Atencia, E. and Martin-Reyes, F. J. (1983). The maximal ergodic Hilbert transform with weights, *Pacific J. Math.* **108**, 257–263.

Atencia, E. and Martin-Reyes, F. J. (1984). Weak type inequalities for the maximal ergodic function and the maximal ergodic Hilbert transform in weighted spaces, *Studia Math.* **78**, 231–244.

Atkinson, K. (1972). The numerical evaluation of the Cauchy transform on simple closed curves, *SIAM J. Numer. Analysis* **9**, 284–299.

Audoin, B. and Roux, J. (1996). An innovative application of the Hilbert transform to time delay estimation of overlapped ultrasonic echoes, *Ultrasonics* **34**, 25–33.

Babenko, K. I. (1948). On conjugate functions, *Doklady Akademii Nauk SSSR* **62**, 157–160. In Russian.

Bachmann, P. (1894). *Die Analytische Zahlentheorie* (Stuttgart: B. G. Teubner). There is a 1968 reprint of this book by Johnson Reprint Corporation, New York.

Baernstein, II., A. (1978). Some sharp inequalities for conjugate functions, *Indiana Math. J.* **27**, 833–852.

Baernstein, II., A. (1979). Some sharp inequalities for conjugate functions, *Proc. Symp. Pure Math.* **35**, part (1), 409–416.

Baird, L. C., Sancaktar, S., and Zweifel, P. F. (1977). The Poincaré-Bertrand formula: A derivation and generalization, *SIAM J. Math. Analysis* **8**, 580–591.

Baker, G. R., Li, X., and Morlet, A. C. (1996). Analytic structure of two 1D-transport equations with nonlocal fluxes, *Physica D* **91**, 349–375.

Balzarotti, A., Colavita, E., Gentile, S., and Rosei, R. (1975). Kramers-Krönig analysis of modulated reflectance data investigation of errors, *Appl. Opt.* **14**, 2412–2417.

Bampi, F., and Zordan, C. (1992). Hilbert transforms in wave propagation theory, *Bollettino Uni. Mat. Ital.* **6-B**(7), 177–191.

Bañuelos, R. and Wang, G. (1995). Sharp inequalities for martingales with applications to the Beurling-Ahlfors and Riesz transforms, *Duke Math. J.* **80**, 575–600.

Bardaro, C., Butzer, P. L., Stens, R. L., and Vinti, G. (2006). Approximation error of the Whittaker cardinal series in terms of an averaged modulus of smoothness covering discontinuous signals, *J. Math. Analysis Appl.* **316**, 269–306.

Bareiss, E. H. and Neuman, C. P. (1965). *Singular integrals and singular integral equations with a Cauchy kernel and the method of symmetric pairing*. Argonne National Laboratory Technical Report 6988.

Barnes, A. E. (1996). Theory of 2-D complex seismic trace analysis, *Geophys.* **61**, 264–272.

Barone, A. and Paternò, G. (1982). *Physics and Applications of the Josephson Effect* (New York: John Wiley).

Barros-Neto, J. (1973). *An Introduction to the Theory of Distributions* (New York: Marcel Dekker).

Bartholdi, E. and Ernst, R. R. (1973). Fourier spectroscopy and the causality principle, *J. Mag. Res.* **11**, 9–19.

Barton, G. (1965). *Dispersion Techniques in Field Theory* (New York: Benjamin).

Bary, N. K. (1964). *A Treatise on Trigonometric Series*, Vols. I and II (Oxford: Pergamon Press).

Basinger, J. O. (1976). Cesàro summability of the conjugate series and the double Hilbert transform, *Proc. Am. Math. Soc.* **56**, 177–182.

Bass, M., Franken, P. A., Hill, A. E., Peters, C. W., and Weinreich, G. (1962a). Optical mixing, *Phys. Rev. Lett.* **8**, 18.

Bass, M., Franken, P. A., Ward, J. F., and Weinreich, G. (1962b). Optical rectification, *Phys. Rev. Lett.* **9**, 446–448.

Bassani, F. and Altarelli, M. (1983). Interaction of radiation with condensed matter. In *Handbook on Synchrotron Radiation*, E.-E. Koch, ed. (Amsterdam: North-Holland), pp. 463–605.

Bassani, F. and Lucarini, V. (1998). General properties of optical harmonic generation from a simple oscillator model, *Nuovo Cimento* **20D**, 1117–1125.

Bassani, F. and Lucarini, V. (1999). Pump and probe nonlinear processes: new modified sum rules and a simple oscillator model, *Eur. Phys. J. B* **12**, 323–330.

Bassani, F. and Lucarini, V. (2000). Asymptotic behavior and general properties of harmonic generation susceptibilities, *Eur. Phys. J. B* **17**, 563–573.

Bassani, F. and Scandolo, S. (1991). Dispersion relations and sum rules in nonlinear optics, *Phys. Rev. B* **44**, 8446–8453.

Bassani, F. and Scandolo, S. (1992). Sum rules for nonlinear optical susceptibilities, *Phys. Stat. Sol.* (b) **173**, 263–270.

Bedrosian, E. (1963). A product theorem for Hilbert transforms, *Proc. IEEE* **51**, 868–869.

Bedrosian, E. (1966a). Author's reply, *Proc. IEEE* **54**, 1459.

Bedrosian, E. (1966b). Author's reply, *Proc. IEEE* **54**, 435.

Bedrosian, E. (1972). Comments on "An extension of the Hilbert transform product theorem," *Proc. IEEE* **60**, 228.

Bell, S. R. (1992). *The Cauchy Transform, Potential Theory, and Conformal Mapping* (Boca Raton, FL: CRC Press) .

Beltrami, E. J. (1967). Linear dissipative systems, nonnegative definite distributional kernels, and the boundary values of bounded-real and positive-real matrices, *J. Math. Analysis Appl.* **19**, 231–246.

Beltrami, E. J. and Wohlers, M. R. (1965). Distributional boundary value theorems and Hilbert transforms, *Arch. Ration. Mech. Analysis* **18**, 304–309.

Beltrami, E. J. and Wohlers, M. R. (1966a). *Distributions and the Boundary Values of Analytic Functions* (New York: Academic Press).

Beltrami, E. J. and Wohlers, M. R. (1966b). Distributional boundary values of functions holomorphic in a half plane, *J. Math. Mech.* **15**, 137–145.

Beltrami, E. J. and Wohlers, M. R. (1967). The Cauchy integral of tempered distributions and some theorems on analytic continuation, *SIAM J. Appl. Math.* **15**, 1077–1087.

Beltzer, A. I. (1983). Kramers-Kronig relationships and wave propagation in composites, *J. Acoust. Soc. Am.* **73**, 355–356.

Bendat, J. S. (1985). *The Hilbert transform and applications to correlation measurements*. Technical Report, Brüel & Kjaer, Naerum, Denmark.

Bendat, J. S. (1990). *Nonlinear System Analysis and Identification from Random Data* (New York: John Wiley).

Bendat, J. S. and Piersol, A. G. (2000). *Random Data Analysis and Measurement Procedures*, 3rd edn (New York: John Wiley).

Benedek, A. and Panzone, R. (1971). Continuity properties of the Hilbert transform, *J. Funct. Analysis* **7**, 217–234.

Benedetto, J. J. (1997). *Harmonic Analysis and Applications* (Boca Raton, FL: CRC Press).

Benjamin, T. B. (1967). Internal waves of permanent form in fluids of great depth, *J. Fluid Mech.* **29**, 559–592.

Bennett, C. (1976). A best constant for Zygmund's conjugate function inequality, *Proc. Am. Math. Soc.* **56**, 256–260.

Bennett, C. and Sharpley, R. (1988). *Interpolation of Operators* (Boston, MA: Academic Press).

Bennett, C. L. (1987a). Evidence for microscopic causality violation, *Phys. Rev. A* **35**, 2409–2419.

Bennett, C. L. (1987b). Further evidence for causality violation, *Phys. Rev. A* **35**, 2420–2428.

Berkson, E. and Gillespie, T. A. (1985). The generalized M. Riesz theorem and transference, *Pacific J. Math.* **120**, 279–288.

Berreman, D. W. (1967). Kramers-Kronig analysis of reflectance measured at oblique incidence, *Appl. Opt.* **9**, 1519–1521.

Berthelot, Y. H. (2001). Surface acoustic impedance and causality, *J. Acoust. Soc. Am.* **109**, 1736–1739.

Bertie, J. E. and Eysel, H. H. (1985). Infrared intensities of liquids I: Determination of infrared optical and dielectric constants by FT-IR using the circle ATR cell, *Appl. Spectroscopy.* **39**, 392–401.

Bertie, J. E. and Lan, Z. (1996). An accurate modified Kramers-Kronig transformation from reflectance to phase shift on attenuated total reflection, *J. Chem. Phys.* **105**, 8502–8514.

Bertie, J. E. and Zhang, S. L. (1992). Infrared intensities of liquids. IX. The Kramers-Kronig transform, and its approximation by the finite Hilbert transform via fast Fourier transforms, *Can. J. Chem.* **70**, 520–531.

Bertin, J. J. (2002). *Aerodynamics for Engineers*, 4th edn (Upper Saddle River, NJ: Prentice Hall).

Bertrand, G. (1922). Equations de Fredholm à intégrales principales au sens de Cauchy, *Comptes Rendus* **172**, 1458–1461.

Bertrand, G. (1923a). Le Problème de Dirichlet et le potentiel de simple couche, *Bull. Sci. Math.* **47**, 282–288.

Bertrand, G. (1923b). Le Problème de Dirichlet et le potentiel de simple couche, *Bull. Sci. Math.* **47**, 298–307.

Besicovitch, A. S. (1926). On a general metric property of summable functions, *J. Lond. Math. Soc.* **1**, 120–128.

Besikovitch, A. (1923). Sur la nature des fonctions à carré sommable et des ensembles mesurables, *Fund. Math.* **4**, 172–195.

Bessis, D., Haffad, A., and Msezane, A. Z. (1994). Momentum-transfer dispersion relations for electron-atom cross sections, *Phys. Rev. A* **49**, 3366–3375.

Bessis, D. and Temkin, A. (2000). Partial-wave dispersion relations: Exact left-hand E-plane discontinuity computed from the Born series, *Phys. Rev. A* **61**, 032702.

Bialecki, B. and Keast, P. (1999). A sinc quadrature subroutine for Cauchy principal value integrals, *J. Comput. Appl. Math.* **112**, 3–20.

Bierens de Haan, D. (1867). *Nouvelles Tables D'Intégrales Définies*, corrected reprinting (New York: Hafner Publishing).

Bierman, G. J. (1971). A particular class of singular integral equations, *SIAM J. Appl. Math.* **20**, 99–109.

Birkhoff, G. (1973). *A Source Book in Classical Mathematics* (Cambridge, MA: Harvard University Press).

Bitsadze, A. V. (1987). The multidimensional Hilbert transform, *Sov. Math. Dokl.* **35**, 390–392.

Bland, S. R. (1970). The two-dimensional oscillating airfoil in a wind tunnel in subsonic flow, *SIAM J. Appl. Math.* **18**, 830–848.

Blasco, O. and Villarroya, P. (2003). Commutators of linear and bilinear Hilbert transforms, *Proc. Am. Math. Soc.* **132**, 1997–2004.

Bleistein, N. and Handelsman, R. A. (1986). *Asymptotic Expansions of Integrals* (New York: Dover).

Bloembergen, N. (1965). *Nonlinear Optics* (New York: Benjamin).

Blue, J. L. (1979). A Legendre polynomial integral, *Math. Comp.* **33**, 739–741.

Blum, K. and Burke, P. G. (1977). Validity of dispersion relations for electron-atom scattering, *Phys. Rev. A* **16**, 163–168.

Blyumin, S. L. and Trakhtman, A. M. (1977). Discrete Hilbert transform on finite intervals, *Radio Eng. Electron. Phys.* **22**, 50–56.

Boas, Jr., R. P. (1936). Some theorems on Fourier transforms and conjugate trigonometric integrals, *Trans. Am. Math. Soc.* **40**, 287–308.

Boas, Jr., R. P. (1954). *Entire Functions* (New York: Academic Press).

Boche, H. and Protzmann, M. (1997). A new algorithm for the reconstruction of bandlimited functions and their Hilbert transform, *IEEE Trans. Instrum. Meas.* **46**, 442–444.

Bochner, S. (1939). Additive set functions on groups, *Ann. Math.* **40**, 760–799.

Bochner, S. (1959a). *Lectures on Fourier Integrals* (Princeton, NJ: Princeton University Press).

Bochner, S. (1959b). Generalized conjugate and analytic functions without expansions, *Proc. Natl. Acad. Sci. USA* **45**, 855–857.

Bochner, S. and Chandrasekharan, K. (1949). *Fourier Transforms*, Annals of Mathematics Studies Number 19 (Princeton, NJ: Princeton University Press). Reprinted Kraus Reprint Corporation, New York (1965).

Bode, H. W. (1945). *Network Analysis and Feedback Amplifier Design* (Princeton, NJ: D. Van Nostrand).

Bogner, R. E. (2001). Hilbert transforms. In *Wiley Encyclopedia of Electrical and Electronics Engineering Online*, J. Webster, ed. (John Wiley & Sons).

Bogolyubov, N. N., Medvedev, B. V., and Polivanov, M. K. (1958). *Theory of Dispersion Relations*. Translation of *Voprosy Teorii Dispersionnykh Sootnoshenii* (Moscow: Moscow State Technical Publishing House).

Bokut, B. V., Penyaz, V. A., and Serdyukov, A. N. (1981). Dispersion sum rules in the optics of naturally gyrotropic media, *Opt. Spectrosc. (USSR)* **50**, 511–513.

Bolton, H. C. (1969a). The modification of the Kronig-Kramers relations under saturation, *Phil. Mag.* **19**, 477–485.

Bolton, H. C. (1969b). Some practical properties of Kronig-Kramers transforms, *Phil. Mag.* **19**, 487–499.

Bolton, H. C., Troup, G. J., and Wilson, G. V. H. (1964). A new use of the Kronig-Kramers relations in nuclear magnetic resonance, *Phil. Mag.* **9**, 591–605.

Bona, J. L. and Kalisch, H. (2004). Singularity formation in the generalized Benjamin-Ono equation, *Disc. Cont. Dyn. Syst.* **11**, 27–45.

Bonzanigo, F. (1972). A note on "Discrete Hilbert transform," *IEEE Trans. Audio & Electroacoustics* **20**, 99–100.

Booij, H. C. and Thoone, G. P. J. M. (1982). Generalization of Kramers-Kronig transforms and some approximations of relations between viscoelastic quantities, *Rheologica Acta* **21**, 15–24.

Boole, G. (1857). XXXVI. On the comparison of transcendents, with certain applications to the theory of definite integrals, *Phil. Trans. Roy. Soc. Lond.* **147**, 745–803.

Born, M. and Wolf, E. (1999). *Principles of Optics* (Cambridge: Cambridge University Press).

Bortz, M. L. and French, R. H. (1989). Quantitative, FFT-based Kramers-Krönig analysis for reflectance data, *Appl. Spectroscopy* **43**, 1498–1501.

Bose, N. K. and Prabhu, K. A. (1979). Two-dimensional discrete Hilbert transform and computational complexity aspects in its implementation, *IEEE Trans. Acoustics Speech Signal Process.* **27**, 356–360.

Boswarva, I. M., Howard, R. E., and Lidiard, A. B. (1962). Faraday effect in semiconductors, *Proc. Roy. Soc. Lond. A* **269**, 125–141.

Bowlden, H. J. and Wilmshurst, J. K. (1963). Evaluation of the one-angle reflection technique for the determination of optical constants, *J. Opt. Soc. Am.* **53**, 1073–1078.

Boyd, D. W. (1967). The Hilbert transform on rearrangement-invariant spaces, *Can. J. Math.* **19**, 599–616.

Boyd, R. W. (1992). *Nonlinear Optics* (Boston, MA: Academic Press).

Bracewell, R. N. (1986). *The Hartley Transform* (New York: Oxford University Press).

Brachman, M. K. (1955). Note on the Kramers-Kronig relations, *J. Appl. Phys.* **26**, 497–498.

Brachman, M. K. and MacDonald, J. R. (1954). Relaxation-time distribution functions and the Kramers-Kronig relations, *Physica* **20**, 1266–1270.

Brachman, M. K. and MacDonald, J. R. (1956). Generalized immittance kernels and the Kramers-Kronig relations, *Physica* **22**, 141–148.

Brackx, F. and De Schepper, H. (2005a). Convolution kernels in Clifford analysis: Old and new, *Math. Meth. Appl. Sci.* **28**, 2173–2182.

Brackx, F. and De Schepper, H. (2005b). Hilbert-Dirac operators in Clifford analysis, *Chinese Ann. Math.* **26B**, 1–14.

Brackx, F., De Knock, B., and De Schepper, H. (2006a). Generalized multidimensional Hilbert transforms in Clifford analysis, *Int. J. Math. Math. Sci.* **2006**, Art. ID 98145, 19 pp.

Brackx, F., De Schepper, H., and Eelbode, D. (2006b). A new Hilbert transform on the unit sphere in $\mathbb{R}^m$, *Complex Variables & Elliptic Eqns.* **51**, 453–462.

Brackx, F., De Knock, B., De Schepper, H., and Eelbode, D. (2006c). On the interplay between the Hilbert transform and conjugate harmonic functions, *Math. Meth. Appl. Sci.* **29**, 1435–1450.

Brau, C. A. (2004). *Modern Problems in Classical Electrodynamics* (New York: Oxford University Press).

Braun, S. and Feldman, M. (1997). Time-frequency characteristics of non-linear systems, *Mech. Syst. Signal Process.* **11,** 611–620.

Brauner, N. and Beltzer, A. I. (1985). The Kramers-Kronig relations method and wave propagation in porous elastic media, *Int. J. Engng Sci.* **23**, 1151–1162.

Bremermann, H. (1965a). *Distributions, Complex Variables, and Fourier Transforms* (Reading, MA: Addison-Wesley).

Bremermann, H. J. (1965b). Several complex variables. In *Studies in Mathematics* Vol. 3 *Studies in Real and Complex Analysis*, I. I. Hirschman, Jr., ed. (Buffalo, NY: Mathematical Association of America; Englewood Cliffs, NJ: Prentice-Hall) pp. 3–33.

Bremermann, H. J. (1967). Some remarks on analytic representations and products of distributions, *SIAM J. Appl. Math.* **15**, 929–943.

Brezinski, C. and Redivo Zaglia, M. (1991). *Extrapolation Methods Theory and Practice* (Amsterdam: North-Holland).

Brillouin, L. (1914). 2. Über die fortpflanzung des lichtes in dispergierenden medien, *Ann. Phys. (Leipzig)* **44**, 203–240.

Brillouin, L. (1960). *Wave Propagation and Group Velocity* (New York: Academic Press).

Brockman, M. W. and Moscowitz, A. (1981). Macroscopic sum rules in natural optical activity, *Mol. Phys.* **43**, 1385–1393.

Bronzan, J. B., Kane, G. L., and Sukhatme, U. P. (1974). Obtaining real parts of scattering amplitudes directly from cross section data using derivative analyticity relations, *Phys. Lett.* **49B**, 272–276.

Brown, B. M. (1961). *The Mathematical Theory of Linear Systems* (New York: Wiley).

Brown, F. C. and Laramore, G. (1967). Magnetooptical experiments on broad absorption bands in solids, *Appl. Opt.* **6**, 669–673.

Brown, Jr., J. L. (1974). Analytic signals and product theorems for Hilbert transforms, *IEEE Trans. Circuits Syst.* **21**, 790–792.

Brown, Jr., J. L. (1986). A Hilbert transform product theorem, *Proc. IEEE* **74**, 520–521.

Brychkov, Yu. A. and Prudnikov, A. P. (1989). *Integral Transforms of Generalized Functions* (New York: Gordon & Breach).

Buchkovska, A. L. and Pilipović, S. (2002). Bilinear Hilbert transform of ultradistributions, *Integral Transforms Spec. Funct.* **13**, 211–221.

Buckingham, A. D. and Stephens, P. J. (1966). Magnetic optical activity. In *Annual Review of Physical Chemistry*, Vol. 17, H. Eyring, C. J. Christensen, and H. S. Johnston, eds. (Palo Alto, CA: Annual Review, Inc.), pp 399–432.

Bueckner, H. F. (1966). On a class of singular integral equations, *J. Math. Analysis Appl.* **14**, 392–426.

Buffoni, B. (2004). Existence by minimization of solitary water waves on an ocean of infinite depth, *Ann. Inst. H. Poincaré Analyse Non Linéaire* **21**, 503–516.

Bülow, T. and Sommer, G. (2001). Hypercomplex signals – A novel extension of the analytical signal to the multidimensional case, *IEEE Trans. Signal Process.* **49**, 2844–2852.

Burge, R. E., Fiddy, M. A., Greenaway, A. H., and Ross, G. (1974). The application of dispersion relations (Hilbert transforms) to phase retrieval, *J. Phys. D: Appl. Phys.* **7**, L65–L68.

Burge, R. E., Fiddy, M. A., Greenaway, A. H., and Ross, G. (1976). The phase problem, *Proc. Roy. Soc. Lond. A* **350**, 191–212.

Burk, F. (1998). *Lebesgue Measure and Integration* (New York: John Wiley).

Burkhardt, H. (1969). *Dispersion Relation Dynamics* (Amsterdam: North-Holland).

Burkholder, D. L., Gundy, R. F., and Silverstein, M. L. (1971). A maximal function characterization of the class H^p, *Trans. Am. Math. Soc.* **157**, 137–153.

Burris, F. E. (1975). Matrix formulation of the discrete Hilbert transform, *IEEE Trans. Circuits & Systems* **22**, 836–838.

Butcher, P. N. and Cotter, D. (1990). *The Elements of Nonlinear Optics* (Cambridge: Cambridge University Press).

Butzer, P. L. and Nessel, R. J. (1971). *Fourier Analysis and Approximation, Vol. 1 One-Dimensional Theory*, (New York: Academic Press).

Butzer, P. L. and Trebels, W. (1968). *Hilberttransformation, gebrochene Integration und Differentiation* (Köln: Westdeutscher Verlag).

Butzer, P. L., Higgins, J. R., and Stens, R. L. (2000). Sampling theory of signal analysis. In *Development of Mathematics 1950–2000*, J.-P. Pier, ed. (Basel: Birkhäuser Verlag), pp. 193–234.

Butzer, P. L., Splettstösser, W., and Stens, R. L. (1988). The sampling theorem and linear prediction in signal analysis, *Jber. d. Dt. Math. Verein.* **90**, 1–70.

Byron, Jr., F. W. and Joachain, C. J. (1977). Exchange amplitudes and dispersion relations for electron-atom scattering, *Phys. Lett. A* **62**, 217–219.

Byron, Jr., F. W. and Joachain, C. J. (1978). Exchange amplitudes and forward dispersion relations for electron-atomic hydrogen scattering, *J. Phys. B: Atom. Mol. Phys.* **11**, 2533–2546.

Byron, Jr., F. W., De Heer, F. J., and Joachain, C. J. (1975). Derivation of dispersion relations for atomic scattering processes, *Phys. Rev. Lett.* **35**, 1147–1150.

Cable, J. R. and Albrecht, A. C. (1986). A direct inverse transform for resonance Raman scattering, *J. Chem. Phys.* **84**, 4745–4754.

Cain, G. D. (1972). Hilbert-transform description of linear filtering, *Electron. Lett.* **8**, 380–382.

Cain, G. D. (1973). Hilbert transform relations for products, *Proc. IEEE* **61**, 663–664.

Calderón, A. P. (1950). On theorems of M. Riesz and Zygmund, *Proc. Am. Math. Soc.* **1**, 533–535.

Calderón, A. P. (1965). Commutators of singular integral operators, *Proc. Natl Acad. Sci. USA* **53**, 1092–1099.

Calderón, A. P. (1966). Singular integrals, *Bull. Am. Math. Soc.* **72**, 427–465.

Calderón, A. P. and Sagher, Y. (1991). The Hilbert transform of the Gaussian. In *Almost Everywhere Convergence II*, A. Bellow and R. L. Jones, eds. (Boston, MA: Academic Press), pp. 109–112.

Calderón, A. P. and Zygmund, A. (1952). On the existence of certain singular integrals, *Acta Math.* **88**, 85–139.

Calderón, A. P. and Zygmund, A. (1954). Singular integrals and periodic functions, *Studia Math.* **14**, 249–271.

Calderón, A. P. and Zygmund, A. (1955). On a problem of Mihlin, *Trans. Am. Math. Soc.* **78**, 209–224. Addenda to the paper *On a problem of Mihlin*, *Trans. Am. Math. Soc.* **84**, 559–560.

Calderón, A. P. and Zygmund, A. (1956a). On singular integrals, *Am. J. Math.* **78**, 289–309.

Calderón, A. P. and Zygmund, A. (1956b). Algebras of certain singular operators, *Am. J. Math.* **78**, 310–320.

Calderón, A. P. and Zygmund, A. (1957). Singular integral operators and differential equations, *Am. J. Math.* **79**, 901–921.

Calderón, A. P., Weiss, M., and Zygmund, A. (1967). On the existence of singular integrals. In *Proceedings of Symposia in Pure Mathematics*, Vol. 10 (Providence, RI: American Mathematical Society), pp. 56–73.

Caldwell, D. J. and Eyring, H. (1971). *The Theory of Optical Activity* (New York: Wiley-Interscience).

Campbell, G. A. and Foster, R. M. (1948). *Fourier Integrals for Practical Applications* (Princeton, NJ: D. Van Nostrand Company).

Campbell, J. and Petersen, K. (1989). The spectral measure and Hilbert transform of a measure-preserving transformation, *Trans. Am. Math. Soc.* **313**, 121–129.

Campbell, L. L. (1968). Sampling theorem for the Fourier transform of a distribution with bounded support, *SIAM J. Appl. Math.* **16**, 626–636.

Candan, Ç., Kutay, M. A., and Ozaktas, H. M. (2000). The discrete fractional Fourier transform, *IEEE Trans. Signal Process.* **48**, 1329–1337.

Capobianco, M. R. (1993). The stability and the convergence of a collocation method for a class of Cauchy singular integral equations, *Math. Nachr.* **162**, 45–58.

Capobianco, M. R., Criscuolo, G., and Giova, R. (2001). A stable and convergent algorithm to evaluate Hilbert transform, *Numer. Algorithms* **28**, 11–26.

Carbery, A., Ricci, F., and Wright, J. (1998). Maximal functions and Hilbert transforms associated to polynomials, *Rev. Mat. Iberoamericana* **14**, 117–144.

Carbery, A., Christ, M., Vance, J., Wainger, S., and Watson, D. K. (1989). Operators associated to flat curves: L^p estimates via dilation methods, *Duke Math. J.* **59**, 675–700.

Carbery, A., Vance, J., Wainger, S., Watson, D., and Wright, J. (1995). L^p estimates for operators associated to flat curves without the Fourier transform, *Pacific J. Math.* **167**, 243–262.

Carcaterra, A. and Sestieri, A. (1997). Complex envelope displacement analysis: A quasi-static approach to vibrations, *J. Sound Vib.* **201**, 205–233.

Cardona, M. (1969). Optical constants of insulators: dispersion relations. In *Optical Properties of Solids*, S. Nudelman and S. S. Mitra, eds. (New York: Plenum Press), pp. 137–151.

Carleman, T. (1922). Sur la résolution de certaines équations intégrales, *Arkiv Mat. Astron. Fysik* **16**, 1–19.

Carleson, L. (1966). On convergence and growth of partial sums of Fourier series, *Acta Math.* **116**, 135–157.

Carrier, G. F., Krook, M., and Pearson, C. E. (1983). *Functions of a Complex Variable* (Ithaca, NY: Hod Books).

Carrington, A. and McLachlan, A. D. (1967). *Introduction to Magnetic Resonance* (New York: Harper and Row).

Carro, M. J. (1998). Discretization of linear operators on $L^p(\mathbb{R}^N)$, *Illinois J. Math.* **42**, 1–18.

Carson, J. R. (1926). *Electric Circuit Theory and the Operational Calculus* (New York: McGraw-Hill).

Cartan, H. (1963). *Elementary Theory of Analytic Functions of One or Several Complex Variables* (Paris: Hermann).

Carton-Lebrun, C. (1977). Product properties of Hilbert transforms, *J. Approx. Theory* **21**, 356–360.

Carton-Lebrun, C. (1979). Remarques sur les relations de dispersion, *Bull. Soc. Math. Belgique B* **31**, 209–213.

Carton-Lebrun, C. (1987). An extension to *BMO* functions of some product properties of Hilbert transforms, *J. Approx. Theory* **49**, 75–78.

Carton-Lebrun, C. (1988). A real variable definition of the Hilbert transform on D′ and S′, *Appl. Analysis* **29**, 235–251.

Carton-Lebrun, C. (1991). Continuity and inversion of the Hilbert transform of distributions, *J. Math. Analysis Appl.* **161**, 274–283.

Carton-Lebrun, C. (2005). Problems related to the analytic representation of tempered distributions. In *Advances in Analysis*, H. G. W. Begehr, R. P. Gilbert, M. E. Muldoon, and M. W. Wong, eds. (Hackensack, NJ: World Scientific), pp. 245–253.

Cartwright, M. L. (1962). *Integral Functions* (London: Cambridge University Press).

Cartwright, M. L. (1982). Manuscripts of Hardy, Littlewood, Marcel Riesz and Titchmarsh, *Bull. Lond. Math. Soc.* **14**, 472–532.

Case, K. M. (1979). Properties of the Benjamin-Ono equation, *J. Math. Phys.* **20**, 972–977.

Caspers, W. J. (1964). Dispersion relations for nonlinear response, *Phys. Rev.* **133**, A1249–A1251.

Castaño González, O. D., de Dios Leyva, M., and Pérez Alvarez, R. (1975). The β parameter and Velický's approximation in the calculation of the optical properties of GaAs, *Phys. Stat. Sol. b* **71**, 111–116.

Castro, F. and Nabet, B. (1999). Numerical computation of the complex dielectric permittivity using Hilbert transform and FFT techniques, *J. Franklin Inst.* **336B**, 53–64.

Cataliotti, F. S., Fort, C., Hänsch, T. W., Inguscio, M., and Prevedelli, M. (1997). Electromagnetically induced transparency in cold atoms: Test of a sum rule for nonlinear optics, *Phys. Rev. A* **56**, 2221–2224.

Cauchy, A. (1822). Mémoire sur les intégrales définies, où l'on fixe le nombre et la nature des constantes arbitraires et des fonctions arbitraires que peuvent comporter les valeurs de ces mêmes intégrales quand elles deviennent indéterminées, *Bull. des Sci., La Soc. Philomatique* (*Paris*), pp. 161–174.

Cauchy, A. (1958). *Œuvres Complètes D'Augustin Cauchy*, Vol. 2, Series 2 (Paris: Gauthier-Villars), pp. 283–299.

Cauer, W. (1932). The Poisson integral for functions with positive real part, *Bull. Am. Math. Soc.* **38**, 713–717.

Červený, V. and Zahradník, J. (1975). Hilbert transform and its geophysical applications, *Acta Univ. Carolinae – Math. Physica* **16**, 67–81.

Cesari, L. (1938). Sulle serie di Fourier delle funzioni lipschitziane di più variabili, *Ann. Scuola Norm. Super. Pisa* **7**, 279–295.

Chambers, W. G. (1975). Failures in the Kramers-Krönig analysis of power-reflectivity, *Infrared Phys.* **15**, 139–141.

Champeney, D. C. (1987). *A Handbook of Fourier Theorems* (Cambridge: Cambridge University Press).

Chandarana, S. (1996). L^p-bounds for hypersingular integral operators along curves, *Pacific J. Math.* **175**, 389–416.

Chandler, G. A. and Graham, I. G. (1993). The computation of water waves modelled by Nekrasov's equation, *SIAM J. Numer. Analysis* **30**, 1041–1065.

Chatfield, C. (1989). *The Analysis of Time Series*, 4th edn (London: Chapman and Hall).

Chaudhry, M. A. and Pandey, J. N. (1985). Generalized n-dimensional Hilbert transforms and applications, *Appl. Analysis* **20**, 221–235.

Chaudhry, M. A. and Pandey, J. N. (1987). The Hilbert transform of Schwartz distributions. II, *Math. Proc. Camb. Phil. Soc.* **102**, 553–559.

Chawla, M. M. and Kumar, S. (1979). Convergence of quadratures for Cauchy principal value integrals, *Computing* **23**, 67–72.

Chen, C.-H., Li, C.-P., and Teng, T.-L. (2002). Surface-wave dispersion measurements using Hilbert-Huang transform, *TAO* **13**, 171–184.

Chen, K.-K. (1944). On the absolute Cesaro summability of negative order for a Fourier series at a given point, *Am. J. Math.* **66**, 299–312.

Chen, K. Y., Yeh, H. C., Su, S. Y., Liu, C. H., and Huang, N. E. (2001). Anatomy of plasma structures in an equatorial spread F event, *Geophys. Res. Lett.* **28**, 3107–3110.

Chen, Q., Huang, N., Riemenschneider, S., and Xu, Y. (2006). A B-spline approach for empirical mode decompositions, *Adv. Comput. Math.* **24**, 171–195.

Chen, Y.-M. (1960). Theorems of asymptotic approximation, *Math. Ann.* **140**, 360–407.

Chen, Y.-M. (1963). On conjugate functions, *Can. J. Math.* **15**, 486–494.

Chernyak, V. and Mukamel, S. (1995). Generalized sum rules for optical nonlinearities of many-electron systems, *J. Chem. Phys.* **103**, 7640–7644.

Chinsky, L., Laigle, A., Peticolis, W. L., and Turpin, P.-Y. (1982). Excited state geometry of uracil from the resonant Raman overtone spectrum using a Kramers-Kronig technique, *J. Chem. Phys.* **76**, 1–5.

Christ, M. (1985). Hilbert transforms along curves I. Nilpotent groups, *Ann. of Math.* **122**, 575–596.

Christ, M. (1990). *Lectures on Singular Integral Operators* (Providence, RI: American Mathematical Society).

Chung, J. (2001). Hilbert transform of generalized functions of L^p-growth, *Integ. Trans. & Special Funct.* **12**, 149–160.

Churchill. R. V. and Brown, J. W. (1984). *Complex Variables and Applications*, 4th edn (New York: McGraw-Hill).

Cima, J. A., Matheson, A. L., and Ross, W. T. (2006). *The Cauchy Transform* (Providence, RI: American Mathematical Society).

Čížek, V. (1970). Discrete Hilbert transform, *IEEE Trans. Audio Electroacoust.* **18**, 340–343.

Čížek, V. (1986). *Discrete Fourier Transforms and Their Applications* (Bristol: Adam Hilger Ltd.).

Clancey, K. F. (1975). On finite Hilbert transforms, *Trans. Am. Math. Soc.* **212**, 347–354.

Cody, W. J., Paciorek, K. A., and Thacher, H. C. Jr. (1970). Chebyshev approximations for Dawson's integral, *Math. Comp.* **24**, 171–178.

Cohen, G. and Lin, M. (2003). Laws of large numbers with rates and the one-sided ergodic Hilbert transform, *Illinois J. Math.* **47**, 997–1031.

Cohn, D. L. (1980). *Measure Theory* (Boston, MA: Birkhäuser).

Coifman, R. R. and Fefferman, C. (1974). Weighted norm inequalities for maximal functions and singular integrals, *Studia Math.* **51**, 241–250.

Cole, K. S. and Cole, R. H. (1941). Dispersion and absorption in dielectrics I. Alternating current characteristics, *J. Chem. Phys.* **9**, 341–351.

Cole, K. S. and Cole, R. H. (1942). Dispersion and absorption in dielectrics II. Direct current characteristics, *J. Chem. Phys.* **10**, 98–105.

Cole, R. H. (1941). Correlations in dielectric data, *Phys. Rev.* **60**, 172.

Collocott, S. J. (1977). Numerical solution of Kramers-Kronig transforms by a Fourier method, *Comput. Phys. Commun.* **13**, 203–206.

Collocott, S. J. and Troup, G. J. (1979). Adaptation: numerical solution of the Kramers-Kronig transforms by trapezoidal summation as compared to a Fourier method, *Comput. Phys. Commun.* **17**, 393–395.

Colwell, P. (1985). *Blaschke Products* (Ann Arbor, MI: The University of Michigan Press).

Condon, E. U. (1937a). Theories of optical rotatory power, *Rev. Mod. Phys.* **9**, 432–457.

Condon, E. U. (1937b). Immersion of the Fourier transform in a continuous group of functional transformations, *Proc. Natl. Acad. Sci. USA* **23**, 158–164.

Constantin, P., Lax, P. D., and Majda, A. (1985). A simple one-dimensional model for the three-dimensional vorticity equation, *Commun. Pure Appl. Math.* **38**, 715–724.

Cooke, J. C. (1970). The solution of some integral equations and their connection with dual integral equations and series, *Glasgow Math. J.* **11**, 9–20.

Coppens, P. (1997). *X-Ray Charge Densities and Chemical Bonding* (New York: Oxford University Press).

Córdoba, A. and Fefferman R. (1977). On the equivalence between the boundedness of certain classes of maximal and multiplier operators in Fourier analysis, *Proc. Natl. Acad. Sci. USA* **74**, 423–425.

Córdoba, A., Córdoba, D., and Fontelos M. A. (2006). Integral inequalities for the Hilbert transform applied to a nonlocal transport equation, *J. Math. Pures Appl.* **86**, 529–540.

Córdoba, A., Nagel, A., Vance, J., Wainger, S., and Weinberg D. (1986). L^p bounds for Hilbert transforms along convex curves, *Invent. Math.* **83**, 59–71.

Corinaldesi, E. (1959). An introduction to dispersion relations, *Nuovo Cimento Supp.* **14**, 369–384.

Cormack, A. M. (1963). Representation of a function by its line integrals, with some radiological applications, *J. Appl. Phys.* **34**, 2722–2727.

Cossar, J. (1939). On conjugate functions, *Proc. Lond. Math. Soc.* **45**, 369–381.

Cossar, J. (1960). Hilbert transforms and almost periodic functions, *Proc. Camb. Phil. Soc.* **56**, 354–366.

Cotlar, M. (1955). A unified theory of Hilbert transforms and ergodic theorems, *Rev. Mat. Cuyana* **1**, 105–167.

Cotlar, M. and Sadosky, C. (1975). A moment theory approach to the Riesz theorem on the conjugate function with general measures, *Studia Math.* **53**, 75–101.

Courant, R. (1936). *Differential and Integral Calculus*, Vol. 2 (London: Blackie & Son).

Craig, M. (1996). Analytic signals for multivariate data, *Math. Geol.* **28**, 315–329.

Criscuolo, G. and Giova, R. (1999). On the evaluation of the finite Hilbert transform by a procedure of interpolatory type, *Bull. Allahabad Soc.* **14**, 21–33.

Criscuolo, G. and Mastroianni, G. (1987). On the convergence of an interpolatory product rule for evaluating Cauchy principal value integrals, *Math. Comp.* **48**, 725–735.

Criscuolo, G. and Mastroianni, G. (1989). Mean convergence of a product rule to evaluate the finite Hilbert transform. In *Numerical and Applied Mathematics*, Vol. 1.1, W. F. Ames, ed. (Basel: J. C. Baltzer AG, Scientific Publishing Company), pp. 43–45.

Criscuolo, G. and Mastroianni, G. (1991). Mean and uniform convergence of quadrature rules for evaluating the finite Hilbert transform. In *Progress in Approximation Theory*, P. Nevai and A. Pinkus, eds. (Boston: Academic Press), pp. 141–175.

Criscuolo, G. and Scuderi, L. (1998). The numerical evaluation of Cauchy principal value integrals with non-standard weight functions, *BIT* **38**, 256–274.

Criscuolo, G., Della Vecchia, B., Lubinsky, D. S., and Mastroianni, G. (1995). Functions of the second kind for Freud weights and series expansions of Hilbert transforms, *J. Math. Analysis Appl.* **189**, 256–296.

Crystal, T. H. (1968). The Hilbert transform as an iterated Fourier transform: Comment on "The Hilbert transform as an iterated Laplace transform," *IEEE Trans. Aerospace Electron. Syst.* **4**, 315.

Cusmariu, A. (2002). Fractional analytic signals, *Signal Process.* **82**, 267–272.

Damelin, S. B. (2003). Marcinkiewicz-Zygmund inequalities and the numerical approximation of singular integrals for exponent weights: methods, results and open problems, some new, some old, *J. Complexity* **19**, 406–415.

Damelin, S. B. and Diethelm, K. (2001). Boundedness and uniform numerical approximation of the weighted Hilbert transform on the real line, *Numer. Funct. Analysis Optim.* **22**, 13–54.

Damera-Venkata, N., Evans, B. L. and McCaslin, S. R. (2000). Design of optimal minimum-phase digital FIR filters using discrete Hilbert transforms, *IEEE Trans. Signal Process.* **48**, 1491–1495.

Danloy, B. (1973). Numerical construction of Gaussian quadrature formulas for $\int_0^1 (-\mathrm{Log}\, x) \cdot x^\alpha \cdot f(x) \cdot \mathrm{d}x$ and $\int_0^1 E_m(x) \cdot f(x) \cdot \mathrm{d}x$, *Math. Comp.* **27**, 861–869.

Davies, B. (1978). *Integral Transforms and Their Applications* (New York: Springer-Verlag).

Davies, K. T. R. and Davies, R. W. (1989). Evaluation of a class of integrals occurring in mathematical physics via a higher order generalization of the principal value, *Can. J. Phys.* **67**, 759–765.

Davies, K. T. R., Davies, R. W., and White, G. D. (1990). Dispersion relations for causal Green's functions: Derivations using the Poincaré-Bertrand theorem and its generalizations, *J. Math. Phys.* **31**, 1356–1373. Erratum: *J. Math. Phys.* **32**, 1651 (1991).

Davies, K. T. R., Glasser, M. L., and Davies, R. W. (1992). Quadrature relations for finite-limit, principal value integrals, *Can. J. Phys.* **70**, 656–666.

Davies, K. T. R., Glasser, M. L., Protopopescu, V., and Tabakin, F. (1996). The mathematics of principal value integrals and applications to nuclear physics, transport theory, and condensed matter physics, *Math. Models Meth. Appl. Sci.* **6**, 833–885.

Davis, B. (1973). On the distributions of conjugate functions of nonnegative measures, *Duke Math. J.* **40**, 695–700.

Davis, B. (1974). On the weak type (1,1) inequality for conjugate functions, *Proc. Am. Math. Soc.* **44**, 307–311.

Davis, B. (1976). On Kolmogorov's inequalities $\left\|\tilde{f}\right\| \leq C_p \|f\|_1, 0 < p < 1$, *Trans. Am. Math. Soc.* **222**, 179–192.

Davis, B. (1979). Application of the conformal invariance of Brownian motion. In *Harmonic Analysis in Euclidean Spaces*, *Proceedings of Symposia in Pure Mathematics*, Vol. 35 part 2, G. Weiss and S. Wainger, eds. (Providence, RI: American Mathematical Society), pp. 303–310.

Davis, J. A. and Nowak, M. D. (2002). Selective edge enhancement of images with an acousto-optic light modulator, *Appl. Opt.* **41**, 4835–4839.

Davis, J. A., McNamara, D. E., and Cottrell, D. M. (1998). Analysis of the fractional Hilbert transform, *Appl. Opt.* **37**, 6911–6913.

Davis, J. A., McNamara, D. E., and Cottrell, D. M. (2000). Image processing with the radial Hilbert transform: theory and experiments, *Opt. Lett.* **25**, 99–101.

Davis, J. A., Smith, D. A., McNamara, D. E., Cottrell, D. M., and Campos, J. (2001). Fractional derivatives – analysis and experimental implementation, *Appl. Opt.* **40**, 5943–5948.

Davis, K. M. and Chang, Y.-C. (1987). *Lectures on Bochner-Riesz Means* (Cambridge: Cambridge University Press).

Davis, P. J. and Rabinowitz, P. (1975). *Methods of Numerical Integration*, (New York: Academic Press).

Davydov, A. S. (1963). Dispersion relations for the refractive index and absorption coefficient in media with exciton absorption, *Sov. Phys. JETP* **16**, 1293–1298.

De Alfaro, V., Fubini, S., Rossetti, G., and Furlan, G., (1966). Sum rules for strong interactions, *Phys. Lett.* **21**, 576–579.

De Bonis, M. C. and Mastroianni, G. (2003). Some simple quadrature rules for evaluating the Hilbert transform on the real line, *Arch. Inequal. Appl.* **1**, 475–494.

De Carli, L. and Laeng, E. (2000). Sharp L^p estimates for the segment multiplier, *Collect. Math.* **51**, 309–326.

De Heer, F. J., Wagenaar, R. W., Blaauw, H. J., and Tip, A. (1976). A dispersion relation for forward scattering, *J. Phys. B: Atom. Mol. Phys.* **9**, L269–L274.

Deans, S. R. (1983). *The Radon Transform and Some of Its Applications* (New York: John Wiley).

Deans, S. R. (1996). Radon and Abel transforms. In *The Transforms and Applications Handbook*, A. D. Poularikas, ed. (Boca Raton, FL: CRC Press), pp. 631–717.

Debiais, G. (2002). Hilbert transform or Kramers-Kronig relations applied to some aspects of linear and non-linear physics. In *Contemporary Problems in Mathematical Physics*, J. Govaerts, M. N. Hounkonnou, and A. Z. Msezane, eds. (River Edge, NJ: World Scientific), pp. 233–258.

Debnath, L. (1990). The Hilbert transform and its applications to fluid dynamics, *Prog. Math.* **24**, 19–40.

Debnath, L. and Bhatta, D. (2007). *Integral Transforms and Their Applications*, 2nd edn (Boca Raton, FL: Chapman & Hall/CRC).

Degasperis, A. and Santini, P. M. (1983). Linear operator and conservation laws for a class of nonlinear integro-differential evolution equations, *Phys. Lett. A* **98**, 240–244.

Degasperis, A., Santini, P. M. and Ablowitz, M. J. (1985). Nonlinear evolution equations associated with a Riemann–Hilbert scattering problem, *J. Math. Phys.* **26**, 2469–2472.

Del Pace, C. and Venturi, A. (1981). Su una equazione di Wiener-Hopf a nucleo singolare [A Wiener–Hopf equation with singular kernel], *Matematiche* (*Catania*) **33**, 333–347.

Delanghe, R (2004). On some properties of the Hilbert transform in Euclidean space, *Bull. Belg. Math. Soc.* **11**, 163–180.

Deléchelle, E., Lemoine, J., and Niang, O. (2005). Empirical mode decomposition: An analytical approach for sifting process, *IEEE Signal Process. Lett.* **12**, 764–767.

Della Vecchia, B. (1994). Two new formulas for the numerical evaluation of the Hilbert transform, *BIT* **34**, 346–360.

Delves, L. M. (1967). The numerical evaluation of principal value integrals, *Comput. J.* **10**, 389–391.

Derriennic, Y. and Lin, M. (2001). Fractional Poisson equations and ergodic theorems for fractional coboundaries, *Israel J. Math.* **123**, 93–130.

Dickinson, B. W. and Steiglitz, K. (1982). Eigenvectors and functions of the discrete Fourier transform, *IEEE Trans. Acoustics Speech Signal Process.* **30**, 25–31

Dienstfrey, A. and Greengard, L. (2001). Analytic continuation, singular-value expansions, and Kramers–Kronig analysis, *Inverse Problems* **17**, 1307–1320.

Diethelm, K. (1994a). Error estimates for a quadrature rule for Cauchy principal value integrals, *Proc. Sympos. Appl. Math.* **48**, 287–291.

Diethelm, K. (1994b). Non-optimality of certain quadrature rules for Cauchy principal value integrals, *Z. Angew. Math. Mech.* **74**, T689–T690.

Diethelm, K. (1994c). Modified compound quadrature rules for strongly singular integrals, *Computing* **52**, 337–354.

Diethelm, K. (1994d). Uniform convergence of optimal order quadrature rules for Cauchy principal value integrals, *J. Comput. Appl. Math.* **56**, 321–329.

Diethelm, K. (1995a). Asymptotically sharp error bounds for a quadrature rule for Cauchy principal value integrals based on piecewise linear interpolation, *Approx. Theory Appl.* **11**, 78–89.

Diethelm, K. (1995b). Gaussian quadrature formulae of the third kind for Cauchy principal value integrals: Basic properties and error estimates, *J. Comput. Appl. Math.* **65**, 97–114.

Diethelm, K. (1995c). The order of convergence of modified interpolatory quadratures for singular integrals of Cauchy type, *Z. Angew. Math. Mech.* **75**, S621–S622.

Diethelm, K. (1996a). A definiteness criterion for linear functionals and its application to Cauchy principal value quadrature, *J. Comput. Appl. Math.* **66**, 167–176.

Diethelm, K. (1996b). Definite quadrature formulae for Cauchy principal value integrals, *Bolyai Soc. Math. Stud.* **5**, 175–186.

Diethelm, K. (1996c). Peano kernels and bounds for the error constants of Gaussian and related quadrature rules for Cauchy principal value integrals, *Numer. Math.* **73**, 53–63.

Diethelm, K. (1997). New error bounds for modified quadrature formulas for Cauchy principal value integrals, *J. Comput. Appl. Math.* **82**, 93–104.

Diethelm, K. (1999). A method for the practical evaluation of the Hilbert transform on the real line, *J. Comput. Appl. Math.* **112**, 45–53.

Diethelm, K. and Köhler, P. (2000). Asymptotic behaviour of fixed-order error constants of modified quadrature formulae for Cauchy principal value integrals, *J. Inequal. Appl.* **5**, 167–190.

Dirac, P. A. M. (1927). The physical interpretation of the quantum dynamics, *Proc. Roy. Soc. Lond. A* **113**, 621–641.

Dirac, P. A. M. (1939). A new notation for quantum mechanics, *Proc. Camb. Phil. Soc.* **35**, 416–418.

Divin, Y., Volkov, O., Pavlovskii, V., Poppe, U., and Urban, K. (2001). Terahertz spectral analysis by ac Josephson effect in high-T_c bicrystal junctions, *IEEE Trans. Appl. Supercond.* **11**, 582–585.

Divin, Y. Y., Schulz, H., Poppe, U., Klein, N., Urban, K., and Stepantsov, E. A. (1995). $YbBa_2Cu_3O_{7-x}$ grain boundary Josephson junctions for Hilbert-transform spectroscopy, *Inst. Phys. Conf. Ser.* **148**, 1645–1648.

Divin, Y. Y., Kotelyanskii, I. M., Shadrin, P. M., Volkov, O. Y., Shirotov, V. V., Gubankov, V. N., Schulz, H., and Poppe, U. (1997a). $YBa_2Cu_3O_{7-x}$ Josephson junctions on $NdGaO_3$ bicrystal substrates, *Inst. Phys. Conf. Ser.* **158**, 467–470.

Divin, Y. Y., Pavlovskii, V. V., Volkov, O. Y., Schulz, H., Poppe, U., Klein, N., and Urban, K. (1997b). Hilbert-transform spectral analysis of millimeter- and submillimeter-wave radiation with high T_c Josephson junctions, *IEEE Trans. Appl. Supercond.* **7**, 3426–3429.

Divin, Yu. Ya., Polyanskii, O. Yu., and Shul'man, A. Ya. (1980). Incoherent radiation spectroscopy by means of the Josephson effect, *Sov. Tech. Phys. Lett.* **6**, 454–455.

Divin, Y. Y., Poppe, U., Urban, K., Volkov, O. Y., Shirotov, V. V., Pavlovskii, V. V., Schmueser, P., Hanke, K. and Geitz, M. (1999). Hilbert-transform spectroscopy with high-T_c Josephson junctions: First spectrometers and first applications, *IEEE Trans. Appl. Supercond.* **9**, 3346–3349.

Divin, Yu. Ya., Polyanski, O. Yu., and Shul'man, A. Ya. (1983). Incoherent radiation spectroscopy based on AC Josephson effect, *IEEE Trans. Magnetics* **19**, 613–615.

Divin, Yu. Ya., Larkin, S. Y., Anischenko, S. E., Khabayev, P. V., and Korsunsky, S. V. (1993). Millimeter-wave Hilbert-transform spectrum analyzer based on Josephson junction, *Int. J. Infrared Milli. Waves* **14**, 1367–1373.

Divin, Yu. Ya., Schulz, H., Poppe, U., Klein, N., Urban, K., and Pavlovskii, V. V. (1996). Millimeter-wave Hilbert-transform spectroscopy with high-T_c Josephson junctions, *Appl. Phys. Lett.* **68**, 1561–1563.

Dolgov, O. V., Kirzhnits, D. A., and Maksimov, E. G. (1981). On an admissible sign of the static dielectric function of matter, *Rev. Mod. Phys.* **53**, 81–93.

Domínguez, M. (1990a). Weighted inequalities for the Hilbert transform and the adjoint operator in the continuous case, *Studia Math.* **95**, 229–236.

Dominguez, M. (1990b). Multivariate prediction theory and weighted inequalities. In *Signal Processing, Scattering and Operator Theory, and Numerical Methods,* M. A. Kaashoek, J. H. Van Schuppen, and A. C. M. Ran, eds. (Boston, MA: Birkhäuser), pp. 123–130.

Dorf, R. C. (1993). *The Electrical Engineering Handbook* (Boca Raton, FL: CRC Press).

Dow, M. L. and Elliott, D. (1979). The numerical solution of singular integral equations over $(-1, 1)$, *SIAM J. Numer. Analysis* **16**, 115–134.

Drago, R. S. (1992). *Physical Methods for Chemists* (Fort Worth, TX: Saunders).

Dragomir, N. M., Dragomir, S. S., and Farrell, P. (2002). Approximating the finite Hilbert transform via trapezoid type inequalities, *Computers Math. Appl.* **43**, 1359–1369.

Dragomir, S. S. (2002a). Inequalities for the discrete Hilbert transform of functions whose derivatives are convex, *J. Korean Math. Soc.* **39**, 709–729.

Dragomir, S. S. (2002b). Approximating the finite Hilbert transform via an Ostrowski type inequality for functions of bounded variation, *J. Inequal. Pure Appl. Math.* **3** (article 51), 19 pp.

Drazin, P. G. and Johnson, R. S. (1989). *Solitons: An Introduction*, (Cambridge: Cambridge University Press).

Duan, J. and Ervin, V. J. (1998). Dynamics of a nonlocal Kuramoto-Sivashinsky equation, *J. Differential Equations* **143**, 243–266.

Duffin, R. J. (1956). Basic properties of discrete analytic functions, *Duke Math. J.* **23**, 335–363.

Duffin, R. J. (1957). Two-dimensional Hilbert transforms, *Proc. Am. Math. Soc.* **8**, 239–245.

Duffin, R. J. (1972). Hilbert transforms in Yukawan potential theory, *Proc. Natl. Acad. Sci. USA* **69**, 3677–3679.

Duggal, B. P. (1980). On the spectrum of a class of integral transforms, *J. Math. Analysis Appl.* **78**, 41–48.

Dugundji, J. (1958). Envelopes and pre-envelopes of real waveforms, *IRE Trans. Inf. Theory* **4**, 53–57.

Dumbrajs, O. and Martinis, M. (1981). Dispersion relations in atomic physics, *J. Phys. B: Atom. Mol. Phys.* **15**, 961–975.

Dunkl, C. F. (1985). Orthogonal polynomials and a Dirichlet problem related to the Hilbert transform, *Indag. Math.* **88**, 147–171.

Duoandikoetxea, J. (2001). *Fourier Analysis* (Providence, RI: American Mathematical Society).

Duoandikoetxea, J. and Vargas, A. (1995) Directional operators and radial functions on the plane, *Ark. Mat.* **33**, 281–291.

Duren, P. L. (2000). *Theory of H^p Spaces* (Mineola, NY: Dover).

Dutta Roy, S. C. and Agrawal, A. (1978). Digital low-pass filtering using the discrete Hilbert transform, *IEEE Trans. Acoustics Speech Signal Process.* **26**, 465–467.

Dyn'kin, E. M. (1991). Methods of the theory of singular integrals: Hilbert transform and Calderón-Zygmund theory. In *Commutative Harmonic Analysis*, V. P. Khavin and N. K. Nikol'skii, eds. (Berlin: Springer-Verlag), pp. 167–259.

Dyn'kin, E. M. and Osilenker, B. P. (1985). Weighted estimates of singular integrals and their applications, *J. Sov. Math.* **30**, 2094–2154. Translation from *Itogi Nauki I Tekhniki, Seriya Matematicheskii Analiz* **21**, 42–129 (1983).

Dzhuraev, A. (1992). *Methods of Singular Integral Equations* (Harlow: Longman Scientific & Technical).

Eastham, M. S. P. (1962). On Hilbert transforms, *Quart. J. Math.* **13**, 247–251.

Echeverría, J. C., Crowe, J. A., Woolfson, M. S., and Hayes-Gill, B. R. (2001). Application of empirical mode decomposition to heart rate variability analysis, *Med. Biol. Eng. Computing* **39**, 471–479.

Edmunds, D. E. and Kokilashvili, V. M. (1995). Two-weighted inequalities for singular integrals, *Can. Math. Bull.* **38**, 295–303.

Edwards, R. E. (1982). *Fourier Series*, Vol. 2, 2nd edn (New York: Springer-Verlag).

Ehlers, J. F. (2001). *Rocket Science for Traders* (New York: John Wiley).

Elliott, D. (1982). The classical collocation method for singular integral equations, *SIAM J. Numer. Analysis* **19**, 816–832.

Elliott, D. and Okada, S. (2004). The finite Hilbert transform and weighted Sobolev spaces, *Math. Nachr.* **266**, 34–47.

Elliott, D. and Paget, D. F. (1975). On the convergence of a quadrature rule for evaluating certain Cauchy principal value integrals, *Numer. Math.* **23**, 311–319.

Elliott, D. and Paget, D. F. (1976). On the convergence of a quadrature rule for evaluating certain Cauchy principal value integrals: an addendum, *Numer. Math.* **25**, 287–289.

Elliott, D. and Paget, D. F. (1979). Gauss type quadrature rules for Cauchy principal value integrals, *Math. Comp.* **33**, 301–309.

Emeis, C. A., Oosterhoff, L. J., and De Vries, G. (1967). Numerical evaluation of Kramers-Kronig relations, *Proc. Roy. Soc. Lond. A* **297**, 54–65.

Ephremidze, L. (1998). On the integrability of the ergodic Hilbert transform for a class of functions with equal absolute values, *Georgian Math. J.* **5**, 101–106.

Ephremidze, L. (2003). On the generalization of the Stein-Weiss theorem for the ergodic Hilbert transform, *Studia Math.* **155**, 67–75.

Ephremidze, L. (2004). The Stein-Weiss theorem for the ergodic Hilbert transform, *Studia Math.* **165**, 61–71.

Erdélyi, A., Magnus, W., Oberhettinger, F., and Tricomi, F. G. eds. (1953). *Higher Transcendental Functions*, Vols. I, II, and III (New York: McGraw-Hill).

Erdélyi, A., Magnus, W., Oberhettinger, F., and Tricomi, F. G. eds. (1954). *Tables of Integral Transforms*, Vols. I and II (New York: McGraw-Hill).

Ernst, R. R. (1969). Numerical Hilbert transform and automatic phase correction in magnetic resonance spectroscopy, *J. Mag. Res.* **1**, 7–26.

Ernst, R. R., Bodenhausen, G., and Wokaun, A. (1987). *Principles of Nuclear Magnetic Resonance in One and Two Dimensions*, (Oxford: Clarendon Press).

Essén, M. (1984). A superharmonic proof of the M. Riesz conjugate function theorem *Arkiv Mat.* **22**, 241–249.

Essén, M. (1992). Some best constant inequalities for conjugate functions, *Int. Series Numer. Math.* **103**, 129–140.

Essén, M., Shea, D., and Stanton, C. (1999). Best constant inequalities for conjugate functions, *J. Comput. Appl. Math.* **105**, 257–264.

Essén, M., Shea, D. F., and Stanton, C. S. (2002). Sharp $L\log^{\alpha} L$ inequalities for conjugate functions, *Ann. Inst. Fourier* **52**, 623–659.

Estrada, R. and Kanwal, R. P. (1985). Distributional solutions of singular integral equations, *J. Integral Equations* **8**, 41–85.

Estrada, R. and Kanwal, R. P. (1987). The Carleman type singular integral equations, *SIAM Rev.* **29**, 263–290.

Eu, J. K. T. and Lohmann, A. W. (1973). Isotropic Hilbert spatial filtering, *Opt. Commun.* **9**, 257–262.

Fabes, E. B. and Rivière, N. M. (1966). Singular integrals with mixed homogeneity, *Studia Math.* **27**, 19–38.

Fang, P. H. (1961). Complex conductivity of some plasmas and semiconductors, *Appl. Sci. Res. B* **9**, 51–64.

Fang, P. H. (1965). Covariance in the Kramers-Kronig relation, *Physica* **31**, 1792–1795.

Fannin, P. C., Molina, A., and Charles, S. W. (1993). On the generation of complex susceptibility data through the use of the Hilbert transform, *J. Phys. D: Appl. Phys.* **26**, 2006–2009.

Fano, U. (1956). Atomic theory of electromagnetic interactions in dense materials, *Phys. Rev.* **103**, 1202–1218.

Fatou, P. (1906). Séries trigonométriques et séries de Taylor, *Acta Math.* **30**, 335–400.

Fefferman, C. (1971). The multiplier problem for the ball, *Ann. of Math.* **94**, 330–336.

Fefferman, C. (1972). Estimates for double Hilbert transforms, *Studia Math.* **44**, 1–15.

Feinberg, G., Sucher, J., and Au, C.-K. (1989). The dispersion theory of dispersion forces, *Phys. Rep.* **180**, 83–157.

Feldman, M. (1997). Non-linear free vibration identification via the Hilbert transform, *J. Sound Vib.* **208**, 475–489.

Feldman, M. (2001). Hilbert transforms. In *Encyclopedia of Vibration* (New York: Academic Press), pp. 642–648.

Felsberg, M. and Sommer, G. (2001). The monogenic signal, *IEEE Trans. Signal Process.* **49**, 3136–3144.

Feng, B.-F. and Kawahara, T. (2000). Multi-hump stationary waves for a Korteweg-de Vries equation with nonlocal perturbations, *Physica D* **137**, 237–246.

Fernández-Cabrera, L. M. and Torrea, J. L. (1993). Vector-valued inequalities with weights, *Publ. Mat.* **37**, 177–208.

Fernández-Cabrera, L. M., Martín-Reyes, F. J., and Torrea, J. L. (1995). On the ergodic averages and the ergodic Hilbert transform, *Can. J. Math.* **47**, 330–343.

Ferrand, J. (1944). Fonctions préharmoniques et fontions préholmorphes, *Bull. Sci. Math.* (*2nd series*) **68**, 152–180.

Ferrando, S., Jones, R. L., and Reinhold, K. (1996). On approach regions for the conjugate Poisson integral and singular integrals, *Studia Math.* **120**, 169–182.

Ferro Fontán, C., Queen, N. M., and Violini, G. (1972). Dispersion sum rules for strong and electromagnetic interactions, *Riv. Nuovo Cimento* **2**, 357–497.

Ferry, J. D. (1970). *Viscoelastic Properties of Polymers*, 2nd edn (New York: John Wiley).

Fessler, T., Ford, W. F., and Smith, D. A. (1983). HURRY: an acceleration algorithm for scalar sequences and series, *ACM Trans. Math. Soft.* **9**, 346–354.

Fink, A. M. (2000). An essay on the history of inequalities, *J. Math. Analysis Appl.* **249**, 118–134.

Fischer, J. and Kolář, P. (1987). Differential forms of the dispersion integral, *Czech. J. Phys. B* **37**, 297–311.

Fischer, J., Pišút, J., Prešnajder, P., and Šebesta, J. (1969). Modified dispersion relations and sum rules, *Czech. J. Phys. B* **19**, 1486–1499.

Flett, T. M. (1958). Some theorems on odd and even functions, *Proc. Lond. Math. Soc.* (3) **8**, 135–148.

Folland, G. B. (1999). *Real Analysis*, 2nd edn (New York: John Wiley).

Forelli, F. (1963). The Marcel Riesz theorem on conjugate functions, *Trans. Am. Math. Soc.* **106**, 369–390.

Foreman, A. J., Jaswon, M. A., and Wood, J. K. (1951). Factors controlling dislocation widths, *Proc. Phys. Soc. Lond.* **64**, 156–163.

Fourès, Y. and Segal, I. E. (1955). Causality and analyticity, *Trans. Am. Math. Soc.* **78**, 385–405.

Franken, P. A., Hill, A. E., Peters, C. W., and Weinreich, G. (1961). Generation of optical harmonics, *Phys. Rev. Lett.* **7**, 118–119.

Fratila, R., ed. (1982). *Theory of Linear Systems* (New York: Research and Education Association).

Freedman, M. I., Falb, P. L., and Anton, J. (1969). A note on causality and analyticity, *SIAM J. Control* **7**, 472–478.

Fried, B. D. and Conte, S. D. (1961). *The Plasma Dispersion Function, The Hilbert Transform of the Gaussian* (New York: Academic Press).

Fröhlich, H. (1958). *Theory of Dielectrics*, 2nd edn (London: Oxford University Press).

Fromme, J. A. and Golberg, M. A. (1979). Numerical solution of a class of integral equations arising in two-dimensional aerodynamics. In *Solution Methods for Integral Equations Theory and Applications*, M. A. Golberg, ed. (New York: Plenum Press), pp. 109–146.

Frye, G. and Warnock, R. L. (1963). Analysis of partial-wave dispersion relations, *Phys. Rev.* **130**, 478–494.

Fuchssteiner, B. and Schulze, T. (1995). A new integrable system: The interacting soliton of the BO, *Phys. Lett. A* **204**, 336–342.

Fuks, B. A. (1963). *Introduction to the Theory of Analytic Functions of Several Complex Variables* (Providence, RI: American Mathematical Society).

Funk, P. (1916). Über eine geometrische anwendung der Abelchen integralgleichung, *Math. Ann.* **77**, 129–135.

Furuya, K. and Zimerman, A. H. (1976). Sum rules for the conductivity of superconducting films, *Phys. Rev. B* **13**, 1357–1358.

Furuya, K., Villani, A., and Zimerman, A. H. (1977). Superconvergent sum rules for the normal reflectivity, *J. Phys. C: Solid State Phys.* **10**, 3189–3198.

Furuya, K., Zimerman, A. H., and Villani, A. (1976). Superconvergent sum rules for the Voigt effect, *J. Phys. C: Solid State Phys.* **9**, 4329–4333.

Futterman, W. I. (1962). Dispersive wave bodies, *J. Geophys. Res.* **67**, 5279–5291.

Gabisonija, I. and Meskhi, A. (1998). Two-weighted inequalities for a discrete Hilbert transform, *Proc. A. Razmadze Math. Inst.* **116**, 107–122.

Gabor, D. (1946). Theory of communication, *J. Inst. Elect. Eng.* **93**, 429–457.

Gakhov, F. D. (1966). *Boundary Value Problems* (New York: Dover).

Gallardo, D. (1989). Weighted integral inequalities for the ergodic maximal operator and other sublinear operators. Convergence of the averages and the ergodic Hilbert transform, *Studia Math.* **94**, 121–147.

Gallardo, D. and Martín-Reyes, F. J. (1989). On the almost everywhere existence of the ergodic Hilbert transform, *Proc. Am. Math. Soc.* **105**, 636–643.

Gamelin, T. W. (1978). *Uniform Algebras and Jensen Measures* (Cambridge: Cambridge University Press).

Gao, M.-Z. (1996). A note on the Hardy-Hilbert inequality, *J. Math. Analysis Appl.* **204**, 346–351.

Gao, M.-Z. (1997). On Hilbert's inequality and its applications, *J. Math. Analysis Appl.* **212**, 316–323.

Gaposhkin, V. F. (1958). A generalization of M. Riesz's theorem on conjugate functions, *Mat. Sbornik* **46**, 359–372.

Garcia-Cuerva, J. and Rubio de Francia, J. L. (1985). *Weighted Norm Inequalities and Related Topics* (Amsterdam: North-Holland).

Gårding, L. (1970). Marcel Riesz in memoriam, *Acta Math.* **124**, I–XI.

Garnett, J. B. (1981). *Bounded Analytic Functions* (New York: Academic Press).

Garsia, A. M. (1970). *Topics in Almost Everywhere Convergence* (Chicago, IL: Markham Publishing Com.).

Gasquet, C. and Witomski, P. (1999). *Fourier Analysis and Applications* (New York: Springer-Verlag).

Gautschi, W. (1970). On the construction of Gaussian quadrature rules from modified moments, *Math. Comp.* **24**, 245–260.

Gautschi, W. (1979). On the preceding paper "A Legendre polynomial integral" by James L. Blue, *Math. Comp.* **33**, 742–743.

Gautschi, W. (1981). A survey of Gauss-Christoffel quadrature formulae. In *E. B. Christoffel*, P. L. Butzer and F. Fehér, eds. (Basel: Birkhäuser Verlag), pp. 72–147.

Gautschi, W. (1990). Computational aspects of orthogonal polynomials. In *Orthogonal Polynomials: Theory and Practice*, P. Nevai, ed. (Dordrecht: Kluwer Academic Publishers), pp. 181–216.

Gautschi, W. and Waldvogel, J. (2001). Computing the Hilbert transform of the generalized Laguerre and Hermite weight functions, *BIT* **41**, 490–503.

Gautschi, W. and Wimp, J. (1987). Computing the Hilbert transform of a Jacobi weight function, *BIT* **27**, 203–215.

Gazonas, G. A. (1986). The numerical evaluation of Cauchy principal value integrals via the fast Fourier transform, *Int. J. Comput. Math.* **18**, 277–288.

Gel'fand, I. M. and Shilov, G. E. (1964). *Generalized Functions*, Vol. 1 (New York: Academic Press).

Gel'fand, I. M. and Shilov, G. E. (1968). *Generalized Functions*, Vol. 2 (New York: Academic Press). (Available as a German translation in Gelfand and Schilow (1962).)

Gel'fand, I. M. and Schilow, G. E. (1962). *Verallgemeinerte Funktionen (Distributionen)* II (Berlin: Deutscher Verlag Wissenschaften).

Gell-Mann, M., Goldberger, M. L., and Thirring, W. E. (1954). Use of causality conditions in quantum theory, *Phys. Rev.* **95**, 1612–1627.

Gera, A. E. (1986). Singular integral equations with a Cauchy kernel, *J. Comput. Appl. Math.* **14**, 311–318.

Gerjuoy, E. and Krall, N. A. (1960). Dispersion relations in atomic scattering problems, *Phys. Rev.* **119**, 705–711.

Gerjuoy, E. and Krall, N. A. (1962). Dispersion relations for electron scattering from atomic helium, *Phys. Rev.* **127**, 2105–2113.

Gerjuoy, E. and Lee, C. M. (1978). On the dispersion relations for electron-atom scattering, *J. Phys. B: Atom. Mol. Phys.* **11**, 1137–1155.

Giacovazzo, C., Siliqi, D., and Fernández-Castaño, C. (1999). The joint probability distribution function of structure factors with rational indices. IV. The $P1$ case, *Acta Cryst.* A **55**, 512–524.

Giambiagi, J. J. and Saavedra, I. (1963). Causality and analyticity in non-relativistic scattering, *Nucl. Phys.* **46**, 413–416.

Gindikin, S. and Michor, P. (1994). *75 Years of Radon Transform* (Boston, MA: International Press).

Ginzberg, V. L. (NB: The usual English spelling is Ginzburg) (1955). Concerning the general relationship between absorption and dispersion of sound waves, *Sov. Phys. Acoust.* **1**, 32–41.

Ginzburg, V. L. (1956). On the macroscopic theory of superconductivity, *Sov. Phys. JETP* **2**, 589–600.

Ginzburg, V. L. and Meĭman, N. N. (1964). Dispersion relations for the indices of refraction and absorption, *Sov. Phys. JETP* **19**, 169–175.

Girard, A. (1629). *Invention Nouvelle en L'algebre* (Amsterdam: Chez Guillaume Iansson Blaeuw). Reprinted by Bierens De Haan, Leiden (1884).

Giraud, G. (1934). Équations à integrals principales, *Ann. Sci. École Norm. Sup.* **51**, 251–372.

Giraud, G. (1936). Sur une classe générale d'équations à integrals principales, *Comptes Rendus Acad. Sci., Series I Math.* **202**, 2124–2127.

Glaeske, H.-J. and Tuan, V. K. (1995). Some applications of the convolution theorem of the Hilbert transform, *Integral Transforms Spec. Funct.* **4**, 263–268.

Glasser, M. L. (1983). A remarkable property of definte integrals, *Math. Comp.* **40**, 561–563.

Glasser, M. L. (1984). Some useful properties of the Hilbert transform, *SIAM J. Math. Analysis* **15**, 1228–1230.

Glover, III, R. E. and Tinkham, M. (1957). Conductivity of superconducting films for photon energies between 0.3 and 40 kT_c, *Phys. Rev.* **108**, 243–256.

Goedecke, G. H. (1975a). Dispersion relations and complex reflectivity, *J. Opt. Soc. Am.* **65**, 146–149.

Goedecke, G. H. (1975b). Complex-reflectivity dispersion relation and radiation reaction, *J. Opt. Soc. Am.* **65**, 1075–1076.

Gohberg, I. and Krupnik, N. (1992a). *One-dimensional Linear Singular Integral Equations, Volume I, Introduction* (Basel: Birkhäuser).

Gohberg, I. and Krupnik, N. (1992b). *One-dimensional Linear Singular Integral Equations, Volume II, General Theory and Applications* (Basel: Birkhäuser).

Gokhberg, I. Ts. and Krupnik, N. Ya. (1968). Norm of the Hilbert transformation in the L_p space, *Funct. Analysis Appl.* **2**, 180–181.

Golberg, M. A. (1990). Introduction to the numerical solution of Cauchy singular integral equations. In *Numerical Solution of Integral Equations*, M. A. Golberg, ed. (New York: Plenum Press), pp. 183–308.

Gold, B., Oppenheim, A. V. and Radar, C. M. (1970). Theory and implementation of the discrete Hilbert transform. In *Proceedings of the Symposium on Computer Processing in Communications, 1969*, Vol. 19 (Brooklyn, NY: Polytechnic Press), pp. 235–250.

Goldberg, R. R. (1960). An integral transform related to the Hilbert transform, *J. Lond. Math. Soc.* **35**, 200–204.

Goldberger, M. L. (1960). Introduction to the theory and applications of dispersion relations. In *Relations de dispersion et particules élémentaires*, C. De Witt and R. Omnes, eds. (Paris: Hermann), pp. 1–157.

Goldberger, M. L. and Watson, K. M. (1964). *Collision Theory* (New York: John Wiley).

Goodspeed, F. M. (1939). Some generalizations of a formula of Ramanujan, *Quart. J. Math.* **10**, 210–218.

Gopinath, D. V. and Nayak, A. R. (1996). Chebyshev-Hilbert transform method for the solution of singular integrals and singular integral equations, *Trans. Theory Stat. Phys.* **25**, 635–657.

Gori, L. and Santi, E. (1995). On the evaluation of Hilbert transforms by means of a particular class of Turán quadrature rules, *Numer. Algorithms* **10**, 27–39.

Gornov, E., Peiponen, K.-E., Svirko, Y., Ino, Y., and Kuwata-Gonokami, M. (2006). Efficient dispersion relations for terahertz spectroscopy, *Appl. Phys. Lett.* **89**, 142903.

Gorter, C. J. (1938). Parmagnetische dispersion und absorption, *Phys. Zeitschr.* **39**, 815–823.

Gorter, C. J. and Kronig, R. De L. (1936). On the theory of absorption and dispersion in paramagnetic and dielectric media, *Physica* **3**, 1009–1020.

Gottlieb, M. (1960). Optical properties of lithium fluoride in the infrared, *J. Opt. Soc. Am.* **50**, 343–349.

Gottlieb, O. and Feldman, M. (1997). Application of a Hilbert transform-based algorithm for parameter estimation of a nonlinear ocean system roll model, *J. Offshore Mech. Arctic Eng.* **119**, 239–243.

Gradshteyn, I. S. and Ryzhik, I. M. (1965). *Tables of Integrals, Series, and Products* (New York: Academic Press).

Grafakos, L. (1994). An elementary proof of the square summability of the discrete Hilbert transform, *Am. Math. Monthly* **101**, 456–458.

Grafakos, L. (1996/7). Linear truncations of the Hilbert transform, *Real Analysis Exch.* **22**, 413–427.

Grafakos, L. (1997). Best bounds for the Hilbert transform on $L^p(\mathbb{R}^1)$, *Math. Res. Lett.* **4**, 469–471.

Grafakos, L. and Li, X. (2004). Uniform bounds for the bilinear Hilbert transforms, I, *Ann. of Math.* **159**, 889–933.

Grafakos, L. and Li, X. (2006). The disc as a bilinear multiplier, *Am. J. Math.* **128**, 91–119.

Gray, A., Mathews, G. B., and MacRobert, T. M. (1966). *A Treatise on Bessel Functions and Their Applications to Physics*, 2nd edn (New York: Dover).

Greenaway, D. L. and Harbeke, G. (1968). *Optical Properties and Band Structure of Semiconductors* (Oxford: Pergamon Press).

Greene, R. E. and Krantz, S. G. (1997). *Function Theory of One Complex Variable* (New York: Wiley).

Gregor, J. (1961). O aproximaci obrazu v Hilbertově transformaci ortogonálními řadami racionálních lomených funkcí, *Apl. Mat.* **6**, 214–240.

Grennberg, A. and Sandell, M. (1994). Estimation of subsample time delay differences in narrowband ultrasonic echoes using the Hilbert transform correlation, *IEEE Trans. Ultrasonics, Ferroelect, Freq. Control* **41**, 588–595.

Gross, B. (1941). On the theory of dielectric loss, *Phys. Rev.* **59**, 748–750.

Gross, B. (1943). On an integral transformation of general circuit theory, *Am. Math. Monthly*, **50**, 90–93.

Gross, B. (1948). On creep and relaxation. II. *J. Appl. Phys.* **19**, 257–264.

Gross, B. (1956). Lineare systeme, *Supp. Nuovo Cimento* **3**, 235–296.

Gross, B. (1968). *Mathematical Structure of the Theories of Viscoelasticity* (Paris: Hermann).

Gross, B. (1975). Applications of the Kramers-Kronig relations, *J. Phys. C: Solid State Phys.* **8**, L226–L227.

Gross, B. and Braga, E. P. (1961). *Singularities of Linear Systems Functions* (Amsterdam: Elsevier Publishing Company).

Grosse, P. and Offermann, V. (1991). Analysis of reflectance data using the Kramers-Kronig relations, *Appl. Phys. A* **52**, 138–142.

Gründler, R. (1983). Superconvergent sum rules for metals and insulators, *Physica Status Solidi* **115**, K147–K150.

Grushevskii, S. P. (1986). Kolmogorov inequality and estimates of boundary value distribution for analytic functions, *Sov. Math.* **30**, 64–71. Translation of *Izvestiya Vysshikh Uchebnykh Zavedeniui Matematika* **30**, 47–52.

Guillemin, E. A. (1949). *The Mathematics of Circuit Analysis* (Cambridge, MA: Technology Press, MIT; New York: John Wiley).

Guillemin, E. A. (1963). *Theory of Linear Physical Systems* (New York: John Wiley).

Gurielashvili, R. I. (1987). On the Hilbert transform, *Analysis Math.* **13**, 121–137.

Güttinger, W. (1966). Generalized functions and dispersion relations in physics, *Fortsch. Physik* **14**, 483–602.

Güttinger, W. (1967). Generalized functions in elementary particle physics and passive system theory: recent trends and problems, *SIAM J. Appl. Math.* **15**, 964–1000.

Hahn, S. L. (1996a). *Hilbert Transforms in Signal Processing* (Boston, MA: Artech House).

Hahn, S. L. (1996b). Hilbert transforms. In *The Transforms and Applications Handbook*, A. D. Poularikas, ed. (Boca Raton, FL: CRC Press), pp. 463–629.

Hahn, S. L. (1996c). The Hilbert transform of the product $a(t)\cos(\omega_0 t + \varphi)$, *Bull. Pol. Acad. Sci. Tech. Sci.* **44**, 75–80.

Hamel, G. (1937). *Integralgleichungen* (Berlin: Springer).

Hamilton, J. (1960). Dispersion relations for elementary particles, *Prog. Nucl. Phys.* **8**, 143–194.

Hansen, E. (1975). *A Table of Series and Products* (Englewood Cliffs, NJ: Prentice-Hall).

Harbecke, B. (1986). Application of Fourier's allied integrals to the Kramers-Kronig transformation of reflectance data, *Appl. Phys. A* **40**, 151–158.

Hardy, G. H. (1901). General theorems in contour integration: with some applications, *Quart. J. Math.* **32**, 369–384.

Hardy, G. H. (1902). The Theory of Cauchy's principal values. (Third paper: Differentiation and integration of principal values), *Proc. Lond. Math. Soc.* **35**, 81–107.

Hardy, G. H. (1908). The Theory of Cauchy's principal values. Fourth paper: The integration of principal values – continued – with applications to the inversion of definite integrals, *Proc. Lond. Math. Soc.* **7**(2), 181–208.

Hardy, G. H. (1924a). Notes on some points in the integral calculus. LVIII. On Hilbert transforms, *Mess. Math.* **54**, 20–27.

Hardy, G. H. (1924b). Notes on some points in the integral calculus. LIX. On Hilbert transforms (continued), *Mess. Math.* **54**, 81–88.

Hardy, G. H. (1926). Notes on some points in the integral calculus. LXII. A singular integral, *Mess. Math.* **56**, 10–16.

Hardy, G. H. (1928a). Notes on some points in the integral calculus. LXVII. On the repeated integral which occurs in the theory of conjugate functions, *Mess. Math.* **58**, 53–58.

Hardy, G. H. (1928b). Remarks on three recent notes in the Journal, *J. Lond. Math. Soc.* **3**, 166–169.

Hardy, G. H. (1932). On Hilbert transforms, *Quart. J. Math. (Oxford)* **3**, 102–112.

Hardy, G. H. (1937). On a theorem of Paley and Wiener, *Proc. Camb. Phil. Soc.* **33**, 1–5.

Hardy, G. H. (1966). *Collected Papers of G. H. Hardy*, 7 vols. (London: Oxford University Press).

Hardy, G. H. and Littlewood, J. E. (1926). The allied series of a Fourier series, *Proc. Lond. Math. Soc.* **24**, 211–246.

Hardy, G. H. and Littlewood, J. E. (1928). A convergence criterion for Fourier series, *Math. Zeitschr.* **28**, 612–634.

Hardy, G. H. and Littlewood, J. E. (1929). A point in the theory of conjugate functions, *J. Lond. Math. Soc.* **4**, 242–245.

Hardy, G. H. and Littlewood, J. E. (1930). A maximal theorem with function-theoretic applications, *Acta Math.* **54**, 81–116.

Hardy, G. H. and Littlewood, J. E. (1932). Some properties of conjugate functions, *J. Reine Angew. Math.* **167**, 405–423.

Hardy, G. H. and Littlewood, J. E. (1936). Some more theorems concerning Fourier series and Fourier power series, *Duke Math. J.* **2**, 354–382.

Hardy, G. H., Littlewood, J. E., and Pólya, G. (1952). *Inequalities*, 2nd edn (London: Cambridge University Press).

Hartley, R. V. L. (1942). A more symmetric Fourier analysis applied to transmission problems, *Proc. IRE* **30**, 144–150.

Hasegawa, T. (2004). Uniform approximations to finite Hilbert transform and its derivative, *J. Comput. Appl. Math.* **163**, 127–138.

Hasegawa, T. and Torii, T. (1991). An automatic quadrature for Cauchy principal value integrals, *Math. Comp.* **56**, 741–754.

Hasegawa, T. and Torii, T. (1994). Hilbert and Hadamard transforms by generalized Chebyshev expansion, *J. Comput. Appl. Math.* **51**, 71–83.

Hassani, S. (1999). *Mathematical Physics A Modern Introduction to Its Foundations* (New York: Springer).

Hauss, M. (1997). An Euler-Maclaurin-type formula involving conjugate Bernoulli polynomials and an application to $\varsigma(2m + 1)$, *Commun. Appl. Analysis* **1**, 15–32.

Hauss, M. (1998). A Boole-type formula involving conjugate Euler polynomials. In *Charlemagne and His Heritage: 1200 Years of Civilization and Science in Europe, Vol. 2: Mathematical Arts*, P. L. Butzer, H. Th. Jongen, and W. Oberschelp, eds. (Turnhout: Brepols), pp. 361–375.

Hayashi, N. and Naumkin, P. I. (1999). Large time asymptotics of solutions to the generalized Benjamin-Ono equation, *Trans. Am. Math. Soc.* **351**, 109–130.

He, P. (1998). Simulation of ultrasound pulse propagation in lossy media obeying a frequency power law, *IEEE Trans. Ultrasonics, Ferroelect., Freq. Control* **45**, 114–125.

Healy, W. P. (1974). Higher multipole moments and dispersion relations in optical activity, *J. Phys. B: Atom. Mol. Phys.* **7**, 1633–1648.

Healy, W. P. (1976). Primitive causality and optical active molecules, *J. Phys. B: Atom. Mol. Phys.* **9**, 2499–2510.

Healy, W. P. and Power, E. A. (1974). Dispersion relations for optically active media, *Am. J. Phys.* **42**, 1070–1074.

Heinig, H. P. (1979). On a Hilbert-type transform, *Portugaliae Math.* **38**, 217–223.

Helgason, S. (1980). *The Radon Transform* (Boston, MA: Birkhäuser).

Hellinger, E. (1935). Hilberts arbeiten über integralgleichungen und unendliche gleichungssysteme. In *David Hilbert Gesammelte Abhandlungen*, Vol. 3 (Springer: Berlin), pp. 94–145.

Helson, H. (1958). Conjugate series and a theorem of Paley, *Pacific J. Math.* **8**, 437–446.

Helson, H. (1959). Conjugate series in several variables, *Pacific J. Math.* **9**, 513–523.

Helson, H. and Szegö, G. (1960). A problem in prediction theory, *Ann. Mat. Pura Appl.* **51**, 107–138.

Henery, R. J. (1984). Hilbert transforms using fast Fourier transforms, *J. Phys. A: Math. Gen.* **17**, 3415–3423.

Henrici, P. (1986). *Applied and Computational Complex Analysis*, Vol. 3, (New York: John Wiley & Sons).

Henry, C. H., Schnatterly, S. E., and Slichter, C. P. (1965). Effect of applied fields on the optical properties of color centers, *Phys. Rev.* **137**, A583–A602.

Herdman, T. L. and Turi, J. (1991). An application of finite Hilbert transforms in the derivation of a state space model for an aeroelastic system, *J. Integral Equations Appl.* **3**, 271–287.

Herglotz, G. (1911). Über potenzreihen mit positivem, reellen Teil im einheitskreis, *Berichte über die Verhandlungen der Königlich Sächsischen Gessellschaft der Wissenschaften zu Leipzig. Mathematisch-Physische Klasse* **63**, 501–511.

Herman, G. T. (1980). *Image Reconstruction from Projections* (New York: Academic Press).

Herring, F. G., Marshall, A. G., Phillips, P. S., and Roe, D. C. (1980). Dispersion versus absorption (DISPA): Modulation broadening and instrumental distortions in electron paramagnetic resonance lineshapes, *J. Mag. Res.* **37**, 293–303.

Hewitt, E. and Ritter, G. (1983). Conjugate Fourier series of certain solenoids, *Trans. Am. Math. Soc.* **276**, 817–840.

Heywood, P. (1963). On a transform discussed by Goldberg, *J. Lond. Math. Soc.* **38**, 162–168.

Heywood, P. (1967). On a modification of the Hilbert transform, *J. Lond. Math. Soc.* **42**, 641–645.

Heywood, P. and Rooney, P. G. (1988). On the inversion of the even and odd Hilbert transformations, *Proc. Roy. Soc. Edinburgh* **109A**, 201–211.

Higgins, J. R. (1977). *Completeness and Basis Properties of Sets of Special Functions* (Cambridge: Cambridge University Press).

Higgins, J. R. (1985). Five short stories about the cardinal series, *Bull. Am. Math. Soc.* **12**, 45–89.

Hilb, E. and Riesz, M. (1924). Neuere untersuchungen über trigonometrische reihen, *Enzyklopädie der Mathematischen Wissenschaften II C10* (Leipzig: B. G. Teubner), pp. 1189–1228.

Hilbert, D. (1904). Grundzüge einer allgemeinen theorie der linearen integralgleichungen, *Nach. Akad. Wissensch. Gottingen. Math.-phys. Klasse*, **3**, 213–259.

Hilbert, D. (1905). Über eine anwendung der integralgleichungen auf ein problem der funktionentheorie. In *Verhandlungen des dritten Internationalen Mathematiker Kongresses* (Heidelberg 8–13 August, 1904), A. Krazer, ed. (Leipzig: B. G. Teubner), pp. 233–240.

Hilbert, D. (1912). *Grundzüge einer Allgemeinen Theorie der Linearen Integralgleichungen* (Leipzig: B. G. Teubner).

Hilgevoord, J. (1962). *Dispersion Relations and Causal Description* (Amsterdam: North-Holland).

Hille, E. and Tamarkin, J. D. (1933). On a theorem of Paley and Wiener, *Ann. Math.* **34**, 606–614.

Hille, E. and Tamarkin, J. D. (1934). A remark on Fourier transforms and functions analytic in a half-plane, *Compos. Math.* **1**, 98–102.

Hille, E. and Tamarkin, J. D. (1935). On the absolute integrability of Fourier transforms, *Fund. Math.* **25**, 329–352.

Hinich, M. J. and Weber, W. E. (1984). A Hilbert transform method for estimating distributed lag models with randomly missed or distorted observations. In *Time Series Analysis of Irregularly Observed Data* (Lecture Notes in Statistics), E. Parzen, ed. (New York: Springer-Verlag), pp. 134–151.

Hinich, M. J. and Weber, W. E. (1994). Estimating linear filters with errors in variables using the Hilbert transform, *Signal Process.* **37**, 215–228.

Hinojosa, J. H. and Mickus, K. L. (2002). Hilbert transform of gravity gradient profiles: Special cases of the general gravity-gradient tensor in the Fourier transform domain, *Geophys.* **67**, 766–769.

Hirschman, I. I. (1955). The decomposition of Walsh and Fourier Series, *Mem. Am. Math. Soc.* No. 15, pp. 1–65.

Hobson, E. W. (1926). *The Theory of Functions of a Real Variable and the Theory of Fourier's Series* Vol. II, 2nd edn (London: Cambridge University Press).

Hochstadt, H. (1973). *Integral Equations* (New York: Wiley).

Hoenders, B. J. (1975). On the solution of the phase retrieval problem, *J. Math. Phys.* **16**, 1719–1725.

Holbrow, C. H. and Davidon, W. C. (1964). An introduction to dispersion relations, *Am. J. Phys.* **32**, 762–774.

Hölder, O. (1889). Ueber einen mittelwerthssatz, *Nachr. Königliche Gesellscahft Wissenschaft. Göttingen*, pp. 38–47.

Holland, A. S. B. (1973). *Introduction to the Theory of Entire Functions* (New York: Academic Press).

Hollenbeck, B. and Verbitsky, I. E. (2000). Best constants for the Riesz projection, *J. Funct. Analysis* **175**, 370–392.

Hollenbeck, B., Kalton, N. J., and Verbitsky, I. E. (2003). Best constants for some operators associated with the Fourier and Hilbert transforms, *Studia Math.* **157**, 237–278.

Hopfield, J. J. (1970). Sum rule relating optical properties to the charge distribution, *Phys. Rev. B* **2**, 973–979.

Horton, Sr., C. W. (1974). Dispersion relationships in sediments and sea water, *J. Acoust. Soc. Am.* **55**, 547–549. Erratum: *J. Acoust. Soc. Am.* **70**, 1182 (1981).

Horváth, J. (1953a). Sur l'itération de la transformée de Hilbert d'une distribution complexe, *C. R. Acad. Sci. Paris* **237**, 1480–1482.

Horváth, J. (1953b). Sur les fonctions conjuguées à plusieurs variables, *Indag. Math.* **15**, 17–29.

Horváth, J. (1956). Singular integral operators and spherical harmonics, *Trans. Am. Math. Soc.* **82**, 52–63.

Horváth, J. (1959). On some composition formulas, *Proc. Am. Math. Soc.* **10**, 433–437.

Horváth, J., Ortner, N., and Wagner, P. (1987). Analytic continuation and convolution of hypersingular higher Hilbert-Riesz kernels, *J. Math. Analysis Appl.* **123**, 429–447.

Houghton, E. L. and Carruthers, N. B. (1982). *Aerodynamics for Engineering Students*, 3rd edn (London: Edward Arnold).

Howell, K. B. (1996). Fourier Transforms. In *The Transforms and Applications Handbook*, A. D. Poularikas, ed. (Boca Raton, FL: CRC Press), pp. 95–225.

Howell, K. B. (2001). *Principles of Fourier Analysis* (Boca Raton, FL: Chapman & Hall/CRC Press).

Huang, J. B. and Urban, M. W. (1992). Evaluation and analysis of attenuated total reflectance FT-IR spectra using Kramers-Kronig transforms, *Appl. Spectrosc.* **46**, 1666–1672.

Huang, N. E., Long, S. R., and Shen, Z. (1996). The mechanism for frequency downshift in nonlinear wave evolution. In *Advances in Applied Mechanics*, Vol. 32, J. W. Hutchinson and T. Y. Wu, eds. (Boston, MA: Academic Press), pp. 51–117.

Huang, N. E. and Shen, S. S. P. ed. (2005). *Hilbert-Huang Transform and Its Applications* (Singapore: World Scientific Publishing Co.).

Huang, N. E., Shen, Z., and Long, S. R. (1999). A new view of nonlinear water waves: The Hilbert spectrum, *Annu. Rev. Fluid Mech.* **31**, 417–457.

Huang, N. E., Shen, Z., Long, S. R., *et al.* (1998). The empirical mode decomposition and the Hilbert spectrum for nonlinear and non-stationary time series analysis, *Proc. Roy. Soc. Lond. A* **454**, 903–995.

Huang, N. E., Wu, M. C., Long, S. R., Shen, S. S. P., Qu, W., Gloersen, P., and Fan, K. L. (2003a). A confidence limit for the empirical mode decomposition and Hilbert spectral analysis, *Proc. Roy. Soc. Lond. A* **459**, 2317–2345.

Huang, N. E., Wu, M.-L., Qu, W., Long, S. R., Shen, S. S. P., and Zhang, J. E. (2003b). Applications of Hilbert-Huang transform to non-stationary financial time series analysis, *Appl. Stochastic Models Business Industry* **19**, 245–268.

Huang, W., Shen, Z., Huang, N. E. and Fung, Y. C. (1998). Engineering analysis of biological variables: An example of blood pressure over 1 day, *Proc. Natl. Acad. Sci. USA* **95**, 4816–4821.

Huang, W., Shen, Z., Huang, N. E., and Fung, Y. C. (1999). Nonlinear indicial response of complex nonstationary oscillations as pulmonary hypertension responding to step hypoxia, *Proc. Natl Acad. Sci. USA* **96**, 1834–1839.

Hulthén, R. (1982). Kramers-Kronig relations generalized: on dispersion relations for finite frequency intervals. A spectrum-restoring filter, *J. Opt. Soc. Am.* **72**, 794–803.

Hunt, R. A. (1972). An estimate of the conjugate function, *Studia Math.* **44**, 371–377.

Hunt, R. A. (1974). Comments on Lusin's conjecture and Carleson's proof for L^2 Fourier series. In *Linear Operators and Approximation II*, P. L. Butzer and B. Szökefalvi-Nagy, eds. (Basel: Birkhäuser Verlag), pp. 235–245.

Hunt, R., Muckenhoupt, B., and Wheeden, R. (1973). Weighted norm inequalities for the conjugate function and Hilbert transform, *Trans. Am. Math. Soc.* **176**, 227–251.

Hunter, D. B. (1972). Some Gauss-type formulae for the evaluation of Cauchy principal values of integrals, *Numer. Math.* **19**, 419–424.

Hunter, D. B. (1973). The numerical evaluation of Cauchy principal values of integrals by Romberg integration, *Numer. Math.* **21**, 185–192.

Huntington, H. B. (1955). Modification of the Peierls-Nabarro model for edge dislocation core, *Proc. Phys. Soc. Lond.* **68**, 1043–1048.

Hutchings, D. C., Sheik-Bahae, M., Hagan, D. J., and Van Stryland, E. W. (1992). Kramers-Krönig relations in nonlinear optics, *Opt. Quant. Electron.* **24**, 1–30.

Hutchinson, D. A. (1968). Hameka theory of optical rotation, *Can. J. Chem.* **46**, 599–604.

Hutt, P. K., Islam, M. M., Rabheru, A., and McDowell, M. R. C. (1976). Empirical tests of the electron-atom dispersion relation, *J. Phys. B: Atom. Mol. Phys.* **9**, 2447–2460.

Igari, S. (1962). Transformations of conjugate functions II, *Tôhoku Math. J.* **14**, 121–126.

Inagaki, T., Kuwata, H., and Ueda, A. (1979). Phase-shift sum rules for the complex polarization amplitude of photon reflection, *Phys. Rev. B* **19**, 2400–2403.

Inagaki, T., Kuwata, H., and Ueda, A. (1980). Dispersion relations and sum rules for the ellipsometric function, *Surf. Sci.* **96**, 54–66.

Inagaki, T., Ueda, A., and Kuwata, H. (1978). Sum rules for complex reflection amplitudes for photons, *Phys. Lett. A* **66**, 329–331.

Ishikawa, S. (1987). Generalized Hilbert transforms in tempered distributions, *Tokyo J. Math.* **10**, 119–132.

Ivanov, I. A. and Dubau, J. (1998). Dispersion relation for the ground-state energy of a two-electron atom, *Phys. Rev. A* **57**, 1516–1518.

Ivić, A. (1985). *The Riemann Zeta-Function* (New York: Wiley).

Iwaniec, T. (1987). Hilbert transform in the complex plane and area inequalities for certain quadratic differentials, *Michigan Math. J.* **34**, 407–434.

Iwaniec, T. and Martin, G. (1996). Riesz transforms and related singular integrals, *J. Reine Angew. Math.* **473**, 25–57.

Jackson, J. D. (1999). *Classical Electrodynamics*, 3rd edn (New York: John Wiley).

Jackson, L. B. (1989). *Digital Filters and Signal Processing* (Boston, MA: Kluwer Academic Publishers).

Jahoda, F. C. (1957). Fundamental absorption of barium oxide from its reflectivity spectrum, *Phys. Rev.* **107**, 1261–1265.

Jajte, R. (1987). On the existence of the ergodic Hilbert transform, *Ann. Prob.* **15**, 831–835.

Jeffrey, A. (1992). *Complex Analysis and Applications* (Boca Raton, FL: CRC Press).

Jiang, H.-M. (1991). Weighted-BMO and the Hilbert transform, *Studia Math.* **100**, 75–80.

Jiang, S. and Rokhlin, V. (2003). Second kind integral equations for the classical potential theory on open surfaces I: analytical apparatus, *J. Comput. Phys.* **191**, 40–74.

Jichang, K. and Debnath, L. (2000). On new generalizations of Hilbert's inequality and their applications, *J. Math. Analysis Appl.* **245**, 248–265.

Jin, Y. S. and Martin, A. (1964). Connection between the asymptotic behavior and the sign of the discontinuity in one-dimensional dispersion relations, *Phys. Rev.* **135**, B1369–B1374.

Jodeit, Jr., M., Kenig, C. E., and Shaw, R. K. (1983). "Integration" of singular integrals of integrable functions and extensions of Stein's $L \log L$ results, *Indiana U. Math. J.* **32**, 859–877.

Johnson, D. W. (1975). A Fourier series method for numerical Kramers-Kronig analysis, *J. Phys. A: Math. Gen.* **8**, 490–495.

Jolly, L. B. W. (1961). *Summation of Series* (New York: Dover).

Jones, D. S. (1965). Some remarks on Hilbert transforms, *J. Inst. Math. Appl.* **1**, 226–240.

Jones, D. S. (1982). *The Theory of Generalized Functions* (Cambridge: Cambridge University Press).

Jones, P. W. (1980). Factorization of A_p weights, *Ann. of Math.* **111**, 511–530.

Jones, P. W. (1994). Bilinear singular integrals and maximal functions. In *Linear and Complex Analysis Problem,* Book 3, Part I, V. P. Havin and N. K. Nikolski (Nikol'skiĭ), eds. (Berlin: Springer-Verlag), p. 414.

Jones, R. L. (1992). A remark on singular integrals with complex homogeneity, *Proc. Am. Math. Soc.* **114**, 763–768.

Jones, W. and March, N. H. (1973). *Theoretical Solid State Physics,* Vol. 2 (New York: Dover).

Joo, T. and Albrecht, A. C. (1993). Inverse transform in resonance Raman scattering: An iterative approach, *J. Phys. Chem.* **97**, 1262–1264.

Journé, J.-L. (1983). *Calderón-Zygmund Operators, Pseudo-Differential Operators and the Cauchy Integral of Calderón* (Berlin: Springer-Verlag).

Kaczmarz, S. (1931). Integrale vom Dini'schen typus, *Studia Math.* **3**, 189–199.

Kaczmarz, S. (1932). The divergence of certain integrals, *J. Lond. Math. Soc.* **7**, 218–222.

Kador, L. (1995). Kramers-Kronig relations in nonlinear optics, *Appl. Phys. Lett.* **66**, 2938–2939.

Kahanpää, L. and Mejlbro, L. (1984). Some new results on the Muckenhoupt conjecture concerning weighted norm inequalities connecting the Hilbert transform with the maximal function. In *Proceedings of the Second Finnish-Polish Summer School in Complex Analysis at Jyväskylä*, J. Ławrynowicz and O. Martio, eds. (Jyväskylä: Universität Jyväskylä, Mathematisches Institut).

Kak, S. C. (1968). Fourier, Laplace and Hilbert transforms, *Electron. Lett.* **4**, 396.

Kak, S. C. (1970). The discrete Hilbert transform, *Proc. IEEE Lett.* **58**, 585–586.

Kak, S. C. (1972). The finite discrete Hilbert transformation. In *Proceedings Fifth Hawaii International Conference on System Sciences*, A. Lew, ed. (North Hollywood, CA: Western Periodicals Company), pp. 124–126.

Kak, S. C. (1973). Hilbert transformation for discrete data, *Int. J. Electron.* **34**, 177–183.

Kak, S. (1977). The discrete finite Hilbert transform, *Indian J. Pure Appl. Math.* **8**, 1385–1390.

Kaneko, M. (1970). Estimates for the maximal function of Hardy-Littlewood and the maximal Hilbert transform with weighted measures, *Tôhoku Math. J.* **22**, 130–137.

Kaneko, M. and Yano, S. (1975). Weighted norm inequalities for singular integrals, *J. Math. Soc. Japan* **27**, 570–588.

Kanter, H. and Vernon, Jr. F. L. (1972). High-frequency response of Josephson point contacts, *J. Appl. Phys.* **43**, 3174–3183.

Kantorovich, L. V. and Akilov, G. P. (1982). *Functional Analysis*, 2nd edn (Oxford: Pergamon Press).

Kanwal, R. P. (1998). *Generalized Functions Theory and Technique*, 2nd edn (Boston, MA: Birkhäuser).

Karl, J. H. (1989). *An Introduction to Digital Signal Processing* (San Diego, CA: Academic Press).

Katz, N. H. and Pereyra, C. (1997). On the two weights problem for the Hilbert transform, *Rev. Mat. Iberoamericana* **13**, 211–243.

Katznelson, Y. (1976). *An Introduction to Harmonic Analysis* (New York: Dover).

Kaufmann, B. (1985). Dispersion relations and the phase problem for the scattering of electromagnetic radiation, *Acta Cryst.* **A41**, 152–155.

Kawasaki, T. (1971). Formal theory of nonlinear response to pulse fields, *Prog. Theor. Phys.* **46**, 1323–1336.

Kawata, T. (1936). On analytic functions regular in the half-plane (I), *Japan. J. Math.* **13**, 421–430.

Keefe, C. D. (2001). Curvefitting imaginary components of optical properties: restrictions on the lineshape due to causality, *J. Mol. Spectrosc.* **205**, 261–268.

Kellogg, O. (1904). Unstetigkeiten in den linear integralgleichungen, *Math. Ann.* **58**, 441–456.

Kenig, C. E., Ponce, G., and Vega, L. (1994). On the generalized Benjamin-Ono equation, *Trans. Am. Math. Soc.* **342**, 155–172.

Kerr, F. H. (1988). Namias' fractional Fourier transforms on L^2 and applications to differential equations, *J. Math. Analysis Appl.* **136**, 404–418.

Kestelman, H. (1960). *Modern Theories of Integration*, 2nd edn (New York: Dover).

Ketolainen, P., Peiponen, K.-E., and Karttunen, K. (1991). Refractive-indcx change due to F color centers in KCl_xBr_{1-x} mixed crystals, *Phys. Rev. B* **43**, 4492–4494.

Khuri, N. N. (1957). Analyticity of the Schrödinger scattering amplitude and nonrelativistic dispersion relations, *Phys. Rev.* **107**, 1148–1156.

Khvedelidze, B. V. (1977). The method of Cauchy-type integrals in the discontinuous boundary-value problems of the theory of holomorphic functions of a complex variable, *J. Sov. Math.* **7**, 309–317.

Kierat, W. and Sztaba, U. (2003). *Distributions, Integral Transforms and Applications* (London: Taylor & Francis).

Kikuchi, M. and Fukao, Y. (1976). Seismic return motion, *Phys. Earth Planet. Interiors* **12**, 343–349.

Kilbas, A. A. and Saigo, M. (2004). H-*Transforms Theory and Applications* (Boca Raton, FL: Chapman & Hall/CRC).

Kim, W.-J. (1995). L^p bounds for Hilbert transforms along convex curves, *J. Math. Analysis Appl.* **193**, 60–95.

Kimel, I. (1982a). Gaussian sum rules for optical rotation, *Phys. Rev. B* **25**, 6561–6569.

Kimel, I. (1982b). Sum rules for the reflectivity coefficients, *Phys. Lett. A* **88**, 62–64.

King, F. W. (1976a). Sum rules for the optical constants, *J. Math. Phys.* **17**, 1509–1514.

King, F. W. (1976b). Sum rules for the forward elastic low energy scattering of light, *J. Phys. B: Atom. Mol. Phys.* **9**, 147–156.

King, F. W. (1977). A Fourier series algorithm for the analysis of reflectance data, *J. Phys. C: Solid State Phys.* **10**, 3199–3204.

King, F. W. (1978a). Analysis of optical data by the conjugate Fourier-series approach, *J. Opt. Soc. Am.* **68**, 994–997.

King, F. W. (1978b). Inverse first moment for the weighted magnetoreflection, *Chem. Phys. Lett.* **56**, 568–570.

King, F. W. (1979). Dispersion relations and sum rules for the normal reflectance of conductors and insulators, *J. Chem. Phys.* **71**, 4726–4733.

King, F. W. (1980). Moments of the optical rotatory power and circular dichroism for an isotropic medium, *Phys. Rev. B* **21**, 4466–4469.

King, F. W. (1981). Integral inequalities for optical constants, *J. Math. Phys.* **22**, 1321–1323.

King, F. W. (1982). Some bounds for the absorption coefficient of an isotropic nonconducting medium, *Phys. Rev. B* **25**, 1381–1383.

King, F. W. (1991). Analysis of some integrals arising in the atomic three-electron problem, *Phys. Rev. A* **44**, 7108–7133.

King, F. W. (1999). Convergence accelerator approach for the evaluation of some three-electron integrals containing explicit r_{ij} factors, *Int. J. Quantum Chem.* **72**, 93–99.

King, F. W. (2002). Efficient numerical approach to the evaluation of Kramers–Kronig transforms, *J. Opt. Soc. Am. B* **19**, 2427–2436.

King, F. W. (2006). Alternative approach to the derivation of dispersion relations for optical constants, *J. Phys. A: Math. Gen.* **39**, 10 427–10 435.

King, F. W. (2007). Numerical evaluation of truncated Kramers-Kronig transforms, *J. Opt. Soc. Am. B* **24**, 1589–1595.

King, F. W., Dykema, K. J., and Lund, A. D. (1992). Calculation of some integrals for the atomic three-electron problem, *Phys. Rev. A* **46**, 5406–5416.

King, F. W., Smethells, G. J. Helleloid, G. T., and Pelzl, P. J. (2002). Numerical evaluation of Hilbert transforms for oscillatory functions: A convergence accelerator approach, *Comput. Phys. Commun.* **145**, 256–266.

Kinukawa, M. and Igari, S. (1961). Transformations of conjugate functions, *Tôhoku Math. J.* **13**, 274–280.

Kircheva, P. P. and Hadjichristov, G. B. (1994). Kramers-Kronig relations in FWM spectroscopy, *J. Phys. B: Atom. Mol. Opt. Phys.* **27**, 3781–3793.

Kishida, H., Hasegawa, T., Iwasa, Y., Koda, T., and Tokura, Y. (1993). Dispersion relation in the third-order electric susceptibility for polysilane film, *Phys. Rev. Lett.* **70**, 3724–3727.

Kittel, C. (1976). *Introduction to Solid State Physics*, 5th edn (New York: John Wiley).

Klein, A. and Zemach, C. (1959). Analytic properties of the amplitude for the scattering of a particle by a central potential, *Ann. Phys. (N.Y.)* **7**, 440–455.

Klucker, R. and Nielsen, U. (1973). Kramers-Kronig analysis of reflection data, *Comput. Phys. Commun.* **6**, 187–193.

Kober, H. (1942). A note on Hilbert's operator, *Bull. Am. Math. Soc.* **48**, 421–427.

Kober, H. (1943a). A note on Hilbert transforms, *J. Lond. Math. Soc.* **18**, 66–71.

Kober, H. (1943b). A note on Hilbert transforms, *Quart. J. Math. (Oxford)* **14**, 49–54.

Kober, H. (1957). *Dictionary of Conformal Representations* (New York: Dover).

Kober, H. (1964). An operator related to Hilbert transforms and to Dirichlet's integral, *J. Lond. Math. Soc.* **39**, 649–656.

Kober, H. (1967). A modification of Hilbert transforms, the Weyl integral and functional equations, *J. Lond. Math. Soc.* **42**, 42–50.

Kober, H. (1971). Some new properties of the Poisson operator, *J. Lond. Math. Soc.* (2) **3**, 640–644.

Kochneff, E. (1992). An analytic family of solutions to the heat equation, *Bull. Lond. Math. Soc.* **24**, 575–586.

Kochneff, E. (1995). The Riesz transforms of the Gaussian, *Illinois J. Math.* **39**, 140–142.

Kochneff, E., Sagher, Y., and Tan, R. (1992). On Bernstein's inequality, *Illinois J. Math.* **36**, 297–309.

Kochneff, E., Sagher, Y., and Tan, R. (1993). On a theorem of Akhiezer, *Illinois J. Math.* **37**, 489–501.

Kochneff, E., Sagher, Y., and Zhou, K. (1992). Homogeneous solutions of the heat equation, *J. Approx. Theory* **69**, 35–47.

Kogan, Sh. M. (1963). On the electrodynamics of weakly nonlinear media, *Sov. Phys. JETP* **16**, 217–219.

Kohlenberg, A. (1953). Exact interpolation of band-limited functions, *J. Appl. Phys.* **24**, 1432–1436.

Kohlmann, K. (1996). Corner detection in natural images based on the 2-D Hilbert transform, *Signal Process.* **48**, 225–234.

Koizumi, S. (1958a). On the singular integrals. I, *Proc. Japan Acad.* **34**, 193–198.

Koizumi, S. (1958b). On the singular integrals. III, *Proc. Japan Acad.* **34**, 594–598.

Koizumi, S. (1959a). On the singular integrals. V, *Proc. Japan Acad.* **35**, 1–6.

Koizumi, S. (1959b). On the Hilbert transform I, *J. Fac. Sci. Hokkaido Univ.* **14**, 153–224. Errataum: see p. 128 of the following reference.

Koizumi, S. (1960). On the Hilbert transform II, *J. Fac. Sci. Hokkaido Univ.* **15**, 93–130.

Kokilashvili, V. M. (1980). Singular integral operators in weighted spaces, *Colloq. Math. Soc. János Bolyai* **35**, 707–714.

Kokilashvili, V. and Krbec, M. (1991). *Weighted Inequalities in Lorentz and Orlicz Spaces* (Singapore: World Scientific).

Kokilashvili, V. and Meskhi, A. (1997). Weighted inequalities for Hilbert transforms and multiplicators of Fourier transforms, *J. Inequal. Appl.* **1**, 239–252.

Kolář, P. and Fischer, J. (1984). On the validity and practical applicability of derivative analyticity relations, *J. Math. Phys.* **25**, 2538–2544.

Kolm, P. and Rokhlin, V. (2001). Numerical quadratures for singular and hypersingular integrals, *Computers Math. Appl.* **41**, 327–352.

Kolmogoroff, A. (1925). Sur les fonctions harmoniques conjuguées les séries de Fourier, *Fund. Math.* **7**, 24–29.

Komori, Y. (2001). Weak l^1 estimates for the generalized discrete Hilbert transforms, *Far East J. Math. Sci.* **3**, 331–338.

Koosis, P. (1998). *Introduction to H_p Spaces*, 2nd edn, (Cambridge: Cambridge University Press).

Koppelman, W. and Pinus, J. D. (1959). Spectral representations for finite Hilbert transformations, *Math. Zeitschr.* **71**, 399–407.

Kozima, K., Suëtaka, W., and Schatz, P. N. (1966). Optical constants of thin films by a Kramers-Kronig method, *J. Opt. Soc. Am.* **56**, 181–184.

Krall, N. A. and Gerjuoy, E. (1960). Upper bound on total electron scattering cross sections in hydrogen, *Phys. Rev.* **120**, 143–144.

Kramer, H. P. (1973). The digital form of operators on band-limited functions, *J. Math. Analysis Appl.* **44**, 275–287.

Kramers, H. A. (1926). *Nature* **117**, 775 (untitled abstract).

Kramers, H. A. (1927). La diffusion de la lumière par les atomes, *Atti. del Congresso Internazionale dei Fisici*, **2**, 545–557. An English translation of this paper can be found in Ter Haar (1998).

Kramers, H. A. (1929). Die dispersion und absorption von Röntgenstrahlen, *Phys. Zeit.* **30**, 522–523.

Kramers, H. A. (1956). *Collected Scientific Papers* (Amsterdam: North-Holland).

Krantz, S. G. (1982). *Function Theory of Several Complex Variables* (New York: Wiley).

Krantz, S. G. (1987). What is several complex variables?, *Am. Math. Monthly* **94**, 236–256.

Krantz, S. G. (1990). *Complex Analysis: The Geometric Viewpoint* (Washington, D.C.: The Mathematical Association of America)

Krantz, S. G. (1999a). *A Panorama of Harmonic Analysis* (Washington, D.C.: The Mathematical Association of America).

Krantz, S. G. (1999b). *Handbook of Complex Variables* (Boston, MA: Birkhäuser).

Krantz, S. G. (2006). *Geometric Function Theory* (Boston, MA: Birkhäuser).

Kress, R. (1999). *Linear Integral Equations*, 2nd edn (Berlin: Springer-Verlag).

Kress, R. and Martensen, E. (1970). Anwendung der Rechteckregel auf die relle Hilberttransformation mit unendlichem intervall, *Zeit. Ang. Math. Mech.* **50**, T61–T64.

Kröger, K. (1975). Dispersion relation for complex reflectivity, *J. Opt. Soc. Am.* **65**, 1075.

Kronig, R. de L. (1926). On the theory of dispersion of X-rays, *J. Opt. Soc. Am. Rev. Sci. Instrum.* **12**, 547–557.

Kronig, R. de L. (1929). Dispersionstheorie im Röntgengebiet, *Phys. Zeit.* **30**, 521–522.

Kronig, R. de L. (1938). Zur theorie der relaxationserscheinungen, *Zeit. Tech. Physik* **19**, 509–516.

Kronig, R. (1942). Algemeene theorie der diëlectrische en magnetische verliezen, *Ned. T. Natuurk.* **9**, 402–409.

Kronig, R. (1946). A supplementary condition in Heisenberg's theory of elementary particles, *Physica* **12**, 543–544.

Krupnik, N. Ya. (1987). *Banach Algebras with Symbol and Singular Operators* (Basel: Birkhäuser Verlag).

Krylov, V. I. and Pal′cev, A. A. (1963). Numerical integration of functions having logarithmic and power singularities, *Vesci Akad. Navuk BSSR. Ser. Fiz. Tehn. Navuk*, (1), 14–23.

Kubo, R. (1957). Statistical-mechanical theory of irreversible processes. I. General theory and simple applications to magnetic and conduction problems, *J. Phys. Soc. Japan* **12**, 570–586.

Kubo, R. and Ichimura, M. (1972). Kramers-Kronig relations and sum rules, *J. Math. Phys.* **13**, 1454–1461.

Kuchiev, M. Yu. (1985). The power-type singularity in the electron-atom scattering, *J. Phys. B: Atom. Mol. Phys.* **18**, L579–L584.

Kuchiev, M. Yu. and Amusia, M. Ya. (1978). Dispersion relation for elastic-hydrogen atom forward scattering amplitude, *Phys. Lett. A* **66**, 195–197.

Kuethe, A. M. and Chow, C.-Y. (1986). *Foundations of Aerodynamics* (New York: John Wiley).

Kuijlaars, A. B. J. (1998). Best constants in one-sided weak-type inequalities, *Methods Appl. Analysis* **5**, 95–108.

Kuipers, L. and Robin, L. (1961). Sur quelques résultats concernant les polynômes de Jacobi et les fonctions de Legendre généralisées, *Nederl. Akad. Wetensch. Proc. Ser. A* (*Indag. Math.*) **23**, 256–259.

Kumar, S. (1980). A note on quadrature formulae for Cauchy principal value integrals, *J. Inst. Maths. Appl.* **26**, 447–451.

Kuzmenko, A. B. (2005). Kramers-Kronig constrained variational analysis of optical spectra, *Rev. Sci. Instrum.* **76**, 083108.

Lacey, M. T. (1998). On the bilinear Hilbert transform, *Doc. Math. J. Deut. Math. Ver.* (*Extra volume ICM*) **2**, 647–656.

Lacey, M. and Thiele, C. (1997a). L^p estimates for the bilinear Hilbert transform, *Proc. Natl. Acad. Sci. USA* **94**, 33–35.

Lacey, M. and Thiele, C. (1997b). L^p estimates on the bilinear Hilbert transform for $2 < p < \infty$, *Ann. of Math.* **146**, 693–724.

Lacey, M. and Thiele, C. (1998). On Calderón's conjecture for the bilinear Hilbert transform, *Proc. Natl. Acad. Sci. USA* **95**, 4828–4830.

Lacey, M. and Thiele C. (1999). On Calderón's conjecture, *Ann. of Math.* **149**, 475–496.

Ladopoulos, E. G. (2005). Nonlinear singular integral equations in elastodynamics by using Hilbert transformations, *Nonlinear Analysis Real World Appl.* **6**, 531–536.

Laeng, E. (2007). Remarks on the Hilbert transform and on some families of multiplier operators related to it, *Collect. Math.* **58**, 25–44.

Lamb, Jr., G. L. (1962). The attenuation of waves in a dispersive medium, *J. Geophys. Res.* **67**, 5273–5277.

Lamperti, J. (1959). A note on conjugate functions, *Proc. Am. Math. Soc.* **10**, 71–76.

Landau, E. (1909). *Handbuch der Lehre von der Verteilung der Primzahlen* (Leipzig: B. G. Teubner).

Landau, L. D. and Lifshitz, E. M. (1960). *Electrodynamics of Continuous Media* (Oxford: Pergamon Press).

Landau, L. D. and Lifshitz, E. M. (1965). *Quantum Mechanics*, 2nd edn (Oxford: Pergamon Press).

Landau, L. D. and Lifshitz, E. M. (1969). *Statistical Physics*, 2nd edn (Reading, MA: Addison-Wesley).

Landau, L. D., Lifshitz, E. M., and Pitaevskiĭ, L. P. (1984). *Electrodynamics of Continuous Media*, 2nd edn (Oxford: Pergamon Press).

Langley, R. S. (1986). On various definitions of the envelope of a random process, *J. Sound Vib.* **105**, 503–512.

Larkin, S., Anischenko, S. Kamychine, V., and Cowalenko, N. (1997). Use of autocorrelation function of the Josephson junction response signal for microwave signal spectrometry. In *The 1997 International Workshop on Superconductivity*. Hawaii (Battimore, MD: NASA Center for Aesospace Information), pp. 349–350.

Larkin, S., Anischenko, S. E., and Khabayev, P. V. (1994). Josephson junction spectrum analyzer for millimeter and submillimeter wavelengths. In *Proceedings of the Fourth International Conference and Exhibition: World Congress on Superconductivity*, Vol. II, Florida (Houston, TX: Johnson Space Center, NASA), pp. 585–593.

Larsen, R. (1971). *An Introduction to the Theory of Multipliers* (New York: Springer-Verlag).

Lasser, R. (1996). *Introduction to Fourier Series* (New York: Marcel Dekker).

Lauwerier, H. A. (1963). The Hilbert problem for generalized functions, *Arch. Ration. Mech. Analysis* **13**, 157–166.

Le Van Quyen, M., Foucher, J., Lachaux, J.-P., Rodriguez, E., Lutz, A., Martinerie, J., and Varela, F. J. (2001). Comparison of Hilbert transform and wavelet methods for the analysis of neuronal synchrony, *J. Neurosci. Methods* **111**, 83–98.

Lebedev, V. I. and Baburin, O. V. (1965). Calculation of the principal values, weights and nodes of the Gauss quadrature formulae of integrals, *USSR Comp. Math. & Math. Phys.* **5**, 81–92.

Lee, C. C., Lahham, M., and Martin, B. G. (1990). Experimental verification of the Kramers-Kronig relationship for acoustic waves, *IEEE Trans. Ultrasonics, Ferroelect., Freq. Control* **37**, 286–294.

Lee, M. H. (1996). Solving certain principal value integrals by reduction to the dilogarithm, *Physica A* **234**, 581–588.

Lee, M. H. and Sindoni, O. I. (1992). Dynamic response and bounds of the susceptibility of a semiclassical gas and Kramers-Kronig relations in optic-data inversion, *Phys. Rev. A* **46**, 3028–3036.

Lee, M. H. and Sindoni, O. I. (1997). Kramers-Kronig relations with logarithmic kernel and application to the phase spectrum in the Drude model, *Phys. Rev. E* **56**, 3891–3896.

Lee, S.-Y. (1995). Phase recovery and reconstruction of the Raman amplitude from the Raman excitation profile, *Chem. Phys. Lett.* **245**, 620–628.

Lee, S.-Y. and Yeo, R. C. K. (1994). A transform from absorption to Raman excitation profile. A time-dependent approach, *Chem. Phys. Lett.* **221**, 459–466.

Lee, S.-Y., Feng, Z. W., and Yeo, R. C. K. (1997). Phase recovery from the Raman excitation profile, time domain information and transform theory, *J. Raman Spectrosc.* **28**, 411–425.

Lee, Y. W. (1932). Synthesis of electronic networks by means of the Fourier transforms of Laguerre's functions, *J. Math. Phys. (MIT)* **11**, 83–113.

Lekishvili, M. M. (1978). Conjugate functions of several variables in the class Lip α, *Math. Notes* **23**, 196–203.

Leontovich, M. (1961). Generalization of the Kramers-Kronig formulas to media with spatial dispersion, *Sov. Phys. JETP* **13**, 634–637.

Lerche, I. (1986). Some singular, nonlinear, integral problems arising in physical problems, *Quart. Appl. Math.* **44**, 319–326.

Lerner, R. M. (1960). A matched filter detection system for complicated Doppler shifted signals, *IEEE Trans. Inform. Theory* **6**, 373–385.

Leuthold, P. E. (1974). Die bedeutung der Hilbert-transformation in der Nachrichtentechnik, *Sci. Electr.* **20**, 127–157.

Leveque, G. (1977). Reflectivity extrapolations in Kramers-Kronig analysis, *J. Phys. C: Solid State Phys.* **10**, 4877–4888.

Lévêque, G. (1986). Augmented partial sum rules for the analysis of optical data, *Phys. Rev. B* **34**, 5070–5072.

Levin, B. Ja. (1964). *Distribution of Zeros of Entire functions* (Providence, RI: American Mathematical Society).

Levin, D. (1973). Development of non-linear transformations for improving convergence of sequences, *Int. J. Comput. Math. B* **3**, 371–388.

Levinson, N. (1965). Simplified treatment of integrals of Cauchy type, the Hilbert problem and singular integral equations. Appendix: Poincaré-Bertrand formula, *SIAM Rev.* **7**, 474–502.

Levinson, N. and Redheffer, R. M. (1970). *Complex Variables* (San Francisco, CA: Holden-Day).

Lewin, L. (1961). On the resolution of a class of waveguide discontinuity problems by the use of singular integral equations, *IRE Trans. Microwave Theory Tech.* **MTT9**, 321–332.

L'Huillier, A. and Balcou, Ph. (1993). High-order harmonic generation in rare gases with a 1-ps 1053-nm laser, *Phys. Rev. Lett.* **70**, 774–777.

Liang, Z. and Marshall, A. G. (1990). Time domain (interferogram) and frequency-domain (absorption-mode and magnitude-mode) noise and precision in Fourier transform spectrometry, *Appl. Spectrosc.* **44**, 766–775.

Lichvár, P., Liška, M., and Galusek, D. (2002). What is the true Kramers-Kronig transform?, *Ceramics – Silikáty* **46**, 25–27.

Lifanov, I. K. (1996). *Singular Integral Equations and Discrete Vortices* (Utrecht: VSP)

Lighthill, M. J. (1958). *Introduction to Fourier Analysis and Generalized Functions* (Cambridge: Cambridge University Press).

Likharev, K. K. (1986). *Dynamics of Josephson Junctions and Circuits* (New York: Gordon and Breach Science Publishers).

Lindsey, III, J. F. and Doelitzsch, D. F. (1993). Standard broadcasting (amplitude modulation). In *The Electrical Engineering Handbook*, R. C. Dorf, ed. (Boca Raton, FL: CRC Press), pp. 1367–1379.

Littlewood, J. E. (1926). Mathematical notes (2): On a theorem of Kolmogoroff, *J. Lond. Math. Soc.* **1**, 229–231.

Littlewood, J. E. (1929). Mathematical notes (10): On a theorem of Zygmund, *J. Lond. Math. Soc.* **4**, 305–307.

Liu, H.-P. and Kosloff, D. D. (1981). Numerical evaluation of the Hilbert transform by the fast Fourier transform (FFT) technique, *Geophys. J. Roy. Astron. Soc.* **67**, 791–799.

Liu, Y.-C. and Okubo, S. (1967). Tests of new π-N superconvergent dispersion relations, *Phys. Rev. Lett.* **19**, 190–192.

Liu, Y.-C. and Okubo, S. (1968). New πN superconvergent sum rules and determination of P'-trajectory parameters, *Phys. Rev.* **168**, 1712–1714.

Löfström, J. (1983). A non-existence theorem for translation invariant operators on weighted L_p-spaces, *Math. Scand.* **53**, 88–96.

Logan, Jr., B. F. (1978). Theory of analytic modulation syst, *Bell Syst. Tech. J.* **57**, 491–576.

Logan, B. F. (1983a). Hilbert transform of a function having a bounded integral and a bounded derivative, *SIAM J. Math. Analysis* **14**, 247–248. Erratum: *SIAM J. Math. Analysis.* **14**, 845 (1983).

Logan, B. F. (1983b). An integral equation connected with the Jacobi polynomials, *SIAM J. Math. Analysis* **14**, 269–322.

Logan, B. F. (1984). On the eigenvaules of a certain integral equation, *SIAM J. Math. Analysis* **15**, 712–717.

Lohmann, A. W., Mendlovic, D., and Zalevsky, Z. (1996a). Fractional Hilbert transform, *Opt. Lett.* **21**, 281–283.

Lohmann, A. W., Mendlovic, D., and Zalevsky, Z. (1998). Fractional transformations in optics. In *Progress in Optics*, Vol. 38, E. Wolf, ed. (Amsterdam: Elsevier), pp. 263–342,

Lohmann, A. W., Ojeda-Castañeda, J., and Diaz-Santana, L. (1996b). Fractional Hilbert transform: optical implementation for 1D objects, *Opt. Mem. Neural Networks* **5**, 131–135.

Lohmann, A. W., Tepichín, E., and Ramírez, J. G. (1997). Optical implementation of the fractional Hilbert transform for two-dimensional objects, *Appl. Opt.* **36**, 6620–6626.

Longman, I. M. (1958). On the numerical evaluation of Cauchy principal values of integrals, *Math. Tables Aids Comp.* **12**, 205–207.

Loomis, L. H. (1946). A note on the Hilbert transform, *Bull. Am. Math. Soc.* **52**, 1082–1086.

Lorentz, H. A. (1952). *The Theory of Electrons* (New York: Dover).

Loughlin, P. J. (1998). Do bounded signals have bounded amplitudes?, *Multidimensional Syst. Signal Process.* **9**, 419–424.

Loughlin, P. J. and Tacer, B. (1997). Comments on the interpretation of instantaneous frequency, *IEEE Signal Process. Lett.* **4**, 123–125.

Love, E. R. (1977). Repeated singular integrals, *J. Lond. Math. Soc.* (2) **15**, 99–102.

Lovell, R. (1974). Application of Kramers-Kronig relations to the interpretation of dielectric data, *J. Phys. C: Solid State Phys.* **7**, 4378–4384.

Lovell, R. (1975) A reply to 'Applications of Kramers-Kronig relations,' *J. Phys. C: Solid State Phys.* **8**, L227.

Lowenthal, S. and Belvaux, Y. (1967). Observation of phase objects by optically processed Hilbert transform, *Appl. Phys. Lett.* **11**, 49–51.

Lubinsky, D. S. and Rabinowitz, P. (1984). Rates of convergence of Gaussian quadrature for singular integrands, *Math. Comp.* **43**, 219–242.

Lucarini, V. and Peiponen, K.-E. (2003). Verification of generalized Kramers-Kronig relations and sum rules on experimental data of third harmonic generation susceptibility on polymer, *J. Chem. Phys.* **119**, 620–627.

Lucarini, V., Saarinen, J. J., and Peiponen, K.-E. (2003a). Multiply subtractive Kramers-Krönig relations for arbitrary-order harmonic generation susceptibilities, *Optics Commun.* **218**, 409–414.

Lucarini, V., Saarinen, J. J., and Peiponen, K.-E. (2003b). Multiply subtractive generalized Kramers-Kronig relations: Application on third-harmonic generation susceptibility on polysilane, *J. Chem. Phys.* **119**, 11 095–11 098.

Lucarini, V., Bassani, F., Peiponen, K.-E., and Saarinen, J. J. (2003c). Dispersion theory and sum rules in linear and nonlinear optics, *La Rivista Del Nuovo Cimento* **26**, 1–120.

Lucarini, V., Ino, Y., Peiponen, K.-E., and Kuwata-Gonokami, M. (2005a). Detection and correction of the misplacement error in terahertz spectroscopy by application of singly subtractive Kramers-Kronig relations, *Phys. Rev. B* **72**, 125107.

Lucarini, V., Saarinen, J. J., Peiponen, K.-E., and Vartiainen, E. M. (2005b). *Kramers-Kronig Relations in Optical Materials Research* (Berlin: Springer-Verlag).

Ludwig, F., Menzel, J., Kaestner, A., Volk, M., and Schilling, M. (2001). THz-spectroscopy with $YBa_2Cu_3O_7$-Josephson junctions on $LaAlO_3$-bicrystals, *IEEE Trans. Appl. Supercond.* **11**, 586–588.

Lukashenko, T. P. (2006). Monotonicity and convexity of a conjugate function, *Moscow Univ. Math. Bull.* **61**, 12–17.

Lusin, N. (1913). Sur la convergence des séries trigonométriques de Fourier, *Comptes Rendus Acad. Sci.* **156**, 1655–1658.

Lusin, N. N. (1915). *Integral i Trigonometrich – eskii riâd* (*Integral and Trigonometric Series*) (Moscow: Tip. Lissnera i D. Sobko.).

Lyubarskii, Y. I. and Seip, K. (1997). Complete interpolating sequences for Paley-Wiener spaces and Muckenhoupt's (A_p) condition, *Rev. Mat. Iberoamericana* **13**, 361–376.

McBride, A. C. and Kerr, F. H. (1987). On Namias's fractional Fourier transforms, *IMA J. Appl. Math.* **39**, 159–175.

McCabe, J. H. (1974). A continued fraction expansion, with a truncation error estimate, for Dawson's integral, *Math. Comp.* **28**, 811–816.

McClellan, J. H. and Parks, T. W. (1972). Eigenvalue and eigenvector decomposition of the discrete Fourier transform, *IEEE Trans. Audio Electroacoustics* **20**, 66–74.

McClure, J. P. and Wong, R. (1978). Explicit error terms for asymptotic expansions of Stieltjes transforms, *Inst. J. Math. Appl.* **22**, 129–145.

McCormick, B. W. (1979). *Aerodynamics, Aeronautics, and Flight Mechanics* (New York: John Wiley).

MacDonald, J. R. (1952). Dielectric dispersion in materials having a distribution of relaxation times, *J. Chem. Phys.* **20**, 1107–1111.

MacDonald, J. R. and Brachman, M. K. (1956). Linear-system integral transform relations, *Rev. Mod. Phys.* **28**, 393–422.

McDowell, M. R. C. and Farmer, C. M. (1977). A note on the exchange amplitude for elastic forward electron-hydrogen-atom scattering. *J. Phys. B: Atom. Mol. Phys.* **10**, L565–L567.

McLean, W. and Elliott, D. (1988). On the *p*-norm of the truncated Hilbert transform, *Bull. Austral. Math. Soc.* **38**, 413–420.

McQuarrie, D. A. (2003). *Mathematical Methods for Scientists and Engineers* (Sausalito, CA: University Science Books).

Madych, W. R. (1990). On the correctness of the problem of inverting the finite Hilbert transform in certain aeroelastic models, *J. Integral Equations Appl.* **2**, 263–267.

Mandal, B. N. (2001). Integral equations and applications to water wave scattering problems. In *Geometry, Analysis and Applications*, R. S. Pathak, ed. (Singapore: World Scientific), pp. 389–404.

Mangulis, V. (1964). Kramers–Kronig or dispersion relations in acoustics, *J. Acoust. Soc. Am.* **36**, 211–212.

Mansfeld, F. and Kendig, M. W. (1999). Discussion of the paper "The application of the transform of Kramers-Kronig for computing the polarization resistance", *Materials Corros.* **50**, 475.

Marcinkiewicz, J. (1936). Sur les séries de Fourier, *Fund. Math.* **27**, 38–69.

Marcinkiewicz, J. (1939). Sur l'interpolation d'opérations, *Comptes Rendus Acad. Sci.* **208**, 1272–1273.

Marichev, O. I. (1983). *Handbook of Integral Transforms of Higher Transcendental Functions* (Chichester: Ellis Horward Ltd.).

Markushevich, A. I. (1966). *Entire Functions* (New York: American Elsevier Publishing)

Marsden, M. J., Richards, F. B., and Riemenschneider, S. D. (1975). Cardinal spline interpolation operators on l^p data, *Indiana Univ. Math. J.* **24**, 677–689. Erratum: *Indiana Univ. Math. J.* **25**, 919 (1976).

Marshall, A. G. (1982). Dispersion versus absorption (DISPA): Hilbert transforms in spectral line shape analysis. In *Fourier, Hadamard, and Hilbert Transforms in Chemistry*, A. G. Marshall, ed. (New York: Plenum Press), pp. 99–123.

Marshall, A. G. and Roe, D. C. (1978). Dispersion versus absorption: spectral line shape analysis for radiofrequency and microwave spectroscopy, *Analytical Chem.* **50**, 756–763.

Martin, P. C. (1967). Sum rules, Kramers–Kronig relations, and transport coefficients in charged systems, *Phys. Rev.* **161**, 143–155.

Martini, A. F., Menon, M. J., Paes, J. T. S., and Silva Neto, M. J. (1999). Differential dispersion relations and elementary amplitudes in a multiple diffraction model, *Phys. Rev. D* **59**, 116006.

Mashreghi, J. (2001). Hilbert transform of $\log|f|$, *Proc. Am. Math. Soc.* **130**, 683–688.

Mastroianni, G. and Occorsio, D. (1995). Interlacing properties of the zeros of the orthogonal polynomials and approximation of the Hilbert transform, *Computers Math. Appl.* **30**, 155–168.

Mastroianni, G. and Occorsio, D. (1996). Legendre polynomials of the second kind, Fourier series and Lagrange interpolation, *J. Comput. Appl. Math.* **75**, 305–327.

Mastroianni, G. and Occorsio, M. R. (1990). A method for computing the Tricomi transform, *Facta Universit. Ser. Math. Inform.* **5**, 109–114.

Mastroianni, G. and Themistoclakis, W. (2005). A numerical method for the generalized airfoil equation based on de la Vallée Poussin interpolation, *J. Comput. Appl. Math.* **180**, 71–105.

Matsuno, Y. (1980). Interaction of the Benjamin-Ono solitons, *J. Phys. A: Math. Gen.* **13**, 1519–1536.

Matsuno, Y. (1984). *Bilinear Transformation Method* (Orlando, FL: Academic Press).

Matsuno, Y. (1986). N-Soliton solutions for the sine-Hilbert equation, *Phys. Lett. A* **119**, 229–233.

Matsuno, Y. (1987a). Periodic problem for the sine-Hilbert equation, *Phys. Lett. A* **120**, 187–190.

Matsuno, Y. (1987b). Kinks of the sine-Hilbert equation and their dynamical motions, *J. Phys. A: Math. Gen.* **20**, 3587–3606.

Matsuno, Y. (1990). Linearization of a novel nonlinear integro-differential evolution equation and its exact solutions, *J. Phys. Soc. Japan* **59**, 1523–1525.

Matsuno, Y. (1991). Linearization of novel nonlinear diffusion equations with the Hilbert kernel and their exact solutions, *J. Math. Phys.* **32**, 120–126.

Matsuno, Y. (1995). Dynamics of interacting algebraic solitons, *Int. J. Mod. Phys B* **9**, 1985–2081.

Matsuno, Y. (1996). Stochastic Benjamin-Ono equation and its application to the dynamics of nonlinear random waves, *Phys. Rev. E* **54**, 6313–6322.

Matsuno, Y. and Kaup, D. J. (1997). Initial value problem of the linearized Benjamin-Ono equation and its applications, *J. Math. Phys.* **38**, 5198–5224.

Matsuoka, K. (1991). On the generalized Hilbert transforms in R^2 and the generalized harmonic analysis, *Tokyo J. Math.* **14**, 395–405.

Maximon, L. C. and O'Connell J. S. (1974). Sum rules for forward elastic photon scattering, *Phys. Lett. B* **48**, 399–402.

Melrose, D. B. and Stoneham, R. J. (1977). Generalized Kramers–Kronig formula for spatially dispersive media, *J. Phys. A: Math. Gen.* **10**, L17–L20.

Menon, M. J., Motter, A. E., and Pimentel, B. M. (1999). Differential dispersion relations with an arbitrary number of subtractions: a recursive approach, *Phys. Lett. B* **451**, 207–210.

Merzbacher, E. (1970). *Quantum Mechanics*, 2nd edn (New York: John Wiley).

Meyer, R. M. (1979). *Essential Mathematics for Applied Fields* (New York: Springer-Verlag).

Meyer, Y. and Coifman, R. (1997). *Wavelets Calderón-Zygmund and Multilinear Operators* (Cambridge: Cambridge University Press).

Mezincescu, G. A. (1985). Dispersion relations and sum rules for the dielectric function when $\epsilon(k, 0) < 0$, *Phys. Lett. A* **108**, 477–479.

Mihlin, S. G. (1950). Singular integral equations, *Am. Math. Soc. Trans. Number* **24**, 1–116.

Mikhlin, S. G. (1965). *Multidimensional Singular Integrals and Integral Equations* (Oxford: Pergamon Press).

Mikhlin, S. G. and Prössdorf, S. (1986). *Singular Integral Operators* (Berlin: Springer-Verlag).

Millane, R. P. (1994). Analytic properties of the Hartley transform and their applications, *Proc. IEEE* **82**, 413–428.

Miller, D. A. B., Seaton, C. T., Prise, M. E. and Smith, S. D. (1981). Band-gap-resonant nonlinear refraction in III-V semiconductors, *Phys. Rev. Lett.* **47**, 197–200.

Miller, K. S. (1970). *An Introduction to Advanced Complex Analysis* (New York: Dover).

Milton, G. W., Eyre, D. J., and Mantese, J. V. (1997). Finite frequency range Kramers Kronig relations: Bounds on the dispersion, *Phys. Rev. Lett.* **79**, 3062–3065.

Minerbo, G. (1971). Causality and analyticity in formal scattering theory, *Phys. Rev. D* **3**, 928–932.

Misell, D. L. and Greenaway, A. H. (1974a). An application of the Hilbert transform in electron microscopy: I. Bright-field microscopy, *J. Phys. D: Appl. Phys.* **7**, 832–855.

Misell, D. L. and Greenaway, A. H. (1974b). An application of the Hilbert transform in electron microscopy: II. Non-iterative solution in bright-field microscopy and the dark-field problem, *J. Phys. D: Appl. Phys.* **7**, 1660–1669.

Misell, D. L., Burge, R. E., and Greenaway, A. H. (1974). Phase determination from image intensity measurements in bright-field optics, *J. Phys. D: Appl. Phys.* **7**, L27–L30.

Mishnev, A. F. (1993). Discrete Hilbert transforms in crystallography, *Acta Cryst.* A**49**, 159–161.

Mishnev, A. F. (1996). Discrete Hilbert transforms and the phase problem in crystallography, *Acta Cryst.* A**52**, 629–633.

Mitrinović, D. S. (1970). *Analytic Inequalities* (New York: Springer-Verlag)

Mitrinović, D. S. and Kečkić, J. D. (1984). *The Cauchy Method of Residues* (Dordrecht: D. Reidel).

Mitrinović, D. S., Pečarić, J. E., and Fink, A. M. (1991). *Inequalities Involving Functions and Their Integrals and Derivatives* (Dordrecht: Kluwer).

Mobley, J., Waters, K. R., and Miller, J. G. (2005). Causal determination of acoustic group velocity and frequency derivative of attenuation with finite-bandwidth Kramers-Kronig relations, *Phys. Rev. E* **72**, 016604.

Moffitt, W. and Moscowitz, A. (1959). Optical activity in absorbing media, *J. Chem. Phys.* **30**, 648–660.

Mojahedi, M., Malloy, K. J., Eleftheriades, G. V., Woodley, J., and Chiao, R. Y. (2003). Abnormal wave propagation in passive media, *IEEE J. Selected Topics Quantum Electron.* **9**, 30–39.

Monegato, G. (1982). The numerical evaluation of one-dimensional Cauchy principal value integrals, *Computing* **29**, 337–354.

Monegato, G. (1998). Numerical resolution of the generalized airfoil equation with Possio kernel. In Tricomi's Ideas and Contemporary Applied Mathematics. Convegno Internazionale in Occasione del Centenario della Nascita di Francesco G. Tricomi (Rome, 28–29 November and Torino, 1–2 December, 1997), Atti Convegni Lincei **147** (Rome: Accademia Nazionale dei Lincei), pp. 103–121.

Monegato, G. and Sloan, I. H. (1997). Numerical solution of the generalized airfoil equation with a flap, *SIAM J. Numer. Analysis* **34**, 2288–2305.

Montgomery, H. L. and Vaughan, R. C. (1974). Hilbert's inequality, *J. Lond. Math. Soc.* (2) **8**, 73–82.

Moore, D. W. (1983). Resonances introduced by discretization, *IMA J. Appl. Math.* **31**, 1–11.

Morán, M. D. and Urbina, W. (1998). Invariant subspaces and commutant for the Gaussian Hilbert transform, *Acta Científica Venezolana* **49**, 102–105.

Morawitz, H. (1970). A numerical approach to principal value integrals in dispersion relations, *J. Comput. Phys.* **6**, 120–123.

Morlet, A. C. (1997). Some further results for a one-dimensional transport equation with nonlocal flux, *Commun. Appl. Analysis* **1**, 315–336.

Morse, P. M. and Feshbach, H. (1953). *Methods of Theoretical Physics*, Part I (New York: McGraw-Hill).

Moscowitz, A. (1962). Theoretical aspects of optical activity part one: small molecules. In *Advances in Chemical Physics*, Vol. 4, I. Prigogine, ed. (New York: Wiley Interscience) pp. 67–112.

Moscowitz, A. (1965). Two theorems useful for optical activity calculations. In *Modern Quantum Chemistry Istanbul Lectures, Part III*: *Action of Light and Organic Crystals*, O. Sinanoğlu, ed. (New York: Academic Press), pp. 31–44.

Moss, T. S. (1961). *Optical Properties of Semi-conductors*, 2nd edn (London: Butterworths).

Moss, W. F. (1983). The two-dimensional oscillating airfoil: A new implementation of the Galerkin method, *SIAM J. Numer. Analysis* **20**, 391–399.

Muckenhoupt, B. (1969). Mean convergence of Jacobi series, *Proc. Am. Math. Soc.* **23**, 306–310.

Muckenhoupt, B. (1970). Mean convergence of Hermite and Laguerre series. II, *Trans. Am. Math. Soc.* **147**, 433–460.

Muckenhoupt, B. (1972). Weighted norm inequalities for the Hardy maximal function, *Trans. Am. Math. Soc.* **165**, 207–226.

Muckenhoupt, B. (1974a). The equivalence of two conditions for weight functions, *Studia Math.* **49**, 101–106.

Muckenhoupt, B. (1974b). Weighted norm inequalities for classical operators. In *Linear Operators and Approximation II*, P. L. Butzer and B. Szökefalvi-Nagy, eds. (Basel: Birkhäuser Verlag), pp. 265–283.

Muckenhoupt, B. (1979). Weighted norm inequalities for classical operators. In *Harmonic Analysis in Euclidean Spaces, Proceedings of Symposia in Pure Mathematics*, Vol. 35, part 1, G. Weiss and S. Wainger, eds. (Providence, RI: American Mathematical Society), pp. 69–83.

Muckenhoupt, B. and Wheeden, R. L. (1976a). Weighted bounded mean oscillation and the Hilbert transform, *Studia Math. T.* **54**, 221–237.

Muckenhoupt, B. and Wheeden, R. L. (1976b). Two weight function norm inequalities for the Hardy-Littlewood maximal function and the Hilbert transform, *Studia Math. T.* **55**, 279–294.

Muckenhoupt, B. and Wheeden, R. L. (1977). Some weighted weak-type inequalities for the Hardy-Littlewood maximal function and the Hilbert transform, *Indiana Univ. Math. J.* **26**, 801–816.

Mukamel, S. (1995). *Principles of Nonlinear Optical Spectroscopy*. (New York: Oxford University Press).

Mukhtarov, Ch. K. (1979). Integral dispersion relations for the real part of the polarizability, the refractive index, and magneto-optical characteristics, *Sov. Phys. Dokl.* **24**, 991–993.

Murakami, T. and Corrington, M. S. (1948). Relation between amplitude and phase in electrical networks, *RCA Rev.* **9**, 602–631.

Musienko, T., Rudakov, V., and Solov'ev, L. (1989). On the application of Kramers-Kronig relations to media with spatial dispersion, *J. Phys.: Condens. Matter* **1**, 6745–6753.

Muskhelishvili, N. I. (1992). *Singular Integral Equations*, 2nd edn (New York: Dover).

Nabarro, F. R. N. (1947). Dislocations in a simple cubic lattice, *Proc. Phys. Soc. Lond.* **59**, 256–272.

Nabarro, F. R. N. (1997). Fifty-year study of the Peierls-Nabarro stress, *Mat. Sci. Eng. A* **234**, 67–76.

Nabighian, M. N. (1984). Towards a three-dimensional automatic interpretation of potential field data via generalized Hilbert transforms: Fundamental relations, *Geophys.* **49**, 780–786.

Nadir, M. (2003). On the approximation of the Hilbert transform, *Far East J. Appl. Math.* **12**, 71–78.

Nagel, A. and Wainger, S. (1976). Hilbert transforms associated with plane curves, *Trans. Am. Math. Soc.* **223**, 235–252.

Nagel, A. and Wainger, S. (1977). L^2 boundedness of Hilbert transforms along surfaces and convolution operators homogeneous with respect to a parameter group, *Am. J. Math.* **99**, 761–785.

Nagel, A., Rivière, N., and Wainger, S. (1974). On Hilbert transforms along curves, *Bull. Am. Math. Soc.* **80**, 106–108.

Nagel, A., Riviere, N. M., and Wainger, S. (1976). On Hilbert transforms along curves. II, *Am. J. Math.* **98**, 395–403.

Nagel, A., Stein, E. M., and Wainger, S. (1979). Hilbert transforms and maximal functions related to variable curves. In *Harmonic Analysis in Euclidean Spaces, Proceedings of Symposia in Pure Mathematics*, Vol. 35, part 1, G. Weiss and S. Wainger, eds. (Provides, RI: American Mathematical Society), pp. 95–98.

Nagel, A., Vance, J. Wainger, S., and Weinberg, D. (1983). Hilbert transforms for convex curves, *Duke Math. J.* **50**, 735–744.

Nagel, A., Vance, J. Wainger, S., and Weinberg D. (1986). The Hilbert transform for convex curves in R^n, *Am. J. Math.* **108**, 485–504.

Nahin, P. J. (2006). *Dr. Euler's Fabulous Formula Cures Many Mathematical Ills* (Princeton, NJ: Princeton University Press).

Nakajima, N. (1988). Phase retrieval using the logarithmic Hilbert transform and the Fourier-series expansion, *J. Opt. Soc. Am. A* **5**, 257–262.

Nakano, J. and Tagami, S. (1988). Delay estimation by a Hilbert transform method, *Austral. J. Stat.* **30**, 217–227.

Namias, V. (1980). The fractional order Fourier transform and its application to quantum mechanics, *J. Inst. Math. Appl.* **25**, 241–265.

Nash, P. L., Bell, R. J., and Alexander R. (1995). On the Kramers-Kronig relation for the phase spectrum, *J. Mod. Optics* **42**, 1837–1842.

Natarajan, A. and Mohankumar, N. (1995). A comparison of some quadrature methods for approximating Cauchy principal value integrals, *J. Comput. Phys.* **116**, 365–368.

Nayfeh, A. H. and Mook, D. T. (1979). *Nonlinear Oscillations* (New York: John Wiley).

Nazarov, F. and Treil, S. (1996). The weighted norm inequalities for Hilbert transform are now trivial, *C. R. Acad. Sci. Paris* (*Série I*), **323**, 717–722.

Needham, T. (1997). *Visual Complex Analysis* (Oxford: Clarendon Press).

Nehari, Z. (1975). *Conformal Mapping* (New York: Dover).

Neri, U. (1971). *Singular Integrals*, Lecture Notes in Mathematics, Vol. 200, (Berlin: Springer-Verlag).

Neufeld, J. D. and Andermann, G. (1972). Kramers-Kronig dispersion-analysis method for treating infrared transmittance data, *J. Opt. Soc. Am.* **62**, 1156–1162.

Nevai, P. (1990). Hilbert transforms and Lagrange interpolation, *J. Approx. Theory* **60**, 360–363.

Nevanlinna, R. (1922). Asymptotische entwicklungen beschränkter funktionen und das Stieltjessche momentenproblem, *Ann. Acad. Sci. Fenn. A* **18**, 1–53.

Nevanlinna, R. (1970). *Analytic Functions* (New York: Springer-Verlag).

Newell, A. C. (1968). An alternative proof of the Poincaré-Bertrand formula for real integrals and its generalization to *n* dimensions, *J. Math. Analysis Appl.* **24**, 149–155.

Nho, W. and Loughlin, P. J. (1999). When is instantaneous frequency the average frequency at each time?, *IEEE Signal Process. Lett.* **6**, 78–80.

Nicholson, J. W. (1911). Notes on Bessel functions, *Quart. J. Pure Appl. Math.* **42**, 216–224.

Nickel, K. (1951). Lösung eines integralgleichungssystems aus der tragflügeltheorie, *Math. Zeitschr.* **54**, 81–96.

Nickel, K. (1953). Lösung von zwei verwandten integralgleichungssystemen, *Math. Zeitschr.* **58**, 49–62.

Nieto-Vesperinas, M. (1980). Dispersion relations in two-dimensions: application to the phase problem, *Optik* **56**, 377–384.

Nikulin, A. Yu. (1997). Unambiguous phase determination using a new X-ray interferometer on separated crystals, *Phys. Lett. A* **229**, 387–391.

Nikulin, A. Yu., Zaumseil, P., and Petrashen, P. V. (1996). Unambiguous determination of crystal-lattice strains in epitaxially grown SiGe/Si multilayers, *J. Appl. Phys.* **80**, 6683–6688.

Nishijima, K. (1974). *Fields and Particles* (New York: Benjamin).

Nitsche, R. and Fritz, T. (2004). Determination of model-free Kramers-Kronig consistent optical constants of thin absorbing films from just one spectral measurement: Application to organic semiconductors, *Phys. Rev. B* **70**, 195432.

Noda, I. (2000). Determination of the two-dimensional correlation spectra using the Hilbert transform, *Appl. Spectrosc.* **54**, 994–999.

Noether, F. (1921). Über eine klasse singulärer integralgleichungen, *Math. Ann.* **82**, 42–63.

Nozières, P. and Pines, D. (1958). A dielectric function of the many body problem: application to the free electron gas, *Nuovo Cimento* **9**, 470–490.

Nozières, P. and Pines, D. (1959). Electron interaction in solids. Characteristic energy loss spectrum, *Phys. Rev.* **113**, 1254–1267.

Nozières, P. and Pines, D. (1999). *The Theory of Quantum Liquids* (Cambridge MA: Persus).

Nussenzveig, H. M. (1972). *Causality and Dispersion Relations* (New York: Academic Press).

Nuttall, A. H. (1966). On the quadrature approximation to the Hilbert transform of modulated signals, *Proc. IEEE* **54**, 1458–1459.

Oden, J. T. (1979). *Applied Functional Analysis* (Englewood Cliffs, NJ: Prentice-Hall).

O'Donnell, M., Jaynes, E. T., and Miller, J. G. (1978). General relationships between ultrasonic attenuation and dispersion, *J. Acoust. Soc. Am.* **63**, 1935–1937.

O'Donnell, M., Jaynes, E. T., and Miller, J. G. (1981). Kramers-Kronig relationship between ultrasonic attenuation and phase velocity, *J. Acoust. Soc. Am.* **69**, 696–701.

Ohta, K. and Ishida, H. (1988). Comparison among several numerical integration methods for Kramers-Kronig transformation, *Appl. Spectrosc.* **42**, 952–957.

Okada, S. (1992a). The Poincaré-Bertrand formula for the Hilbert transform. *Miniconference on Probality and Analysis* (Sydney, 24–26 July, 1991), Proceedings of the Centre for Mathematics and its Applications, Australian National University **29**, I. Doust and B. Jefferies, eds. (Canberra: Australian National University).

Okada, S. (1992b). Finite Hilbert transforms and compactness, *Bull. Austral. Math. Soc.* **46**, 475–478.

Okada, S. and Elliott, D. (1991). The finite Hilbert transform in $\mathscr{L}^2$, *Math. Nachr.* **153**, 43–56.

Okada, S. and Elliott, D. (1994). Hölder continuous functions and the finite Hilbert transform, *Math. Nachr.* **169**, 219–233.

Okikiolu, G. O. (1965a). On integral operators associated with the Hilbert transform, *Math. Zeitschr.* **90**, 54–70.

Okikiolu, G. O. (1965b). A generalization of the Hilbert transform, *J. Lond. Math. Soc.* **40**, 27–30.

Okikiolu, G. O. (1966). *n*th order integral operators associated with the Hilbert transforms, *Pacific J. Math.* **18**, 343–360.

Okikiolu, G. O. (1967a). A class of integral operators associated with the Hilbert transform, *Proc. Lon. Math. Soc. (Series 3)* **17**, 342–354.

Okikiolu, G. O. (1967b). On certain extensions of the Hilbert operator, *Math. Ann.* **169**, 315–327.

Okikiolu, G. O. (1971). *Aspects of the Theory of Bounded Integral Operators in L^p-Spaces* (London: Academic Press).

Okubo, S. (1974). Theory of self-reproducing kernel and dispersion inequalities, *J. Math. Phys.* **15**, 963–973.

Olejniczak, K. J. (1996). The Hartley transform. In *The Transforms and Applications Handbook*, A. D. Poularikas, ed. (Boca Raton, FL: CRC Press), pp. 463–629.

Oleszkiewicz, K. (1993). An elementary proof of Hilbert's inequality, *Am. Math. Monthly* **100**, 276–280.

Olhede, S. and Walden, A. T. (2004). The Hilbert spectrum via wavelet projections, *Proc. Roy. Soc. Lond. A* **460**, 955–975.

Olkkonen, H. (1990). Fast sinc interpolation of digitized signals using the Hilbert transform, *J. Biomed. Eng.* **12**, 531–532.

Olmstead, W. E. and Raynor, S. (1964). Depression of an infinite liquid surface by an incompressible gas jet, *J. Fluid Mech.* **19**, 561–576.

Olver, F. W. J. (1997). *Asymptotics and Special Functions* (Wellesley, MA: A K Peters).

O'Neil, R. and Weiss, G. (1963). The Hilbert transform and rearrangement of functions, *Studia Math.* **23**, 189–198.

Onishchuk, O. V., Popov, G. Ya., and Farshait, P. G. (1986). On the singularities of contact forces in the bending of plates with fine inclusions, *Appl. Math. Mech. (USSR)* **50**, 219–226.

Ono, H. (1975). Algebraic solitary waves in stratified fluids, *J. Phys. Soc. Japan* **39**, 1082–1091.

Oosterhoff, L. J. (1965). Interaction of light with matter and theory of optical rotatory power. In *Modern Quantum Chemistry Istanbul Lectures, Part III: Action of Light and Organic Crystals*, O. Sinanoğlu, ed. (New York: Academic Press), pp. 5–29.

Oppenheim, A. V., Schafer, R. W., and Buck J. R. (1999). *Discrete-Time Signal Processing* (Upper Saddle River, NJ: Prentice-Hall).

Oppenheim, A. V., Willsky, A. S., and Nawab, S. H. (1997). *Signals and Systems* (Upper Saddle River, NJ: Prentice Hall).

Ortner, N. (2004). On some contributions of John Horváth to the theory of distributions, *J. Math. Analysis Appl.* **297**, 353–383.

Orton, M. (1973). Hilbert transforms, Plemelj relations, and Fourier transforms of distributions, *SIAM J. Math. Analysis* **4**, 656–670.

Orton, M. (1977). Hilbert boundary value problems — a distributional approach, *Proc. Roy. Soc. Edinburgh* **76A**, 193–208.

Orton, M. (1979). Distributions and singular integral equations of Carleman type, *Appl. Analysis* **9**, 219–231.

Osgood, W. F. (1966). *Topics in the Theory of Functions of Several Complex Variables* (New York: Dover).

Oskolkov, K. (1998). Schrödinger equation and oscillatory Hilbert transforms of second degree, *J. Fourier Analysis Appl.* **4**, 341–356.

Oswald, J. R. V. (1956). The theory of analytic band-limited signals applied to carrier systems, *IRE Trans. Circuit Theory* **3**, 244–251.

Ozaktas, H. M., Arikan, O., and Kutay, M. A. (1996). Digital computation of the fractional Fourier transform, *IEEE Trans. Signal Process.* **44**, 2141–2150.

Ozaktas, H. M., Zalevsky, Z., and Kutay, M. A. (2001). *The Fractional Fourier Transform* (Chichester: John Wiley).

Padala, S. K. and Prabhu, K. M. M. (1997). Systolic arrays for the discrete Hilbert transform, *IEE Proc.-Circuits Devices Syst.* **144**, 259–264.

Page, C. H. (1955). *Physical Mathematics* (Princeton, NJ: D. Van Nostrand).

Paget, D. F. and Elliott, D. (1972). An algorithm for the numerical evaluation of certain Cauchy principal value integrals, *Numer. Math.* **19**, 373–385.

Pake, G. E. and Purcell, E. M. (1948). Line shapes in nuclear paramagnetism, *Phys. Rev.* **74**, 1184–1188. Erratum: *Phys. Rev.* **75**, 534 (1949).

Paley, R. E. A. C. and Wiener, N. (1933). Notes on the theory and applications of Fourier transforms. I-II, *Trans. Am. Math. Soc.* **35**, 348–355.

Paley, R. E. A. C. and Wiener, N. (1934). *Fourier Transforms in the Complex Domain* (Providence, RI: American Mathematical Society).

Palmer, K. F., Williams, M. Z., and Budde, B. A. (1998). Multiple subtractive Kramers-Kronig analysis of optical data, *Appl. Opt.* **37**, 2660–2673.

Pandey, J. N. (1982). An extension of the Gel'fand-Shilov technique for Hilbert transforms, *Appl. Analysis* **13**, 279–290.

Pandey, J. N. (1983). The Hilbert transform of Schwartz distributions, *Proc. Am. Math. Soc.* **89**, 86–90.

Pandey, J. N. (1996). *The Hilbert Transform of Schwartz Distributions and Applications* (New York: John Wiley & Sons).

Pandey, J. N. (1997). The Hilbert transform of periodic distributions, *Integral Transforms Spec. Funct.* **5**, 117–142.

Pandey, J. N. (2001). The Hilbert transform of almost periodic functions and distributions. In *Geometry, Analysis and Applications*, R. S. Pathak, ed. (Singapore: World Scientific). 227–243.

Pandey, J. N. (2004). The Hilbert transform of almost periodic functions and distributions, *J. Computat. Analysis Appl.* **6**, 199–210.

Pandey, J. N. and Chaudhry, M. A. (1983). The Hilbert transform of generalized functions and applications, *Can. J. Math.* **35**, 478–495.

Pandey, J. N. and Hughes, E. (1976). An approximate Hilbert transform and its inversion, *Tôhoku Math. J.* **28**, 497–509.

Pandey, J. N. and Singh, O. P. (1991). On the p-norm of the truncated n-dimensional Hilbert transform, *Bull. Austral. Math. Soc.* **43**, 241–250.

Panman, P. (1999). The ergodic Hilbert transform on the weighted spaces $L_p(G,w)$, *Rocky Mountain J. Math.* **29**, 317–334.

Papadopoulos, S. F. (1999). A note on the M. Riesz theorem for the conjugate function, *Bull. Pol. Acad. Sci. Math.* **47**, 283–288.

Papoulis, A. (1962). *The Fourier Integral and its Applications* (New York: McGraw-Hill)

Papoulis, A. (1965). *Probability, Random Variables, and Stochastic Processes* (New York: McGraw-Hill).

Papoulis, A. (1973). Hilbert transform of periodic functions, *IEEE Trans. Circuit Theory* **20**, 454.

Parke, S. (1965). The Kramers-Kronig relations and their application to the dielectric properties of glasses. In *Physics of Non-Crystalline Solids*, J. A. Prins, ed. (Amsterdam: North-Holland), pp. 417–425.

Parker, P. J. and Anderson, B. D. O. (1990). Hilbert transforms from interpolation data, *Math. Control Signals Syst.* **3**, 97–124.

Parris, D. and Van Der Walt, S. J. (1975). A new numerical method for evaluating the Kramers-Kronig transformation, *Analytical Biochem.* **68**, 321–327.

Patapoff, T. W., Turpin, P.-Y., and Peticolis, W. L. (1986). The use of a Kramers-Kronig transform technique to calculate the resonance Raman excitation profiles of anharmonic molecular vibrations, *J. Phys. Chem.* **90**, 2347–2351.

Pathak, R. S. (1997). *Integral Transforms of Generalized Functions and Their Applications* (Amsterdam: Gordon and Breach).

Paveri-Fontana, S. L. and Zweifel, P. F. (1994). The half-Hartley and the half-Hilbert transform, *J. Math. Phys.* **35**, 2648–2656. Erratum: *J. Math. Phys.* **35**, 6226 (1994).

Pei, S.-C. and Ding, J.-J. (2001). Relations between fractional operations and time-frequency distributions, and their applications, *IEEE Trans. Signal Process* **49**, 1638–1655.

Pei, S.-C. and Jaw, S.-B. (1989). Computation of discrete Hilbert transform through fast Hartley transform, *IEEE Trans. Circuits Syst.* **36**, 1251–1252.

Pei, S.-C. and Jaw, S.-B. (1990). The analytic signal in the Hartley transform domain, *IEEE Trans. Circuits Syst.* **37**, 1546–1548.

Pei, S.-C. and Wang, P.-H. (2001). Analytical design of maximally flat FIR fractional Hilbert transformers, *Signal Process.* **81**, 643–661.

Pei, S.-C. and Yeh, M.-H. (1997). Improved discrete fractional Fourier transform, *Opt. Lett.* **22**, 1047–1049.

Pei, S.-C. and Yeh, M.-H. (1998). Discrete fractional Hilbert transform, *Proceedings of the 1998 IEEE International Symposium on Circuits and Systems, ISCAS '98*, **4**, 506–509.

Pei, S.-C. and Yeh, M.-H. (2000). Discrete fractional Hilbert transform, *IEEE Trans. Circuits Systems II Analog Digital Signal Process.* **47**, 1307–1311.

Pei, S.-C., Yeh, M.-H., and Tseng, C.-C. (1999). Discrete fractional Fourier transform based on orthogonal projections, *IEEE Trans. Signal Process.* **47**, 1335–1348.

Pei, S.-C., Tseng, C.-C., Yeh, M.-H., and Shyu, J.-J. (1998). Discrete fractional Hartley and Fourier transforms, *IEEE Trans. Circuits Systems II: Analog Digital Signal Process.* **45**, 665–675.

Peierls, R. (1940). The size of a dislocation, *Proc. Phys. Soc. Lond.* **52**, 34–37.

Peiponen, K.-E. (1985). A simple derivation of some sum rules for the powers of the optical constants of insulators, *Lett. Nuovo Cimento* **44**, 445–448.

Peiponen, K.-E. (1987a). On the derivation of sum rules for physical quantities and their applications for linear and non-linear optical constants, *J. Phys. C: Solid State Phys.* **20**, 2785–2788.

Peiponen, K.-E. (1987b). Sum rules for the nonlinear susceptibilities in the case of sum frequency generation, *Phys. Rev. B* **35**, 4116–4117.

Peiponen, K.-E. (1987c). Sum rules for non-linear susceptibilities in the case of difference-frequency generation, *J. Phys. C: Solid State Phys.* **20**, L285–L286.

Peiponen, K.-E. (1988). Nonlinear susceptibilities as a function of several complex angular-frequency variables, *Phys. Rev. B* **37**, 6463–6467.

Peiponen, K.-E. (2001). On the dispersion theory of meromorphic optical constants, *J. Phys. A: Math. Gen.* **34**, 6525–6530.

Peiponen, K.-E. (2005). Multiply subtractive Kramers-Kronig relations for impedance function of concrete, *Cement Conrete Res.* **35**, 1435–1437.

Peiponen, K.-E. and Hämäläinen, R. M. K. (1986). On the effective field and optical constants of alkali halides containing *F* colour centers, *Nuovo Cimento* **7 D**, 371–378.

Peiponen, K.-E. and Saarinen, J. J. (2002). Dispersion theory of meromorphic total reflectivity, *Phys. Rev. A* **65**, 063810.

Peiponen, K.-E. and Vartiainen, E. M. (1991). Kramers-Kronig relations in optical data inversion, *Phys. Rev. B* **44**, 8301–8303.

Peiponen, K.-E. and Vartiainen, E. M. (2006). Dispersion theory of the reflectivity of *s*-polarized and *p*-polarized light, *J. Opt. Soc. Am. B* **23**, 114–119.

Peiponen, K.-E., Saarinen, J. J., and Svirko, Y. (2004). Derivation of general dispersion relations and sum rules for meromorphic nonlinear optical spectroscopy, *Phys. Rev. A* **65**, 043818.

Peiponen, K.-E., Vartiainen, E. M., and Asakura, T. (1992). Sum rules for non-linear optical constants, *J. Phys.: Condens. Matter* **4**, L299–L301.

Peiponen, K.-E., Vartiainen, E. M., and Asakura, T. (1996). Phase retrieval in optical spectroscopy: resolving optical constants from power spectra, *Appl. Spectrosc.* **50**, 1283–1289.

Peiponen, K.-E., Vartiainen, E. M., and Asakura, T. (1997a). Complex analysis in dispersion theory, *Opt. Rev.* **4**, 433–441.

Peiponen, K.-E., Vartiainen, E. M., and Asakura, T. (1997b). Dispersion relations and phase retrieval in optical spectroscopy. In *Progress in Optics* **37**, 57–94.

Peiponen, K.-E., Vartiainen, E. M., and Asakura, T. (1997c). Dispersion theory and phase retrieval of meromorphic total susceptibility, *J. Phys.: Condens. Matter* **9**, 8937–8943.

Peiponen, K.-E., Vartiainen, E. M., and Asakura, T. (1998). Dispersion theory of effective meromorphic nonlinear susceptibilities of nanocomposites, *J. Phys.: Condens. Matter* **10**, 2483–2488.

Peiponen, K.-E., Vartiainen, E. M., and Asakura, T. (1999). *Dispersion, Complex Analysis and Optical Spectroscopy* (Berlin: Springer).

Peiponen, K.-E., Vartiainen, E. M., and Tsuboi, T. (1990). Sum rules for the squared modulus of the nonlinear Raman susceptibility, *Phys. Rev. A* **41**, 527–530.

Peiponen, K.-E., Lucarini, V., Saarinen, J. J., and Vartiainen, E. (2004a). Kramers-Kronig relations and sum rules in nonlinear optical spectroscopy, *Appl. Spectrosc.* **58**, 499–509.

Peiponen, K.-E., Lucarini, V., Vartiainen, E. M., and Saarinen, J. J. (2004b). Kramers-Kronig relations and sum rules of negative refractive index media, *Eur. Phys. J. B* **41**, 61–65.

Peiponen, K.-E., Gornov, E., Svirko, Y., Ino, Y., Kuwata-Gonokami, M., and Lucarini, V. (2005). Testing the validity of terahertz reflection spectra by dispersion relations, *Phys. Rev. B* **72**, 245109.

Pełczynski, A. (1985). Norms of classical operators in function spaces, *Astérisque* No. 131, 137–162.

Pelzl, P. J. and King, F. W. (1998). Convergence accelerator approach for the high-precision evaluation of three-electron correlated integrals, *Phys. Rev. E*, **57**, 7268–7273.

Pelzl, P. J., Smethells, G. J., and King, F. W. (2002). Improvements on the application of convergence accelerators for the evaluation of some three-electron atomic integrals, *Phys. Rev. E*, **65**, 036707.

Peřina, J. (1985). *Coherence of Light*, 2nd edn (Dordrecht: D. Reidel Publishing Company).

Perry, P. A. and Brazil, T. J. (1998). Forcing causality on S-parameter data using the Hilbert transform, *IEEE Microwave Guided Wave Lett.* **8**, 378–380.

Petermichl, S. and Wittwer, J. (2002). A sharp estimate for the weighted Hilbert transform via Bellman functions, *Michigan Math. J.* **50**, 71–87.

Peters, A. S. (1963). A note on the integral equation of the first kind with a Cauchy kernel, *Comm. Pure Appl. Math.* **16**, 57–61.

Peters, A. S. (1968). Abel's equation and the Cauchy integral equation of the second kind, *Comm. Pure Appl. Math.* **21**, 51–65.

Peters, J. M. H. (1995). A beginner's guide to the Hilbert transform, *Int. J. Math. Educ. Sci. Technol.* **26**, 89–106.

Petersen, K. (1983a). Another proof of the existence of the ergodic Hilbert transform, *Proc. Am. Math. Soc.* **88**, 39–43.

Petersen, K. (1983b). *Ergodic Theory*. (Cambridge: Cambridge University Press).

Peterson, C. W. and Knight, B. W. (1973). Causality calculations in the time domain: an efficient alternative to the Kramers-Kronig method, *J. Opt. Soc. Am.* **63**, 1238–1242.

Pfaffelhuber, E. (1971a). Generalized impulse response and causality, *IEEE Trans. Circuit Theory* **CT-18**, 218–223.

Pfaffelhuber, E. (1971b). Sampling series for band-limited generalized functions, *IEEE Trans. Inform. Theory* **IT-17**, 650–654.

Philipp, H. R. and Ehrenreich, H. (1964). Optical constants in the X-ray range, *J. Appl. Phys.* **35**, 1416–1419.

Philipp, H. R. and Taft, E. A. (1959). Optical constants of germanium in the region 1 to 10 ev, *Phys. Rev.* **113**, 1002–1005.

Philipp, H. R. and Taft, E. A. (1964). Kramers-Kronig analysis of reflectance data for diamond, *Phys. Rev.* **136**, A1445–A1448. The lead author's name is misspelled on the title page as Phillip.

Phillips, K. (1967). The maximal theorems of Hardy and Littlewood, *Am. Math. Monthly* **74**, 648–660.

Phillips, S. C., Gledhill, R. J., Essex, J. W., and Edge, C. M. (2003). Application of the Hilbert-Huang transform to the analysis of molecular dynamics simulations, *J. Phys. Chem. A* **107**, 4869–4876.

Phong, D. H. and Stein, E. M. (1986). Hilbert integrals, singular integrals, and Radon transforms II, *Invent. Math.* **86**, 75–113.

Pichorides, S. K. (1972). On the best values of the constants in the theorems of M. Riesz, Zygmund and Kolmogorov, *Studia Math.* **46**, 165–179.

Pichorides, S. K. (1975a). Une propriété de la transformée de Hilbert, *C. R. Acad. Sci. Paris* **280**, 1197–1199.

Pichorides, S. K. (1975b). On the conjugate of bounded functions, *Bull. Am. Math. Soc.* **81**, 143–144.

Pick, L. (1994). Weighted estimates for the Hilbert transform of odd functions, *Georgian Math. J.* **1**, 77–97.

Piessens, R. (1970). Numerical evaluation of Cauchy principal values of integrals, *BIT* **10**, 476–480.

Pilipović, S. (1987). Hilbert transformation of Beurling ultradistributions, *Rend. Sem. Mat. Univ. Padov* **77**, 1–13.

Pilipović, S. and Sad, N. (1990). A topology of the Hilbert transform of Beurling ultradifferentiable functions, *Glasnik Mat.* **25**, 95–102.

Pinsky, M. (2001). A generalized Kolmogorov inequality for the Hilbert transform, *Proc. Am. Math. Soc.* **130**, 753–758.

Pipkin, A. C. (1991). *A Course on Integral Equations*. (New York: Springer-Verlag).

Pippard, A. B. (1986). *Response and Stability: An Introduction to the Physical Theory*. (Cambridge: Cambridge University Press).

Plaskett, J. S. and Schatz, P. N. (1963). On the Robinson and Price (Kramers-Kronig) method of interpreting reflection data taken through a transparent window, *J. Chem. Phys.* **38**, 612–617.

Plemelj, J. (1908a). Ein ergänzungssatz zur Cauchyschen integraldarstellung analytischer funktionen, randwerte betreffend, *Monat. Math.* **19**, 205–210.

Plemelj, J. (1908b). Riemannsche funktionenscharen mit gegebener monodromiegruppe, *Monat. Math.* **19**, 211–245.

Plessner, A. (1923). Zur theorie der konjugierten trigonometrischen reihen. In *Mitteilungen des Mathematischen Seminars der Universität Giessen* Im selbsterverlag des Mathematischen seminars (Giessen: Universität Giessen).

Plieth, W. J. and Naegele, K. (1975). Kramers-Kronig analysis for the determination of the optical constants of thin surface films, *Surf. Sci.* **50**, 53–63.

Podolsky, B. and Kunz, K. S. (1969). *Fundamentals of Electrodynamics*. (New York: Marcel Dekker)

Poincaré, H. (1910). *Leçons de Mécanique Céleste* Vol. 3 (Paris: Gauthier-Villars).

Pollard, S. (1926). Extension to Stieltjes integrals of a theorem due to Plessner, *J. Lond. Math. Soc.* **2**, 37–41.

Polyakov, P. L. (2007). On a boundary value problem in subsonic aeroelasticity and the cofinite Hilbert transform, *J. Integral Equations Appl.* **19**, 33–70.

Porras, I. and King, F. W. (1994). Evaluation of some integrals for the atomic three-electron problem using convergence accelerators, *Phys. Rev. A* **49**, 1637–1645.

Portis, A. M. (1953). Electronic structure of *F* centers: saturation of the electron spin resonance, *Phys. Rev.* **91**, 1071–1078.

Potamianos, A. and Maragos, P. (1994). A comparison of the energy operator and the Hilbert transform approach to signal and speech demodulation, *Signal Process.* **37**, 95–120.

Poularikas, A. D., ed. (1996a). *The Transforms and Applications Handbook* (Boca Raton, FL: CRC Press).

Poularikas, A. D. (1996b). The Z-transform. In *The Transforms and Applications Handbook*, A. D. Poularikas, ed. (Boca Raton, FL: CRC Press), pp. 387–461.

Poularikas, A. D. (1999). *The Handbook of Formulas and Tables for Signal Processing* (Boca Raton, FL: CRC Press).

Press, W. H., Teukolsky, S. A., Vetterling, W. T., and Flannery, B. P. (1992). *Numerical Recipes in Fortran 77*, 2nd edn (Cambridge: Cambridge University Press).

Press, W. H., Teukolsky, S. A., Vetterling, W. T., and Flannery, B. P. (2002). *Numerical Recipes in C++*, 2nd edn (Cambridge: Cambridge University Press).

Prestini, E. (1985). Variants of the maximal double Hilbert transform, *Trans. Am. Math. Soc.* **290**, 761–771.

Price, P. J. (1963). Theory of quadratic response functions, *Phys. Rev.* **130**, 1792–1797.

Priestley, H. A. (1997). *Introduction to Integration* (Oxford: Oxford University Press).

Priestley, M. B. (1988). *Non-linear and Non-stationary Time Series Analysis* (London: Academic Press).

Pringsheim, A. (1900). Ueber das verhalten von potenzeihen auf dem convergenzkreise, *Sitzungsberichte Math.-Physik. Akad. Wissenschaften München* **30**, 37–100.

Pritz, T. (1999). Verification of local Kramers-Kronig relations for complex modulus by means of fractional derivative model, *J. Sound Vib.* **228**, 1145–1165.

Privalov, I. I. (1919). *Cauchy's Integral* (in Russian) (Saratov).

Priwaloff, I. (1916a). Sur la convergence des séries trigonométriques conjuguées, *Comptes Rendus Acad. Sci.* **162**, 123–126.

Priwaloff, I. (1916b). Sur la convergence des séries trigonométriques conjuguées, *Comptes Rendus Acad. Sci.* **165**, 96–99.

Priwaloff, I. (1916c). Sur les fonctions conjuguées, *Bull. Soc. Math. France* **44**, 100–103.

Pykhteev, G. N. (1959). On the evaluation of certain singular integrals with a kernel of the Cauchy type, *J. Appl. Math. Mech.* **23**, 1536–1548.

Qian, T. (2006). Analytic signals and harmonic measures, *J. Math. Analysis Appl.* **314**, 526–536.

Qian, T., Chen, Q., and Li, L. (2005). Analytic unit quadrature signals with nonlinear phase, *Physica D* **203**, 80–87.

Rabinowitz, P. (1978). The numerical evaluation of Cauchy principal value integrals. In *Symposium on Numerical Mathematics* (Durban: University of Natal), pp. 53–82.

Rabinowitz, P. (1983). Gauss-Kronrod integration rules for Cauchy principal value integrals, *Math. Comp.* **41**, 63–78.

Rabinowitz, P. (1986). Some practical aspects in the numerical evaluation of Cauchy principal value integrals, *Int. J. Comput. Math.* **20**, 283–298.

Rabinowitz, P. and Lubinsky, D. S. (1989). Noninterpolatory integration rules for Cauchy principal value integrals, *Math. Comp.* **53**, 279–295.

Radon, J. (1917). Über die bestimmung von funktionen durch ihre integralwerte läangs gewisser mannigfaltigkeiten, *Sitzungsbericht. Sachsischen Akad. Wissenschaft. Leipzig Math.-Naturwissenschaft. Klasse* **69**, 262–277.

Radon, J. (1986). On the determination of functions from their integral values along certain manifolds (P. C. Parks, translator), *IEEE Trans. Medical Imaging* **MI-5**, 170–176.

Rakotondratsimba, Y. (1999). Two-weight inequality for the discrete Hilbert transform, *Soochow J. Math.* **25**, 353–373.

Ramachandran, G. N. (1969). Application of Hilbert transforms to the phase problem in crystallography, *Mat. Res. Bull.* **4**, 525–533.

Rao, A. and Kumaresan, R. (1998). A parametric modeling approach to Hilbert transformation, *IEEE Signal Process. Lett.* **5**, 15–17.

Rapapa, N. P. and Scandolo, S. (1996). Universal constraints for the third-harmonic generation susceptibility, *J. Phys.: Condens. Matter* **8**, 6997–7004.

Rasigni, M. and Rasigni, G. (1977). Optical constants of lithium deposits as detected from the Kramers-Kronig analysis, *J. Opt. Soc. Am.* **67**, 54–59.

Räty, J., Peiponen, K.-E., and Asakura, T. (2004). *UV-Visible Reflection Spectroscopy of Liquids* (Berlin: Springer).

Read, R. R. and Treitel, S. (1973). The stabilization of two-dimensional recursive filters via the discrete Hilbert transform, *IEEE Trans. Geosci. Electron.* **GE11**, 153–160.

Reddy, G. R., Sathyanarayana, P., and Swamy, M. N. S. (1991a). CAS-CAS transform for 2-D signals: a few applications, *Circuits Syst. Signal Process.* **10**, 163–173.

Reddy, G. R., Sathyanarayana, P., and Swamy, M. N. S. (1991b). The stabilization of two-dimensional non-symmetric half-plane recursive filters via the discrete Hilbert transformations, *Int. J. Circuit Theory Appl.* **19**, 1–8.

Redheffer, R. (1968). An inequality for the Hilbert transform, *Proc. Natl. Acad. Sci. USA* **61**, 810–811.

Redheffer, R. (1971). An inequality for the Hilbert transform, *Pacific J. Math.* **37**, 181–211.

Reich, E. (1967). Some estimates for the two-dimensional Hilbert transform, *J. Analyse Math.* **18**, 279–293.

Reick, C. H. (2002). Linear response of the Lorenz system, *Phys. Rev. E* **66**, 036103.

Reid, C. (1996). *Hilbert* (New York: Copernicus (Springer-Verlag)).

Remacle, F. and Levine, R. D. (1993). Time domain information from resonant Raman excitation profiles: a direct inversion by maximum entropy, *J. Chem. Phys.* **99,** 4908–4925.

Rhodes, D. R. (1977). A reactance theorem, *Proc. Roy. Soc. Lond. A* **353**, 1–10.

Richardson, L. F. (1927). The deferred approach to the limit. Part I. – Single lattice, *Phil. Trans. Roy. Soc. Lond. Ser. A* **226**, 299–349.

Ridener, Jr., F. L. and Good, Jr., R. H. (1974). Dispersion relations for third-degree nonlinear systems, *Phys. Rev. B* **10**, 4980–4987.

Ridener, Jr., F. L. and Good, Jr., R. H. (1975). Dispersion relations for nonlinear systems of arbitrary degree, *Phys. Rev. B* **11**, 2768–2770.

Riesz, M. (1924). Les fonctions conjuguées et les séries de Fourier, *Comptes Rendus* **178**, 1464–1467.

Riesz, M. (1927). Sur les fonctions conjugées, *Math. Zeitschr.* **27**, 218–244.

Riesz, M. (1988). *Marcel Riesz Collected Papers*, L. Gårding and L. Hörmander, eds. (Berlin: Springer-Verlag).

Rihacek, A. W. (1966). Hilbert transforms and the complex representation of real signals, *Proc. IEEE* **54**, 434–435.

Rimmer, M. P. and Dexter, D. L. (1960). Optical constants of germanium in the region 0–10 ev, *J. Appl. Phys.* **31**, 775–777.

Robinson, T. S. (1952). Optical constants by reflection, *Proc. Phys. Soc. Lond.* B **65**, 910–911.

Robinson, T. S. and Price, W. C. (1953). The determination of infra-red absorption spectra from reflection measurements, *Proc. Phys. Soc. Lond.* B **66**, 969–974.

Roe, D. C., Marshall, A. G., and Smallcombe, S. H. (1978). Dispersion versus absorption: analysis of line-broadening mechanisms in nuclear magnetic resonance spectrometry, *Analytical Chem.* **50**, 764–767.

Roessler, D. M. (1965a). Kramers-Kronig analysis of reflection data, *Brit. J. Appl. Phys.* **16**, 1119–1123. Corrigendum, *Brit. J. Appl. Phys.* **16**, 1777.

Roessler, D. M. (1965b). Kramers-Kronig analysis of non-normal incidence reflection, *Brit. J. Appl. Phys.* **16**, 1359–1366.

Roessler, D. M. (1966). Kramers-Kronig analysis of reflection data III. Approximations, with reference to sodium iodide, *Brit. J. Appl. Phys.* **17**, 1313–1317.

Rogers, L. J. (1888). An extension of a certain theorem in inequalities, *Mess. Math.* **17**, 145–150.

Roman, P. (1965). *Advanced Quantum Theory* (Reading, MA: Addison-Wesley).

Roman, P. and Marathay, A. S. (1963). Analyticity and phase retrieval, *Nuovo Cimento* **30**, 1452–1464.

Rooney, P. G. (1972). On the ranges of certain fractional integrals, *Can. J. Math.* **24**, 1198–1216.

Rooney, P. G. (1975). On Tricomi's relation for the Hilbert transformation, *Glasgow Math. J.* **16**, 52–56.

Rooney, P. G. (1980). On the Y_v and H_v transformations, *Can. J. Math.* **32**, 1021–1044.

Rooney, P. G. (1986). On the spectrum of an integral operator, *Glasgow Math. J.* **28**, 5–9.

Roos, B. W. (1969). *Analytic Functions and Distributions in Physics and Engineering* (New York: John Wiley).

Rosenblum, M. and Rovnyak, J. (1974). Two theorems on finite Hilbert transforms, *J. Math. Analysis Appl.* **48**, 708–720.

Rosenblum, M. G., Pikovsky, A. S., and Kurths, J. (1996). Phase synchronization of chaotic oscillators, *Phys. Rev. Lett.* **76**, 1804–1807.

Rosenfeld, L. (1928). Quantenmechanische theorie der natürlichen optischen aktivität von flüssigkeiten und gasen, *Zeit. Phys.* **52**, 161–174.

Rosenfeld, L. (1965). *Theory of Electrons* (New York: Dover).

Ross, G., Fiddy, M. A., and Moezzi, H. (1980). The solution to the inverse scattering problem, based on fast zero location from two measurements, *Optica Acta* **27**, 1433–1444.

Ross, G., Fiddy, M. A. Nieto-Vesperinas, M. and Wheeler, M. W. L. (1978). The phase problem in scattering phenomena: the zeros of entire functions and their significance, *Proc. Roy. Soc. Lond. A* **360**, 25–45.

Roth, J., Rao, B., and Dignam, M. J. (1975). Application of the causality condition to thin film spectroscopy, *J. Chem. Soc. Faraday Trans. II* **71**, 86–94.

Rubio de Francia, J. L. (1982). Factorization and extrapolation of weights, *Bull. Am. Math. Soc.* **7**, 393–395.

Saal, L. (1990). A result about the Hilbert transform along curves, *Proc. Am. Math. Soc.* **110**, 905–914.

Saal, L. and Urciuolo, M. (1993). The Hilbert transform along curves that are analytic at infinity, *Illinois J. Math.* **37**, 561–574.

Saarinen, J. J. (2002). Sum rules for arbitrary-order harmonic generation susceptibilities, *Eur. Phys. J. B* **30**, 551–557.

Sabri, M. S. and Steenaart, W. (1974). A new approach to digital filter realization using the Hilbert transform. In*Canadian Communications & Power Conference*, (New York: IEEE), pp. 89–90.

Sabri, M. S. and Steenaart, W. (1975). Discrete Hilbert transform filtering. In *Proceeding of the Eighteenth Midwest Symposium on Circuits & Systems*, M. N. S Swamy, ed. (North Hollywood, CA: Western Periodicals Co.), pp. 449–455.

Sabri, M. S. and Steenaart, W. (1976). Discrete Hilbert transform filtering, in *1976 IEEE International Conference on Acoustics, Speech, and Signal Processing*, (Rome, NY: Canterbury Press), pp. 116–119.

Sabri, M. S. and Steenaart, W. (1977). Discrete Hilbert transform filtering, *IEEE Trans. Acoustics Speech Signal Process.* **ASSP-25**, 452–454.

Šachl, V. (1963). Dispersion relations as a criterion for the imaginary part of a spectral function, *Czech. J. Phys. B* **13**, 792–799.

Sack, R. A. and Donovan, A. F. (1972). An algorithm for Gaussian quadrature given modified moments, *Numer. Math.* **18**, 465–478.

Sadosky, C. (1979). *Interpolation of Operators and Singular Integrals*, (New York: Marcel Dekker).

Sagher, Y. and Xiang, N. (1996). Complex methods in the calculation of some distribution functions. In *Convergence in Ergodic Theory and Probability* (Berlin: de Gruyter), pp. 381–387.

Saito, M. (1974). Hilbert transforms for sampled data, *J. Phys. Earth* **22**, 313–324.

Sakai, H. and Vanasse, G. A. (1966). Hilbert transform in Fourier spectroscopy, *J. Opt. Soc. Am.* **56**, 131–132.

Samotij, K. (1991). An example of a Hilbert transform, *Proc. Am. Math. Soc.* **112**, 965–972.

Santhanam, B. and McClellan, J. H. (1996). The discrete rotational Fourier transform, *IEEE Trans. Signal Process.* **44**, 994–998.

Santini, P. M., Ablowitz, M. J., and Fokas, A. S. (1987). On the initial value problem for a class of nonlinear integral evolution equations including the sine-Hilbert equation, *J. Math. Phys.* **28**, 2310–2316.

Saslow, W. M. (1970). Two classes of Kramers-Kronig sum rules, *Phys. Lett. A* **33**, 157–158.

Sato, R. (1986). A remark on the ergodic Hilbert transform, *Math. J. Okayama Univ.* **28**, 159–163.

Sato, R. (1987a). On the ergodic Hilbert transform for Lamperti operators, *Proc. Am. Math. Soc.* **99**, 484–488.

Sato, R. (1987b). On the ergodic Hilbert transform for operators in L_p, $1 < p < \infty$, *Can. Math. Bull.* **30**, 210–214.

Sato, R. (1998 [2000]). On the weighted ergodic properties of invertible Lamperti operators, *Math. J. Okayama Univ.* **40**, 147–176.

Saxton, W. O. (1974). Phase determination in bright-field electron microscopy using complementary half-plane apertures, *J. Phys. D: Appl. Phys.* **7**, L63–L64.

Scaife, B. K. P. (1971). The theory of dispersion in polar dielectrics. In *Complex Permittivity*, B. K. P. Scaife, compiler (London: The English Universities Press Ltd.), pp. 3–17.

Scaife, B. K. P. (1972). The theory of the macroscopic properties of isotropic dielectrics. In *Dielectric and Related Molecular Processes,* Vol. 1 (London: The Chemical Society).

Scandolo, S. and Bassani, F. (1992a). A nonlinear sum rule for atomic hydrogen, *Nuovo Cimento D* **14**, 873–880.

Scandolo, S. and Bassani, F. (1992b). Nonlinear sum rules: The three-level and the anharmonic-oscillator models, *Phys. Rev. B* **45**, 13 257–13 261.

Scandolo, S. and Bassani, F. (1995a). Kramers-Kronig relations and sum rules for the second-harmonic susceptibility, *Phys. Rev. B* **51**, 6925–6927.

Scandolo, S. and Bassani, F. (1995b). Miller's rule and the static limit for second-harmonic generation, *Phys. Rev. B* **51**, 6928–6931.

Scandolo, S., Bassani, F., and Lucarini, V. (2001). Spatial-dispersion and relativistic effects in the optical sum rules, *Eur. Phys. J. B* **23**, 319–323.

Scarfone, L. M. and Chlipala, J. D. (1975). Single-band model of substitutional disordered ternary alloys in the coherent-potential approximation, *Phys. Rev. B* **11**, 4960–4979.

Schatz, P. N., Maeda, S., and Kozima, K. (1963). Determination of optical constants from reflection bands using dispersion relations, *J. Chem. Phys.* **38**, 2658–2661.

Schiff, L. I. (1955). *Quantum Mechanics*, 2nd edn (New York: McGraw-Hill).

Schlömilch, O. (1848). *Analytische Studien*, Vols. 1&2, (Leipzig: W. Engelmann).

Schlurmann, T. (2002). Spectral analysis of nonlinear water waves based on the Hilbert-Huang transformation, *Trans. ASME: J. Offshore Mech. Arctic Eng.* **124**, 22–27

Schmidt, R. and Black, E. (trans.) (1986). *Three Treatises, by Viète, Girard & de Beaune* (Annapolis, MD: Golden Hind Press).

Schnatterly, S. E. (1969). Magnetoreflection measurements on the noble metals, *Phys. Rev.* **183**, 664–667.

Schneider, C. B. (1998). Inversion formulas for the discretized Hilbert transform on the unit circle, *SIAM J. Num. Analysis* **35**, 71–77.

Schochet, S. (1986). Explicit solutions of the viscous model vorticity equation, *Commun. Pure Appl. Math.* **39**, 531–537.

Schoeck, G. (1994). The generalized Peierls-Nabarro model, *Phil. Mag. A* **69**, 1085–1095.

Schützer, W. and Tiomno, J. (1951). On the connection of the scattering and derivative matrices with causality, *Phys. Rev.* **83**, 249–251.

Schwartz, L. (1962). Causalité et analyticité, *An. Acad. Brasileira Ciências*, **34**, 13–21.

Schwartz, L. (1966a). *Théorie des distributions* (Paris: Hermann).

Schwartz, L. (1966b). *Mathematics for the Physical Sciences* (Paris: Hermann).

Schwarzl, F. R. and Struik, L. C. E. (1967–1968). Analysis of relaxation measurements, *Advan. Mol. Relax. Processes* **1**, 201–255.

Scouler, W. J. (1969). Optical properties of Mg_2Si, Mg_2Ge, and Mg_2Sn from 0.6 to 11.0 eV at 77°K, *Phys. Rev.* **178**, 1353–1357.

Segovia, C. and Torrea, J. L. (1990). A note on the commutator of the Hilbert transform, *Rev. Unión Mat. Argentina* **35**, 259–264.

Segovia, C. and Wheeden, R. L. (1971). Fractional differentiation of the commutator of the Hilbert transform, *J. Funct. Analysis* **8**, 341–359.

Selesnick, I. W. (2001). Hilbert transform pairs of wavelet bases, *IEEE Signal Process. Lett.* **8**, 170–173.

Selesnick, I. W. (2002). The design of approximate Hilbert transform pairs of wavelet bases, *IEEE Trans. Signal Process.* **50**, 1144–1152.

Sen, P. and Sen, P. K. (1987). Kramers-Kronig type of dispersion relation in nonlinear optics, *Pramana J. Phys.* **28**, 661–667.

Shaik, J. S. and Iftekharuddin, K. M. (2003). Detection and tracking of rotated and scaled targets by use of Hilbert-wavelet transform, *Appl. Opt.* **42**, 4718–4735.

Shamir, E. (1964). Reduced Hilbert transforms and singular integral equations, *J. Analyse Math.* **12**, 277–305.

Shannon, C. E. (1949). Communication in the presence of noise, *Proc. Inst. Radio Eng.* **37**, 10–21.

Sharnoff, M. (1964). Validity conditions for the Kramers-Kronig relations, *Am. J. Phys.* **32**, 40–44.

Sharpley, R. C. and Vatchev, V. (2006). Analysis of the intrinsic mode functions, *Constr. Approx.* **24**, 17–47.

Shen, Y. R. (1984). *The Principles of Nonlinear Optics* (New York: John Wiley).

Shiles, E., Sasaki, T., Inokuti, M., and Smith, D. Y. (1980). Self-consistency and sum-rule tests in the Kramers-Kronig analysis of optical data: applications to aluminum, *Phys. Rev. B* **22**, 1612–1628.

Shohat, J. A. and Tamarkin, J. D. (1970). *The Problem of Moments*, (Providence, RI: American Mathematical Society).

Shore, K. A. and Chan, D. A. S. (1990). Kramers-Kronig relations for nonlinear optics, *Electron. Lett.* **26**, 1206–1207.

Shtrauss, V. (2006). FIR Kramers-Kronig transformers for relaxation data conversion, *Signal Process.* **86**, 2887–2900.

Shul'man, A. Ya., Kosarev, E. L., and Tarasov, M. A. (2003). Hilbert-transform spectroscopy based on the a.c. Josephson effect. Theory and computational technique, *J. Commun. Tech. Electron.* **48**, 1124–1136.

Sidky, E. Y. and Pan, X. (2005). Recovering a compactly supported function from knowledge of its Hilbert transform on a finite interval, *IEEE Signal Process. Lett.* **12**, 97–100.

Silva, H. and Gross, B. (1941). Some measurements on the validity of the principle of superposition in solid dielectrics, *Phys. Rev.* **60**, 684–687.

Simon, M. and Tomlinson, G. R. (1984). Use of the Hilbert transform in modal analysis of linear and non-linear structures, *J. Sound Vib.* **96**, 421–436.

Simpson, R. S. and Blackwell, C. A. (1966). Hilbert transform analysis of the output noise spectrum of an ideal AM demodulator and FM discriminator, *Proc. IEEE* **54**, 444–445.

Singh, O. P. and Pandey, J. N. (1990a). The *n*-dimensional Hilbert transform of distributions, its inversion and applications, *Can. J. Math.* **42**, 239–258.

Singh, O. P. and Pandey, J. N. (1990b). The *n*-dimensional Hilbert transform of distributions, *Progr. Math.* (*Varanasi*) **24**, 95–105.

Skwarek, V. and Hans, V. (2001). An improved method for hardware-based complex demodulation, *Measurement* **29**, 87–93.

Slichter, C. P. (1963). *Principles of Magnetic Resonance* (New York: Harper and Row).

Sloan, I. H. (1968). The numerical evaluation of principal value integrals, *J. Comput. Phys.* **3**, 332–333.

Smet, F. and Smet, P. (1974). Relations de dispersion pour les phénomènes non linéaires d'ordre 2. *Nuovo Cimento* **20**, 273–280.

Smet, F. and van Groenendael, A. (1979). Dispersion relations for *n*-order nonlinear phenomena, *Phys. Rev. A* **19**, 334–337.

Smirnoff, V. (1929). Sur les valeurs limites des fonctions analytiques, *Comptes Rendus Acad. Sci.* (*Series* I) **188**, 131–133.

Smith, D. A. and Ford, W. F. (1979). Acceleration of linear and logarithmic convergence, *SIAM J. Num. Analysis* **16**, 223–240.

Smith, D. A. and Ford, W. F. (1982). Numerical comparisons of nonlinear convergence accelerators, *Math. Comp.* **38**, 481–499.

Smith, D. Y. (1976a). Comments on the dispersion relations for the complex refractive index of circularly and elliptically polarized light, *J. Opt. Soc. Am.* **66**, 454–460.

Smith, D. Y. (1976b). Dispersion relations and sum rules for magnetoreflectivity, *J. Opt. Soc. Am.* **66**, 547–554.

Smith, D. Y. (1976c). Superconvergence and sum rules for the optical constants: Natural and magneto-optical activity, *Phys. Rev. B* **13**, 5303–5315.

Smith, D. Y. (1977). Dispersion relations for complex reflectivities, *J. Opt. Soc. Am.* **67**, 570–571.

Smith, D. Y. (1980). Dispersion theory and moments relations in magneto-optics. In *Theoretical Aspects and New Developments in Magneto-Optics*, J. T. Devreese, ed. (New York: Plenum Press). pp. 133–182.

Smith, D. Y. (1985). Dispersion theory, sum rules, and their application to the analysis of optical data. In *Handbook of Optical Constants of Solids*, E. D. Palik, ed. (Orlando, FL: Academic Press), pp. 35–68.

Smith, D. Y. and Graham, G. (1980). Oscillator strengths of defects in insulators: The generalization of Smakula's equation, *J. Phys. Colloq. C6* **41**, C6–80–C6–83.

Smith, D. Y. and Manogue, C. A. (1981). Superconvergence relations and sum rules for reflection spectroscopy, *J. Opt. Soc. Am.* **71**, 935–947.

Smith, D. Y. and Shiles, E. (1978). Finite-energy f-sum rules for valence electrons, *Phys. Rev. B* **17**, 4689–4694.

Smith, D. Y., Inokuti, M., and Karstens, W. (2001). A generalized Cauchy dispersion formula and the refractivity of elemental semiconductors, *J. Phys.: Condens. Matter* **13**, 3883–3893.

Smith, E. (1969). A pile-up of dislocations in a bi-metallic solid, *Scripta Metallurgica* **3**, 415–418.

Smith, W. E. and Lyness, J. N. (1969). Application of Hilbert transform theory to numerical quadrature, *Math. Comp.* **23**, 231–252.

Smithies, F. (1997). *Cauchy and the Creation of Complex Function Theory* (Cambridge: Cambridge University Press).

Sneddon, I. N. (1972). *The Use of Integral Transforms* (New York: McGraw-Hill).

Söhngen, H. (1939). Die Lösungen der integralgleichung $g(x) = (1/2\pi)\oint_{-a}^{a} f(\xi)\mathrm{d}\xi/(x-\xi)$ und deren anwendung in der tragflügeltheorie, *Math. Zeitschr.* **45**, 245–264.

Söhngen, H. (1954). Zur theorie der endlichen Hilbert-transformation, *Math. Zeitschr.* **60**, 31–51.

Sokół-Sokółski, K. (1947). On trigonometric series conjugate to Fourier series of two variables, *Fund. Math.* **34**, 166–182.

Sommerfeld, A. (1914). 1. Über die fortpflanzung des lichtes in dispergierenden medien, *Ann. Phys. (Leipzig)* **44**, 177–202.

Spanos, P. D. and Miller, S. M. (1994). Hilbert transform generalization of a classical random vibration integral, *J. Appl. Mech.* **61**, 575–581.

Srivastav, R. P. (1997). A hybrid Fourier transform, *Appl. Math. Lett.* **10**, 1–5.

Srivastava, H. M. and Tuan, V. K. (1995). A new convolution theorem for the Stieltjes transform and its application to a class of singular integral equations, *Arch. Math.* **64**, 144–149.

Stallard, B. R., Champion, P. M., Callis, P. R. and Albrecht, A. C. (1983). Advances in calculating Raman excitation profiles by means of the transform theory, *J. Chem. Phys.* **78**, 712–722.

Stanković, B. (1988). Abelian theorems for the Stieltjes-Hilbert transform of distributions, *Univ. u Novom Sadu Zb. Rad. Prirod.-Mat. Fak. Ser. Mat.* **18**, 89–101.

Stanomir, D., Negrescu, C., and Pârvu, V. (1997). Hilbert transforms for discrete-time signals revisited, *Rev. Roum. Sci. Techn. Électrotech. et Énerg.* (*Électronique trans. l'inform.*) **42**, 255–269.

Stark, H. (1971). An extension of the Hilbert transform product theorem, *Proc. IEEE* **59**, 1359–1360.

Stark, H. (1972). Author's reply, *Proc. IEEE* **60**, 228.

Stark, H., Bennett, W. R., and Arm, M. (1969). Design considerations in power spectra measurements by diffraction of coherent light, *Appl. Opt.* **8**, 2165–2172.

Steele, J. M. (2004). *The Cauchy-Schwarz Master Class An Introduction to the Art of Mathematical Inequalities* (Cambridge: Cambridge University Press).

Stein, E. M. (1957). Note on singular integrals, *Proc. Am. Math. Soc.* **8**, 250–254.

Stein, E. M. (1970). *Singular Integrals and Differentiability Properties of Functions* (Princeton, NJ: Princeton University Press).

Stein, E. M. (1993). *Harmonic Analysis: Real-Variable Methods, Orthogonality, and Oscillatory Integrals* (Princeton, NJ: Princeton University Press).

Stein, E. M. (1998). Singular integrals: the roles of Calderón and Zygmund, *Notices AMS* **45**, 1130–1140.

Stein, E. M. and Wainger, S. (1970). The estimation of an integral arising in multiplier transformations, *Studia Math.* **35**, 101–104.

Stein, E. M. and Wainger, S. (1978). Problems in harmonic analysis related to curvature, *Bull. Am. Math. Soc.* **84**, 1239–1295.

Stein, E. M. and Weiss, G. (1959). An extension of a theorem of Marcinkiewicz and some of its applications, *J. Math. Mech.* **8**, 263–284.

Stein, E. M. and Weiss, G. (1971). *Introduction to Fourier Analysis on Euclidean Spaces* (Princeton, NJ: Princeton University Press).

Stein, P. (1933). On a theorem of M. Riesz, *J. Lond. Math. Soc.* **8**, 242–247.

Stenger, F. (1976). Approximation via Whittaker's cardinal function, *J. Approx. Theory* **17**, 222–240.

Stenger, F. (2000). Summary of sinc numerical methods, *J. Comput. Appl. Math.* **121**, 379–420.

Stens, R. L. (1983). A unified approach to sampling theorems for derivatives and Hilbert transforms, *Signal Process.* **5**, 139–151.

Stens, R. L. (1984). Approximation of functions by Whittaker's cardinal series. In *General Inequalities 4*, W. Walter, ed. (Basel: Birkhäuser Verlag).

Stern, E. A., McGroddy, J. C., and Harte, W. E. (1964). Polar reflection Faraday effect in metals, *Phys. Rev.* **135**, A1306–A1314.

Stern, F. (1963). Elementary theory of the optical properties of solids. In *Solid State Physics*, Vol. 15, F. Seitz and D. Turnbull, eds. (New York: Academic Press). pp. 299–408.

Stewart, C. E. (1960). On the numerical evaluation of singular integrals of Cauchy type, *J. Soc. Indust. Appl. Math.* **8**, 342–353.

Stoer, J. and Bulirsch, R. (1980). *Introduction to Numerical Analysis* (New York: Springer-Verlag).

Stout, G. H. and Jensen, L. H. (1989). *X-Ray Structure Determination*, 2nd edn (New York: John Wiley).

Stroud, A. H. and Secrest, D. (1966). *Gaussian Quadrature Formulas* (Englewood Cliffs, NJ: Prentice-Hall).

Stroud, D. (1979). Percolation effects and sum rules in the optical properties of composites, *Phys. Rev. B* **19**, 1783–1791.

Sucher, J. (2002). Higher-order poles and mass-shell singularities in electron-hydrogen scattering, *Phys. Rev. A* **66**, 042706.

Sugawara, M. and Kanazawa, A. (1961). Subtractions in dispersion relations, *Phys. Rev.* **123**, 1895–1902.

Sugiyama, A. (1992). Pseudopotential theory for interionic interaction potentials in metals at non-zero temperature. I. Elementary derivation of asymptotic expressions on basis of Hilbert transforms, *J. Phys. Soc. Japan* **61**, 4061–4084.

Sukhatme U., Kane, G. L., Blankenbecler, R., and Davier, M. (1975). Extensions of the derivative dispersion relations for amplitude analyses, *Phys. Rev. D* **12**, 3431–3440.

Suzuki, N. (1976). The finite Hilbert transform on $L_2(0, \pi)$ is a shift, *Proc. Japan Acad.* **52**, 544–547.

Symanzik, K. (1960). On the many-particle structure of Green's functions in quantum field theory, *J. Math. Phys.* **1**, 249–273.

Szabo, T. L. (1994). Time domain wave equations for lossy media obeying a frequency power law, *J. Acoust. Soc. Am.* **96**, 491–500.

Szabo, T. L. (1995). Causal theories and data for acoustic attenuation obeying a frequency power law, *J. Acoust. Soc. Am.* **97**, 14–24.

Tamarkin, J. D. (1931). Remarks on the theory of conjugate functions, *Proc. Lond. Math. Soc.* **34**, 379–391.

Tang, M.-T. and Chang, S.-L. (1990). Kramers-Kronig relations in three-beam X-ray diffraction: application to phase determination, *Phys. Lett. A* **143**, 405–408.

Tarasov, M. A., Shul'man, A. Ya., Prokopenko, G. V., Koshelets, V. P., Polyanski, O. Yu., Lapitskaya, I. L., and Vystavkin, A. N. (1995). Quasioptical Hilbert transform spectrometer, *IEEE Trans. Appl. Supercond.* **5**, 2686–2689.

Tauber, A. (1891). Über den zusammenhang des reellen und imaginären theiles einer potenzreihe, *Monat. Math. Physik* **2**, 79–118.

Taurian, O. E. (1980). A method and a program for the numerical evaluation of the Hilbert transform of a real function, *Comput. Phys. Commun.* **20**, 291–307.

Taylor, J. G. (1958). Dispersion relations and Schwartz's distributions, *Ann. Phys. (NY)* **5**, 391–398.

Taylor, L. S. (1981). The phase retrieval problem, *IEEE Trans. Antennas Prop.* **AP-29**, 386–391.

Temkin, A. and Drachman, R. J. (2000). Partial wave dispersion relations: application to electron-atom scattering, *J. Phys. B: Atom. Mol. Opt. Phys.* **33**, L505–L511.

Temkin, A., Bhatia, A. K., and Kim, Y. S. (1986). A new dispersion relation for electron-atom scattering, *J. Phys. B: Atom. Mol. Phys.* **19**, L701–L706.

Ter Haar, D. (1998). *Masters of Modern Physics The Scientific Contributions of H. A. Kramers* (Princeton, NJ: Princeton University Press).

Terhune, R. W., Maker, P. D., and Savage, C. M. (1962). Optical harmonic generation in calcite, *Phys. Rev. Lett.* **8**, 404–406.

Thirring, W. (1959). Dispersion relations, *Nuovo Cimento Supp.* **14**, 385–400.

Thomaz, M. T. and Nussenzveig, H. M. (1982). Dispersion relations and sum rules for natural optical activity, *Ann. Phys. (NY)* **139**, 14–35.

Thomée, V. and Murthy, A. S. V. (1998). A numerical method for the Benjamin-Ono equation, *BIT* **38**, 597–611.

Tickanen, L. D., Tejedor-Tejedor, M. I., and Anderson, M. A. (1992). Concurrent determination of optical constants and the Kramers-Kronig integration constant (anchor point) using variable-angle ATR/FT-IR spectroscopy, *Appl. Spectrosc.* **46**, 1848–1858.

Tickanen, L. D., Tejedor-Tejedor, M. I., and M. A. Anderson (1997). Quantitative characterization of aqueous suspensions using variable-angle ATR/FT-IR spectroscopy: determination of optical constants and absorption coefficient spectra, *Langmuir* **13**, 4829–4836.

Tinkham, M. (1956). Energy gap interpretation of experiments on infrared transmittance through superconducting films, *Phys. Rev.* **104**, 845–846.

Tinkham, M. and Ferrell, R. A. (1959). Determination of the superconducting skin depth from the energy gap and sum rule, *Phys. Rev. Lett.* **2**, 331–333.

Tip, A. (1977). The analytical structure of the elastic forward electron-atom scattering amplitude, *J. Phys. B: Atom. Mol. Phys.* **10**, L11–L16.

Titchmarsh, E. C. (1925a). Conjugate trigonometric integrals, *Proc. Lond. Math. Soc.* **24**, 109–130.

Titchmarsh, E. C. (1925b). The convergence of certain integrals, *Proc. Lond. Math. Soc.* **24**, 347–358.

Titchmarsh, E. C. (1926). Reciprocal formulae involving series and integrals, *Math. Zeitschr.* **25**, 321–347. Erratum: *Math. Zeitschr.* **26**, 496 (1927).

Titchmarsh, E. C. (1929). On conjugate functions, *Proc. Lond. Math. Soc.* (2) **29**, 49–80.

Titchmarsh, E. C. (1930a). Additional note on conjugate functions, *J. Lond. Math. Soc.* **4**, 204–206.

Titchmarsh, E. C. (1930b). A theorem on conjugate functions, *J. Lond. Math. Soc.* **5**, 89–91.

Titchmarsh, E. C. (1939). *The Theory of Functions*, 2nd edn (London: Oxford University Press).

Titchmarsh, E. C. (1948). *Introduction to the Theory of Fourier Integrals*, 2nd edn (London: Oxford University Press).

Toland, J. F. (1997a). The Peierls-Nabarro and Benjamin-Ono equations, *J. Funct. Analysis* **145**, 136–150.

Toland, J. F. (1997b). A few remarks about the Hilbert transform, *J. Funct. Analysis* **145**, 151–174.

Toll, J. S. (1956). Causality and the dispersion relation: logical foundations, *Phys. Rev.* **104**, 1760–1770.

Tomiyama, K., Clough, S. A., and Kneizys, F. X. (1987). Complex susceptibility for collisional broadening, *Appl. Opt.* **26**, 2020–2028.

Tomlinson, G. R. (1987). Developments in the use of the Hilbert transform for detecting and quantifying non-linearity associated with frequency response functions, *Mech. Syst. Signal Process.* **1**, 151–171.

Torchinsky, A. (1986). *Real-Variable Methods in Harmonic Analysis* (Orlando, FL: Academic Press).

Treil, S. R. (1983). A geometric approach to the weighted estimates of Hilbert transforms, *Funct. Analysis Appl.* **17**, 319–321.

Treil, S. R. and Volberg, A. L. (1995). Weighted embeddings and weighted norm inequalities for the Hilbert transform and the maximal operator, *Algebra i Analiz* **7**, 205–226.

Treil, S. and Volberg, A. (1997). Wavelets and the angle between past and future, *J. Funct. Analysis* **143**, 269–308.

Treil, S. Volberg, A., and Zheng, D. (1997). Hilbert transform, Toeplitz operators and Hankel operators, and invariant A_∞ weights, *Rev. Mat. Iberoamericana* **13**, 319–360.

Tricomi, F. (1926). Formula d'inversione dell'ordine di due integrazioni doppie, *Rend. Accad. Naz. Lincei* **3**, 535–539.

Tricomi, F. G. (1951a). On the finite Hilbert transformation, *Quart. J. Math.* (2)**2**, 199–211.

Tricomi, F. G. (1951b). The airfoil equation for a double interval, *Zeit. Angew. Math. Physik* **2**, 402–406.

Tricomi, F. G. (1955). Sull'inversione dell'ordine di integrali principali nel senso di Cauchy, *Rend. Accad. Naz. Lincei* **18**, 3–7.

Tricomi, F. G. (1985). *Integral Equations* (New York: Dover)

Troup, G. J. (1971). The failure of the modified Kronig-Kramers relations for second harmonic generation, *Phil. Mag.* **24**, 613–617.

Troup, G. J. and Bambini, A. (1973). The use of the modified Kramers-Kronig relation in the rate equation approach of laser theory, *Phys. Lett. A* **45**, 393–394.

Tschoegl, N. W. (1989). *The Phenomenological Theory of Linear Viscoelastic Behavior* (Berlin: Springer-Verlag).

Tseng, C.-C. and Pei, S.-C. (2000). Design and application of discrete-time fractional Hilbert transformer, *IEEE Trans. Systems Circuits II Analog Digital Signal Process.* **47**, 1529–1533.

Tsereteli, O. D. (1977). Metric properties of conjugate functions, *J. Sov. Math.* **7**, 317–343.

Tuttle, Jr. D. F. (1958). *Network Synthesis* (New York: John Wiley).

Urkowitz, H. (1962). Hilbert transforms of bandpass functions, *Proc. IRE* **50**, 2143.

Urkowitz, H. (1964). Pre-envelopes of nonstationary bandpass processes, *J. Franklin Inst.* **277**, 31–36.

Ursell, F. (1983). Integrals with a large parameter: Hilbert transforms, *Math. Proc. Camb. Phil. Soc.* **93**, 141–149.

Vakman, D. (1994). Computer measuring of frequency stability and the analytic signal, *IEEE Trans. Instrum. Meas.* **43**, 668–671.

Vakman, D. (1996). On the analytic signal, the Teager-Kaiser energy algorithm, and other methods for defining amplitude and frequency, *IEEE Trans. Signal Process.* **44**, 791–797.

Vakman, D. (1997). Analytic waves, *Int. J. Theor. Phys.* **36**, 227–247.

Vakman, D. and Vaĭnshteĭn, L. A. (1977). Amplitude, phase, frequency – fundamental concepts of oscillation theory, *Sov. Phys. Usp.* **20**, 1002–1016.

Van Der Pol, B. and Bremmer, H. (1987). *Operational Calculus*, 3rd edn (New York: Chelsea Publishing Co.).

Van Kampen, N. G. (1953a). *S*-matrix and causality condition. I Maxwell field, *Phys. Rev.* **89**, 1072–1079.

Van Kampen, N. G. (1953b). *S*-matrix and causality condition. II Nonrelativistic particles, *Phys. Rev.* **91**, 1267–1276.

Van Kampen, N. G. (1957). Can the *S*-matrix be generated from its lowest-order term?, *Physica* **23**, 157–163.

Van Kampen, N. G. (1958a). Causaliteit en Kramers-Kronigrelaties I, *Ned. T. Natuurk.* **24**, 1–14.

Van Kampen, N. G. (1958b). Causaliteit en Kramers-Kronigrelaties II, *Ned. T. Natuurk.* **24**, 29–42. For a French translation of this and the preceding reference, see Van Kampen (1961).

Van Kampen, N. G. (1961). Causalité et relations de Kramers-Kronig, *J. Physique* **22**, 179–191.

Van Vleck, J. H. (1945). The relation between absorption and the frequency dependence of refraction, *Massachusetts Institute of Technology Radiation Laboratory Report* 735.

Van Wijngaarden, L. (1963). On the Kramers-Kronig relations, with special reference to gravity waves, *Konink. Nederland. Akad. Wetenschappen Proc. Phys. Sci.* **66**, 339–347.

VanderNoot, T. J. (1992). Hilbert transformation of immittance data using the fast Fourier transform, *J. Electroanalytical Chem.* **322**, 9–24.

Varsavsky, O. A. (1949). Sobre la transformacion de Hilbert, *Revista Union Mat. Argentina* **14**, 20–37.

Vartiainen, E. M. and Peiponen, K.-E. (1994). Meromorphic degenerate nonlinear susceptibility: Phase retrieval from the amplitude spectrum, *Phys. Rev. B* **50**, 1941–1944.

Vartiainen, E. M., Peiponen, K.-E., and Asakura, T. (1993a). Comparison between the optical constants obtained by the Kramers-Kronig analysis and the maximum entropy: infrared optical properties of orthorhombic sulfur, *Appl. Opt.* **32**, 1126–1129.

Vartiainen, E. M., Peiponen, K.-E., and Asakura, T. (1993b). Sum rules in testing non-linear susceptibility obtained using the maximum entropy model, *J. Phys.: Condens. Matter* **5**, L113-L116.

Vartiainen, E. M., Peiponen, K.-E., Kishida, H., and Koda, T. (1996). Phase retrieval in nonlinear optical spectroscopy by the maximum-entropy method: an application to the $|\chi^{(3)}|$ spectra of polysilane, *J. Opt. Soc. Am. B* **13**, 2106–2114.

Veal, B. W. and Paulikas, A. P. (1974). Optical properties of molybdenum. I. Experiment and Kramers-Kronig analysis, *Phys. Rev. B* **10**, 1280–1289.

Velický, B. (1961a). Dispersion relation for complex reflectivity, *Czech. J. Phys. B* **11**, 541–543.

Velický, B. (1961b). The use of the Kramers-Kronig relations in determining optical constants, *Czech. J. Phys. B* **11**, 787–798.

Veltcheva, A. D. (2002). Wave and group transformation by a Hilbert spectrum, *Coastal Eng. J.* **44**, 283–300.

Verbitskiĭ, I. È. (1984). An estimate of the norm of a function in a Hardy space in terms of the norms of its real and imaginary parts, *Am. Math. Soc. Transl.* **124**, 11–15. Translation of *Mat. Issled. Vyp.* **54**, 16–20 (1980).

Verbitsky, I. È. and Krupnik, N. Ya. (1994). The norm of the Riesz projection. In *Linear and Complex Analysis Problem Book 3, Part I*, V. P. Havin and N. K. Nikolski, eds. (Berlin: Springer-Verlag). pp. 422–423.

Verdun, F. R., Giancaspro, C., and Marshall, A. G. (1988). Effects of noise, time-domain damping, zero filling and the FFT algorithm on the "exact" interpolation of fast Fourier transform spectra, *Appl. Spectrosc.* **42**, 715–721.

Verleur, H. W. (1968). Determination of the optical constants from reflectance or transmittance measurements on bulk crystals or thin films, *J. Opt. Soc. Am.* **58**, 1356–1364.

Veselago, V. G. (1967). Properties of materials having simultaneously negative values of the dielectric (ε) and the magnetic (μ) susceptibilities, *Sov. Phys. – Solid State* **8**, 2854–2856.

Veselago, V. G. (1968). The electrodynamics of substances with simultaneously negative values of ε and μ, *Sov. Phys. Usp.* **10**, 509–514.

Viète, F. (1646). *Opera Mathematica*, Lvgdvni Batavorvm, ex officinâ Bonaventurae & Abrahami Elzeviriorum, Leiden. (Reprinted by Georg Olms Verlag, Hildesheim, 1970).

Villani, A. and Zimerman, A. H. (1973a). Generalized f-sum rules for the optical constants, *Phys. Lett. A* **44**, 295–297.

Villani, A. and Zimerman, A. H. (1973b). Superconvergent sum rules for the optical constants, *Phys. Rev. B* **8**, 3914–3916.

Ville, J. (1948). Théorie et applications de la notion de signal analytique, *Cables Transmission A* **2**, 61–74.

Vladimirov, V. S. (1979). *Generalized Functions in Mathematical Physics* (Moscow: Mir).

Voelcker, H. B. (1966a). Toward a unified theory of modulation Part I: Phase-envelope relationships, *Proc. IEEE* **54**, 340–353.

Voelcker, H. B. (1966b). Toward a unified theory of modulation - Part II: Zero manipulation, *Proc. IEEE* **54**, 735–755.

Volkov, O. Y., Pavlovskii, V. V., Divin, Y. Y., and Poppe, U. (1999). Far-infrared Hilbert-transform spectrometer based on Stirling cooler. In *Proceedings of EUCAS 1999,* The Fourth European Conference on Applied Superconductivity (Sitges, Spain, 14–17 September, 1999), X. Obradocs, F. Sandiumenge, and J. Fontcubertadet, eds., pp. 623–626.

Vrinceanu, D., Msezane, A. Z., Bessis, D., and Temkin, A. (2001). Exchange forces in dispersion relations investigated using circuit relations, *Phys. Rev. Lett.* **86**, 3256–3259.

Walker, J. S. (1988). *Fourier Analysis* (New York: Oxford University Press).

Walker, J. S. (1996). *Fast Fourier Transforms*, 2nd edn (Boca Raton, FL: CRC Press).

Wall, H. S. (1948). *Analytic Theory of Continued Fractions*, (Bronx, NY: Chelsea Publishing Company).

Wallis, D. H. and Wickramasinghe, N. C. (1999). Determination of optical spectra by a modified Kramers Kronig integral, *Astrophys. Space Sci.* **262**, 193–213.

Walther, A. (1962). The question of phase retrieval in optics, *Opt. Acta* **10**, 41–49.

Wang, L. J. (2002). Causal "all-pass" filters and Kramers-Kronig relations, *Opt. Commun.* **213**, 27–32.

Wang, S.-L. (1965). On some theorems on conjugate functions, *Scientia Sinica* **14**, 507–521.

Wang, T.-C. L. and Marshall, A. G. (1983). Plots of dispersion vs. absorption for detection of multiple positions or widths of Gaussian spectral signals, *Analytical Chem.* **55**, 2348–2353.

Wang, T.-X. (1990). Three-dimensional filtering using Hilbert transform, *Chinese Sci. Bull.* **35**, 123–127.

Waters, K. R., Mobley, J., and Miller, J. G. (2005). Causality-imposed (Kramers-Kronig) relationships between attenuation and dispersion, *IEEE Trans. Ultrasonics, Ferroelect., Freq. Control* **52**, 822–833.

Waters, K. R., Hughes, M. S., Brandenburger, G. H., and Miller, J. G. (2000a). On a time-domain representation of the Kramers-Krönig dispersion relations, *J. Acoust. Soc. Am.* **108**, 2114–2119.

Waters, K. R., Hughes, M. S., Mobley, J., Brandenburger, G. H., and Miller, J. G. (2000b). On the applicability of Kramers-Krönig relations for ultrasonic attenuation obeying a frequency power law, *J. Acoust. Soc. Am.* **108**, 556–563.

Waters, K. R., Hughes, M. S., Mobley, J., and Miller, J. G. (2003). Differential forms of the Kramers-Krönig dispersion relations, *IEEE Trans. Ultrasonics, Ferroelect., Freq. Control* **50**, 68–76.

Watson, G. N. (1944). *A Treatise on the Theory of Bessel Functions*, 2nd edn (Cambridge: Cambridge University Press).

Weaver, R. L. (1986). Causality and theories of multiple scattering in random media, *Wave Motion* **8**, 473–483.

Weaver, R. L. and Pao, Y.-H. (1981). Dispersion relations for linear wave propagation in homogeneous and inhomogeneous media, *J. Math. Phys.* **22**, 1909–1918.

Wegert, E. and Wolfersdorf, L. v. (1988). A uniform inequality for the finite Hilbert transform in weighted Lebesgue spaces, *Demonstratio Math.* **21**, 529–538.

Weideman, J. A. C. (1994). Computation of the complex error function, *SIAM J. Numer. Analysis* **31**, 1497–1518.

Weideman, J. A. C. (1995). Computing the Hilbert transform on the real line, *Math. Comp.* **64**, 745–762.

Weinberg, D. A. (1981). The Hilbert transform and maximal function for approximately homogeneous curves, *Trans. Am. Math. Soc.* **267**, 295–306.

Weir, A. J. (1973). *Lebesgue Integration and Measure* (London: Cambridge University Press).

Weiss, G. (1965). Harmonic analysis. In *Studies In Real and Complex Analysis*, Studies in Mathematics Vol. 3, I. I. Hirschman, Jr., ed. (New York: Mathematical Association of America), pp. 124–178.

Weniger, E. J. (1989). Nonlinear sequence transformations for the acceleration of convergence and the summation of divergent series, *Comput. Phys. Rep.* **10**, 189–371.

Weniger, E. J. (1991). On the derivation of iterated sequence transformations for the acceleration of convergence and the summation of divergent series, *Comput. Phys. Commun.* **64**, 19–45.

Weniger, E. J. (1996). Nonlinear sequence transformations: a computational tool for quantum mechanical and quantum chemical calculations, *Int. J. Quantum Chem.* **57**, 265–280. Erratum: *Int. J. Quantum Chem.* **58**, 319–321 (1996).

Wheeden, R. L. and Zygmund, A. (1977). *Measure and Integral* (New York: Marcel Dekker).

Wheeler, J. C. (1974). Modified moments and Gaussian quadratures, *Rocky Mountain J. Math.* **4**, 287–296.

Wheelon, A. D. (1968). *Tables of Summable Series and Integrals Involving Bessel Functions* (San Francisco, CA: Holden-Day).

Whittaker, E. T. (1915). XVIII.-On the functions which are represented by the expansions of the interpolation-theory, *Proc. Roy. Soc. Edinburgh* **35**, 181–194.

Whittaker, E. T. and Watson, G. N. (1927). *A Course of Modern Analysis* (London: Cambridge University Press).

Widom, H. (1960). Singular integral equations in L_p, *Trans. Am. Math. Soc.* **97**, 131–160.

Wiener, N. (1929). Hermitian polynomials and Fourier series, *J. Math. Phys.* (*MIT*) **8**, 70–73.

Wiener, N. (1933). *The Fourier Integral and Certain of its Applications* (New York: Dover).

Wiener, N. (1960). *Extrapolation, Interpolation, and Smoothing of Stationary Time Series* (New York: Technology Press, MIT, & John Wiley)

Wightman, A. S. (1960). Analytic functions of several complex variables. In *Relations de dispersion et particules élémentaires*, C. De Witt and R. Omnes, eds. (Paris: Hermann), pp. 227–315.

Wigner, E. P. (1964). Causality, *R*-matrix, and collision matrix. In *Dispersion Relations and their Connection with Causality*, E. P. Wigner, ed. (New York: Academic Press), pp. 40–67.

Williams, C. P. and Marshall, A. G. (1992). Hartley/Hilbert transform spectroscopy: absorption-mode resolution with magnitude-mode precision, *Analytical Chem.* **64**, 916–923.

Williams, J. R. (1980). Fourier efficiency using analytic translation and Hilbert samples, *J. Acoust. Soc. Am.* **67**, 581–588.

Williams, W. E. (1978). Note on a singular integral equation, *J. Inst. Math. Appl.* **22**, 211–214.

Wimp, J. (1981). *Sequence Transformations and their Applications* (New York: Academic Press).

Witte, H., Eiselt, M., Patakova, I., Petranek, S., Griessbach, G., Krajca, V., and Rother, M. (1991). Use of discrete Hilbert transformation for automatic spike mapping: a methodological investigation, *Med. Biol. Eng. Comput.* **29**, 242–248.

Wolf, A. A. (1967). The Hilbert transform as an iterated Laplace transform, *IEEE Trans. Aerospace Elect. Syst.* **3**, 728–729.

Wolf, E. (1958). Reciprocity inequalities, coherence time and bandwidth in signal analysis and optics, *Proc. Phys. Soc. Lond.* **71**, 257–269.

Wolfersdorf, L. v. (1985). On the theory of nonlinear singular integral equations of the Cauchy type, *Math. Methods Appl. Sci.* **7**, 493–517.

Wong, D. Y. (1964). Dispersion relations and applications. In *Dispersion Relations and Their Connection with Causality*, E. P. Wigner, ed. (New York: Academic Press), pp. 68–96.

Wong, R. (1980a). Error bounds for asymptotic expansions of integrals, *SIAM Rev.* **22**, 401–435.

Wong, R. (1980b). Asymptotic expansion of the Hilbert transform, *SIAM J. Math. Analysis* **11**, 92–99.

Wong, R. (1989). *Asymptotic Approximations of Integrals* (Boston, MA: Academic Press).

Wood, F. M. (1929). Reciprocal integral formulae, *Proc. Lond. Math. Soc.* (2) **29**, 29–48.

Woolcock, W. S. (1967). Asymptotic behavior of Stieltjes transforms. I, *J. Math. Phys.* **8**, 1270–1275.

Woolcock, W. S. (1968). Asymptotic behavior of Stieltjes transforms. II, *J. Math. Phys.* **9**, 1350–1356.

Wooten, F. (1972). *Optical Properties of Solids* (New York: Academic Press).

Wright, K. R. and Hutchinson, J. S. (1999). Oscillator phase and the reaction dynamics of HN_3: A model for correlated motion, *Phys. Chem. Chem. Phys.* **1**, 1299–1309.

Wu, C.-H., Mahler, G., and Birman, J. L. (1978). Electric quadrupole sum rules in solids, *Phys. Rev. B* **18**, 4221–4224.

Wu, C.-K. and Andermann, G. (1968). Improved Kramers-Kronig dispersion analysis of infrared reflectance data for lithium fluoride, *J. Opt. Soc. Am.* **58**, 519–525.

Wu, T. T. (1962). Some properties of impedance as a causal operator, *J. Math. Phys.* **3**, 262–271.

Wu, T.-Y. and Ohmura, T. (1962). *Quantum Theory of Scattering* (Englewood Cliffs, NJ: Prentice-Hall).

Xu, Y. and Yan, D. (2006). The Bedrosian identity for the Hilbert transform of product functions, *Proc. Am. Math. Soc.* **134**, 2719–2728.

Yamamoto, K. and Ishida, H. (1997). Complex refractive index determination for uniaxial anisotropy with the use of the Kramers-Kronig analysis, *Appl. Spectrosc.* **51**, 1287–1293.

Yamamoto, T. (2006). Approximation of the Hilbert transform via use of sinc convolution, *Electron. Trans. Numer. Analysis* **23**, 320–328.

Yang, B.-C. (2000). On new generalizations of Hilbert's inequality, *J. Math. Analysis Appl.* **248**, 29–40.

Yang, J. N., Lei, Y., Pan, S., and Huang, N. (2003a). System identification of linear structures based on Hilbert-Huang spectral analysis. Part 1: normal modes, *Earthquake Engng. Struct. Dyn.* **32**, 1443–1467.

Yang, J. N., Lei, Y., Pan, S., and Huang, N. (2003b). System identification of linear structures based on Hilbert-Huang spectral analysis. Part 2: complex modes, *Earthquake Engng. Struct. Dyn.* **32**, 1533–1554.

Yariv, A. (1989). *Quantum Electronics* (New York: Wiley).

Youla, D. C., Castriota, L. J., and Carlin, H. J. (1959). Bounded real scattering matrices and the foundations of linear passive network theory, *IRE Trans. Circuit Theory* **6**, 102–124.

Young, R. H. (1976). Validity of the Kramers-Kronig transformation used in reflection spectroscopy, *J. Opt. Soc. Am.* **67**, 520–523.

Young, W. H. (1911). On the convergence of a Fourier series and of its allied series, *Proc. Lond. Math. Soc.* (2) **10**, 254–272.

Young, W. H. (1912). Note on a certain functional reciprocity in the theory of Fourier series, *Mess. Math.* **41**, 161–166.

Zabusky, N. J. and Kruskal, M. D. (1965). Interaction of "solitons" in a collisionless plasma and the recurrence of initial states, *Phys. Rev. Lett.* **15**, 240–243.

Zadeh, L. A. and Desoer, C. A. (1963). *Linear System Theory* (New York: McGraw-Hill).

Zaidi, N. H. (1976). On a transform discussed by Goldberg, *J. Lond. Math. Soc.* (2) **14**, 240–244.

Zaidi, N. H. (1977). The Boas transform and its inversion, *Karachi Univ. J. Sci.* **5**, 7–19.

Žak, I. E. (1950). On some properties of conjugate double trigonometric series, *Dokl. Akad. Nauk SSSR* (*NS*) **73**, 5–8.

Žak, I. E. (1952). Concerning a theorem of L. Cesari on conjugate functions of two variables, *Dokl. Akad. Nauk SSSR* (*NS*) **87**, 877–880.

Zanotti, G., Fogale, F., and Capitani, G. (1996). The use of discrete Hilbert transforms in phase extension and improvement, *Acta Cryst. A* **52**, 757–765.

Zayed, A. I. (1993). *Advances in Shannon's Sampling Theory* (Boca Raton, FL: CRC Press)

Zayed, A. I. (1996). *Handbook of Function and Generalized Function Transformations* (Boca Raton, FL: CRC Press).

Zayed, A. I. (1998). Hilbert transform associated with the fractional Fourier transform, *IEEE Signal Process. Lett.* **5**, 206–208.

Zayed, A. I. and García, A. G. (1999). New sampling formulae for the fractional Fourier transform, *Signal Process.* **77**, 111–114.

Zemanian, A. H. (1965). *Distribution Theory and Transform Analysis* (New York: Dover)

Zhechev, B. (2005). Hilbert transform relations, *Cybernetics Inform. Tech.* **5**, 3–13.

Zhizhiashvili, L. V. (1983). Validity of A. N. Kolmogorov's theorem for conjugate functions of several variables, *Math. Notes* **32**, 486–490. Translated from *Mat. Zametki* **32**, 13–21 (1982).

Zhizhiashvili, L. V. (1996). *Trigonometric Fourier Series and Their Conjugates* (Dordrecht: Kluwer).

Zhu, Y. M., Peyrin, F., and Goutte, R. (1990). The use of a two-dimensional Hilbert transform for Wigner analysis of 2-dimensional real signals, *Signal Process.* **19**, 205–220.

Ziesler, S. (1994). L^p-boundedness of the Hilbert transform and maximal function associated to flat plane curves, *Proc. Am. Math. Soc.* **122**, 1035–1043.

Ziesler, S. (1995). L^p-boundedness of the Hilbert transform and maximal function along flat curves in $\mathbb{R}^n$, *Pacific J. Math.* **168**, 383–405.

Zimering, S. (1969). Some asymptotic behavior of Stieltjes transforms, *J. Math. Phys.* **10**, 181–183.

Zygmund, A. (1929). Sur les fonctions conjuguées, *Fund. Math.* **13**, 284–303. Erratum: *Fund. Math.* **18**, 312 (1932).

Zygmund, A. (1932). A remark on conjugate series, *Proc. Lond. Math. Soc.* **34**, 392–400.

Zygmund, A. (1934). Some points in the theory of trigonometric and power series, *Trans. Am. Math. Soc.* **36**, 586–617.

Zygmund, A. (1949). On the boundary values of functions of several complex variables, I., *Fund. Math.* **36**, 207–235.

Zygmund, A. (1955). *Trigonometrical Series* (New York: Dover).

Zygmund, A. (1956a). On a theorem of Marcinkiewicz concerning interpolation of operations, *J. Math. Pures Appl.* **35**, 223–248.

Zygmund, A. (1956b). Hilbert transforms in E^n. In *Proceedings of the International Congress of Mathematicians 1954*, Vol. 3 (Amsterdam: North-Holland), pp. 140–151.

Zygmund, A. (1957). On singular integrals, *Rend. Mat. Appl.* **16**, 468–505.

Zygmund, A. (1968). *Trigonometric Series*, Vols. I & II (Cambridge: Cambridge University Press).

Zygmund, A. (1971). *Intégrales Singuliéres* (Berlin: Springer-Verlag).

Author index

Only a limited number of citations are included for Hilbert, Cauchy, and Fourier, as these authors occur very prominently throughout the book. Authors cited implicitly (as part of *et al.*) are indicated by italic font. Page numbers for volume 2 are indicated by (2).

Subject index

Page numbers for volume 2 are indicated by (2).